Sets of Numbers

Natural numbers $\quad N = \{1, 2, 3, 4, 5, \ldots\}$ (counting numbers)

Whole numbers $\quad W = \{0, 1, 2, 3, 4, 5, \ldots\}$

Integers $\quad Z = \{\ldots, -3, -2, -1, 0, 1, 2, 3, \ldots\}$

Rational numbers $\quad Q = \left\{ \dfrac{p}{q} \;\middle|\; p, q \in Z, q \neq 0 \right\}$

Irrational numbers $\quad I = \{x \mid x \in R, x \text{ is not a rational number}\}$

Real numbers $\quad R = \{x \mid x \text{ corresponds to a point on the number line}\}$

Complex numbers $\quad C = \{a + bi \mid a, b \in R\}$

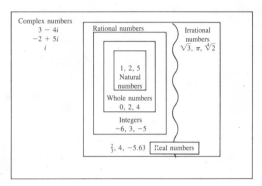

$$N \subset W \subset Z \subset Q \subset \mathcal{R} \subset C$$

Properties of Real Numbers

Commutative Properties

$a + b = b + a$

$a \cdot b = b \cdot a$

Associative Properties

$a + (b + c) = (a + b) + c$

$a \cdot (b \cdot c) = (a \cdot b) \cdot c$

Distributive Property

$a \cdot (b + c) = a \cdot b + a \cdot c$ or $(a + b) \cdot c = a \cdot c + b \cdot c$

Identities

There is a unique real number called the **additive identity**, represented by 0, which has the property that $a + 0 = 0 + a = a$ for all $a \in R$.

There is a unique real number called the **multiplicative identity**, represented by 1, which has the property that $a \cdot 1 = 1 \cdot a = a$ for all $a \in R$.

Inverses

Each real number a has a unique **additive inverse**, represented by $-a$, which has the property that $a + (-a) = (-a) + (a) = 0$.

Each real number a, except Q, has a unique **multiplicative inverse**, represented by $\dfrac{1}{a}$, which has the property that $a\left(\dfrac{1}{a}\right) = \left(\dfrac{1}{a}\right)a = 1$.

Closure Properties

The sum of two real numbers is a real number.

The product of two real numbers is a real number.

Understanding
Intermediate
Algebra

Second Edition

Reviewing the material before doing exercises makes each solution you go through more meaningful. The better you understand the concepts underlying the exercise, the easier the material becomes, and the less likely you are to confuse examples or forget steps.

When reviewing the material, take the time to *think about what you are reading*. Try not to get frustrated if it takes you an hour or so to read and understand a few pages of a math text—that time will be well spent. As you read your text and your notes, think about the concepts being discussed: **(a)** how they relate to previous concepts covered, and **(b)** how the examples illustrate the concepts being discussed. More than likely, worked-out examples will follow verbal material, so look carefully at these examples and try to understand why each step in the solution is taken. When you finish reading, take a few minutes and think about what you have just read.

Doing Exercises

After you have finished reviewing the appropriate material, you should be ready to do the relevant exercises. Although your ultimate goal is to be able to work out the exercises accurately *and* quickly, when you are working out exercises on a topic which is new to you it is a good idea to take your time and think about what you are doing while you are doing it.

Think about how the exercises you are doing illustrate the concepts you have reviewed. Think about the steps you are taking and ask yourself why you are proceeding in this particular way and not some other: Why this technique or step and not a different one?

Do not worry about speed now. If you take the time at home to think about what you are doing, the material becomes more understandable and easier to remember. You will then be less likely to "do the wrong thing" in an exercise. The more complex-looking exercises are less likely to throw you. In addition, if you think about these things in advance, you will need much less time to think about them during an exam, and so you will have more time to work out the problems.

Once you believe you thoroughly understand what you are doing and why, you may work on increasing your speed.

Reading Directions

One important but frequently overlooked aspect of an algebraic problem is the verbal instructions. Sometimes these instructions are given in a single word, such as "simplify" or "solve" (occasionally it takes more time to understand the instructions than it takes to do the exercise). The verbal instructions tell us what we are expected to do, so make sure you read the instructions carefully and understand what is being asked.

Two examples may look the same, but the instructions may be asking you to do two different things. For example,

Identify the following property:

$$a + (b + c) = (a + b) + c$$

versus

Verify the following property by replacing the variables with numbers:

$$a + (b + c) = (a + b) + c$$

Strengthening Your Study Skills

Studying Algebra— How Often?

In most college courses, you are typically expected to spend 2–4 hours studying outside of class for every hour spent in class.

It is especially important that you spend this amount of time studying algebra, since you must both acquire and *perfect* skills; and, as most of you who play a musical instrument or participate seriously in athletics already should know, it takes time and lots of practice to develop and perfect a skill.

It is also important that you distribute your studying over time. That is, do not try to do all your studying in 1, 2, or even 3 days, and then skip studying the other days. You will find that understanding algebra and acquiring the necessary skills are much easier if you spread your studying out over the week, doing a little each day. If you study in this way, you will need less time to study just before exams.

In addition, if your study sessions are more than 1 hour long, it is a good idea to take a 10-minute break within every hour you spend reading math or working exercises. The break helps to clear your mind, and allows you to think more clearly.

Previewing Material

Before you attend your next class, preview the material to be covered beforehand. First, skim the section to be covered, look at the headings, and try to guess what the sections will be about. Then read the material over carefully.

You will find that when you read over the material before you go to class, you will be able to follow the instructor more easily, things will make more sense, and you will learn the material more quickly. Now, if there was something you did not understand when you previewed the material, the teacher will be able to answer your questions *before* you work your assignment at home.

What to Do First

Before you attempt algebra exercises, either for homework or for practicing your skills, it is important to review the relevant portions of your notes and text.

Memorizing a bunch of seemingly unrelated algebraic steps to follow in an example may serve you initially, but in the long run (most likely before Chapter 3), your memory will be overburdened—you will tend to confuse examples and/or forget steps.

On the other hand, two different examples may have the same instructions but require you to do different things. For example,

Evaluate: $2(3 + 8)$ versus   Evaluate: $2 + (3 + 8)$

You are asked to evaluate both expressions, but the solutions require different steps.

It is a good idea to familiarize yourself with the various ways the same basic instructions can be worded. In any case, always look at an example carefully and ask yourself what is being asked and what needs to be done, *before you do it*.

Preface

This text is designed to help you understand algebra. We are convinced that if you understand what you are doing and why, you will be a much better algebra student. (Our students who have used this book in its preliminary form seem to agree with us.) This does not mean that after reading each section you will understand all the concepts clearly. Much of what you learn comes through the course of doing lots and lots of exercises and seeing for yourself exactly what is involved in completing an exercise. However, if you read the textbook carefully and take good notes in class you will find algebra not quite so menacing.

Here are a few suggestions for using this textbook:

- Always read the textbook with a pencil and paper in hand. Reading mathematics is not like reading other subjects. *You* must be involved in the learning process. Work out the examples along with the textbook and *think* about what you are reading. Make sure you understand what is being done and why.

- You must work homework exercises on a daily basis. While attending class and listening to your instructor are important, do not mistake understanding someone else's work for the ability to do the work yourself. (Think about watching someone else driving a car, as opposed to driving yourself.) Make sure *you* know how to do the exercises.

- Read the Study Skills which appear in the preface and at the end of each section in the first four chapters. They discuss the best ways to use the textbook and your notes. They also offer a variety of suggestions on how to study, do homework, and prepare for and take tests. If you want more information on improving your algebra study skills, we direct you to the book *Studying Mathematics*, by Mary Catherine Hudspeth and Lewis R. Hirsch (1982, Kendall/Hunt Publishing Company, Dubuque, Iowa).

- Do not get discouraged if you have difficulty with some topics. Certain topics may not be absolutely clear the first time you see them. Be persistent. We all need time to absorb new ideas and become familiar with them. What was initially difficult will become less so as you spend more time with a subject. Keep at it and you will see that you are making steady progress.

A series of videotapes which reviews the text material one chapter at a time is also available.

More detailed information on all of the supplements is available from the publisher.

ACKNOWLEDGMENTS   The authors would like to acknowledge the many users, both instructors and students, of the first edition whose invaluable comments and suggestions helped shape this second edition.

The authors sincerely thank the following reviewers of this revised edition for their thoughtful comments and numerous suggestions: Neil Aiken, Milwaukee Area Technical College; Harold Bennett, Texas Technical University; Benjamin Bockstege, Broward Community College; Harvey Braverman, New York City Technical College; Helen Burrier, Kirkwood Community College; William Chatfield, University of Wisconsin at Platteville; W. D. Clark, Stephen F. Austin State University; Edward Davenport, Central Missouri State University; Lloyd Davis, College of San Mateo; Bob Denton, Orange Coast College; Max Ellis, Oklahoma State University; Mike Emerick, Parkman College; Thomas Englert, Southwest Texas State University; Aparna Ganguli, University of Minnesota; William Grimes, Central Missouri State University; Linda Hall, Pan American University; Lou Hoelze, Bucks County Community College; Mark King, University of New Orleans; John Manon, University of Delaware; Susan McCory, San Jose State University; Peggy Miller, Kearney State College; Fern Mizell, Austin Community College; Wendell Motler, Florida A & M University; Nancy Petty, Utah State University; A. Allan Riveland, Washburn University; Doug Robertson, University of Minnesota; Donald Rossi, DeAnza College; Dorothy Schellenback, Hartnell Community College; Edith Silver, Mercer Community College; Elizabeth Sirjani, Washington State University; Gerald Skidmore, Alvin Community College; Don Skow, Pan American University; Bruce Teague, Santa Fe Community College; Francis Ventola, Brookdale Community College; Mary Voxman, University of Idaho; and Barbara Worley, Des Moines Area Community College.

The authors would also like to thank Susan Lewis for helping to prepare the original manuscript for class testing at Rutgers University and Queens College, and Patrick Fitzgerald for his support on the first edition. Obviously, the production of a textbook is a collaborative effort and we must thank our editor, Jay Ricci, for his constant support; Susan Reiland for her expert supervision of the entire production; and Valerie DeBellis, Nancy Little, and Beverly Stevens for their assistance in checking solutions. Of course, any errors that remain are the sole responsibility of the authors, and we would greatly appreciate their being called to our attention.

Finally, we would like to thank our wives, Cindy and Sora, and our families for their unwavering encouragement.

□ The answer section contains answers to all the odd-numbered exercises, as well as to all review exercises and practice test problems. The answer to each verbal problem contains a description of what the variable(s) represent and the equation (or system of equations) used to solve it. In addition, the answers to the cumulative review exercises and cumulative practice tests contain a reference to the section in which the relevant material is covered.

□ Throughout the text a second color has been used to highlight important ideas, definitions, and procedural outlines.

NEW TO THE SECOND EDITION   Many of the comments and suggestions made by the users of the first edition have been incorporated into the second edition. Among these are:

1. To increase the flexibility of the text for many users, we have interchanged (and modified) the chapters on systems of linear equations and conic sections. Now systems of linear equations can be treated before conic sections.

2. A section has been added to Chapter 9 (Systems of Linear Equations) covering matrix methods of solving linear systems. This section includes both Gaussian and Gauss–Jordan elimination.

3. More verbal problems have been integrated throughout the text.

4. Most sections contain a **Mini-Review** which consists of exercises that allow students to periodically review important topics as well as help them prepare for the material to come. These Mini-Reviews afford the student additional opportunity to see new topics within the framework of what they have already learned.

SUPPLEMENTS   A student's study guide is available which contains the worked-out solutions to many odd-numbered exercises, the answers to the Questions for Thought, two additional practice tests for each chapter, and three additional cumulative practice tests.

An instructor's manual contains the answers to the even-numbered exercises, answers to the Questions for Thought, five additional chapter tests for each chapter, two additional cumulative tests for every three chapters, and two final exams. The instructor's guide also contains some additional suggestions on using the Study Skills.

An extensive computer software package, consisting of two parts, is available. The first type is *diagnostic software*, which offers the student the opportunity to work out a variety of multiple-choice exercises and receive diagnostic computer responses keyed to the incorrect answer the student chooses. The second type of software is new to the second edition. It is *tutorial software*, which generates problems on a specific topic, at varying degrees of difficulty, and guides the student through a step-by-step solution.

Two supplements are available to accompany *Understanding Intermediate Algebra*: (1) A **geometry supplement** covering basic vocabulary and important facts and formulas relating to parallel lines, angles, triangles, quadrilaterals, and circles; and (2) A **supplemental chapter on sequences and series**.

Basic rules and/or procedures are highlighted so that students can find important ideas quickly and easily.

FEATURES   The various steps in the solutions to examples are explained in detail. Many steps appear with annotations (highlighted in a second color) which involve the student in the solution. These comments explain how and why a solution is proceeding in a certain way.

☐ There are over 5,200 exercises, including calculator exercises. Not only have the exercise sets been matched odd/even, but they have also been designed so that, in many situations, successive odd-numbered exercises compare and contrast subtle differences in applying the concepts covered in the section. Additionally, variety has been added to the exercise sets so that the student must be alert as to what the problem is asking. For example, the exercise set in Section 4.5, which deals primarily with solving rational equations, also contains some exercises on adding rational expressions.

☐ One of the main sources of students' difficulties is that they do not know how to study algebra. In this regard we offer a totally unique feature. The preface and each section in the first four chapters conclude with a **Study Skill**. This is a brief paragraph discussing some aspect of studying algebra, doing homework, or preparing for or taking exams. Our students who have used the preliminary version of this book indicated that they found the Study Skills very helpful. The study skills appearing in this text are part of a collection of general mathematics study skills developed by Lewis R. Hirsch and Mary C. Hudspeth at the Pennsylvania State University. For a more detailed discussion of how to study mathematics we refer you to the book *Studying Mathematics* by M. C. Hudspeth and L. R. Hirsch (1982, Kendall/Hunt Publishing Company, Dubuque, Iowa).

☐ Almost every exercise set contains **Questions for Thought**, which offer the student an opportunity to *think* about various algebraic ideas. They may be asked to compare and contrast related ideas, or examine an incorrect solution and explain why the solution is wrong. The Questions for Thought are intended to be answered in complete sentences and in grammatically correct English. The Questions for Thought were originally designed for having students write across the curriculum, and can be used by instructors for this purpose.

☐ Each chapter contains a chapter summary describing the basic concepts in the chapter. Each point listed in the summary is accompanied by an example illustrating the concept or procedure.

☐ Each chapter contains a set of chapter review exercises and a chapter practice test. Additionally, there are four cumulative review exercise sets and four cumulative practice tests, following Chapters 3, 6, 9, and 12. These offer the student more opportunities to practice choosing the appropriate procedure in a variety of situations.

Preface

To the Instructor

This second edition of *Understanding Intermediate Algebra* retains the same basic structure and philosophy as the first edition.

PURPOSE   *Understanding Intermediate Algebra*, 2nd edition, is our attempt to offer an intermediate algebra textbook which reflects our philosophy—that students can *understand* what they are doing in algebra and why.

We offer a view of algebra that takes every opportunity to explain why things are done in a certain way, to show how concepts or topics are related, and to show how supposedly "new" topics are actually just new applications of concepts already learned.

This text assumes a knowledge of elementary algebra. However, we realize that students arrive in intermediate algebra with a diversity of previous mathematical experiences. Some may have recently learned elementary algebra but to varying degrees; some may have learned elementary algebra well but have been away from it for awhile. In this spirit, we have tried to develop a text flexible enough to meet the needs of this heterogeneous group by providing explanations for what we regard as key elementary concepts that are crucial to the understanding of the more difficult intermediate algebra concepts. The less well prepared students certainly benefit from these explanations; the better prepared students appreciate this second look as a way to help them fit the seemingly unrelated parts of algebra together. The instructor has the choice of discussing this material or leaving it for students to review on their own.

PEDAGOGY   We believe that a student can successfully learn and understand algebra by mastering the basic concepts and being able to apply them to a wide variety of situations. Thus, each section begins by relating the topic to be discussed to material previously learned. In this way the students can see algebra as a whole rather than as a series of isolated ideas.

Basic concepts, rules, and definitions are motivated and explained via numerical and algebraic examples.

Concepts are developed in a series of carefully constructed illustrative examples. Through the course of these examples we compare and contrast related ideas, helping the student to understand the sometimes subtle distinctions among various situations. In addition, these examples strengthen a student's understanding of how this "new" idea fits into the overall picture.

Every opportunity has been taken to point out common errors often made by students and to explain the misconception that often leads to a particular error.

Table of Contents

Production service: Susan L. Reiland

Interior design: James Chadwick

Artwork: Ben Turner Graphics

Composition: Jonathan Peck Typographers, Ltd.

Cover design: Pollock Design Group

Cover: Miklos Pogany, *KLARIKA MARCH*. Courtesy of Victoria Munroe
Gallery, New York City.

Printed in the United States of America
97 96 95 94 93 92 91 90   8 7 6 5 4 3 2 1 0

Library of Congress Cataloging-in-Publication Data

Hirsch, Lewis.
    Understanding intermediate algebra / Lewis Hirsch, Arthur Goodman.
    —2nd ed.
       p.   cm.
    ISBN 0-314-48124-9
    1. Algebra.   I. Goodman, Arthur.   II. Title.
QA152.2.H57   1990   89-22746
512.9--dc20   CIP

Understanding Intermediate Algebra

A Course for College Students

Second Edition

Lewis Hirsch
Rutgers University

Arthur Goodman
Queens College of the City University of New York

WEST PUBLISHING COMPANY

St. Paul   New York
Los Angeles   San Francisco

Understanding Intermediate Algebra

Second Edition

1

The Fundamental Concepts

Study Skills

In elementary algebra we developed certain skills, learned a number of basic concepts and principles, and learned how to apply them in various situations.

In intermediate algebra we will develop our skills to a higher level and, although we will learn new concepts, to a great extent we will draw upon the same basic notions covered in elementary algebra. This requires that we understand the fundamental principles of elementary algebra and how to recognize and apply these principles. Therefore, we will begin with a review of the elementary concepts and will eventually generalize them to the more complex cases in later chapters. We begin with the notion of sets.

1.1

Basic Definitions

Sets

A *set* is simply a *well-defined* collection of objects. The phrase *well-defined* means that there are clearly determined criteria for membership in the set. The criteria can be a list of objects in the set, called the ***elements*** or members of the set, or it can be a description of the objects in the set.

For example, it is not sufficient to say "the set of all tall people in our class." "Tall" is not well defined; we may not agree on whether a $5'11''$ person belongs in the set or not. It is possible to say "the set of people in our class over 6 feet tall."

One way to represent a set is to list the elements of the set, and to enclose the list in *set brackets,* which look like { }.

We often designate sets by using capital letters such as A, B, C. For example,

$$A = \{1, 6, 17, 45\} \qquad B = \{k, l, m, n, o\}$$

We say that the set A is the set consisting of the numerals 1, 6, 17, and 45.

The symbol we use to indicate that an object is a member of a particular set is "$\in$". Thus, $x \in S$ is a symbolic way of writing that x is a member or an element of the set S.

We use the symbol "$\notin$" to indicate that an object is *not* an element of a set. (In general, when we put a "/" through a math symbol it means *not*; thus, "$\neq$" means "not equal to.") For example, using the sets A and B listed above, we have:

$17 \in A$ 17 is an element of A.

$p \notin B$ p is not an element of B.

Sometimes, in order to exhibit a set that contains many elements or a set that contains an infinite number of elements, we use a variation on the listing method. We list a few elements followed by a comma and three dots. For example, two sets of numbers to which we frequently refer are N and W. The set of numbers we use for counting is called the set of ***natural numbers,*** and is usually denoted by the letter N:

$$N = \{1, 2, 3, 4, 5, \ldots\}$$

If we include the number 0 with this set it is called the set of **whole numbers** and is denoted by the letter W:

$$W = \{0, 1, 2, 3, 4, 5, \ldots\}$$

Of course, this method of listing a set can be used only when the first few elements clearly show the pattern for *all* the elements in the set.

Often when we describe a set we use the word *between,* which can be ambiguous. When we say "the numbers between 7 and 15" do we mean including or excluding 7 and 15 themselves? Therefore, let's agree that unless we specifically indicate otherwise, when we say *between* and *in between,* we do *not* include the first and last numbers.

EXAMPLE 1

List the elements of the following sets:

(a) The set A of even numbers between 6 and 54

(b) The set B of odd numbers greater than 19

Solution

(a) An even number is (an integer) exactly divisible by 2.

$$A = \{8, 10, 12, 14, \ldots, 52\}$$ *

Note that 6 and 54 are not included. Also note that if the set is large but not infinite, we use the three-dot notation, but list the last element of the set preceded by a comma.

(b) Since no upper limit to this set is given, our answer is

$$B = \{21, 23, 25, \ldots\}$$

Note that 19 is not included. ■

Another way of writing a set is by using set-builder notation. **Set-builder notation** consists of the set braces, a variable that acts as a place holder, a vertical bar (|) which is read "such that," and a sentence that describes what the variable can be. This last part is called the **condition** on the variable. For example,

$$\{ \quad x \quad | \quad x \text{ is an odd number greater than 1 and less than 15}\}$$

$\uparrow$ $\uparrow$ $\uparrow$

Variable Such that   Condition on the variable

This is read "the set of all x such that x is an odd number greater than 1 and less than 15," which is the set $\{3, 5, 7, 9, 11, 13\}$.

It is possible to place a condition on the variable that cannot be satisfied, as for example

$$F = \{x \mid x \text{ is a whole number less than 0}\}$$

Since it is impossible for a whole number to be less than 0, this set F has no members. A set with no members is called the **empty set** or the **null set** and is symbolized by "$\varnothing$". Thus, we have $F = \varnothing$.

---

* Throughout the text, we will use a box to indicate our final answer.

In dealing with numbers, we use the equals symbol "=" to indicate that two expressions represent the same number. For example, $6 + 2 = 8$ since both $6 + 2$ and 8 represent the same number. For sets, the same symbol of equality is used to indicate that two sets are identical (they contain the same elements). Hence,

$$\{1, 3, 8\} = \{3, 8, 1\}$$

because both sets contain the same elements. Note that the order of the elements is unimportant.

DEFINITION

The set B is a *subset* of A, written $B \subset A$, if all elements of B are also contained in A.

Hence, if $B = \{3, 7, 9\}$ and $A = \{2, 3, 4, 5, 7, 9\}$ then B is a subset of A, since all elements of B (3, 7, and 9) are contained in A.

Note that $N \subset W$ since all natural numbers are contained in the set of whole numbers.

We can also create new sets from old sets by certain *operations* on sets.

DEFINITION

The *union* of two sets A and B, written $A \cup B$, is the set made up of elements in A or B, or in both A and B.

For example, if $S = \{a, b, c\}$ and $T = \{\text{red, white, blue}\}$, then $S \cup T$ is the new set formed by combining elements of S and T:

$$S \cup T = \{a, b, c, \text{red, white, blue}\}$$

We often see $A \cup B$ defined symbolically as follows:

$$A \cup B = \{x \mid x \in A \quad \text{or} \quad x \in B\}$$

(In mathematics, the word *or* includes the possibility that x is an element of *both* A and B.)

DEFINITION

The *intersection* of two sets A and B, written $A \cap B$, is the set made up of all elements common to both A and B.

For example, if $S = \{1, 3, 8, 9, 15\}$ and $T = \{3, 7, 9, 11\}$, then $S \cap T$ is the new set made up of elements that are common to both sets:

$$S \cap T = \{3, 9\}$$

We often see $A \cap B$ defined symbolically as follows:

$$A \cap B = \{x \mid x \in A \quad \text{and} \quad x \in B\}$$

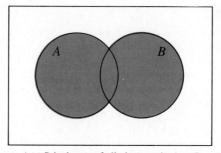 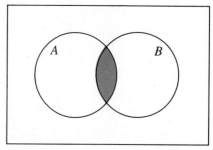

Figure 1.1
Union and intersection

a. $A \cup B$ is the set of all elements in A or B, or in both A and B.

b. $A \cap B$ is the set of elements common to both A and B.

The shaded areas in Figure 1.1 represent the union and intersection of sets A and B, respectively.

Before we continue to the next section, we should review some basic terminology and notation.

Sums, Terms, Products, and Factors

Sum is the word we use for addition. In an expression involving a sum, the quantities to be added in the sum are called the **terms.** The symbol used to indicate a sum is the familiar "$+$" sign.

Product is the word we use for multiplication. In an expression involving a product, the numbers being multiplied are called **factors.**

The most frequent error made by students in algebra is that of confusing terms with factors and factors with terms. There are things that can be done with factors that cannot be done with terms, and vice versa. For example,

$$\frac{\cancel{x}y}{\cancel{x}}$$ *The x is a **factor** of the numerator and therefore **can** be cancelled with the factor of x in the denominator.*

but

$$\frac{x + y}{x}$$ *The x is a **term** of the numerator and therefore **cannot** be cancelled with the factor of x in the denominator.*

Saying that a is a **multiple** of b is equivalent to saying that b is a **factor** of a.

30 is a multiple of 6   because 6 is a factor of 30.
63 is a multiple of 7   because 7 is a factor of 63.

Thus, a factor of n is a number that divides exactly into n, whereas a multiple of n is a number that is exactly divisible by n.

EXAMPLE 2

List the elements of the following sets:

(a) $A = \{y \mid y \in W, y \text{ is a multiple of 6, and } y \text{ is less than 30}\}$
(b) $B = \{s \mid s \in W, s \text{ is a multiple of 4, and } s \text{ is less than 28}\}$
(c) $A \cap B$
(d) $A \cup B$

Solution

(a) $A = \{y \mid y \in W, y$ is a multiple of 6, and y is less than 30$\}$. Hence,

$A =$ ┃ $\{0, 6, 12, 18, 24\}$ ┃   *Note that 0 is included as a whole-number multiple of 6 ($0 = 6 \cdot 0$).*

(b) $B = \{s \mid s \in W, s$ is a multiple of 4, and s is less than 28$\}$. Hence,

$$B = \boxed{\{0, 4, 8, 12, 16, 20, 24\}}$$

(c) $A \cap B$: Looking at sets A and B, we have

$$A \cap B = \boxed{\{0, 12, 24\}}$$

(d) $A \cup B$: Looking at sets A and B, we have

$$A \cup B = \boxed{\{0, 4, 6, 8, 12, 16, 18, 20, 24\}}$$

Note that in $A \cup B$, even though the elements 0, 12, and 24 appear in both sets, we write them only once.   ■

There is another subset of the natural numbers that is very important. It is called the set of *prime numbers*.

DEFINITION

A *prime number* is a natural number, greater than 1, which is divisible only by itself and 1. In other words, a prime number is a number whose only factors are itself and 1.
   A natural number greater than 1 which is not prime is called *composite*.

For example, the numbers 7 and 11 are prime numbers because they are not divisible by any number other than themselves and 1. The number 12 is composite (not prime) because it is divisible by other numbers, such as 3 and 4.
   Every composite number can be uniquely written as a product of prime factors. For example,

$$70 = 2 \cdot 5 \cdot 7$$

EXAMPLE 3

Write the number 54 as a product of prime factors.

Solution

We start by writing the number as a product of two factors (any two factors we recognize) and continue factoring until *all* the factors are prime numbers.

We have many choices for our first two factors. We will illustrate just two of the possible paths to the answer.

$54 = 2 \cdot 27$ *Factor 27.* $54 = 9 \cdot 6$ *Factor 9 and factor 6.*

    $= 2 \cdot 3 \cdot 9$ *Factor 9.*   **OR** $= 3 \cdot 3 \cdot 2 \cdot 3$

    $= \boxed{2 \cdot 3 \cdot 3 \cdot 3}$ $= \boxed{2 \cdot 3 \cdot 3 \cdot 3}$

No matter which factors you decide to start with, the final answer (since it involves *prime* factors only) will be the same.   ∎

STUDY SKILLS 1.1   **Comparing and Contrasting Examples**

When learning most things for the first time, it is very easy to get confused and treat things which are different as though they were the same because they "look" similar. Algebraic notation can be especially confusing because of the detail involved. Move or change one symbol in an expression and the entire example is different; change one word in a verbal problem and the whole problem has a new meaning.

It is important that you become capable of making these distinctions. The best way to do this is by comparing and contrasting examples and concepts that look almost identical, but are not. It is also important that you ask yourself in what ways these things are similar and in what ways they differ. For example, as

you may remember from elementary algebra, the commutative property of addition is similar in some respects to the commutative property of multiplication, but different from it in others. Also, the two expressions $3 + 2 \cdot 4$ and $3 \cdot 2 + 4$ look similar, but are actually very different.

When you are working out exercises (or reading a concept), ask yourself, "What examples or concepts are similar to those which I am now doing? In what ways are they similar? How do I recognize the differences?" Doing this while you are working the exercises will help to prevent you from making careless errors later on.

Exercises 1.1

In Exercises 1–12, indicate whether the statement is true or false.

1. $3 \in N$

2. $-3 \in N$

3. $0 \notin W$

4. $0 \in N$

5. $\{a, b\} \subset \{a, b, c\}$

6. $e \in \{a, b, c, d\}$

7. $\{1, 2, 3\} = \{3, 2, 1\}$

8. $12 \in \{1, 2, 3\}$

9. $a \in \{x \mid x \text{ is a letter of the alphabet}\}$

10. $\{a, b, d\} \subset \{a, b, c, d\}$

11. $\{a \mid a \text{ is a multiple of } 6\} \subset \{a \mid a \text{ is a multiple of } 3\}$

12. $\{x \mid x \text{ is a multiple of } 3\} \subset \{x \mid x \text{ is a multiple of } 6\}$

In Exercises 13–28, list the elements in the specified set.

13. $\{n \mid n \in N, n \text{ is less than } 12\}$

14. $\{x \mid x \in W, x \text{ is less than } 12\}$

15. $\{m \mid m \in N, m \text{ is greater than } 6 \text{ and less than } 13\}$

16. $\{m \mid m \in N, m \text{ is greater than } 2 \text{ and less than or equal to } 8\}$

17. $\{a \mid a \in W, a \text{ is greater than or equal to } 3 \text{ and less than } 14\}$

18. $\{t \mid t \in W, t \text{ is greater than } 30 \text{ and less than or equal to } 42\}$

19. $\{n \mid n \in W, n$ is greater than 8 and less than 6$\}$ **20.** $\{n \mid n \in W, n$ is greater than 5 and less than 5$\}$

21. $\{n \mid n$ is a prime number between 40 and 50, $n \in N\}$ **22.** $\{n \mid n$ is a composite number between 40 and 50, $n \in N\}$

23. $\{m \mid m$ is both a prime and a composite number, $m \in N\}$ **24.** $\{m \mid m \in N$ and m is neither a prime nor a composite number$\}$ $\{1\}$

25. $\{t \mid t \in W$ and t is a multiple of 6$\}$ **26.** $\{t \mid t \in W$ and t is a multiple of 3 but not of 6$\}$

27. $\{n \mid n \in W$ and n is a multiple of 8$\}$ **28.** $\{n \mid n \in W$ and n is a multiple of 8 but not of 4$\}$ $\emptyset$

29. $\{n \mid n$ is a factor of 36$\}$ **30.** $\{t \mid t$ is a factor of 54$\}$

In Exercises 31–38, let

$$A = \{0, 1, 2, 3, 4, 5, 6\}$$
$$B = \{x \mid x \in W, x \text{ is a multiple of 3, and } x \text{ is less than 36}\}$$
$$C = \{p \mid p \text{ is a prime number}\}$$
$$D = \{x \mid x \text{ is between 6 and 14, } x \in N\}$$
$$E = \{x \mid x \in N \text{ and } x \text{ is less than or equal to 12}\}$$

List the elements in the indicated set.

31. $A \cap B$ **32.** $A \cap C$

33. $A \cup B$ **34.** $B \cap C$

35. $D \cap C$ **36.** $D \cap E$

37. $A \cap D$ **38.** $A \cap E$

In Exercises 39–46, write the number as a product of prime factors. If the number is prime, say so.

39. 66   **40.** 78

41. 128   **42.** 144

43. 61   **44.** 51

45. 91   **46.** 73

1.2

The Real Numbers: Order and the Number Line

Until now we have been working within the framework of the set of whole numbers. However, we all recognize that the set of whole numbers is not sufficient to supply us with all the numbers we need to describe various situations.

We can represent whole numbers on the number line as follows:

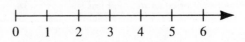

As we remember from elementary algebra, we cannot find the solution to the equation

$$x + 2 = 1$$

if we are restricted to whole number solutions, because there is no whole number which, when increased by 2, will yield 1. Thus, we introduce the **integers,** the set $\{. . . , -3, -2, -1, 0, 1, 2, 3, . . .\}$, designated by Z, in order to find the solution, -1.

The integers can be represented on the number line by extending the whole number line to the left of 0 and labeling the units as follows:

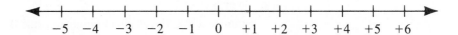

The integers seem to be a satisfactory system until we try to solve the equation

$$3x = 1$$

which has no solution in the integers (what integer multiplied by 3 will yield 1?). This requires us to create new numbers called **rational numbers** or *fractions* in order to find the solution, $\frac{1}{3}$.

The set of rational numbers is usually designated by the letter Q. This set is difficult to list, primarily because no matter where you start, there is no *next* rational number. (What is the "first" fraction after 0?) Instead, we use set-builder notation to describe the set of rational numbers:

$$Q = \left\{ \frac{p}{q} \;\middle|\; p, q \in Z, \quad \text{and} \quad q \neq 0 \right\}$$

In words, this says that the set of rational numbers, Q, is the set of all numbers that can be represented as fractions, or quotients of integers, provided the denominator is not equal to 0.

Thus, the following are all rational numbers:

$$8 = \frac{8}{1} \qquad \frac{-3}{7} \qquad 0 = \frac{0}{3} \qquad 0.73 = \frac{73}{100}$$

(Notice that integers are also rational numbers since they *can* be represented as fractions.)

Now we associate the rational numbers with units on the number line and with points *between* the units as follows:

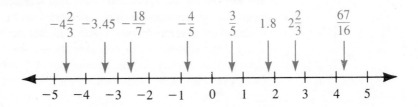

But we will find that there are still points on the number line which cannot be labeled with a rational number. The numbers associated with these points are called *irrational numbers*.

To get a better idea of what irrational numbers look like, let's first examine rational numbers in decimal form (often called *decimal fractions*). In order to convert a fraction into its decimal form we divide the numerator by the denominator. However, converting from the decimal form to a fraction is not quite so straightforward. In the previous example, we recognized that the decimal 0.73 is equal to $\frac{73}{100}$. Similarly, we recognize that the decimal $0.333\overline{3}$ (where the bar above the last 3 indicates that the 3 repeats forever) is equal to the fraction $\frac{1}{3}$.

On the other hand, it is highly unlikely that we would recognize the decimal $0.407407\overline{407}$ as being equal to the fraction $\frac{11}{27}$. (Divide 27 into 11 and verify that you get $0.407407\overline{407}$.)

It turns out that if a decimal terminates (the decimal ends or at some point is followed by zeros), or if a decimal is repeating (the same group of digits in the decimal is repeated infinitely), then the decimal represents a rational number.

This leaves us with nonterminating and nonrepeating decimals, which are *not* rational numbers. In other words, such a decimal cannot be represented as the quotient of two integers. This set of nonrepeating, nonterminating decimals is called the set of **irrational numbers** and we shall designate the set by the letter *I*.

It is necessary for us to consider irrational numbers because just as the whole numbers were insufficient to fill all our needs, so too the rational numbers do not quite do the job, either. When we try to solve the equation

$$x^2 = 9$$

(we are looking for a number which when multiplied by itself gives a product of 9) we will fairly quickly come up with two answers:

$$3 \quad (since\ 3 \cdot 3 = 9) \quad\text{and}\quad -3 \quad [since\ (-3)(-3) = 9]$$

These are called **square roots** of 9. Thus, we see that both 3 and -3 are square roots of 9.

When we try to solve

$$x^2 = 2$$

however, the answer does not come as easily. We could say our answer is $\sqrt{2}$ or $-\sqrt{2}$, but what are those numbers exactly? What decimal number, when squared, will give us 2? It turns out that, if we try to find the answer by trial and error (using a calculator to do the multiplication would help), we can get closer and closer to 2 but we will never get 2 exactly.

For example, if we try $(1.4)(1.4)$, we get 1.96, so we see that 1.4 is too small. If we try $(1.5)(1.5)$, we get 2.25, and we see that 1.5 is too big. If we continue in this way we can get better and better approximations to a square root of 2. We might reach the approximate answer 1.414235, but $(1.414235)(1.414235) = 2.0000606$ (rounded off to seven decimal places). In fact, no matter how many places we get for x, the decimal never stops (because the square of x is never exactly 2) and never repeats. This implies that the square root of 2 (written $\sqrt{2}$) is an irrational number. Other examples of irrational numbers are $\sqrt{7}$, $\sqrt{15}$, $-\sqrt{5}$, π.

The important thing for us to recognize is that the irrational numbers also represent points on the number line. If we take all the rational numbers together with all the irrational numbers (both positive and negative), we get all the points on the number line. This set is called the set of **real numbers,** and is usually designated by the letter R:

$$R = \{x \mid x \text{ corresponds to a point on the number line}\}$$

The real number line is shown in Figure 1.2.

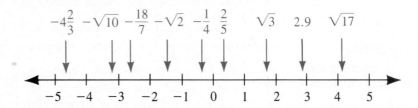

Figure 1.2
Real numbers

We use set notation to describe the relationship among the sets discussed above as shown in Figure 1.3. From now on unless we are told otherwise, we will assume that the set of real numbers serves as our basic frame of reference.

$$N \subset W \subset Z \subset Q \subset R$$

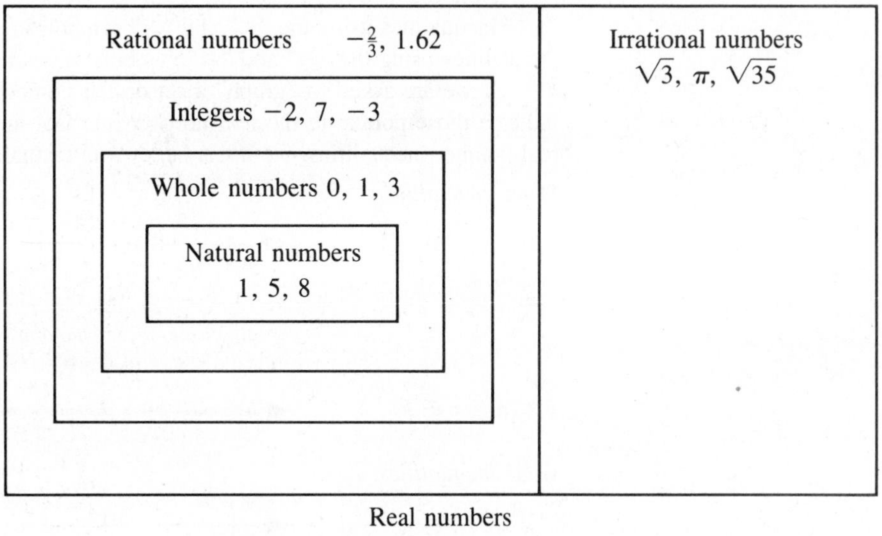

Rational numbers $-\frac{2}{3}$, 1.62

Integers −2, 7, −3

Whole numbers 0, 1, 3

Natural numbers
1, 5, 8

Irrational numbers
$\sqrt{3}$, π, $\sqrt{35}$

Real numbers

Figure 1.3
Relationships among subsets
of real numbers

Order and the Real Number Line

The number line provides us with a simple way of defining the idea of numerical "order." For example, 5 is less than 7 because 5 is to the left of 7 on the number line.

In general, we define a to be "less than" b if a is to the left of b on the number line. The symbol we use for "less than" is "$<$".

> $a < b$ means that a is to the left of b on the number line.

$2 < 6$ is the symbolic statement for "2 is less than 6," which *means* that 2 is to the left of 6 on the number line.

Similarly, the symbol ">" is used for the expression "greater than." Thus, $a > b$ means a is to the right of b on the number line. The symbols "<" and ">" are called ***inequality symbols.***

The accompanying box contains a list of all of our equality and inequality symbols, what each means, and an example of a *true* statement using each symbol.

**Equality and
Inequality Symbols**

$a = b$	a "is equal to" b	$8 + 6 = 9 + 5$
$a \neq b$	a "is not equal to" b	$4 + 5 \neq 10$
$a < b$	a "is less than" b	$3 < 8$
$a \leq b$	a "is less than *or* equal to" b	$3 \leq 8$
$a > b$	a "is greater than" b	$7 - 2 > 4$
$a \geq b$	a "is greater than *or* equal to" b	$6 \geq 6$

Note that $a \leq b$ means that either $a < b$ or $a = b$, and similarly for $a \geq b$.

Inequalities using the "<" and ">" symbols are called ***strict inequalities,*** while inequalities using the "≤" and "≥" symbols are called ***weak inequalities.***

If we are asked to "graph" a set on the number line, it means that we want to indicate those points on the real number line that are in the set. If we are graphing real-number inequalities, we use a heavy *line* (actually half-line or ray) as follows:

Strict inequalities:

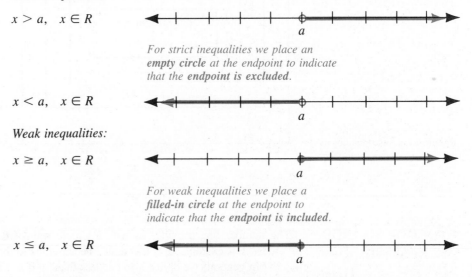

$x > a, \quad x \in R$

*For strict inequalities we place an **empty circle** at the endpoint to indicate that the **endpoint is excluded**.*

$x < a, \quad x \in R$

Weak inequalities:

$x \geq a, \quad x \in R$

*For weak inequalities we place a **filled-in circle** at the endpoint to indicate that the **endpoint is included**.*

$x \leq a, \quad x \in R$

In all four cases above, the set of concern is the real numbers. The heavy line indicates that starting with a, all points on the line are included.

EXAMPLE 1

Graph the following sets on the number line.

(a) $A = \{x \mid x > -3\}$ (b) $B = \{s \mid s \leq 4\}$

(c) $A \cap B$ (d) $C = \{a \mid a < 7, a \in N\}$

Solution

We always assume that we are dealing with the set of real numbers unless otherwise noted.

(a) $\{x \mid x > -3\}$

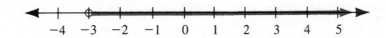

*Note the empty circle to indicate the point −3 is **excluded**.*

(b) $\{s \mid s \leq 4\}$

*Note the filled-in circle to indicate that the point 4 is **included**.*

(c) $A \cap B$ is the set consisting of the points A and B have in common. Looking at the previous graphs of A and B, we get

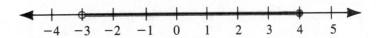

(d) $C = \{a \mid a < 7, a \in N\}$ We are limited to the *natural numbers* rather than the real numbers and we indicate this set by putting a filled-in circle at those points in the given set.

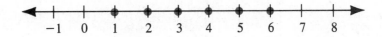

*Note that we **exclude 7** since we have a strict inequality. Also notice that we **exclude 0** since 0 is not a natural number.*   ▪

The ***double inequality***, $a < x < b$, is used to indicate "betweenness." For example,

$-3 < x < 6$ means that x is between −3 and 6 and is read "x is greater than −3 and less than 6."

We read the middle variable first, then the left-hand number, and then the right-hand number.

The double inequality is actually a combination of two inequalities that must be satisfied simultaneously. That is, $a < x < b$ is actually a combination of the two inequalities

$$a < x \qquad \text{and} \qquad x < b$$

where x satisfies **both** inequalities at the same time.

For example, $3 < x < 7$ means that $3 < x$ and $x < 7$.

Obviously, for the double inequality $a < x < b$ to make sense, a must be less than b.

Unlike single inequalities which have one endpoint (they begin at a point and go off forever in one direction), double inequalities have two endpoints.

With double inequalities we use the same convention for representing endpoints on the number line as with single inequalities: an empty circle at the endpoint for strict inequalities ("<" and ">"), and a filled-in circle for weak inequalities ("≤" and "≥"). Thus, the graph of $-2 \le x < 5$ would look like this:

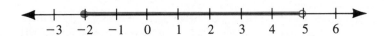

EXAMPLE 2

Graph the following double inequalities on the number line.

(a) $\{s \mid -2 \le s \le 6\}$ **(b)** $\{s \mid -2 < s < 6\}$

(c) $\{x \mid 3 \le x < 8\}$ **(d)** $\{z \mid -5 \le z < 3, z \in \mathbf{Z}\}$

Solution

(a) $\{s \mid -2 \le s \le 6\}$ *This is the set of all (real) numbers lying between -2 and 6, including -2 and 6.*

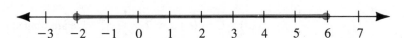

Note that both -2 and 6 are included (filled-in circles at -2 and 6).

(b) $\{s \mid -2 < s < 6\}$

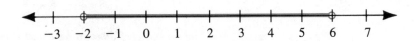

Note that both -2 and 6 are excluded (empty circles at -2 and 6).

(c) $\{x \mid 3 \le x < 8\}$

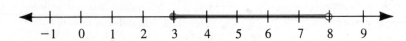

*3 is **included** and 8 is **excluded**.*

(d) $\{z \mid -5 \le z < 3, z \in Z\}$

Note that we are limited to the integers rather than the real numbers.   ■

STUDY SKILLS 1.2   **Coping with Getting Stuck**

All of us have had the frustrating experience of getting stuck on a problem; sometimes even the simple problems can give us difficulty.

Perhaps you do not know how to begin; or you are stuck halfway through an exercise and are at a loss as to how to continue; or your answer and the book's answer do not seem to match. (Do not assume the book's solutions are 100% correct—we are only human even if we are math teachers. But do be sure to check that you have copied the problem accurately.)

Assuming you have reviewed all the relevant material beforehand, be sure you have spent enough time on the problem. Some people take one look at a problem and simply give up without giving the problem much thought. This is not what we regard as "getting stuck," since it is giving up before even having started.

If you find after a reasonable amount of time,

effort, and *thought*, that you are still not getting anywhere, and if you have looked back through your notes and textbook and still have no clue as to what to do, try to find exercises similar to the one you are stuck on (with answers in the back) that you can do. Analyze what you did to arrive at the solution and try to apply those principles to the problem you are finding difficult. If you have difficulty with those similar problems as well, you may have missed something in your notes or in the textbook. Reread the material and try again. If you are still not successful, go on to different problems or take a break and come back to it later.

If you are still stuck, wait until the next day. Sometimes a good night's rest is helpful. Finally, if you are still stuck after rereading the material, see your teacher (or tutor) as soon as possible.

Exercises 1.2

In Exercises 1–18, *indicate whether the statement is true or false.*

1. $5 \in N$

2. $.3 \in N$

3. $-8 \in Z$

4. $0 \in Q$

5. $\dfrac{8}{4} \in Z$

6. $\dfrac{3}{4} \in N$

7. $\dfrac{2}{3} \notin Q$

8. $-\dfrac{4}{7} \in Q$

9. $1.8 \in Q$

10. $10 \in R$

11. $\sqrt{19} \in Q$

12. $\sqrt{19} \in R$

13. $\sqrt{9} \in Q$

14. $.474747 \in R$

15. $W \subset N$

16. $Z \subset Q$

17. $Q \subset R$

18. $R \subset Q$

*In Exercises 19–24, list as many ordering symbols as are appropriate. Choose from $<$, $>$,
$\leq$, $\geq$, $=$, and $\neq$.*

19. -3 _____ -2

20. -5 _____ -7

21. $6 \cdot 2$ _____ $6 + 2$

22. $4 + 1$ _____ $4 \cdot 1$

23. $3 \cdot 0$ _____ $3 + 0$

24. $18 - 6$ _____ $6 \cdot 2$

*In Exercises 25–48, graph the set on the number line. Unless otherwise indicated, all
numbers are real numbers.*

25. $\{x \mid x < 4\}$

26. $\{x \mid x > 4\}$

27. $\{a \mid a \leq -3\}$

28. $\{a \mid a \geq -3\}$

29. $\{b \mid b > 0\}$

30. $\{b \mid b < 0\}$

31. $\{y \mid y \geq -4\}$

32. $\{y \mid y \leq -4\}$

33. $\{c \mid c \geq 0\}$

34. $\{c \mid c \leq 0\}$

35. $\{y \mid -2 < y < 5, y \in Z\}$

36. $\{y \mid -5 < y < 2, y \in W\}$

37. $\{r \mid 5 \leq r \leq 12\}$

38. $\{r \mid -12 \leq r \leq 5\}$

39. $\{z \mid -3 < z \leq 0, z \in Z\}$

40. $\{n \mid 0 < n \leq 3, n \in N\}$

41. $\{t \mid -7 \leq t < 0\}$

42. $\{t \mid 0 \leq t < 7\}$

43. $\{x \mid -4 < x \leq 4, x \in Z\}$

44. $\{y \mid -2 \leq y < 5, y \in W\}$

45. $\{x \mid -4 < x \leq -7\}$

46. $\{y \mid -5 \leq y \leq 8\}$

47. $\{a \mid -4 < a < 4\}$

48. $\{a \mid -4 \leq a \leq 4\}$

In Exercises 49–54, let

$$A = \{x \mid x \geq -3, x \in Z\} \qquad B = \{x \mid x < -3, x \in Z\}$$

$$C = \{x \mid -4 \leq x \leq 6, x \in Z\} \qquad D = \{x \mid 1 \leq x < 9, x \in Z\}$$

Graph the set on the number line and then describe it using set notation.

49. $C \cup D$

50. $A \cap B$

51. $A \cap D$

52. $B \cap C$

53. $C \cap D$

54. $A \cap C$

In Exercises 55–62, let

$$A = \{x \mid x \geq -3\} \qquad B = \{x \mid x < -3\}$$

$$C = \{x \mid -4 \leq x \leq 6\} \qquad D = \{x \mid 1 \leq x < 9\}$$

Graph the set on the number line and then describe it using set notation.

55. $A \cup B$

56. $A \cap B$

57. $C \cap D$

58. $C \cup D$

59. $A \cup C$

60. $A \cap C$

61. $B \cap D$

62. $B \cup D$

? QUESTION FOR THOUGHT

63. The following is a method of representing an infinitely-repeating decimal as a fraction. We will show that $0.578578\overline{578}$ is a rational number.

Let $x = 0.578578\overline{578}$.   *Multiply both sides of the equation by* $1,000$.

Then $1,000x = 578.578\overline{578}$

Now subtract $x = 0.578\overline{578}$ from $1,000x = 578.578\overline{578}$:

$$1,000x = 578.578\overline{578}$$
$$-\quad x = \quad\ 0.578\overline{578}$$

Notice that the decimal portions of the numbers match up exactly.

Therefore, $999x = 578$

$$x = \frac{578}{999}$$

Our decimal $x = 0.578578\overline{578}$ is $\frac{578}{999}$ and is therefore a rational number.

Try this method of representing an infinitely-repeating decimal as a fraction for $0.674674\overline{674}$. Try it for $0.9292\overline{92}$.

1.3

Properties of Real Numbers

Real numbers, along with the operations of addition $(+)$ and multiplication $(\cdot)$ obey eleven properties listed in this section. Most of these properties are straightforward and may seem trivial. We shall see later on that these eleven basic properties are quite powerful in that they allow us to do much in simplifying algebraic expressions.

The Commutative Properties

1. *For addition* $a + b = b + a$
2. *For multiplication* $a \cdot b = b \cdot a$

Observe first that there are two commutative properties: one for each operation. Note the similarities and differences between them. Essentially, the commutative properties indicate that in adding (or multiplying) a pair of numbers, order is unimportant, as the following examples illustrate.

*Commutative property of **addition***

(a) $x + 3 = 3 + x$
(b) $(a + 8) + 4 = 4 + (a + 8)$ *Note $(a + 8)$ is interchanged with 4.*

*Commutative property of **multiplication***

(c) $(y + 4) \cdot 3 = 3 \cdot (y + 4)$

(d) $(x + 5)(x - 3) = (x - 3)(x + 5)$

The Associative Properties

3. *For addition* $a + (b + c) = (a + b) + c$

4. *For multiplication* $a \cdot (b \cdot c) = (a \cdot b) \cdot c$

For the associative properties, the order of the variables remains the same, but the grouping changes. The associative properties indicate that how you *group* terms for addition or factors for multiplication is unimportant. Consider the following examples:

*Associative property of **addition***

(a) $(x + 3) + 4 = x + (3 + 4)$ *Note that the order of the terms remains the same; only the grouping is changed.*

(b) $(x + y) + (z + q) = x + (y + z) + q$ *The associative property can be generalized to include more than one pair of grouping symbols.*

*Associative property of **multiplication***

(c) $5 \cdot (3 \cdot 4) \cdot (2x) = (5 \cdot 3 \cdot 4 \cdot 2)(x)$

It is very important to note that subtraction and division are neither commutative nor associative. For example,

$$5 - 4 \overset{?}{=} 4 - 5 \qquad (12 \div 2) \div 2 \overset{?}{=} 12 \div (2 \div 2)$$
$$1 \neq -1 \qquad\qquad 6 \div 2 \overset{?}{=} 12 \div 1$$
$$3 \neq 12$$

The associative and commutative properties involve a single operation; that is, addition and multiplication have their own associative and commutative properties. The distributive property involves both operations together, as shown in the accompanying box.

The Distributive Property

5. $a \cdot (b + c) = a \cdot b + a \cdot c$ or $(a + b) \cdot c = a \cdot c + b \cdot c$

For example,

(a) $3(x + 2) = 3 \cdot x + 3 \cdot 2$

(b) $xy(a + b + c) = xy \cdot a + xy \cdot b + xy \cdot c$ *Note that we can generalize the distributive property to more than two terms.*

(c) $4x + 8 = 4(x + 2)$ *The distributive property also yields a method for factoring as well as multiplying.*

(d) $(3x + 2)(x + 4) = (3x + 2)x + (3x + 2)4$

(e) $a(b - c) = ab - ac$ *This is another variation of the distributive property using subtraction.*

Example **(d)** illustrates that the variables in the properties above may stand for more complex expressions such as $3x + 2$, as well as a single letter or number.

$$a \quad \cdot (b + c) = \quad a \quad \cdot b + \quad a \quad \cdot c$$
$$(3x + 2) \cdot (x + 4) = (3x + 2) \cdot x + (3x + 2) \cdot 4$$

Identities

6. *For addition* There is a unique real number called the ***additive identity***, represented by 0, which has the property that

$$a + 0 = 0 + a = a \quad \text{for all } a \in R$$

7. *For multiplication* There is a unique real number called the ***multiplicative identity***, represented by 1, which has the property that

$$a \cdot 1 = 1 \cdot a = a \quad \text{for all } a \in R$$

Zero is the additive identity; in other words, *adding* 0 to any number will not change the number. The multiplicative identity is 1; thus, *multiplying* any number by 1 will not change the number. Note the similarities and differences between definitions. Each identity serves the same function with respect to its operation.

Inverses

8. *For addition* Each real number a has a unique ***additive inverse***, represented by $-a$, which has the property that

$$a + (-a) = (-a) + (a) = 0$$

9. *For multiplication* Each real number a, except 0, has a unique ***multiplicative inverse***, represented by $\frac{1}{a}$, which has the property that

$$a \cdot \frac{1}{a} = \frac{1}{a} \cdot a = 1$$

Additive inverses are often called **opposites** (or **negatives**) and multiplicative inverses are often called **reciprocals.** Again, note the similarities and differences between definitions:

The *additive* inverse of x is that number which, when *added* to x, yields the *additive* identity. The *multiplicative* inverse of x is that number which, when *multiplied* by x, yields the *multiplicative* identity.

Closure Properties	**10.** *For addition* The sum of two real numbers is a real number.
	11. *For multiplication* The product of two real numbers is a real number.

While the closure property may seem like another unimpressive, common-sense property, it plays an important part in the development of number systems. For example, we could start with whole numbers and develop a system of whole numbers that have the associative, commutative, distributive, and identity properties (there are no inverses). This system is *closed* under addition and multiplication; that is, the sum and product of two whole numbers will be whole numbers. However, *the whole number system is not closed under the operation of subtraction;* the difference between two whole numbers will not necessarily be a whole number. (For example, $7 - 9$ is not a whole number.)

Thus, we create the integer system, which is closed under subtraction (differences between integers are still integers). When we define division, however, we find that the integers are not closed under division and thus we create the rational number system.

EXAMPLE 1

If the statement is true, name the property which the statement illustrates. If the statement is not true, write "false."

(a) $(x + 7) + 8 = x + (7 + 8)$

(b) $(2 + x) \cdot 1 = 2 + x$

(c) $3 + (x \cdot y) = (3 + x) \cdot (3 + y)$

(d) $5 \cdot a + 0 = 5 \cdot a$

(e) $[3(xy)z] = [(3x)(yz)]$

(f) $y + (5 + x) = y + (x + 5)$

(g) $(x - y + z)(a + b) = (x - y + z)a + (x - y + z)b$

(h) $(x + 4) + [-(x + 4)] = 0$

Solution

(a) $(x + 7) + 8 = x + (7 + 8)$ Associative property of addition

(b) $(2 + x) \cdot 1 = 2 + x$ Multiplicative identity

(c) $3 + (x \cdot y) = (3 + x) \cdot (3 + y)$ False

(d) $5 \cdot a + 0 = 5 \cdot a$ Additive identity

(e) $[3(xy)z] = [(3x)(yz)]$ Associative property of multiplication

(f) $y + (5 + x) = y + (x + 5)$ Commutative property of addition

(g) $(x - y + z)(a + b) = (x - y + z)a + (x - y + z)b$
Distributive property

(h) $(x + 4) + [-(x + 4)] = 0$ Additive inverse property ∎

As we stated at the beginning of this section, the eleven basic properties are powerful in that many of our techniques for simplifying expressions can be derived from these properties. For example, the following theorem can be derived from the eleven properties just discussed. (A theorem is a statement of an important fact that can be proven.)

THEOREM

The Multiplication Property of Zero

For all real numbers a,

$$a \cdot 0 = 0 \cdot a = 0$$

The proof of this theorem is discussed in the Questions for Thought at the end of Exercises 1.3. We will discuss more theorems that can be derived from the properties above when we cover integers in the next section.

STUDY SKILLS 1.3 **Reviewing Old Material**

One of the most difficult aspects of learning algebra is that each skill and concept depend on those previously learned. If you have not acquired a certain skill or learned a particular concept well enough, this will more than likely affect your ability to learn the next skill or concept.

Thus, even though you have finished a topic that was particularly difficult for you, you should not breathe too big a sigh of relief. Eventually you will have to learn that topic well in order to understand subsequent topics. It is important that you try to master all skills and understand all concepts.

Whether or not you have had difficulty with a topic, you should be constantly reviewing previous material as you continue to learn new subject matter. Reviewing helps to give you a perspective on the material you have covered. It helps you to tie the different topics together and makes them *all* more

meaningful. Some statement you read 3 weeks ago, and which may have seemed very abstract then suddenly becomes simple and obvious in the light of all you now know.

Since many problems require you to draw on the skills you have developed previously, it is important for you to review so that you will not forget or confuse them. You will be surprised to find how much constant reviewing aids in learning new material.

When working the exercises, always try to work out some exercises from earlier chapters or sections. Try to include some review exercises at every study session, or at least at every other session. Take the time to reread the text material in previous chapters. When you review, think about how the material you are reviewing relates to the topic you are presently learning.

Exercises 1.3

In Exercises 1–34, if the statement is true, name the property which the statement illustrates; if the statement is false, write "false."

1. $(5 + 3) + 7 = 5 + (3 + 7)$

2. $(5 + 3) + 7 = 7 + (5 + 3)$

3. $(x + y)z = (y + x)z$

4. $(x + y)z = z(x + y)$

5. $(x + y)z = xz + yz$

6. $(xy)z = x(yz)$

7. $10 - (7 - 3) = (10 - 7) - 3$

8. $a - (b - c) = (a - b) - c$

9. $(x + 3)(x + 2) = (x + 3)x + (x + 3)2$

10. $(x + 3)(x + 2) = x(x + 2) + 3(x + 2)$

11. $(a + b) \cdot 1 = a + b$

12. $(a + b) - (a + b) = 0$

13. $(a + b) \cdot 0 = a + b$

14. $(a + b) + 0 = a + b$

15. $x[(x + 3)(x + 2)] = [x(x + 3)](x + 2)$

16. $x\left(\dfrac{1}{x}\right) = 1 \qquad (x \neq 0)$

17. $36 \cdot (12 \cdot 3) = (36 \cdot 12) \cdot 3$

18. $(m + 3)\left(\dfrac{1}{m + 3}\right) = 1 \qquad (m \neq -3)$

19. $3 \cdot x + y = 3x + 3y$

20. $3(xy) = (3x)(3y)$

21. $(r + s) + [-(r + s)] = 0$

22. $(a + b)[-(a + b)] = 0$

23. $(x - y) + (z + 3) + a = x - (y + z) + (3 + a)$

24. $(x - y)(z + 3)(a) = (z + 3)(x - y)(a)$

25. $(a + b)(e + f) = (a + b)(f + e)$

26. $abef = abfe$

27. $ef(a + b) = efa + efb$

28. $e(f)(a + b) = e(fa + fb)$

29. $2x + 3x = (2 + 3)x$

30. $(a + b) + (-a + b) = 0$

31. The product of two integers is an integer.

32. The difference of two whole numbers is a whole number.

33. The quotient of two integers is an integer.

34. The difference of two integers is an integer.

? QUESTIONS FOR THOUGHT

35. Are both forms of the distributive property necessary? Could we derive one form from the other by using the other properties?

36. Is there a commutative or associative property of subtraction? Explain your answer.

37. Is there a commutative or associative property of division? Explain your answer.

38. Supply the reason for each step in the following proof for the multiplication property of zero. (We have already supplied the property for step 3, which will be discussed in Chapter 2.)

1. $\qquad a \cdot 0 = a \cdot (0 + 0)$

2. $\qquad a \cdot 0 = a \cdot 0 + a \cdot 0$

3. $-(a \cdot 0) + a \cdot 0 = -(a \cdot 0) + (a \cdot 0 + a \cdot 0)$ Addition property of equality

4. $\qquad 0 = -(a \cdot 0) + (a \cdot 0 + a \cdot 0)$

5. $\qquad 0 = [-(a \cdot 0) + (a \cdot 0)] + a \cdot 0$

6. $\qquad 0 = 0 + a \cdot 0$

7. $\qquad 0 = a \cdot 0$

1.4

Operations with Real Numbers

In Section 1.3 we discussed the property of additive inverses. We mentioned that the additive inverse of a number is also called the opposite or negative of the number. We represent an additive inverse by putting a negative sign before a number (hence the term *negative*). Thus, $-x$ is the additive inverse of x; by definition, it is the number which, when added to x, will yield 0.

$$-3 \quad \text{is the additive inverse of 3} \quad \text{because} \quad (-3) + 3 = 0$$
$$6 \quad \text{is the additive inverse of } -6 \quad \text{because} \quad 6 + (-6) = 0$$

In the last example, note that 6 is the additive inverse of -6. This illustrates that the additive inverse of a number is not necessarily negative. *The additive inverse of a number yields a number opposite in sign* (hence the term *opposite*). Symbolically we have

THEOREM

$$-(-x) = x$$

The following theorem can be derived from the real number properties and will be useful to us later on.

THEOREM

$$(-1)x = -x$$

Absolute Value

Geometrically, the **absolute value** of a number is its distance from 0 on the number line. The absolute value of x is symbolized as follows: $|x|$. Hence,

$$|-4| = 4 \quad \text{since } -4 \text{ is 4 units away from 0 on the number line.}$$
$$|4| = 4 \quad \text{since } 4 \text{ is 4 units away from 0 on the number line.}$$

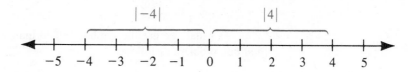

Algebraically, we define absolute value as follows:

$$|x| = \begin{cases} x & \text{if } x \geq 0 \\ -x & \text{if } x < 0 \end{cases}$$

If x is positive or 0, it is left alone. If x is less than 0, or negative, then the absolute value of x is $-x$, which is positive. Consequently, $|x|$ *can never be negative.* For example, by the algebraic definition of absolute value, we have

$$|-8| = -(-8) \quad \textit{Since } -8 < 0$$
$$ = 8 \quad \textit{By theorem above: } -(-x) = x$$
$$|0| = 0 \quad \textit{Since } 0 \geq 0$$
$$|6| = 6 \quad \textit{Since } 6 \geq 0$$

We introduce the concept of absolute values here so that we may refer back to whole numbers in defining operations with signed numbers.

Addition of Real Numbers

Recall from elementary algebra that we defined addition of signed numbers as moving a specified distance either right (if positive) or left (if negative) on the number line. Hence, $(-8) + (+3)$ can be represented on the number line as follows:

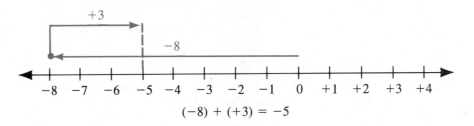

$$(-8) + (+3) = -5$$

As a result, we developed the following rules for adding signed numbers:

Addition of Signed Numbers	1. When adding two numbers with the same sign, add their absolute values and keep their common sign.
	2. When adding two numbers with opposite signs, subtract the smaller absolute value from the larger, and keep the sign of the number with the larger absolute value.

As usual, we use the symbol "+" to indicate addition, and would represent an addition problem as follows:

$(-4) + (+3)$ *Find the sum of -4 and $+3$.*

EXAMPLE 1   Find the following sums.

(a) $(-6) + (-12)$ **(b)** $(+7) + (-15)$

(c) $(+9) + (-3)$ **(d)** $(-4) + (+8) + (-2) + (-12)$

Solution

(a) $(-6) + (-12)$

$$= -(6 + 12) = \boxed{-18}$$
$\uparrow$
and keep the negative sign.

Both numbers have the same signs. Since $|-6| = 6$ and $|-12| = 12$, we add 6 and 12 to get 18

(b) $(+7) + (-15)$

$$= -(15 - 7) = \boxed{-8}$$
$\uparrow$
and keep the negative sign since $|-15| > |+7|$.

The two numbers have opposite signs. Since $|+7| = 7$ and $|-15| = 15$, we find the difference in absolute values

(c) $(+9) + (-3) = +(9 - 3) = \boxed{+6}$

(d) $(-4) + (+8) + (-2) + (-12)$

$$= (+4) + (-2) + (-12)$$

$$= (+2) + (-12) = \boxed{-10}$$

Actually, we could derive the rules for adding signed numbers from the eleven properties given in Section 1.3. For example, to add -4 and -6:

$(-4) + (-6) = (-1)4 + (-1)6$	*By theorem: $-x = (-1)x$*
$= -1(4 + 6)$	*Distributive property*
$= -1(10)$	*Addition of whole numbers*
$= -10$	*Again, since $-1(x) = -x$*

Of course, the rules for addition were developed to shortcut this process and should be used instead. The above steps do demonstrate the power of the eleven real number properties in that they can be used to perform addition of signed numbers.

Subtraction

Just as with whole numbers, subtraction is defined in terms of addition. To find $a - b$, you are looking for a number which, when added to b, will yield a.

We define subtraction as follows:

$$\boxed{a - b = a + (-b)}$$

Thus, if we add $a - b$ to b we get $(a - b) + b = a + (-b) + b = a$, as required. Hence, subtracting b *is* adding the additive inverse of b. Since the additive inverse changes the sign of a number, **to subtract b we change the sign of b and add.**

EXAMPLE 2

Perform the indicated operations.

(a) $(-6) - (+4)$ **(b)** $(+7) - (-2)$

(c) $(-13) - (-7)$ **(d)** $(-17) + (-8) - (-4) - (-9)$

(e) $-6 + 3 - 5 - 4 + 7$

Solution

(a) $(-6) - (+4) = (-6) + (-4)$ *Change sign of +4 and add.*

$$= \boxed{-10}$$

(b) $(+7) - (-2) = (+7) + (+2)$ *Change −2 to + 2 and add.*

$$= \boxed{+9}$$

(c) $(-13) - (-7) = (-13) + (+7)$ *Change −7 to +7 and add.*

$$= \boxed{-6}$$

(d) $(-17) + (-8) - (-4) - (-9)$

$$= (-17) + (-8) + (+4) + (+9)$$

$$= \boxed{-12}$$

Here we would change all subtraction to addition (and change the signs of the numbers being subtracted) and add.

(e) $-6 + 3 - 5 - 4 + 7$

We often drop the addition symbol between signed numbers we are adding. In this case addition is understood: we are adding (-6), $(+3)$, (-5), (-4), and $(+7)$.

$$= -6 + 3 - 5 - 4 + 7$$

$$= -3 - 5 - 4 + 7 \qquad Since\ (-6) + (+3) = -3$$

$$= -8 - 4 + 7 \qquad Since\ (-3) + (-5) = -8$$

$$= -12 + 7 \qquad Since\ (-8) + (-4) = -12$$

$$= \boxed{-5}$$

Multiplication

In elementary algebra, we derived the rules for multiplying signed numbers by viewing multiplication as repeated addition. Thus, for example, $3(-2) = (-2) + (-2) + (-2) = -6$; hence, the product of a positive and a negative number is negative.

Rules for Multiplying Two Signed Numbers	**1.** If the signs are the same, their product is positive. **2.** If the signs are opposite, their product is negative.

Again, we can use the properties to derive these rules (see the Questions for Thought at the end of Exercises 1.4).

In general, we can derive the following theorems:

Theorems	**1.** $\quad (-a)b = a(-b)$
	2. $\quad a(-b) = -(ab)$
	3. $\quad (-a)(-b) = ab$

EXAMPLE 3

Perform the operations.

(a) $\quad (-3)(+4)$ **(b)** $\quad (-6)(-9)$ **(c)** $\quad (-7)(+2)(-3)(-1)$

Solution

(a) $\quad (-3)(+4) = \boxed{-12}$

(b) $\quad (-6)(-9) = \boxed{+54}$

(c) $\quad (-7)(+2)(-3)(-1) = (-14)(-3)(-1)$

$$= (+42)(-1) = \boxed{-42} \qquad \blacksquare$$

When multiplying more than two signed numbers together, we can figure out the sign of the product quickly as follows:

If there is an even number of negative factors in a *product*, the product will be positive; otherwise, the product will be negative.

For example, $(-2)(-3)(+1)(-2)(-4)$ will be *positive* 48 since there are four (an even number of) negative factors in the product.

Division

Division is defined in terms of multiplication. To find $\dfrac{a}{b}$, we are looking for the number which, when multiplied by b, will yield a. For example, $\dfrac{6}{3} = 2$ because $2 \cdot 3 = 6$.

Just as we defined subtraction as adding the additive inverse, we define division as multiplication by the multiplicative inverse:

$$\boxed{\dfrac{a}{b} = a\left(\dfrac{1}{b}\right)}$$

As a result, the sign rules for division are the same as for multiplication.

| **Rules for Division of One Signed Number by Another** | **1.** If the signs are the same, the quotient is positive. |
| | **2.** If the signs are different, the quotient is negative. |

What happens if we try to divide a nonzero number by 0? Let's assume $\frac{6}{0} = x$; then this means $x \cdot 0 = 6$. But multiplication by 0 will only yield 0. We end up with $0 = 6$, which is of course false.

Thus, for any *nonzero* number a, $\frac{a}{0}$ will yield the contradiction $a = 0$. Therefore, division by 0 is *undefined*. On the other hand, $\frac{0}{a} = 0$ since $0 \cdot a = 0$.

If we try to divide 0 by 0, we run into problems for another reason. Let's suppose we let $\frac{0}{0}$ be equal to some number, r. Then this means $r \cdot 0 = 0$. But this is true for all numbers r. This means that any number will work; that is, $\frac{0}{0}$ is not unique. Therefore, we say that $\frac{0}{0}$ is *indeterminate* (there is no unique answer).

If a is any nonzero number, then

$\frac{a}{0}$ is undefined

$\frac{0}{a} = 0$

$\frac{0}{0}$ is indeterminate

EXAMPLE 4

Perform the operations.

(a) $(-6) \div (-3)$ **(b)** $\frac{-21}{+7}$ **(c)** $\frac{-5}{0}$

Solution

(a) $(-6) \div (-3) = \boxed{+2}$

(b) $\frac{-21}{+7} = \boxed{-3}$

(c) $\frac{-5}{0}$ is $\boxed{\text{undefined}}$

Exponents

Multiplication by a positive integer is shorthand for repeated addition (for example, $5 + 5 + 5 + 5 = 4 \cdot 5$). Similarly, exponential notation is shorthand for repeated multiplication. For example, $5 \cdot 5 \cdot 5 \cdot 5 = 5^4$. We make the definition given in the box.

Exponential Notation

$$x^n = x \cdot x \cdot x \cdot x \cdot \cdots \cdot x \qquad \text{where the factor } x \text{ occurs } n \text{ times.}$$

In x^n, x is called the **base** and n is called the **exponent.** A natural number exponent tells how many times x occurs as a factor in the product. The expression x^n is called a **power,** the nth power of x.

Note that $x = x^1$, that is, a missing exponent is assumed to be 1.

To compute or evaluate an expression with a numerical base means to multiply out the expression.

EXAMPLE 5

Write the following without exponents and then evaluate.

(a) 3^5 **(b)** $(-2)^4$

Solution

(a) $3^5 = 3 \cdot 3 \cdot 3 \cdot 3 \cdot 3$ *Written without exponents*

$= \boxed{243}$ *Evaluated*

(b) $(-2)^4 = (-2)(-2)(-2)(-2)$ *Written without exponents*

$= \boxed{+16}$ *Evaluated*

■

Multiple Operations

If there is more than one operation to perform in an expression, there must be agreement as to which operation should be performed first: an order of operations. For example, given the expression $3 + 2 \cdot 4$, if addition were performed first, the value would be 20; if multiplication were performed first, the value would be 11. To rectify this ambiguity, the order of operations given in the next box is agreed upon.

Order of Operations

1. Start by performing operations in the grouping symbols, innermost first.
2. Calculate powers (and roots) in any order.
3. Perform multiplication and division, working left to right.
4. Perform addition and subtraction, working left to right.

EXAMPLE 6

Perform the operations.

(a) $38 - 5^2$ **(b)** $7^2 + 3^2 - (2 + 3)^2$ **(c)** $3 + 7 - 2[5 - 3(9 - 8)]$

Solution

(a) $38 - 5^2$ *Powers first*

     $= 38 - 25$ *You are subtracting 5^2.*

     $=$ $\boxed{13}$

(b) $7^2 + 3^2 - (2 + 3)^2$ *Addition in parentheses*

     $= 7^2 + 3^2 - (5)^2$ *Then powers*

     $= 49 + 9 - 25$ *Next we add.*

     $= 58 - 25$

     $=$ $\boxed{33}$

(c) $3 + 7 - 2[5 - 3(9 - 8)]$ *Innermost grouping symbol first*

     $= 3 + 7 - 2[5 - 3(1)]$ *Then multiplication in brackets*

     $= 3 + 7 - 2[5 - 3]$ *Then subtraction in brackets*

     $= 3 + 7 - 2[2]$ *Next we multiply.*

     $= 3 + 7 - 4$

     $=$ $\boxed{6}$

           ■

Multiple operations with integers can be confusing. For example,

      $(-3)(-2)$ is a multiplication problem.

      $(-3) - (2)$ is a subtraction problem.

Keep in mind that grouping symbols do not represent operations in themselves. It is the symbol between *pairs* of grouping symbols that indicates the operation to be performed (the lack of a symbol between pairs of grouping symbols indicates multiplication). This may seem obvious to you as presented above in its simplified form, but is not as obvious in an expression such as

$$7 - (-7)(-7) = 7 - (49) = -42$$

Two other expressions which are distinct and yet frequently confused concern integers raised to powers. For example,

$$-2^4 = -(2 \cdot 2 \cdot 2 \cdot 2) = -16 \quad \text{and} \quad (-2)^4 = (-2)(-2)(-2)(-2) = +16$$

Keep in mind that the exponent applies only to that which is to its immediate left. In the expression $3 \cdot 2^4$, we raise 2 to the fourth power and then multiply by 3 to get 48. Similarly, in -2^4, we raise 2 to the fourth power and then bring in the minus sign to get -16.

EXAMPLE 7

Perform the operations.

(a) $\left(-\dfrac{3}{4}\right)(-8) - (-5)(-1)$

(b) $-6(-3 - 2) - 3^2$

(c) $-6(-3 - 2)(-3)^2$

(d) $\dfrac{4[-5 - 3(-2)^2]}{-6 - 7(-3 - 2) + 5}$

(e) $|-5 - 7| - |7 - 5|$

(f) $-9 - \{6 - 2[12 - (8 - 15)] - 4\}$

Solution

(a) $\left(-\dfrac{3}{4}\right)(-8) - (-5)(-1)$ *Multiplication*

$= (+6) - (+5)$ *Then subtraction*

$= 6 - 5$

$= \boxed{1}$

(b) $-6(-3 - 2) - 3^2$ *Operations in parentheses first*

$= -6(-5) - 3^2$ *Then powers*

$= -6(-5) - 9$ *Next multiplication*

$= 30 - 9$ *Subtraction*

$= \boxed{21}$

(c) $-6(-3 - 2)(-3)^2$ *Parentheses first*

$= -6(-5)(-3)^2$ *Then powers*

$= -6(-5)(+9)$ *Multiply*

$= \boxed{270}$

Note the differences between parts **(b)** *and* **(c).**

(d) $\dfrac{4[-5 - 3(-2)^2]}{-6 - 7(-3 - 2) + 5} = \dfrac{4[-5 - 3(+4)]}{-6 - 7(-5) + 5}$

$= \dfrac{4[-5 - 12]}{-6 + 35 + 5}$

$= \dfrac{4(-17)}{34}$

$= \dfrac{-68}{34}$

$= \boxed{-2}$

(e) $|-5 - 7| - |7 - 5|$ *Perform operations in absolute value symbols first.*

$= |-12| - |2|$ *Then find absolute values.*

$= 12 - 2$ *Then subtract.*

$= \boxed{10}$

(f) $-9 - \{6 - 2[12 - (8 - 15)] - 4\} = -9 - \{6 - 2[12 - (-7)] - 4\}$

$= -9 - \{6 - 2[12 + 7] - 4\}$

$= -9 - \{6 - 2[19] - 4\}$

$= -9 - \{6 - 38 - 4\}$

$= -9 - \{-36\}$

$= -9 + 36$

$= \boxed{27}$ ∎

Substitution

Applications of algebra very often require us to substitute numerical values for variables and then to perform the operations with the substituted values. In such a situation we will be asked to evaluate an expression given the values for the variables.

EXAMPLE 8

Given $a = -3$, $b = -2$, $c = 4$, and $d = 0$, evaluate the following.

(a) $a^2 - 2ab + b^2$ **(b)** $|a - b| - |a|$ **(c)** $\dfrac{(4a^2 + 3c)d}{6}$

Solution

It is usually a good idea to enclose the values being substituted in parentheses. This helps us to avoid confusing the original operations.

(a) $a^2 - 2ab + b^2 = a^2 - 2 \cdot a \cdot b + b^2$

$= (-3)^2 - 2 \cdot (-3) \cdot (-2) + (-2)^2$

$= 9 - 2(6) + 4$

$= 9 - 12 + 4$

$= \boxed{1}$

(b) $|a - b| - |a| = |(-3) - (-2)| - |(-3)|$ *We perform the operations in abso-*
*lute values **first**; then take the abso-*
$= |-3 + 2| - |-3|$ *lute value of the result.*

$= |-1| - |-3|$

$= 1 - 3$

$= \boxed{-2}$

(c) $\dfrac{(4a^2 + 3c)d}{6} = \dfrac{[4 \cdot a^2 + 3 \cdot c] \cdot d}{6}$

$= \dfrac{[4(-3)^2 + 3(4)](0)}{6}$ *Multiplication by 0 yields 0.*

$= \dfrac{0}{6}$

$= \boxed{0}$ ∎

Exercises 1.4

In Exercises 1–78, evaluate each of the expressions, if possible.

1. $-3 + 8$

2. $-8 + 3$

3. $-3 - 8$

4. $8 - (-3)$

5. $-3(-8)$

6. $-8(3)$

7. $-3 - 4 - 5$

8. $7 - 5 - 6$

9. $-3 - 4(5)$

10. $7 - 5(-6)$

11. $-3(-4)(-5)$

12. $7(-5)(-6)$

13. $3(-4 - 5)$

14. $7(-5 - 6)$

15. $8 - 4 \cdot 3 - 7$

16. $(9 - 6)(2 - 5)$

17. $8 - (4 \cdot 3 - 7)$

18. $9 - (6 \cdot 2 - 5)$

19. $(8 - 4)(3 - 7)$

20. $9 - 6(2 - 5)$

21. $\dfrac{-20}{-5}$

22. $\dfrac{-28}{7}$

23. $\dfrac{-5 - 11}{-9 + 4}$

24. $\dfrac{-8 - 7}{-2 - 7}$

25. $\dfrac{-10 - 2 - 4}{-2}$

26. $\dfrac{-9 - 3 - 6}{-3}$

27. $\dfrac{-10 - (2 - 4)}{-2}$

28. $\dfrac{-9 - (3 - 6)}{-3}$

29. $\dfrac{-10 - 2(-4)}{-2}$

30. $\dfrac{-9 - 3(-6)}{-3}$

31. $\dfrac{-4(-3)(-6)}{-4(-3) - 6}$

32. $\dfrac{-2(-4)(-6)}{-2(-4) - 6}$

33. $8 - 3(5 - 1)$

34. $9 - 5(4 - 1)$

35. $-7 - 2(4 - 6)$

36. $-6 - 3(2 - 5)$

37. $7 + 2[4 + 3(4 + 1)]$

38. $6 + 3[1 + 2(3 + 1)]$

39. $7 - 2[4 - 3(4 - 1)]$

40. $6 - 3[1 - 2(3 - 1)]$

41. $8 - \left(\dfrac{10}{-5}\right)$

42. $-6 - \left(\dfrac{-8}{-4}\right)$

43. $8\left(\dfrac{10}{-5}\right)$

44. $-6\left(\dfrac{-8}{-4}\right)$

45. $\dfrac{12}{-4} - \left(\dfrac{10}{-2}\right)$

46. $\dfrac{18}{-3} - \left(\dfrac{12}{-6}\right)$

47. $\dfrac{-8 + 2}{4 - 6} - \dfrac{6 - 11}{-3 - 2}$

48. $\dfrac{13 - (-5)}{-3 - 3} - \dfrac{13 - (-5)}{-3(-3)}$

49. $-12 - \dfrac{6 - 2(-3)}{-3}$

50. $18 + \dfrac{7 - 4(-2)}{-5}$

51. $(-6)^2$

52. -6^2

53. $2^2 + 3^2 + 4^2$

54. $(2 + 3 + 4)^2$

55. $2(5)^2$

56. $(2 \cdot 5)^2$

57. -3^4

58. $(-3)^4$

59. $-8 - 2(-4)^2$

60. $10 - 3(2)^3$

61. $2(-5)(-6)^2$

62. $3(-4)(-5)^2$

63. $2(-5) - 6^2$

64. $3(-4) - 5^2$

65. $\dfrac{5[-8 - 3(-2)^2]}{-6 - 6 - 2}$

66. $\dfrac{5[-12 - 2(-3)^2]}{-4 - 6 - 2}$

67. $-3 - 2[-4 - 3(-2 - 1)]$

68. $-5 - 4[3 - 4(-3 - 2)]$

69. $-3\{5 - 3[2 - 6(3 - 5)]\}$

70. $-5\{6 - 2[3 - 2(-2 - 2)]\}$

71. $|3 - 8| - |3| - |-8|$

72. $|2 - 9| - |-2| - |9|$

73. $|-3 - 2 - 4| - (2 - 3)^3$

74. $|-4 - 5 - 1| - (5 - 6)^3$

75. $(1 - 4)^2 - (4 - 1)^2$

76. $(2 - 6)^2 - (6 - 2)^2$

77. $(-3 - 2)^2 - (-3 + 2)^2$

78. $(-5 - 4)^2 - (-5 + 1)^2$

In Exercises 79–88, evaluate the expressions for $x = -2$, $y = -3$, and $z = 5$.

79. $x + y + z$

80. $x(y + z)$

81. xyz

82. $xy + z$

83. $-x^2 - 4x + 2$

84. $-z^2 + 3z + 1$

85. $|xy - z|$

86. $|x - y - z|$

87. $\dfrac{3x^2y - x^3y^2}{3x - 2y}$

88. $\dfrac{3x^2y + x^2y^2}{3x + 2y}$

🔲 CALCULATOR EXERCISES

In Exercises 89–94, evaluate each of the expressions.

89. $1.692 - 3.965 + 8.754$

90. $-8.236 - 12.257 + 4.3$

91. $(2.6)^2 - |7.8 - 13.69|$

92. $|-2.94 - 3.8| + (3.9)^2$

93. $\dfrac{42.2}{1.63 - (2.1)(5.8)}$

94. $\dfrac{-9.6}{8.49 - (2.3)^2}$

? QUESTION FOR THOUGHT

95. Supply the reason for each of the following steps:

$$(-3)(4) = [(-1)(3)](4)$$
$$= (-1)[(3)(4)]$$
$$= (-1)[12]$$
$$= -12$$

1.5

Algebraic Expressions

A *variable* is a symbol that stands for a number (or numbers). In algebra, there are primarily two ways variables are used. One use of a variable is to hold the place of a particular number (or numbers) which has not yet been identified but which needs to be found. Equations are examples of this type of variable use.

A second use of variables is to describe a general relationship between numbers and/or arithmetic operations, such as in the commutative property:

$$a + b = b + a$$

While variables represent unknown quantities, a *constant,* on the other hand, is a symbol whose value is fixed, such as 2, -6, 8.33, or π.

DEFINITION

An *algebraic expression* is an expression consisting of constants, variables, grouping symbols, and symbols of operations put together according to the rules of algebra.

Our goal in this section is to review how to simplify algebraic expressions. That is, given a basic set of guidelines and the real number properties, we will take algebraic expressions and change them into simpler equivalent expressions. (We will postpone working with roots until Chapter 6.) By *equivalent expressions,* we mean expressions that represent the same number for all valid replacements of the variables in the expression. For example, the expression $3x + 7x$ is equivalent to the expression $10x$ because, when we substitute any number for x, the computed value of $3x + 7x$ will be the same as the computed value of $10x$.

Products

Recall the definition of exponential notation:

$$x^n = x \cdot x \cdot \cdots \cdot x \qquad \textit{where the factor x occurs n times}$$

The natural number n is called the *exponent*; x is called the *base*, and x^n is called the *power*. Thus,

$$5a^4b^2 = 5aaaabb \qquad (-3x)^3 = (-3x)(-3x)(-3x)$$

DEFINITION

The numerical factor of a term is called the *numerical coefficient* or simply the *coefficient*.

For example,

The coefficient of $3x^2y$ is 3.

The coefficient of $-5a^4b^2$ is -5.

The coefficient of z^3y is understood to be 1.

We will call the nonnumerical factors in a term the *literal part*.

For example, the literal part of $3x^2y$ is x^2y.

When we multiply two powers with the same base, such as $x^5 \cdot x^4$, we can write out the following:

$$x^5 \cdot x^4 = (x \cdot x \cdot x \cdot x \cdot x)(x \cdot x \cdot x \cdot x) \qquad \textit{Write } x^5 \textit{ and } x^4 \textit{ without exponents.}$$
$$= x \cdot x \cdot x \cdot x \cdot x \cdot x \cdot x \cdot x \cdot x \qquad \textit{Then count the x's to get}$$
$$= x^9$$

We can see that multiplying two powers with the same base is a matter of counting up the number of times x appears as a factor. This gives us the first rule of exponents.

The First Rule of Exponents	$x^n x^m = x^{n+m}$

Thus, to *multiply* two powers of the *same base,* we simply keep the base and *add* the exponents.

The first rule can be generalized to include more than two expressions. For example,

$$x^p x^q x^r x^m = x^{p+q+r+m}$$

When asked to *simplify* an expression involving exponents, we should write the expression with bases and exponents occurring as few times as possible.

EXAMPLE 1

Simplify each of the following.

(a) $x^2 x^7 x$ **(b)** $(-2)^3(-2)^2$

Solution

(a) $x^2x^7x = x^{2+7+1}$ *Remember that $x = x^1$.*

   $= \boxed{x^{10}}$

(b) $(-2)^3(-2)^2 = (-2)^{3+2}$

   $= \boxed{(-2)^5}$ *Evaluated, this answer is* $\boxed{-32}$ ∎

To find a product such as $(3x^3y)(-4xy^2)$, we *could* proceed as follows:

$(3x^3y)(-4xy^2)$

$= (3)[(x^3y)(-4x)](y^2)$ *Associative property of multiplication*
   (Notice that only the grouping symbols were changed.)

$= (3)[(-4x)(x^3y)](y^2)$ *Commutative property of multiplication*

$= (3)(-4)(x \cdot x^3)(y \cdot y^2)$ *Associative property of multiplication*

$= -12x^4y^3$ *Multiplication and the first rule of exponents*

We stated above that we *could* proceed in this way; however, we usually do *not* proceed in this manner unless we need to *prove* that $(3x^3y)(-4xy^2)$ is equal to $-12x^4y^3$.

Actually, to find a *product* such as $(3x^3y)(-4xy^2)$, we can proceed by ignoring the original order and grouping of the variables and constants. Therefore, we usually multiply the coefficients first; then we multiply the powers of each variable using the first rule of exponents.

$(3x^3y)(-4xy^2)$ *Reorder and regroup.*

$= 3(-4)x^3 \cdot x \cdot y \cdot y^2$ *Multiply.*

$= -12x^4y^3$

The statement that "we can ignore the original order and grouping of the variables and constants" is, in fact, a restatement of the commutative and associative properties of multiplication.

EXAMPLE 2

Multiply the following.

(a) $(-7a^2b^3c^2)(-2a^3b^4)(-3ac^5)$ (b) $-(3x)^4$ (c) $(-3x)^4$

Solution

(a) $(-7a^2b^3c^2)(-2a^3b^4)(-3ac^5)$ *First we reorder and regroup.*

   $= (-7)(-2)(-3)(a^2a^3a)(b^3b^4)(c^2c^5)$ *Then multiply.*

   $= \boxed{-42a^6b^7c^7}$

(b) $-(3x)^4 = -(3x)(3x)(3x)(3x)$

   $= -(3 \cdot 3 \cdot 3 \cdot 3 \cdot x \cdot x \cdot x \cdot x)$

   $= \boxed{-81x^4}$

(c) $(-3x)^4 = (-3x)(-3x)(-3x)(-3x)$

$\qquad\quad = (-3)(-3)(-3)(-3)x \cdot x \cdot x \cdot x$

$\qquad\quad = \boxed{+81x^4}$

Note the difference between parts **(b)** *and* **(c)**.   ■

Combining Terms

We discussed the product of expressions; now we will discuss how we *combine* or add and subtract terms. You probably remember from elementary algebra that you can add "*like terms*," terms with identical literal parts, but you cannot add "*unlike terms*." What allows us to add like terms is the distributive property:

$$ba + ca = (b + c)a$$

For example,

$5x + 7x = (5 + 7)x$ *Distributive property*

$\qquad\;\; = 12x$

$8ab - 17ab + 4ab = (8 - 17 + 4)ab$ *Distributive property*

$\qquad\qquad\qquad\;\; = -5ab$

EXAMPLE 3

Combine the following.

(a) $2x^2 - 7x^2$ **(b)** $5x - 2y - 4x - 5y$ **(c)** $4xy^2 - 4y^2x + 3x^2y$

Solution

We will use the distributive property to demonstrate how it is used in combining terms. This step, however, should be done mentally.

(a) $2x^2 - 7x^2 = (2 - 7)x^2$ *This step is usually done mentally.*

$\qquad\qquad\;\; = \boxed{-5x^2}$

(b) $5x - 2y - 4x - 5y = (5 - 4)x + (-2 - 5)y$ *Note that we combine only "like" terms.*

$\qquad\qquad\qquad\qquad\;\; = 1x - 7y$

$\qquad\qquad\qquad\qquad\;\; = \boxed{x - 7y}$ *A coefficient of 1 is understood.*

(c) $4xy^2 - 4y^2x + 3x^2y = (4 - 4)xy^2 + 3x^2y$ *Note that $xy^2 = y^2x$ by the commutative property and they are therefore "like" terms.*

$\qquad\qquad\qquad\qquad\;\; = 0xy^2 + 3x^2y$

$\qquad\qquad\qquad\qquad\;\; = \boxed{3x^2y}$ *Since $0xy^2 = 0$*   ■

In summary, to combine "like" terms, add their numerical coefficients.

Removing Grouping Symbols

If we wanted to compute a numerical expression such as $7(9 + 4)$, we would evaluate this as follows:

$$7(9 + 4) = 7(13) = 91$$

We perform the operations within parentheses first.

However, in multiplying algebraic expressions with variables, as in the product $3x(x + 4)$, we cannot simplify within the parentheses. Instead, we multiply using the distributive property:

$$(a + b)c = ac + bc \qquad \text{or} \qquad a(b + c) = ab + ac \cdot$$

Verbally stated, multiplication distributes over addition.

Multiplication distributes over subtraction as well:

$$a(b - c) = ab - ac$$

For example,

$$7(9 + 4) = 7 \cdot 9 + 7 \cdot 4 = 63 + 28 = 91 \qquad \text{\textit{Note we still arrive at the same answer as } 7(13).}$$

$$3x(x + 4) = (3x)(x) + (3x)(4) = 3x^2 + 12x$$

$$5(x - y - 3) = 5x - 5y + 5(-3) = 5x - 5y - 15$$

$$-2(x + y - 3) = (-2)x + (-2)y + (-2)(-3) = -2x - 2y + 6$$

When a negative sign immediately precedes a grouping symbol as in

$$-(x + 4 - 3y)$$

we may interpret this as subtracting the *quantity* $x + 4 - 3y$ from zero or as the negative of the quantity $x + 4 - 3y$. In the previous section, we found that we can rewrite $-a$ as $(-1)a$. We do the same to $-(x + 4 - 3y)$ in order to remove the grouping symbol:

$$-(x + 4 - 3y) = (-1)(x + 4 - 3y) \qquad \text{\textit{Since } } -a = (-1)a$$
$$\text{\textit{Then we use the distributive property.}}$$
$$= (-1)(x) + (-1)(4) + (-1)(-3y)$$
$$= -x - 4 + 3y$$

Thus, we can also interpret $-(x + 4 - 3y)$ as multiplying the quantity $x + 4 - 3y$ by -1.

When subtracting a quantity within a pair of grouping symbols, change the sign of each *term* within the grouping symbols.

If a positive sign precedes a grouping symbol as in

$$+(x + 4 - 7y)$$

we interpret it as $+1(x + 4 - 7y)$. Hence,

$$+(x + 4 - 7y) = +1(x + 4 - 7y)$$
$$= (+1)x + (+1)(+4) + (+1)(-7y)$$
$$= x + 4 - 7y$$

Note that the signs of the terms within the grouping symbols remain unchanged.

EXAMPLE 4

Perform the following operations.

(a) $-5(2x - 3y + 5)$ **(b)** $-3x(-a + b)$

(c) $-3x - (a + b)$ **(d)** $-2x - [y + (3 - a)]$

Solution

(a) $-5(2x - 3y + 5) = -5(2x) - 5(-3y) - 5(5)$ *Distributive property*

$$= \boxed{-10x + 15y - 25}$$

(b) $-3x(-a + b) = (-3x)(-a) - 3x(b)$ *Distributive property*

$$= \boxed{3ax - 3bx}$$

(We prefer to write factors of terms in alphabetical order; hence, we write 3ax rather than 3xa.)

(c) $-3x - (a + b) = \boxed{-3x - a - b}$ *Since you are subtracting $a + b$ change the sign of each term.*

*Compare parts **(b)** and **(c)** of this example.*

(d) $-2x - [y + (3 - a)]$ *Follow the order of operations.*

$$= -2x - [y + 3 - a]$$ *Work inside the brackets; remove parentheses.*

$$= \boxed{-2x - y - 3 + a}$$ *In subtracting $y + 3 - a$, change the sign of each term.* ∎

At this point we will look at examples requiring us to remove grouping symbols and then combine terms where possible.

EXAMPLE 5

Perform the operations and simplify.

(a) $6x - 3(x - 4)$ **(b)** $3a - [5a - 3(2a - 1)]$

(c) $3 - \{2y - 5[2 - 3y + (5 - 8y)]\}$

Solution

We follow the order of operations discussed in the last section.

(a) $6x - 3(x - 4) = 6x - 3x + 12$ *Multiply $x - 4$ by -3 and combine where possible.*

$$= \boxed{3x + 12}$$

(b) $3a - [5a - 3(2a - 1)] = 3a - [5a - 6a + 3]$ *Simplify in brackets first (multiply $2a - 1$ by 3). Then combine terms in brackets.*

$$= 3a - [-a + 3]$$ *Next, remove brackets.*

$$= 3a + a - 3$$ *Then combine terms.*

$$= \boxed{4a - 3}$$

(c) $3 - \{2y - 5[2 - 3y + (5 - 8y)]\}$ *Work in brackets first (remove parentheses).*

$$= 3 - \{2y - 5[2 - 3y + 5 - 8y]\}$$ *Next, combine terms in brackets.*

$$= 3 - \{2y - 5[7 - 11y]\}$$ *Then multiply $7 - 11y$ by -5.*

$$= 3 - \{2y - 35 + 55y\} \qquad \textit{Combine terms in braces.}$$
$$= 3 - \{57y - 35\} \qquad \textit{Subtract } 57y - 35.$$
$$= 3 - 57y + 35$$
$$= \boxed{38 - 57y}$$

STUDY SKILLS 1.5   **Checking Your Work**

We develop confidence in what we do by knowing that we are right. One way to check to see if we are right is to look at the answers usually provided in the back of the book. However, few algebra texts provide *all* the answers. And of course, answers are not provided during exams, when we need confidence most.

It is frustrating to find that you incorrectly worked a problem on an exam, and then discover that you would easily have seen your error had you just taken the time to check your work. Therefore, you should know how to check your answers.

The method of checking work should be different from the method used in the solution. In this way you are more likely to discover any errors you might have made. If you simply rework the problem the same way, you cannot be sure you did not make the same mistake twice.

Ideally, the checking method should be quicker than the method for solving the problem (although this is not always possible).

Learn how to check your answers, and practice checking your homework exercises as you do them.

Exercises 1.5

In Exercises 1–68, simplify the expression as completely as possible.

1. $6x + 2x$

2. $5a + 4a$

3. $6x(2x)$

4. $5a(4a)$

5. $2x - 6x$

6. $5a - 4a$

7. $2x(-6x)$

8. $5a(-4a)$

9. $3m - 4m - 5m$

10. $-8y(-3y)(-2y)$

11. $3m(-4m)(-5m)$

12. $-8y - 3y - 2y$

13. $-2t^2 - 3t^2 - 4t^2$

14. $-5z^3 - 2z^3 - 4z^3$

15. $-2t^2(-3t^2)(-4t^2)$

16. $-5z^3(-2z^3)(-4z^3)$

17. $2x + 3y + 5z$

18. $3s + 5t + 6u$

19. $2x(3y)(5z)$

20. $3s(5t)(6u)$

21. $x^3 + x^2 + 2x$

22. $a^5 + 4a^3 + a^2$

23. $x^3(x^2)(2x)$

24. $a^5(4a^3)(a^2)$

25. $-5x(3xy) - 2x^2y$

26. $6r(-2r^2)(-4r^3t)$

27. $-5x(3xy)(-2x^2y)$

28. $6r(-2r^2t) - 4r^3t$

29. $2x^2 + 3x - 5 - x^2 - x - 1$

30. $-7t^3 - 4t^2 - 8 - t^3 - 5t^2 + 2$

31. $10x^2y - 6xy^2 + x^2y - xy^2$

32. $8a^2b^2 - 5a^2b^3 - a^2b^3 + a^2b^2$

33. $3(m + 3n) + 3(2m + n)$

34. $5(2u + 3w) + 4(u + 3w)$

35. $6(a - 2b) - 4(a + b)$

36. $7(3p - q) - 3(p + 2q)$

37. $8(2c - d) - (10c + 8d)$

38. $10(y - z) - (4y + 10z)$

39. $x(x - y) + y(y - x)$

40. $w(w^2 - 4) + w^2(w + 3)$

41. $a^2(a + 3b) - a(a^2 + 3ab)$

42. $t^3(t^2 - 3t) - t^2(t^3 - 3t)$

43. $5a^2bc(-2ab^2)(-4bc^2)$

44. $-2xyz(-4x^3y)(6yz^2)$

45. $(2x)^3(3x)^2$

46. $(5a)^2(2a)^4$

47. $2x^3(3x)^2$

48. $5a^2(2a)^4$

49. $(-2x)^5(x^6)$

50. $(-3a)^4a^8$

51. $(-2x)^4 - (2x)^4$

52. $(-5a)^2 - (5a)^2$

53. $4b - 5(b - 2)$

54. $8z - 9(z - 3)$

55. $8t - 3[t - 4(t + 1)]$

56. $7y - 5[y - 6(y - 1)]$

57. $a - 4[a - 4(a - 4)]$

58. $c - 6[c - 6(c - 6)]$

59. $5 + 2[b - 5(2 - b)]$

60. $4 + 3[d - 5(4 - d)]$

61. $x + x[x + 3(x - 3)]$

62. $y^2 + y[y + y(y - 4)]$

63. $x - \{y - 3[x - 2(y - x)]\}$

64. $y - 2\{x - [z - (x - y)]\}$

65. $3x + 2y[x + y(x - 3y) - y^2]$

66. $6r + 2t[r - t(r - 5t) - r^2]$

67. $6s^2 - [st - s(t + 5s) - s^2]$

68. $9z^3 - [wz - z(w - 4z^2) - z^3]$

CALCULATOR EXERCISES

In Exercises 69–72, simplify the expression as completely as possible.

69. $3.6x - 15.9y - 11.2x$

70. $-2.35x + 2.8y - 7.94x - 2.5y$

71. $2.3(4.1x - 3) - 5.8(5 - 3.2x)$

72. $5.63(4.2x - 3) - 7.58(x - 3)$

1.6

Translating Phrases and Sentences into Algebraic Form

In this section we will focus our attention on translating phrases and sentences into their algebraic form, and leave the formulation and solution of entire problems to Chapter 2.

Many students look for *key words* to help them quickly determine which algebraic symbols to use in translating an English phrase or sentence. However, the key words do not usually indicate how the symbols should be put together to yield an accurate translation of the words. In order to be able to put the symbols together in a meaningful way, we must understand how the key words are being used in the context of the given verbal expression.

Let's look at two verbal expressions that use the same key words and yet do not have the same meaning:

1. Five times the sum of six and four

2. The sum of five times six and four

Both expressions contain the following key words:

"five"	the number 5	(5)
"times"	meaning multiply	(·)
"sum"	meaning addition	(+)
"six"	the number 6	(6)
"four"	the number 4	(4)

How should the symbols be put together?

Expression 1 says: five times . . . .
   Five times what? . . . *the sum*
   Thus, we must *first* find the sum of 6 and 4, before multiplying by 5.
Expression 2 says: the sum of . . . .
   The sum of what *and* what? . . . *the sum of* 5 *times* 6 **and** 4
   Thus, we must *first* compute 5 times 6 before we can determine what the sum is.

Translating the two verbal expressions into algebraic form, we get:

1. $5(6 + 4) = 5 \cdot 10 = 50$
2. $5 \cdot 6 + 4 = 30 + 4 = 34$

Thus, in comparing expressions 1 and 2, we can see that the simple change of word positions in a phrase can change the meaning quite a bit.

EXAMPLE 1

Translate each of the following into algebraic form.
(a)   Nine more than 5 times a number
(b)   Seven less than 5 times a number is 26.
(c)   The sum of two numbers is 6 less than their product.

Solution

(a)   "Nine more than" means we are going to add 9 to something. To what? To "5 *times a number*."
   If we represent the number by n (you are free to choose the letter you like), we get

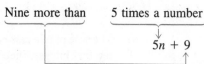

Thus, our answer is

$$5n + 9$$

Of course, $9 + 5n$ is also correct.

(b) "Seven less than" means we are going to subtract 7 from something. From what? From "5 *times a number*."

The word *is* translates to "=."

If we represent the number by *s*, we get

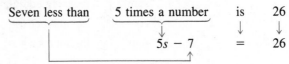

Thus, our final answer is

$$5s - 7 = 26$$

It is very important to note here that $7 - 5s = 26$ *is **not** a correct translation.*

(c) The sentence mentions two numbers, so let's call them *x* and *y*.

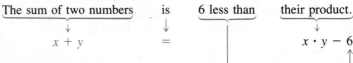

Thus, our translation is

$$x + y = xy - 6$$

    ■

EXAMPLE 2

Translate each of the following into algebraic form.

(a) The sum of two consecutive integers

(b) The sum of two consecutive even integers

(c) The sum of two consecutive odd integers

Solution

This example illustrates how important it is to clearly state what the variable you are using represents. Let's begin by looking at some numerical examples for guidance.

Examples of two consecutive integers	Examples of two consecutive even integers	Examples of two consecutive odd integers
2 and 3	4 and 6	7 and 9
25 and 26	28 and 30	31 and 33
x and $x + 1$	x and $x + 2$	x and $x + 2$

(a) The case of two consecutive integers is fairly straightforward. If we let *x* represent the first integer, then the next consecutive integer is $x + 1$. Therefore, the sum of two consecutive integers can be represented as

$$x + (x + 1) = \boxed{2x + 1}$$

Based on the numerical examples given above, we notice that the two cases of consecutive even or odd integers are basically the same. In order to get from one even integer to the next we have to add 2, and to get from one odd integer to the next we have to add 2.

It is not the adding of 2 which makes the numbers even or odd, but rather whether the *first* number is even or odd. If x is even, then so is $x + 2$. If x is odd, then so is $x + 2$.

(b) If we let x represent the first even integer, then the sum of two consecutive even integers can be represented as

$$x + (x + 2) = \boxed{2x + 2}$$

(c) If we let x represent the first odd integer, then the sum of two consecutive odd integers can be represented as

$$x + (x + 2) = \boxed{2x + 2}$$

It all depends on how we designate x.   ∎

The next example again illustrates the need to clearly understand the meaning of the words in an example in order to be able to put them together properly.

EXAMPLE 3

Translate each of the following into algebraic form.
(a) The sum of the squares of two consecutive even integers
(b) The square of the sum of two consecutive odd integers

Solution

(a) Based on our work in the last example, let x represent the first of the two consecutive even integers. Then $x + 2$ represents the second of the two consecutive even integers.

The "sum of the squares" means that we must *first* square each one of the two consecutive even integers and *then* add the results. Thus, our translation becomes

$$\boxed{x^2 + (x + 2)^2}$$

(b) We let x represent the first of the two consecutive odd integers. Then $x + 2$ represents the second of the two consecutive odd integers.

The "square of the sum" means that we must *first* add the two numbers and *then* square the result. Thus, our translation becomes

$$(x + x + 2)^2 \quad \text{or} \quad \boxed{(2x + 2)^2}$$ ∎

Even though both phrases in Example 2 contained the words "square" and "sum," those words do not give us the meaning of the entire phrase. We still need to analyze how the words fit together.

While it is necessary to recognize that particular words and phrases imply specific arithmetic operations, and to understand what the problem means as a whole, there is still one more important factor necessary to successfully translate from verbal expressions to algebraic ones. This additional factor is the ability to apply your basic general knowledge and common sense to a particular problem.

Let's look at how these various components blend together in the formulation of a problem.

EXAMPLE 4

Suppose Harry averages $10 an hour working as a programmer for his company and $25 an hour as a consultant in his spare time.

(a) How much does Harry make working 40 hours for his company?

(b) In terms of x, how much does Harry make working x hours for his company?

(c) How much does Harry make if he does 40 hours of consulting?

(d) If Harry works x hours for the company and y hours consulting, how much does he make all together, in terms of x and y?

(e) If Harry still works x hours for the company and he works a total of 20 hours in both jobs, how much does he make all together, in terms of x?

Solution

(a) How much does Harry make working 40 hours for the company?

The problem assumes that you understand that if a person is paid a certain amount per hour (pay rate), and works for a number of hours (# of hours), then the total pay would be found by multiplying the pay rate by the # of hours:

$$\text{Total pay} = (\text{Pay rate per hour}) \cdot (\text{\# of hours})$$

Hence, Harry makes $(10)(40) = \boxed{\$400}$ by working 40 hours for the company at $10 per hour.

(b) In terms of x, how much does Harry make working x hours for his company?

Again, we find the total pay by multiplying the rate per hour by the number of hours. Hence,

$$\text{Harry makes } (10) \cdot x = \boxed{10x \text{ dollars}}$$

(c) How much does Harry make if he does 40 hours of consulting?

$$\text{Harry makes } (25)(40) = \boxed{\$1,000}$$

(d) If Harry works x hours for the company and y hours consulting, how much does he make all together, in terms of x and y?

Now we are required to find Harry's total pay for the two jobs. His total is the sum of the incomes from each job, or,

$$\begin{pmatrix} \text{Pay from} \\ \text{company} \end{pmatrix} + \begin{pmatrix} \text{Pay from} \\ \text{consulting} \end{pmatrix} = \text{Total pay}$$

which is found by

$$\begin{pmatrix} \text{Rate from} \\ \text{company} \end{pmatrix} \cdot \begin{pmatrix} \text{\# of hours} \\ \text{working for} \\ \text{company} \end{pmatrix} + \begin{pmatrix} \text{Rate from} \\ \text{consulting} \end{pmatrix} \cdot \begin{pmatrix} \text{\# of hours} \\ \text{consulting} \end{pmatrix} = \text{Total pay}$$

Hence, we have

$$10 \cdot x \qquad + \qquad 25 \cdot y \qquad = \text{Total pay}$$

Thus, his total pay, in terms of x and y, is

$$\boxed{10x + 25y \text{ dollars}}$$

(e)   This part is similar to part (d), except that we are given only one of the number of hours worked. Let's use the same analysis as in part (d) and see what we can fill in at this point.

$$\begin{pmatrix} \text{Pay from} \\ \text{company} \end{pmatrix} + \begin{pmatrix} \text{Pay from} \\ \text{consulting} \end{pmatrix} = \text{Total pay}$$

which is found by

$$\begin{pmatrix} \text{Rate from} \\ \text{company} \end{pmatrix} \cdot \begin{pmatrix} \text{\# of hours} \\ \text{working for} \\ \text{company} \end{pmatrix} + \begin{pmatrix} \text{Rate from} \\ \text{consulting} \end{pmatrix} \cdot \begin{pmatrix} \text{\# of hours} \\ \text{consulting} \end{pmatrix} = \text{Total pay}$$

$$10 \cdot x \qquad + \qquad 25 \cdot ? \qquad = \text{Total pay}$$

What remains is to fill in the question mark.

Although we are not given the amount of hours consulting, we are given that Harry worked a total of 20 hours. Therefore, if he worked a total of 20 hours and x of those hours are spent working for the company, then the remaining hours must have been spent consulting. Hence,

$$(\text{Total hours}) - \begin{pmatrix} \text{\# of hours} \\ \text{working for} \\ \text{company} \end{pmatrix} = \begin{pmatrix} \text{\# of hours} \\ \text{consulting} \end{pmatrix}$$

$$20 \qquad - \qquad x \qquad = \begin{pmatrix} \text{\# of hours} \\ \text{consulting} \end{pmatrix}$$

Since the number of hours consulting is $20 - x$, we can fill in the rest of the previous expression:

$$\begin{pmatrix} \text{Rate from} \\ \text{company} \end{pmatrix} \cdot \begin{pmatrix} \text{\# of hours} \\ \text{working for} \\ \text{company} \end{pmatrix} + \begin{pmatrix} \text{Rate from} \\ \text{consulting} \end{pmatrix} \cdot \begin{pmatrix} \text{\# of hours} \\ \text{consulting} \end{pmatrix} = \text{Total pay}$$

$$10 \cdot x \qquad + \qquad 25 \cdot ? \qquad = \text{Total pay}$$

$$10 \cdot x \qquad + \qquad 25 \cdot (20 - x) \qquad = \text{Total pay}$$

Harry's total pay is $10x + 25(20 - x)$, which we usually simplify:

$$10x + 25(20 - x) = 10x + 500 - 25x$$
$$= \boxed{500 - 15x \text{ dollars}} \qquad \blacksquare$$

EXAMPLE 5

A collection of coins consists of nickels, dimes, and quarters. If there are three times as many nickels as dimes, and seven more quarters than dimes, represent the total value of the coins using one variable.

Solution

Here, as with Example 4, we have to distinguish between two different numerical concepts. In Example 4 we had to distinguish between the quantity of hours worked and the pay Harry received for working these hours.

In this coin problem we will have to distinguish between quantity and value: the number of coins and the value of the coins. We must start out by trying to describe the number of coins in terms of one variable before we can determine the values.

If you read the problem carefully, you will notice that the number of quarters and nickels is given in terms of the number of dimes. It would therefore be reasonable to let x represent the number of dimes.

Let # of dimes = x.

Then # of nickels = $3x$.   *(Three times as many nickels as dimes)*

And # of quarters = $x + 7$.   *(7 more quarters than dimes)*

Now that we have expressed the number of each coin in terms of one variable, we can now look at the values of each type of coin. To simplify our arithmetic, we will express our values in terms of cents rather than dollars.

How do we determine the value of x dimes?

The value of 1 dime is 10 cents.

The value of 2 dimes is 20 cents.

The value of 5 dimes is 50 cents.

We can see that we find the value of dimes by multiplying the *number* of dimes by 10, the *value of* 1 *dime*.

In general,

$$\begin{pmatrix} \text{Total value of} \\ \text{one type of item} \end{pmatrix} = \begin{pmatrix} \text{Value of} \\ \text{one item} \end{pmatrix} \cdot \begin{pmatrix} \text{\# of items} \\ \text{of that value} \end{pmatrix}$$

Thus, the value of the dimes is $10 \cdot x = 10x$. Hence, we have

$$\begin{pmatrix}\text{Value of}\\ \text{all dimes}\end{pmatrix} + \begin{pmatrix}\text{Value of}\\ \text{all nickels}\end{pmatrix} + \begin{pmatrix}\text{Value of}\\ \text{all quarters}\end{pmatrix} = \begin{pmatrix}\text{Total value}\\ \text{of all coins}\end{pmatrix}$$

which is found by

$$\begin{pmatrix}\text{Value of}\\ \text{one dime}\end{pmatrix} \cdot \begin{pmatrix}\text{\# of}\\ \text{dimes}\end{pmatrix} + \begin{pmatrix}\text{Value of}\\ \text{one nickel}\end{pmatrix} \cdot \begin{pmatrix}\text{\# of}\\ \text{nickels}\end{pmatrix}$$

$$+ \begin{pmatrix}\text{Value of}\\ \text{one quarter}\end{pmatrix} \cdot \begin{pmatrix}\text{\# of}\\ \text{quarters}\end{pmatrix} = \begin{pmatrix}\text{Total value}\\ \text{of all}\\ \text{coins}\end{pmatrix}$$

$$10 \cdot (x) + 5 \cdot (3x) + 25 \cdot (x + 7)$$

Hence, the total *value* of all coins (in cents) is:

$$10x + 5(3x) + 25(x + 7) = 10x + 15x + 25x + 175$$

$$= \boxed{50x + 175 \text{ cents}}$$

■

EXAMPLE 6

The length of a rectangle is 5 more than twice its width.

(a) Express the *area* of the rectangle in terms of one variable.

(b) Express the *perimeter* of the rectangle in terms of one variable.

Solution

We draw a picture of a rectangle. Since the length is expressed in terms of the width,

We let the width $= x$.

Then the length $= 2x + 5$ *(5 more than twice the width)*

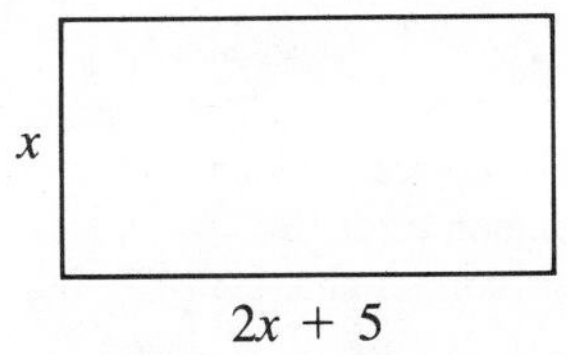

Figure 1.4
Rectangle for Example 6

Then we label the rectangle as shown in Figure 1.4.

(a) The area, A, of a rectangle, is given by

$$\begin{aligned} A &= (\text{Length})(\text{Width}) \\ &= (2x + 5)(x) \\ &= \boxed{2x^2 + 5x} \end{aligned}$$

(b) The perimeter, P, of a rectangle is the distance around the rectangle. Hence,

$P = 2W + 2L$ *Where W = width and L = length*

Therefore, in terms of x,

$$\begin{aligned} P &= 2 \cdot x + 2(2x + 5) \\ &= 2x + 4x + 10 \\ &= \boxed{6x + 10} \end{aligned}$$

■

STUDY SKILLS 1.6   **Making Study Cards**

Study cards are 3″ × 5″ or 5″ × 8″ index cards that contain summary information needed for convenient review. The process of making study cards is a learning experience in itself. We will discuss how to use study cards in the next chapter. For now, we will cover three types of cards: the definition/principle card, the warning card, and the quiz card.

The *definition/principle* (**D/P**) *cards* contain a single definition, concept, or rule for a particular topic.

Here is an example of a D/P card.

The front of each D/P card should contain the following:

1.   A heading of a few words

2.   The definition, concept, or rule accurately recorded

3.   If possible, a restatement of the definition, concept, or rule in your own words

The back of the card should contain examples illustrating the idea on the front of the card.

FRONT

$$\underline{\text{The Distributive Property}}$$
$$a(b+c) = ab + ac$$
$$\text{or}$$
$$(b+c)a = ba + ca$$
Multiply each \underline{term} by a.

BACK

(1) $3x(x+2y) = 3x(x) + 3x(2y)$
$$= 3x^2 + 6xy$$
(2) $-2x(3x-y) = -2x(3x) - 2x(-y)$
$$= -6x^2 + 2xy$$
(3) $(2x+y)(x+y) = 2x(x+y) + y(x+y)$

Warning (**W**) *cards* contain errors that you may be making consistently on homework, quizzes, or exams, or those common errors pointed out by your teacher or your text. The front of the warning card should contain the word WARNING; the back of the card

should contain an example of both the correct way an example should be done and the common error. Be sure to label clearly which solution is correct and which is not.

For example:

FRONT

WARNING
\underline{EXPONENTS}
    An exponent refers only to the factor immediately to the left of the exponent.

BACK

EXAMPLES
$2 \cdot 3^2 = 2 \cdot 3 \cdot 3$ *NOT* $(2 \cdot 3)(2 \cdot 3)$
$(-3)^2 = (-3)(-3) = 9$
↑ Parentheses mean -3 is the factor to be squared.
BUT
$-3^2 = -3 \cdot 3 = -9$
↑ The factor being squared here is 3, not -3.

Quiz cards are another type of study card. They will be used to help us construct practice tests. For now, go through your text and pick out a few of the odd-numbered exercises (just the problem) from each section, putting one or two problems on one side of each card. Make sure that you copy the *instructions* as well as the problem accurately. On the back of the card, write down the exercise number and section of the book where the problem was found. For example:

FRONT

Translate the given sentence. Indicate clearly what each variable represents.

Four more than three times a number is seven less than the number.

BACK

Exercise 5
Section 1.6

Exercises 1.6

In Exercises 1–8, translate the given phrase or sentence. Indicate clearly what each variable represents.

1. Eight more than a number

2. Eight less than a number

3. Three less than twice a number

4. Three more than twice a number

5. Four more than three times a number is seven less than the number.

6. Nine less than twice a number is six more than twice the number.

7. The sum of two numbers is one more than their product.

8. The quotient of two numbers is five less than their sum.

In Exercises 9–18, represent all the numbers in the exercise in terms of one variable.

9. There are two numbers, the larger of which is 5 more than twice the smaller.

10. There are two numbers, the smaller of which is 12 less than 3 times the larger.

11. There are three numbers. The middle number is 3 times the smallest, and the largest number is 12 more than the middle number.

12. There are three numbers. The largest number is 5 more than 8 times the smallest, and the middle number is 10 less than the largest number.

13. Represent two consecutive integers.

14. Represent two consecutive odd integers.

15. Represent three consecutive even integers.

16. Represent three consecutive odd integers.

17. Represent the sum of the cubes of two consecutive even integers.

18. Represent the cube of the sum of two consecutive odd integers.

In Exercises 19–22, represent the numbers in terms of one variable.

19. The sum of two numbers is 40.

20. The sum of two numbers is 29.

21. The sum of three numbers is 100. One of the numbers is twice one of the other numbers.

22. The sum of three numbers is 82. One of the numbers is 9 more than one of the others.

23. The length of a rectangle is 3 times its width. Represent its area and perimeter in terms of one variable.

24. The length of a rectangle is 6 less than 4 times its width. Represent its area and perimeter in terms of one variable.

25. The length of the first side of a triangle is twice the length of its second side, and the length of its third side is 4 more than the length of its second side. Express its perimeter in terms of one variable.

26. The length of the first side of a triangle is 2 more than 3 times the length of its third side, and the length of its second side is 5 more than the length of its third side. Express its perimeter in terms of one variable.

27. A collection of coins consists of twelve nickels, nine dimes, and ten quarters.

    (a) How many coins are there?

    (b) What is the value of the coins?

28. The length of a rectangular plot of land is 20 meters and its width is 12 meters. Heavy-duty fence for the length costs $5 per meter, while regular fence for the width costs $2 per meter.

    (a) How much regular fence is needed?

    (b) How much will the regular fence cost?

    (c) How much heavy-duty fence is needed?

    (d) How much will the heavy-duty fence cost?

    (e) What will the total cost for fencing the plot of land be?

29. Repeat Exercise 27 for n nickels, d dimes, and q quarters.

30. Flight attendant A serves six meals per minute for 15 minutes while flight attendant B serves eight meals per minute for 20 minutes.

    (a) How many meals does flight attendant A serve?

    (b) How many meals does flight attendant B serve?

    (c) How many meals do they serve all together?

31. Repeat Exercise 28 if the width is w meters and the length is 3 times the width.

32. Repeat Exercise 30 if flight attendant A serves n meals per minute for 18 minutes while flight attendant B serves nine meals per minute for t minutes.

33. A collection of twenty coins consists of nickels and dimes. Represent the value of all the coins using one variable.

34. A collection of thirty-seven coins consists of nickels, dimes, and quarters. There are ten more dimes than nickels. Represent the value of all the coins using one variable.

CHAPTER 1   Summary

After having completed this chapter, you should be able to:

1. Understand set notation and identify the sets N and W (Section 1.1).

2. Given sets A and B, list the elements of $A \cap B$ and $A \cup B$ (Section 1.1).

    For example: Given

    $$A = \{x \mid x \in N \text{ and } x \text{ is less than } 15\}$$

    $$B = \{y \mid y \text{ is a whole-number multiple of 3 and is less than } 12\}$$

Then
$$A = \{1, 2, 3, 4, 5, 6, 7, 8, 9, 10, 11, 12, 13, 14\}$$
$$B = \{0, 3, 6, 9\}$$

and
$$A \cap B = \{3, 6, 9\}$$
$$A \cup B = \{0, 1, 2, 3, 4, 5, 6, 7, 8, 9, 10, 11, 12, 13, 14\}$$

3. Distinguish between integers (Z), the rational (Q) and irrational (I) numbers, and understand the makeup of the set of real numbers (R) (Section 1.2).

4. Graph inequalities on the number line (Section 1.2).

   For example:

   (a) On the number line, the graph of $\{x \mid -5 < x \le 3, x \in Z\}$ is:

   (b) On the number line, the graph of $\{x \mid -2 < x \le 3\}$ is:

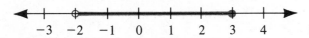

   Note the empty circle at -2 and the filled-in circle at 3.

5. Identify the properties of the real numbers (Section 1.3).

   For example: The statement

   $$(x + 4) + (x - 6) = (4 + x) + (x - 6)$$

   illustrates the **commutative property of addition.**

6. Perform operations on integers (Section 1.4).

   For example: Perform the given operations.

   (a) $-6 - 2[5 - (7 - 8)]$ *Perform operations inside parentheses first.*

   $= -6 - 2[5 - (-1)]$ *Then simplify in brackets.*

   $= -6 - 2[5 + 1]$

   $= -6 - 2[6]$

   $= -6 - 12$

   $= -18$

   (b) $\dfrac{-5 + 4(2 - 3)}{-5(3 - 3)}$ *Perform operations inside parentheses.*

   $= \dfrac{-5 + 4(-1)}{-5(0)}$

   $= \dfrac{-5 + 4(-1)}{0}$

   Undefined   *Since division by 0 is undefined*

7. Perform operations on and simplify algebraic expressions (Section 1.5).

For example: Perform the operations and simplify the following.

(a) $3x(2x - y) - 2x(x + y)$ *First use the distributive property.*

$\quad\quad = 6x^2 - 3xy - 2x^2 - 2xy$ *Combine like terms.*

$\quad\quad = 4x^2 - 5xy$

(b) $2a - [3a - 5(2a - 1)]$ *Perform operations inside brackets.*

$\quad\quad = 2a - [3a - 10a + 5]$

$\quad\quad = 2a - [-7a + 5]$ *Then subtract (change the sign of each term in*

$\quad\quad = 2a + 7a - 5$ *brackets).*

$\quad\quad = 9a - 5$

8. Translate verbal sentences into algebraic statements (Section 1.6).

For example: If Jack has n nickels, q quarters, and d dimes, express the number of coins and the value of his coins in terms of n, q, and d.

Solution: If Jack has n nickels, q quarters, and d dimes, then he has $n + q + d$ coins.

The value of his nickels is $5n$ cents.

The value of his quarters is $25q$ cents.

The value of his dimes is $10d$ cents.

So the total value of all his coins is

$$5n + 25q + 10d \text{ cents}$$

CHAPTER 1   Review Exercises

In Exercises 1–6, you are given

$$A = \{x \mid x \le 4, x \in N\} \quad\quad B = \{a \mid a > 5, a \in W\}$$

$$C = \{a, r, b, s, e\} \quad\quad\quad D = \{b, c, f, g\}$$

List the elements in the following sets.

1. A **2.** B

3. $C \cap D$ **4.** $C \cup D$

5. $A \cup B$ **6.** $A \cap B$

In Exercises 7–12, you are given

$$A = \{a \mid a \in W, \text{ and } a \text{ is a factor of } 12\}$$

$$B = \{b \mid b \in W, \text{ and } b \text{ is a multiple of } 12\}$$

$$C = \{c \mid c \in W, \text{ and } c \text{ is a multiple of } 6\}$$

List the elements in the following sets.

7. A

8. B

9. $A \cap B$

10. $A \cup B$

11. $B \cap C$

12. $A \cap C$

In Exercises 13–14, you are given

$$A = \{x \mid -2 < x \le 4, x \in Z\}$$

$$B = \{y \mid 3 \le y \le 12, y \in Z\}$$

Describe the sets using set notation.

13. $A \cap B$

14. $A \cup B$

In Exercises 15–20, graph the sets on the real number line.

15. $\{x \mid x \le 4\}$

16. $\{x \mid x > 4\}$

17. $\{b \mid -8 \le b \le 5\}$

18. $\{b \mid -8 < b < 5\}$

19. $\{a \mid -2 < a \le 4\}$

20. $\{b \mid -2 \le b < 4\}$

In Exercises 21–28, determine if the statement is true or false.

21. $\frac{1}{2} \in Z$

22. $\sqrt{9} \in Z$

23. $.47\overline{6476} \in Q$

24. $\pi \in R$

25. $\pi \in Q$

26. $6\frac{2}{7} \in Q$

27. $I \subset R$

28. $Q \subset I$

In Exercises 29–36, if the statement is true, state the property illustrated by the statement. If the statement is not true, write "False."

29. $(x - 4) + 3 = 3 + (x - 4)$

30. $(5 - x)7 = 7(5 - x)$

31. $(x - 4) \cdot 3 = x \cdot 3 - 4 \cdot 3$

32. $(x + 5)(y - 2) = x(y - 2) + 5(y - 2)$

33. $\left(\dfrac{1}{2x}\right)2x = 1 \qquad (x \ne 0)$

34. $(5x + 2) + 0 = 5x + 2$

35. $(3x - 2) + (4x - 1) = 3x - (2 + 4x) - 1$

36. $3 \cdot 2(x - 1) = 6(2x - 2)$

In Exercises 37–54, perform the operations.

37. $(-2) + (-3) - (-4) + (-5)$

38. $(-6) - (-2) + (-3) - (+4)$

39. $6 - 2 + 5 - 8 - 9$

40. $-7 - 2 - 3 + 8 - 5$

41. $(-2)(-3)(-5)$

42. $(-7)(-3)(-5)(+2)$

43. $(-2)^6$

44. -2^6

45. $(-2) - (-3)^2$

46. $-5 - (-2)^3$

47. $(-6 - 3)(-2 - 5)$

48. $(-4 + 7)(-3 - 2)$

49. $5 - 3[2 - (4 - 8) + 7]$

50. $-6 + 2[5 - (3 - 9)]$

51. $5 - \{2 + 3[6 - 4(5 - 9)] - 2\}$

52. $-7 + \{3 - 6[12 - 5(9 - 11)]\}$

53. $\dfrac{4[5 - 3(8 - 12)]}{-6 - 2(5 - 6)}$

54. $\dfrac{6 - [2 - (3 - 5)]}{7 + [5 + 6(6 - 8)]}$

In Exercises 55–60, evaluate, given $x = -2$, $y = -1$, and $z = 0$.

55. $x^2 - 2xy + y^2$

56. $2x^2 + 3xy - 3y^2$

57. $|x - y| - (|x| - |y|)$

58. $|x + y| + |x| - |y|$

59. $\dfrac{2x^2y^3 + 3y^2}{zx^2y}$

60. $\dfrac{(5xy^3 + x^4)z}{x^2y}$

In Exercises 61–78, perform the operations and express your answer in simplest form.

61. $(2x + y)(-3x^2y)$

62. $(-6ab^2)(-3a^2b)(-2b)$

63. $(2xy^2)^2(-3x)^2$

64. $(-3r^2s^2)^3(-2r^2)$

65. $3x - 2y - 4x + 5y - 3x$

66. $-6a + 2b + 3a - 4b - 5a$

67. $-2r^2s + 5rs^2 - 3sr^2 - 4s^2r$

68. $-5x^2y^3 + 6xy^2 + 2x^2y^3 - 8y^2x$

69. $-2x - (3 - x)$

70. $(y - 5) - (y - 4)$

71. $3a(a - b + c)$

72. $5rs(2r + 3s)$

73. $2x - 3(x - 4)$

74. $7y - 9(2x + 1)$

75. $3a - [5 - (a - 4)]$

76. $-2r + 3[s - 2(s - 6)]$

77. $5x - \{3x + 2[x - 3(5 - x)]\}$

78. $7 - 3\{y - 2[y - 4(y - 1)]\}$

In Exercises 79–86, translate the statements algebraically.

79. Five less than the product of two numbers is 3 more than their sum.

80. Eight more than the sum of a number and itself is 3 less than the product of the number and itself.

81. The sum of the first two of three consecutive odd integers is 5 less than the third.

82. The product of the last two of three consecutive even integers is 8 more than 10 times the first.

83. The length of a rectangle is 5 less than 4 times its width. Express its area and perimeter in terms of one variable.

84. The length of a rectangle is 5 more than 3 times its width. Express its area and perimeter in terms of one variable.

85. The sum of the squares of two numbers is 8 more than the product of the two numbers.

86. The square of the sum of two numbers is 8 more than the product of the two numbers.

87. Xavier has forty coins in nickels and dimes. Express the total value of his coins in terms of the number of dimes.

88. Cassandra makes $12 an hour as a garage mechanic and $8 an hour as a typist. If she works a total of 30 hours, and she works x of these hours as a mechanic, express the total amount she makes in terms of x.

CHAPTER 1   Practice Test

1. Given

$$A = \{x \mid x \text{ is a prime factor of } 210\}$$

$$B = \{y \mid y \text{ is a prime number less than } 25\}$$

  (a) List the elements in $A \cap B$.

  (b) List the elements in $A \cup B$.

2. Indicate whether each of the following is true or false:

  (a) $\sqrt{16} \in I$

  (b) $2 \in Q$

  (c) $-\dfrac{3}{4} \in Z$

3. Graph the following sets on the real number line.

   (a) $\{a \mid a > 4\}$

   (b) $\{x \mid -3 \leq x < 10\}$

4. If the statement is true, state the property which the given statement illustrates. If the statement is not true, write "False."

   (a) $7(xy + z) = 7xy + z$

   (b) $(x + y) + 3 = 3 + (x + y)$

5. Evaluate the following:

   (a) $-3 - (-6) + (-4) - (-9)$

   (b) $(-7)^2 - (-6)(-2)(-3)$

   (c) $|3 - 8| - |5 - 9|$

   (d) $6 - 5[-2 - (7 - 9)]$

6. Evaluate the following, given $x = -2$ and $y = -3$:

   (a) $(x - y)^2$

   (b) $\dfrac{x^2 - y^2}{x^2 - 2xy - y^2}$

7. Simplify the following:

   (a) $(5x^3y^2)(-2x^2y)(-xy^2)$

   (b) $3rs^2 - 5r^2s - 4rs^2 - 7rs$

   (c) $2a - 3(a - 2) - (6 - a)$

   (d) $7r - \{3 + 2[s - (r - 2s)]\}$

8. The length of a rectangle is 8 less than 3 times its width. Express its perimeter in terms of one variable.

9. Wallace has thirty-four coins in nickels and dimes. If x represents the number of dimes, express the number of nickels, in terms of x. In terms of x, what is the total value of his coins?

2

Equations and Inequalities

Study Skills

Undoubtedly the most useful and most used skill in all of algebra is the ability to solve different types of equations and inequalities. In this chapter we begin a process, which will continue throughout the text, of developing methods to solve various types of equations and inequalities.

2.1

First-Degree Equations

An *equation* is a mathematical statement that two algebraic expressions are equal. Let's begin by examining the three equations

1. $5(x + 2) - 5x - 15 = 0$　　　2.　$5(x + 2) - 5x - 10 = 0$
3. $5(x + 2) - 15 = 0$

These equations may look very similar at first glance. However, if we examine them more carefully, we see that there are very fundamental differences. If we simplify the left-hand side of each equation, we obtain the following:

1. $5(x + 2) - 5x - 15 = 0$　　　2.　$5(x + 2) - 5x - 10 = 0$
    $5x + 10 - 5x - 15 = 0$　　　　　$5x + 10 - 5x - 10 = 0$
    $-5 = 0$　　　　　　　　　　　　　$0 = 0$

3. $5(x + 2) - 15 = 0$
    $5x + 10 - 15 = 0$
    $5x - 5 = 0$

After simplifying each equation, we can see that:

Equation 1 is false.

DEFINITION

An equation which is always false, regardless of the value of the variable, is called a *contradiction*.

Equation 2 is true.

DEFINITION

An equation which is always true, regardless of the value of the variable, is called an *identity*.

Equation 3 is neither true nor false. Its truth depends on the value of the variable. If x is equal to 1, the equation is true; otherwise, it is false.

DEFINITION

An equation whose truth depends on the value of the variable is called a *conditional equation*.

EXAMPLE 1

Determine whether each of the following is an identity or a contradiction.

(a) $x(x - 3) - (5x + 7) = 2x(x + 3) - 7(2x - 1) - x^2 - 14$

(b) $3x + 6 - (3x - 2) = 5 + 4x + 4(3 - x)$

Solution

Clearly, we cannot tell what types of equations we have in their present form. Our first step in dealing with an equation is always to simplify each side as completely as possible using the methods outlined in Chapter 1.

(a) $x(x - 3) - (5x + 7) = 2x(x + 3) - 7(2x - 1) - x^2 - 14$

Multiply out using the distributive property; combine like terms.

$$x^2 - 3x - 5x - 7 = 2x^2 + 6x - 14x + 7 - x^2 - 14$$

$$x^2 - 8x - 7 = x^2 - 8x - 7$$

Thus, we see that the equation is an $\boxed{\text{identity}}$

(b) $3x + 6 - (3x - 2) = 5 + 4x + 4(3 - x)$

$$3x + 6 - 3x + 2 = 5 + 4x + 12 - 4x$$

$$8 = 17$$

Thus, we see that the equation is a $\boxed{\text{contradiction}}$ ■

In the case of a conditional equation, we are usually interested in finding those values which make it true.

DEFINITION

A value that makes an equation true is called a *solution* of the equation. We also say that the value *satisfies* the equation.

EXAMPLE 2

Determine whether each of the values listed satisfies the given equation.

$$4 + 3(w - 1) = 5(w + 1) - (1 - w) \quad \text{for} \quad w = 1, -1$$

Solution

To check whether a particular value satisfies an equation, we replace each occurrence of the variable, on both sides, by the proposed value.

CHECK $w = 1$: $4 + 3(w - 1) = 5(w + 1) - (1 - w)$

$$4 + 3(1 - 1) \stackrel{?}{=} 5(1 + 1) - (1 - 1)$$

$$4 + 3(0) \stackrel{?}{=} 5(2) - 0$$

$$4 + 0 \stackrel{?}{=} 10 - 0$$

$$4 \neq 10$$

Therefore, $w = 1$ is *not* a solution.

CHECK $w = -1$: $\quad 4 + 3(w - 1) = 5(w + 1) - (1 - w)$

$$4 + 3(-1 - 1) \stackrel{2}{=} 5(-1 + 1) - [1 - (-1)]$$

$$4 + 3(-2) \stackrel{2}{=} 5(0) - (1 + 1)$$

$$4 - 6 \stackrel{2}{=} 0 - 2$$

$$-2 \stackrel{\checkmark}{=} -2$$

Therefore, $w = -1$ *is* a solution.

Note that the equation is conditional. It is neither always true nor always false. ∎

Properties of Equality

If we consider the following four equations:

$$5a - 2(a - 3) = 2a + 10$$
$$5a - 2a + 6 = 2a + 10$$
$$3a + 6 = 2a + 10$$
$$a = 4$$

we can easily check that $a = 4$ is a solution to each of them. (Check this!) In fact, as we shall soon see, $a = 4$ is the only solution to these equations.

DEFINITION

Equations which have exactly the same set of solutions are called *equivalent equations*.

Of the four equations above, the solution to the last one was the most obvious. Consequently, an equation of this form (that is, $a = 4$) is often called an ***obvious equation***.

In order to develop a systematic method for solving equations we would like to be able to transform an equation into successively simpler *equivalent* ones, finally resulting in an obvious equation. Developing such a method depends on recognizing the basic properties of equations.

If $a = b$, then a and b are interchangeable. Any expression containing a can be rewritten with b replacing a. Thus, if $a = b$, the expression $a + 7$ can be rewritten as $b + 7$ and we have $a + 7 = b + 7$. Another way of saying this is that if $a = b$, we can "add 7 to both sides of the equation," obtaining $a + 7 = b + 7$, without disturbing the equality. Consequently, recognizing what the equal sign means, we can see that equalities have the properties listed in the next box.

Properties of Equality

1. If $a = b$, then $a + c = b + c$.

   Addition Property of Equality

2. If $a = b$, then $a - c = b - c$.

   Subtraction Property of Equality

3. If $a = b$, then $a \cdot c = b \cdot c$.

   Multiplication Property of Equality

4. If $a = b$, then $\dfrac{a}{c} = \dfrac{b}{c}$, $c \neq 0$.

   Division Property of Equality

A comment about the multiplication property is in order here. While property 3 as stated is correct, if we apply the multiplication property with $c = 0$, we may not obtain an equivalent equation. For example, the equation $x + 3 = 5$ is a conditional equation whose only solution is $x = 2$. If we apply the multiplication property with $c = 0$ (that is, we multiply both sides of the equation by 0), we get

$$x + 3 = 5$$
$$0(x + 3) = 0(5)$$
$$0 = 0$$

which is an identity.

Multiplying both sides of the equation by 0 has not given us an equivalent equation. In order to ensure that whenever we apply property 3 we will obtain an equivalent equation, we must restrict the use of property 3 to the case when $c \neq 0$.

Solving First-Degree Equations

We mentioned earlier that as we progress through this text we will develop methods to solve various kinds of equations. As we shall see, *the method of solution we choose will very much depend on the kind of equation we are trying to solve.* Therefore, when we begin to solve an equation, we must first try to determine what kind of equation we are trying to solve.

In this section we are going to restrict our attention to a specific kind of equation.

DEFINITION

A *first-degree equation* is an equation that can be put in the form $ax + b = 0$, where a and b are constants ($a \neq 0$).

In other words, a first-degree equation is an equation in which the variable appears to the first power only. For example,

$$5x + 3 = 0, \quad 3x + 7 = x - 2, \quad \text{and} \quad 8 - 2x = 1$$

are all first-degree equations, while

$$2x^2 = 4x + 3$$

is a second-degree equation. (The degree of an equation is not always obvious and we will have more to say about this a bit later on in this section.)

Referring back to the properties of equality, we can verbally summarize these properties with regard to equations as shown in the box.

An equation can be transformed into an equivalent equation by adding or subtracting the same quantity on both sides of the equation, or by multiplying or dividing both sides of the equation by the same *nonzero* quantity.

Our method of solution for a first-degree equation in one variable consists of applying the properties of equality to transform the original equation into progressively simpler *equivalent* equations until we eventually obtain an obvious equation. Recall that an obvious equation is an equation of the form $x = \#$ or $\# = x$. In other words, in solving a *first-degree* equation,

> Our goal is to isolate the variable on one side of the equation.

EXAMPLE 3

Solve for a. $3a - 5 = -11$

Solution

In order to isolate a, we must deal with the 3 multiplying a and the 5 being subtracted. Since the 3 is multiplying the a, in order to get rid of it we use the inverse operation and divide by 3. However, dividing by 3 means we must divide both sides of the equation by 3, which forces us to work with fractions. Consequently, it is a better strategy to leave the division for last.

We proceed as follows:

$$3a - 5 = -11 \qquad \textit{First we add } +5 \textit{ to both sides of the equation.}$$
$$3a - 5 \; \boxed{+ 5} = -11 \; \boxed{+ 5}$$
$$3a = -6 \qquad \textit{Then divide both sides of the equation by 3.}$$
$$\frac{3a}{3} = \frac{-6}{3}$$
$$\boxed{a = -2}$$

CHECK $a = -2$: $3a - 5 = -11$
$$3(-2) - 5 \overset{?}{=} -11$$
$$-6 - 5 \overset{?}{=} -11$$
$$-11 \overset{\checkmark}{=} -11 \qquad \blacksquare$$

EXAMPLE 4

Solution

Solve for x. $12 - 4x = x - 3$

We want to isolate x on one side of the equation. In order to do this, we collect all the x terms on one side of the equation and all the non-x terms on the other side. The side on which we choose to isolate x is irrelevant.

$$12 - 4x = x - 3$$ *We decide to isolate x on the right.*
Add 4x to both sides of the equation.

$$12 - 4x + 4x = x - 3 + 4x$$ *Combine like terms.*

$$12 = 5x - 3$$ *Next, add +3 to both sides of the equation.*

$$12 + 3 = 5x - 3 + 3$$

$$15 = 5x$$ *Finally, divide both sides of the equation by 5.*

$$\frac{15}{5} = \frac{5x}{5}$$

$$\boxed{3 = x}$$

CHECK $x = 3$: $12 - 4x = x - 3$

$$12 - 4(3) \overset{?}{=} 3 - 3$$

$$12 - 12 \overset{\checkmark}{=} 0$$ ∎

Very often it is necessary to simplify an equation before proceeding to the actual solution.

EXAMPLE 5

Solution

Solve for t. $3(t - 4) - 4(t - 3) = t + 3 - (t - 2)$

We begin by simplifying each side of the equation as completely as possible.

$$3(t - 4) - 4(t - 3) = t + 3 - (t - 2)$$ *Remove parentheses.*

$$3t - 12 - 4t + 12 = t + 3 - t + 2$$ *Then combine like terms.*

$$-t = 5$$ *We are not finished yet. We want t alone so we divide both sides of the equation by the coefficient of t, which is −1.*

$$\frac{-t}{-1} = \frac{5}{-1}$$

$$\boxed{t = -5}$$

CHECK $t = -5$: $3(t - 4) - 4(t - 3) = t + 3 - (t - 2)$

$$3(-5 - 4) - 4(-5 - 3) \overset{?}{=} -5 + 3 - (-5 - 2)$$

$$3(-9) - 4(-8) \overset{?}{=} -2 - (-7)$$

$$-27 + 32 \overset{?}{=} -2 + 7$$

$$5 \overset{\checkmark}{=} 5$$ ∎

We can summarize the approach we have used in the last few examples into the outline given in the next box.

Strategy for Solving First-Degree Equations	**1.** *Simplify both sides* of the equation as completely as possible.
	2. By using the addition and subtraction properties of equality, *collect the variable terms* on one side of the equation and the numerical terms on the other side.
	3. By using the division property of equality, make the coefficient of the variable 1. That is, *isolate the variable* on one side of the equation.
	4. *Check* the solution in the original equation.

EXAMPLE 6 Solve for x. $x(x + 12) + 36 = x^2 - 3(5x - 12)$

Solution Following our outline, we first simplify both sides of the equation to get

$x^2 + 12x + 36 = x^2 - 15x + 36$ *This may look like a second-degree equation but it is not.*

Now apply the addition property of equality.   *First add $-x^2$ to both sides.*

$x^2 + 12x + 36 - x^2 = x^2 - 15x + 36 - x^2$

$12x + 36 = -15x + 36$ *Add $+15x$ to both sides of the equation.*

$12x + 36 + 15x = -15x + 36 + 15x$

$27x + 36 = 36$ *Then add -36 to both sides.*

$27x = 0$ *Treat 0 like any other numerical term. Divide both sides of the equation by 27.*

$\dfrac{27x}{27} = \dfrac{0}{27}$

$\boxed{x = 0}$

The check is left for the student.   ■

EXAMPLE 7 Solve for y. $3(y - 1) - 4(y + 2) = 11 - y$

Solution $3(y - 1) - 4(y + 2) = 11 - y$

$3y - 3 - 4y - 8 = 11 - y$

$-y - 11 = 11 - y$ *Add $+y$ to both sides.*

$-11 = 11$

Since $-11 = 11$ is always false, and is equivalent to our original equation, our original equation is a contradiction (this means there is no value of y that makes the equation true) and therefore there is no solution .

When we solve an equation which is in fact a contradiction, the variable drops out entirely and we get an equation which says that two unequal numbers are equal—which is always false.   ■

EXAMPLE 8

Solve for s. $8 - (s - 1) = -s + 9$

Solution

$$8 - (s - 1) = -s + 9$$
$$8 - s + 1 = -s + 9$$
$$9 - s = -s + 9 \qquad \textit{Add } +s \textit{ to both sides.}$$
$$9 = 9$$

Since $9 = 9$ is always true and is equivalent to our original equation, our original equation is an $\boxed{\text{identity}}$ (this means every value of s will make the equation true) and we say the equation is

$$\boxed{\text{true for all real numbers}}$$

Actually, you might already have recognized the third line of the solution as an identity.

When we solve an equation which is in fact an identity, the variable drops out entirely and we get an equation which says that a number is equal to itself—which is always true. ∎

We have seen that if a first-degree equation is conditional, then it has a unique solution which we can find by using the outline given previously. Otherwise, it must be either an identity or a contradiction.

STUDY SKILLS 2.1   **Preparing for Exams: Is Doing Homework Enough?**

At some point in time you will probably take an algebra exam. More than likely, your time will be limited and you will not be allowed to refer to any books or notes during the test.

Working problems at home may help you to develop your skills and to better understand the material, but there is still no guarantee that you can demonstrate the same high level of performance during a test as you may be showing on your homework. There are several reasons for this:

1. Unlike homework, exams must be completed during a limited time.

2. The fact that you are being assigned a grade may make you anxious and therefore more prone to careless errors.

3. Homework usually covers a limited amount of material, while exams usually cover much more material. This increases the chance of confusing or forgetting skills, concepts, and rules.

4. Your books and notes *are* available as a guide while working homework exercises.

Even if you do not deliberately go through your textbook or notes while working on homework exercises, the fact that you know what sections the exercises are from, and what your last lecture was about, cues you in on how the exercises are to be solved. You may not realize how much you depend on these cues and may be at a loss when an exam does not provide them for you.

If you believe that you understand the material and/or you do well on your homework, but your exam grades just do not seem to be as high as you think they should be, then the study skills discussed in this chapter should be helpful.

Exercises 2.1

In Exercises 1–8, determine whether the equation is an identity or contradiction.

1. $x - (x - 3) = 4$

2. $3a - 3(a - 2) = -6$

3. $4(w - 2) = 4w - 8$

4. $4(w - 2) = 4w - 2$

5. $2(y + 1) - (y - 5) = 4(y - 2) - 3(y - 5)$

6. $3(y - 1) - (3y - 1) = 2y - 1 - (2y + 1)$

7. $6(2y + 1) - 4(3y - 1) = y - (y + 10)$

8. $2(y - 4) + 2(4 - y) = 5(7 - y) + 5(y - 7)$

In Exercises 9–16, determine whether the listed values of the variable satisfy the equation.

9. $5(x - 3) - (x + 2) = 8 - x$; $x = 0, 5$

10. $7(x + 1) - (x - 2) = x - 1$; $x = 2, -2$

11. $3(a - 5) + 2(1 - a) = a - 8 - (a - 5)$; $a = -3, 0$

12. $4(t + 3) - (7 - t) = 3(t - 1) + 2(t + 4)$; $t = -1, 4$

13. $x^2 - 5x = 5 - x$; $x = -1, 5$

14. $6x - x^2 = x - 6$; $x = -4, 3$

15. $a(a + 8) = (a + 2)^2$; $a = -1, 1$

16. $(a + 4)^2 = a(a + 6)$; $a = -8, 8$

In Exercises 17–60, solve the equation.

17. $2x - 7 = 5$

18. $3a - 5 = 19$

19. $4y + 2 = -1$

20. $7z + 6 = -2$

21. $m + 3 = 3 - m$

22. $m - 3 = 3 - m$

23. $m + 3 = 3 + m$

24. $m - 3 = 3 + m$

25. $3t - 5 = 5t - 13$

26. $8w - 9 = 13w + 21$

27. $11 - 3y = 38$

28. $9 - 7y = 16$

29. $5s + 2 = 3s - 7$

30. $40t + 15 = 22t - 3$

31. $5 - 21x = 3x - 16$

32. $4 - 3b = 5b + 7$

33. $3x - 12 = 12 - 3x$

34. $5y - 7 = 7 - 5y$

35. $3x - 12 = -12 - 3x$

36. $5y - 7 = -7 - 5y$

37. $2t + 1 = 7 - t$

38. $6 - 5t = t - 12$

39. $2(x - 1) = 2(x + 1)$

40. $3(a + 5) = 4(a + 5) - (a + 5)$

41. $3(x + 1) - x = 2(x + 3)$

42. $4(x - 3) - 2x = 4(x - 1)$

43. $5(2t - 1) - 3t = 5 - 7t$

44. $6(3q - 4) + 5q - 2 = 5(4q + 1) + 3q - 7$

45. $2(x - 1) + 3(x + 1) = x - 3$

46. $3(x - 2) + 5(x - 3) = 15x$

47. $6(3 - z) + 2(4z - 5) = z - (3 - z)$

48. $9 - 5(x + 4) = -4(x + 1) - (x + 7)$

49. $4(3x - 1) - 5(3x - 2) = 2(x + 3) - 5x$

50. $7z - 3(4 - z) = 5(2z) + 12$

51. $x(x - 2) - 15 = x(x + 5) - 3(x + 5)$

52. $a(a + 4) - 6(a + 4) = a(a - 4) + 6(a - 4)$

53. $6x - [2 - (x - 1)] = x - 5(x + 1)$

54. $3x - [5 - 2(x + 1)] = x - 7(x + 2)$

55. $5x - 2[x - 3(7 - x)] = 3 - 2(x - 8)$

56. $3a - [5 - 2a + 3(a - 6)] = 2 + 3(4 - a)$

57. $3 - [4t + 5(2t - 1) - 3t] = 3(4 - t)$

58. $5d - [6d + 4(5 - d)] = 3 - [2d - (2d - 1)]$

59. $4 - 3\{2x - 5[3 + (x - 1)]\} = 4 - [2 - (x - 5)]$

60. $5 - \{4 - 2x + 2[2 - (x - 1)]\} = 3 - x$

CALCULATOR EXERCISES

In Exercises 61–64, solve the equation.

61. $3.24x - 5.2 = 7.74x + 0.3$

62. $7.24x - 12.3 = 0.5 + 4.84x$

63. $5.85(3.5x - 4) = 7.2(2.75x + 1.5)$

64. $6.5(2.5x - 4) = -2.5(5x + 5.34)$

? QUESTIONS FOR THOUGHT

65. What is the difference between a conditional equation and an identity?

66. What is the difference between a contradiction and an identity?

67. What does it mean to say that two equations are equivalent?

2.2

Verbal Problems

Virtually all of the work we have done so far to acquire algebraic skills, such as the ability to simplify expressions and solve equations, is designed to prepare us for applying these skills in the "real world." This normally involves translating a problem, whether it is from physics, chemistry, psychology, statistics, or some other field, into mathematical language, getting an equation or inequality out of this translation, and then solving it.

While you may think that some of the verbal problems in this book are somewhat artificial or contrived and have no connection with "reality," they do offer us an opportunity to practice the skills necessary to cope with a wide variety of problems.

In solving the problems that follow, we will always be finding an *algebraic solution*. By this we mean that we are using a systematic approach which we will outline a bit later in this section. We will not use a trial-and-error procedure because it is a very risky strategy for problem-solving. What if the problem has more than one solution, or has no solution, or the trial-and-error process is very long? Therefore, we will not accept a trial-and-error approach to solving a verbal problem.

Let's begin by looking at an example. (We will, of course, rely heavily on the material on translating verbal expressions which was covered in Section 1.6.)

EXAMPLE 1

The length of a rectangle is 3 less than 4 times its width. If the perimeter is 54 cm, what are the dimensions of the rectangle?

Solution

A diagram is especially useful in a problem like this. Note that the length is described in terms of the width.

Let w = Width.

Then $4w - 3$ = Length.   (The problem tells us that the length is "3 less than 4 times the width.")

Our rectangle is shown in Figure 2.1.

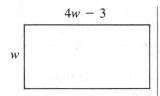

$4w - 3$

w

Figure 2.1
Rectangle for Example 1

Recall from Section 1.6, Example 6, that the formula for the perimeter, P, of a rectangle is

$$P = 2W + 2L \qquad \textit{Where W = width and L = length}$$

Our equation is therefore

$$2w + 2(4w - 3) = 54$$
We will indicate the equation we obtain by enclosing it in a shaded rectangular panel.

Now we solve the equation:

$$2w + 8w - 6 = 54$$
$$10w - 6 = 54$$
$$10w = 60$$
$$w = 6$$

The width, w, is 6 and therefore the length is $4w - 3 = 4(6) - 3 = 24 - 3 = 21$. The rectangle is

$$\boxed{6 \text{ cm by } 21 \text{ cm}}$$

CHECK: 21 is 3 less than 4 times 6.
$$2(6) + 2(21) = 12 + 42 = 54$$ ∎

The basic elements of the solution to Example 1 can be outlined as follows.

Outline of Strategy for Solving Verbal Problems	
1.	*Read the problem carefully*, as many times as is necessary to understand what the problem is saying and what it is asking.
2.	*Use diagrams* whenever you think it will make the given information clearer.
3.	*Find the underlying relationship or formula* relevant to the given problem. Ask if there is some underlying relationship or formula you need to know. If not, then the words of the problem themselves give the required relationship.
4.	Clearly *identify the unknown quantity* (or quantities) in the problem, and label it (them) using one variable.
5.	By using the underlying formula or relationship in the problem, *write an equation* involving the unknown quantity (or quantities). (This is the *crucial* and probably most difficult step.)
6.	*Solve* the equation.
7.	*Answer the question*. Make sure you have answered the question that was asked.
8.	*Check* the answer(s) in the original words of the problem.

EXAMPLE 2

A discount shoe outlet collected $763 on the sale of 62 pairs of sneakers. Some of the sneakers sold were women's sneakers selling for $14 per pair and the remainder were girls' sneakers selling for $11 per pair. How many of each type were sold?

Solution

At this point in our development we will restrict ourselves to solutions involving one variable. In Chapter 9 we will discuss a two-variable approach to this type of problem.

Let x = # of pairs of women's sneakers sold.

Then $62 - x$ = # of pairs of girls' sneakers sold because there are 62 pairs sold all together.   (*Total minus part equals remainder*)

Our equation involves the amount of money collected on the sale of the sneakers (that is, the *value* of the sneakers sold). We compute the amount as follows:

$$\left(\begin{array}{c} \text{Amount collected from} \\ \text{sale on all women's sneakers} \end{array} \right) + \left(\begin{array}{c} \text{Amount collected from} \\ \text{sale on all girls' sneakers} \end{array} \right) = \left(\begin{array}{c} \text{Total} \\ \text{collected} \end{array} \right)$$

$$\left(\begin{array}{c} \text{\# of women's} \\ \text{sneakers sold} \end{array} \right) \cdot \left(\begin{array}{c} \text{Cost of 1 pair} \\ \text{of women's sneakers} \end{array} \right)$$

$$+ \left(\begin{array}{c} \text{\# of girls'} \\ \text{sneakers sold} \end{array} \right) \cdot \left(\begin{array}{c} \text{Cost of 1 pair} \\ \text{of girls' sneakers} \end{array} \right) = 763$$

$$x \cdot 14 + (62 - x) \cdot 11 = 763$$

Then our equation is

$$14x + 11(62 - x) = 763$$

We now solve the equation:

$$14x + 682 - 11x = 763$$
$$3x + 682 = 763$$
$$3x = 81$$
$$x = 27$$

Thus, there were x = | 27 pairs of women's sneakers sold
and $62 - 27$ = | 35 pairs of girls' sneakers sold

CHECK: $27 + 35 = 62$
$14(27) + 11(35) = 378 + 385 = 763$ ∎

EXAMPLE 3

A printer is producing a pamphlet that will cost $3.45 per copy. There will be a certain number of color pages which cost 13¢ each, and 28 more than that number of black and white pages which cost 5¢ each. If, in addition, there is a 25¢ per copy cover charge, how many pages are there in the pamphlet all together?

Solution

Let c = # of color pages.

Then $c + 28$ = # of black and white pages.   (*There are 28 more black and white than color pages.*)

Our equation expresses the cost of printing each pamphlet. (We write our equations in cents to simplify the arithmetic.)

$$\begin{pmatrix} \text{Cost of printing} \\ \text{color pages} \end{pmatrix} + \begin{pmatrix} \text{Cost of printing} \\ \text{black and white pages} \end{pmatrix} + \begin{pmatrix} \text{Cover} \\ \text{charge} \end{pmatrix} = \begin{pmatrix} \text{Total} \\ \text{cost} \end{pmatrix}$$

$$\begin{pmatrix} \text{\# color} \\ \text{pages} \end{pmatrix} \cdot \begin{pmatrix} \text{cost of} \\ \text{1 color} \\ \text{page} \end{pmatrix} + \begin{pmatrix} \text{\# black} \\ \text{and white} \\ \text{pages} \end{pmatrix} \cdot \begin{pmatrix} \text{cost of 1} \\ \text{black and} \\ \text{white page} \end{pmatrix} + \begin{pmatrix} \text{Cover} \\ \text{charge} \end{pmatrix} = \begin{pmatrix} \text{Total} \\ \text{cost} \end{pmatrix}$$

$$c \quad\cdot\quad 13 \quad+\quad (c+28) \quad\cdot\quad 5 \quad+\quad 25 \quad=\quad 345$$

$$13c + 5(c + 28) + 25 = 345$$

Now we solve the equation:

$$13c + 5c + 140 + 25 = 345$$
$$18c + 165 = 345$$
$$18c = 180$$
$$c = 10$$

There were 10 color pages and 38 black and white pages (why 38?). Do not forget to answer the question. The problem asks for the total number of pages; therefore, the answer is

$$10 + 38 = \boxed{48 \text{ pages}}$$

The check is left to the student. ∎

If you look back at Examples 2 and 3 you will probably notice that while on the surface they seem to have nothing to do with each other, the equations we used to solve them are structurally very similar. Such problems are often called *value problems*. Example 4 has the same structure.

EXAMPLE 4

Jenna had 68 coins in nickels, dimes, and quarters totaling $7.30. If she had 3 more dimes than quarters, how many of each coin did she have?

Solution

$$\text{Let } x = \text{\# of quarters.}$$
$$\text{Then } x + 3 = \text{\# of dimes.}$$

What remains is to express the number of nickels in terms of x. If we subtract the number of quarters and dimes from the total number of coins, this will leave us with the remaining number of coins, the nickels. Hence,

$$\begin{pmatrix} \text{Total \#} \\ \text{of coins} \end{pmatrix} - \begin{pmatrix} \text{\# of} \\ \text{quarters} \end{pmatrix} - \begin{pmatrix} \text{\# of} \\ \text{dimes} \end{pmatrix} = \begin{pmatrix} \text{\# of} \\ \text{nickels} \end{pmatrix}$$

$$68 \quad-\quad x \quad-\quad (x+3) = \text{\# of nickels}$$

$$\text{The number of nickels} = 68 - x - (x + 3)$$
$$= 68 - x - x - 3$$
$$= 65 - 2x$$

Now we construct an equation involving the *value* of each type of coin. We again set up the problem in cents rather than dollars.

$$\begin{pmatrix}\text{Value of} \\ \text{all quarters}\end{pmatrix} + \begin{pmatrix}\text{Value of} \\ \text{all dimes}\end{pmatrix} + \begin{pmatrix}\text{Value of} \\ \text{all nickels}\end{pmatrix} = \begin{pmatrix}\text{Total value} \\ \text{of all coins}\end{pmatrix}$$

$$\begin{pmatrix}\text{\# of} \\ \text{quarters}\end{pmatrix} \cdot \begin{pmatrix}\text{Value of} \\ \text{a quarter}\end{pmatrix} + \begin{pmatrix}\text{\# of} \\ \text{dimes}\end{pmatrix} \cdot \begin{pmatrix}\text{Value of} \\ \text{a dime}\end{pmatrix}$$

$$+ \begin{pmatrix}\text{\# of} \\ \text{nickels}\end{pmatrix} \cdot \begin{pmatrix}\text{Value of} \\ \text{a nickel}\end{pmatrix} = \begin{pmatrix}\text{Total value} \\ \text{of all coins}\end{pmatrix}$$

$$x \cdot 25 + (x + 3) \cdot 10 + (65 - 2x) \cdot 5 = 730$$

Our equation is

$$25x + 10(x + 3) + 5(65 - 2x) = 730$$

We now solve the equation:

$$25x + 10x + 30 + 325 - 10x = 730$$
$$25x + 355 = 730$$
$$25x = 375$$
$$x = 15$$

Since $x = 15$, then

The number of quarters is 15;

the number of dimes is $15 + 3 = 18$; and

the number of nickels is $65 - 2(15) = 35$.

Our answer is

$$\boxed{15 \text{ quarters, 18 dimes, and 35 nickels}}$$

CHECK: 15 quarters $\doteq$ 15(.25) = $3.75

                18 dimes   = 18(.10) = $1.80   *Note that there are 3 more*

                35 nickels = 35(.05) = $1.75   *dimes than quarters.*

Total # of coins 68   Total value $7.30   ∎

EXAMPLE 5   Jane begins a 20-mile race at 9:00 A.M., running at an average speed of 10 mph. One hour later her brother leaves the starting line on a motorbike and follows her route at the rate of 40 mph. At what time does he catch up to her?

Solution

A simple diagram may help us visualize the problem. Figure 2.2 emphasizes the fact that when Jane's brother overtakes her, they have both travelled the *same distance*.

Figure 2.2
Diagram for Example 5

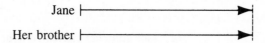

The basic idea in the problem is that if you travel at a constant rate of speed, then

$$\text{Distance} = (\text{Rate})(\text{Time})$$

or

$$d = rt$$

As indicated above, we have

$$d_{\text{Jane}} = d_{\text{brother}}$$

(d_{Jane} *is called a* **subscripted variable** *and is used to indicate the distance travelled by Jane.* d_{Jane} *and* d_{brother} *are two different variables, just like x and y.*)

Using the formula $d = rt$, we get

$$r_{\text{Jane}} \cdot t_{\text{Jane}} = r_{\text{brother}} \cdot t_{\text{brother}}$$

The information given in the problem tells us $r_{\text{Jane}} = 10$ and $r_{\text{brother}} = 40$, so our problem becomes

$$10t_{\text{Jane}} = 40t_{\text{brother}}$$

Since we are told that Jane's brother left 1 hour later,

Let $t = \#$ of hours Jane runs until her brother catches her.

Then $t - 1 = \#$ of hours Jane's brother rides until he catches up. (He leaves 1 hour later so he travels 1 hour *less*.)

Thus, our equation becomes

$$10t = 40(t - 1)$$
$$10t = 40t - 40$$
$$-30t = -40$$
$$t = \frac{-40}{-30} = \frac{4}{3} = 1\frac{1}{3} \text{ hours} = 1 \text{ hr } 20 \text{ min}$$

Jane's brother caught her 1 hour and 20 minutes after *she* began the race, at

$$\boxed{10:20}$$

CHECK: Jane started at 9:00 and met her brother at 10:20. Thus, she ran for 1 hour and 20 minutes ($1\frac{1}{3}$ hours). Travelling for $1\frac{1}{3}$ hours at 10 mph, Jane travels

$$\left(1\frac{1}{3}\right)10 = \left(\frac{4}{3}\right)10 = \frac{40}{3} = 13\frac{1}{3} \text{ miles}$$

Jane's brother started an hour after Jane, at 10:00, and met his sister at 10:20. Therefore, her brother travelled for 20 minutes ($\frac{1}{3}$ hour). Travelling for $\frac{1}{3}$ hour at 40 mph, her brother travels

$$\left(\frac{1}{3}\right)40 = \frac{40}{3} = 13\frac{1}{3} \text{ miles}$$

Since the distances are the same, the answer checks out.

Note that we could have let t = time for Jane's brother to catch up, and $t + 1$ = time Jane runs until her brother catches up. While the value for t will not be the same, the answer to the problem remains unchanged.   ∎

EXAMPLE 6

Jean can stuff 30 envelopes per hour and Bob can stuff 45 envelopes per hour.

(a)   How long would it take them to stuff 1,650 envelopes?

(b)   How long would it take if Bob started an hour after Jean?

Solution

As with Example 5, we are dealing with a rate (a quantity per unit of time). The basic relationship between the amount of work done and the rate at which the work is being done is as follows:

Amount of work done = (Work rate) · (Time)

or

$$W = r \cdot t$$

Thus, if Jean can stuff 30 envelopes per hour, her rate is 30 envelopes/hr.

In 4 hours, Jean can stuff: $W = (30)(4) = 120$ envelopes.

In 4 hours, Bob can stuff: $W = (45)(4) = 180$ envelopes.

Now let us proceed to answer the questions.

(a)   How long will it take Jean and Bob to stuff 1,650 envelopes? We can approach this problem in two ways.

METHOD 1:   First, we let x = the number of hours it takes for them to stuff 1,650 envelopes together. Then we can say that since Jean can stuff 30 envelopes per hour, the amount she can stuff in x hours is $30x$. Since Bob can stuff 45 envelopes in an hour, the amount he can stuff in x hours is $45x$. Together they stuff

$$\left(\begin{array}{c}\text{Amount of work} \\ \text{done by Jean}\end{array}\right) + \left(\begin{array}{c}\text{Amount of work} \\ \text{done by Bob}\end{array}\right) = \left(\begin{array}{c}\text{Total amount of} \\ \text{work done}\end{array}\right)$$

(Jean's rate) · (Jean's time) + (Bob's rate) · (Bob's time) = 1,650

30 · x + 45 · x = 1,650

Hence, our equation is

$$30x + 45x = 1,650$$

We solve the equation:

$$30x + 45x = 1,650$$
$$75x = 1,650$$
$$x = 22$$

Hence, it takes $\boxed{22 \text{ hours}}$ for Bob and Jean to stuff 1,650 envelopes.

METHOD 2:   The second way to approach the same problem is similar to the first method, but is more intuitive. If Jean can stuff 30 envelopes an hour and Bob can stuff 45 envelopes an hour, then together they can stuff $30 + 45 = 75$ envelopes in an hour. Our equation becomes

$$75x = 1,650$$

which is equivalent to the equation we found before.

Why bother with Method 1 at all? It will become apparent with part **(b)** of the example.

(b)   How long would it take if Bob started an hour after Jean? Because we are starting with different rates at different hours, we cannot simply add the rates as we did in Method 2 above. Instead, we will apply the logic of Method 1.

First, we let $x = $ # hours Jean spends stuffing envelopes.

Then $x - 1 = $ # hours Bob spends stuffing envelopes.   *(Since he starts an hour later, he is working an hour less than Jean.)*

Thus, we have

$$\left(\begin{array}{c}\text{Amount of work} \\ \text{done by Jean}\end{array}\right) + \left(\begin{array}{c}\text{Amount of work} \\ \text{done by Bob}\end{array}\right) = \left(\begin{array}{c}\text{Total amount of} \\ \text{work done}\end{array}\right)$$

(Jean's rate) · (Jean's time) + (Bob's rate) · (Bob's time) = 1,650

30 · x + 45 · $(x - 1)$ = 1,650

Hence, our equation is

$$30x + 45(x - 1) = 1,650$$

We solve as follows:

$$30x + 45(x - 1) = 1,650$$
$$30x + 45x - 45 = 1,650$$
$$75x - 45 = 1,650$$
$$75x = 1,695$$
$$x = 22\tfrac{3}{5}$$

So it takes them $\boxed{22\frac{3}{5} \text{ hours} \quad \text{or} \quad 22 \text{ hr } 36 \text{ min}}$ to stuff 1,650 envelopes if Bob starts an hour later.

The check is left to the student. ∎

One final comment: Do not get discouraged if you do not always get a complete solution to every problem. As you do more problems, you will get better at solving problems. Make an honest attempt to solve the problems and keep a written record of your work so that you can go back over your work to see exactly where you got stuck.

STUDY SKILLS 2.2 **Preparing for Exams: When to Study**

When to start studying and how to distribute your time studying for exams is as important as how to study. To begin with, "pulling all-nighters" (staying up all night to study just prior to an exam) seldom works. As with athletic or musical skills, algebraic skills cannot be developed overnight. In addition, without an adequate amount of rest, you will not have the clear head you need to work on an algebra exam. It is usually best to start studying early—from $1\frac{1}{2}$ to 2 weeks before the exam. In this way you have the time to perfect your skills and, if you run into a problem, you can consult your teacher to get an answer in time to include it as part of your studying.

It is also a good idea to distribute your study sessions over a period of time. That is, instead of putting in 6 hours in one day and none the next two days, put in 2 hours each day over the three days. You will find that not only will your studying be less boring, but also you will retain more with less effort.

As we mentioned before, your study activity should be varied during a study session. It is also a good idea to take short breaks and relax. A study "hour" could consist of 50 minutes of studying and a 10-minute break.

Exercises 2.2

Solve each of the following problems algebraically. Be sure to clearly label what the variable represents.

1. One number is 3 more than 4 times another and their sum is 43. What are the numbers?

2. One number is 5 more than twice another and their sum is 23. What are the numbers?

3. One number is 8 less than 5 times another. If the sum of the two numbers is −20, find the numbers.

4. One number is 6 less than 7 times another. If the sum of the two numbers is −6, find the numbers.

5. The sum of three consecutive integers is 66. Find them.

6. The sum of four consecutive integers is 1 more than the third. Find them.

7. Three consecutive even integers are such that the sum of the second and third is 6 more than twice the first. Find them.

8. The sum of three consecutive even integers is 36. Find them.

9. The sum of four consecutive odd integers is 56. Find them.

10. Four consecutive odd integers are such that the product of the first and third is 48 less than the product of the second and fourth. Find them.

11. The length of a rectangle is twice its width. If the perimeter is 42 meters, what are its dimensions?

12. The width of a rectangle is 8 cm less than its length. If the perimeter is 72 cm, what are its dimensions?

13. The length of a rectangle is 7 more than 3 times the width. If the perimeter is 54 cm, find the dimensions of the rectangle.

14. The length of a rectangle is 5 less than 4 times the width. If the perimeter is 80 cm, find the dimensions of the rectangle.

15. The first side of a triangle is 5 cm less than the second side, and the third side is twice as long as the first side. If the perimeter of the triangle is 33 cm, find the lengths of the sides of the triangle.

16. The shortest side of a triangle is 8 inches less than the medium side, and the longest side is 4 inches longer than 3 times the shortest side. If the perimeter is 27 inches, find the length of the shortest side.

17. The length of a rectangle is 1 more than 3 times the width. If the width is increased by 2 and the length is doubled, the new perimeter is 3 less than 5 times the original length. Find the original dimensions.

18. The width of a rectangle is 12 less than twice the length. If the length is increased by 3 and the width is doubled, the new perimeter is 6 less than 6 times the original length. What are the original dimensions?

19. A collection of nickels, dimes, and quarters is worth $5.30. There are two more dimes than nickels and four more quarters than dimes. How many of each type of coin are there?

20. A collection of nickels, dimes, and quarters is worth $21.75. There are twelve fewer nickels than dimes and as many quarters as there are nickels and dimes together. How many of each type of coin are there?

21. Bernard has $164 in $1, $5, and $10 bills. If he has twenty-five bills all together and he has one more $10 bill than $5 bills, how many of each type of bill does he have?

22. A collection of thirty-six coins consisting of nickels, dimes, and quarters is worth $4.00. If there are twice as many dimes as nickels, how many of each type of coin are there?

23. Barbie buys seventy-four stamps at the post office for $6.70. If she bought only 5¢, 10¢, and 15¢ stamps, and there were twice as many 15¢ stamps as 10¢ stamps, how many of each did she buy?

24. Ken buys a total of 100 stamps at the post office for $11.42. If he bought 5¢, 12¢, and 15¢ stamps only, and there were ten more 15¢ stamps than 5¢ stamps, how many of each did he buy?

25. A truck carries a load of fifty packages; some are 20-lb packages and the rest are 25-lb packages. If the total weight of all packages is 1,075 lb, how many of each type are on the truck?

26. A truck carries a load of eighty-five boxes; some are 30-lb boxes, some are 25-lb boxes, and some are 20-lb boxes. If there are three times as many 30-lb. boxes as 20-lb boxes, and the total weight of all the boxes is 2,245 lb, how many of each type of box are in the truck?

27. Marge works part-time in a shoe store. She earns a salary of $90 per week plus a commission of $2 on each pair of boots and $1 on each pair of shoes she sells. During a certain week she made thirty sales and earned a total income of $137. How many pairs of shoes did she sell?

28. A merchant wishes to purchase a shipment of clock radios. Simple AM models cost $35 each, while AM/FM models cost $50 each. In addition, there is a delivery charge of $70. If he spends $1,000 on twenty-four clock radios, how many of each type did he buy?

29. A gourmet food shop wants to mix some $2-per-pound coffee beans with 30 pounds of $3-per-pound coffee beans to produce a mixture which will sell for $2.60 per pound. How much $2-per-pound coffee should be used?

30. A toy store sells puzzles for $3 each and games for $7 each. A man spends $220 buying a certain number of puzzles and ten more than that many games. How many puzzles were bought?

31. Orchestra seats to a certain Broadway show are $24 each and balcony seats are $14 each. If a theater club spends $1,164 on the purchase of fifty-six seats, how many orchestra seats were purchased?

32. Orchestra seats to the Broadway show *Iwog* are $15 each and balcony seats are $7 each. If 156 tickets were sold for the matinee performance grossing $1,204, how many of each type of ticket were sold?

33. A plumber charges $22 per hour for her time and $13 per hour for her assistant's time. On a certain job the assistant works alone for 2 hours doing preparatory work, then the plumber and her assistant complete the job together. If the total bill for the job was $236, how many hours does the plumber work?

34. The plumber and assistant of Exercise 33 complete another job in which the assistant does the preparatory work alone and then the plumber completes the job alone. If the job took 9 hours and the bill came to $171, how many hours did each work?

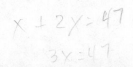

35. At 3:00 P.M., two cars are 345 miles apart and are travelling toward each other. If one car is travelling at 55 mph and the other at 60 mph, at what time will they meet?

36. Two cars leave from the same spot at the same time and travel in opposite directions. If one car travels at 50 mph and the other travels at 60 mph, how long will it be until they are 495 miles apart?

37. Two cars leave the same spot at the same time and travel in opposite directions. If one car is travelling at 35 kph and the other at 50 kph, how long will it take for them to be 595 kilometers apart?

38. Alice left her house at 7:00 A.M. for work. Her husband found her pocketbook at home at 8:00 A.M., so he jumped into his car to try to catch up with her. If Alice travels to work at 30 mph and her husband is travelling at 75 mph, can he catch up to her before she gets to work if her work is 60 miles from her house? If he does, what times does he catch up to her?

39. How long would it take someone jogging at 17 kph to overtake someone walking at 7 kph with a 3-hour head start?

40. Repeat Exercise 39 if the walker walks at the rate of 6 kph for 3 hours and then walks at 7 kph.

41. A person drives to a convention at the rate of 48 kph and returns home at the rate of 54 kph. If the total driving time is 17 hours, how far away is the convention?

42. A person can drive from town A to town B at a certain rate of speed in 6 hours. If he increases his speed by 20 kph he can make the trip in 4 hours. How far is it from town A to town B?

43. A manager needs 5,125 copies of a document. A new duplicating machine can make 50 copies per minute while an older model can make 35 copies per minute. The older machine begins making copies but breaks down before completing the job and is replaced by the new machine, which completes the job. If the total time for the job is 1 hour and 50 minutes, how many copies did the older machine make?

44. A new computer printer can print 30 pages per minute while an older model can print 20 pages per minute. The older printer begins printing a 1,190-page report but breaks down before completing the job. The newer printer then completes the job. If the total printing time for the report was 47 minutes, how many pages did the older printer print?

45. An experienced worker can process 60 items per hour, while a new worker can process 30 per hour. How many hours will it take to complete a job of processing 6,750 items if they work together?

46. Repeat Exercise 45 if the experienced worker works alone for 12 hours and then the new worker completes the job alone.

47. Repeat Exercise 45 if the experienced worker begins the job of processing the 6,750 items 3 hours before being joined by the new worker.

48. Repeat Exercise 45 if the new worker begins the job of processing the 6,750 items 3 hours before being joined by the experienced worker.

49. An experienced assembly-line worker can package 80 items per hour, while a new trainee can package 48 items per hour. A trainee begins packaging a stack of 656 items at 10:00 A.M. and is then joined by the experienced worker 3 hours later. At what time will they finish packaging the stack?

50. A college professor is grading eighty exam papers. He grades the papers at a certain rate for 3 hours. Then fatigue sets in and he completes the grading in 2 more hours at a rate of 5 exam papers per hour slower. How fast was he grading the papers for the first 3 hours?

2.3

First-Degree Inequalities

In Section 2.1 we started by discussing the various types of first-degree equations and discovered that not all first-degree equations have unique solutions.

A similar analysis applies to inequalities.

An inequality may be a ***contradiction,*** such as $x^2 < 0$. (Why is this always false?)

Or it may be an *identity*, such as $x + 1 > x$. (Why is this always true?)

Or it may be *conditional*, such as $x + 1 > 3$.

EXAMPLE 1 | Determine whether the following is an identity or a contradiction:

$$7x - (3x - 4) < 5x + (4 - x)$$

Solution | As with some equations, we may not be able to tell what type of inequality we have in its present form. Thus, our first step is again to simplify each side as completely as possible.

$7x - (3x - 4) < 5x + (4 - x)$ *Remove parentheses by the distributive*

$\quad 7x - 3x + 4 < 5x + 4 - x$ *property and combine terms.*

$\qquad\quad 4x + 4 < 4x + 4$

Thus, we see that the inequality is a ⎡contradiction⎤ (Why?) ∎

As with equations, the values that make an inequality true are called *solutions* of the inequality, and we say that those values *satisfy* the inequality. But while conditional linear equations have single-value solutions, the solutions to conditional linear inequalities are infinite sets. Even the solution to the double inequality $0 < x < 1$ is the infinite set of numbers which lie between 0 and 1.

EXAMPLE 2 | Determine whether each of the values listed satisfies the given inequality.

$$8x + 2(x - 4) \geq 12 \qquad \text{for} \quad x = -1, 2, 4$$

Solution | We replace each occurrence of the variable, on both sides of the inequality, by the proposed value.

CHECK $x = -1$: $8x + 2(x - 4) \geq 12$

$\qquad\qquad\qquad 8(-1) + 2(-1 - 4) \overset{?}{\geq} 12$

$\qquad\qquad\qquad\qquad -8 + 2(-5) \overset{?}{\geq} 12$

$\qquad\qquad\qquad\qquad\qquad -18 \ngeq 12$

Therefore, ⎡ $x = -1$ is *not* a solution ⎤

CHECK $x = 2$: $8x + 2(x - 4) \geq 12$

$\qquad\qquad\qquad 8(2) + 2(2 - 4) \overset{?}{\geq} 12$

$\qquad\qquad\qquad\quad 16 + 2(-2) \overset{?}{\geq} 12$

$\qquad\qquad\qquad\qquad\quad 12 \overset{\checkmark}{\geq} 12$

Therefore, ⎡ 2 is a solution ⎤

*Remember: A number is equal to itself and is therefore greater than **or** equal to itself.*

CHECK $x = 4$: $8x + 2(x - 4) \geq 12$

$$8(4) + 2(4 - 4) \overset{?}{\geq} 12$$

$$32 + 2(0) \overset{?}{\geq} 12$$

$$32 \overset{\checkmark}{\geq} 12$$

Therefore, | $x = 4$ does satisfy the inequality | ∎

Properties of Inequalities

Inequalities that have exactly the same set of solutions are called *equivalent inequalities*. As with equations, it will be our goal to take an inequality and transform it into successively simpler *equivalent* inequalities, finally resulting in an *obvious inequality*—one that has the variable isolated on one side of the inequality (or in the middle of a double inequality).

The properties of inequalities will help us to systematically achieve our goal. These properties are a bit different from the properties of equalities. Keeping in mind that $a < b$ means that a is to the left of b on the real number line, we can see that adding or subtracting the same quantity on both sides of an inequality *preserves* the direction (or sense) of the inequality. For example,

$$-2 < 1$$
$$-2 + 4 < 1 + 4 \qquad \text{\textit{Add 4 to both sides.}}$$
$$2 < 5$$

$$-2 < 0$$
$$-2 - 4 < 0 - 4 \qquad \text{\textit{Subtract 4 from both sides.}}$$
$$-6 < -4$$

However, multiplication and division are somewhat different.

Multiplying by a positive number

$$-2 < 5$$
$$3(-2) \; ? \; 3(5) \qquad \text{\textit{Multiply both sides by 3.}}$$
$$-6 < 15$$

Multiplying by a negative number

$$-2 < 5$$
$$-3(-2) \; ? \; -3(5) \qquad \text{\textit{Multiply both sides by -3.}}$$
$$+6 > -15 \qquad \text{\textit{The inequality symbol has reversed.}}$$

These numerical examples illustrate what the situation is in general. We also need to point out that the division and multiplication properties must be the same. After all, $\dfrac{a}{c} = \left(\dfrac{1}{c}\right)a$, so that dividing by c is the same as multiplying by $\dfrac{1}{c}$. If c is positive so is $\dfrac{1}{c}$, and if c is negative so is $\dfrac{1}{c}$.

Properties of Inequalities	In order to obtain an equivalent inequality:
	1. We may add or subtract the same quantity on each side of an inequality and the inequality symbol remains the same.
	2. We may multiply or divide each member of an inequality by a *positive* quantity and the inequality symbol remains the same.
	3. We may multiply or divide each member of an inequality by a *negative* quantity and the inequality symbol is *reversed*.

Algebraically, we can write these properties as follows (for properties, definitions, and theorems throughout this book, the "<" symbol can be replaced by ">," "$\geq$," or "$\leq$"):

Properties of Inequalities	If $a < b$, then
	$$a + c < b + c$$
	$$a - c < b - c$$
	$ac < bc$ when c is positive; $\qquad ac > bc$ when c is negative
	$\dfrac{a}{c} < \dfrac{b}{c}$ when c is positive; $\qquad \dfrac{a}{c} > \dfrac{b}{c}$ when c is negative

Solving First-Degree Inequalities

DEFINITION

A ***first-degree inequality*** is an inequality that can be put in the form $ax + b < 0$, where a and b are constants and $a \neq 0$. (Recall that the "<" symbol can be replaced with ">", "$\geq$", or "$\leq$".)

The procedure we have outlined for solving first-degree equations works equally well for first-degree inequalities. However, we must exercise a bit more care as Example 3 illustrates.

EXAMPLE 3

Solve for x. $\quad 9 - 5(x - 3) \geq 14$

Solution

As with an equation, our goal is to isolate the variable on one side of the inequality.

$$9 - 5(x - 3) \geq 14 \qquad \text{\textit{First simplify the left side of the inequality.}}$$
$$9 - 5x + 15 \geq 14$$
$$24 - 5x \geq 14 \qquad \text{\textit{Then apply the inequality properties.}}$$
$$24 - 5x - 24 \geq 14 - 24 \qquad \text{\textit{Note that the inequality symbol does not change.}}$$

$$-5x \geq -10$$

$$\frac{-5x}{-5} \leq \frac{-10}{-5}$$

$$\boxed{x \leq 2}$$

*The inequality symbol does not reverse when we subtract from both sides. However, when we **divide** both sides of an inequality by a **negative** number we must **reverse** the inequality symbol.*

Visualizing the solutions of an inequality on the real number line can be helpful. To sketch the graph of the solution set for this example (see Figure 2.3), we use the notation introduced in Chapter 1.

Figure 2.3
Solution set for Example 3

$$-3 \quad -2 \quad -1 \quad 0 \quad 1 \quad 2 \quad 3 \quad 4 \quad 5 \quad 6$$

*Remember, a **filled-in** circle indicates that the point 2 is to be included as part of the solution.*

This solution means that any number less than or equal to 2 should satisfy the original inequality. How do we check this? It is impossible to check every number less than or equal to 2.

We would like to check that the endpoint, 2, is correct and that the inequality symbol is facing in the right direction. In general, the endpoint of our answer is the solution to the equation corresponding to the original inequality. That is, the endpoint, 2, is the solution to the *equation,* $9 - 5(x - 3) = 14$.

Thus, we check by replacing x with 2 in the original inequality to determine if both sides of the inequality yield the same value. This tells us that our endpoint, 2, is correct.

To determine if our inequality symbol is facing the right direction, we check by using a number less than 2 to see that it does satisfy the inequality. (This you can do mentally.)

CHECKS:

We check to see what happens when $x = 2$:

$$9 - 5(x - 3) = 14$$
$$9 - 5(2 - 3) \stackrel{?}{=} 14$$
$$9 - 5(-1) \stackrel{?}{=} 14$$
$$9 + 5 \stackrel{\checkmark}{=} 14$$

Therefore, our *endpoint,* 2, is correct.

We can choose any number less than 2, say $x = 0$. This should satisfy the inequality.

$$9 - 5(x - 3) \geq 14$$
$$9 - 5(0 - 3) \stackrel{?}{\geq} 14$$
$$9 - 5(-3) \stackrel{?}{\geq} 14$$
$$9 + 15 \stackrel{\checkmark}{\geq} 14$$

Therefore, the direction of our inequality symbol is correct. ∎

EXAMPLE 4

Solve for a and sketch the solution set on the real number line.

$$5a - 3 > -9 - (4 - 3a)$$

Solution

$$5a - 3 > -9 - (4 - 3a)$$
$$5a - 3 > -9 - 4 + 3a \qquad \text{\textit{Simplify the right side of the inequality.}}$$
$$5a - 3 > -13 + 3a \qquad \text{\textit{Then apply the inequality properties.}}$$
$$5a - 3 + 3 > -13 + 3a + 3$$
$$5a > -10 + 3a$$
$$5a - 3a > -10 + 3a - 3a$$
$$2a > -10$$
$$\frac{2a}{2} > \frac{-10}{2}$$
$$\boxed{a > -5}$$

We graph the solution set on the number line as shown in Figure 2.4.

Figure 2.4
Solution set for Example 4

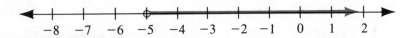

*Note that an **empty** circle indicates that the point −5 is to be excluded.*

The check is left to the student. ∎

As we discussed in Chapter 1, the **double inequality,** $a < x < b$, is used to indicate "betweenness." For example,

$$-7 < x < 1 \quad \text{means that } x \text{ is between } -7 \text{ and } 1$$

and is read

"x is greater than −7 and less than 1"

Note that we read the middle variable first, then the left-hand number, and then the right-hand number.

The inequality symbols separate the double inequality into three parts called ***members:***

$$a \qquad < \qquad x \qquad < \qquad b$$

Left member   Middle member   Right member

Recall that the double inequality is actually a combination of two inequalities that must be satisfied simultaneously. That is,

$$a < x < b$$

is actually a combination of the two inequalities

$$a < x \qquad \text{and} \qquad x < b$$

where x satisfies **both** inequalities at the same time. In other words, x must be *both* greater than a *and* less than b.

We have to be careful to avoid putting two single inequalities together into one double inequality where either it is not required, or it does not make sense. For example, if we have a region such as that shown in Figure 2.5, where $x < -3$ *or* $x > 5$, we cannot use the double inequality notation because x *cannot* satisfy both inequalities simultaneously (x is not "between" -3 and 5).

Figure 2.5
$x < -3$ **or** $x > 5$

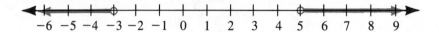

When writing a double inequality in its obvious form ($a < x < b$, where a and b are constants and x is the variable), the left-hand constant, a, should be the smaller of the two constants; this way the double inequality reflects the relative positions occupied by the members on the number line (see Figure 2.6).

Figure 2.6
$4 < x \leq 8$

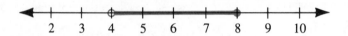

EXAMPLE 5

Which of the following do not make sense?

(a) $4 < x < 9$ **(b)** $-5 < x < -7$ **(c)** $-5 < z > 4$

(d) $-2 \geq y > -5$

Solution

(a) $4 < x < 9$ is satisfied by $x = 5$ and therefore

$$\boxed{\text{makes sense}}$$

(b) $-5 < x < -7$

$$\boxed{\text{does not make sense}}$$

since there is no number which is both less than -7 and, at the same time, greater than -5.

(c) $-5 < z > 4$ This is

$$\boxed{\text{not acceptable notation}}$$

Remember that double inequalities are used to indicate "betweenness."

(d) $-2 \geq y > -5$ is satisfied by $y = -3$ and

$$\boxed{\text{does make sense}}$$

However, the standard practice is to put the *smaller number to the left* of the variable. Hence, it should be rewritten as

$$-5 < y \le -2$$ ∎

As with obvious double inequalities, the double inequality

$$-7 \le 2x + 3 < 9$$

is actually two inequalities:

$$-7 \le 2x + 3 \quad \text{and} \quad 2x + 3 < 9$$

We *could* solve them both, proceeding as follows:

$$-7 - 3 \le 2x + 3 - 3 \quad \text{and} \quad 2x + 3 - 3 < 9 - 3$$
$$-10 \le 2x \quad \text{and} \quad 2x < 6$$
$$\frac{-10}{2} \le \frac{2x}{2} \quad \text{and} \quad \frac{2x}{2} < \frac{6}{2}$$
$$-5 \le x \quad \text{and} \quad x < 3$$

Since x has to satisfy both inequalities simultaneously, we would rewrite the two inequalities as one double inequality:

$$-5 \le x < 3$$

On the other hand, notice that we applied exactly the same transformation to each inequality at the same time. So it would be more convenient to leave the double inequality as it is and to *apply the same transformation to each **member*** of the inequality. Again the goal is to isolate x, but for double inequalities, x will be the middle member.

$$-7 \le 2x + 3 < 9$$
$$-7 - 3 \le 2x + 3 - 3 < 9 - 3$$
$$-10 \le 2x < 6$$
$$\frac{-10}{2} \le \frac{2x}{2} < \frac{6}{2}$$
$$-5 \le x < 3$$

To isolate x in the middle, first subtract 3 from all three members.

Then divide all three members by 2.

We should check that the endpoints, -5 and 3, are correct, and then check a value between the endpoints. We leave this to the student.

EXAMPLE 6   Solve for x and sketch the solution set on the real number line:

$$-1 < 7 - 4x \le 9$$

Solution   In solving a double inequality, we want to isolate the variable as the *middle member* of the inequality.

$$-1 \boxed{-7} < 7 - 4x \boxed{-7} \le 9 \boxed{-7}$$ *Subtract 7 from each member of the inequality.*

$$-8 < -4x \le 2$$

$$\frac{-8}{-4} > \frac{-4x}{-4} \ge \frac{2}{-4}$$ *Divide each member of the inequality by -4. Note that both inequality symbols reverse when we divide by a negative number.*

$$2 > x \ge -\frac{1}{2}$$ *Simplify fractions.*

$$\boxed{-\frac{1}{2} \le x < 2}$$ *Rewrite the double inequality in standard form.*

The sketch of this solution set is shown in Figure 2.7.

Figure 2.7
Solution set for Example 6

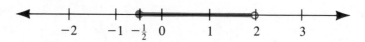

$$-2 \quad -1 \ -\tfrac{1}{2} \ 0 \quad 1 \quad 2 \quad 3$$

Note the filled-in circle at $-\frac{1}{2}$ and the empty circle at 2.

We leave the check to the student.   ■

As is the case with equations, if we find that in the process of solving an inequality the variable drops out entirely, then we have either an identity or a contradiction.

STUDY SKILLS 2.3   Preparing for Exams: Study Activities

If you are going to learn algebra well enough to be able to demonstrate high levels of performance on exams, then you must concern yourself with developing your skills in algebraic manipulation and understanding what you are doing and why you are doing it.

Many students concentrate only on skills and resort to memorizing the procedures for algebraic manipulations. This may work for quizzes or a test covering just a few topics. For exams covering a chapter or more this can be quite a burden on your memory. Eventually interference occurs and problems and procedures get confused. If you find yourself doing well on quizzes but not on longer exams, this may be your problem.

Concentrating on understanding what a method is and why it works is important. Neither the teacher nor the textbook can cover every possible way in which a particular concept may present itself in a problem. If you understand the concept, you should be able to recognize it in any problem. But again, if you concentrate only on understanding concepts and not on developing skills, you may find yourself prone to making careless and costly errors under the pressure of an exam.

In order to achieve both skill development and conceptual understanding, your studying should include four activities: (**1**) practicing problems, (**2**) reviewing your notes and textbook, (**3**) drilling with study cards, and (**4**) reflecting on the material and exercises.

Rather than doing any one of these activities over a long period of time, it is best to do a little of the first three activities during a study session and save some time for reflection at the end of the session.

Exercises 2.3

In Exercises 1–8, determine whether the inequality is an identity or a contradiction.

1. $x + 1 > x - 1$

2. $a - 7 \leq a + 3$

3. $2x - (x - 3) \geq x + 5$

4. $2x - (x - 3) < x + 5$

5. $-6 \leq r - (r + 6) < 3$

6. $-6 < r - (r + 6) < 3$

7. $0 < x - (x - 3) \leq 5$

8. $-7 \leq a - (a + 3) < -3$

In Exercises 9–14, determine whether the listed values of the variable satisfy the inequality.

9. $4 + 3u > 10;\quad u = 2, 3$

10. $7 + 2u \leq 15;\quad u = 4, 5$

11. $6 - 4y \leq 8;\quad y = -1, -2$

12. $8 - 5y \geq -3;\quad y = -1, 1$

13. $-3 \leq 3 - (4 - z) < 3;\quad z = -2, 4$

14. $-4 < 8 - 2(1 - r) \leq 4;\quad r = 5, -1$

In Exercises 15–24, determine whether the inequality makes sense. Put those which do make sense into standard form if they are not already in standard form.

15. $5 < x < 8$

16. $-5 < z < -8$

17. $-6 > w \geq -8$

18. $-6 \leq a < -5$

19. $-3 > x > 2$ $2 < x < -3$

20. $2 < a < -4$

21. $-3 < a < -2$

22. $-4 < b < 3$

23. $2 > x > -3$

24. $2 < x < -3$

In Exercises 25–34, solve each inequality.

25. $5x - 2 < 13$

26. $3x + 7 \geq 4$

27. $5y \geq 2 - 3y$

28. $5y < 7y - 2$

29. $3x + 4 \leq 2x - 5$

30. $5x - 7 < 2x + 6$

31. $3x + 4 \geq 2x - 5$

32. $5x - 7 > 2x + 6$

33. $11 - 3y < 38$

34. $9 - 7y \geq 16$

In Exercises 35–58, solve each inequality and sketch the graph of the solution set on the real number line.

35. $3a - 8 < 8 - 3a$

36. $5y - 4 \geq 4 - 5y$

37. $3a - 8 < -8 - 3a$

38. $5y - 4 \geq -4 - 5y$

39. $3t - 5 > 5t - 13$

40. $8w - 9 < 13w + 21$

41. $2y - 3 \leq 5y + 7$

42. $4 - 3t \geq 13 - t$

43. $5(a - 2) - 7a > 2a + 10$

44. $4(m + 7) - 5 < 20$

45. $5(a + 1) - 6(a - 2) < 10$

46. $4(t - 3) - 5(t + 2) > 0$

47. $7 - 5(t - 2) \leq 2$

48. $4 + 3(1 - a) \geq -8$

49. $2(y + 3) + 3(y - 4) < 3(y + 2) + 2(y + 1)$

50. $5(x - 2) - 3(x + 4) \geq 2x - 22$

51. $5 - 2[3x - 2(2x - 3)] \geq 3x - [2 - 2(x - 1)]$

52. $3a - \{2 - 5[3 + 2(a - 4)]\} \geq a - 6$

53. $1 \leq c + 3 < 5$

54. $-3 < r - 3 \leq 1$

55. $6 < 2k - 1 \leq 11$

56. $11 \leq 3z + 2 < 15$

57. $-1 < 4 - t < 3$

58. $-3 \leq 6 - t \leq -1$

In Exercises 59–68, solve each inequality.

59. $1 \le 8 - 3t \le 12$

60. $0 < 9 - 5t < 29$

61. $2 < 5 - 3(x + 1) < 17$

62. $13 < 7 - 2(x - 3) \le 31$

63. $-2 < x - 5 \le 1$

64. $-4 < k - 3 < 2$

65. $-2 \le 5 - x < 1$

66. $-4 < 3 - k < 2$

67. $1 \le 1 - (5z - 2) \le 11$

68. $0 < 2 - (3w - 4) < 6$

▫ CALCULATOR EXERCISES

In Exercises 69–70, solve each inequality and sketch the graph of the solution set on the real number line.

69. $0.39 \le 0.72x - 1.5 < 8.1$

70. $7.55 < 0.75x - 2.5 \le 8.5$

? QUESTIONS FOR THOUGHT

71. Describe what is *wrong* (if anything) with the following:

(a) $3x < -6$

$\dfrac{3x}{3} \overset{?}{<} \dfrac{-6}{3}$

$x \overset{?}{>} -2$

(b) $-2x > 4$

$\dfrac{-2x}{-2} \overset{?}{>} \dfrac{4}{-2}$

$x \overset{?}{<} 2$

(c) $-2x > 4$

$\dfrac{-2x}{-2} \overset{?}{>} \dfrac{4}{-2}$

$x \overset{?}{>} -2$

(d) $x + 2 \le 5$

$x + 2 - 2 \overset{?}{\le} 5 - 2$

$x \overset{?}{\ge} 3$

72. What is the difference between a conditional inequality and an identity?

73. What is the difference between a contradiction and an identity?

74. Discuss the differences between the properties of equations and the properties of inequalities.

75. What does it mean to say that two inequalities are equivalent?

76. Describe the difference between the method of checking an equation and that of checking an inequality.

2.4

Verbal Problems Involving Inequalities

The outline suggested in Section 2.2 for solving verbal problems applies equally well to problems which give rise to inequalities. A few examples will serve to illustrate this idea.

EXAMPLE 1

What numbers satisfy the condition that "3 less than 4 times the number is less than 29"?

Solution

In translating this problem algebraically we must be careful to distinguish the phrase "3 less than" from the phrase "3 is less than." The phrase "3 less than" indicates that we are subtracting 3 from some quantity—it is not a statement of inequality. An example of this would be the phrase "3 less than 8," which gets translated as $8 - 3$.

On the other hand, the phrase "3 is less than 8" is a statement of inequality, which is translated as $3 < 8$.

Let $x = $ a number satisfying the condition given in the example.

The given condition can be translated as follows:

$$\underbrace{\text{3 less than 4 times the number}}_{4x - 3} \quad \underbrace{\text{is less than}}_{<} \quad \underbrace{29}_{29}$$

Thus, our inequality is

$$\boxed{4x - 3 < 29}$$

Now we solve this inequality:

$$4x - 3 \;\boxed{+ 3} < 29 \;\boxed{+ 3}$$
$$4x < 32$$
$$\frac{4x}{4} < \frac{32}{4}$$
$$x < 8$$

CHECK:

Check the endpoint $x = 8$:

$$4x - 3 = 29$$
$$4(8) - 3 \overset{?}{=} 29$$
$$32 - 3 \overset{?}{=} 29$$
$$29 \overset{\checkmark}{=} 29$$

The endpoint, 8, is correct.

Check any value less than 8, say 7:

$$4x - 3 < 29$$
$$4(7) - 3 \overset{?}{<} 29$$
$$28 - 3 \overset{?}{<} 29$$
$$25 \overset{\checkmark}{<} 29$$

The direction of the inequality symbol is correct.

Thus,

$$\boxed{\text{any number less than 8}}$$

satisfies the condition of this problem. ■

EXAMPLE 2

The perimeter of a rectangle is to be at least 48 cm and no greater than 72 cm. If the width is 9 cm, what is the range of possible values for the length?

Solution

Our rectangle, with length L, is shown in Figure 2.8. The perimeter, P, is $2L + 2(9)$ or

$$P = 2L + 18$$

Figure 2.8
Rectangle for Example 2

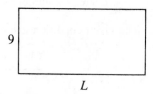

The problem tells us that we require

$$48 \leq \text{Perimeter} \leq 72$$

Since the perimeter is $2L + 18$, our inequality becomes

$$48 \leq 2L + 18 \leq 72$$

Now we solve the inequality:

$$48 - 18 \leq 2L + 18 - 18 \leq 72 - 18$$
$$30 \leq 2L \leq 54$$
$$\frac{30}{2} \leq \frac{2L}{2} \leq \frac{54}{2}$$
$$15 \leq L \leq 27$$

The length can range between

$$\boxed{15 \text{ cm and } 27 \text{ cm, inclusively}} \quad \text{(including 15 and 27)}.$$

The check is left to the student. ∎

EXAMPLE 3

An aide in the mathematics department gets paid $3 per hour for clerical work and $8 per hour for tutoring. If she wants to work a total of 20 hours and earn at least $135, what is the maximum number of hours she can spend on clerical work?

Solution

This problem is similar to the value problems covered in Section 2.2, except that it requires an inequality. Let's take a closer look at the problem.

If the aide works the whole 20 hours on clerical work, then she would be paid $20(3) = \$60$, certainly less than her goal. So we know that she has to spend less than 20 hours on clerical work.

On the other hand, if she does not spend any hours on clerical work, then she would spend all 20 hours tutoring and would make $20(8) = \$160$, more than enough for her goal. Thus, the maximum number of hours to be spent doing clerical work must lie between 0 and 20 for the aide to realize her goal of earning at least $135.

Now let's set up the problem:

Let x = # of hours spent on clerical work.

Then $20 - x$ = # of hours spent tutoring.

$$\begin{pmatrix} \text{Amount earned} \\ \text{from clerical work} \end{pmatrix} + \begin{pmatrix} \text{Amount earned} \\ \text{from tutoring} \end{pmatrix} = \begin{pmatrix} \text{Total amount} \\ \text{earned} \end{pmatrix} \geq 135$$

$$\begin{pmatrix} \text{Hourly rate} \\ \text{for} \\ \text{clerical} \end{pmatrix} \cdot \begin{pmatrix} \text{\# hours} \\ \text{spent on} \\ \text{clerical} \end{pmatrix} + \begin{pmatrix} \text{Hourly rate} \\ \text{for} \\ \text{tutoring} \end{pmatrix} \cdot \begin{pmatrix} \text{\# hours} \\ \text{spent on} \\ \text{tutoring} \end{pmatrix} \geq 135$$

$$3 \quad \cdot \quad x \quad + \quad 8 \quad \cdot \quad (20 - x) \quad \geq 135$$

Hence, our inequality is

$$3x + 8(20 - x) \geq 135$$

Now we solve our inequality:

$$3x + 8(20 - x) \geq 135$$
$$3x + 160 - 8x \geq 135$$
$$160 - 5x \geq 135$$
$$-5x \geq -25$$
$$x \leq 5 \qquad \text{\textit{Note that we are dividing by} } -5 \text{ \textit{and must therefore}}$$
$$\text{\textit{reverse the inequality symbol.}}$$

Thus, the number of hours spent on clerical work must be less than or equal to 5, or *the maximum number of hours the aide must spend on clerical work is*

$$\boxed{5 \text{ hours}}$$

if she is to make at least $135.

The process of making up study cards is a learning experience in itself. Study cards are convenient to use—you can carry them along with you and use them for review in between classes or as you wait for a bus. Use the (D/P and W) cards as follows:

1. Look at the heading of a card and, covering the rest of the card, see if you can remember what the rest of the card says.

2. Continue this process with the remaining cards. Pull out cards you know well and put them aside, but review them from time to time. Study cards you do not know.

3. Shuffle the cards so that they are in random order and repeat the process again from the beginning.

4. As you go through the cards, ask yourself the following questions (where appropriate):

(a) When do I use this rule, method, or principle?

(b) What are the differences and similarities between problems?

(c) What are some examples of the definitions or concepts?

(d) What concept is illustrated by the problem?

(e) Why does this process work?

(f) Is there a way to check this problem?

Exercises 2.4

Solve each of the following problems algebraically. Be sure to clearly label what each variable represents.

1. What numbers satisfy the condition "4 less than 3 times the number is less than 17"?

2. What numbers satisfy the condition "5 less than 4 times the number is greater than 19"?

3. If 12 more than 6 times a number is greater than 3 times the number, how large must the number be?

4. If 9 less than 5 times a number is less than twice the number, how small must the number be?

5. If 3 less than 4 times a number is greater than or equal to 3 less than 7 times the number, what is the largest the number can be?

6. If 7 more than twice a number is less than 6 more than twice the number, what can the number be?

7. What is the maximum length of the side of a square which has a maximum perimeter of 72 feet?

8. What is the minimum length of the side of a square which has a minimum perimeter of 94 feet?

9. The width of a rectangle is 8 cm. If the perimeter is to be at least 80 cm, what must the length be?

10. If the width of a rectangle is 10 meters and the perimeter is not to exceed 120 meters, how large can the length be?

11. The length of a rectangle is 18 in. If the perimeter is to be at least 50 in. but not greater than 70 in., what is the range of values for the width?

12. The length of a rectangle is twice the width. If the perimeter is to be at least 80 feet but not more than 100 feet, what is the range of values for the width of the rectangle?

13. The length of a rectangle is 3 times the width. If the perimeter is between 100 and 200 feet, what is the range of values for the width?

14. The length of a rectangle is 4 times its width. If the width varies from 8 to 12 feet, what is the range of values for the perimeter?

15. The length of a rectangle is 5 inches more than 3 times the width. If the width varies from 12 inches to 16 inches, what is the range of values for the perimeter?

16. The length of a rectangle is 3 feet less than twice its width. If the width varies from 6 to 13 feet, what is the range of values for the length? What is the range of values for the perimeter?

17. The medium side of a triangle is 2 centimeters longer than the shortest side and the longest side is twice as long as the shortest side. If the perimeter of the triangle is to be at least 30 cm and no more than 50 cm, what is the range of values for the shortest side?

18. The first side of a triangle is 3 times the second side and the third side is 5 m more than the second side. If the perimeter is to be between 35 and 50 m, what is the range of values for each side?

19. An organization wants to sell tickets to a concert. It plans on selling 300 reserved-seat tickets and 150 tickets at the door. The price of a reserved-seat ticket is to be $2 more than a ticket at the door. If the organization wants to collect at least $3,750, what is the minimum price it can charge for a reserved-seat ticket?

20. Repeat Exercise 19 if the price of a reserved-seat ticket is to be $3 more than a ticket at the door.

21. Jerome has fifty coins in dimes and quarters in his pockets. What is the minimum number of quarters he must have in order to have at least $9.80?

22. Allen has forty coins in dimes and quarters in his pockets. What is the minimum number of dimes he must have in order to have at most $8.95?

23. Xerxes has forty coins in nickels and dimes. What is the maximum number of dimes he should have in order to have no more than $2.85?

24. Bridgett has fifty coins in nickels and dimes. What is the minimum number of dimes she should have in order to have at least $3.45?

25. Carolyn makes two types of stuffed animals: stuffed panda bears and stuffed elephants. She charges $20 for the bears and $25 for the elephants. If she decides to make twenty-four stuffed animals this week, what is the minimum number of elephants she should make if she is to gross at least $575 for the twenty-four animals?

26. Jules enjoys spending time as a teacher, although he is paid only $32 a day. On the other hand, he receives $85 a day for what he considers to be a less enjoyable job as a car mechanic. If he plans to work 25 days this month, what is the minimum number of days he should work as a mechanic if he is to make at least $1,171?

27. Ian makes $6 an hour when he tutors and $4 an hour when he works at a local restaurant. Out of 30 hours working time, what is the minimum number of hours he should tutor if he is to make at least $190?

28. Barry makes $5 for each case of chocolate bars he sells and $7 for each case of hard candy. If he is supposed to sell seventy-five cases of either type, what is the minimum number of cases of hard candy he should sell in order to make at least $530?

29. Joe makes 25¢ on every regular hot dog he sells from his stand and 45¢ on every super-dog he sells. Joe just bought a new sign for his stand for $38.60. What is the minimum number of super-dogs he must sell in order to make enough to pay for his new sign by the 100th sale of either dog?

30. Repeat Exercise 29 and determine the minimum number of super-dogs Joe must sell by his 100th sale (of either dog), if his new sign cost him $52.50.

31. Zach wants to invest his money in the stock market. He decides on two stock issues which both sell at the same price. Stock A is a "safe," conservative stock which pays an annual dividend of $2 per share. Stock B, on the other hand, is a more risky investment but pays an annual dividend of $3 per share. Zach wants to invest enough money to buy 1,000 shares of either stock, but he does not want to take the risk of buying all stock B. He decides that he would like to get at least $2,400 annual dividends from his stock investment. What is the minimum number of shares of stock B he should buy in order to assure a $2,400 dividend annually?

2.5

Absolute-Value Equations and Inequalities

In Chapter 1 we defined the absolute value of x, $|x|$, *geometrically* as follows:

$|x|$ is the distance on the real number line between x and 0.

In this section we will draw on this idea of the absolute value of an expression being its distance to 0 to solve first-degree equations and inequalities involving absolute values.

EXAMPLE 1

Solve for x. $|x| = 4$

Solution

We will begin solving absolute-value equations and inequalities by using the geometric definition of absolute value to interpret the equation or inequality on the real number line. We will then replace the absolute-value equation or inequality with one *without* absolute values.

The equation $|x| = 4$ means that the distance from x to 0 is 4. In other words, x is located 4 units away from 0. Therefore, x must be 4 or -4, as illustrated in Figure 2.9.

Figure 2.9
$|x| = 4$

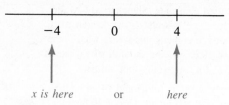

Therefore, $x = -4$ or $x = 4$

EXAMPLE 2

Solve for a. $|3a - 1| = 5$

Solution

$3a - 1$ must be 5 units away from 0, as shown in Figure 2.10.

Figure 2.10
$|3a - 1| = 5$

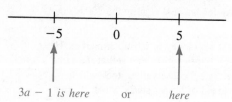

Therefore,

$$3a - 1 = -5 \qquad \text{or} \qquad 3a - 1 = 5$$
$$3a = -4 \qquad\qquad\qquad 3a = 6$$
$$\boxed{a = \frac{-4}{3}} \qquad\qquad\qquad \boxed{a = 2}$$

CHECK $a = \dfrac{-4}{3}$: $|3a - 1| = 5$

$$|3(\tfrac{-4}{3}) - 1| \overset{?}{=} 5$$

$$|-4 - 1| \overset{?}{=} 5$$

$$|-5| \overset{\checkmark}{=} 5$$

CHECK $a = 2$: $|3a - 1| = 5$

$$|3(2) - 1| \overset{?}{=} 5$$

$$|6 - 1| \overset{?}{=} 5$$

$$|5| \overset{\checkmark}{=} 5 \quad \blacksquare$$

Now that we have seen the geometric (number-line) interpretation of absolute-value equations, we can see that:

For $a \geq 0$,

$|x| = a$ *is equivalent to* $x = a$ or $x = -a$.

Hence,

$|x| = 7$ yields two equations (and therefore two answers):

$$x = 7 \quad \text{or} \quad x = -7$$

$|2t - 5| = 7$ yields two equations:

$$2t - 5 = 7 \quad \text{or} \quad 2t - 5 = -7$$

We would solve for t to find the two answers: $t = 6$ or $t = -1$.

EXAMPLE 3

Solve for t. $|2t + 3| - 4 = 3$

Solution

In order to apply the analysis we have used thus far, we would like to have the absolute value isolated on one side of the equation.

$|2t + 3| - 4 + 4 = 3 + 4$ *Add 4 to both sides.*

$\quad |2t + 3| = 7$

This means that the expression $2t + 3$ must be 7 units away from 0, or, equivalently,

$$2t + 3 = 7 \quad \text{or} \quad 2t + 3 = -7$$
$$2t = 4 \qquad\qquad 2t = -10$$
$$\boxed{t = 2} \quad \text{or} \quad \boxed{t = -5}$$

CHECK $t = 2$: $|2t + 3| - 4 = 3$

$$|2(2) + 3| - 4 \overset{?}{=} 3$$

$$|4 + 3| - 4 \overset{?}{=} 3$$

$$|7| - 4 \overset{?}{=} 3$$

$$7 - 4 \overset{\checkmark}{=} 3$$

CHECK $t = -5$: $|2t + 3| - 4 = 3$

$$|2(-5) + 3| - 4 \overset{?}{=} 3$$

$$|-10 + 3| - 4 \overset{?}{=} 3$$

$$|-7| - 4 \overset{?}{=} 3$$

$$7 - 4 \overset{\checkmark}{=} 3 \quad \blacksquare$$

EXAMPLE 4   *Solve for y.* $|y - 3| = -2$

Solution   Be careful! Do not blindly apply a procedure where it is not appropriate. This example requires that the absolute value, which is a distance, be negative. But distance is always measured in *positive* units. The absolute value of an expression can never be negative; therefore, there are

$$\boxed{\text{no solutions}}$$

■

EXAMPLE 5   *Solve for x.* $|2x + 5| = |x - 2|$

Solution   In order for the absolute value of two expressions to be equal, they must both be the same distance from 0. This can happen if the two expressions are equal *or* if they are negatives of each other. For instance, if $|a| = |b|$ then a and b can both be 6, $|6| = |6|$, or a can be 6 and b can be -6 (or vice versa), $|6| = |-6|$.

Thus, our absolute-value equation will be true if

$2x + 5 = x - 2$ or $2x + 5 = -(x - 2)$ *Notice that we must take the*

$$\boxed{x = -7}$$

$$2x + 5 = -x + 2$$

negative of the entire expression.

$$3x = -3$$

$$\boxed{x = -1}$$

The check is left to the student.   ■

The same kind of analysis allows us to solve absolute-value inequalities as well. Again, our goal is to replace the absolute-value inequality with an inequality (or inequalities) *without* absolute values.

EXAMPLE 6   *Solve for x.* $|x| < 3$

Solution   This inequality means that x must be less than 3 units away from 0. Therefore, x must fall between -3 and 3, as illustrated in Figure 2.11.

Figure 2.11
$|x| < 3$

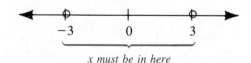

x must be in here

We can write the solution set as the double inequality

$$\boxed{-3 < x < 3}$$

■

EXAMPLE 7   Solve for n and sketch the solution on the real number line:

$$|2n - 1| < 5$$

Solution

Proceeding as we did in the last example, we require that $2n - 1$ be less than 5 units away from 0. Therefore, $2n - 1$ must lie between -5 and 5, as shown in Figure 2.12.

Figure 2.12
$|2n - 1| < 5$

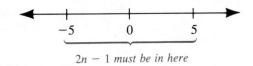

$2n - 1$ *must be in here*

We write this as the double inequality:

$$-5 < 2n - 1 < 5 \qquad \textit{Now solve for n.}$$
$$-5 + 1 < 2n - 1 + 1 < 5 + 1$$
$$-4 < 2n < 6$$
$$\frac{-4}{2} < \frac{2n}{2} < \frac{6}{2}$$
$$\boxed{-2 < n < 3}$$

Keep in mind that this solution means that in order for the expression $2n - 1$ to be less than 5 units away from 0, n must be between -2 and 3, as shown in Figure 2.13.

Figure 2.13
Solution set for Example 7

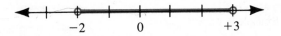

Note that the empty circles at the endpoints indicate a strict inequality.

The check is left to the student. ∎

EXAMPLE 8

Solve for y. $|y| > 2$

Solution

This inequality means that y must be more than 2 units away from 0. Therefore, y must be to the left of -2, or to the right of 2, as indicated in Figure 2.14.

Figure 2.14
$|y| > 2$

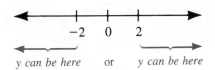

y can be here   or   *y can be here*

We must write the solution set as *two* separate inequalities:

$$\boxed{y < -2 \quad \text{or} \quad y > 2}$$

∎

EXAMPLE 9

Solve for w and sketch the solution on the real number line:

$$|3w - 2| > 4$$

Solution

This inequality says that the expression $3w - 2$ must be more than 4 units away from 0. Therefore, $3w - 2$ must be to the left of -4 or to the right of 4, as shown in Figure 2.15.

Figure 2.15
$|3w - 2| > 4$

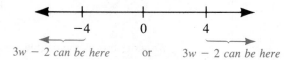

We can write the solution set as *two* separate inequalities:

$$
\begin{array}{ccc}
3w - 2 < -4 & \text{or} & 3w - 2 > 4 \\
3w - 2 + 2 < -4 + 2 & \text{or} & 3w - 2 + 2 > 4 + 2 \\
3w < -2 & & 3w > 6
\end{array}
$$

$$\boxed{w < \dfrac{-2}{3} \quad \text{or} \quad w > 2}$$

On the number line the solution set appears as shown in Figure 2.16.

Figure 2.16
Solution set for Example 9

With regard to solving absolute-value *inequalities,* we can summarize the previous discussions as follows:

For $a > 0$,

$$|x| < a \quad \text{is equivalent to} \quad -a < x < a$$

whereas

$$|x| > a \quad \text{is equivalent to} \quad x < -a \text{ or } x > a$$

Note the differences between the two types of inequalities and how they should be handled.

EXAMPLE 10

Solve each of the following for x and sketch the solution on the real number line:
(a) $|5x - 3| \le 7$ **(b)** $|5x - 3| > 7$

Solution

(a) $|5x - 3| \leq 7$ is equivalent to $-7 \leq 5x - 3 \leq 7$.

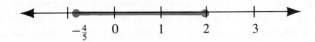

$$-7 + 3 \leq 5x - 3 + 3 \leq 7 + 3$$
$$-4 \leq 5x \leq 10$$
$$-\frac{4}{5} \leq \frac{5x}{5} \leq \frac{10}{5}$$

$$\boxed{-\frac{4}{5} \leq x \leq 2}$$

The solution set is shown in Figure 2.17.

Figure 2.17
Solution set for
Example 10(a)

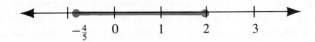

(b) $|5x - 3| > 7$ is equivalent to

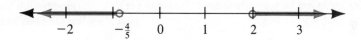

$$5x - 3 < -7 \quad \text{or} \quad 5x - 3 > 7$$
$$5x - 3 + 3 < -7 + 3 \quad \text{or} \quad 5x - 3 + 3 > 7 + 3$$
$$5x < -4 \quad \text{or} \quad 5x > 10$$

$$\boxed{x < -\frac{4}{5} \quad \text{or} \quad x > 2}$$

The solution set is shown in Figure 2.18.

Figure 2.18
Solution set for
Example 10(b)

Notice the similarities and the differences between parts (a) and (b). We cannot use a double inequality to describe the solution in part (b). ∎

EXAMPLE 11

Solve for t:
(a) $|t + 3| < -2$ (b) $|t - 5| > -1$

Solution

As we have mentioned before, look at a problem carefully before you blindly plunge ahead.

(a) Since absolute value means distance from 0, this part of this example is asking that a distance be less than -2, which is impossible. (Distance is always measured in positive units.) Thus, the inequality in part (a) is a

$$\boxed{\text{contradiction}}$$

and has no solutions.

(b) Part **(b)** is asking that a distance be greater than -1, which is always true (a distance is *always* greater than or equal to 0). Therefore, part **(b)** is an

$$\boxed{\text{identity}}$$

and all values of t satisfy the inequality. ∎

STUDY SKILLS 2.5   **Preparing for Exams: Reviewing Your Notes and Text; Reflecting**

Another activity we suggested as an important facet of studying for exams is to review your notes and text. Your notes are a summary of the information you believed was important at the time you wrote them down. In the process of reviewing your notes and text you may turn up something you missed. A gap in your understanding may get filled and consequently give more meaning to some of the definitions, rules, and concepts on your study cards (and make them easier to remember). Perhaps you will understand a shortcut that you missed the first time around.

Reviewing the explanations or problems in the text *and* your notes gives you a better perspective and helps to tie the material together. Concepts will begin to make more sense when you review and think about how they are interrelated. It is also important to practice review problems so that you will not forget those skills you have already learned. Do not forget to

review old homework exercises, quizzes, and exams—especially those problems which were incorrectly done. Review problems also offer an excellent opportunity to work on your speed as well as your accuracy.

We discussed the importance of reflecting on the material you are reading and the exercises you are doing. Your thinking time is usually limited during an exam, and you want to be able to anticipate variations in problems and to make sure that your careless errors will be minimized at that time. For this reason it is a good idea to try to think about possible problems ahead of time. In areas where you tend to get confused, make the distinctions that exist as clear as possible.

As you review material, ask yourself the study questions given in Study Skills 1.1. Also look at the Questions for Thought at the end of most of the exercise sets and ask yourself those questions as well.

Exercises 2.5

In Exercises 1–42, solve the absolute-value equation or inequality.

1. $|x| = 4$

2. $|a| = 6$

3. $|x| < 4$

4. $|a| > 6$

5. $|x| > 4$

6. $|a| < 6$

7. $|x| \leq 4$

8. $|a| \geq 6$

9. $|x| \geq 4$

10. $|a| \leq 6$

11. $|x| = -4$

12. $|a| = -6$

13. $|x| > -4$

14. $|a| \leq -6$

15. $|x| \leq -4$

16. $|a| > -6$

17. $|t - 3| = 2$

18. $|y| + 2 = 7$

19. $|t| - 3 = 2$

20. $|y + 2| = 7$

21. $|5 - n| = 1$

22. $|3 - y| = 2$

23. $|a - 5| < 3$

24. $|a + 1| > 4$

25. $|a - 1| \geq 2$

26. $|a + 7| \leq 3$

27. $|2a - 5| < -1$

28. $|2a - 9| \geq -5$

29. $|2a| - 5 < -1$

30. $|2a| - 9 \geq -5$

31. $|3x - 2| - 3 = 1$

32. $|5t - 4| + 4 = 3$

33. $|3x - 2| + 3 = 1$

34. $|5t - 4| - 4 = 3$

35. $|2(x - 1) + 7| = 5$

36. $|3(y + 2) - 10| = 4$

37. $|3 - a| \leq 2$

38. $|6 - m| > 5$

39. $|5 - 2a| > 1$

40. $|7 - 2a| \leq 3$

41. $|4(x - 1) - 5x| < 4$

42. $|3(y + 2) - 7y| \geq 6$

In Exercises 43–52, sketch the solution set of the equation or inequality on the real number line.

43. $|x - 1| = 5$

44. $|a + 3| = 4$

45. $|3 - x| \leq 2$

46. $|5 - a| \geq 3$

47. $|2x + 7| > 1$

48. $|2a - 9| < 5$

49. $|4x - 5| < 3$

50. $|5a - 4| < 3$

51. $|3 - 4x| < 1$

52. $|4 - 3x| < 1$

In Exercises 53–62, solve the absolute-value equation.

53. $|5t - 1| = |4t + 3|$

54. $|3t + 2| = |2t - 5|$

55. $|4r - 3| = |2r + 9|$

56. $|5r + 6| = |3r - 12|$

57. $|a - 5| = |2 - a|$

58. $|4 - n| = |n - 6|$

59. $|3x - 4| = |4x - 3|$

60. $|2x - 5| = |5 - 2x|$

61. $|x + 1| = |x - 1|$

62. $|t - 3| = |t + 3|$

 ? QUESTION FOR THOUGHT

63. Verbally describe the differences between the following two inequalities:

$$|x| < a \quad \text{and} \quad |x| > a$$

Describe the approach you would use in solving them.

CHAPTER 2   Summary

After having completed this chapter, you should be able to:

1. Understand and recognize the basic types of equations and inequalities: conditional, identity, and contradiction; and the properties of equations and inequalities (Sections 2.1 and 2.3).

2. Solve and check a first-degree equation in one variable (Section 2.1).

For example: *Solve for x.* $5(x - 3) - (8x - 1) = 3(x - 2) - 8(x + 2)$

Solution: $5(x - 3) - (8x - 1) = 3(x - 2) - 8(x + 2)$ *Simplify both sides of*
$$5x - 15 - 8x + 1 = 3x - 6 - 8x - 16$$ *the equation, then apply equality properties.*

$$-3x - 14 = -5x - 22$$ *Add 5x to both sides of the equation.*

$$2x - 14 = -22$$ *Add 14 to both sides of the equation.*

$$2x = -8$$
$$x = -4$$ *Divide both sides of the equation by 2.*

CHECK $x = -4$: $5(-4 - 3) - [8(-4) - 1] \stackrel{?}{=} 3(-4 - 2) - 8(-4 + 2)$
$$5(-7) - (-32 - 1) \stackrel{?}{=} 3(-6) - 8(-2)$$
$$-35 - (-33) \stackrel{?}{=} -18 + 16$$
$$-2 \stackrel{\checkmark}{=} -2$$

3. Solve, check, and sketch the solution set of a first-degree inequality in one variable (Section 2.3).

For example: *Solve for x.* $7 - 3(x + 2) < 22$

Solution: $7 - 3(x + 2) < 22$ *Simplify where necessary.*
$$7 - 3x - 6 < 22$$
$$1 - 3x < 22$$ *Then apply inequality properties. Add −1 to both sides of the inequality.*
$$-3x < 21$$
$$\frac{-3x}{-3} > \frac{21}{-3}$$ *Dividing by a negative number reverses the inequality symbol.*
$$x > -7$$

CHECK: **Check the endpoint** $x = -7$:   Pick a value of x greater than −7. Let $x = -2$ (why −2?):

$$7 - 3(x + 2) = 22 \qquad\qquad 7 - 3(x + 2) < 22$$
$$7 - 3(-7 + 2) \stackrel{?}{=} 22 \qquad\qquad 7 - 3(-2 + 2) \stackrel{?}{<} 22$$
$$7 - 3(-5) \stackrel{?}{=} 22 \qquad\qquad 7 - 3(0) \stackrel{?}{<} 22$$
$$7 + 15 \stackrel{?}{=} 22 \qquad\qquad 7 - 0 \stackrel{\checkmark}{<} 22$$
$$22 \stackrel{\checkmark}{=} 22$$

The solution set on the real number line is shown here:

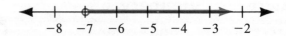

4. Solve verbal problems algebraically by using first-degree equations or inequalities in one variable (Sections 2.2 and 2.4).

For example: John spends a total of $19 on twelve batteries. If alkaline batteries cost $2 each and regular batteries cost $1 each, how many of each type did he buy?

Solution:

Let x = # of alkaline batteries purchased.

Then $12 - x$ = # of regular batteries purchased.

$$2x \quad + \quad 1(12 - x) \quad = \quad 19$$

Cost of   *Cost of*   *Total cost*
alkaline batteries   *regular batteries*

Our equation is

$$2x + (12 - x) = 19$$
$$2x + 12 - x = 19$$
$$x + 12 = 19$$
$$x = 7$$

John purchased 7 alkaline batteries and $12 - 7 = 5$ regular batteries.

5. Solve absolute-value equations and inequalities of the first degree (Section 2.5).

For example:

(a) *Solve for t.* $|3t - 5| = 8$

Solution: $|3t - 5| = 8$ means that $3t - 5$ is 8 units away from 0. Therefore,

$$3t - 5 = 8 \quad \text{or} \quad 3t - 5 = -8$$
$$3t = 13 \qquad\qquad 3t = -3$$
$$t = \frac{13}{3} \quad \text{or} \quad t = -1$$

(b) *Solve for a.* $|2a - 3| \leq 7$

Solution: $|2a - 3| \leq 7$ means that $2a - 3$ is less than or equal to 7 units away from 0 on the real number line. Therefore,

$2a - 3$ is between -7 and 7, inclusively.

Hence, our inequality:

$$-7 \leq 2a - 3 \leq 7 \qquad \text{\textit{Add }} +3 \text{ \textit{to each member.}}$$
$$-7 + 3 \leq 2a - 3 + 3 \leq 7 + 3$$
$$-4 \leq 2a \leq 10 \qquad \text{\textit{Divide each member by 2.}}$$
$$\frac{-4}{2} \leq \frac{2a}{2} \leq \frac{10}{2}$$
$$-2 \leq a \leq 5$$

(c) *Solve for x.* $|8 - 5x| > 2$

Solution: $|8 - 5x| > 2$ means that $8 - 5x$ is more than 2 units away from 0 on the real number line. Therefore,

$$8 - 5x < -2 \quad \text{or} \quad 8 - 5x > 2$$

Solve each inequality.

$$8 - 5x < -2 \quad \text{or} \quad 8 - 5x > 2 \qquad \textit{First add } -8 \textit{ to both sides of each inequality.}$$

$$-5x < -10 \quad \text{or} \quad -5x > -6 \qquad \textit{Divide each side of each inequality by } -2.$$

$$x > 2 \qquad \text{or} \qquad x < \frac{6}{5} \qquad \textit{Note that we reverse both inequality symbols.}$$

CHAPTER 2   Review Exercises

In Exercises 1–34, solve the equation:

1. $5x - 2 = -2$

2. $3x - 4 = 8$

3. $3x - 5 = 2x + 6$

4. $3y - 5 = 8y - 9$

5. $11x + 2 = 6x - 3$

6. $7x + 8 = 3x$

7. $5(a - 3) = 2(a - 4)$

8. $7 - (2 - b) = 3b - 4(b - 1)$

9. $6(q - 4) + 2(q + 5) = 8q - 19$

10. $3(z - 7) + 2z - 4 = 5(z - 5)$

11. $3x - 5x = 7x - 4x$

12. $2a + 5a = 7a - 3a$

13. $6(8 - a) = 3(a - 4) + 2(a - 1)$

14. $3(a - 4) - 2(a - 4) = a - 4$

15. $7[x - 3(x - 3)] = 4x - 2$

16. $5[x - 2(x + 5) + 3] = 3x - 5$

17. $5\{x - [2 - (x - 3)]\} = x - 2$

18. $2 - \{x - 6[x - (4 - x)]\} = 3(x + 2)$

19. $|x| = 3$

20. $|a| = -5$

21. $|x| = -3$

22. $|y| = 7$

23. $|2x| = 8$

24. $|-3x| = 5$

25. $|a + 1| = 4$

26. $|a - 3| = 5$

27. $|4z + 5| = 0$

28. $|7x + 2| = 6$

29. $|2y - 5| - 8 = -3$

30. $|4x - 9| + 5 = 7$

31. $|x - 5| = |x - 1|$

32. $|a + 3| = |a + 1|$

33. $|2t - 4| = |t - 2|$

34. $|2a - 5| = |a - 3|$

In Exercises 35–50, solve the inequality.

35. $3x - 4 \leq 5$

36. $2x \leq x - 3$

37. $5x - 4 \leq 2x$

38. $3x - 2 > 5x - 6$

39. $5z + 4 > 2z - 1$

40. $3y - 2 > -y + 3$

41. $5(s - 4) < 3(s - 4)$

42. $2(t - 3) + 4 \leq 2t - 7$

43. $3(r - 2) - 5(r - 1) \geq -2r - 1$

44. $5(x - 7) - 2(x - 1) < 3(x - 11)$

45. $-3 \leq 2x - 1 \leq 5$

46. $5 \leq 3x + 2 \leq 20$

47. $-5 \leq 6 - 2x \leq 7$

48. $4 \leq 3 - x \leq 12$

49. $-3 \leq 3(x - 4) \leq 6$

50. $-5 \leq 5(x - 3) \leq 6$

In Exercises 51–72, solve the inequality and sketch a graph of its solution set if it is conditional.

51. $3[a - 3(a + 2)] > 0$

52. $2 - (b - 4) \leq 5 - b$

53. $5 - [2 - 3(2 - x)] < -3x + 1$

54. $5 - [2 - 3(x - 2)] > -3x + 1$

55. $3 - \{2 - 5[q - (2 - q)]\} \leq 3[q - (q - 5)]$

56. $2\{x - [3 - (5 - x)]\} > 2 - x$

57. $|x| < 4$

58. $|x| > 4$

59. $|s| \geq 5$

60. $|x| \leq 5$

61. $|t| < 0$

62. $|t| \leq 0$

63. $|t - 1| < 2$

64. $|t - 1| > 2$

65. $|a - 6| \geq 3$

66. $|a - 6| \leq 3$

67. $|r + 9| \leq 4$

68. $|r - 9| \geq 4$

69. $|2x - 1| \geq 2$

70. $|2x - 1| > 2$

71. $|3x - 2| < 4$

72. $|3x - 2| > 4$

In Exercises 73–78, solve the inequality.

73. $|2x - 3| \leq 5$

74. $|2x - 3| \geq 5$

75. $|3x + 5| + 2 > 7$

76. $|3x + 5| + 2 < 7$

77. $|3 - 2x| + 8 \leq 4$

78. $|3 - 2x| - 8 \leq 4$

79. If 3 times a number is 4 less than 4 times the number, what is the number?

80. Five less than 5 times a number is 3 greater than the number. What is the number?

81. If 5 times the sum of a number and 6 is 2 less than the number, what is the number?

82. If the sum of 5 times a number and 6 is 2 less than the number, what is the number?

83. Harold has nickels and dimes in his pockets, thirty-five coins in all, totaling $3.45. How many of each coin does he have?

84. Sarah has seventy coins in nickels and quarters. If they total $12.50, how many of each coin does she have?

85. Alex has forty-two coins in nickels and quarters totaling $7.50. How many of each coin does he have?

86. Earl has forty-five coins in quarters and dimes totaling $9.15. How many of each coin does he have?

87. Cindy has seventy coins in quarters, dimes, and nickels totaling $7.15. If she has 3 more nickels than quarters, how many of each type does she have?

88. Lew has eighty-two coins in quarters, dimes, and nickels totaling $9.85. If he has 7 more nickels than quarters, how many of each coin does he have?

89. Thirty packages were delivered. Some were 8-lb packages and the rest were 5-lb packages. If the total weight of all the packages was 186 lb, how many of each type of package were delivered?

90. Forty-five packages were in a storeroom. Some weighed 8 lb and the rest weighed 5 lb. If the total weight of all the packages was 276 lb, how many of each type of package were in the storeroom?

91. A TV repairman charges $15 to fix a television and $10 to fix a radio. Yesterday he fixed a total of twenty-three radios and televisions. If he collected $295, how many of each did he fix?

92. A plumber charges $25 an hour for his services and $10 an hour for his assistant's services. On a certain job the assistant did the preparatory work alone and then the plumber completed the job alone. The total bill for the job came to $134.50: $27 for parts and the rest for 7 hours total labor. How many hours did each work on the job?

93. Bob averages $12 an hour editing textbooks and $20 an hour tutoring. If he is working 30 hours this week, what is the minimum number of hours he must tutor in order for him to make at least $456?

94. Refer to Exercise 93. If Bob works 40 hours, what is the maximum number of hours he can tutor if he is to make no more than $680?

95. Jerry has eighty coins in nickels, dimes, and quarters. He has twice as many dimes as nickels. What is the minimum number of quarters he needs if he is to have at least $25?

CHAPTER 2 Practice Test

Solve the following equations.

1. $7x - 9 = 5x + 2$

2. $3x - 4(x - 5) = 7(x - 2) - 8(x - 1)$

3. $6a - [2 - 3(a - 4)] = 5(a - 10)$

Solve the following inequalities.

4. $3x + 5 \leq 5x - 3$

5. $3x - 2[x - 3(1 - x)] > 5$

6. $-3 < 5 - 2x < 4$

Solve the following and graph the solution set on the real number line.

7. $|2x - 7| + 2 = 9$

8. $|3x - 4| < 5$

9. $|5x - 7| \geq 2$

10. A truck carries a load of ninety-three boxes; some weigh 35 kg and the rest weigh 45 kg. If the total weight of all the boxes is 3,465 kg, how many of each type are in the truck?

11. How long would it take a jogger travelling at 8 mph to catch up with a person walking 3 mph with a 2-hour head start?

12. Ken has thirty-two coins in his pocket in nickels and dimes. What is the minimum number of dimes he must have in order to have at least $2.65?

 omit

STUDY SKILLS 2.6   **Preparing for Exams: Using Quiz Cards**

A few days before the exam, select an appropriate number of problems from the quiz cards, old exams, or quizzes, and make up a practice test for yourself. You may need the advice of your teacher as to the number of problems and the amount of time to allow yourself for the test. If they are available, old quizzes and exams may help guide you.

Now find a quiet, well-lit place with no distractions, set your clock for the appropriate time limit (the same as your class exam will be) and take the test. Pretend it is a real test; that is, do not leave your seat or look at your notes, books, or answers until your time is up. (Before giving yourself a test you may want to refer to next chapter's Study Skills discussions on taking exams.)

When your time is up, stop; you may now look up the answers and grade yourself. If you are making errors, check over what you are doing wrong. Find the section where those problem types are covered, review the material, and try more problems of that type.

If you do not finish your practice test on time, you should definitely work on your speed. Remember that speed, as well as accuracy, is important on most exams.

Think about what you were doing as you took your test. You may want to change your test-taking strategy or reread the next chapter's discussion on taking exams. If you were not satisfied with your performance and you have time after the review, give yourself another practice test.

3

Polynomials

3.1

Polynomials: Sums, Differences, and Products

In Chapter 1 we discussed the real number system and demonstrated how the real number properties can be used to simplify various algebraic expressions. Most of the algebraic expressions covered in that chapter were of a particular variety: they were polynomial expressions.

In this chapter we will take a closer look at polynomials and polynomial operations. We will begin with the basic algebraic expression called the *monomial*.

DEFINITION

A *monomial* is an algebraic expression that is either a constant or a product of a constant and one or more variables raised to whole-number powers.

The following are examples of monomials:

$$3x^2y = 3xxy \qquad 5xy^4 = 5xyyyy \qquad \frac{x}{4} = \frac{1}{4}x \qquad 7$$

Note that each is a product of constants and/or variables. On the other hand,

$$3x + 2y \qquad 5x^{1/2} \qquad \frac{3}{x+2} \qquad x - 4$$

cannot be represented as products of constants and/or variables with whole-number exponents, and are therefore *not* monomials.

DEFINITION

A *polynomial* is a finite sum of monomials.

We may name polynomials by the number of monomials or terms making up the polynomial: a *binomial* is a polynomial consisting of two terms and a *trinomial* is a polynomial consisting of three terms.

Besides the number of terms, we can also classify a polynomial by its degree. Before defining the degree of a polynomial, however, we first define the degree of a monomial.

DEFINITION

The *degree of a monomial* is the sum of the exponents of its variables. The degree of a nonzero constant is zero. The degree of the *number* 0 is undefined.

For example:

$3x^4$ has degree 4 since the exponent of its variable is 4.

$5x^3y^6$ has degree 9 since the sum of the exponents of the variables is 9.

$-3x^2y^3z$ has degree 6 since $2 + 3 + 1 = 6$ (remember $z = z^1$).

4   has degree 0   (as we will see in Chapter 5, we can write 4 as $4x^0$).

Now we can define the degree of a polynomial:

DEFINITION

The *degree of a polynomial* is the highest degree of any monomial in it.

EXAMPLE 1

Identify the degree of each of the following:
(a) $3x^5 + 2x^3 + 4$ (b) $7x^2y^4 - 3x^7y^5z^2$ (c) -7

Solution

(a) $\underbrace{3x^5}_{Degree\ 5}$ $+$ $\underbrace{2x^3}_{Degree\ 3}$ $+$ $\underbrace{4}_{Degree\ 0}$

$3x^5 + 2x^3 + 4$ has $\boxed{\text{degree } 5}$ *since the highest monomial degree is* 5.

(b) $\underbrace{7x^2y^4}_{Degree\ 6}$ $-$ $\underbrace{3x^7y^5z^2}_{Degree\ 14}$

$7x^2y^4 - 3x^7y^5z^2$ has $\boxed{\text{degree } 14}$ *since the highest degree of any monomial is* 14.

(c) -7 *is a constant. We can rewrite it as* $-7x^0$.

Therefore, it has $\boxed{\text{degree } 0.}$ ∎

When the number 0 is considered as a polynomial, we call it the *zero polynomial.* Since we can write 0 as $0x^4$ or as $0x^{10}$, you can see that we would have difficulty assigning a degree to the zero polynomial. Hence, *the zero polynomial has no degree.* Thus, while constants other than 0 have zero degree, 0 has no degree.

A third way to classify polynomials is by the number of variables. For example, $3x^2y^2 - 2xz$ is a fourth-degree binomial in three variables: x, y, and z; $5x^3 - 2x^2 + 3x$ is a third-degree trinomial in one variable.

We usually write polynomials in descending powers; that is, the highest-degree term is written first, followed by the next highest-degree term, and so on.

Symbolically, we often see polynomials in one variable defined as follows:

DEFINITION

A *polynomial in one variable* is an expression of the form

$$a_n x^n + a_{n-1} x^{n-1} + a_{n-2} x^{n-2} + \cdots + a_2 x^2 + a_1 x + a_0, \qquad a_n \neq 0$$

where the a_i's are real numbers, x is a variable, and n is a positive integer called the *degree* of the polynomial.

Let's examine this definition, especially the notation, more closely. To begin with, $a_0, a_1, a_2, a_3, \ldots$ are simply real-number coefficients: a_2 is the coefficient of x^2, a_4 is the coefficient of x^4, a_n is the coefficient of x^n. The subscript numbers are used to conveniently differentiate the constants from one another. We could have used letters of the alphabet instead, but then we would have to stop at 26 since there are only 26 letters in our alphabet.

When a polynomial is written with *descending powers*, it is said to be in **standard form.** A polynomial such as $5 + 3x - 4x^2$ would be rewritten as $-4x^2 + 3x + 5$. In this form, by the definition above, n would be 2, $a_2 = -4$, $a_1 = 3$, and $a_0 = 5$. Note how the subscripts of a conveniently match the exponents of x.

When a polynomial is written in descending powers, we assume that missing powers of the variable have coefficients of 0. For example:

$$3x^5 - 2x + 3 \quad \text{can be rewritten as} \quad 3x^5 + 0x^4 + 0x^3 + 0x^2 - 2x + 3$$

Then, $n = 5$, $a_5 = 3$, $a_4 = 0$, $a_3 = 0$, $a_2 = 0$, $a_1 = -2$, and $a_0 = 3$.

Notice how the meaning of degree is worked into the preceding definition of a polynomial in one variable: the highest power of x is n and therefore the degree of the polynomial is n.

Adding and Subtracting Polynomials

Addition and subtraction of polynomials is simply a matter of grouping with parentheses (if not supplied in the problem), removing grouping symbols, and combining like terms.

EXAMPLE 2

(a) Add the polynomials $3x^2 - 2x + 5$, $5x^3 - 4$, and $3x + 2$.

(b) Subtract $3x^3 + 5x^2 - 2x - 3$ from $7x - 4$.

Solution

(a) We fill in the grouping symbols to differentiate the given polynomials:

$$(3x^2 - 2x + 5) + (5x^3 - 4) + (3x + 2) \qquad \textit{Remove the grouping symbols.}$$

$$= 3x^2 - 2x + 5 + 5x^3 - 4 + 3x + 2 \qquad \textit{Then combine like terms.}$$

$$= \boxed{5x^3 + 3x^2 + x + 3}$$

(b) Subtraction can be tricky. Make sure you understand what is being subtracted:

$$(7x - 4) - (3x^3 + 5x^2 - 2x - 3)$$

Note how the polynomials are placed. Remove grouping symbols by distributing -1 (*observe the sign of each term*).

$$= 7x - 4 - 3x^3 - 5x^2 + 2x + 3$$ *Combine like terms.*

$$= \boxed{-3x^3 - 5x^2 + 9x - 1}$$ ∎

EXAMPLE 3 Perform the given operations and simplify.

(a) $(3x^2 + 2x - 4) + (5x^2 - 3x + 2) - (3x^2 - 2x + 1)$

(b) Subtract $3x^2 - 2x + 5$ from the sum of $2x^3 - 3x$ and $5x^2 - x + 4$.

(c) Subtract the sum of $2x^3 - 3x$ and $5x^2 - x + 4$ from $3x^2 - 2x + 5$.

Solution **(a)** $(3x^2 + 2x - 4) + (5x^2 - 3x + 2) - (3x^2 - 2x + 1)$

Remove grouping symbols.

$$= 3x^2 + 2x - 4 + 5x^2 - 3x + 2 - 3x^2 + 2x - 1$$

Then combine like terms.

$$= \boxed{5x^2 + x - 3}$$

(b) We write this problem in one step as follows:

$$[(2x^3 - 3x) + (5x^2 - x + 4)] - (3x^2 - 2x + 5)$$

First remove parentheses in brackets.

$$= [2x^3 - 3x + 5x^2 - x + 4] - (3x^2 - 2x + 5)$$

Then combine like terms in brackets.

$$= [2x^3 + 5x^2 - 4x + 4] - (3x^2 - 2x + 5)$$

Remove parentheses.

$$= 2x^3 + 5x^2 - 4x + 4 - 3x^2 + 2x - 5$$

Then combine like terms.

$$= \boxed{2x^3 + 2x^2 - 2x - 1}$$

(c) We translate the problem as follows:

$$(3x^2 - 2x + 5) - [(2x^3 - 3x) + (5x^2 - x + 4)]$$

Remove parentheses in brackets.

$$= (3x^2 - 2x + 5) - [2x^3 - 3x + 5x^2 - x + 4]$$

Then combine terms in brackets.

$$= (3x^2 - 2x + 5) - [2x^3 + 5x^2 - 4x + 4]$$

Remove remaining grouping symbols.

$$= 3x^2 - 2x + 5 - 2x^3 - 5x^2 + 4x - 4$$

We write the answer in standard form.

$$= \boxed{-2x^3 - 2x^2 + 2x + 1}$$

Note the differences between parts **(b)** and **(c)** of this example. ∎

Products of Polynomials

The distributive property gives us a procedure for multiplying polynomials. Recall that the property states that multiplication distributes over addition, that is:

$$a(b + c) = a \cdot b + a \cdot c$$

or

$$(b + c)a = b \cdot a + c \cdot a$$

As we discussed in Chapter 1, the real number properties, along with the first rule of exponents, give us a procedure for multiplying monomials by polynomials.

EXAMPLE 4

Perform the operations for each of the following:

(a) $5a^2b(4ab + 3ab^2)$ **(b)** $3x^3y^2(2x^2 - 7y^3)$

Solution

(a) $5a^2b(4ab + 3ab^2)$ *Apply the distributive property.*

   $= (5a^2b)(4ab) + (5a^2b)(3ab^2)$ *Then use the first rule of exponents.*

   $= \boxed{20a^3b^2 + 15a^3b^3}$

(b) $3x^3y^2(2x^2 - 7y^3) = 3x^3y^2(2x^2) - 3x^3y^2(7y^3)$

                   $= \boxed{6x^5y^2 - 21x^3y^5}$ ∎

Our next step is to demonstrate how to multiply two polynomials of more than one term. We can distribute any polynomial in the same way that we distribute a monomial. For example, in multiplying $(2x + 1)(x + 3)$, we can distribute $(x + 3)$ as follows:

$$(2x + 1)(x + 3) = 2x(x + 3) + 1\,(x + 3)$$
$$(B \ + C) \cdot \ A \ \ = B \cdot \ A \ \ + C \cdot \ A$$

It is important for you to understand that variables in equivalent expressions may stand not only for numbers, but also for other variables, expressions, or polynomials as well. Hence, we can let A stand for the binomial $x + 3$ and apply the distributive property above. The problem is still unfinished, for we must now apply the distributive property again and then combine terms:

$(2x + 1)(x + 3)$ *Distribute $x + 3$.*

$= (2x)(x + 3) + 1(x + 3)$ *Apply the distributive property again.*

$= (2x)x + (2x)3 + (1)x + (1)3$

$= 2x^2 + 6x + x + 3$ *Combine like terms.*

$= \boxed{2x^2 + 7x + 3}$

You may have learned to multiply binomials by some memorized procedure such as **FOIL**, or to multiply polynomials by the vertical method. However, our factoring problems will require that you thoroughly understand the logic underlying polynomial multiplication. For this reason, it is important that you realize that polynomial multiplication is derived from the *distributive property*.

EXAMPLE 5

Multiply the following:

(a) $(3a - 2)(2a + 5)$ (b) $(2y^2 + 3y + 1)(y - 5)$

Solution

(a) $(3a - 2)(2a + 5) = (3a)(2a + 5) - 2(2a + 5)$
$$= 6a^2 + 15a - 4a - 10$$
$$= \boxed{6a^2 + 11a - 10}$$

(b) $(2y^2 + 3y + 1)(y - 5) = (2y^2)(y - 5) + 3y(y - 5) + 1(y - 5)$
$$= 2y^3 - 10y^2 + 3y^2 - 15y + y - 5$$
$$= \boxed{2y^3 - 7y^2 - 14y - 5}$$

■

We can find the product of two polynomials by the **vertical method** as well; for example, $(x^2 - 3x + 4)(2x^2 - 3)$ can be multiplied as follows:

$$
\begin{array}{r}
x^2 - 3x + 4 \\
2x^2 \quad\quad - 3 \\
\hline
-3x^2 + 9x - 12 \quad \leftarrow (-3)(x^2 - 3x + 4) \\
2x^4 - 6x^3 + 8x^2 \quad\quad\quad\quad \leftarrow (2x^2)(x^2 - 3x + 4) \\
\hline
2x^4 - 6x^3 + 5x^2 + 9x - 12 \\
\uparrow \quad \uparrow \quad \uparrow \quad \uparrow \quad \uparrow
\end{array}
$$

Like terms lined up in columns

You have probably already recognized the rule given in the next box.

When two polynomials are multiplied together, each term of one polynomial is multiplied by each term of the other.

We will finish this section with a few more examples.

EXAMPLE 6

Perform the indicated operations.

(a) $(3x - 7)(2x + 5)$ (b) $(3x - 5)^2$
(c) $(x + y + z)(x + y - z)$ (d) $(3x^2 - 2y)(2x + 5y^2)$

Solution

(a) $(3x - 7)(2x + 5)$ *First distribute $2x + 5$.*
$$= 3x(2x + 5) - 7(2x + 5)$$
$$= 6x^2 + 15x - 14x - 35$$
$$= \boxed{6x^2 + x - 35}$$

(b) $(3x - 5)^2 = (3x - 5)(3x - 5)$ *First distribute $3x - 5$.*

$\qquad\qquad = 3x(3x - 5) - 5(3x - 5)$

$\qquad\qquad = 9x^2 - 15x - 15x + 25$

$\qquad\qquad = \boxed{9x^2 - 30x + 25}$ *Note that the answer is **not** $9x^2 - 25$.*

(c) $(x + y + z)(x + y - z) = x(x + y - z) + y(x + y - z) + z(x + y - z)$

$\qquad\qquad\qquad\qquad = x^2 + xy - xz + yx + y^2 - yz + zx + zy - z^2$

$\qquad\qquad\qquad\qquad = \boxed{x^2 + 2xy + y^2 - z^2}$

(d) $(3x^2 - 2y)(2x + 5y^2) = 3x^2(2x + 5y^2) - 2y(2x + 5y^2)$

$\qquad\qquad\qquad\qquad = \boxed{6x^3 + 15x^2y^2 - 4xy - 10y^3}$ ■

EXAMPLE 7

A circular garden has a radius of 10 feet. If a walk with a uniform width of x feet is to be built surrounding the garden, express the area of the walk in terms of x.

Solution

We draw a diagram of the circular garden and surrounding walkway, labelling the radius of the garden, 10 feet, and the width of the walkway, x feet, as shown in Figure 3.1. Notice that the area of the walk is the area of the shaded region between the two circles.

Figure 3.1

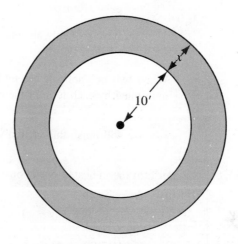

In order to find the area of the shaded region, we have to find the area of the inner and outer circles, and then subtract the area of the inner circle from the area of the outer circle. Since the area of a circle is πr^2, where r is the radius of the circle, we need to find the radius of each circle.

The radius of the inner circle is given as 10 feet and, as we can see by the figure, the radius of the outer circle must be $10 + x$ feet. Thus,

$$\text{Area}_{\text{shaded region}} = \text{Area}_{\text{outer circle}} - \text{Area}_{\text{inner circle}}$$

$$= \pi R^2 \qquad\quad - \pi r^2 \qquad\qquad \textit{where R is the radius of the outer circle and r is the radius of the inner circle.}$$

$$= \pi(10 + x)^2 - \pi(10^2) \qquad \textit{Now we perform the operations.}$$

$$= \pi(10 + x)(10 + x) - \pi(10)^2$$

$$= \pi(100 + 20x + x^2) - 100\pi$$

$$= 100\pi + 20\pi x + \pi x^2 - 100\pi \qquad \textit{Simplify.}$$

$$= \pi x^2 + 20\pi x \text{ square feet}$$

STUDY SKILLS 3.1   **Taking an Algebra Exam: Just Before the Exam**

You will need to concentrate and think clearly during the exam. For this reason it is important that you get plenty of rest the night before the exam, and that you have adequate nourishment.

It is *not* a good idea to study up until the last possible moment. You may find something that you missed and become anxious because there is not enough time to learn it. Then rather than simply missing a problem or two on the exam, the anxiety

may affect your performance on the entire exam. It is better to stop studying some time before the exam and do something else. You could, however, review formulas you need to remember and warnings (common errors you want to avoid) just before the exam.

Also, be sure to give yourself plenty of time to get to the exam.

Exercises 3.1

In Exercises 1–12, determine if the expression is a polynomial. If it is a polynomial, classify it by the following:

(a) *the number of terms (monomial, binomial, trinomial, polynomial)*
(b) *the degree of the polynomial*
(c) *the number of variables*

1. $3x^2 + 2$

2. $2x^3 - 3xy^4 + 5$

3. 73

4. -5

5. $-17x^{1/2} + 8x^5$

6. $3xy^4 + 2x^3 + 5$

7. $-17x^2y^4z^5$

8. $2x^4y^8 - 9x^2yz^2 + 6x^2 - 2s$

9. $\dfrac{1}{x}$

10. $\dfrac{-12b^2}{3b^3 - 2b + 2}$

11. $3x^2y - 7xy^4z^7 + 3$

12. 0

In Exercises 13–66, perform the indicated operations.

13. $(3x^2 - 2x + 5) + (2x^2 - 7x + 4)$

14. $(7xy^2 - 3x^2y) - (4xy^2 - 3x)$

15. $(15x^2 - 3xy - 4y^2) - (16x^2 - 3x + 2)$

16. $(2x^2 - 3x) - (-4x + 5) - (3x^2 - 2x + 1)$

17. $(3a^2 - 2ab + 4b^2) + [(3a^2 - b^2) - (3ab)]$

18. $[(5s^2 - 3) - (2st - 4t^2)] - 7s^2 - 3st + 2t^2$

19. $(3a^2 - 2ab + 4b^2) - [(3a^2 - b^2) - (3ab)]$

20. $7s^2 - 3st + 2t^2 - [(5s^2 - 3) - (2st - 4t^2)]$

21. $[(3a^2 - b^2) - (3ab)] - (3a^2 - 2ab + 4b^2)$

22. Subtract $5y^2 - 3y + 2$ from $y - 1$.

23. Add $3xy + 2y^2$ to $-7y^2 - 3y + 4$.

24. Subtract $y - 1$ from $5y^2 - 3y + 2$.

25. Subtract $3x^2 - 2xy + 7y^2$ from $-8x^2 - 5xy + 9y^2$.

26. Find the sum of $3a^2 - 6ab + 16b^2$, $3ab - 6b^2$, and $-5a^2 - 3ab$.

27. Subtract the sum of $2a^2 - 3ab + 4b^2$ and $6a^2 + 2ab - 2b^2$ from $3a^2 - 2b^2$.

28. Subtract the sum of $3x + 2y$ and $3y - 2x$ from the sum of $x - y$ and $x + y$.

29. $x^2y(3x^2 - 2xy + 2y^2)$

30. $-3a^2b^2(4a^2b - 3ab + 2a)$

31. $(3x^2 - 2xy + 2y^2)(x^2y)$

32. $-13a^3b^2(5a^2b - 3a^2 + 2)$

33. $(x + 3)(x + 4)$

34. $(y - 4)(y - 8)$

35. $(3x - 1)(2x + 1)$

36. $(2x - 3)(3x + 5)$

37. $(3x + 1)(2x - 1)$

38. $(2x + 3)(3x - 5)$

39. $(3a - b)(a + b)$

40. $(2a - 5b)(a + 2b)$

41. $(2r - s)^2$

42. $(3b - 2)^2$

43. $(2r + s)(2r - s)$

44. $(3b - 2)(3b + 2)$

45. $(3x - 2y)(3x + 2y)$

46. $(2x^2 + y)(2x^2 - y)$

47. $(3x - 2y)^2$

48. $(2x^2 - y)^2$

49. $(x - y)(a - b)$

50. $(a + b)(x + y)$

51. $(2r + 3s)(2a + 3b)$

52. $(4x^2 - 5y)(5x - 3y^2)$

53. $(2y^2 - 3)(y^2 + 1)$

54. $(3a^2 - x^3)(3a^2 + x^3)$

55. $(5x^2 - 3x + 4)(x + 3)$

56. $(5x^2 - 3x + 4)(x - 3)$

57. $(9a^2 + 3a - 5)(3a + 1)$

58. $(7x^2 + 4x - 7)(5x - 2)$

59. $(a^2 - ab + b^2)(a + b)$

60. $(a^2 + ab + b^2)(a - b)$

61. $(x - y - z)^2$

62. $(x + y + z)^2$

63. $(x^2 - 2x + 4)(x + 2)$

64. $(x^2 + 2x + 4)(x - 2)$

65. $(a + b + c + d)(a + b)$

66. $(x^3 + x^2 + x + 1)(x - 1)$

67. The length of a rectangle is 5 more than three times its width. If the width of the rectangle is w, express the area and perimeter of the rectangle in terms of w.

68. The length of a rectangle is 1 less than twice its width. If the width of the rectangle is w, express the area and perimeter of the rectangle in terms of w.

69. A circular garden has a radius of 8 feet. If a walk with a uniform width of x feet is to be built surrounding the garden, express the area of the walk in terms of x.

70. A circular swimming pool has a radius of 50 feet. If a concrete walkway with a uniform width of x feet surrounds the pool, express the area of the walkway in terms of x.

71. A 20 ft by 60 ft rectangular swimming pool is surrounded by a path of uniform width. If the width of the walk is x feet, express the area of the walk in terms of x (see Figure 7.2 on page 317).

72. A 30 ft by 30 ft square swimming pool is surrounded by a concrete walkway of uniform width. If the width of the walkway is x feet, express the total area of the pool and walkway in terms of x (see Figure 7.2 on page 317).

CALCULATOR EXERCISES

In Exercises 73–74, perform the indicated operations.

73. $6.24x(3.1x^2 - 5.2x + 6)$

74. $(2.6x + 3)(1.2x - 3.5)$

? QUESTIONS FOR THOUGHT

75. Supply the real number property for each of the following steps:

$$(x + y)(3x - 2y) = (x + y)(3x) + (x + y)(-2y)$$
$$= x(3x) + y(3x) + x(-2y) + y(-2y)$$
$$= (3x)x + (3x)y + (-2y)x + (-2y)y$$
$$= 3(xx) + 3(xy) - 2(yx) - 2(yy)$$
$$= 3x^2 + 3xy - 2yx - 2y^2$$
$$= 3x^2 + 3xy - 2xy - 2y^2$$
$$= 3x^2 + (3 - 2)xy - 2y^2$$
$$= 3x^2 + xy - 2y^2$$

76. Discuss what is **wrong** (if anything) with the following:

(a) $(3x^2)(2x^3) \overset{?}{=} 6x^6$ **(b)** $2x - \{3x - 2[4 - x]\} \overset{?}{=} 2x - \{3x - 8 + 2x\}$
$$\overset{?}{=} 2x - \{5x - 8\}$$
$$\overset{?}{=} -10x^2 + 16x$$

3.2

General Forms and Special Products

We can now develop some procedures for multiplying polynomials quickly as long as we keep in mind *why* the process works. Our emphasis will be on quick procedures for *binomial* multiplication since such products occur frequently in algebra. Later on in this section, we will generalize the procedures to more complex cases.

General Forms

If we multiplied $(x + a)(x + b)$, we would get the following:

$(x + a)(x + b)$ *Distribute $x + b$.*
$= x(x + b) + a(x + b)$ *Distribute x and a.*
$= x^2 + bx + ax + ab$ *Factor x from the middle terms.*
$= x^2 + (a + b)x + ab$

Thus, we have:

General Form #1	$(x + a)(x + b) = x^2 + (a + b)x + ab$

Verbally stated, when two binomials of the form (Variable + Constant) are multiplied, the product will be a trinomial, where the coefficient of the first-degree term will be the *sum* of the binomial constants and the numerical term will be the *product* of the binomial constants. Knowing this form allows us to quickly multiply binomials of this type mentally.

EXAMPLE 1

(a) $(x + 3)(x + 5)$ **(b)** $(y - 7)(y + 4)$ **(c)** $(a - 6)(a - 5)$

Solution

$(x + a)(x + b) = x^2 + (a + b)x + ab$

(a) $(x + 3)(x + 5) = x^2 + (3 + 5)x + 5 \cdot 3$

$\qquad = \boxed{x^2 + 8x + 15}$

(b) $(y - 7)(y + 4) = y^2 + (-7 + 4)y + (-7)(4)$

$\qquad = \boxed{y^2 - 3y - 28}$

(c) $(a - 6)(a - 5) = a^2 + (-6 - 5)a + (-6)(-5)$

$\qquad = \boxed{a^2 - 11a + 30}$

$\blacksquare$

If we multiplied out $(ax + b)(cx + d)$, we would get:

$(ax + b)(cx + d)$ *Distribute cx + d.*

$= ax(cx + d) + b(cx + d)$ *Distribute ax and b.*

$= acx^2 + adx + bcx + bd$ *Factor x from the middle terms.*

$= acx^2 + (ad + bc)x + bd$

Thus, we have:

General Form #2	$(ax + b)(cx + d) = acx^2 + (ad + bc)x + bd$

This form is certainly more complicated than the previous general form. Therefore, rather than trying to memorize this form to perform quick multiplication of binomials, we will present another view of the same procedure for multiplying binomials, called the **FOIL method.**

    FOIL stands for

First,   Outer,   Inner,   Last

It is a systematic way to keep track of the terms to be multiplied, demonstrated as follows:

$$(3x + 4)(2x + 5) = (3x)(2x) + (3x)(5) + (4)(2x) + (4)(5)$$
$$= 6x^2 + 15x + 8x + 20$$
$$= 6x^2 + 23x + 20$$

Keep in mind that FOIL is only a name to help us to be systematic in carrying out *binomial* multiplication; *each term of one binomial is still being multiplied by each term of the other*.

EXAMPLE 2

Perform the indicated operations.

(a) $(5x - 4)(2x + 1)$ (b) $(3a + 4)(3a - 4)$ (c) $(3x + 1)^2$

Solution

(a) $(5x - 4)(2x + 1) = (5x)(2x) + (5x)(1) - 4(2x) + (-4)(1)$
$$= 10x^2 + 5x - 8x - 4$$
$$= \boxed{10x^2 - 3x - 4}$$

(b) $(3a + 4)(3a - 4) = (3a)(3a) - (3a)4 + 4(3a) - 4(4)$
$$= 9a^2 - 12a + 12a - 16$$
$$= \boxed{9a^2 - 16}$$

(c) $(3x + 1)^2 = (3x + 1)(3x + 1)$
$$= (3x)(3x) + (3x)(1) + 1(3x) + 1(1)$$
$$= 9x^2 + 3x + 3x + 1$$
$$= \boxed{9x^2 + 6x + 1}$$

You should practice these until you can do them flawlessly in your head.

Special Products

Special products are specific products of binomials which can be derived from the general forms given previously (which, in turn, are derived from the distributive properties).

Special Products		
1.	$(a + b)(a - b) = a^2 - b^2$	*Difference of two squares*
2.	$(a + b)^2 = a^2 + 2ab + b^2$	*Perfect square of sum*
3.	$(a - b)^2 = a^2 - 2ab + b^2$	*Perfect square of difference*

PROOF:

1. $(a + b)(a - b) = a(a - b) + b(a - b)$
$$= a^2 - ab + ba - b^2$$
$$= a^2 - b^2$$

2. $(a + b)^2 = (a + b)(a + b)$
$$= a(a + b) + b(a + b)$$
$$= a^2 + ab + ba + b^2$$
$$= a^2 + 2ab + b^2$$

Proof of special product #3 is left as an exercise.

Special product #1 is known as the ***difference of two squares,*** while special products #2 and #3 are known as ***perfect squares.*** Note the differences between the special products, especially between special products #1 and #3, which are often confused with each other.

The expressions $a + b$ and $a - b$ are called ***conjugates of each other.*** A *conjugate* of a binomial has identical terms, except the sign of *one* of its terms is opposite. Special product #1 tells us that the product of conjugates yields the difference of two squares.

Conjugates will be important to us in subsequent chapters, but for now you should understand that $-a - b$ is *not* the conjugate of $a + b$, but rather the *negative* of $a + b$, since

$$-(a + b) = -a - b$$

A *conjugate* of $x - 4$ is $x + 4$, but the *negative* of $x - 4$ is $-(x - 4) = -x + 4$, usually written as $4 - x$.

Notice that in the negative of a binomial, the signs of *both* terms are changed, while in a conjugate only one sign is changed.

Special products are important in factoring; in many cases, the quickest way to factor an expression is to recognize it as a special product. In addition, recognizing and using special products can reduce the time needed for multiplication.

EXAMPLE 3

Perform the indicated operations.

(a) $(3x + 5)^2$ **(b)** $(2a - 7b)(2a + 7b)$ **(c)** $(2a - 7b)^2$

Solution

All these problems can be worked by using the general forms, or more slowly by using the distributive property. The quickest way, however, is to use the special products.

(a) $(3x + 5)^2$ *Special product #2*

$$= (3x)^2 + 2(3x)(5) + (5)^2$$

$$= \boxed{9x^2 + 30x + 25}$$

(b) $(2a - 7b)(2a + 7b)$ *Special product #1*

$= (2a)^2 - (7b)^2$

$= \boxed{4a^2 - 49b^2}$

(c) $(2a - 7b)^2$

$= (2a)^2 - 2(2a)(7b) + (7b)^2$ *Special product #3*

$= \boxed{4a^2 - 28ab + 49b^2}$ ∎

Study the differences between parts **(b)** and **(c)** of Example 3; note their similarities. Keep in mind that when you square a binomial, you will get a middle term in the product. Again, practice these so that you can quickly do them mentally.

Multiple Operations

Let's put together what we have learned so far and examine how to simplify expressions requiring multiple operations with polynomials. We follow the same order of operations discussed in Chapter 1 (that is, parentheses, exponents, multiplication and division, addition and subtraction).

EXAMPLE 4

Perform the indicated operations.

(a) $2x - 3x(x - 5)$ **(b)** $(x - y)^3$

(c) $2x(x - 3)(3x + 1)$ **(d)** $5x^3 - 2x(x - 4)(2x + 3)$

(e) $(a - b)^2 - (a - b)(a + b)$

Solution

(a) $2x - 3x(x - 5)$ *First distribute $-3x$.*

$= 2x - 3x^2 + 15x$ *Then combine terms.*

$= \boxed{-3x^2 + 17x}$

(b) $(x - y)^3$ *Expand $(x - y)^3$ first.*

$= (x - y)(x - y)(x - y)$ *Then multiply two binomials together.*

$= (x - y)(x^2 - 2xy + y^2)$ *Multiply the result by the third binomial.*

$= x^3 - 2x^2y + xy^2 - x^2y + 2xy^2 - y^3$ *Combine like terms.*

$= \boxed{x^3 - 3x^2y + 3xy^2 - y^3}$

(c) $2x(x - 3)(3x + 1)$ *There is less chance of your making an error if you multiply the binomials first.*

$= 2x(3x^2 - 8x - 3)$

$= \boxed{6x^3 - 16x^2 - 6x}$

(d) $5x^3 - 2x(x - 4)(2x + 3)$ *Multiply binomials first.*

$= 5x^3 - 2x(2x^2 - 5x - 12)$ *Distribute* $-2x$.

$= 5x^3 - 4x^3 + 10x^2 + 24x$ *Combine "like" terms.*

$= \boxed{x^3 + 10x^2 + 24x}$

(e) $(a - b)^2 - (a - b)(a + b)$ $(a - b)^2$ *is a perfect square.*

$= a^2 - 2ab + b^2 - (a - b)(a + b)$ $(a + b)(a - b)$ *is the difference of two squares.*

$= a^2 - 2ab + b^2 - (a^2 - b^2)$ *Remove parentheses.*

$= a^2 - 2ab + b^2 - a^2 + b^2$

$= \boxed{-2ab + 2b^2}$

Note that in part **(e)**, we multiplied $(a + b)(a - b)$ before we subtracted. It is a good habit to retain parentheses to remind us that we are subtracting the entire expression $(a^2 - b^2)$. ∎

Applying Special Products

Of course, we can apply the special products to binomials containing more complex monomials, as shown in the next example.

EXAMPLE 5

Perform the operations using special products or general forms.

(a) $(2x^2 - 3y)(2x^2 + 3y)$ **(b)** $(3r^3 - 5x^2)^2$

(c) $(x^2y - 3s^2)(x^2y - 4s^2)$

Solution

(a) $(2x^2 - 3y)(2x^2 + 3y) = (2x^2)^2 - (3y)^2$ *Difference of two squares*

$= \boxed{4x^4 - 9y^2}$

(b) $(3r^3 - 5x^2)^2 = (3r^3)^2 - 2(3r^3)(5x^2) + (5x^2)^2$ *Perfect square*

$= \boxed{9r^6 - 30r^3x^2 + 25x^4}$

(c) $(x^2y - 3s^2)(x^2y - 4s^2) = (x^2y)^2 + (-3 - 4)(x^2y)(s^2) + (-3)(-4)(s^2)^2$

$= \boxed{x^4y^2 - 7x^2ys^2 + 12s^4}$ ∎

General forms and special products allow us to multiply quickly without going through intermediate steps. They can also be applied to more complex expressions. For example, to multiply

$$[x - (r + s)][x + (r + s)]$$

you could simplify the expressions within the brackets and multiply out using the horizontal or vertical method. On the other hand, if you recognize this problem as a form of special product #1 (the difference of two squares), and apply what you already know about special products, you could reduce your labor for this problem. Again, keep in mind that in equivalent expressions such as special products, the variables can represent polynomials as well as other letters and numbers.

$$(a - b) \cdot (a + b) = a^2 - b^2$$
$$[x - (r + s)] \cdot [x + (r + s)] = x^2 - (r + s)^2$$

We finish the problem by applying special product #2 to $(r + s)^2$:

$$x^2 - (r + s)^2 = x^2 - (r^2 + 2rs + s^2)$$
$$= \boxed{x^2 - r^2 - 2rs - s^2}$$

EXAMPLE 6

Perform the operations using special products:

$$[(x + y) + z]^2$$

Solution

We can view this problem as the square of the sum of two terms: $(x + y)$ and z, and use special product #2:

$$(a + b)^2 = a^2 + 2 \cdot a \cdot b + b^2$$
$$[(x + y) + z]^2 = (x + y)^2 + 2(x + y)z + z^2 \qquad \text{\textit{Apply special product #2 to}}$$
$$\text{\textit{(x + y)}}^2.$$
$$= x^2 + 2xy + y^2 + 2(x + y)z + z^2$$
$$= \boxed{x^2 + 2xy + y^2 + 2xz + 2yz + z^2}$$

The key is to recognize complex expressions as being in special product form. ∎

EXAMPLE 7

Perform the operations using special products or general forms.
(a) $(x + y - 3)(x + y + 3)$ **(b)** $(x + y + z)^2$
(c) $(a + b + 3)(a + b - 4)$

Solution

At first glance, these problems do not seem to be in special product or general form. However, we can regroup some of the expressions within parentheses to get them into special product or general form.

(a) $(x + y - 3)(x + y + 3)$ can be rewritten as

$$[(x + y) - 3][(x + y) + 3]$$

Note that it is a difference of two squares with $x + y$ as the first term and 3 as the second term.

$$[(x + y) - 3][(x + y) + 3] \qquad \text{\textit{Apply special product #1.}}$$
$$= (x + y)^2 - 3^2 \qquad \text{\textit{Then apply special product #2 to (x + y)}}^2.$$
$$= \boxed{x^2 + 2xy + y^2 - 9}$$

Note that if we had rewritten $(x + y - 3)(x + y + 3)$ as

$$[x + (y - 3)][x + (y + 3)]$$

we could not have used the special products, since $y - 3$ and $y + 3$ are not identical.

(b) $(x + y + z)^2$ was done before. This time, we will regroup it differently to demonstrate that we still arrive at the same answer.

$$(x + y + z)^2 = [x + (y + z)][x + (y + z)]$$

Apply special product #2.

$$= x^2 + 2x(y + z) + (y + z)^2$$

Then apply special product #2 to $(y + z)^2$.

$$= x^2 + 2x(y + z) + (y^2 + 2yz + z^2)$$

$$= \boxed{x^2 + 2xy + 2xz + y^2 + 2yz + z^2}$$

(c) $(a + b + 3)(a + b - 4)$ can be rewritten as

$$[(a + b) + 3][(a + b) - 4]$$

Now it has the form of the problem $(x + 3)(x - 4)$:

$$(x \quad + 3) \quad (x \quad - 4) = \quad x^2 \quad - \quad x \quad - 12$$

$$[(a + b) + 3][(a + b) - 4] = (a + b)^2 - (a + b) - 12$$

$$= a^2 + 2ab + b^2 - (a + b) - 12$$

$$= \boxed{a^2 + 2ab + b^2 - a - b - 12} \qquad \blacksquare$$

Multiplication in the manner illustrated in Example 7 is much faster than using the horizontal or vertical method. Again, you should practice these problems for we will return to these forms again in the next sections.

STUDY SKILLS 3.2   **Taking an Algebra Exam: Beginning the Exam**

At the exam, make sure that you listen carefully to the instructions given by your instructor or the proctor.

As soon as you are allowed to begin, jot down the formulas you think you might need, and write some key words (warnings) to remind you to avoid common errors or errors you have made previously. Writing down the formulas first will relieve you of the burden of worrying about whether you will remember them when you need to, thus allowing you to concentrate more.

You should refer back to the relevant warnings as you go through the exam to make sure you avoid these errors.

Remember to read the directions carefully.

Exercises 3.2

Perform the operations and simplify. Use special products wherever possible.

1. $(x + 4)(x + 5)$

2. $(a - 2)(a + 4)$

3. $(x + 3)(x - 7)$

4. $(y - 3)(y - 9)$

5. $(x - 8)(x - 11)$

6. $(x - 21)(x + 20)$

7. $(2x + 1)(x - 4)$

8. $(3a + 2b)(a - b)$

9. $(5a - 4)(5a + 4)$

10. $(6s - 4)(6s + 4)$

11. $(3z + 5)^2$

12. $(7r - 3t)(7r + 3t)$

13. $(3z - 5)^2$

14. $(7r - 3t)^2$

15. $(3z - 5)(3z + 5)$

16. $(7r + 3t)^2$

17. $(5r^2 + 3s)(3r + 5s)$

18. $(4x^2 - 5)(5x - 4)$

19. $(3s - 2y)^2$

20. $(7x - y)(7x - y)$

21. $(3y + 10z)^2$

22. $(y - 2x)(y + 2x)$

23. $(3y + 10z)(3y - 10z)$

24. $(y - 2x)^2$

25. $(3a + 2b)(3a + 4b)$

26. $(4x - y)(4x + y)$

27. $(8a - 1)^2$

28. $(8a - 1)(8a + 1)$

29. $(5r - 2s)(5r - 2s)$

30. $(2r - 3x)(2r + 7x)$

31. $(7x - 8)(8x - 7)$

32. $(5x + 9y)(9x - 5y)$

33. $(5t - 3s^2)(5t + 3s)$

34. $(2x^2 - 3)(2x + 3)$

35. $(3y^3 - 4x)^2$

36. $(2x^2 - 3)^2$

37. $(3y^3 - 4x)(3y^3 + 4x)$

38. $(7r^2s^2 - 5g)(7r^2s^2 + 5g)$

39. $(2rst - 7xyz)^2$

40. $(7r^2 - t^3)^2$

41. $(2rst - 7xyz)(2rst + 7xyz)$

42. $(7r^2 + t^3)(7r^2 - t^3)$

43. $(a^6 - 3b^3)(2a^6 - b^3)$

44. $(3xy - 2a^2)(5xy + 3a^2)$

45. $(2x^2 - 3y^2)(x^2 - 3y^2)$

46. $(5a^2 - 2b^2)(3a^2 + b^2)$

47. $(x - 4) - (3x + 1)^2$

48. $5x^2 - (2x + 1)(3x + 2)$

49. $(a - b)^2 - (b + a)^2$

50. $(3x - 1)(2x + 3) - (x - 4)(x + 1)$

51. $(3a + 2)(-5a)(2a - 1)$

52. $(3x - 1)(2x + 3) - x - 4x - 1$

53. $(3a + 2) - 5a(2a - 1)$

54. $b^2 - 2b(3b + 2)(2b - 3)$

55. $3r^3 - 3r(r - s)(r + s)$

56. $(5r - 2)3r - (5r - 2)(r - 1)$

57. $(3y + 1)(y + 2) - (2y - 3)^2$

58. $(x - 2y)^2 - (x - 2y)^2$

59. $(5a - 3b)^3$

60. $(x - 3y)^3$

61. $3x(x + 1) - 2x(x + 1)^2$

62. $[x - (a + b)][x + (a + b)]$

63. $[(a + b) + 1][(a + b) - 1]$

64. $[(x + 2y) - 3z]^2$

65. $[(a - 2b) + 5z]^2$

66. $[(x + 2y) - 3z][(x + 2y) + 3z]$

67. $[a - 2b + 5z][a - 2b - 5z]$

68. $[2(a + b) + 1][3(a + b) - 5)]$

69. $[(a + b) + (2x + 1)][(a + b) - (2x + 1)]$

70. $(a - b - 3)(a + b + 3)$

71. $(a^n - 3)(a^n + 3)$

72. $(x^n - y^n)(x^n + y^n)$

? QUESTIONS FOR THOUGHT

73. *Verbally* describe how to find the product $(x + a)(x + b)$.

74. Discuss what is *wrong* (if anything) with each of the following.

(a) $(x + y)^2 - (x + y)(x - y) \overset{?}{=} x^2 + 2xy + y^2 - x^2 - y^2$

$\overset{?}{=} 2xy$

(b) $3x(x - 5)(x + 5) \overset{?}{=} (3x^2 - 15x)(3x^2 + 15x)$

$\overset{?}{=} 9x^4 - 225x^2$

75. Discuss the differences between the expression $(x + y)^2$ and the expression $x^2 + y^2$.

3.3

Factoring Out the Greatest Common Factor

We can view polynomial multiplication as changing products into sums. In the same way, we can view factoring as changing sums into products. The distributive property gives us the method for multiplying polynomials and it also gives us a method for factoring polynomials:

$$\overset{\textit{Multiplying}}{\xrightarrow{\hspace{3cm}}}$$
$$(a + b)c = a \cdot c + b \cdot c$$
$$\underset{\textit{Factoring}}{\xleftarrow{\hspace{3cm}}}$$

Before we begin to factor polynomials, we should mention that in factoring whole numbers, we were interested in a representation as a product of *whole numbers*. Thus, 12 could be factored as $4 \cdot 3$, $6 \cdot 2$, $3 \cdot 2 \cdot 2$, or $12 \cdot 1$. Even though 12 could be represented as a product of $\frac{1}{3}$ and 36, we did not consider factors that are not whole numbers.

In the same way, we will not consider factors of polynomials with fractional or irrational coefficients; *all polynomial factors should have integer coefficients*. Hence, although $x - 1$ can be represented as the product $\frac{1}{4}(4x - 4)$, we will not consider $\frac{1}{4}$ as a factor of $x - 1$ since it is not an integer.

The first type of factoring we will consider is called ***common monomial factoring***. Essentially, we are interested in factoring out the greatest monomial common to each term in a polynomial.

For example, the greatest common monomial factor of $6x^2 + 8x$ is $2x$ because $2x$ is the greatest common factor of *both* terms $6x^2$ and $8x$. Therefore, we can rewrite $6x^2 + 8x$ as $2x \cdot 3x + 2x \cdot 4$ and then factor out $2x$ by the distributive property to get $2x(3x + 4)$. Thus,

$$6x^2 + 8x = 2x \cdot 3x + 2x \cdot 4 = 2x(3x + 4)$$

Keep in mind that this is a two-step process: first you find the greatest common factor, then you determine what remains within the parentheses.

EXAMPLE 1

Factor the following as completely as possible.

(a) $3x^2 + 2x^3 - 4x^5$ **(b)** $24x^2y^3 - 16xy^3 - 8y^4$

(c) $3x^3y^3 - 9xy^4 + 3$ **(d)** $2a^2b - 8ab^2c + 5c^2$

Solution

In general, it is probably easiest to begin by first determining the greatest common numerical factor, then the greatest common x factor, then the greatest common y factor, and so forth.

(a) $3x^2 + 2x^3 - 4x^5$ *We factor out x^2 from each term since it is the*
                                           greatest common factor of $3x^2$, $2x^3$, and $4x^5$.

$\qquad = \boxed{x^2(3 + 2x - 4x^3)}$

Always check your answer by multiplying (you can check each term mentally).

CHECK: $x^2(3 + 2x - 4x^3) = 3x^2 + 2x^3 - 4x^5$

(b) $24x^2y^3 - 16xy^3 - 8y^4$ *The greatest common factor of $24x^2y^3$, $16xy^3$, and $8y^4$ is $8y^3$.*

$= \boxed{8y^3(3x^2 - 2x - y)}$ *Check this answer.*

(c) $3x^3y^3 - 9xy^4 + 3$ *The greatest common factor is 3.*

$= \boxed{3(x^3y^3 - 3xy^4 + 1)}$ *Do not forget to include $+ 1$ to hold a place for the 3.*

CHECK: $3(x^3y^3 - 3xy^4 + 1) = 3x^3y^3 - 9xy^4 + 3$

(d) $2a^2b - 8ab^2c + 5c^2$ *Since there is no common factor of $2a^2b$, $8ab^2c$, and $5c^2$ other than 1, we write:* $\boxed{\text{not factorable}}$

If we had factored part **(b)** as $4y(6x^2y^2 - 4xy^2 - 2y^3)$, it would check when we multiplied it out. But it is not factored *completely*; we have not factored out the *greatest* common factor. We could still factor $2y^2$ from the trinomial. Thus,

$4y(6x^2y^2 - 4xy^2 - 2y^3)$ *Factor $2y^2$ from $6x^2y^2 - 4xy^2 - 2y^3$.*

$= 4y[2y^2(3x^2 - 2x - y)]$

$= 8y^3(3x^2 - 2x - y)$

Always check that there are no more common factors in the parentheses.   ■

We can generalize common factoring to factoring *polynomials* of more than one term from expressions as follows:

Factor the following: $3x(y - 4) + 2(y - 4)$

Note that $y - 4$ is common to both expressions, $3x(y - 4)$ and $2(y - 4)$, and therefore can be factored out, just as we would factor A from $3xA + 2A$.

$$3x \cdot \quad A \quad + 2 \cdot \quad A \quad = \quad A \quad \cdot (3x + 2)$$
$$3x(y - 4) + 2(y - 4) = (y - 4)(3x + 2)$$

If we read the equation from right to left, we see multiplication by the distributive property.

EXAMPLE 2   Factor the following completely:

(a) $5a(x - 2y) - 3b(x - 2y)$ **(b)** $3x(x - 3) + 5(x - 3)$

(c) $(x + 2)^2 + (x + 2)$

Solution   **(a)** $5a(x - 2y) - 3b(x - 2y)$ *Since $x - 2y$ is common to both terms, $5a(x - 2y)$ and $-3b(x - 2y)$, we can factor $x - 2y$ out, and are left with $5a$ and $-3b$.*

$= \boxed{(x - 2y)(5a - 3b)}$

(b) $\quad 3x(x - 3) + 5(x - 3)$ $\qquad$ *Factor out* $x - 3$ *and we are left with* $3x$ *and* $+ 5.$

$\qquad = \boxed{(x - 3)(3x + 5)}$

(c) $\quad (x + 2)^2 + (x + 2)$ $\qquad$ $x + 2$ *is the factor common to both* $(x + 2)^2$ *and* $x + 2$. *This is like factoring* $A^2 + A$ *to get* $A(A + 1).$

$$A^2 \quad + \quad A \quad = \quad A \cdot (\quad A \quad + 1)$$
$$(x + 2)^2 + (x + 2) = (x + 2)[(x + 2) + 1]$$
$$= (x + 2)[x + 2 + 1]$$
$$= \boxed{(x + 2)(x + 3)} \qquad \blacksquare$$

Not all polynomials are conveniently grouped as in Example 2. We often have to take a step or two to put the polynomial in factorable form. For example, in factoring

$$ax + ay + bx + by$$

you can see that there is no common *monomial* we can factor from *all four terms*. However, if we group the terms by pairs and factor the pairs, we can then factor the *binomial* from each group as follows:

$$ax + ay + bx + by \qquad \textit{Group the pairs together.}$$
$$= ax + ay \quad + \quad bx + by \qquad \textit{Then factor each pair.}$$
$$= a(x + y) + b(x + y) \qquad \textit{Factor } x + y \textit{ from each group.}$$
$$= (x + y)(a + b)$$

Grouping and factoring *parts* of a polynomial in order to factor the polynomial itself is called ***factoring by grouping***.

EXAMPLE 3

Factor the following completely.

(a) $\quad 7xy + 2y + 7xa + 2a$ $\qquad$ **(b)** $\quad 3xb - 2b + 15x - 10$

Solution

(a) $\quad 7xy + 2y + 7xa + 2a$ $\qquad$ *Since there is no common factor of all terms, we group in pairs.*

$\qquad = 7xy + 2y \quad + \quad 7xa + 2a$ $\qquad$ *Factor each pair.*

$\qquad = y(7x + 2) + a(7x + 2)$ $\qquad$ *Factor* $7x + 2$ *from each group.*

$\qquad = \boxed{(7x + 2)(y + a)}$

(b) $\quad 3xb - 2b + 15x - 10$ $\qquad$ *There is no common factor so group each pair.*

$\qquad = 3xb - 2b \quad + \quad 15x - 10$ $\qquad$ *Factor each pair.*

$\qquad = b(3x - 2) + 5(3x - 2)$ $\qquad$ *Factor* $3x - 2$ *from each group.*

$\qquad = \boxed{(3x - 2)(b + 5)}$ $\qquad \blacksquare$

A word of caution: Factoring parts of a polynomial, as in

$$3xb - 2b + 15x - 10 = b(3x - 2) + 5(3x - 2)$$

is an intermediate step to guide us in factoring by grouping. It is *not* the factored form of the polynomial. (You would not say 13 is factored as $3 \cdot 3 + 2 \cdot 2$, would you?) Thus, keep in mind that when you are asked to factor a polynomial, the whole polynomial must be represented as a *product* of polynomials.

When a negative sign appears between the pairs of binomials we intend to group, we occasionally have to factor out a negative factor first, to see if each group contains the same binomial factor. In factoring $5ac + 20c - 3a - 12$, for example, we would have to factor out -3 in $-3a - 12$:

$5ac + 20c - 3a - 12$ *First separate the pairs.*

$= 5ac + 20c \quad - 3a - 12$

              ↑   ↑

[*Be careful. Check with multiplication.*]

            ↓   *Factor out 5c and -3.*

$= 5c(a + 4) - 3(a + 4)$ *Factor out $a + 4$.*

$= \boxed{(a + 4)(5c - 3)}$

EXAMPLE 4

Factor the following completely.

(a) $10x^2 - 2x + 5x - 1$ **(b)** $6x^2 + 2x - 9x - 3$

(c) $x^3 + x^2 + x + 1$ **(d)** $x^3 + x^2 - x + 1$

Solution

(a) $10x^2 - 2x + 5x - 1$ *Group each pair.*

      $= 10x^2 - 2x \quad + \quad 5x - 1$ *Then factor each pair.*

      $= 2x(5x - 1) + 1(5x - 1)$

                  Note: It is helpful to hold this place with the understood factor of 1.

      $= \boxed{(5x - 1)(2x + 1)}$

(b) $6x^2 + 2x - 9x - 3 = 6x^2 + 2x \quad - 9x - 3$ *Factor out 2x from the first*

      $= 2x(3x + 1) - 3(3x + 1)$ *pair, -3 from the second pair.*

      $= \boxed{(3x + 1)(2x - 3)}$

(c) $x^3 + x^2 + x + 1 = x^3 + x^2 \quad + \quad x + 1$

             $= x^2(x + 1) + 1(x + 1)$

                $= \boxed{(x + 1)(x^2 + 1)}$

(d) $x^3 + x^2 - x + 1 = x^2(x + 1) - 1(x - 1)$ *Note that the binomials $x - 1$ and $x + 1$ are not identical and therefore this expression cannot be factored by grouping.*

Answer: $\boxed{\text{Not factorable}}$

Note the differences between parts (c) and (d). Even if we tried grouping in part (d) without factoring out -1, we would get $x^2(x + 1) + 1(-x + 1)$, which is still not factorable by grouping since $x + 1$ and $-x + 1$ are not identical.   ■

STUDY SKILLS 3.3   **Taking an Algebra Exam: What to Do First**

Not all exams are arranged in ascending order of difficulty (from easiest to most difficult). Since time is usually an important factor, you do not want to spend so much time working on a few problems that you find difficult and then find that you do not have enough time to solve the problems that are easier for you. Therefore, it is strongly recommended that you first look over the exam and then follow the order given below:

1.   Start with the problems which you know how to solve quickly.

2.   Then go back and work on problems which you know how to solve but take longer.

3.   Then work on those problems which you find more difficult, but for which you have a general idea of how to proceed.

4.   Finally, divide the remaining time between doing the problems you find most difficult and checking your solutions. Do not forget to check the warnings you wrote down at the beginning of the exam.

You probably should not be spending a lot of time on any single problem. To determine the average amount of time you should be spending on a problem, divide the amount of time given for the exam by the number of problems on the exam. For example, if the exam lasts for 50 minutes and there are 20 problems, you should spend an average of $\frac{50}{20} = 2\frac{1}{2}$ minutes per problem. Remember, this is just an estimate. You should spend less time on "quick" problems (or those worth fewer points), and more time on the more difficult problems (or those worth more points). As you work the problems be aware of the time; if half the time is gone you should have completed about half of the exam.

Exercises 3.3

Factor the following completely.

1. $4x^2 + 2x$

2. $5ab + ab^2$

3. $x^2 + x$

4. $3x^3 + 9$

5. $3xy^2 - 6x^2y^3$

6. $7x^2 - 3xy + 2x$

7. $6x^4 + 9x^3 - 21x^2$

8. $12x^2y^3 - 18x^2$

9. $35x^4y^4z - 15x^3y^5z + 10x^2y^3z^2$

10. $15a^2b - 10ab^2 - 10$

11. $24r^3s^4 - 18r^3s^5 - 6r^2s^3$

12. $15a^2b - 10ab^2$

13. $35a^2b^3 - 21a^3b^2 + 7ab$

14. $15rs^4t - 15r^2s^3 + 10rs^2$

15. $3x(x + 2) + 5(x + 2)$

16. $7r(2s + 1) + 3(2s + 1)$

17. $3x(x + 2) - 5(x + 2)$

18. $7r(2s + 1) - 3(2s + 1)$

19. $2x(x + 3y) + 5y(x + 3y)$

20. $2s(a - b) + 3(a - b)$

21. $3x(x + 4y) - 5y(x + 4y)$

22. $2x(x - 5) + (x - 5)$

23. $(2r + 1)(a - 2) + 5(a - 2)$

24. $(3a + 4)(x - 1) + 2x(x - 1)$

25. $2x(a - 3)^2 + 2(a - 3)^2$

26. $5y(a - b)^2 - 5(a - b)^2$

27. $16a^2(b - 4)^2 - 4a(b - 4)$

28. $3y(y - 2)^2 + (y - 2)$

29. $2x^2 - 8x + 3x - 12$

30. $5a^2 + 10a + 7a + 14$

31. $3x^2 - 12xy + 5xy - 20y^2$

32. $2a^2 + 2ab + 3ab + 3b^2$

33. $7ax - 7bx + 3ay - 3by$

34. $5ra - 5rb + 3sa - 3sb$

35. $7ax + 7bx - 3ay - 3by$

36. $5ra + 5rb - 3sa - 3sb$

37. $7ax - 7bx - 3ay + 3by$ **38.** $5ra - 5rb - 3sa - 3sb$ **39.** $7ax - 7bx - 3ay - 3by$

40. $5ra - 5rb - 3sa + 3sb$ **41.** $2r^2 + 2rs - sr - s^2$ **42.** $3x^2 + 6x - 4x - 8$

43. $2r^2 + 2rs - sr + s^2$ **44.** $3x^2 - 4x + 6x - 8$ **45.** $5a^2 - 5ab - 2ab + 2b^2$

46. $2x^2 + 10x - x - 5$ **47.** $3a^2 - 6a - a + 2$ **48.** $5x^2 + 20x + x + 4$

49. $3a^2 - 6a + a - 2$ **50.** $5x^2 - 20x + x - 4$ **51.** $3a^2 + 6a - a - 2$

52. $5x^2 + 20x - x + 4$ **53.** $a^3 + 2a^2 + 4a + 8$ **54.** $a^3 - 3a^2 + a - 3$

$7x(a-6)\ /\ 3y\ (a-6)$

? QUESTIONS FOR THOUGHT

55. Verbally describe what is **wrong** (if anything) with the following:

    (a) $9x^2 + 15x + 3 \overset{?}{=} 3(3x^2 + 5x)$

    (b) $(x - 2)^2 + (x - 2) \overset{?}{=} (x - 2)^3$

56. Factoring by grouping illustrates what property?

57. Complete the following expressions so that they can be factored by grouping:

    (a) $5x + 10y + 3x + ?$

    (b) $4a - 12y - 5a^2 + ?$

3.4

Factoring Trinomials

Factoring $x^2 + qx + p$

When we multiplied $x + a$ by $x + b$ in Section 3.2 we found that

$$(x + a)(x + b) = x^2 + (a + b)x + ab$$

Note the relationship between the constants a and b in the binomials to be multiplied, and the coefficients of the trinomial product: The x term coefficient is the sum of a and b; the numerical term is the product of a and b. Therefore, if the trinomial $x^2 + qx + p$ could be factored into two binomials, $(x + a)$ and $(x + b)$, then q is the sum of a and b and p is the product of a and b. Thus, all we need to find are two factors of p which sum to q. For example,

To factor $x^2 + 8x + 12$, we need to find two factors of $+12$ which sum to $+8$.

We first determine the signs of the two factors, arriving at the answer by logical deduction as follows:

The two factors must have the same signs since their product, $+12$, is positive.

Both signs must be positive since the sum $+8$ is positive.

If we systematically check the factors of 12 $(1 \cdot 12; 2 \cdot 6; 3 \cdot 4)$ we arrive at $+6$ and $+2$ as the factors of $+12$ which sum to $+8$. Thus,

$$x^2 + 8x + 12 = (x + 6)(x + 2)$$

EXAMPLE 1 | Factor each of the following.

(a) $x^2 - 4x - 12$ **(b)** $x^2 - 21x + 54$

(c) $a^2 - 10a + 25$ **(d)** $a^2 - 25$

(e) $3y^3 - 6y^2 - 105y$

Solution | **(a)** $x^2 - 4x - 12$ Find factors of -12 which sum to -4. First consider the signs of the two factors:

The signs must be opposite since the product is negative.

Since we know the signs are *opposite,* we ignore the signs and look for two factors of 12 whose *difference* is 4.

Possible candidates are 12 and 1, 6 and 2, or 3 and 4. The pair 6 and 2 yield a difference of 4 and are therefore the two factors. Now, looking at the signs, we must have -6 and $+2$ in order to sum to -4 as required. Hence,

$$\boxed{x^2 - 4x - 12 = (x - 6)(x + 2)}$$

Always take the time to check your answer by multiplying the factors. Be careful with the signs.

(b) $x^2 - 21x + 54$ Find the two factors of $+54$ which sum to -21. The signs of the two factors must be the same (since $+54$ is positive):

Both factors must be negative since their sum, -21, is negative.

Since we know the signs are the same, we ignore the signs and look for two factors of 54 whose *sum* is 21.

Possible candidates are 1 and 54, 2 and 27, 3 and 18, or 6 and 9. The answer is 3 and 18; considering the signs, we must have -3 and -18. Hence,

$$\boxed{x^2 - 21x + 54 = (x - 18)(x - 3)}$$ Check the answer.

(c) $a^2 - 10a + 25$ Find two factors of $+25$ which sum to -10.

Since $+25$ is positive, the signs must be the same.

Both factors must be negative since their sum is -10.

The factors must be -5 and -5. Hence,

$$\boxed{a^2 - 10a + 25 = (a - 5)(a - 5)}$$

(d) $a^2 - 25$ We can rewrite this as $a^2 + 0a - 25$ to determine that we are seeking two factors of -25 which sum to 0.

The signs of the two factors must be opposite (since -25 is negative).

Ignoring the signs for the moment, we are searching for two factors of 25 whose *difference* is 0.

The factors must be 5 and 5. It obviously does not matter which factor has the negative sign:

$$a^2 - 25 = (a - 5)(a + 5)$$ *Note this is a difference of two squares.*

Compare this problem with the one in part **(c).**

(e) $3y^3 - 6y^2 - 105y$ Do not forget always to factor the greatest common factor first.

$$3y^3 - 6y^2 - 105y = 3y(y^2 - 2y - 35)$$

Now try to factor $y^2 - 2y - 35$:

The signs of two factors of -35 are opposite. Therefore, ignoring the signs, what two factors of 35 yield a *difference* of 2? Answer: 7 and 5. Including signs: -7 and $+5$, since they must sum to -2.

Thus, $y^2 - 2y - 35 = (y - 7)(y + 5)$ and

$$3y^3 - 6y^2 - 105y = 3y(y^2 - 2y - 35)$$
$$= 3y(y - 7)(y + 5)$$ ■

Factoring $Ax^2 + Bx + C$

We saw in Section 3.2 that

$$(ax + b)(cx + d) = acx^2 + (ad + bc)x + bd$$

which is obviously more complex than the previous case; it is complicated by the coefficients of the x terms in the binomials.

Note the relationships between the constants a, b, c, and d, and the coefficients of the trinomial product. Therefore, if $Ax^2 + Bx + C$ were to factor into two binomials, A would be the product of the x coefficients in the binomials ($a \cdot c$), C would be the product of the numerical term coefficients in the binomials ($b \cdot d$), and B would be the interaction (inner + outer) of the four coefficients a, b, c, and d. We will demonstrate the trial-and-error process by example.

EXAMPLE 2

Factor $2x^2 + 5x + 3$.

Solution

Note first that there are no common factors.

The only possible factorization of 2 is $2 \cdot 1$. These must be the binomial x term coefficients.

The only possible factorization of 3 is $3 \cdot 1$. These must be the binomial numerical term coefficients.

There are two possible answers:

$$(2x + 1)(x + 3) \qquad \text{and} \qquad (2x + 3)(x + 1)$$

Multiplying out, we get

$$2x^2 + 7x + 3 \qquad \text{and} \qquad 2x^2 + 5x + 3$$

Note that both first and last terms are identical. The middle term indicates that $(2x + 3)(x + 1)$ is the answer. Hence,

$$2x^2 + 5x + 3 = \boxed{(2x + 3)(x + 1)}$$

■

EXAMPLE 3

Factor $6x^2 + 19x + 10$ completely.

Solution

Note that there is no common monomial to factor from $6x^2 + 19x + 10$. Since there seem to be many possible combinations, we check out the possibilities as follows:

Possible factorizations of 6 are $6 \cdot 1$ and $2 \cdot 3$.
Possible factorizations of 10 are $10 \cdot 1$ and $5 \cdot 2$.

The possible factorizations of $6x^2 + 19x + 10$ are

$$(6x + 5)(x + 2) = 6x^2 + 17x + 10$$
$$(6x + 2)(x + 5) = 6x^2 + 32x + 10*$$
$$(6x + 10)(x + 1) = 6x^2 + 16x + 10*$$
$$(6x + 1)(x + 10) = 6x^2 + 61x + 10$$
$$(2x + 5)(3x + 2) = 6x^2 + 19x + 10$$
$$(2x + 2)(3x + 5) = 6x^2 + 16x + 10*$$
$$(2x + 10)(3x + 1) = 6x^2 + 32x + 10*$$
$$(2x + 1)(3x + 10) = 6x^2 + 23x + 10$$

Each pair of binomials will yield the same first and last term, but only one combination will yield the correct middle term, $19x$:

$$6x^2 + 19x + 10 = \boxed{(2x + 5)(3x + 2)}$$

Note the following:

1. You should stop when you hit the right combination and then check your answer by multiplying.

2. Take a close look at the second possibility, $(6x + 2)(x + 5)$. This possibility can be factored further into $2(3x + 1)(x + 5)$ by fac-

toring 2 from $6x + 2$. If this possibility were the answer, it would imply that 2 is a common factor of $6x^2 + 19x + 10$ (why?). But clearly, there is no common factor of $6x^2 + 19x + 10$, so we can eliminate $(6x + 2)(x + 5)$ as a possibility. For the same reason, we can eliminate the other possibilities that appear with an asterisk (*). ∎

EXAMPLE 4

Factor $10a^2 + 19a - 15$ completely.

Solution

There is no common monomial to factor from $10a^2 + 19a - 15$.

Possible factorizations of 10 are $10 \cdot 1$ and $5 \cdot 2$.
Possible factorizations of 15 are $15 \cdot 1$ and $5 \cdot 3$.

The possible factorizations of $10a^2 + 19a - 15$ are

$$(10a + 3)(a - 5) = 10a^2 - 47a - 15$$
$$(10a + 5)(a - 3) = 10a^2 - 25a - 15*$$
$$(10a + 1)(a - 15) = 10a^2 - 149a - 15$$
$$(10a + 15)(a - 1) = 10a^2 + 5a - 15*$$
$$(5a + 3)(2a - 5) = 10a^2 - 19a - 15 \qquad \textit{Stop here.}$$

Although $(5a + 3)(2a - 5)$ is not the correct factorization (the sign of the middle term should be positive), switching the signs will switch the sign of the middle term of its product:

$$10a^2 + 19a - 15 = \boxed{(5a - 3)(2a + 5)} \ .$$

Note again that since the polynomial contains no common factor, we could have automatically eliminated factorizations with an asterisk since they contain common factors. ∎

EXAMPLE 5

Factor $12a^3 + 2a^2 - 4a$.

Solution

$12a^3 + 2a^2 - 4a$ *Factor out the common monomial, 2a, first.*

$= 2a(6a^2 + a - 2)$ *Factor $6a^2 + a - 2$ into $(2a - 1)(3a + 2)$.*

$= \boxed{2a(2a - 1)(3a + 2)}$ ∎

This is a laborious trial-and-error process—especially when the numbers have numerous factors. If you do not go about checking the factors in a systematic manner, it is very easy to miss the correct factors (and their relative positions) and assume the expression cannot be factored. For this reason, we recommend the *factoring by grouping* method, which we now describe.

Factoring Trinomials by Grouping

Another way to factor trinomials is to use grouping. To factor $Ax^2 + Bx + C$ we would:

1. Find the product, AC.
2. Find two factors of AC which sum to B.
3. Rewrite the middle term as a sum of terms whose coefficients are the factors found in step 2.
4. Factor by grouping.

EXAMPLE 6

Factor $12x^2 - 17x - 5$ by grouping.

Solution

1. Find the product, AC:

$$12(-5) = -60$$

2. Find two factors of -60 which add up to -17:

-20 and $+3$, since $(-20)(+3) = -60$ and $-20 + 3 = -17$

3. Rewrite the middle term, $-17x$, as $-20x + 3x$. Hence,

$$12x^2 - 17x - 5 = 12x^2 - 20x + 3x - 5$$

4. Factor by grouping:

$$12x^2 - 20x + 3x - 5$$
$$= 4x(3x - 5) + 1(3x - 5)$$
$$= \boxed{(3x - 5)(4x + 1)}$$

Note that we could also write this as $12x^2 + 3x - 20x - 5$.

Although we still have to look for factors (of a number larger than A or C), the process of factoring by grouping is usually a bit more efficient than the trial-and-error process.

EXAMPLE 7

Factor $20a^2 - 7ab - 6b^2$ completely.

Solution

Find the product, $20(-6) = -120$.
Which two factors of -120 will yield -7 when added? Answer: -15 and $+8$.
Rewrite $-7ab$ as $-15ab + 8ab$.

Thus,

$$20a^2 - 7ab - 6b^2 = 20a^2 - 15ab + 8ab - 6b^2$$
$$= 5a(4a - 3b) + 2b(4a - 3b)$$
$$= \boxed{(4a - 3b)(5a + 2b)}$$

EXAMPLE 8

Factor the following completely.

(a) $5x^2 - 13x - 6$ **(b)** $6a^2 + ab - 15b^2$ **(C)** $8x^4 - 10x^2y - 3y^2$

Solution

(a) Since $(5)(-6) = -30$, we want factors of -30 which sum to -13.

Answer: -15 and $+2$.

Hence,

$5x^2 - 13x - 6$ *Rewrite $-13x$ as $-15x + 2x$.*

$= 5x^2 - 15x + 2x - 6$ *Factor by grouping.*

$= 5x(x - 3) + 2(x - 3)$

$= \boxed{(x - 3)(5x + 2)}$

(b) The factors of $(6)(-15) = -90$ which sum to $+1$ are $+10$ and -9. Hence,

$6a^2 + ab - 15b^2$ *We rewrite the middle term, $+ab$, as $+10ab - 9ab$.*

$= 6a^2 + 10ab - 9ab - 15b^2$ *Factor by grouping.*

$= 2a(3a + 5b) - 3b(3a + 5b)$

$= \boxed{(3a + 5b)(2a - 3b)}$

(c) Factors of $8(-3) = -24$ which sum to -10 are $+2$ and -12. Therefore,

$8x^4 - 10x^2y - 3y^2 = 8x^4 + 2x^2y - 12x^2y - 3y^2$

$= 2x^2(4x^2 + y) - 3y(4x^2 + y)$

$= \boxed{(4x^2 + y)(2x^2 - 3y)}$ ∎

If there are no pairs of factors of *AC* which sum to the middle term coefficient, then the trinomial does not factor into two binomials.

EXAMPLE 9

Factor the following completely.

(a) $5r^2 - 2rs + 10s^2$ **(b)** $18x^2y - 57xy^2 + 30y^3$

(c) $-10z^2 + 11z - 3$ **(d)** $x(x - 3) + (x + 3)(x - 1)$

Solution

(a) The factors of $5(10) = 50$ are $1 \cdot 50$, $2 \cdot 25$, and $5 \cdot 10$. Since there are no pairs of factors of $10(5) = 50$ which sum to -2, the polynomial is

$$\boxed{\text{not factorable}}$$

(b) $18x^2y - 57xy^2 + 30y^3$ *Factor the common monomial, 3y, first.*

$$= 3y(6x^2 - 19xy + 10y^2)$$

$$= 3y[6x^2 - 15xy - 4xy + 10y^2]$$

$$= 3y[3x(2x - 5y) - 2y(2x - 5y)]$$

$$= 3y[(2x - 5y)(3x - 2y)]$$

Now factor $6x^2 - 19xy + 10y^2$. The factors of $6 \cdot 10 = 60$ which sum to -19 are -15 and -4.

or simply

$$= \boxed{3y(2x - 5y)(3x - 2y)}$$

(c) The factors of $(-10)(-3) = 30$ which sum to 11 are $+6$ and $+5$. Hence,

$$-10z^2 + 11z - 3 = -10z^2 + 6z + 5z - 3$$

$$= -2z(5z - 3) + 1(5z - 3)$$

$$= \boxed{(5z - 3)(-2z + 1)}$$

Another way to solve this problem is to factor -1 from the trinomial first:

$$-10z^2 + 11z - 3 = -[10z^2 - 11z + 3] \quad \text{\textit{Factor by grouping.}}$$

$$= -[10z^2 - 6z - 5z + 3]$$

$$= -[2z(5z - 3) - 1(5z - 3)]$$

$$= -[(5z - 3)(2z - 1)]$$

or simply

$$= \boxed{-(5z - 3)(2z - 1)}$$

Look at the two answers given for this same problem. They actually are the same. If you multiply out the second factor of the second answer, $-(2z - 1)$, you will get $-2z + 1$, which is the second factor of the first answer. [Remember that $-(ab) = (-a)(b)$.]

(d) At first glance you may be tempted to factor $x - 3$ or $x + 3$ from each term. But since they are not identical, we cannot factor either. Hence, we have no recourse but to multiply out the expression first, and then see if the simplified polynomial factors.

$$x(x - 3) + (x + 3)(x - 1) = x^2 - 3x + (x^2 + 2x - 3)$$

$$= x^2 - 3x + x^2 + 2x - 3$$

$$= 2x^2 - x - 3 \quad \text{\textit{which factors into}}$$

$$= \boxed{(2x - 3)(x + 1)} \quad \blacksquare$$

STUDY SKILLS 3.4   **Taking an Algebra Exam: Dealing with Panic**

In the first two chapters of this text we have given you advice on how to learn algebra. In the last chapter we discussed how to prepare for an algebra exam. If you followed this advice and put the proper amount of time to good use you should feel fairly confident and less anxious about the exam. But you may still find during the course of the exam that you are suddenly stuck or you "draw a blank." This may lead you to panic and say irrational things like "I'm stuck. . . . I can't do this problem. . . . I can't do any of these problems. . . . I'm going to fail this test." Your heart may start to beat faster and your breath may quicken. You are entering a panic cycle.

These statements are irrational. Getting stuck on a few problems does not mean that you cannot do any algebra. These statements only serve to interfere with your concentrating on the exam itself. How can you think about solving a problem while you are telling yourself that you cannot? The increased heart and breath rate are part of this cycle.

What we would like to do is to break this cycle. What we recommend that you do is first put aside the exam and silently say to yourself **STOP!** Then try to relax, clear your mind, and encourage yourself by saying to yourself things such as "This is only one (or a few) problems, not the whole test" or "I've done problems like this before, so I'll get the solution soon." (Haven't you ever talked to yourself this way before?)

Now take some slow deep breaths and search for some problems that you know how to solve and start with those. Build your concentration and confidence slowly with more problems. When you are through with the problems you can complete go back to the ones you were stuck on. If you have the time, take a few minutes and rest your head on your desk, and then try again. But make sure you have checked the problems you have completed.

Exercises 3.4

Factor the following completely.

1. $x^2 + 4x - 45$
2. $x^2 - 10x + 24$
3. $y^2 - 3y - 10$
4. $a^2 - 8ab - 20b^2$
5. $x^2 - 2xy - 15y^2$
6. $r^2 - rs - 2s^2$
7. $r^2 - 81$
8. $a^2 - 49$
9. $x^2 - xy - 6y^2$
10. $a^2 + 4ab - 21b^2$
11. $x^2 - xy + 6y^2$
12. $a^2 + 4ab + 21b^2$
13. $r^2 - 7rs + 12s^2$
14. $a^2 - 12ab + 20b^2$
15. $r^2 - rs - 12s^2$
16. $a^2 + 12ab - 20b^2$
17. $9x^2 - 49y^2$
18. $4a^2 - 81b^2$
19. $15x^2 + 17x + 4$
20. $9a^2 - 6a - 8$
21. $15x^2 - xy - 6y^2$
22. $10a^2 + 23ab - 12b^2$
23. $15x^2 + xy - 6y^2$
24. $10a^2 - 23ab + 12b^2$
25. $10xy + 3x^2 - 25y^2$
26. $13xy + 6y^2 + 6x^2$
27. $21x^2 - 41xy - 10y^2$
28. $10x^2 - 27xy + 18y^2$
29. $10a^2 - 21ab - 10b^2$
30. $7x^2 - 20xy + 12y^2$
31. $10a^2 + 21ab - 10b^2$
32. $7x^2 + 20xy + 12y^2$
33. $25 - 5y - 2y^2$
34. $21 + 8a - 4a^2$
35. $25 - 5y - y^2$
36. $21 - 8a - 4a^2$
37. $2x^3 + 4x^2 - 16x$
38. $3y^3 + 30y^2 + 63y$
39. $3x^3y + 6x^2y - 45xy$
40. $2a^3b + 2a^2b - 24ab$
41. $18abx^2 + 15abx - 18ab$
42. $24ab - 2a^2b - 2a^3b$
43. $x^3 + 3x^2 - 28x$
44. $y^3 - 5y^2 - 36y$
45. $2x(x - 3) + (x - 1)(x + 2)$
46. $2u(u + 1) + (u + 2)(u - 5)$
47. $18ab - 15abx - 18abx^2$
48. $30y^3 + 25y^2 + 5y$

49. $90y^4 - 114y^3 + 36y^2$

50. $54xy^4 + 45x^2y^3 + 6x^3y^2$

51. $125a^4b - 25a^3b^2 - 30a^2b^3$

52. $18a^3y^2 + 3a^2y^3 - 3ay^4$

53. $24r^3s + 20r^2s^2 - 24rs^3$

54. $42r^4s + 7r^3s^2 - 84r^2s^3$

55. $6r^4 - r^2 - 2$

56. $6x^4 - x^2 - 12$

57. $3x^2 - 27$

58. $2x^2 - 50$

59. $20xy^5 + 2x^2y^3 - 8x^3y$

60. $6x^2 + 5xy^2 - 6y^4$

61. $108x^3y + 72x^2y^3 - 15xy^5$

62. $25x^4 + 10x^2y^2 + 15y^4$

63. $12a^7b + a^4b - 6ab$

64. $24a^7b - 20a^4b^3 + 4ab^5$

65. $15a^8 + 19a^4 - 10$

66. $4a^4 + 8a^2 + 3$

? QUESTIONS FOR THOUGHT

67. In this section we mentioned that if a polynomial does *not* have a common factor, we can eliminate any binomial factors that *do* have common factors. For example, the trinomial $6x^2 - 5x - 6$ has no common numerical factors. Why does this imply that $3x - 3$ could not be a binomial factor of $6x^2 - 5x - 6$?

68. Find all k such that:

(a) $x^2 + kx + 10$ factors.

(b) $x^2 + kx - 10$ factors.

(c) $2x^2 + kx - 3$ factors.

69. Find all p such that $x^2 + 3x + p$ factors, where $-20 < p < 20$.

70. How would you factor the following?

(a) $3(x + y)^2 - 4(x + y) - 4$

(b) $2a^2 + 4ab + 2b^2 - 7a - 7b + 3$

3.5

Factoring Using Special Products

The factoring discussed in the previous section was that which applied to the general forms discussed in Section 3.2. In this section we will be concerned mainly with factoring the special products.

For example, in the last section we discussed factoring $x^2 - 10x + 25$ and $x^2 - 25$ by using trial and error and it took a few steps to arrive at each solution. However, if we had recognized these polynomials as forms of special products, we could have cut down our labor a bit. In this section, we will discuss how to recognize and factor special products without resorting to trial-and-error factoring or factoring by grouping.

First we relist in the box the special products that we have had thus far.

Factoring Special Products		
1.	$a^2 - b^2 = (a + b)(a - b)$	*Difference of two squares*
2.	$a^2 + 2ab + b^2 = (a + b)^2$	*Perfect square of sum*
3.	$a^2 - 2ab + b^2 = (a - b)^2$	*Perfect square of difference*

The first factorization, $a^2 - b^2$, is called the *difference of two squares*: the difference of squares of two terms can be factored into two binomial conjugates. For example, we could rewrite the binomial

$$4x^2 - 9y^2$$

as the difference of two squares

$$(2x)^2 - (3y)^2$$

which factors into

$$(2x - 3y)(2x + 3y)$$

The other two factorizations are called *perfect squares* (just as $36 = 6^2$ is a perfect square). Notice that the first and last terms of both perfect square trinomials are perfect squares. For example,

$$9x^2 - 12xy + 4y^2$$
$$= (3x)^2 - 2(3x)(2y) + (2y)^2$$
$$= (3x - 2y)^2$$

If a trinomial is factorable, it can be factored by the methods covered in the previous section. However, if a trinomial can be recognized as a special product, it can be factored quickly and with less effort. For example, when we see a trinomial such as

$$16x^2 - 40xy + 25y^2$$

there are quite a few possible combinations of factors to check if we use trial and error or factoring by grouping. Our experience with special products, however, makes us suspicious when we see two perfect square terms, $16x^2$ and $25y^2$, in the trinomial. We would immediately check to see if it is a perfect square by choosing

$$(4x - 5y)(4x - 5y)$$

as the first *possible* factorization. Multiplication will confirm our suspicion that this is indeed the proper factorization of $16x^2 - 40xy + 25y^2$.

With regard to the difference of two squares, the only binomials up to this point which factor into two *binomials* are those which are the difference of two squares. Obviously, it is easier to use the difference of squares factorization than to introduce a middle term coefficient of 0 as we did in the previous section using the other factoring methods.

In order to factor quickly, you must recognize the relationships given in the previous box.

EXAMPLE 1

Factor the following completely.

(a) $x^2 - 6x + 9$ (b) $9a^2 + 30a + 25$

(c) $4x^2 - 25y^2$ (d) $16a^2b^2 - 81c^4$

Solution

(a) $x^2 - 6x + 9 = x^2 - 2(3x) + 3^2$ *Note the relationships between terms in a perfect square trinomial.*

$= \boxed{(x - 3)^2}$

(b) $9a^2 + 30a + 25 = (3a)^2 + 2(3a)(5) + 5^2$

$= \boxed{(3a + 5)^2}$

(c) $4x^2 - 25y^2 = (2x)^2 - (5y)^2$

$= \boxed{(2x - 5y)(2x + 5y)}$

(d) $16a^2b^2 - 81c^4 = (4ab)^2 - (9c^2)^2$

$= \boxed{(4ab - 9c^2)(4ab + 9c^2)}$ ∎

We will now add two more special products.

Factoring the Sum and Difference of Cubes	
4. $a^3 - b^3 = (a - b)(a^2 + ab + b^2)$	*Difference of two cubes*
5. $a^3 + b^3 = (a + b)(a^2 - ab + b^2)$	*Sum of two cubes*

Again, these are forms you should remember if you are to factor quickly.

Note the similarities and differences between factoring the sum and difference of two cubes. Keep in mind that $a^3 - b^3$ is *not* the same as $(a - b)^3$, which when multiplied out will have middle terms. Also note that the right-hand factor, either $a^2 - ab + b^2$ or $a^2 + ab + b^2$, will not have binomial factors. (Do not confuse it with the perfect square $a^2 + 2ab + b^2$, which has binomial factors.)

EXAMPLE 2

Factor the following completely.

(a) $27x^3 - y^3$ (b) $8c^3 + 125b^3$

Solution

(a) $27x^3 - y^3$ *Write as a difference of two cubes.*

$\quad A^3 \quad - \quad B^3$

$= (3x)^3 - y^3$ *Factor.*

$\quad (A - B) \quad (A^2 \;+\; A \cdot B \;+\; B^2)$

$= (3x - y)[(3x)^2 + (3x)(y) + y^2]$ *Simplify inside the brackets.*

$= \boxed{(3x - y)(9x^2 + 3xy + y^2)}$

(b) $8c^3 + 125b^3$ *Rewrite as a sum of two cubes.*

   $= (2c)^3 + (5b)^3$ *Factor.*

   $= (2c + 5b)[(2c)^2 - (2c)(5b) + (5b)^2]$ *Simplify inside the brackets.*

   $= \boxed{(2c + 5b)(4c^2 - 10cb + 25b^2)}$

Note that although the sum of two *cubes* **does** factor, the sum of two *squares* does **not**.   ∎

Some polynomials are not readily recognizable as special products until we factor out the greatest common factor. For example, you may not recognize the expression $3x^3 - 3xy^2$ as a special product, but if you factor $3x$ from it you get

$3x^3 - 3xy^2 = 3x(x^2 - y^2)$ *Now factor* $(x^2 - y^2)$.

   $= 3x(x - y)(x + y)$

Remember, your first step should always be to factor out the greatest common factor; this should always make the factoring process easier.

EXAMPLE 3   Factor the following completely.

   (a) $2a^3b + 8a^2b^2 + 8ab^3$ **(b)** $x^4 - y^4$ **(c)** $27x^6 - 8y^6$

Solution   **(a)** $2a^3b + 8a^2b^2 + 8ab^3$ *Factor out the common monomial first.*

   $= 2ab(a^2 + 4ab + 4b^2)$ $a^2 + 4ab + 4b^2$ *is a perfect square*

   $= \boxed{2ab(a + 2b)^2}$

(b) $x^4 - y^4$ *There is no common monomial to factor. Rewrite as a difference of two squares.*

   $= (x^2)^2 - (y^2)^2$ *Factor.*

   $= (x^2 - y^2)(x^2 + y^2)$ *Do not forget to factor* $x^2 - y^2$.

   $= \boxed{(x - y)(x + y)(x^2 + y^2)}$

Always check to see if there is any more factoring to be done.

(c) $27x^6 - 8y^6$ *There is no common monomial to factor. Rewrite as a difference of two cubes.*

   $= (3x^2)^3 - (2y^2)^3$

   $= (3x^2 - 2y^2)[(3x^2)^2 + (3x^2)(2y^2) + (2y^2)^2]$ *Simplify inside the brackets.*

   $= \boxed{(3x^2 - 2y^2)(9x^4 + 6x^2y^2 + 4y^4)}$ ∎

At the end of Section 3.2 we applied our knowledge of special products to multiply more complex expressions. We will now do the same here. For example, we would factor $(x + y)^2 - (r + s)^2$ by first recognizing that it is a difference of two squares. Hence,

$$
\begin{array}{ccccccc}
a^2 & - & b^2 & = & (a & - & b) & \cdot & (a & + & b) \\
\end{array}
$$
$$(x + y)^2 - (r + s)^2 = [(x + y) - (r + s)][(x + y) + (r + s)]$$
$$= \boxed{(x + y - r - s)(x + y + r + s)}$$

Then simplify inside the brackets.

As before, we can factor a more complex expression by recognizing the form as a special product.

EXAMPLE 4

Factor completely. $(x - 3)^2 - 4$

Solution

$(x - 3)^2 - 4$

$= (x - 3)^2 - 2^2$ *The difference of two squares $(x - 3)^2 - 2^2$*

$= [(x - 3) - 2][(x - 3) + 2]$ *Factor.*

$= \boxed{(x - 5)(x - 1)}$ *Then simplify inside the brackets.*

We could have multiplied the expression out and then factored, but this approach is quicker. ∎

Problems such as those in Example 5 may require grouping before you can recognize them as special products.

EXAMPLE 5

Factor the following completely.

(a) $x^2 - 4xy + 4y^2 - 9z^2$ **(b)** $x^3 + x^2 - x - 1$

Solution

(a)   If we tried to factor in pairs, we would find no common factor. But suppose we group together the first *three* terms:

$$x^2 - 4xy + 4y^2 - 9z^2 = (x^2 - 4xy + 4y^2) - 9z^2$$

Note that $x^2 - 4xy + 4y^2$ is the perfect square $(x - 2y)^2$. Hence,

$$x^2 - 4xy + 4y^2 - 9z^2 = (x - 2y)^2 - (3z)^2$$ *Which is a difference of two squares*

$$= [(x - 2y) - 3z][(x - 2y) + 3z]$$

$$= \boxed{(x - 2y - 3z)(x - 2y + 3z)}$$

(b) $x^3 + x^2 - x - 1$ *Factor by grouping.*

$$= x^2(x + 1) - 1(x + 1)$$ *Then factor out $x + 1$.*

$$= (x + 1)(x^2 - 1)$$ *Do not forget to factor $x^2 - 1$.*

$$= \boxed{(x + 1)(x - 1)(x + 1)}$$ ∎

In general, we offer the following advice for factoring polynomials.

General Advice for Factoring Polynomials	

1. Always factor out the greatest common factor first.
2. If the polynomial to be factored is a binomial, then it may be a difference of two squares, or a sum or difference of two cubes (remember that a sum of two squares does not factor).
3. If the polynomial to be factored is a trinomial, then:
   (a) If two of the three terms are perfect squares, the polynomial may be a perfect square.
   (b) Otherwise, the polynomial may be one of the general forms.
4. If the polynomial to be factored consists of four or more terms, then try factoring by grouping.

STUDY SKILLS 3.5 **Taking an Algebra Exam: A Few Other Comments About Exams**

Do not forget to check over all your work as we have suggested on numerous occasions. Reread all directions and make sure that you have answered all the questions as directed.

If you are required to show your work (such as for partial credit), make sure that your work is neat. Do not forget to put your final answers where directed or at least indicate your answers clearly by putting a box or circle around them. For multiple-choice tests be sure you have filled in the correct space.

One other bit of advice: Some students are unnerved when they see others finishing the exam early. They begin to believe that there may be something wrong with themselves because they are still working on the exam. They should not be concerned that some students can do the work quickly and others leave the exam early, not because the exam was easy for them, but because they give up.

In any case, do not be in a hurry to leave the exam. If you are given 1 hour for the exam then take the entire hour. If you have followed the suggestions in this chapter such as checking your work, etc., and you still have time left over, relax for a few minutes and then go back and check your work again.

Exercises 3.5

Factor the following as completely as possible.

1. $16x^2 - 9y^2$
2. $x^2 - 6x + 9$
3. $x^2 + 4xy + 4y^2$
4. $9a^2 + 30ab + 25b^2$
5. $x^2 - 4xy + 4y^2$
6. $9a^2 - 30ab + 25b^2$
7. $x^2 - 4xy - 4y^2$
8. $9a^2 - 30ab - 25b^2$
9. $6x^2y^2 - 5xy + 1$
10. $9r^2 + 42rs + 49s^2$
11. $81r^2s^2 - 16$
12. $25x^2y^2 - 4$
13. $49x^2y^2 - 42xy + 9$
14. $36a^2b^2 - 60abc + 25c^2$
15. $9x^2y^2 - 4z^2$
16. $9a^2b^2 - 4$
17. $25a^2 - 4a^2b^2$
18. $1 - 9a^2$
19. $64x^4 - 16x^2y + y^2$
20. $4a^2 + 4ab^2 - b^4$
21. $1 - 14a + 49a^2$
22. $125a^3 + b^3$
23. $8a^3 - b^3$
24. $8a^3 - 343b^3$
25. $x^3 + 125y^3$
26. $9x^4 + 24x^2 + 16$
27. $4x^2 + 25y^2$

28. $9x^4 - 24x^2 - 16$

29. $4y^2 - 20y + 10$

30. $45a^3b - 20ab^3$

31. $18a^3 + 24a^2b + 8ab^2$

32. $63r^2 - 28rs^3$

33. $12a^3c + 36a^2c^2 + 27ac^3$

34. $18x^5 + 24x^3 + 8x$

35. $9a^6 - b^6$

36. $245x^5y + 70x^3y + 5xy$

37. $4x^4 - 81y^4$

38. $5r^3s - 45rs^3$

39. $4x^6 + 28x^3y^3 + 49y^6$

40. $x^8 - y^8$

41. $4x^6 - 9y^6$

42. $64x^6 - 16x^3 + 1$

43. $25a^6 + 10a^3b + b^2$

44. $12y^6 - 27y^2$

45. $24x^4y - 54x^2y^3$

46. $16x^4 - 81y^4$

47. $12y^6 + 27y^2$

48. $12x^4 - 12y^4$

49. $(a + b)^2 - 4$

50. $(3x - 2)^2 - (a + b)^2$

51. $(3a + b)^2 - (a + b)^2$

52. $(5x - 3y)^2 - (2x + 1)^2$

53. $8x^4 - 14x^2 + 3$

54. $a^6 + a^3b^3 - 2b^6$

55. $27a^6 + 28a^3b^3 + b^6$

56. $4s^4 - 5s^2y^2 + y^4$

57. $9a^5b - 12a^3b^3 + 3ab^5$

58. $40a^7b - 35a^4b^4 - 5ab^7$

59. $5a^5b - 15a^3b^3 - 20ab^5$

60. $48a^7b + 378a^4b^4 - 48ab^7$

61. $(x + y)^3 - (a + b)^3$

62. $4x^2 - 12x + 9 - a^2 - 2ab - b^2$

63. $a^2 - 2ab + b^2 - 16$

64. $r^2 - 10r + 25 - x^2$

65. $x^2 + 6x + 9 - r^2$

66. $x^2 + 2xy + y^2 - 81$

67. $a^3 + a^2 - 4a - 4$

68. $r^3 - r^2 - 9r + 9$

69. $x^4 + x^3 - x - 1$

70. $8x^4 - 16x^3 - x + 2$

71. $a^5 - a^3 + a^2 - 1$

72. $x^5 - 4x^3 + 8x^2 - 32$

? QUESTIONS FOR THOUGHT

73. Describe what relationship must exist among terms in order for a trinomial to factor as a perfect square.

74. Describe the factors of the sum and the differences of two cubes. How do they differ? How are they the same?

75. Discuss the differences between $a^3 - b^3$ and $(a - b)^3$.

76. Factor $x^6 - y^6$ completely.

   (a) Begin by using the difference of two squares.

   (b) Begin by using the difference of two cubes.

   (c) Should the answers to parts **(a)** and **(b)** be the same? Why or why not?

77. Why is $(a^2b - b)(a + 1)$ *not* a complete factorization of $a^3b + a^2b - ab - b$? What should the answer be?

3.6

Polynomial Division

In this section we will discuss division of polynomials—in particular, division by binomials and trinomials. We will discuss division by monomials in Chapter 4 within the context of fractional expressions.

Recall that when we divide numbers using long division, we use the following procedure:

Which is shorthand for

$$\begin{array}{r} 12 \\ 23\overline{)279} \\ 23 \\ \hline 49 \\ 46 \\ \hline 3 \end{array}$$

$$\begin{array}{r} 10 + 2 \\ 20 + 3\overline{)200 + 70 + 9} \\ 200 + 30 \\ \hline 40 + 9 \\ 40 + 6 \\ \hline 3 \end{array}$$

We divide 20 *into* 200.

We divide 20 *into* 40.

First we focus our attention on dividing by 20, but then we multiply the number in the quotient by $20 + 3$, subtract, bring down the next term, and repeat this process. Polynomial division is handled similarly. We will demonstrate by example.

EXAMPLE 1

$(2x^2 + 3x + 3) \div (x + 4)$

Solution

1. Put both polynomials into standard form (arrange terms in descending order of degree).

$$x + 4\overline{)2x^2 + 3x + 3}$$

2. Divide the highest-degree term of the divisor into the highest-degree term of the dividend:

$$\dfrac{2x^2}{x} = 2x$$

$$\begin{array}{r} 2x \\ x + 4\overline{)2x^2 + 3x + 3} \end{array}$$

3. Multiply the resulting quotient, $2x$, by the whole divisor:

$$2x(x + 4) = 2x^2 + 8x$$

Multiply
$$\begin{array}{r} 2x \\ x + 4\overline{)2x^2 + 3x + 3} \\ 2x^2 + 8x \end{array}$$

4. Subtract the result of step 3 from the dividend:

$$\begin{array}{r} 2x^2 + 3x \\ -(2x^2 + 8x) \end{array} \rightarrow \begin{array}{r} 2x^2 + 3x \\ -2x^2 - 8x \\ \hline -5x \end{array}$$

Change signs of all terms in $2x^2 + 8x$ and add.

$$\begin{array}{r} 2x \\ x + 4\overline{)2x^2 + 3x + 3} \\ -(2x^2 + 8x) \\ \hline -5x \end{array}$$

Subtract

5. Bring down the next term.

$$\begin{array}{r} 2x \\ x + 4\overline{)2x^2 + 3x + 3} \\ -(2x^2 + 8x) \\ \hline -5x + 3 \end{array}$$

6. Repeat steps 2–5 until no more terms can be brought down.

   (a) Divide x into $-5x$ to get -5.

   (b) Multiply -5 by $x + 4$ to get $-5x - 20$.

   (c) Subtract $-5x - 20$ from $-5x + 3$ to get 23.

$$\begin{array}{r} 2x - 5 \\ x + 4 \overline{) 2x^2 + 3x + 3} \\ -(2x^2 + 8x) \\ \hline -5x + 3 \\ -(-5x - 20) \quad \text{\textit{Subtract}} \\ \hline + 23 \end{array}$$

7. The remaining term, $+23$, is the remainder.

$$\begin{array}{r} 2x - 5 \\ x + 4 \overline{) 2x^2 + 3x + 3} \\ -(2x^2 + 8x) \\ \hline -5x + 3 \\ -(-5x - 20) \\ \hline + 23 \end{array}$$

The process ends when the degree of the remainder is less than the degree of the divisor. In this case, the degree of 23 is 0 and the degree of $x + 1$ is 1.

Answer: $\boxed{2x - 5, \quad \text{Rem. } +23}$

Or, we can rewrite this expression as $\boxed{2x - 5 + \dfrac{23}{x + 4}}$

We should check the answer in the same way we check long division problems with numbers:

$$\text{Dividend} = (\text{Quotient}) \cdot (\text{Divisor}) + \text{Remainder}$$
$$2x^2 + 3x + 3 \overset{?}{=} (2x - 5)(x + 4) + 23 \quad \text{\textit{Multiply first.}}$$
$$\overset{?}{=} 2x^2 + 3x - 20 + 23$$
$$\overset{\checkmark}{=} 2x^2 + 3x + 3 \qquad\qquad \blacksquare$$

When writing the dividend in standard form, it is a good idea to leave a space between terms wherever a consecutive power is missing, or fill in the power with a coefficient of 0. This stops you from making the error of combining two different powers of x in a subtraction step.

EXAMPLE 2

$(2x^3 + x - 18) \div (x - 2)$

Solution

$$\begin{array}{r} 2x^2 + 4x + 9 \\ x - 2 \overline{) 2x^3 + 0x^2 + x - 18} \\ -(2x^3 - 4x^2) \\ \hline 4x^2 + x \\ -(4x^2 - 8x) \\ \hline 9x - 18 \\ -(9x - 18) \\ \hline 0 \end{array}$$

Missing power of x in dividend. Write in power with 0 coefficient.

Answer: $2x^2 + 4x + 9$

CHECK: $2x^3 + x - 18 \overset{?}{=} (2x^2 + 4x + 9)(x - 2)$
$$\overset{?}{=} 2x^3 + 4x^2 + 9x - 4x^2 - 8x - 18$$
$$\overset{\checkmark}{=} 2x^3 + x - 18$$

EXAMPLE 3 $(a^5 + 3a^3 + a^2 + 3a + 7) \div (a^3 + a + 1)$

Solution

$$
\begin{array}{r}
a^2 \qquad\quad + 2 \\
a^3 + a + 1\,)\overline{a^5 \qquad + 3a^3 + a^2 + 3a + 7} \\
-(a^5 \qquad + a^3 + a^2) \\
\hline
2a^3 \qquad\quad + 3a + 7 \\
-(2a^3 \qquad\quad + 2a + 2) \\
\hline
a + 5
\end{array}
$$

These last two terms must be brought down.

Answer: $a^2 + 2$, Rem. $a + 5$ or $a^2 + 2 + \dfrac{a + 5}{a^3 + a + 1}$

Note that the degree of the remainder must be less than the degree of the divisor.

Exercises 3.6

Divide the following using polynomial long division.

1. $(x^2 - x - 20) \div (x - 5)$

2. $(3y^2 + 5y + 7) \div (y + 1)$

3. $(3a^2 + 10a + 8) \div (3a + 4)$

4. $(2x^2 + x - 14) \div (2x - 3)$

5. $(21z^2 + z - 16) \div (3z + 1)$

6. $(5a^2 + 16a + 11) \div (5a + 1)$

7. $(a^3 + a^2 + a - 8) \div (a - 1)$

8. $(x^3 + 2x^2 + 2x + 1) \div (x + 1)$

9. $(3x^3 - 11x^2 - 5x + 12) \div (x - 4)$

10. $(y^3 - 2y^2 + 3y - 2) \div (y - 2)$

11. $(4z^2 - 15) \div (2z + 3)$

12. $(4a^2 - 35) \div (2a - 7)$

13. $(x^4 + x^3 - 4x^2 - 3x - 2) \div (x - 2)$

14. $(y^4 + 4y^3 - 2y^2 - 2y + 3) \div (y + 1)$

15. $(-10 + 2y^2 + 5y - 5y^3 + 3y^4) \div (y - 1)$

16. $(-10x^2 + 12 + 9x - 3x^3 + 6x^4) \div (2x - 1)$

17. $(6 - 5a + 2a^4 + 4a^3) \div (a + 2)$

18. $(10x^4 - 6x - 15 + 5x^3) \div (2x + 1)$

19. $(y^3 - 1) \div (y + 1)$

20. $(8a^3 - 1) \div (2a - 1)$

21. $(8a^3 + 1) \div (2a - 1)$

22. $(y^3 + 1) \div (y + 1)$

23. $(2y^5 + 4 - 3y^4 - y^2 + y) \div (y^2 + 1)$

24. $(5x^4 + 1) \div (x - 1)$

25. $(-10 + 12z + 3z^5 + 3z^2 - 4z^3) \div (z^2 + 2z - 1)$

26. $(5y^6 + 17y^2 - 14 + 7y - 9y^3 - 28y^4 + 15y^5) \div (y^2 + 3y - 5)$

27. $(7x^6 + 1 + x - 3x^2 - 14x^4) \div (x^2 - 2)$

28. $(x^4 - y^4) \div (x - y)$

CHAPTER 3 Summary

After having completed this chapter, you should be able to:

1. Identify the degree and coefficients of the terms in a polynomial, as well as the degree of the polynomial itself (Section 3.1).

   For example: The polynomial $5x^6 - 3x + 6$ has *three* terms:

   The first term, $5x^6$, has coefficient 5 and degree 6.

   The second term, $-3x$, has coefficient -3 and degree 1.

   The constant term, 6, has degree 0.

   The polynomial has degree 6.

2. Combine and simplify polynomial expressions (Section 3.1).

   For example:

   (a) $2(3x^2 - x + 3y) - 3(y - x) - (2x^2 - 3y)$
   $$= 6x^2 - 2x + 6y - 3y + 3x - 2x^2 + 3y$$
   $$= 4x^2 + x + 6y$$

   (b) $2x^2y(3xy^2 - 6x^3y^2) - 3x(-2xy^2)(-4x^3y) = 6x^3y^3 - 12x^5y^3 - 24x^5y^3$
   $$= 6x^3y^3 - 36x^5y^3$$

3. Multiply polynomials (Section 3.1).

   For example:

   $$(2x - y)(x^2 - 3xy - y^2) = 2x^3 - 6x^2y - 2xy^2 - x^2y + 3xy^2 + y^3$$
   $$= 2x^3 - 7x^2y + xy^2 + y^3$$

4. Multiply polynomials using general forms and special products (Section 3.2).

   For example:

   (a) $(3x - 4)^2 = (3x)^2 - 2(3x)(4) + 4^2$
   $$= 9x^2 - 24x + 16$$

   (b) $[(x - y) - 7][(x - y) + 7] = (x - y)^2 - 7^2$
   $$= x^2 - 2xy + y^2 - 49$$

5. Factor various types of polynomials (Sections 3.3–3.5).

   For example:

   Common monomial factoring: **(a)** $2x^2 + 7x = x(2x + 7)$

   Factoring by grouping: **(b)** $x^2 - 3x - xy + 3y$
   $$= x(x - 3) - y(x - 3)$$
   $$= (x - 3)(x - y)$$

   General forms: **(c)** $y^2 - 10y + 21 = (y - 7)(y - 3)$

   (d) $2x^2 + 7x + 6 = (2x + 3)(x + 2)$

   (e) $6x^5 - 15x^3y + 9xy^2$
   $$= 3x(2x^4 - 5x^2y + 3y^2)$$
   $$= 3x(2x^2 - 3y)(x^2 - y)$$

Using special products:

(f) $16m^2 - 9n^2 = (4m - 3n)(4m + 3n)$ *Difference of two squares*

(g) $25x^2 - 60x + 36$

$\quad = (5x - 6)(5x - 6)$ *Perfect square*

(h) $x^3 - 8 = x^3 - 2^3$ *Difference of two cubes*

$\quad = (x - 2)(x^2 + 2x + 4)$

(i) $t^2 + 6t + 9 - u^2 = (t + 3)^2 - u^2$ *Grouping*

$\quad = (t + 3 - u)(t + 3 + u)$ *Difference of two squares*

6. Use long division to divide polynomials (Section 3.6).

For example: $(8x^3 - 28x + 15) \div (2x - 3)$

Solution:

$$
\begin{array}{r}
4x^2 + 6x - 5 \\
2x - 3 \overline{)8x^3 + 0x^2 - 28x + 15} \\
\underline{-(8x^3 - 12x^2)} \\
12x^2 - 28x \\
\underline{-(12x^2 - 18x)} \\
-10x + 15 \\
\underline{-(-10x + 15)} \\
0
\end{array}
$$

Answer: $4x^2 + 6x - 5$

CHAPTER 3   Review Exercises

In Exercises 1–26, perform the operations and simplify.

1. Subtract the sum of $2x^2 - 3x + 4$ and $5x^2 - 3$ from $2x^3 - 4$.

2. Subtract $2x^3 - 3x^2 - 2$ from the sum of $2x^3 - 3x + 4$ and $3x^2 - 5x + 9$.

3. $3x^2(2xy - 3y + 1)$

4. $5rs^2(3r^2s - 2rs^2 - 3)$

5. $3ab(2a - 3b) - 2a(3ab - 4b^2)$

6. $3xy(2x - 3y + 4) - 2x(5xy - 3y) - 2xy^2$

7. $3x - 2(x - 3) - [x - 2(5 - x)]$

8. $5x - 4\{3 - 2[x - (2 - x)]\}$

9. $(x - 3)(x + 2)$

10. $(x - 5)(x - 8)$

11. $(3y - 2)(2y - 1)$

12. $(5y - 3a)(5y + 3a)$

13. $(3x - 4y)^2$

14. $(5a + 3b)(a - 2b)$

15. $(4x^2 - 5y)^2$

16. $(4a - 3b^2)(4a + 3b^2)$

17. $(7x^2 - 5y^3)(7x^2 + 5y^3)$

18. $(2x - 5)(3x^2 - 2x + 1)$

19. $(5y^2 - 3y + 7)(2y - 3)$

20. $(7x^2 + 3xy - 2y^2)(3x - 2y)$

21. $(2x + 3y + 4)(2x + 3y - 5)$

22. $[(a + b) + (x - y)][(a + b) - (x - y)]$

23. $[(x - y) + 5]^2$

24. $(x - 3 + y)(x - 3 - y)$

25. $(a + b - 4)^2$

26. $(x + y + 5)^2$

In Exercises 27–72, factor the expression completely.

27. $6x^2y - 12xy^2 + 9xy$

28. $15a^2b^2 - 10ab^4 + 5ab^2$

29. $3x(a + b) - 2(a + b)$

30. $5y(y - 1) + 3(y - 1)$

31. $2(a - b)^2 + 3(a - b)$

32. $4(a - b)^2 + 7(a - b)$

33. $5ax - 5a + 3bx - 3b$

34. $2xa + 2xb + 3a + 3b$

35. $5y^2 - 5y + 3y - 3$

36. $14a^2 - 7ab + 6ab - 3b^2$

37. $a^2 + a(a - 10)$

38. $t^3 + t(t - 6)$

39. $2t(t + 2) - (t - 2)(t + 4)$

40. $3m(m - 1) - (2m + 1)(m - 2)$

41. $x^2 - 2x - 35$

42. $a^2 + 5a - 36$

43. $a^2 + 5ab - 14b^2$

44. $y^4 + x^2y^2 - 6x^4$

45. $35a^2 + 17ab + 2b^2$

46. $x^2 - 6xy + 9y^2$

47. $a^2 - 6ab^2 + 9b^4$

48. $2x^3 + 6x^2 - 54x$

49. $3a^3 - 21a^2 + 30a$

50. $5a^3b + 5a^2b^2 - 30ab^3$

51. $2x^3 - 50xy^2$

52. $3y^3 + 24y^2 + 48y$

53. $6x^2 + 5x - 6$

54. $25y^2 - 5y - 12$

55. $8x^3 + 125y^3$

56. $x^3 - 27$

57. $6a^2 - 17ab - 3b^2$

58. $12x^2 + 16xy^2 + 5y^4$

59. $21a^4 + 41a^2b^2 + 10b^4$

60. $54a^4 - 16ab^3$

61. $25x^4 - 40x^2y^2 + 16y^4$

62. $18x^3 + 15x^2y - 18xy^2$

63. $20x^3y - 60x^2y^2 + 45xy^3$

64. $4a^4 - 9b^4$

65. $6a^4b - 8a^3b^2 - 8a^2b^3$

66. $28x^4y - 63x^2y^3$

67. $6x^5 - 10x^3 - 4x$

68. $30x^5y - 85x^3y^2 + 25xy^3$

69. $(a - b)^2 - 4$

70. $(3x - 2)^2 - (5y + 3)^2$

71. $9y^2 + 30y + 25 - 9x^2$

72. $25x^4 - y^2 - 8y - 16$

In Exercises 73–76, find the quotient using long division.

73. $(3x^3 - 4x^2 + 7x - 5) \div (x - 2)$

74. $(x^4 - x - 1) \div (x + 1)$

75. $(8a^3 - 27) \div (2a - 3)$

76. $(y^6 + y^5 + y^4 + y^3 + y^2 + y + 1) \div (y^2 + y + 1)$

CHAPTER 3 Practice Test

Perform the operations and simplify the following.

1. $(9a^2 + 6ab + 4b^2)(3a - 2b)$

2. $(x - 3)(x + 3) - (x - 3)^2$

3. $(x - y - 2)(x - y + 2)$

4. $(3x + y^2)^2$

Factor the following completely:

5. $10x^3y^2 - 6x^2y^3 + 2xy$

6. $6ax + 15a - 2bx - 5b$

7. $x^2 + 11xy - 60y^2$

8. $10x^2 - 11x - 6$

9. $2x^2 - 3x + 7$

10. $3a^3b - 3ab^3$

11. $r^3s - 10r^2s^2 + 25rs^3$

12. $8a^3 - 1$

13. $(2x + y)^2 - 64$

14. $a^3 + 4a^2 - a - 4$

15. Find the quotient using long division: $(x^3 + 7x - 8) \div (x - 3)$

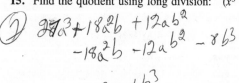

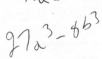

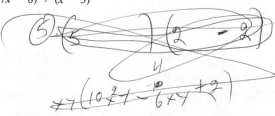

In Exercises 1–4,

$A = \{x \mid x \text{ is a prime number}; \quad 7 < x < 41\}$

$B = \{y \mid y \text{ is a multiple of 3}; \quad 6 \leq y \leq 42\}$

$C = \{z \mid z \text{ is a multiple of 5}; \quad 6 < z < 40\}$

List the elements in the following sets:

1. A

2. $A \cap B$

3. $A \cap C$

4. $A \cup C$

In Exercises 5–6, answer true or false.

5. $\dfrac{6}{2} \in Q$

6. $\sqrt{7} \in Q$

In Exercises 7–8, graph the following sets:

7. $\{x \mid -2 \leq x < 8\}$

8. $\{x \mid 2 \leq x < 8, \quad x \in Z\}$

In Exercises 9–10, list the property illustrated by the statement. If the statement is false, state so.

9. The sum of two real numbers is a real number.

10. $\left(\dfrac{1}{x-3}\right)(x-3) = 1 \quad (x \neq 3)$

In Exercises 11–14, perform the operations.

11. $-2 + 3 - 4 - 8 + 5$

12. $(-2)(4) - (-3)(-3)$

13. $(-9 + 3)(-4 - 2)$

14. $6 - \{3 - 5[2 - (3 - 9)]\}$

In Exercises 15–16, evaluate the expressions given $x = -4$, $y = 2$, and $z = 0$.

15. $|y - x| - (y - x)$

16. $\dfrac{3x^2y - 2x}{3xz}$

In Exercises 17–18, perform the operations and express your answer in simplest form.

17. $-3x(x^2 - 2x + 3)$

18. $(-2xy)^2(-3x^2y)$

In Exercises 19–22, translate the statements algebraically.

19. Three more than the product of a number and 8

20. The product of the first two of three consecutive integers less the third consecutive integer

21. The length of a rectangle is 3 less than 4 times its width. Express its area and perimeter in terms of its width.

22. Harold has thirty-five coins in dimes and nickels; express the value of his coins in terms of the number of nickels.

In Exercises 23–30, solve the equation.

23. $4x - 3 = 5x + 8$

24. $5(a - 1) = 5(a + 2)$

25. $2x + 3 - (x - 2) = 5 - (x - 4)$

26. $3x - 2 + 2(x - 1) = 1 - 2(3 - x)$

27. $3(x - 4) + 2(3 - x) = 5[x - (2 - 3x)]$

28. $5(x - 2) - 2(3x + 1) = 2(x - 4)$

29. $|2x + 1| = 9$

30. $|3x - 2| = |x - 1|$

(continued)

In Exercises 31–38, solve the inequalities and graph the solutions on the number line.

31. $3x - 5 \leq x - 8$

32. $3 - 5x > 5 + 3x$

33. $6(x - 2) + (2x - 3) < 2(4x + 1)$

34. $-5 < 3 - 2x \leq 9$

35. $|x - 5| < 4$

36. $|2x + 1| > 7$

37. $|5 - 4x| > 7$

38. $|8 - 3x| \leq 8$

In Exercises 39–46, solve the problem algebraically.

39. If 4 times a number is 3 less than twice the number, what is the number?

40. Cindy has twenty-five coins in quarters and dimes, totaling $3.25. How many of each coin does she have?

41. A man went to a store and bought 28 packages of paper plates for $33. Some of the packages contained 300 plates and cost $2 per package, and the rest of the packages contained 100 plates and cost $1 per package. How many plates did he buy all together?

42. A painter charges $9 an hour to paint a house and his assistant gets $6 an hour. If the bill for labor for painting a house comes to $417, how many hours did the assistant paint if all together they worked a total of 58 hours?

43. A GL-70 printer can print 80 characters per second, while a VF-44 printer can print 120 characters per second. How long would it take a VF-44 printer to print a document which takes 30 minutes for the GL-70 to print?

44. A company installed a photocopy machine which can make 90 copies per minute. For a certain job it had to make 9,750 copies. The job was begun on the new machine, but after awhile the machine broke down. The remaining copies were made by the old machine, which makes 60 copies per minute. If the total job took 2 hours, how many copies did the new machine make?

45. The length of a rectangle is 2 more than 3 times its width. If the width varies from 5 to 12 feet, what is the range of values for the perimeter?

46. Charles gets $6 an hour as a tutor and $10 an hour as a mechanic. If he is to work a total of 30 hours, what is the minimum number of hours he should work as a mechanic if he wants to make at least $232?

In Exercises 47–48, identify the degree of the polynomial.

47. $5xy^2 - 2x^2y^3 + 2$

48. 3

In Exercises 49–60, perform the operations and express your answer in simplest form.

49. $(3x^2 + 2x - 4) - (2x^2 - 3x + 5)$

50. $-3x(2x - 5y + 4)$

51. $(x - y)(x - 3y)$

52. $(3x - 2y)(x - y)$

53. $(x + y - 2)(x + y)$

54. $(2a + b)(4a^2 - 2ab + b^2)$

55. $(3x - 5y)^2$

56. $(2m + 3n)(2m - 3n)$

57. $(3x - 5y)(3x + 5y)$

58. $(2m + 3n)^2$

59. $(2x + y - 3)(2x + y + 3)$

60. $(x - 2y + 3)^2$

In Exercises 61–82, factor as completely as possible.

61. $x^2 - 5x - 24$

62. $y^2 - 4x^2$

63. $y^2 - 12xy + 35x^2$

64. $10a^2 - 3ab - b^2$

65. $4y^2 + 16yz + 15z^2$

66. $9x^2 - 25z^2$

67. $25a^2 + 20ab + 4b^2$

68. $4x^2 - 12x - 9$

69. $36x^2 - 9$

70. $25x^2 - 30xz + 9z^2$

71. $25a^2 + 20ab - 4b^2$

72. $12y^3 - 16y^2 - 3y$

73. $3y^3 + 5y^2 - 2y$

74. $25a^4 + 10a^2b^2 - 8b^4$

75. $49a^4 - 14a^2z - 3z^2$

76. $18a^5b - 9a^3b^2 - 2ab^3$

77. $(x - y)^2 - 16$

78. $(a - 2b)^2 - 25$

79. $16a^4 - b^4$

80. $(x + y)^2 + 3(x + y) + 2$

81. $8a^3 + 125b^3$

82. $x^3 - 25x - x^2 + 25$

In Exercises 83–86, find the quotients.

83. $(x^2 + 2x + 3) \div (x + 4)$

84. $(2x^2 + 7x - 1) \div (2x + 1)$

85. $(4x^3 + 3x + 1) \div (2x + 3)$

86. $(x^4 + 2) \div (x^2 + 2x + 1)$

CHAPTERS 1–3   CUMULATIVE PRACTICE TEST

1. Given the sets

$$A = \{a \mid a \text{ is a multiple of } 4, \quad 3 < a < 21\}$$
$$B = \{x \mid x \text{ is a multiple of } 2, \quad 0 < x < 24\}$$

find:

(a) $A \cap B$

(b) $A \cup B$

2. What real number property is illustrated by the following statement?

$$(x + 2y)(x + 3y) = x(x + 3y) + 2y(x + 3y)$$

3. Evaluate the following:

(a) $(-2)(-2) - (-2)^2(2)$

(b) $5 - \{6 + [2 - 3(4 - 9)]\}$

4. Given $x = -4$ and $y = -5$, evaluate $\dfrac{x^2 - 3xy + 2y^2}{x - y}$.

5. Solve the following equations:

(a) $3x - 2 = 5x + 4$

(b) $3(x - 5) - 2(x - 5) = 3 - (5 - x)$

(c) $|x - 3| = 8$

(*continued*)

6. Solve each of the following inequalities and graph the solution set on the number line.

    (a) $3 - (2x - 4) < 5 - (7 - 2x)$ **(b)** $|3x - 2| < 5$

    (c) $|5 - 4x| \geq 1$

7. A truck is carrying 170 packages weighing a total of 3,140 lb. If each package weighs either 10 lb or 30 lb, how many of each weight package are on the truck?

8. How long would it take a car travelling at 55 mph to overtake a car travelling at 40 mph with 1 hour head start?

9. Perform the operations and simplify:

    (a) $(2x^2 - 3xy + 4y^2) - (5x^2 - 2xy + y^2)$ **(b)** $(3a - 2b)(5a + 3b)$

    (c) $(2x^2 - y)(2x^2 + y)$ **(d)** $(3y - 2z)^2$

    (e) $(x + y - 3)^2$

10. Factor the following completely:

    (a) $a^2 - 9a + 14$ **(b)** $6r^2 - 5rs - 6s^2$

    (c) $10a^4 - 9a^2y - 9y^2$ **(d)** $4a^4 - 12a^2b^2 + 9b^4$

    (e) $(a + 2b)^2 - 25$ **(f)** $125x^3 - 1$

11. Find the quotient: $(2x^3 + 6x + 5) \div (x + 2)$

4

Rational Expressions

Study Skills

4.1

Equivalent Fractions

In Chapter 1 we defined the set of rational numbers, Q, in the following way:

$$Q = \left\{ \frac{a}{b} \ \middle| \ a, b \in Z, \quad b \neq 0 \right\}$$

which means that a rational number is any number which can be represented as a quotient of two integers, provided the denominator is not 0.

Similarly, we can define a **rational expression** or an **algebraic fraction** as a quotient of two polynomials, provided the denominator is not the *zero polynomial*. (Remember, the zero polynomial is simply 0.)

But even if the denominator is not the zero polynomial, we must still be careful about division by 0; a nonzero polynomial could have a value of 0 when you substitute certain values for the variable. For example, $\dfrac{3x - 4}{x - 5}$ is a rational expression with the nonzero polynomial $x - 5$ in the denominator. However, since $x - 5$ is 0 when $x = 5$,

$$\frac{3x - 4}{x - 5} \quad \text{is not defined for } x = 5.$$

EXAMPLE 1

What value(s) of the variable must be excluded for each of the following?

(a) $\dfrac{5x - 1}{2x - 1}$ (b) $\dfrac{5}{2x}$ (c) $\dfrac{3x + 7}{5}$ (d) $\dfrac{x + y}{x - y}$

Solution

Essentially we need to eliminate those values which make the denominator 0 and hence the fraction undefined.

(a) $\dfrac{5x - 1}{2x - 1}$ will be undefined when $2x - 1 = 0$. Hence, $\boxed{x = \frac{1}{2} \text{ is excluded}}$

(b) $\dfrac{5}{2x}$ will be undefined when $2x = 0$. Hence, we can write $\boxed{x = 0 \text{ is excluded}}$

or simply $\boxed{x \neq 0}$

(c) $\dfrac{3x + 7}{5}$ $\boxed{\text{is defined for all values of } x}$ since the *denominator* is never 0.

(d) $\dfrac{x + y}{x - y}$ is undefined when $x - y = 0$. Hence, $\boxed{x \neq y}$ ■

Just as the integer p may be regarded as a rational number by rewriting it as $\dfrac{p}{1}$, polynomials may also be considered as rational expressions since they can be represented as a quotient of the polynomial and 1.

We know from previous experience with rational expressions that two expressions may look different but may actually be equivalent or represent the same amount. With numerical fractions we can draw pictures to demonstrate that $\frac{2}{3}$ and $\frac{4}{6}$ are equivalent. On the other hand, how do we determine if $\frac{35x^2}{14x}$, $\frac{5x}{2}$, and $\frac{5x^4}{2x^3}$ are equivalent? With variables involved, we primarily use the fundamental principle of fractions.

The Fundamental Principle of Fractions	$$\frac{a \cdot k}{b \cdot k} = \frac{a}{b} \qquad b, k \neq 0$$

This principle says that if we *divide* or *multiply* the numerator and denominator of a fraction by the same nonzero expression, we obtain an equivalent fraction. In the boxed equation, moving from left to right is called ***reducing fractions to lower terms***; moving from right to left is called ***building fractions to higher terms***.

Reducing Fractions

A fraction ***reduced to lowest terms*** or ***written in simplest form*** is a fraction that has no factors (other than 1) common to both its numerator and denominator. This requires us to factor both numerator and denominator and then divide out or "cancel" factors common to the numerator and denominator. It is usually quicker to factor the greatest common factor first.

EXAMPLE 2

Express the following in simplest form:

(a) $\dfrac{56}{98}$ **(b)** $\dfrac{33x^4y^2}{15xy^5}$ **(c)** $\dfrac{x+y}{x-y}$

Solution

(a) $\dfrac{56}{98} = \dfrac{\cancel{7} \cdot \cancel{2} \cdot 2 \cdot 2}{\cancel{7} \cdot 7 \cdot \cancel{2}}$ *For convenience, we cross out the factors common to the numerator and denominator and then rewrite the fraction in reduced form.*

$\qquad = \boxed{\dfrac{4}{7}}$

Instead of factoring the numerator and denominator completely, we could factor out the greatest common factor and then reduce:

$\dfrac{56}{98} = \dfrac{4 \cdot \cancel{14}}{7 \cdot \cancel{14}}$ *The greatest common factor is 14.*

$\qquad = \dfrac{4}{7}$

(b) $\dfrac{33x^4y^2}{15xy^5} = \dfrac{11 \cdot \cancel{3} \cdot \cancel{x} \cdot x \cdot x \cdot x \cdot \cancel{y} \cdot \cancel{y}}{5 \cdot \cancel{3} \cdot \cancel{x} \cdot \cancel{y} \cdot \cancel{y} \cdot y \cdot y \cdot y} = \dfrac{11x^3}{5y^3}$

or

$\dfrac{33x^4y^2}{15xy^5} = \dfrac{11 \cdot \cancel{3} \cdot \cancel{x} \cdot x^3 \cdot \cancel{y^2}}{5 \cdot \cancel{3} \cdot \cancel{x} \cdot \cancel{y^2} \cdot y^3} = \boxed{\dfrac{11x^3}{5y^3}}$

(c) $\dfrac{x + y}{x - y}$ $\boxed{\text{cannot be reduced}}$

Remember that the fundamental principle of fractions refers only to common factors.

We can cancel common factors but not common terms. ∎

We can now apply the factoring techniques we studied in Chapter 3.

EXAMPLE 3

Express the following in simplest form:

(a) $\dfrac{x^2 - y^2}{(x - y)^2}$ **(b)** $\dfrac{x - 8}{x^2 - 5x - 24}$ **(c)** $\dfrac{4x^2 - 4xy - 3y^2}{2x^2 - xy - 3y^2}$

(d) $\dfrac{16 - x^2}{x^2 - x - 12}$ **(e)** $\dfrac{2ax + bx - 2ay - by}{3ax + 2bx - 3ay - 2by}$

Solution

(a) $\dfrac{x^2 - y^2}{(x - y)^2} = \dfrac{(x - y)(x + y)}{(x - y)(x - y)}$ *First factor; then reduce.*

$= \dfrac{\cancel{(x - y)}(x + y)}{\cancel{(x - y)}(x - y)}$

$= \boxed{\dfrac{x + y}{x - y}}$

(b) $\dfrac{x - 8}{x^2 - 5x - 24} = \dfrac{x - 8}{(x - 8)(x + 3)}$ *First factor; then reduce.*

$= \dfrac{\cancel{x - 8}}{\cancel{(x - 8)}(x + 3)}$

$= \boxed{\dfrac{1}{x + 3}}$

(c) $\dfrac{4x^2 - 4xy - 3y^2}{2x^2 - xy - 3y^2} = \dfrac{(2x - 3y)(2x + y)}{(2x - 3y)(x + y)}$ *Factor and reduce.*

$= \dfrac{\cancel{(2x - 3y)}(2x + y)}{\cancel{(2x - 3y)}(x + y)}$

$= \boxed{\dfrac{2x + y}{x + y}}$

(d) $\dfrac{16 - x^2}{x^2 - x - 12} = \dfrac{(4 - x)(4 + x)}{(x - 4)(x + 3)}$ *Factor.*

At first glance it appears that there are no common factors, but as we discussed in Chapter 3, $4 - x$ is the negative of $x - 4$; that is,

$$-(4 - x) = -4 + x = x - 4$$

Thus, we could factor -1 from *either* $4 - x$ or $x - 4$:

$$\frac{(4 - x)(4 + x)}{(x - 4)(x + 3)} = \frac{(-1)\cancel{(x - 4)}(4 + x)}{\cancel{(x - 4)}(x + 3)}$$

$$= \frac{-1(4 + x)}{x + 3} \quad \text{or} \quad \boxed{\frac{-x - 4}{x + 3}}$$

(e) $\dfrac{2ax + bx - 2ay - by}{3ax + 2bx - 3ay - 2by} = \dfrac{x(2a + b) - y(2a + b)}{x(3a + 2b) - y(3a + 2b)}$ *Factor by grouping. (Do not try to reduce yet.)*

$$= \frac{(x - y)(2a + b)}{(x - y)(3a + 2b)}$$ *Now reduce.*

$$= \frac{(2a + b)\cancel{(x - y)}}{(3a + 2b)\cancel{(x - y)}}$$

$$= \boxed{\frac{2a + b}{3a + 2b}} \qquad\qquad \blacksquare$$

Building Fractions to Higher Terms

In the process of adding fractions with different denominators, we will have to use the fundamental principle of fractions to do the opposite of reducing: **building to higher terms**. We demonstrate by example.

EXAMPLE 4

Fill in the question mark to make the two fractions equivalent.

(a) $\dfrac{3x}{5y^2z} = \dfrac{?}{20xy^4z^2}$ **(b)** $\dfrac{2x - y}{x^2 - y^2} = \dfrac{?}{(x - y)(x + y)^2}$

Solution

The fundamental principle of fractions says that two fractions are equivalent if one is obtained by multiplying the denominator and the numerator of the other by the same (nonzero) expression. Thus, all we need to do is to look at what additional factors appear in the denominator of the second fraction and multiply the numerator of the first fraction by these same factors.

(a) $\qquad \dfrac{3x}{5y^2z} = \dfrac{?}{20xy^4z^2}$ *To make $5y^2z$ into $20xy^4z^2$, we need to multiply by $4xy^2z$. Therefore, we must multiply the **numerator** by $4xy^2z$.*

$$\frac{3x \, (4xy^2z)}{5y^2z \, (4xy^2z)} = \frac{12x^2y^2z}{20xy^4z^2}$$

The answer is $\boxed{12x^2y^2z}$

(b) $\qquad \dfrac{2x - y}{x^2 - y^2} = \dfrac{?}{(x - y)(x + y)^2}$ *Factor the denominator of the left-hand fraction.*

$$\frac{(2x - y)}{(x - y)(x + y)} = \frac{?}{(x - y)(x + y)^2}$$ *Looking at the denominators and moving from left to right, we see that we are missing a factor $x + y$.*

$$\frac{(2x - y)(x + y)}{(x - y)(x + y)(x + y)} = \frac{(2x - y)(x + y)}{(x - y)(x + y)^2}$$ *Thus, we must multiply the **numerator** by $(x + y)$.*

The answer is $\boxed{(2x - y)(x + y) \quad \text{or} \quad 2x^2 + xy - y^2}$ $\qquad \blacksquare$

Signs of Fractions

In a rational expression, a sign may precede any of the following: the numerator, the denominator, or the entire fraction. In general,

$$\frac{a}{b} = -\frac{a}{-b} = -\frac{-a}{b} = \frac{-a}{-b}$$

and

$$-\frac{a}{b} = \frac{-a}{b} = \frac{a}{-b} = -\frac{-a}{-b}$$

You may check that these are equal by letting $a = 6$ and $b = 3$. Note that if you change *exactly* two of the three signs of a fraction, the result will be an equivalent fraction.

Of the three equivalent forms,

$$-\frac{x}{3} = \frac{-x}{3} = \frac{x}{-3}$$

the first two are usually the preferred forms.

STUDY SKILLS 4.1   Reviewing Your Exam: Diagnosing Your Strengths and Weaknesses

Your exam will be a useful tool in helping you to determine what topics, skills, or concepts you need to work on in preparation for the next topic, or in preparation for future exams. After you get your exam back, you should review it carefully: examine what you did correctly and the problems you missed.

Don't quickly gloss over your errors and assume that any errors were minor or careless. Students often mistakenly label many of their errors as "careless," when in fact, they are a result of not clearly understanding a certain concept or procedure. Be honest with yourself. Don't delude yourself into thinking that all errors are careless. Ask yourself the following questions about your errors:

Did I understand the directions or perhaps misunderstand them?

Did I understand the topic the question relates to?

Did I misuse a rule or property?

Did I make an arithmetic error?

Look over the entire exam. Did you consistently make the same type of error throughout the exam? Did you consistently miss problems covering a particular topic or concept? You should try to follow your work and see what you were doing on the exam. If you think you have a problem understanding a concept, topic, or approach to a problem, you should immediately seek help from your teacher or tutor, and reread relevant portions of your text.

Exercises 4.1

In Exercises 1–8, what values of the variable(s) should be excluded for the fraction?

1. $\dfrac{x}{x - 2}$

2. $\dfrac{x + 2}{3y + 2}$

3. $\dfrac{3xy^2}{2}$

4. $\dfrac{3x^2z}{5}$

5. $\dfrac{a + b}{5x}$

6. $\dfrac{x^2}{y}$

7. $\dfrac{3xy}{3x - y}$

8. $\dfrac{5ab}{2a - b}$

In Exercises 9–44, reduce to lowest terms.

9. $\dfrac{3x^2y}{15xy}$

10. $\dfrac{24a^3b^2}{36ab^4}$

11. $\dfrac{16x^2y^3a}{18xy^4a^3}$

12. $\dfrac{87a^2b^3}{57a^5b^9}$

13. $\dfrac{(x-5)(x+4)}{(x-3)(x+4)}$

14. $\dfrac{(x-2)(x+1)}{(x-2)(x+3)}$

15. $\dfrac{(2x-3)(x-5)}{(2x+3)(x+3)}$

16. $\dfrac{(5y-1)(2y-7)}{(5y+1)(3y+2)}$

17. $\dfrac{(2x-3)(x-5)}{(2x-3)(5-x)}$

18. $\dfrac{(5y-1)(3y+2)}{(1-5y)(3y+2)}$

19. $\dfrac{3x^2-3x}{6x^2+18x}$

20. $\dfrac{6a^2-3a}{15a^2-15a}$

21. $\dfrac{x^2-9}{x^2-6x+9}$

22. $\dfrac{x^2-6x+9}{x^2-9}$

23. $\dfrac{w^2-8wz+7z^2}{w^2+8wz+7z^2}$

24. $\dfrac{x^2-4xy+4y^2}{x^2+4xy+4y^2}$

25. $\dfrac{x^2+x-12}{x-3}$

26. $\dfrac{a^2+2a-63}{a-7}$

27. $\dfrac{x^2+x-12}{3-x}$

28. $\dfrac{a^2+2a-63}{7-a}$

29. $\dfrac{3a^2-13a-30}{15a^2+28a+5}$

30. $\dfrac{3a^2-a-2}{2a^2+a-3}$

31. $\dfrac{2x^3-2x^2-12x}{3x^2-6x}$

32. $\dfrac{10a^3-45a^2-25a}{2a^2-10a}$

33. $\dfrac{2x^2}{3x^2-6}$

34. $\dfrac{5a^2}{2a^2-10}$

35. $\dfrac{3a^2-7a-6}{3+5a-2a^2}$

36. $\dfrac{2x^2-3x-2}{2+5x-3x^2}$

37. $\dfrac{6r^2-r-2}{2+r-6r^2}$

38. $\dfrac{15y^2-y-2}{2+y-15y^2}$

39. $\dfrac{a^3+b^3}{(a+b)^3}$

40. $\dfrac{a^3+b^3}{a^2+b^2}$

41. $\dfrac{ax+bx-2ay-2by}{4xy-2x^2}$

42. $\dfrac{4x^2-8x+3}{4x^2-6x+2xy-3y}$

43. $\dfrac{(a+b)^2-(x+y)^2}{a+b+x+y}$

44. $\dfrac{x-1}{x^3+x^2-x-1}$

In Exercises 45–56, fill in the missing expression.

45. $\dfrac{2x}{3y}=\dfrac{?}{9x^2y}$

46. $\dfrac{5a}{3b}=\dfrac{?}{21a^2b}$

47. $\dfrac{5}{3a^2b}=\dfrac{?}{15a^4b^2}$

48. $\dfrac{8}{5x^2y}=\dfrac{?}{20x^3y^3}$

49. $\dfrac{3x}{x-5}=\dfrac{?}{(x-5)(x+5)}$

50. $\dfrac{2z}{(z-6)}=\dfrac{?}{(z-6)^2}$

51. $\dfrac{a-b}{x+y}=\dfrac{?}{x^2+2xy+y^2}$

52. $\dfrac{x-y}{a-b}=\dfrac{?}{a^2-b^2}$

53. $\dfrac{y-2}{2y-3}=\dfrac{?}{12-5y-2y^2}$

54. $\dfrac{x+1}{2x-5}=\dfrac{?}{15-x-2x^2}$

55. $\dfrac{x-y}{a^2-b^2}=\dfrac{?}{a^2x+a^2y-b^2x-b^2y}$

56. $\dfrac{5x+y}{r-s}=\dfrac{?}{r^2a+r^2b-s^2a-s^2b}$

? QUESTIONS FOR THOUGHT

57. Discuss what is *wrong* (if anything) with the following:

(a) $\dfrac{\dfrac{a}{a^2}+\dfrac{b}{b^2}}{a+b}\overset{?}{=}a+b$

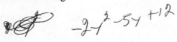

(b) $\dfrac{a^3 + b^3}{(a + b)^3} \overset{?}{=} 1$

(c) $\dfrac{ax + bx + 2ay + 2by}{a + b} \overset{?}{=} \dfrac{(a + b)x + (a + b)2y}{a + b}$

$\overset{?}{=} \dfrac{\cancel{(a + b)}x + (a + b)2y}{\cancel{a + b}}$

$\overset{?}{=} x + (a + b)2y$

 MINI-REVIEW

58. *Evaluate.* $\dfrac{-4 - [4 - 5(2 - 8)]}{4 - 5 \cdot 2 - 8}$

59. *Perform the operations and simplify.* $4x^2 - 9y^2 - (2x - 3y)^2$

60. *Solve for x.* $3 - [x - (2 - x)] = 4 - x$

61. *Factor completely.* $x^3 - x + x^2 - 1$

62. Valerie has 28 coins in nickels and dimes. If their total value is $2.50, how many of each coin does she have?

4.2

Multiplication and Division of Rational Expressions

Multiplication

Now that we have some experience with rational expressions, we can examine the arithmetic operations with these expressions. We begin with multiplication and division since performing these operations is more straightforward than addition and subtraction. We define multiplication as follows:

Multiplication of Fractions	$\dfrac{a}{b} \cdot \dfrac{c}{d} = \dfrac{a \cdot c}{b \cdot d} \qquad b, d \neq 0$

The product of fractions is defined as the product of the numerators divided by the product of the denominators, provided neither denominator is 0. For example:

$$\frac{3x}{y} \cdot \frac{4x^2}{7y^3} = \frac{(3x)(4x^2)}{(y)(7y^3)} = \frac{12x^3}{7y^4}$$

$$\frac{x - y}{2x + y} \cdot \frac{3x - 2y}{x + y} = \frac{(x - y)(3x - 2y)}{(2x + y)(x + y)} = \frac{3x^2 - 5xy + 2y^2}{2x^2 + 3xy + y^2}$$

Now that we have defined multiplication, we can take another look at the fundamental principle of fractions. By the multiplicative identity property,

$$\frac{a}{b} = \frac{a}{b} \cdot 1 \qquad \textit{Multiplication by 1 does not change the value of an expression.}$$

$$= \frac{a}{b} \cdot \frac{k}{k} \qquad \textit{Since } \frac{k}{k} = 1$$

$$= \frac{a \cdot k}{b \cdot k}$$

which is the fundamental principle of fractions. Thus, the fundamental principle of fractions can be viewed as simply stating that multiplying by 1 does not change the value of an expression.

In the last section we required that our final answer be reduced to lowest terms. In the process of multiplication, factors of each numerator remain in the numerator of the product and factors of each denominator remain in the denominator of the product. Therefore, it is much more efficient to reduce any common factors before we actually carry out the multiplication.

EXAMPLE 1

Perform the operations. Express your answer in simplest form.

(a) $\dfrac{24x^2y}{13a^3b^2} \cdot \dfrac{26ab^3}{3xy^2}$ (b) $\dfrac{a^2 - 2ab + b^2}{a^2b - ab^2} \cdot \dfrac{a^3 + a^2b}{a^2 - b^2}$

(c) $\dfrac{x + 2y}{x - y} \cdot \dfrac{y - x}{x - 2y}$

(d) $\left(\dfrac{ax + bx}{2x^2 + 4xy + 2y^2}\right)\left(\dfrac{x + y}{ax - ay + bx - by}\right)(4x^2 - 4y^2)$

Solution

(a) $\dfrac{24x^2y}{13a^3b^2} \cdot \dfrac{26ab^3}{3xy^2} = \dfrac{\overset{8}{24}\overset{x}{x^2}y}{\underset{a^2}{13a^3b^2}} \cdot \dfrac{\overset{2}{26}\overset{b}{ab^3}}{\underset{y}{3xy^2}}$

$$= \boxed{\dfrac{16bx}{a^2y}}$$

(b) $\dfrac{a^2 - 2ab + b^2}{a^2b - ab^2} \cdot \dfrac{a^3 + a^2b}{a^2 - b^2}$ *Factor.*

$$= \dfrac{(a - b)(a - b)}{ab(a - b)} \cdot \dfrac{a^2(a + b)}{(a - b)(a + b)} \qquad \textit{Then reduce.}$$

$$= \dfrac{(a - b)(a - b)}{ab(a - b)} \cdot \dfrac{\overset{a}{a^2}(a + b)}{(a - b)(a + b)}$$

$$= \boxed{\dfrac{a}{b}}$$

(c) $\dfrac{x + 2y}{x - y} \cdot \dfrac{y - x}{x - 2y} = \dfrac{x + 2y}{x - y} \cdot \dfrac{(-1)(x - y)}{x - 2y}$ *Since $y - x = (-1)(x - y)$*

$\qquad\qquad = \dfrac{x + 2y}{\cancel{x - y}} \cdot \dfrac{-1\cancel{(x - y)}}{x - 2y}$ *Then reduce.*

$\qquad\qquad = \boxed{\dfrac{-1(x + 2y)}{x - 2y}}$ or $\boxed{\dfrac{-x - 2y}{x - 2y}}$

(d) $\left(\dfrac{ax + bx}{2x^2 + 4xy + 2y^2}\right)\left(\dfrac{x + y}{ax - ay + bx - by}\right)(4x^2 - 4y^2)$ *Factor and reduce.*

$\qquad = \dfrac{x\cancel{(a + b)}}{\cancel{2}(x + y)\cancel{(x + y)}} \cdot \dfrac{\cancel{x + y}}{\cancel{(a + b)}\cancel{(x - y)}} \cdot \dfrac{\overset{2}{\cancel{4}}\cancel{(x - y)}(x + y)}{1}$

$\qquad\qquad\qquad\qquad\qquad\qquad\qquad \uparrow$
Factor the second denominator
by grouping.

Note the 1 in the denominator.

$\qquad = \dfrac{2x}{1} = \boxed{2x}$ ∎

Division

We can now proceed to formulate the rule for division of fractions. Just as subtraction is the inverse of addition, division is the inverse operation of multiplication. That is, $38 \div 2$ is 19 only because $19 \cdot 2 = 38$. Thus,

$$38 \div 2 = 19 \qquad\qquad 19 \cdot 2 = 38$$

$$\dfrac{a}{b} \div \dfrac{c}{d} = \dfrac{ad}{bc} \qquad \text{because} \qquad \dfrac{ad}{b\cancel{c}} \cdot \dfrac{\cancel{c}}{d} = \dfrac{a}{b}$$

We rewrite

$$\dfrac{ad}{bc} \quad \text{as} \quad \dfrac{a}{b} \cdot \dfrac{d}{c}$$

Hence, we have the rule for division given in the box.

Division of Fractions	$\dfrac{a}{b} \div \dfrac{c}{d} = \dfrac{a}{b} \cdot \dfrac{d}{c} \qquad b, c, d, \ne 0$

Recall from Chapter 1 that the multiplicative inverse of a number is called its *reciprocal*. We usually define reciprocal as follows:

DEFINITION

If $x \ne 0$, the *reciprocal* of x is $\dfrac{1}{x}$.

Thus, the reciprocal of 5 is $\frac{1}{5}$;

the reciprocal of $x^2 + 2$ is $\frac{1}{x^2 + 2}$.

The reciprocal of $\frac{a}{b}$ is $\frac{1}{\frac{a}{b}}$, which is $1 \div \frac{a}{b}$, and, by definition of division,

$$1 \div \frac{a}{b} = 1 \cdot \frac{b}{a} = \frac{b}{a}$$

Hence, if $a \neq 0$ and $b \neq 0$, the reciprocal of $\frac{a}{b}$ is $\frac{b}{a}$.

We can state the rule for division as follows:

To divide *by* a fraction, multiply by its reciprocal.

EXAMPLE 2

Perform the operations. Express your answer in simplest form.

(a) $\dfrac{16x^3}{9y^4} \div \dfrac{32x^6}{27y^3}$ **(b)** $\dfrac{3x^2 - 5xy + 2y^2}{2x^2 + 3xy + y^2} \div \dfrac{x - y}{x^2 + xy}$

(c) $\dfrac{2a^2 - ab - b^2}{2a^2 - 3ab - 2b^2} \div \left[\dfrac{3a^2 - 2ab - b^2}{2a + b} \cdot (2a + b) \right]$

Solution

(a) $\dfrac{16x^3}{9y^4} \div \dfrac{32x^6}{27y^3}$ *Dividing by $\frac{32x^6}{27y^3}$ means multiplying by $\frac{27y^3}{32x^6}$.*

$= \dfrac{16x^3}{9y^4} \cdot \dfrac{27y^3}{32x^6}$

$= \dfrac{16x^3}{9y^4} \cdot \dfrac{27y^3}{32x^6}$ *Reduce.*

$= \boxed{\dfrac{3}{2x^3y}}$

(b) $\dfrac{3x^2 - 5xy + 2y^2}{2x^2 + 3xy + y^2} \div \dfrac{x - y}{x^2 + xy}$ *Multiply by the reciprocal of the divisor (the expression you are dividing by).*

$= \dfrac{3x^2 - 5xy + 2y^2}{2x^2 + 3xy + y^2} \cdot \dfrac{x^2 + xy}{x - y}$ *Then factor and reduce.*

$= \dfrac{(3x - 2y)(x - y)}{(2x + y)(x + y)} \cdot \dfrac{x(x + y)}{x - y}$

$= \boxed{\dfrac{x(3x - 2y)}{2x + y}}$ or $\boxed{\dfrac{3x^2 - 2xy}{2x + y}}$

(c) $\dfrac{2a^2 - ab - b^2}{2a^2 - 3ab - 2b^2} \div \left[\dfrac{3a^2 - 2ab - b^2}{2a + b} \cdot (2a + b) \right]$ *Multiply inside brackets first.*

$= \dfrac{2a^2 - ab - b^2}{2a^2 - 3ab - 2b^2} \div \left[\dfrac{(3a + b)(a - b)}{2a + b} \cdot \dfrac{(2a + b)}{1} \right]$ *Simplify inside brackets.*

$= \dfrac{2a^2 - ab - b^2}{2a^2 - 3ab - 2b^2} \div \dfrac{(3a + b)(a - b)}{1}$ *Use the rule for division of fractions.*

$= \dfrac{2a^2 - ab - b^2}{2a^2 - 3ab - 2b^2} \cdot \dfrac{1}{(3a + b)(a - b)}$ *Factor and reduce.*

$= \dfrac{(a - b)(2a + b)}{(2a + b)(a - 2b)} \cdot \dfrac{1}{(3a + b)(a - b)}$

$= \boxed{\dfrac{1}{(a - 2b)(3a + b)}}$ or $\boxed{\dfrac{1}{3a^2 - 5ab - 2b^2}}$

Either answer is acceptable.

STUDY SKILLS 4.2   **Reviewing Your Exam: Checking Your Understanding**

If you carefully looked over your exam and you believe that you understand the material and what you did wrong, then do the following:

Copy the problems over on a clean sheet of paper and rework the problems without your text, notes, or exam. When you are finished, check to see if your answers are correct. If your answers are correct, try to find problems in the text similar to those problems and work the new problems on a clean sheet of paper (again without notes, text, or exam). If some of your new answers are incorrect then you may have learned how to solve your test problems, but you probably have not thoroughly learned the topic being tested. You may need to repeat these steps several times until you are confident you understand your errors.

In any case, you should keep your exam (with the correct answers) since they are a good source of information for future studying. You may want to record errors that you consistently made on the exam on your warning cards: types of exam problems can be used when you make up your quiz cards (see Section 1.6).

Exercises 4.2

Perform the indicated operations. Reduce all answers to lowest terms.

1. $\dfrac{10a^2b}{3xy^3} \cdot \dfrac{9xy^4}{5a^4b^7}$

2. $\dfrac{32a^2}{7xb^2} \cdot \dfrac{21x^2b^4}{8a^3}$

3. $\dfrac{17a^2b^3}{18yx} \div \dfrac{34a^2}{9xy}$

4. $\dfrac{38xy^2z}{81a^2b^3} \div \dfrac{19ac^2}{27y^2}$

5. $\dfrac{16x^3b}{9a} \cdot 24a^2b^3$

6. $24a^2b^3 \cdot \dfrac{16x^3b}{9a}$

7. $\dfrac{32r^2s^3}{12a^2b} \cdot \left(\dfrac{15ab^2}{16r} \cdot \dfrac{24rs^2}{9ab}\right)$

8. $\dfrac{32x^2c}{12a^2b} \cdot \left(\dfrac{17a^2x^2}{34b^2c} \cdot \dfrac{16x}{9a}\right)$

9. $\dfrac{32r^2s^3}{12a^2b} \div \left(\dfrac{15ab^2}{16r} \cdot \dfrac{24rs^2}{5ab}\right)$

10. $\dfrac{32x^2c}{12a^2b} \div \left(\dfrac{17a^2x^2}{34b^2c} \cdot \dfrac{16x}{9a}\right)$

11. $\dfrac{x^2 - x - 2}{x + 3} \div \dfrac{3x + 9}{2x + 2}$

12. $\dfrac{4x + 8}{3x - 6} \cdot \dfrac{6x - 12}{8x + 16}$

13. $\dfrac{x^2 - x - 2}{x + 3} \cdot \dfrac{3x + 9}{2x + 2}$

14. $\dfrac{4x + 8}{3x - 6} \div \dfrac{6x - 12}{8x + 16}$

15. $\dfrac{5a^3 - 5a^2b}{3a^2 + 3ab} \cdot (a + b)$

16. $\dfrac{x - 3}{2x^2 - 5x - 3} \cdot (8x + 4)$

17. $\dfrac{5a^3 - 5a^2b}{3a^2 + 3ab} \div (a + b)$

18. $(8x + 4) \div \dfrac{x - 3}{2x^2 - 5x - 3}$

19. $\dfrac{2x^2 + 3x - 5}{2x + 5} \cdot \dfrac{1}{1 - x}$

20. $\dfrac{2y^2 - 9y - 5}{2y + 1} \cdot \dfrac{1}{5 - y}$

21. $\dfrac{2a^2 - 7a + 6}{4a^2 - 9} \cdot \dfrac{4a^2 + 12a + 9}{a^2 - a - 2}$

22. $\dfrac{2x^2 - 5x - 12}{3x^2 - 11x - 4} \cdot \dfrac{3x^2 - 14x - 5}{2x^2 - 7x - 15}$

23. $\dfrac{x^2 - y^2}{(x + y)^3} \cdot \dfrac{(x + y)^2}{(x - y)^2}$

24. $\dfrac{2r^2 + rs - 3s^2}{r^2 - s^2} \cdot \dfrac{2r - 2s}{2r^2 + 5rs + 3s^2}$

25. $\dfrac{9x^2 + 3xy - 2y^2}{6x^2y - 2xy^2} \div \dfrac{3x + 2y}{6x^2y}$

26. $\dfrac{2a^2 - 5ab - 3b^2}{4a^2b + 2ab^2} \div \dfrac{2a + b}{4ab}$

27. $\dfrac{2x^2 + xy - 3y^2}{x^2 - y^2} \div \dfrac{2x^2 + 5xy + 3y^2}{2y - 2x}$

28. $\dfrac{9x^2 - 9xy + 2y^2}{2xy^2 - 6x^2y} \cdot \dfrac{6x^3y}{3x - 2y}$

29. $\dfrac{6q^2 - pq - 2p^2}{8pq^2 + 4p^2q} \cdot \dfrac{8p^2q^2}{6pq^2 - 4p^2q}$

30. $\dfrac{6a^2 - ab - 2b^2}{8ab^2 + 4a^2b} \cdot \dfrac{16a^2b^2}{4a^2b - 6ab^2}$

31. $\dfrac{2ax - 2bx - 3a + 3b}{2a^2b + 2ab^2} \div \dfrac{a - b}{4a^2b - 4ab^2}$

32. $\dfrac{ax + bx - ay - by}{a + b} \div \dfrac{x - y}{r}$

33. $\left(\dfrac{2a^2 + 3ab + b^2}{3a^2 - ab - 2b^2}\right)\left(\dfrac{a - b}{2a - b}\right)\left(\dfrac{3a + b}{a + b}\right)$

34. $\left(\dfrac{6x^2 + xy - 2y^2}{4x^2 - 8xy + 3y^2}\right)\left(\dfrac{x - y}{3x + 2y}\right)\left(\dfrac{8x - 12y}{2x - 2y}\right)$

35. $\dfrac{9a^2 + 9ab + 2b^2}{3a^2 - 2ab - b^2} \cdot \left(\dfrac{a - b}{3a^2 + 4ab + b^2} \div \dfrac{3a + 2b}{a + b}\right)$

36. $\dfrac{9a^2 + 9ab + 2b^2}{3a^2 - 2ab - b^2} \div \left(\dfrac{a - b}{3a^2 + 4ab + b^2} \cdot \dfrac{3a + 2b}{a + b}\right)$

37. $\dfrac{2r^2 - 5rs - 3s^2}{3xyr - xys} \div \left(\dfrac{r - 3s}{x^2y} \div \dfrac{3r - s}{2r + s}\right)$

38. $\left(\dfrac{2r^2 - 5rs - 3s^2}{3xyr - xys} \div \dfrac{r - 3s}{x^2y}\right) \div \dfrac{3r - s}{2r + s}$

39. $\dfrac{4ax + 6x - 2ay - 3y}{x^2 + xy} \div \left(\dfrac{2x^2 - xy - y^2}{3x^3 + 3x^2y} \cdot \dfrac{2a + 3}{x - y}\right)$

40. $\dfrac{2ay - by}{8a^3 - b^3} \div \dfrac{y^2}{4a^2y + 2aby + b^2y}$

🔄 MINI-REVIEW

41. *Evaluate.* $|-6 - 3| - |6| - |3|$

42. Solve for x if $2x - 3 \le 5x + 4$.

43. The statement: $3\left(\dfrac{1}{2}\right) + 4\left(\dfrac{1}{2}\right) = (3 + 4)\left(\dfrac{1}{2}\right)$ illustrates which real number property?

44. *Solve for x.* $|5 - 2x| = 5$

45. How long will it take for Bobby running 6 mph to overtake Linda running 5 mph if Linda had an hour head start?

46. The new DVX model machine can process 40 items per minute, while the old DVX model can process 32 items per minute. How long will it take to complete a job of processing 3,960 items if both machines are operating together?

4.3

Sums and Differences of Rational Expressions

The process of adding and subtracting fractions with the same or *common denominators* can be demonstrated by using the distributive property along with the definition of rational multiplication as follows:

$$\frac{a}{c} + \frac{b}{c} = a\left(\frac{1}{c}\right) + b\left(\frac{1}{c}\right) \qquad \textit{Rational multiplication}$$

$$= (a + b)\left(\frac{1}{c}\right) \qquad \textit{The distributive property}$$

$$= \frac{a + b}{c} \qquad \textit{Rational multiplication}$$

Hence, we have the rules stated in the box.

Addition and Subtraction of Rational Expressions

$$\frac{a}{c} + \frac{b}{c} = \frac{a + b}{c} \qquad \text{and} \qquad \frac{a}{c} - \frac{b}{c} = \frac{a - b}{c}$$

In combining fractions with common denominators, we simply combine the numerators and place this result over the common denominator.

EXAMPLE 1

Perform the operations.

(a) $\dfrac{5y}{x} + \dfrac{3}{x}$ (b) $\dfrac{2x - 3}{x - 2} - \dfrac{x - 1}{x - 2}$

(c) $\dfrac{4x^2 - 33x}{x^2 + x - 2} + \dfrac{2x - 6}{x^2 + x - 2} - \dfrac{(3x - 2)(x - 7)}{x^2 + x - 2}$

Solution

(a) $\dfrac{5y}{x} + \dfrac{3}{x} = \boxed{\dfrac{5y + 3}{x}}$

(b) $\dfrac{2x - 3}{x - 2} - \dfrac{x - 1}{x - 2}$

$\qquad$ *Remember, we are subtracting the **quantity** $x - 1$ and therefore **must** include parentheses.*

$= \dfrac{2x - 3 - (x - 1)}{x - 2}$ $\qquad$ *Remove parentheses.*

$= \dfrac{2x - 3 - x + 1}{x - 2}$ $\qquad$ *Observe the signs of the terms in the numerator. Combine terms in the numerator.*

$= \dfrac{x - 2}{x - 2}$ $\qquad$ *Then reduce.*

$= \boxed{1}$

(c) $\dfrac{4x^2 - 33x}{x^2 + x - 2} + \dfrac{2x - 6}{x^2 + x - 2} - \dfrac{(3x - 2)(x - 7)}{x^2 + x - 2}$ *Combine numerators.*

$= \dfrac{4x^2 - 33x + (2x - 6) - (3x - 2)(x - 7)}{x^2 + x - 2}$ *Remember that you are subtracting a product; find the product first.*

$= \dfrac{4x^2 - 33x + 2x - 6 - (3x^2 - 23x + 14)}{x^2 + x - 2}$ *Then subtract.*

$= \dfrac{4x^2 - 33x + 2x - 6 - 3x^2 + 23x - 14}{x^2 + x - 2}$ *Next, combine terms.*

$= \dfrac{x^2 - 8x - 20}{(x - 1)(x + 2)}$ *Factor.*

$= \dfrac{\cancel{(x + 2)}(x - 10)}{(x - 1)\cancel{(x + 2)}}$ *Reduce.*

$= \boxed{\dfrac{x - 10}{x - 1}}$ ∎

Our last step should be to reduce the answer to lowest terms. Also, keep in mind that we must combine all terms in the numerator *before* reducing.

The next type of problem to consider is one in which the denominators are different. If the denominators are different, we use the fundamental principle of fractions to build new fractions. The new fractions must be equivalent to the original fractions and all have the same denominator. Once the denominators are the same, we can add the fractions as we did before. The idea is to find an expression that can serve as a common denominator.

This new denominator must be divisible by the original denominators and, for convenience, it should be the "smallest" expression divisible by the original denominators. By "smallest" we mean having the least number of factors. Hence, we call the smallest expression divisible by all denominators in question the **least common denominator,** or **LCD.** To find the LCD, we follow the procedure outlined in the accompanying box.

Procedure for Finding the LCD	1. Factor each denominator *completely.*
	2. The LCD consists of the product of each *distinct* factor the *maximum* number of times it appears in any *one* denominator.

EXAMPLE 2

Find the LCD of the following:

(a) $\dfrac{3}{14x^3}, \quad \dfrac{5}{21x^2y^4}, \quad \dfrac{1}{12x^4y^3}$ (b) $\dfrac{3}{x^2 - 9}, \quad \dfrac{6}{x^2 + 6x + 9}$

Solution

(a) The LCD of $\dfrac{3}{14x^3}, \dfrac{5}{21x^2y^4},$ and $\dfrac{1}{12x^4y^3}$ is found in the following way:

1. Factor each denominator as completely as possible.

$$14x^3 = 7 \cdot 2 \cdot x \cdot x \cdot x$$
$$21x^2y^4 = 7 \cdot 3 \cdot x \cdot x \cdot y \cdot y \cdot y \cdot y$$
$$12x^4y^3 = 2 \cdot 2 \cdot 3 \cdot x \cdot x \cdot x \cdot x \cdot y \cdot y \cdot y$$

2. Write each factor the maximum (not the total) number of times it appears in any one denominator.

The LCD is $7 \cdot 2 \cdot 2 \cdot 3 \cdot x \cdot x \cdot x \cdot x \cdot y \cdot y \cdot y \cdot y = \boxed{84x^4y^4}$

7   appears *once* in the first two denominators.

2   appears *twice* in the third denominator, and once in the first denominator.

3   appears *once* in the second and third denominators.

x appears *four* times in the third denominator, three times in the first denominator, and twice in the second denominator.

y appears *four* times in the second denominator, and three times in the third denominator.

Looking at the LCD and the denominators in factored form, note that all three denominators appear in the LCD. This means that the LCD is divisible by all three denominators. Also note that each factor in the LCD is necessary. For example, we cannot drop a factor of y or else $21x^2y^4$ will not divide into the LCD; we cannot drop a factor of x or $12x^4y^3$ will not divide into the LCD. The fact that there are no "extra" factors makes it the *least* common denominator.

(b) The LCD of $\dfrac{3}{x^2 - 9}$ and $\dfrac{6}{x^2 + 6x + 9}$ is found as follows:

1. Factor each denominator completely.

$$x^2 - 9 = (x + 3)(x - 3)$$
$$x^2 + 6x + 9 = (x + 3)(x + 3)$$

2. Write each factor the maximum number of times it appears in any one denominator.

The LCD is $(x - 3)(x + 3)(x + 3) = \boxed{(x - 3)(x + 3)^2}$

$x - 3$ appears once in the first denominator and $x + 3$ appears twice in the second denominator. ∎

Now let's see how we can use the LCD and the fundamental principle of fractions to add or subtract fractions with unlike denominators. We would add fractions with unlike denominators by following three general steps.

To Combine Fractions with Different Denominators	1. Find the LCD.
	2. Build each fraction to higher terms with the LCD in each denominator.
	3. Combine as with fractions with common denominators.

EXAMPLE 3

Perform the indicated operations.

(a) $\dfrac{3}{5x^2y} - \dfrac{2}{3xy^2} + x$

(b) $\dfrac{3x}{x^2 - y^2} + \dfrac{5}{2x^2(x - y)}$

(c) $\dfrac{5x}{x^2 - y^2} - \dfrac{6}{3x^2 - 4xy + y^2}$

(d) $\dfrac{5x - 10}{x - 4} + \dfrac{3x - 2}{4 - x}$

Solution

(a) First find the LCD:

$$\text{LCD:} \quad 5 \cdot 3 \cdot x^2y^2$$

Build fractions to higher terms with the LCD, $5 \cdot 3 \cdot x^2y^2$, in each denominator. Look at $5 \cdot 3 \cdot x^2y^2$ and determine what factors are missing.

$\dfrac{3}{5x^2y} = \dfrac{3\,(3y)}{5x^2y\,(3y)}$ *Multiply numerator and denominator by the missing factors to make $5x^2y$ into $5 \cdot 3 \cdot x^2y^2$.*

$\dfrac{2}{3xy^2} = \dfrac{2\,(5x)}{3xy^2\,(5x)}$ *Multiply numerator and denominator by the missing factors to make $3xy^2$ into $5 \cdot 3 \cdot x^2y^2$.*

$\dfrac{x}{1} = \dfrac{x\,(5 \cdot 3 \cdot x^2y^2)}{1\,(5 \cdot 3 \cdot x^2y^2)}$ *Rewrite x as $\dfrac{x}{1}$ and fill in the missing factors to make 1 into $5 \cdot 3 \cdot x^2y^2$.*

$\dfrac{3}{5x^2y} - \dfrac{2}{3xy^2} + x = \dfrac{3(3y) - 2(5x) + x(5 \cdot 3 \cdot x^2y^2)}{5 \cdot 3 \cdot x^2y^2}$ *Place numerators over the LCD and combine.*

$= \boxed{\dfrac{9y - 10x + 15x^3y^2}{15x^2y^2}}$

Your last step should be to simplify (reduce) your answer. In this case, the answer is already in simplest form.

(b) The LCD is found by factoring the denominators and writing each factor the maximum number of times it appears in any denominator:

$$x^2 - y^2 = (x - y)(x + y)$$
$$2x^2(x - y) = 2x^2(x - y)$$

Thus, the LCD is $2x^2(x - y)(x + y)$.

$$\frac{3x}{(x - y)(x + y)} = \frac{3x(2x^2)}{(x - y)(x + y)(2x^2)} = \frac{6x^3}{2x^2(x - y)(x + y)}$$

Build fractions by filling in missing factors.

$$\frac{5}{2x^2(x - y)} = \frac{5(x + y)}{2x^2(x - y)(x + y)} = \frac{5x + 5y}{2x^2(x - y)(x + y)}$$

Combine as with like fractions.

$$\frac{6x^3}{2x^2(x - y)(x + y)} + \frac{5x + 5y}{2x^2(x - y)(x + y)} = \boxed{\frac{6x^3 + 5x + 5y}{2x^2(x - y)(x + y)}}$$

(c)

$$x^2 - y^2 = (x - y)(x + y)$$
$$3x^2 - 4xy + y^2 = (x - y)(3x - y)$$

Therefore, the LCD is $(x - y)(x + y)(3x - y)$.

Rather than isolate each fraction individually, we will write the new fractions beneath the original fractions with all denominators factored:

$$\frac{5x}{(x - y)(x + y)} - \frac{6}{(x - y)(3x - y)}$$
$$\frac{5x(3x - y)}{(x - y)(x + y)(3x - y)} - \frac{6(x + y)}{(x - y)(x + y)(3x - y)}$$

Fill in missing factors.

Place numerators above the common denominator and combine.

$$= \frac{5x(3x - y) - 6(x + y)}{(x - y)(x + y)(3x - y)}$$

Distribute $5x$ and -6.

$$= \boxed{\frac{15x^2 - 5xy - 6x - 6y}{(x + y)(x - y)(3x - y)}}$$

(d) $\dfrac{5x - 10}{x - 4} + \dfrac{3x - 2}{4 - x}$

Note: Since $x - 4$ is the negative of $4 - x$, we multiply the numerator and denominator of the second fraction by -1.

$$= \frac{5x - 10}{x - 4} + \frac{(3x - 2)(-1)}{(4 - x)(-1)}$$

Now the denominators are the same.

$$= \frac{5x - 10}{x - 4} + \frac{(3x - 2)(-1)}{x - 4}$$

$$= \frac{5x - 10 + (3x - 2)(-1)}{x - 4}$$

Combine numerators; place results above the common denominator.

$$= \frac{5x - 10 - 3x + 2}{x - 4} = \frac{2x - 8}{x - 4} = \frac{2(x - 4)}{x - 4}$$

Factor and reduce.

$$= \boxed{2}$$

Dividing a Polynomial by a Monomial

We discussed polynomial division in the last chapter and postponed discussing division by monomials until now because the process typically requires us to rewrite the quotient as a fraction.

EXAMPLE 4

Find the quotient. $(6x^5y^2 - 9x^3y + 12x^2) \div 6x^2y$

Solution

Division by monomials can be accomplished in two ways, depending on what type of answer is being sought. Both methods require us to rewrite the quotient as a fraction.

METHOD 1: Just as we can combine several fractions which have the same denominator into a single fraction, we can also write a single fraction with several terms in the numerator as a sum or difference of several fractions.

$$\frac{6x^5y^2 - 9x^3y + 12x^2}{6x^2y} = \frac{6x^5y^2}{6x^2y} - \frac{9x^3y}{6x^2y} + \frac{12x^2}{6x^2y}$$

Notice that we are doing the reverse of combining fractions.

$$= \frac{\overset{x^3}{\cancel{6}}\overset{y}{\cancel{x^5}}\overset{}{\cancel{y^2}}}{\cancel{6x^2y}} - \frac{\overset{3}{\cancel{9}}\overset{x}{\cancel{x^3}}\cancel{y}}{\underset{2}{\cancel{6x^2y}}} + \frac{\overset{2}{\cancel{12}}\cancel{x^2}}{\cancel{6x^2y}}$$

Reduce each fraction.

$$= \boxed{x^3y - \frac{3x}{2} + \frac{2}{y}}$$

METHOD 2: An alternative approach is to reduce the fraction as discussed in Section 4.1:

$$\frac{6x^5y^2 - 9x^3y + 12x^2}{6x^2y}$$

Factor and reduce.

$$= \frac{\overset{}{\cancel{3}}\cancel{x^2}(2x^3y^2 - 3xy + 4)}{\underset{2}{\cancel{6x^2y}}}$$

$$= \boxed{\frac{2x^3y^2 - 3xy + 4}{2y}}$$

The answers obtained by the two methods may look different but they are in fact equal. Try combining the fractions in the answer obtained by Method 1 into a single fraction and see if you get the answer produced by Method 2. ■

Exercises 4.3

Perform the indicated operations. Express your answer in simplest form.

1. $\dfrac{x + 2}{x} - \dfrac{2 - x}{x}$

2. $\dfrac{x - 4}{2x} + \dfrac{x + 4}{2x}$

3. $\dfrac{10a}{a - b} - \dfrac{10b}{a - b}$

4. $\dfrac{5x}{x + 2} + \dfrac{10}{x + 2}$

5. $\dfrac{3a}{3a^2 + a - 2} - \dfrac{2}{3a^2 + a - 2}$

6. $\dfrac{x}{3x^2 + x - 2} + \dfrac{1}{3x^2 + x - 2}$

7. $\dfrac{x^2}{x^2 - y^2} + \dfrac{y^2}{x^2 - y^2} - \dfrac{2xy}{x^2 - y^2}$

8. $\dfrac{a^2}{a^2 - b^2} + \dfrac{b^2}{a^2 - b^2} + \dfrac{2ab}{b^2 - a^2}$

9. $\dfrac{5x}{x + 3} + \dfrac{3x}{x + 3} + 4$

10. $\dfrac{3t}{t-2} + \dfrac{2}{t-2} + 3t$

11. $\dfrac{3x}{2y^2} - \dfrac{4x^2}{9y}$

12. $\dfrac{24}{5x^2y} + \dfrac{14}{45xy^3}$

13. $\dfrac{7y}{6x^2} - \dfrac{8x^2}{9y^2}$

14. $\dfrac{5a}{4b^2} - \dfrac{16b}{25a^2}$

15. $\dfrac{7y}{6x^2} \cdot \dfrac{8x^2}{9y^2}$

16. $\dfrac{5a}{4b^2} \div \dfrac{16b}{25a^2}$

17. $\dfrac{36a^2}{b^2c} + \dfrac{24}{bc^3} - \dfrac{3}{7bc}$

18. $\dfrac{3x}{2a^2b} + \dfrac{2y}{3ab^2} - \dfrac{3}{a^3b^2}$

19. $\dfrac{3}{x} + \dfrac{x}{x+2}$

20. $\dfrac{5}{y} + \dfrac{y}{y-3}$

21. $\dfrac{a}{a-b} - \dfrac{b}{a}$

22. $\dfrac{2}{x-7} - \dfrac{7}{x}$

23. $\dfrac{5}{x+7} - \dfrac{2}{x-3}$

24. $\dfrac{3}{x-2} + \dfrac{4}{x+5}$

25. $\dfrac{2}{x-7} + \dfrac{3x+1}{x+2}$

26. $\dfrac{3}{a+4} + \dfrac{5a+1}{a-2}$

27. $\dfrac{2r+s}{r-s} \quad \dfrac{r-2s}{r+s}$ $(r-s)$

28. $\dfrac{3x+y}{x+y} - \dfrac{2x+y}{x-y}$

29. $\dfrac{4}{x-4} + \dfrac{x}{4-x}$

30. $\dfrac{a}{a-b} + \dfrac{b}{b-a}$

31. $\dfrac{7a+3}{2a-1} + \dfrac{5a+4}{1-2a}$

32. $\dfrac{3x+1}{x-2} + \dfrac{2x+3}{2-x}$

33. $\dfrac{a+7}{a-3} + \dfrac{2a+10}{3-a}$

34. $\dfrac{x+1}{x-4} + \dfrac{2x+5}{4-x}$

35. $\dfrac{a}{a-b} - \dfrac{b}{a^2-b^2}$

36. $\dfrac{x}{x-y} + \dfrac{y}{(x-y)^2}$

37. $\dfrac{a}{a-b} \div \dfrac{b}{a^2-b^2}$

38. $\dfrac{x}{x-y} \cdot \dfrac{y}{(x-y)^2}$

39. $\dfrac{2y+1}{y+2} + \dfrac{3y}{y+3} + y$

40. $\dfrac{2x+1}{x+3} + \dfrac{3x}{x-2} + 2x$

41. $\dfrac{5x+1}{x+2} + \dfrac{2x+6}{x^2+5x+6}$

42. $\dfrac{2a+3}{a^2-2a-8} - \dfrac{a-5}{a-4}$

43. $\dfrac{5x+1}{x+2} \div \dfrac{2x+6}{x^2+5x+6}$

44. $\dfrac{2a+3}{a^2-2a-8} \div \dfrac{a-5}{a-4}$

45. $\dfrac{a+3}{a-3} + \dfrac{a}{a+4} - \dfrac{3}{a^2+a-12}$

46. $\dfrac{x+6}{x-3} + \dfrac{x}{x+2} - \dfrac{10}{x^2-x-6}$

47. $\dfrac{y+7}{y+5} - \dfrac{y}{y-3} + \dfrac{16}{y^2+2y-15}$

48. $\dfrac{3x+4}{x+4} + \dfrac{3x}{x-2} + \dfrac{4}{x^2+2x-8}$

49. $\dfrac{3x-4}{x-5} + \dfrac{4x}{10+3x-x^2}$

50. $\dfrac{2r-5}{r-2} + \dfrac{3r}{6-r-r^2}$

51. $\dfrac{2a-1}{a^2+a-6} + \dfrac{a+2}{a^2-2a-15} - \dfrac{a+1}{a^2-7a+10}$

52. $\dfrac{3y+2}{y^2-2y+1} - \dfrac{7y-3}{y^2-1} + \dfrac{5}{y+1}$

53. $\dfrac{5a}{3a-1} + \dfrac{2a+1}{5a+2} + 5a+1$

54. $\dfrac{3x}{x-1} + \dfrac{x+3}{x-2} + 2x-3$

55. $\dfrac{2r+xs}{r+s} + \dfrac{2s+xr}{r+s}$

56. $\dfrac{ax+ay}{a+b} + \dfrac{bx+by}{a+b}$

57. $\dfrac{r^2}{r^3-s^3} + \dfrac{rs}{r^3-s^3} + \dfrac{s^2}{r^3-s^3}$

58. $\dfrac{a^2}{a^3+8b^3} - \dfrac{2ab}{a^3+8b^3} + \dfrac{4b^2}{a^3+8b^3}$

59. $\dfrac{2s+t}{s^3-t^3} + \dfrac{3s}{s^2+st+t^2}$

60. $\dfrac{3x+1}{x^3+8} + \dfrac{2x+1}{x^2-2x+4}$

In Exercises 61–66, find the quotient by each of the following methods:

(a) *Reduce the fraction.*

(b) *Rewrite the fraction as sums or differences of fractions with the same denominator (and then simplify each fraction).*

61. $\dfrac{x^2 + 4x}{4x}$

62. $\dfrac{a^3 - 6a^2}{3a}$

63. $\dfrac{15x^3y^2 - 10x^2y^3}{5x^2y^2}$

64. $\dfrac{18r^2s^6 + 24r^3s^4}{6r^2s^3}$

65. $\dfrac{6m^2n - 4m^3n^2 - 9mn}{15mn^2}$

66. $\dfrac{10u^2v^3 - 15u^4v + 20uv^3}{25u^2v^3}$

 MINI-REVIEW

67. *Divide using polynomial long division.* $(4x^2 - 4x + 2) \div (2x + 1)$

68. A truck carries a load of 60 packages: some are 25-lb packages and the rest are 30-lb packages. If the total weight of all the packages is 1,680 lb, how many packages of each kind are in the truck?

69. Solve for x if $|3 - 2x| \le 5$.

70. Perform the operations and simplify:

$$\frac{2x^3y + x^2y - 6xy}{x^2 - 2x - 8} \cdot \frac{x - 4}{2x^3y^2 - x^2y^2 - 3xy^2}$$

71. Perform the operations and simplify:

$$\frac{2x^3y + x^2y - 6xy}{x^2 - 2x - 8} \div \frac{x - 4}{2x^3y^2 - x^2y^2 - 3xy^2}$$

4.4

Mixed Operations and Complex Fractions

Now we are in the position to perform mixed operations with fractions.

EXAMPLE 1

Perform the indicated operations and simplify.

$$\left(\frac{x + 6}{x + 2} - \frac{4}{x}\right) \div \frac{x^2 - 16}{x^2 - 4}$$

Solution

$$\left(\frac{x+6}{x+2} - \frac{4}{x}\right) \div \frac{x^2 - 16}{x^2 - 4}$$

Combine fractions in parentheses. The LCD is $x(x+2)$.

$$= \left(\frac{x(x+6)}{x(x+2)} - \frac{4(x+2)}{x(x+2)}\right) \div \frac{x^2 - 16}{x^2 - 4}$$

$$= \left(\frac{x(x+6) - 4(x+2)}{x(x+2)}\right) \div \frac{x^2 - 16}{x^2 - 4}$$

$$= \left(\frac{x^2 + 6x - 4x - 8}{x(x+2)}\right) \div \frac{x^2 - 16}{x^2 - 4}$$

$$= \left(\frac{x^2 + 2x - 8}{x(x+2)}\right) \div \frac{x^2 - 16}{x^2 - 4}$$

Next we follow the rule for division.

$$= \frac{x^2 + 2x - 8}{x(x+2)} \cdot \frac{x^2 - 4}{x^2 - 16}$$

Factor and reduce.

$$= \frac{\cancel{(x+4)}(x-2)}{x\cancel{(x+2)}} \cdot \frac{\cancel{(x+2)}(x-2)}{\cancel{(x+4)}(x-4)}$$

$$= \boxed{\frac{(x-2)^2}{x(x-4)}} \quad \text{or} \quad \boxed{\frac{x^2 - 4x + 4}{x^2 - 4x}} \qquad \blacksquare$$

Another way of expressing a quotient of fractions is by using a large fraction bar instead of the quotient sign, $\div$. Thus, we can rewrite

$$\frac{a}{b} \div \frac{c}{d} \quad \text{as} \quad \frac{\dfrac{a}{b}}{\dfrac{c}{d}}$$

Quotients written this way (fractions within fractions) are called **complex fractions**. This notation is often more convenient than using the quotient sign. It allows us to demonstrate that the "multiply by the reciprocal of the divisor" rule is simply an application of the fundamental principle of fractions:

$$\frac{\dfrac{a}{b}}{\dfrac{c}{d}} = \frac{\dfrac{a}{b} \cdot \dfrac{d}{c}}{\dfrac{c}{d} \cdot \dfrac{d}{c}}$$

Multiply numerator and denominator of the complex fraction by $\dfrac{d}{c}$ (fundamental principle of fractions).

$$= \frac{\dfrac{a}{b} \cdot \dfrac{d}{c}}{\dfrac{\cancel{c}}{\cancel{d}} \cdot \dfrac{\cancel{d}}{\cancel{c}}} = \frac{\dfrac{a}{b} \cdot \dfrac{d}{c}}{1}$$

Reduce.

$$= \frac{a}{b} \cdot \frac{d}{c}$$

Thus, a problem such as

$$\frac{\dfrac{3a^2b}{2xy^3}}{\dfrac{9ab^2}{8x^2y}}$$

can be rewritten as

$$\frac{3a^2b}{2xy^3} \div \frac{9ab^2}{8x^2y} = \frac{3a^2b}{2xy^3} \cdot \frac{8x^2y}{9ab^2}$$

$$= \frac{\overset{a}{\cancel{3a^2b}}}{\underset{y^2}{\cancel{2xy^3}}} \cdot \frac{\overset{4x}{\cancel{8x^2y}}}{\underset{3\ \ b}{\cancel{9ab^2}}}$$

$$= \frac{4ax}{3by^2}$$

EXAMPLE 2   Express the following as a simple fraction reduced to lowest terms:

$$\frac{\dfrac{1}{y} - \dfrac{4y}{x^2}}{\dfrac{1}{y^2} - \dfrac{2}{xy}}$$

Solution   We offer two methods of solution. For the first method, we treat the problem as a multiple-operation problem similar to Example 1. That is, we treat the problem as

$$\left(\frac{1}{y} - \frac{4y}{x^2}\right) \div \left(\frac{1}{y^2} - \frac{2}{xy}\right)$$

without rewriting it in this way. Hence, we combine the fractions in the numerator of the complex fraction, combine the fractions in the denominator of the complex fraction, and then we have a quotient of two fractions.

METHOD 1:
$$\frac{\dfrac{1}{y} - \dfrac{4y}{x^2}}{\dfrac{1}{y^2} - \dfrac{2}{xy}} = \frac{\dfrac{x^2}{x^2y} - \dfrac{4y^2}{x^2y}}{\dfrac{x}{xy^2} - \dfrac{2y}{xy^2}}$$

Combine the fractions in the numerator of the complex fraction. The LCD is x^2y.

Do the same for the denominator of the complex fraction. The LCD is xy^2.

$$= \frac{\dfrac{x^2 - 4y^2}{x^2y}}{\dfrac{x - 2y}{xy^2}}$$

This is now a division problem, so we multiply by the reciprocal of the divisor.

$$= \frac{x^2 - 4y^2}{x^2y} \cdot \frac{xy^2}{x - 2y}$$

Factor and reduce.

$$= \frac{(x - 2y)(x + 2y)}{x^2y} \cdot \frac{\overset{y}{\cancel{xy^2}}}{\cancel{x - 2y}}$$

$$= \boxed{\dfrac{y(x + 2y)}{x}} \quad \text{or} \quad \boxed{\dfrac{xy + 2y^2}{x}}$$

An alternative way to clear denominators of fractions within a complex fraction is to apply the fundamental principle by multiplying the numerator and denominator of the complex fraction by the LCD of *all* simple fractions in the complex fraction.

METHOD 2:

$$\dfrac{\dfrac{1}{y} - \dfrac{4y}{x^2}}{\dfrac{1}{y^2} - \dfrac{2}{xy}}$$

Find the LCD of all simple fractions, $\dfrac{1}{y}$, $\dfrac{4y}{x^2}$, $\dfrac{1}{y^2}$, and $\dfrac{2}{xy}$, which is x^2y^2.

Next, multiply the numerator and the denominator of the complex fraction by the LCD, x^2y^2.

$$= \dfrac{\left(\dfrac{1}{y} - \dfrac{4y}{x^2}\right)\left(\dfrac{x^2y^2}{1}\right)}{\left(\dfrac{1}{y^2} - \dfrac{2}{xy}\right)\left(\dfrac{x^2y^2}{1}\right)}$$

Next, use the distributive property in the numerator and denominator.

$$= \dfrac{\left(\dfrac{1}{y}\right)\left(\dfrac{x^2y^2}{1}\right) - \left(\dfrac{4y}{x^2}\right)\left(\dfrac{x^2y^2}{1}\right)}{\left(\dfrac{1}{y^2}\right)\left(\dfrac{x^2y^2}{1}\right) - \left(\dfrac{2}{xy}\right)\left(\dfrac{x^2y^2}{1}\right)}$$

Then reduce where appropriate.

$$= \dfrac{\dfrac{1}{y} \cdot \dfrac{x^2y^2}{1} - \dfrac{4y}{x^2} \cdot \dfrac{x^2y^2}{1}}{\dfrac{1}{y^2} \cdot \dfrac{x^2y^2}{1} - \dfrac{2}{xy} \cdot \dfrac{x^2y^2}{1}}$$

$$= \dfrac{x^2y - 4y^3}{x^2 - 2xy} = \dfrac{y(x - 2y)(x + 2y)}{x(x - 2y)}$$

$$= \dfrac{y(x + 2y)}{x}$$

∎

EXAMPLE 3

Express as a simple fraction reduced to lowest terms:

$$\dfrac{\dfrac{2a}{a - 1} + \dfrac{3}{a + 1}}{\dfrac{1}{a^2 - 1} + 5}$$

Solution

Using Method 2, we first find the LCD of $\dfrac{2a}{a - 1}$, $\dfrac{3}{a + 1}$, and $\dfrac{1}{a^2 - 1}$, which is $(a + 1)(a - 1)$. [Remember that $a^2 - 1 = (a + 1)(a - 1)$.]

*Multiply the numerator **and** the denominator of the complex fraction by the LCD, $(a - 1)(a + 1)$.*

$$\dfrac{\left(\dfrac{2a}{a - 1} + \dfrac{3}{a + 1}\right)\left(\dfrac{(a + 1)(a - 1)}{1}\right)}{\left(\dfrac{1}{(a + 1)(a - 1)} + 5\right)\left(\dfrac{(a + 1)(a - 1)}{1}\right)}$$

Then apply the distributive property, factor if necessary, and reduce where appropriate.

$$= \dfrac{\dfrac{2a}{a - 1} \cdot \dfrac{(a + 1)(a - 1)}{1} + \dfrac{3}{a + 1} \cdot \dfrac{(a + 1)(a - 1)}{1}}{\dfrac{1}{(a + 1)(a - 1)} \cdot \dfrac{(a + 1)(a - 1)}{1} + 5 \cdot \dfrac{(a + 1)(a - 1)}{1}}$$

$$= \frac{2a(a+1) + 3(a-1)}{1 + 5(a+1)(a-1)}$$

At this point you may be tempted to cancel either $a+1$ or $a-1$ since they appear in both the numerator and the denominator. However, since neither of the binomials are *factors* of the numerator (or denominator), they cannot be cancelled.

Our next step is to perform the operations and combine terms where possible.

$$= \frac{2a^2 + 2a + 3a - 3}{1 + 5(a^2 - 1)}$$

$$= \boxed{\frac{2a^2 + 5a - 3}{5a^2 - 4}}$$

When you use this method, keep in mind that it is based on the fundamental principle of fractions, which requires the multiplying factor of the numerator to be the same as the multiplying factor of the denominator. ∎

Exercises 4.4

Express each of the following as a simple fraction reduced to lowest terms.

1. $\dfrac{\dfrac{3}{xy^2}}{\dfrac{15}{x^2y}}$

2. $\dfrac{\dfrac{5}{3a^2b}}{\dfrac{25}{9ab^2}}$

3. $\dfrac{\dfrac{3}{x-y}}{\dfrac{x-y}{3}}$

4. $\dfrac{\dfrac{5}{x+y}}{\dfrac{5}{x-y}}$

5. $\dfrac{a - \dfrac{1}{3}}{\dfrac{9a^2 - 1}{3a}}$

6. $\dfrac{1 - \dfrac{1}{x^2}}{\dfrac{x-1}{x}}$

7. $\dfrac{x + \dfrac{2}{xy^2}}{\dfrac{1}{x} + 2}$

8. $\left(\dfrac{1}{x} + \dfrac{1}{y}\right) \div \dfrac{3}{xy^2}$

9. $\left(1 - \dfrac{4}{x^2}\right) \div \left(\dfrac{1}{x} - \dfrac{2}{x^2}\right)$

10. $\dfrac{x - \dfrac{3}{y^2}}{\dfrac{2}{x} - \dfrac{3}{y^2}}$

11. $\dfrac{1 - \dfrac{5}{y}}{y + 3 - \dfrac{40}{y}}$

12. $\dfrac{6 - \dfrac{8}{x}}{x - 4 - \dfrac{30}{y}}$

13. $\dfrac{1 - \dfrac{4}{z} + \dfrac{4}{z^2}}{\dfrac{1}{z^2} - \dfrac{2}{z^3}}$

14. $\dfrac{2 + \dfrac{3}{x} - \dfrac{2}{x^2}}{\dfrac{2}{x^2} - \dfrac{1}{x^3}}$

15. $\dfrac{\dfrac{4}{y^2} - \dfrac{12}{xy} + \dfrac{9}{x^2}}{\dfrac{4}{y^2} - \dfrac{9}{x^2}}$

16. $\dfrac{\dfrac{9}{t} + \dfrac{12}{s} + \dfrac{4t}{s^2}}{\dfrac{9}{t} + \dfrac{9}{s} + \dfrac{2t}{s^2}}$

17. $\dfrac{\dfrac{2x}{y} + 7 + \dfrac{5y}{x}}{3x + 2y - \dfrac{y^2}{x}}$

18. $\dfrac{\dfrac{6a}{b} - 5 - \dfrac{6b}{a}}{2 - \dfrac{3b}{a}}$

19. $\left(3 + \dfrac{1}{2x-1}\right) \div \left(5 + \dfrac{x}{2x-1}\right)$ **20.** $\left(a + \dfrac{2a}{a-1}\right) \div \left(a - \dfrac{2a}{a-1}\right)$ **21.** $\dfrac{x + \dfrac{2}{x+3}}{x - 5 + \dfrac{12}{x+3}}$

22. $\dfrac{x - \dfrac{4}{2x-7}}{x + 9 + \dfrac{68}{2x-7}}$ **23.** $\dfrac{\dfrac{2}{x} - \dfrac{y}{x-y}}{\dfrac{x}{x-y}}$ **24.** $\dfrac{\dfrac{2y^2}{2y-5} - \dfrac{5y-25}{2y-5}}{y - \dfrac{25}{2y-5}}$

25. $\dfrac{\dfrac{2x}{x-3}}{\dfrac{x}{x+2} + \dfrac{3}{x-3}}$ **26.** $\dfrac{\dfrac{3a+1}{a-4}}{\dfrac{2a+1}{a+2} - \dfrac{a}{a-4}}$ **27.** $\dfrac{\dfrac{x}{3x-5} + \dfrac{8x-6}{3x^2-32x+45}}{\dfrac{x}{3x-5} + \dfrac{2}{x-9}}$

28. $\dfrac{\dfrac{x}{x-3} - \dfrac{4x+20}{x^2-9}}{\dfrac{x+1}{x+3} - \dfrac{12}{x^2-9}}$ **29.** $\dfrac{\dfrac{2a^2-6a}{a+2} - a}{\dfrac{a}{a-3} - \dfrac{4a}{a^2-a-6}}$ **30.** $\dfrac{\dfrac{3x-1}{2x+1} + \dfrac{2x}{4x^2-1}}{\dfrac{3}{2x-1} + 3}$

31. $\dfrac{3a - \dfrac{2}{a}}{a - \dfrac{2}{\dfrac{3}{a+2}}}$ **32.** $\dfrac{\dfrac{x}{x+1} + 3}{\dfrac{x}{x+1} + \dfrac{3}{\dfrac{x}{x+1}}}$ **33.** $3 - \dfrac{x}{3 - \dfrac{x}{3-x}}$

34. $1 - \dfrac{1}{1 - \dfrac{1}{\dfrac{1}{1-x}}}$

? QUESTION FOR THOUGHT

35. Explain what is *wrong* (if anything) with the following:

$$\frac{\dfrac{3}{x} + \dfrac{2}{y}}{\dfrac{4}{x^2} + \dfrac{1}{y^2}} \overset{?}{=} \frac{\left(\dfrac{3}{x} + \dfrac{2}{y}\right)xy}{\left(\dfrac{4}{x^2} + \dfrac{1}{y^2}\right)x^2y^2} \overset{?}{=} \frac{3y + 2x}{4y^2 + x^2}$$

4.5

Fractional Equations and Inequalities

Up to now we have been concerned with performing operations with fractional expressions. In this section we will discuss how to approach fractional equations and inequalities.

In Chapter 2 we developed a strategy for solving first-degree equations. In Section 4.3 we discussed the idea of the least common denominator (LCD). Our method for solving fractional equations and inequalities will use both of these ideas, as illustrated in the following examples.

EXAMPLE 1

Solve for x. $\dfrac{x}{4} - \dfrac{x-3}{3} = \dfrac{x+3}{6}$

Solution

Rather than solve this equation directly, we would much prefer to solve an equivalent equation without fractions. The idea of an LCD that we used in Section 4.3 can be used to accomplish this.

Since we are dealing with an equation, we can multiply both sides of the equation by any nonzero quantity we choose. If we multiply the entire equation by a number that is divisible by 3, 4, and 6, we will "eliminate" the denominators. The LCD is exactly the smallest such quantity. This process is called *clearing the denominators*.

We multiply both sides of the equation by the LCD of all the denominators in the equation, which is 12.

$$\frac{x}{4} - \frac{x-3}{3} = \frac{x+3}{6}$$

$$12\left(\frac{x}{4} - \frac{x-3}{3}\right) = 12\left(\frac{x+3}{6}\right) \qquad \textit{Each fraction is multiplied by } 12.$$

$$\frac{12}{1}\cdot\frac{x}{4} - \frac{12}{1}\cdot\frac{x-3}{3} = \frac{12}{1}\cdot\frac{x+3}{6} \qquad \textit{Reduce.}$$

$$\frac{\overset{3}{\cancel{12}}}{1}\cdot\frac{x}{\cancel{4}} - \frac{\overset{4}{\cancel{12}}}{1}\cdot\frac{x-3}{\cancel{3}} = \frac{\overset{2}{\cancel{12}}}{1}\cdot\frac{x+3}{\cancel{6}}$$

$$3x - 4(x-3) = 2(x+3) \qquad \textit{Note that the parentheses around } x-3 \text{ and}$$
$$3x - 4x + 12 = 2x + 6 \qquad \quad x+3 \textit{ are necessary.}$$
$$-x + 12 = 2x + 6$$
$$6 = 3x$$
$$\boxed{2 = x}$$

CHECK $x = 2$: $\quad \dfrac{x}{4} - \dfrac{x-3}{3} = \dfrac{x+3}{6}$

$$\frac{2}{4} - \frac{2-3}{3} \overset{?}{=} \frac{2+3}{6}$$

$$\frac{1}{2} - \frac{-1}{3} \overset{?}{=} \frac{5}{6}$$

$$\frac{3}{6} + \frac{2}{6} \overset{\checkmark}{=} \frac{5}{6} \qquad\qquad\qquad ■$$

When a variable does not appear in any denominator, clearing the denominators is the first step to take in solving fractional inequalities as well.

EXAMPLE 2

Solve for q. $\quad \dfrac{q}{5} + \dfrac{5-q}{3} \le 2$

Solution

We clear the denominators by multiplying both sides of the inequality by the LCD, which is $5 \cdot 3 = 15$.

$$\frac{15}{1} \cdot \frac{q}{5} + \frac{15}{1} \cdot \frac{5 - q}{3} \le 15 \cdot 2$$

$$\frac{\overset{3}{\cancel{15}}}{1} \cdot \frac{q}{\cancel{5}} + \frac{\overset{5}{\cancel{15}}}{1} \cdot \frac{5 - q}{\cancel{3}} \le 15 \cdot 2$$

$$3q + 5(5 - q) \le 30$$

$$3q + 25 - 5q \le 30$$

$$25 - 2q \le 30$$

$$-2q \le 5$$

$$\boxed{q \ge -\frac{5}{2}}$$ *Do not forget to reverse the inequality symbol when dividing by a negative number.*

CHECK:

Let $q = -\frac{5}{2}$.

Then:

$$\frac{-\frac{5}{2}}{5} + \frac{5 - \left(-\frac{5}{2}\right)}{3} \overset{?}{=} 2$$

$$-\frac{5}{2} \cdot \frac{1}{5} + \frac{\frac{15}{2}}{3} \overset{?}{=} 2$$

$$-\frac{1}{2} + \frac{15}{2} \cdot \frac{1}{3} \overset{?}{=} 2$$

$$-\frac{1}{2} + \frac{5}{2} \overset{\checkmark}{=} 2$$

This tells us that the endpoint,
$-\frac{5}{2}$, *is correct.*

Pick a number greater than $-\frac{5}{2}$. Let $q = 0$.

Then:

$$\frac{0}{5} + \frac{5 - (-0)}{3} \overset{?}{\le} 2$$

$$0 + \frac{5}{3} \overset{\checkmark}{\le} 2$$

Thus, the inequality symbol is pointed in the correct direction.

■

EXAMPLE 3

Solve for a. $\dfrac{5}{2a} + \dfrac{3}{a - 2} = \dfrac{7}{10a}$

Solution

We clear the denominator by multiplying both sides of the equation by the LCD, which is $10a(a - 2)$.

$$\frac{5}{2a} + \frac{3}{a - 2} = \frac{7}{10a}$$

Each fraction is multiplied by the LCD.

$$\left(\frac{10a(a - 2)}{1}\right)\frac{5}{2a} + \left(\frac{10a(a - 2)}{1}\right)\frac{3}{a - 2} = \left(\frac{10a(a - 2)}{1}\right)\frac{7}{10a}$$

Reduce.

$$\overset{5}{\cancel{10a}(a-2)}{1} \cdot \frac{5}{\cancel{2a}} + \frac{10a\cancel{(a-2)}}{1} \cdot \frac{3}{\cancel{a-2}} = \frac{\cancel{10a}(a-2)}{1} \cdot \frac{7}{\cancel{10a}}$$

$$25(a-2) + 30a = 7(a-2)$$

$$25a - 50 + 30a = 7a - 14$$

$$55a - 50 = 7a - 14$$

$$48a = 36$$

$$a = \frac{36}{48} = \boxed{\frac{3}{4}}$$

CHECK $a = \frac{3}{4}$: $\quad \dfrac{5}{2a} + \dfrac{3}{a-2} = \dfrac{7}{10a}$

$$\frac{5}{2\left(\frac{3}{4}\right)} + \frac{3}{\frac{3}{4}-2} \overset{?}{=} \frac{7}{10\left(\frac{3}{4}\right)}$$

$$\frac{5}{\frac{3}{2}} + \frac{3}{-\frac{5}{4}} \overset{?}{=} \frac{7}{\frac{15}{2}}$$

$$\frac{5}{1} \cdot \frac{2}{3} - \frac{3}{1} \cdot \frac{4}{5} \overset{?}{=} \frac{7}{1} \cdot \frac{2}{15}$$

$$\frac{10}{3} - \frac{12}{5} \overset{?}{=} \frac{14}{15}$$

$$\frac{50}{15} - \frac{36}{15} \overset{\checkmark}{=} \frac{14}{15}$$

As the next example shows, solving fractional equations requires some care.

EXAMPLE 4

Solve for x. $\dfrac{7}{x+5} + 2 = \dfrac{2-x}{x+5}$

Solution

Again we clear the denominator by multiplying both sides of the equation by the LCD, which is $x + 5$:

$$\frac{7}{x+5} + 2 = \frac{2-x}{x+5}$$

First multiply both sides of the equation by $x + 5$, and reduce.

$$\cancel{(x+5)}\left(\frac{7}{\cancel{x+5}}\right) + 2(x+5) = \cancel{(x+5)}\left(\frac{2-x}{\cancel{x+5}}\right)$$

$$7 + 2(x+5) = 2 - x$$

$$7 + 2x + 10 = 2 - x$$

$$2x + 17 = 2 - x$$

$$3x = -15$$

$$x = -5$$

CHECK $x = -5$: $\quad \dfrac{7}{x + 5} + 2 = \dfrac{2 - x}{x + 5}$

$$\dfrac{7}{-5 + 5} + 2 \stackrel{?}{=} \dfrac{2 - (-5)}{-5 + 5}$$

$$\dfrac{7}{0} + 2 \neq \dfrac{7}{0}$$

Since we are never allowed to divide by 0, $\frac{7}{0}$ is undefined. Therefore, $x = -5$ is *not* a solution.

Have we made an error? No. As we pointed out in Chapter 2, if we multiply an equation by 0 we may no longer have an equivalent equation (that is, an equation with the same solution set). This is exactly what has happened here.

We multiplied the original equation by $x + 5$ to clear the denominators, but if $x = -5$, then $x + 5$ is equal to 0, and so we have multiplied the original equation by 0. The resulting equation, for which $x = -5$ is a solution, is not equivalent to the original equation, for which $x = -5$ is not even an allowable value.

Our logic tells us that $x = -5$ is the only possible solution. Since $x = -5$ does not satisfy the original equation, the original equation has

$$\boxed{\text{no solutions}}$$

Example 4 shows that when we multiply an equation by a variable quantity which might be equal to 0, we *must* check our answer(s) in the original equation. This check is not optional, but rather a necessary step in the solution. We are not checking for errors—we are checking to see if we have obtained a valid solution.

Another way of saying this is that we look at our original equation and ask, "What are the possible replacement values for x?" We must disqualify $x = -5$ from the outset since it requires division by 0, which is undefined.

EXAMPLE 5

Solve for t. $\quad \dfrac{5}{t + 3} - \dfrac{4}{3t} = \dfrac{7}{t^2 + 3t}$

Solution

We will use the LCD to clear the denominators. In order to find the LCD we want each denominator in factored form.

$$\dfrac{5}{t + 3} - \dfrac{4}{3t} = \dfrac{7}{t(t + 3)} \qquad LCD = 3t(t + 3)$$

$$3t(t + 3) \cdot \dfrac{5}{t + 3} - 3t(t + 3) \cdot \dfrac{4}{3t} = 3t(t + 3) \cdot \dfrac{7}{t(t + 3)}$$

$$15t - 4(t + 3) = 21$$

$$15t - 4t - 12 = 21$$

$$11t - 12 = 21$$

$$11t = 33$$

$$\boxed{t = 3}$$

The check is left to the student.

EXAMPLE 6   *Combine and simplify.* $\dfrac{2}{t} + \dfrac{3}{t + 2} - \dfrac{4}{5t}$

Solution   It is very important to recognize the difference between this example and the previous one. Example 5 was an equation. The multiplication property of *equality* allowed us to "clear the denominators." This example, on the other hand, is an expression, *not* an equation. Therefore, the multiplication property of equality does not apply; we cannot "clear the denominators."

Instead, we are being asked to combine three fractions. We will use the LCD again in this example, but this time to *build* the fractions. Following the outline used in Section 4.3, we proceed as follows:

$$\frac{2}{t} + \frac{3}{t + 2} - \frac{4}{5t} \qquad LCD = 5t(t + 2)$$

$$= \frac{2 \cdot 5(t + 2)}{5t(t + 2)} + \frac{3 \cdot 5t}{5t(t + 2)} - \frac{4(t + 2)}{5t(t + 2)}$$

$$= \frac{10(t + 2) + 15t - 4(t + 2)}{5t(t + 2)}$$

$$= \frac{10t + 20 + 15t - 4t - 8}{5t(t + 2)} = \boxed{\frac{21t + 12}{5t(t + 2)}}$$

Exercises 4.5

In each of the following, if the exercise is an equation or inequality, solve it. If it is an expression, perform the indicated operations and simplify.

1. $\dfrac{x}{3} - \dfrac{x}{2} + \dfrac{x}{4} = 1$

2. $\dfrac{a}{4} - \dfrac{a}{5} - \dfrac{a}{10} = 1$

3. $\dfrac{t}{3} + \dfrac{t}{5} < \dfrac{t}{6} - 11$

4. $\dfrac{n}{6} - \dfrac{n}{9} > \dfrac{n}{3} - 5$

5. $\dfrac{a - 1}{6} + \dfrac{a + 1}{10} = a - 3$

6. $\dfrac{w - 3}{8} + \dfrac{w + 4}{12} = w - 4$

7. $\dfrac{y - 5}{2} = \dfrac{y - 2}{5}$

8. $\dfrac{z - 2}{7} = \dfrac{z - 7}{2}$

9. $\dfrac{y - 5}{2} \le \dfrac{y - 2}{5} + 3$

10. $\dfrac{z - 2}{7} \ge \dfrac{z - 7}{2} - 5$

11. $\dfrac{x - 3}{4} - \dfrac{x - 4}{3} = 2$

12. $\dfrac{x - 3}{5} - \dfrac{x + 2}{6} = 1$

13. $\dfrac{x - 3}{4} - \dfrac{x - 4}{3} \ge 2$

14. $\dfrac{x - 3}{5} - \dfrac{x + 2}{6} \le 1$

15. $\dfrac{3x + 11}{6} - \dfrac{2x + 1}{3} = x + 5$

16. $\dfrac{7x + 1}{4} - \dfrac{3x + 1}{8} = \dfrac{x + 2}{2}$

17. $\dfrac{x - 3}{5} - \dfrac{3x + 1}{4} < 8$

18. $\dfrac{5x + 1}{2} - \dfrac{x + 1}{4} > 2$

19. $\dfrac{5}{x} - \dfrac{1}{2} = \dfrac{3}{x}$

20. $\dfrac{4}{3x} + \dfrac{9}{x} - \dfrac{6}{5}$

21. $\dfrac{4}{x} - \dfrac{1}{5} + \dfrac{7}{2x}$

22. $\dfrac{2}{x} + \dfrac{1}{3} = \dfrac{5}{x}$

23. $\dfrac{1}{t - 3} + \dfrac{2}{t} = \dfrac{5}{3t}$

24. $\dfrac{2}{y + 2} + \dfrac{3}{y} = \dfrac{5}{2y}$

25. $\dfrac{6}{a - 3} - \dfrac{3}{8} = \dfrac{21}{4a - 12}$

26. $\dfrac{5}{z - 1} - \dfrac{7}{z} = \dfrac{3}{2z - 2}$

27. $\dfrac{7}{x - 5} + 2 = \dfrac{x + 2}{x - 5}$

28. $\dfrac{9}{x + 2} - 3 = \dfrac{x + 11}{x + 2}$

29. $\dfrac{4}{y^2 - 2y} - \dfrac{3}{2y} = \dfrac{17}{6y}$

30. $\dfrac{3}{z^2 - 4z} + \dfrac{7}{4z} - \dfrac{3}{8}$

31. $\dfrac{5}{y^2 + 3y} - \dfrac{4}{3y} + \dfrac{1}{2}$

32. $\dfrac{9}{z^2 + 5z} + \dfrac{6}{5z} = \dfrac{3}{10z}$

33. $\dfrac{9}{x^2 + 4x} = \dfrac{6}{x^2 + 2x}$

34. $\dfrac{6}{t^2 - 3t} = \dfrac{12}{t^2 - 9t}$

35. $\dfrac{1}{x^2 - x - 2} + \dfrac{2}{x^2 - 1} = \dfrac{1}{x^2 - 3x + 2}$

36. $\dfrac{3}{a^2 + 3a - 10} + \dfrac{12}{a^2 - 2a} = \dfrac{4}{a^2 + 5a}$

37. $\dfrac{1}{x - 4} - \dfrac{5}{x + 2} = \dfrac{6}{x^2 - 2x - 8}$

38. $\dfrac{3}{3x - 2} - \dfrac{7}{x + 1} = \dfrac{5}{3x^2 + x - 2}$

39. $\dfrac{n}{3n + 2} + \dfrac{6}{9n^2 - 4} - \dfrac{2}{3n - 2}$

40. $\dfrac{n}{2n - 3} + \dfrac{2}{2n + 3} - \dfrac{5n}{4n^2 - 9}$

41. $\dfrac{1}{3n + 4} + \dfrac{8}{9n^2 - 16} = \dfrac{1}{3n - 4}$

42. $\dfrac{5}{2n + 1} + \dfrac{3}{2n - 1} = \dfrac{22}{4n^2 - 1}$

43. $\dfrac{4}{2x - 1} + \dfrac{2}{x + 3} = \dfrac{5}{2x^2 + 5x - 3}$

44. $\dfrac{7}{3a + 2} + \dfrac{4}{a + 5} = \dfrac{8}{3a^2 + 17a + 10}$

45. $\dfrac{6}{3a + 5} - \dfrac{2}{a - 4} = \dfrac{10}{3a^2 - 7a - 20}$

46. $\dfrac{5}{y - 2} - \dfrac{3}{2y - 1} = \dfrac{4}{2y^2 - 5y + 2}$

47. $\dfrac{3}{x^2 - x - 6} + \dfrac{2}{2x^2 - 5x - 3} = \dfrac{5}{2x^2 + 5x + 2}$

48. $\dfrac{10}{2a^2 - a - 15} - \dfrac{5}{3a^2 - 4a - 15} = \dfrac{3}{6a^2 + 25a + 25}$

49. $\dfrac{4}{4x^2 - 9} - \dfrac{5}{4x^2 - 8x + 3} = \dfrac{8}{4x^2 + 4x - 3}$

50. $\dfrac{6}{9y^2 - 1} - \dfrac{4}{9y^2 + 9y + 2} = \dfrac{9}{9y^2 + 3y - 2}$

MINI-REVIEW

51. *Factor completely.* $(x - y)^2 - 9$

52. *Solve for a.* $3a - 8 = 5a - (4 + 2a)$

53. Perform the operations and express your answer in simplest form:

$$\dfrac{3x^2 - 6x}{x + 4} \div \left(\dfrac{2y^2 - xy^2}{x^2 - 9} \cdot \dfrac{x^2}{xy + 4y} \right)$$

54. The sum of three consecutive even integers is 174. Find them.

4.6

Literal Equations

When we solve an equation, it is not always the case that we will get a numerical answer. Solving a conditional equation with one variable will always give us a numerical solution. However, the situation is different in the case of an equation with more than one variable.

An equation that contains more than one variable (letter) is called a ***literal equation***. In certain special cases it is called a *formula*. A literal equation in which one of the variables is totally isolated on one side of the equation is said to be solved ***explicitly*** for that variable.

For example, the following three equations are alternate forms of exactly the same equation:

$$y = -3x + 7 \qquad \text{Solved explicitly for } y$$

$$x = \frac{-y + 7}{3} \qquad \text{Solved explicitly for } x$$

$$3x + y = 7 \qquad \text{Not solved explicitly for either variable}$$

When we are asked to solve a literal equation we must be told which variable we are to solve for explicitly.

EXAMPLE 1

Solve the equation. $4x - 7y = 12$

(a) Explicitly for x **(b)** Explicitly for y

Solution

(a) Solving explicitly for x means we want x isolated on one side of the equation.

$$4x - 7y = 12 \qquad \textit{Add } 7y \textit{ to both sides.}$$

$$4x - 7y + 7y = 12 + 7y$$

$$4x = 7y + 12 \qquad \textit{Then divide both sides by } 4.$$

$$\frac{4x}{4} = \frac{7y + 12}{4}$$

$$\boxed{x = \frac{7y + 12}{4}}$$

(b) Solving explicitly for y means we want y isolated on one side of the equation.

$$4x - 7y = 12 \qquad \textit{Add } -4x \textit{ to both sides.}$$

$$4x - 7y - 4x = 12 - 4x$$

$$-7y = -4x + 12 \qquad \textit{Then divide both sides by } -7.$$

$$\frac{-7y}{-7} = \frac{-4x + 12}{-7}$$

$$y = \frac{-4x + 12}{-7} \qquad \textit{Multiply numerator and denominator by } -1 \textit{ to put into a more preferred form.}$$

$$\boxed{y = \frac{4x - 12}{7}}$$

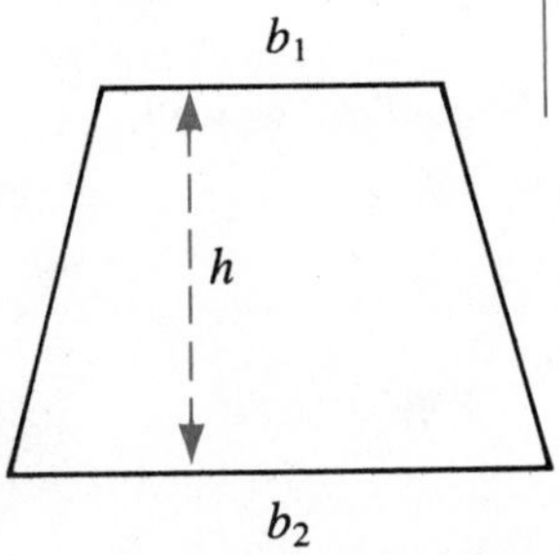

A literal equation that has a "real-life" interpretation is often called a *formula*. For instance, the area, A, of a trapezoid (see Figure 4.1) is given by the formula

$$A = \frac{1}{2}h(b_1 + b_2)$$

This formula is quite useful if we are given h, b_1, and b_2 and want to compute A. However, if we are given A, h, and b_1 and we want to compute b_2, we would much prefer to have a formula solved explicitly for b_2.

Figure 4.1
Trapezoid

EXAMPLE 2

Solve explicitly for b_2. $A = \frac{1}{2}h(b_1 + b_2)$

Solution

We offer two approaches to illustrate a point.

METHOD 1:
$$A = \frac{1}{2}h(b_1 + b_2)$$ *Multiply the equation by 2 to clear the denominator.*

$$2A = h(b_1 + b_2)$$ *Apply the distributive property.*

$$2A = hb_1 + hb_2$$ *Then isolate b_2 (subtract hb_1 from both sides).*

$$2A - hb_1 = hb_2$$ *Divide both sides by h.*

$$\frac{2A - hb_1}{h} = \frac{hb_2}{h}$$

$$\boxed{\frac{2A - hb_1}{h} = b_2}$$

METHOD 2:
$$A = \frac{1}{2}h(b_1 + b_2)$$ *Again clear the denominator.*

$$2A = h(b_1 + b_2)$$ *Divide both sides by h.*

$$\frac{2A}{h} = \frac{h(b_1 + b_2)}{h}$$

$$\frac{2A}{h} = b_1 + b_2$$ *Isolate b_2 (subtract b_1 from both sides).*

$$\boxed{\frac{2A}{h} - b_1 = b_2}$$

Both methods are equally correct. What is important is that you recognize that the two answers are equivalent. If we combine the second answer into a single fraction we get

$$\frac{2A}{h} - b_1 = \frac{2A}{h} - \frac{hb_1}{h} = \frac{2A - hb_1}{h}$$

which is the first answer. Keep this in mind when you check your answers with those in the answer key in the back of the book. ■

The same procedure can be applied to solving literal *inequalities* as well.

EXAMPLE 3

Solve explicitly for a. $a - 2c \leq 6a + 7d$

Solution

$$a - 2c \leq 6a + 7d$$ *Add 2c to both sides of the inequality.*

$$a - 2c + 2c \leq 6a + 7d + 2c$$

$$a \leq 6a + 7d + 2c$$ *Do not stop here! The variable **a** must appear on only one side of the inequality.*

$$a - 6a \leq 6a + 7d + 2c - 6a$$

$$-5a \leq 7d + 2c$$

$$\frac{-5a}{-5} \geq \frac{7d + 2c}{-5}$$

*Remember that when we divide both sides of an inequality by a negative number, we **reverse** the inequality symbol.*

$$\boxed{a \geq \frac{7d + 2c}{-5}}$$

■

EXAMPLE 4

Solve for t. $at + 9 = 3t + b$

Solution

We want to get the t terms on one side of the equation and the non-t terms on the other side.

$$at + 9 = 3t + b$$
$$at + 9 - 3t = 3t + b - 3t$$ *First add $-3t$ to both sides of the equation.*
$$at - 3t + 9 = b$$
$$at - 3t + 9 - 9 = b - 9$$ *Then add -9 to both sides.*
$$at - 3t = b - 9$$ *In order to isolate t, we factor out the t on the left-hand side.*
$$t(a - 3) = b - 9$$ *Then divide both sides by $a - 3$.*
$$\frac{t(a - 3)}{a - 3} = \frac{b - 9}{a - 3}$$
$$\boxed{t = \frac{b - 9}{a - 3}}$$

Since we cannot divide by 0, this solution assumes that $a \neq 3$.   ■

Notice that in Example 4 we had to factor out t from $at - 3t$. This may seem like a new step that we have not seen before, but actually this step was implicit in the equations we solved prior to this section. For example, in solving

$$7t - 9 = 3t + 8$$

we subtract $3t$ from both sides and add 9 to both sides so our equation would look like this:

$$7t - 9 - 3t + 9 = 3t + 8 - 3t + 9$$
$$7t - 3t = 8 + 9$$

We normally combine like terms, $7t - 3t = 4t$ and $8 + 9 = 17$, and simplify our equation:

$$4t = 17$$

If you think back to how we are allowed to combine terms (by the distributive property), you will see that the factoring step is implicit in combining terms:

$$7t - 3t = (7 - 3)t \quad \text{*Distributive property*}$$
$$= 4t$$

So this procedure was not new; you have actually been doing it all along.

EXAMPLE 5 *Solve for u.* $y = \dfrac{u + 1}{u + 2}$

Solution

$$y = \frac{u + 1}{u + 2}$$

Begin by multiplying both sides by $u + 2$ to clear the denominator.

$$(u + 2)y = \frac{u + 1}{u + 2} \cdot \frac{u + 2}{1}$$

$$(u + 2)y = u + 1$$

Apply the distributive property.

$$uy + 2y = u + 1$$

We collect all the u terms on one side, non-u terms on the other side.

$$uy + 2y - u - 2y = u + 1 - u - 2y$$

$$uy - u = 1 - 2y$$

Factor out u on the left side.

$$u(y - 1) = 1 - 2y$$

Then divide both sides of the equation by $y - 1$.

$$\frac{u(y - 1)}{y - 1} = \frac{1 - 2y}{y - 1}$$

$$\boxed{u = \frac{1 - 2y}{y - 1}}$$

This solution assumes that $y \neq 1$ and $u \neq -2$. Why? ∎

Exercises 4.6

Solve each of the following equations or inequalities explicitly for the indicated variable.

1. $5x + 7y = 4$ for x

2. $5x + 7y = 4$ for y

3. $2x - 9y = 11$ for y

4. $2x - 9y = 11$ for x

5. $w + 4z - 1 = 2w - z + 3$ for w

6. $w + 4z - 1 = 2w - z + 3$ for z

7. $2(6r - 5t) > 5(2r + t)$ for r

8. $2(6r - 5t) > 5(2r + t)$ for t

9. $3m - 4n + 6p = 5n + 2p - 8$ for n

10. $3m - 4n + 6p = 5n + 2p - 8$ for m

11. $\dfrac{a}{5} - \dfrac{b}{3} = \dfrac{a}{2} - \dfrac{b}{6}$ for a

12. $\dfrac{c}{12} - \dfrac{d}{8} = \dfrac{c}{4} - \dfrac{d}{6}$ for d

13. $\dfrac{x + y}{3} - \dfrac{x}{2} + \dfrac{y}{6} = 3(x - y)$ for x

14. $\dfrac{w - z}{10} - \dfrac{w}{5} + \dfrac{z}{4} = 2(z + w)$ for z

15. $ax + b = cx + d$ for x

16. $p - rz = q - tz$ for z

17. $3x + 2y - 5 = ax + by + 1$ for x

18. $3x + 2y - 5 = ax + by + 1$ for y

19. $(x + 3)(y + 7) = a$ for x

20. $(3x - 2)(2y - 1) = b$ for y

21. $y = \dfrac{u - 1}{u + 1}$ for u

22. $z = \dfrac{w - 2}{w + 3}$ for w

23. $x = \dfrac{2t - 3}{3t - 2}$ for t

24. $u = \dfrac{3y - 4}{2y - 3}$ for y

Each of the following is a formula from mathematics or the physical or social sciences. Solve each formula for the indicated variable.

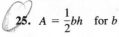

25. $A = \frac{1}{2}bh$ for b

26. $A = \frac{1}{2}bh$ for h

27. $A = \frac{1}{2}h(b_1 + b_2)$ for b_1

28. $A = \frac{1}{2}h(b_1 + b_2)$ for h

29. $A = P(1 + rt)$ for r

30. $A = P(1 + rt)$ for P

31. $C = \frac{5}{9}(F - 32)$ for F

32. $F = \frac{9}{5}C + 32$ for C

33. $\frac{P_1}{V_1} = \frac{P_2}{V_2}$ for P_2

34. $\frac{P_1}{V_1} = \frac{P_2}{V_2}$ for V_2

35. $S = s_0 + v_0 t + \frac{1}{2}gt^2$ for g

36. $S = s_0 + v_0 t + \frac{1}{2}gt^2$ for v_0

37. $\frac{x - \mu}{s} < 1.96$ for x (assume $s > 0$)

38. $\frac{x - \mu}{s} < 1.96$ for μ (assume $s > 0$)

39. $\frac{1}{f} = \frac{1}{f_1} + \frac{1}{f_2}$ for f_1

40. $\frac{1}{f} = \frac{1}{f_1} + \frac{1}{f_2}$ for f

41. $S = 2\pi r^2 + 2\pi rh$ for h

42. $S = 2LH + 2LW + 2WH$ for W

 MINI-REVIEW

43. *Solve for x.* $|3x - 2| > 5$

44. *Factor completely.* $8x^3 - 64y^3$

45. Express as a simple fraction reduced to lowest terms:

$$\frac{1 + \dfrac{2}{x + 3}}{\dfrac{3}{x + 3} + 1}$$

46. Two cars leave from the same spot at the same time and travel in opposite directions. If one car is traveling at 45 mph and the other is traveling at 55 mph, how long will it be until they are 325 miles apart?

47. The length of a rectangle is 5 more than 3 times its width. If its perimeter is 42 inches, find the dimensions of the rectangle.

4.7

More Verbal Problems

The outline suggested in Chapter 2 for solving verbal problems applies equally well to problems which give rise to fractional equations and inequalities. We will repeat this advice in the next box for your reference.

**Outline of Strategy
for Solving Verbal
Problems**

1.   *Read the problem carefully*, as many times as is necessary to understand what the problem is saying and what it is asking.

2.   *Use diagrams* whenever you think it will make the given information clearer.

3.   *Find the underlying relationship or formula* relevant to the given problem. Ask whether there is some underlying relationship or formula you need to know. If not, then the words of the problem themselves give the required relationship.

4.   Clearly *identify the unknown quantity* (or quantities) in the problem, and label it (them) using one variable.

5.   By using the underlying formula or relationship in the problem, *write an equation* involving the unknown quantity (or quantities).

6.   *Solve* the equation.

7.   *Answer the question.* Make sure you have answered the question that was asked.

8.   *Check* the answer(s) in the original words of the problem.

EXAMPLE 1

Gary drove from his house to Elaine's house. He had driven over the speed limit for one-third of the distance, when he was stopped and given a ticket for speeding. He then drove under the speed limit for one-half the total distance when his car ran out of gas. He walked the remaining 5 miles to Elaine's house. What is the distance from Gary's to Elaine's house?

Solution

At first glance, this may look like a distance–rate problem, but actually the problem is simpler. First, draw a picture to represent the distances travelled and label the distances, as shown in Figure 4.2.

Figure 4.2
Diagram for Example 1

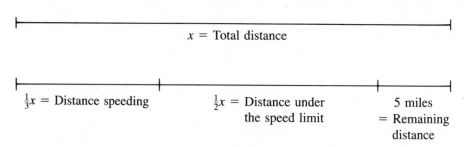

The figure illustrates the relationships among the distances travelled under each condition. It gives us the following equation:

$$\frac{1}{3}x + \frac{1}{2}x + 5 = x$$

We solve this equation by first multiplying the equation by the LCD of the fractions, which is 6:

$$\overset{2}{\cancel{6}}\left(\frac{1}{\cancel{3}}x\right) + \overset{3}{\cancel{6}}\left(\frac{1}{\cancel{2}}x\right) + 6\cdot 5 = 6\cdot x$$

$$2x + 3x + 30 = 6x$$

$$5x + 30 = 6x \qquad \textit{Subtract 5x from both sides.}$$

$$30 = x$$

$$x = \boxed{30 \text{ miles}}$$

CHECK:

$$\text{One-third the distance speeding} = \frac{1}{3}(30) = 10 \text{ miles}$$

$$\text{One-half the distance under the limit} = \frac{1}{2}(30) = 15 \text{ miles}$$

$$\text{Remaining distance} = \underline{5 \text{ miles}}$$

$$\text{Total} = 30 \text{ miles}$$

An alternative, shorter way to solve the same problem is to consider the fact that Gary has already travelled $\frac{1}{2} + \frac{1}{3}$, or $\frac{5}{6}$, of the distance before he runs out of gas. So he still has $\frac{1}{6}$ the total distance to walk. This remaining distance is given as 5 miles. Hence, we have

$$\frac{1}{6} \text{ the total distance} = 5 \text{ miles} \qquad \text{or} \qquad \frac{1}{6}x = 5$$

Multiplying both sides of the equation by 6, we have

$$x = 30 \text{ miles} \qquad\qquad \blacksquare$$

Ratio and proportion problems are fairly straightforward to set up, as long as you remember to match up the corresponding units.

EXAMPLE 2

In a local district, the ratio of Democrats to Republicans is 5 to 7. If there are 2,100 Republicans in the district, how many Democrats are there?

Solution

First, you should realize that the ratio of a to b is simply the fraction $\frac{a}{b}$ reduced to lowest terms. Hence, to *find* the ratio of the number of Democrats to the number of Republicans, we would create the fraction

$$\frac{\text{Number of Democrats}}{\text{Number of Republicans}}$$

and reduce it to get $\frac{5}{7}$.

We let x = # of Democrats and set up our equation with two fractions, being careful to match up the units:

$$\frac{5 \text{ Democrats}}{7 \text{ Republicans}} = \frac{x \text{ Democrats}}{2,100 \text{ Republicans}} \quad \text{or} \quad \frac{5}{7} = \frac{x}{2,100}$$

Multiply both sides of the equation by 2,100 to get

$$1,500 = x$$

Hence, there are $\boxed{1,500 \text{ Democrats}}$ in that district.   ∎

EXAMPLE 3   If 1 kilogram is 2.2 pounds, how many kilograms are in 106 pounds?

Solution   Again, we match up the units to check that the fractions are set up properly. Let x = # kg in 106 lb.

$$\frac{x \text{ kg}}{106 \text{ lb}} = \frac{1 \text{ kg}}{2.2 \text{ lb}}$$

$$\frac{x}{106} = \frac{1}{2.2}$$

Since we are solving for x, we multiply both sides of the equation by 106:

$$x = \frac{106}{2.2} = \boxed{48.18 \text{ kg}}$$ ∎

EXAMPLE 4   If the ratio of Republicans to Democrats in a district is 4 to 9, and there is a total of 2,210 Republicans and Democrats in that district, how many Republicans are there?

Solution   Let x = # of Republicans.

Then $2,210 - x$ = # of Democrats.   *(Total minus part = remainder)*

Hence, our equation becomes

$$\frac{4}{9} = \frac{x}{2,210 - x}$$

Multiplying both sides of the equation by $9(2,210 - x)$, we get

$$9(2,210 - x)\left(\frac{4}{9}\right) = \left(\frac{x}{2,210 - x}\right)(9)(2,210 - x)$$

$$4(2,210 - x) = 9x$$

$$8,840 - 4x = 9x$$

$$8,840 = 13x$$

$$680 = x$$

Thus, there are $\boxed{680 \text{ Republicans}}$ in the district.

We leave the check to the student.   ∎

EXAMPLE 5

Solution

How many liters each of 35% and 60% alcohol solutions must be mixed together to make 40 liters of a 53% alcohol solution?

We can let x = amount of 35% solution.

Then $40 - x$ = amount of 60% solution. Why?

We can visualize this problem as shown in Figure 4.3.

Figure 4.3
Diagram for Example 5

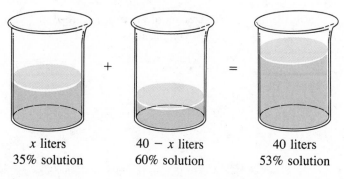

x liters
35% solution

+

$40 - x$ liters
60% solution

=

40 liters
53% solution

In order to write an equation to solve this problem, we need to differentiate between the amount of *solution* we have and the amount of *actual alcohol* in the solution. For instance, if we have 100 liters of a 30% alcohol solution, the amount of actual alcohol is

$$0.30(100) = 30 \text{ liters}$$

The basic relationship we use in this example is that the amounts of alcohol in each solution add up to the total amount of alcohol in the final solution.

Thus, our equation is

$$0.35x \quad + \quad 0.60(40 - x) \quad = \quad 0.53(40)$$

Amount of alcohol   Amount of alcohol   Total amount of alcohol
in the 35% solution + in the 60% solution = in final solution

Since $0.35 = \dfrac{35}{100}$, $0.60 = \dfrac{60}{100}$, and $0.53 = \dfrac{53}{100}$, we multiply both sides of the equation by 100 to clear the decimals.

$$35x + 60(40 - x) = 53(40)$$
$$35x + 2{,}400 - 60x = 2{,}120$$
$$-25x + 2{,}400 = 2{,}120$$
$$-25x = -280$$
$$x = \frac{-280}{-25} = 11.2$$

We must mix
$40 - 11.2 =$

| 11.2 liters of 35% alcohol solution with |
| 28.8 liters of 60% alcohol solution. |

CHECK: $11.2 + 28.8 \overset{\checkmark}{=} 40$ $0.35(11.2) + 0.60(28.8) \overset{?}{=} 0.53(40)$

$3.92 + 17.28 \overset{\checkmark}{=} 21.2$ ∎

EXAMPLE 6

Mrs. Stone invests a certain amount of money in a bank account paying 9.3% interest per year, and $3,500 more than this amount in a mutual fund paying 10.2% interest per year. If the annual income from the two investments is $1,263.75, how much is invested at each rate?

Solution

The amount of interest, I, earned in 1 year if P dollars is invested at a rate of $r\%$ per year is computed as

$$I = P \cdot r \quad \text{Interest} = \text{(Principal)(Rate)}$$

where r is written as a decimal.

For instance, if $1,000 is invested at 8% per year, then the interest earned in 1 year is

$$I = (1000)(0.08) = \$80$$

Based on the information given in this example:

Let x = amount invested at 9.3%.

Then $x + 3,500$ = amount invested at 10.2%.

Our equation is

$$0.093x \quad + 0.102(x + 3,500) = \quad 1,263.75$$

9.3% written as a decimal is 0.093; 10.2% is 0.102.

Interest received		*Interest received*		*Total interest*
for the 9.3%	+	*for the 10.2%*	=	*received for*
investment		*investment*		*both investments*

Multiplying both sides of the equation by 1,000 clears the decimals:

$$93x + 102(x + 3,500) = 1,263,750$$
$$93x + 102x + 357,000 = 1,263,750$$
$$195x = 906,750$$
$$x = 4,650$$

and $3,500 more, or

$4,650 invested at 9.3%
$8,150 invested at 10.2%

CHECK: $8,150 \overset{\checkmark}{=} 4,650 + 3,500$ $0.093(4,650) + 0.102(8,150) \overset{?}{=} 1,263.75$

$432.45 + 831.30 \overset{\checkmark}{=} 1,263.75$ ∎

Notice the striking similarity between Examples 5 and 6 and the value problems we discussed in Chapter 2.

EXAMPLE 7

A man can paint a house in 20 hours, while his son can paint the same house in 40 hours. How long will it take them to paint the house if they work together?

Solution

In order to solve this problem algebraically we need to assume that father and son work in a totally cooperative manner and that they work at a constant rate.

Our basic strategy in a problem like this is to analyze what *part* of a job each person does.

Let x = # of hours it takes the father and son to paint the house working together.

If it takes the father 20 hours to paint the house, then he paints $\frac{1}{20}$ of the house in 1 hour, $\frac{3}{20}$ of the house in 3 hours, and $\frac{7}{20}$ of the house in 7 hours. Thus, he paints $\frac{x}{20}$ of the house in x hours.

Similarly, if his son takes 40 hours to paint the same house, then he paints $\frac{1}{40}$ of the house in 1 hour, or $\frac{x}{40}$ of the house in x hours.

If we add the portions of the job each individual completes, we should get 1 complete job. Thus,

$$\frac{x}{20} \quad + \quad \frac{x}{40} \quad = \quad 1$$

$$\underset{\substack{\textit{Portion completed} \\ \textit{by the father} \\ \textit{in x hours}}}{} \quad + \quad \underset{\substack{\textit{Portion completed} \\ \textit{by his son} \\ \textit{in x hours}}}{} \quad = \quad \underset{\substack{1 \textit{ complete} \\ \textit{job}}}{}$$

Multiply by the LCD, which is 40:

$$\frac{40}{1} \cdot \frac{x}{20} + \frac{40}{1} \cdot \frac{x}{40} = 40 \cdot 1$$

$$2x + x = 40$$

$$3x = 40$$

$$x = \frac{40}{3} = 13\tfrac{1}{3} \text{ hours} = \boxed{13 \text{ hours and } 20 \text{ minutes}}$$

The check is left to the student. ∎

EXAMPLE 8

A train can make a 480-km trip in the same time that a car can make a 320-km trip. If the train travels 40 kph faster than the car, how long does it take the car to make its trip?

Solution

We will use the relationship $d = r \cdot t$ in the form $t = \dfrac{d}{r}$.

Let r = rate for car.

Then $r + 40$ = rate for train.

Since we are told that the times for the train and car are the same, we have

$$t_{train} = t_{car}$$

$$\frac{d_{train}}{r_{train}} = \frac{d_{car}}{r_{car}}$$

Thus, our equation is

$$\frac{480}{r + 40} = \frac{320}{r} \qquad LCD = r(r + 40)$$

$$r(r + 40) \cdot \frac{480}{r + 40} = r(r + 40) \cdot \frac{320}{r}$$

$$480r = 320(r + 40)$$

$$480r = 320r + 12{,}800$$

$$160r = 12{,}800$$

$$r = 80$$

Keep in mind that the problem asks for the *time* it takes the car to make the 320-km trip, which is

$$t = \frac{320}{80} = \boxed{4 \text{ hours}}$$

CHECK: The rate of the train is $80 + 40 = 120$ kph.

The time for the train to make its 480-km trip is

$$t = \frac{480}{120} = 4 \text{ hours} \qquad \blacksquare$$

Exercises 4.7

Solve each of the following problems algebraically. That is, set up an equation or inequality and solve it. Be sure to label clearly what the variable represents.

1. If three-fourths of a number is 7 less than two-fifths of the number, what is the number?

2. If 5 more than five-sixths of a number is 3 less than two-thirds of the number, what is the number?

3. The ratio of men to women in a certain mathematics class is 7 to 9. If there are 810 women in the class, how many men are in the class?

4. In a certain town, the ratio of dogs to cats is 8 to 11. If there is a total of 2,812 cats and dogs, how many cats are there?

5. The ratio of two positive numbers is 5 to 12. Find the two numbers, if one number is 21 less than the other.

6. The ratio of two positive numbers is 7 to 9. Find the two numbers, if one number is 4 more than the other.

7. If 1 inch is 2.54 cm, how many inches are there in 52 cm? (Round your answer to two decimal places.)

8. If 1 kilogram is 2.2 lb, how many pounds are there in 17 kg? (Round your answer to two decimal places.)

9. On planet G, the units of currency are the droogs, the dreeps, and the dribbles. If 5 droogs are equivalent to 4 dreeps, and 7 dreeps are equivalent to 25 dribbles, how many dribbles are in 28 droogs?

10. On planet P, length is measured in wings, wongs, and wytes. If 14 wings are equivalent to 9 wytes, and 9 wongs are equivalent to 7 wings, how many wongs make a wyte?

11. The first side of a triangle is one-half the second side; the third side is two cm more than the second side. If the perimeter is 22 cm, find the lengths of the sides.

12. The width of a rectangle is three-fifths the length. If the perimeter is 80 inches, find the dimensions of the rectangle.

13. The perimeter of a rectangle is 50 cm. If its length is $2\frac{1}{2}$ times its width, find its dimensions.

14. The shortest side of a triangle is two-thirds its medium side, and the longest side is 5 feet longer than the shortest side. If the perimeter is $23\frac{2}{3}$ ft, find the length of each side.

15. Mike was walking home from the ballfield when one-quarter of the way home he decided to take a cab. If he was 6 miles from his home when he decided to take a cab, how far is his home from the ballfield?

16. Half the height of a radio tower is painted blue and $\frac{1}{5}$ the height is painted red. If 22 feet remain unpainted, how tall is the tower?

17. Valerie walks from her home to her school. One-fifth of the way there she finds a nickel. One-quarter of the rest of the way she finds a dime. If she is still 2 blocks from school when she finds the dime, how many blocks did she walk from her home to school? (Assume that all blocks are equal in length.)

18. Mike walks from his home to his school. One-fifth of the way there he finds a nickel. One-quarter of the rest of the way there he finds a dime. If he is still 3 blocks from school when he finds the dime, how far did he walk from his home to his school? (Assume that all blocks are equal in length.)

19. A law of physics states that if an electrical circuit has three resistors in parallel, then the reciprocal of the total resistance of the circuit is the sum of the reciprocals of the individual resistances. As a formula, we have:

$$\frac{1}{R} = \frac{1}{R_1} + \frac{1}{R_2} + \frac{1}{R_3}$$

where R_1, R_2, and R_3 are the individual resistances measured in ohms, and R is the total resistance measured in ohms (see the accompanying figure).

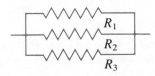

    In such a circuit, the total resistance is $1\frac{1}{4}$ ohms, and two of the resistors are 2 ohms and 5 ohms. What is the third resistance?

20. Einstein's theory of relativity states that velocities must be added according to the formula

$$v = \frac{v_1 + v_2}{1 + \frac{v_1 v_2}{c^2}}$$

where v is the resultant velocity, v_1 and v_2 are the velocities to be added, and c is the speed of light. If v_1 and v_2 are both one-half the speed of light, what is the resultant velocity?

21. Cindy has invested in two interest-bearing investments: a certificate of deposit that yields 11% interest per year, and a bond that yields 12% interest. She invested $3,000 more in the certificate than in the bond. If she receives a total of $905 in annual interest, how much is invested in each venture?

22. Nick invested his money in two money-market certificates: one yields 9% interest per year and the other, higher-risk, certificate yields 14% per year. He invested $3,520 more in the certificate yielding 9% than in the other certificate. How much did he invest in each certificate if he receives a total of $1,936 in annual interest from the two certificates?

23. Arthur has a total of $25,000 saved in two banks: one bank gives $5\frac{1}{2}$% yearly interest, while the other gives 7% interest. If his total yearly interest from both banks is $1,465, how much was saved in each bank?

24. Patrick has invested $15,000 in two bonds: one bond yields 9% annual interest and the other yields $8\frac{3}{4}$% annual interest. How much is invested in each certificiate if the combined yearly interest from both bonds is $1,340?

25. Lew wants to invest $18,000 in two bonds: one yields $8\frac{1}{2}$% annual interest and the other yields 11% annual interest. How much should he invest in each bond if he wants to receive a combined annual interest rate of 10%?

26. Joan saved $21,000 in two banks. One bank gives $7\frac{1}{2}$% annual interest, and the other gives 9% annual interest. How much did she save in each bank if she received a combined annual interest rate of 8% on her savings?

27. It takes Carol 3 hours to paint a small room and it takes Bill 5 hours to paint the same room. Working together, how long would it take for them to finish the room?

28. Pipe A can fill a pool with water in 3 days, and pipe B can fill the same pool in 2 days. If both pipes were used, how long would it take to fill the pool?

29. The Quickie Cleaning Service can clean a certain office building in 30 hours, while the Super-Quickie Cleaning Service can do the same job in 20 hours. If the two services work together, how long will it take to get the building clean?

30. The Echo brand candy corn machine can make enough candy corn to fill a supermarket order in 16 hours. On the other hand, the newer Echo 2 machine can make enough to fill the same order in 13 hours. How many hours would it take to make enough candy corn to fill the supermarket order if both machines were working together?

31. A bricklayer can complete a wall in $2\frac{2}{3}$ days, while his assistant can do the same size wall in 5 days. Working together, how long will it take for them to complete the wall?

32. A clerk can process 500 forms in $1\frac{2}{5}$ days. It takes $2\frac{1}{2}$ days for another clerk to process the same 500 forms. If both clerks work together, how long will it take them to process the 500 forms?

33. Repeat Exercise 29 if the Quickie Cleaning Service works alone for 10 hours and is then replaced by the Super-Quickie Cleaning Service.

34. Repeat Exercise 29 if the Super-Quickie Cleaning Service works alone for 10 hours and is then replaced by the Quickie Cleaning Service.

35. When a bathtub faucet is turned on (and the drain is shut), it can fill a tub in 10 minutes; when the drain is open, a full tub can *empty* in 15 minutes. How long would it take for the bathtub to fill if the water were turned on with the drain left open? [*Hint:* Let the rate at which the bathtub drains be negative.]

36. How long would it take for the bathtub of Exercise 35 to fill if a full tub can empty in 6 minutes when the drain is opened?

37. Bill can ride 10 kilometers in the same time that Jill can ride 15 kilometers. If Jill rides 10 kph faster than Bill, how fast does Bill ride?

38. A car can make a 200-mile trip in the same time that a bike can make a 60-mile trip. If the car travels 35 mph faster than the bike, how long does it take the bike to make its trip?

39. Marla drove 600 miles on an interstate highway. Her speed was 50 mph except for a part of the highway under construction, where her speed was 20 mph. If her total driving time was 14 hours, how many miles did she drive at the slower speed?

40. Susan decided to run to the store, which is 8 km from her house. She ran at a rate of 7 kph for part of the way and then walked at a rate of 3 kph the rest of the way. If the total trip took 2 hours, how many km did she run?

41. How many ounces of a 20% solution of alcohol must be mixed with 5 ounces of a 50% solution in order to get a mixture of 30% alcohol?

42. How many ounces of a 35% solution of sulfuric acid (and distilled water) must be mixed with 12 ounces of a 20% solution in order to get a 30% solution of sulfuric acid?

43. A chemist has a 30% solution of alcohol and a 75% solution of alcohol. How much of each should be mixed together in order to get 80 ml. of a 50% mixture?

44. A chemist has a 45% mixture of hydrochloric acid and an 80% mixture of the same acid. How much of each should be mixed together in order to get 60 ml of a 60% solution of hydrochloric acid?

45. Two alloys, one 40% iron and the other 60% iron, are to be melted down to form another alloy. How much of each alloy should be melted down and mixed to form 80 tons of an alloy which is 55% iron?

46. Two alloys, one 28% iron and the other 82% iron, are to be melted down to form another alloy. How much of each alloy should be melted down and mixed to form 108 tons of an alloy which is 40% iron?

47. Abner has 2 liters of a 60% solution of alcohol. How much pure alcohol must he add in order to have an 80% solution?

48. Susan has 2 liters of a 60% solution of alcohol. How much pure water should she add in order to have a 40% solution?

49. Jack's radiator has a 3-gallon capacity. His radiator is filled to capacity with a 30% mixture of antifreeze and water. He drains off some of the old solution and refills the radiator to capacity with pure water to get a 20% mixture. How much did he drain off?

50. Jack's car radiator has a 3-gallon capacity. His radiator is filled to capacity with a 30% mixture of antifreeze and water. He drains off some of the old solution and refills the radiator to capacity with pure antifreeze to get a 45% mixture. How much did he drain off?

51. Advance tickets were sold at a concert for $7.50 each. At the door, tickets were $10.25 each. How many advance tickets were sold if 3,600 tickets were sold, netting $30,850?

52. Children's tickets at a movie sold for $2.50, while adult tickets sold for $5.25. On a single weekday the movie *Spiders* took in $10.650. If 2,500 tickets were sold that day, how many of each type were sold?

53. Sam had a bunch of nickels, dimes, and quarters totaling $20. If he had five more dimes than nickels and twice as many quarters as nickels, how many of each coin did he have?

54. Jake had 205 coins, consisting of nickels, dimes, and quarters. If the total value of the coins was $22 and there were three times as many dimes as nickels, how many of each coin did he have?

55. Orchestra seating at a play sold for $12.50, balcony tickets sold for $10.25, and general admission tickets sold for $8.00. For a single showing, there were twice as many general admission tickets sold as orchestra tickets. If there were 900 tickets sold and the total gross for the showing was $8,662.50, how many of each ticket were sold?

56. April saved a total of $19,000 in three banks: The First National Bank, the First Federal Bank, and the Fidelity Bank. The First National Bank gave $7\frac{1}{2}\%$ yearly interest, First Federal gave $8\frac{1}{2}\%$ yearly interest, and Fidelity gave 9% yearly interest. April had twice as much money saved in First Federal as she had saved in Fidelity. How much did she have saved in each bank if she received $1,555 in yearly interest for all her savings combined?

57. Gerry's math teacher assigns course grades in the following way: Two exams are given, each worth 20% of the course grade; quizzes and homework combined together are worth 20% of the course grade; and the final exam is worth 40% of the course grade. Gerry received a grade of 85 on the first exam, 65 on the second exam, and had a quiz and homework combined average of 72. What score must he get on the final exam in order to receive a grade of at least 80 for the course?

58. Repeat Exercise 57 if Gerry received the same exam and quiz and homework scores but wanted to find the score he would need on the final exam in order to receive a grade of at least 70 for the course.

? QUESTION FOR THOUGHT

59. The distance from Philadelphia to State College is 200 miles. Joe decides he wants to travel the distance in 4 hours, and therefore, he figures that he must average 50 mph for the trip. His car develops problems, however, and so he could travel only 25 mph. Halfway there, the problem suddenly disappears. Joe now figures that in order for him to make the total trip within the 4-hour limit, he must now average 75 mph for the rest of the trip. What is wrong with this logic?

 MINI-REVIEW

60. *Simplify.* $5 - \{2x - 3[x - 2(x - 3)]\}$

61. *Factor completely.* $3x^3y - 6x^2y^2 + 3xy^3$

62. *Factor completely.* $(x - a)^2 - 9$

63. *Solve for x and graph the solution set.* $|2x + 8| \leq 10$

CHAPTER 4   Summary

After having completed Chapter 4, you should be able to:

1. Apply the fundamental principle of fractions to reduce fractions (Section 4.1).

   For example: $\dfrac{2x^2 - 8x}{x^2 - 16} = \dfrac{2x(x - 4)}{(x + 4)(x - 4)} = \dfrac{2x}{x + 4}$

2. Multiply and divide rational expressions (Section 4.2).

   For example:

   (a) $\dfrac{x^2 - 3x + 2}{4x - 8} \cdot \dfrac{8x}{x^3 - x^2} = \dfrac{(x - 2)(x - 1)}{4(x - 2)} \cdot \dfrac{8x}{x^2(x - 1)}$

   $$= \dfrac{(x - 2)(x - 1)}{4(x - 2)} \cdot \dfrac{\overset{2}{8x}}{\underset{x}{x^2(x - 1)}} = \dfrac{2}{x}$$

   (b) $\dfrac{x^3y - xy^3}{4x + 4y} \div (x^2 - xy) = \dfrac{x^3y - xy^3}{4x + 4y} \cdot \dfrac{1}{x^2 - xy}$ *Use rule for division.*

   $$= \dfrac{xy(x + y)(x - y)}{4(x + y)} \cdot \dfrac{1}{x(x - y)}$$

   $$= \dfrac{xy(x + y)(x - y)}{4(x + y)} \cdot \dfrac{1}{x(x - y)} = \dfrac{y}{4}$$

3. Find the least common denominator of several fractions (Section 4.3).

   For example: Find the LCD of

   $$\dfrac{1}{2x^2}, \qquad \dfrac{2}{x^2 + 2x}, \qquad \dfrac{3}{x^2 - 4}$$

   Factor each denominator: $2x^2$, $x(x + 2)$, $(x + 2)(x - 2)$

   The distinct factors are: 2, x, $x + 2$, $x - 2$

   The LCD is $2x^2(x + 2)(x - 2)$.

4. Use the fundamental principle to combine fractions with unlike denominators (Section 4.3).

   For example:

   $$\dfrac{5}{2x^2} - \dfrac{3}{x^2 + 2x} + \dfrac{2}{x^2 - 4}$$

   $$= \dfrac{5}{2x^2} - \dfrac{3}{x(x + 2)} + \dfrac{2}{(x + 2)(x - 2)} \qquad LCD = 2x^2(x + 2)(x - 2)$$

   $$= \dfrac{5(x + 2)(x - 2)}{2x^2(x + 2)(x - 2)} - \dfrac{3(2x)(x - 2)}{2x^2(x + 2)(x - 2)} + \dfrac{2(2x^2)}{2x^2(x + 2)(x - 2)}$$

   $$= \dfrac{5(x^2 - 4) - 6x(x - 2) + 4x^2}{2x^2(x + 2)(x - 2)}$$

   $$= \dfrac{5x^2 - 20 - 6x^2 + 12x + 4x^2}{2x^2(x + 2)(x - 2)}$$

   $$= \dfrac{3x^2 + 12x - 20}{2x^2(x + 2)(x - 2)}$$

5. Simplify complex fractions (Section 4.4).

For example:

$$\frac{\dfrac{1}{s} + \dfrac{6}{t}}{\dfrac{1}{s^2} - \dfrac{36}{t^2}} = \frac{\left(\dfrac{1}{s} + \dfrac{6}{t}\right)(s^2 t^2)}{\left(\dfrac{1}{s^2} - \dfrac{36}{t^2}\right)(s^2 t^2)}$$

Multiply numerator and denominator by the LCD of all simple fractions: $s^2 t^2$.

$$= \frac{st^2 + 6s^2 t}{t^2 - 36 s^2}$$

$$= \frac{st(t + 6s)}{(t - 6s)(t + 6s)}$$

$$= \frac{st}{t - 6s}$$

6. Use the LCD to solve fractional equations (Section 4.5).

For example: Solve for x.

$$\frac{6}{x + 3} - \frac{3}{x + 6} = \frac{4}{x + 3}$$

Multiply both sides of the equation by $(x + 3)(x + 6)$.

$$(x + 3)(x + 6) \cdot \frac{6}{x + 3} - (x + 3)(x + 6) \cdot \frac{3}{x + 6} = (x + 3)(x + 6) \cdot \frac{4}{x + 3}$$

$$6(x + 6) - 3(x + 3) = 4(x + 6)$$

$$6x + 36 - 3x - 9 = 4x + 24$$

$$3x + 27 = 4x + 24$$

$$3 = x$$

Be sure to check that the solution is valid.

CHECK:
$$\frac{6}{3 + 3} - \frac{3}{3 + 6} \overset{?}{=} \frac{4}{3 + 3}$$

$$\frac{6}{6} - \frac{3}{9} \overset{?}{=} \frac{4}{6}$$

$$1 - \frac{1}{3} \overset{\checkmark}{=} \frac{2}{3}$$

7. Solve a literal equation explicitly for a specified variable (Section 4.6).

For example: Solve for t.

$$3t - 7s = at - xs + 1$$

Collect t terms on one side and non-t terms on the other side of the equation.
*Subtract **at** from both sides.*

$$3t - 7s - at = at - xs + 1 - at$$

$$3t - at - 7s = -xs + 1$$

Add 7s to both sides.

$$3t - at - 7s + 7s = -xs + 1 + 7s$$

$$3t - at = 7s - xs + 1$$

Factor t from the left-hand side.

$$t(3 - a) = 7s - xs + 1$$

Divide both sides of the equation by $3 - a$.

$$t = \frac{7s - xs + 1}{3 - a} \qquad (a \neq 3)$$

8. Solve verbal problems which give rise to fractional equations (Section 4.7).

For example: A clerk can complete a job in 4 hours. If another clerk can complete the same job in 2 hours, how long would it take them to complete the job if they both worked together?

Let x be the number of hours it takes them to complete the job together. Then

$\dfrac{x}{4}$ represents the portion of the job completed by the first clerk in x hours.

$\dfrac{x}{2}$ represents the portion of the job completed by the second clerk in x hours.

Therefore, our equation is:

$\dfrac{x}{4} + \dfrac{x}{2} = 1$ *Multiply both sides of the equation by 4.*

$x + 2x = 4$

$3x = 4$

$x = \dfrac{4}{3} = 1\dfrac{1}{3}$

Hence, working together, it takes them $1\frac{1}{3}$ hours or 1 hour and 20 minutes to complete the job.

CHAPTER 4   Review Exercises

In Exercises 1–8, reduce to lowest terms.

1. $\dfrac{4x^2y^3}{16xy^5}$

2. $\dfrac{12a^2b^4}{16a^5bc^2}$

3. $\dfrac{x^2 + 2x - 8}{x^2 + 3x - 10}$

4. $\dfrac{6x^2 - 7x - 3}{(2x - 3)^2}$

5. $\dfrac{x^4 - 2x^3 + 3x^2}{x^2}$

6. $\dfrac{12a^3b^4 - 16ab^3 + 18a}{4a^2b}$

7. $\dfrac{5xa - 7a + 5xb - 7b}{3xa - 2a + 3xb - 2b}$

8. $\dfrac{8a^3 - b^3}{4a^2 + 2ab + b^2}$

In Exercises 9–50, perform the indicated operations. Express your answer in simplest form.

9. $\dfrac{4x^2y^3z^2}{12xy^4} \cdot \dfrac{24xy^5}{16xy}$

10. $\dfrac{4x^2y^3z^2}{12xy^4} \div \dfrac{24xy^5}{16xy}$

11. $\dfrac{5}{3x^2y} + \dfrac{1}{3x^2y}$

12. $\dfrac{3}{2a^2b} - \dfrac{3a}{2a^3b}$

13. $\dfrac{3x}{x - 1} + \dfrac{3}{x - 1}$

14. $\dfrac{2a}{a - 3} - \dfrac{6}{a - 3}$

15. $\dfrac{2x^2}{x^2 + x - 6} + \dfrac{2x}{x^2 + x - 6} - \dfrac{12}{x^2 + x - 6}$

16. $\dfrac{3a^2}{a^2 - a - 6} - \dfrac{3a}{a^2 - a - 6} - \dfrac{18}{a^2 - a - 6}$

17. $\dfrac{5}{3a^2b} - \dfrac{8}{4ab^4}$

18. $\dfrac{7}{3xy^2} + \dfrac{2}{5xy} - \dfrac{4}{30xy}$

19. $\dfrac{3x + 1}{2x^2} - \dfrac{3x - 2}{5x}$

20. $\dfrac{7x - 1}{3x} + \dfrac{5x + 2}{6x^2}$

21. $\dfrac{x - 7}{5 - x} + \dfrac{3x + 3}{x - 5}$

22. $\dfrac{3a + 1}{a - 4} + \dfrac{2a + 3}{4 - a}$

23. $\dfrac{x^2 + x - 6}{x + 4} \cdot \dfrac{2x^2 + 8x}{x^2 + x - 6}$

24. $\dfrac{a^2 + 2a - 15}{4a^2 + 8a} \cdot \dfrac{2a^2 + 4a}{a + 5}$

25. $\dfrac{a^2 - 2ab + b^2}{a + b} \div \dfrac{(a - b)^3}{a + b}$

26. $\dfrac{r^2 - rs - 2s^2}{2r - s} \div \dfrac{12r^2 - 24rs}{8r^2 - 4rs}$

27. $\dfrac{3x}{2x + 3} - \dfrac{5}{x - 4}$

28. $\dfrac{5a}{a - 1} + \dfrac{3a}{2a + 1}$

29. $\dfrac{3x - 2}{2x - 7} + \dfrac{5x + 2}{2x - 3}$

30. $\dfrac{7x + 1}{x + 3} - \dfrac{2x + 3}{x - 2}$

31. $\dfrac{5a}{a^2 - 3a} + \dfrac{2}{4a^3 + 4a^2}$

32. $\dfrac{3}{x^2 + 2x - 8} - \dfrac{5}{x^2 - x - 2}$

33. $\dfrac{5}{x^2 - 4x + 4} + \dfrac{3}{x^2 - 4}$

34. $\dfrac{3}{a^2 - 6a + 9} - \dfrac{5}{a^2 - 9}$

35. $\dfrac{2x}{7x^2 - 14x - 21} + \dfrac{2x}{14x - 42}$

36. $\dfrac{5}{x - y} + \dfrac{3}{y^2 - x^2}$

37. $\dfrac{5x}{x - 2} + \dfrac{3x}{x + 2} - \dfrac{2x + 3}{x^2 - 4}$

38. $\dfrac{2x}{x - 5} - \dfrac{3x}{x + 5} + \dfrac{x + 1}{x^2 - 25}$

39. $\left(\dfrac{2x + y}{5x^2y - xy^2}\right)\left(\dfrac{25x^2 - y^2}{10x^2 + 3xy - y^2}\right)\left(\dfrac{5x^2 - xy}{5x + y}\right)$

40. $\left(\dfrac{3a + 2b}{2a^2 + 3ab}\right)\left(\dfrac{2a^2 + ab - 3b^2}{9a^2 - 4b^2}\right)\left(\dfrac{3a^2b^2 - 2ab^3}{a - b}\right)$

41. $\dfrac{4x + 11}{x^2 + x - 6} - \dfrac{x + 2}{x^2 + 4x + 3}$

42. $\dfrac{2x + 1}{2x^2 + 13x + 15} - \dfrac{x - 1}{2x^2 - 3x - 9}$

43. $\dfrac{5x}{x^2 - x - 2} + \dfrac{4x + 3}{x^3 + x^2} - \dfrac{x - 6}{x^3 - 2x^2}$

44. $\dfrac{2r}{r^2 - rs - 2s^2} + \dfrac{2s}{r^2 - 3rs + 2s^2} + \dfrac{3}{r^2 - s^2}$

45. $\dfrac{4x^2 + 12x + 9}{8x^3 + 27} \cdot \dfrac{12x^3 - 18x^2 + 27x}{4x^2 - 9}$

46. $\dfrac{4x^2 - 12x + 9}{8x^3 - 27} \div \dfrac{4x^2 - 9}{12x^3 + 18x^2 + 27x}$

47. $4x \div \left(\dfrac{8x^2 - 8xy}{2ax + bx - 2ay - by} \div \dfrac{2ax + 2bx + 3ay + 3by}{2a^2 + 3ab + b^2}\right)$

48. $\left(\dfrac{8x^2 - 8xy}{2ax + bx - 2ay - by} \cdot \dfrac{2a^2 + 3ab + b^2}{2ax + 2bx + 3ay + 3by}\right) \div 4x$

49. $\left(\dfrac{x}{2} + \dfrac{3}{x}\right) \cdot \dfrac{x + 1}{x}$

50. $\left(\dfrac{x}{y} - \dfrac{y}{x}\right) \div \dfrac{x^2 - y^2}{x^2y^2}$

In Exercises 51–56, express the complex fraction as a simple fraction reduced to lowest terms.

51. $\dfrac{\dfrac{3x^2y}{2ab}}{\dfrac{9x}{16a^2}}$

52. $\dfrac{\dfrac{x}{x - y}}{\dfrac{y}{x - y}}$

53. $\dfrac{\dfrac{3}{a} - \dfrac{2}{a}}{\dfrac{5}{a}}$

54. $\dfrac{\dfrac{2}{x^2} + \dfrac{1}{x}}{\dfrac{4}{x^2} - 1}$

55. $\dfrac{\dfrac{3}{b + 1} + 2}{\dfrac{2}{b - 1} + b}$

56. $\dfrac{\dfrac{z}{z - 2} + \dfrac{1}{z + 1}}{\dfrac{3z}{z^2 - z - 2}}$

In Exercises 57–74, solve the equations or inequalities.

57. $\dfrac{x}{3} + \dfrac{x-1}{2} = \dfrac{7}{6}$

58. $\dfrac{x}{2} - \dfrac{1}{3} = \dfrac{2x+3}{6}$

59. $\dfrac{x}{5} - \dfrac{x+1}{3} < \dfrac{1}{3}$

60. $\dfrac{x}{2} - \dfrac{1}{3} \geq \dfrac{2x+3}{6}$

61. $\dfrac{5}{x} - \dfrac{1}{3} = \dfrac{11}{3x}$

62. $\dfrac{7}{x} - \dfrac{3}{2x} = 5$

63. $\dfrac{x+1}{3} - \dfrac{x}{2} > 4$

64. $\dfrac{2-x}{14} \leq \dfrac{x+1}{7} + 2$

65. $-\dfrac{7}{x} + 1 = -13$

66. $\dfrac{3}{x-1} + 4 = \dfrac{5}{x-1}$

67. $\dfrac{5}{x-2} - 1 = 0$

68. $\dfrac{2}{x-3} + 1 = 3$

69. $\dfrac{x-2}{5} - \dfrac{3-x}{15} > \dfrac{1}{9}$

70. $\dfrac{2x+1}{3} + \dfrac{x}{4} \leq \dfrac{5}{6}$

71. $\dfrac{7}{x-1} + 4 = \dfrac{x+6}{x-1}$

72. $\dfrac{3}{x-2} + 5 = \dfrac{1+x}{x-2}$

73. $\dfrac{4x+1}{x^2-x-6} = \dfrac{2}{x-3} + \dfrac{5}{x+2}$

74. $\dfrac{5}{x+3} + \dfrac{2}{x} = \dfrac{x-12}{x^2+3x}$

In Exercises 75–84, solve the equations explicitly for the given variable.

75. $5x - 3y = 2x + 7y$ for x

76. $5x - 3y = 2x + 7y$ for y

77. $3xy = 2xy + 4$ for y

78. $5ab - 2a = 3ab + 2b$ for b

79. $\dfrac{2x+1}{y} = x$ for y

80. $\dfrac{ax+b}{c} = a$ for a

81. $\dfrac{ax+b}{cx+d} = y$ for x

82. $\dfrac{2s+3}{3s-2} = r$ for s

83. $\dfrac{1}{a} + \dfrac{1}{b} + \dfrac{1}{c} = \dfrac{1}{d}$ for b

84. $\dfrac{a}{c} - \dfrac{b}{d} = e$ for c

Solve each of the following problems algebraically:

85. If 1 inch is 2.54 centimeters, how many inches are there in 1 centimeter?

86. The ratio of Democrats to Republicans in a district is 4 to 3. If there is a total of 1,890 Democrats and Republicans, how many Democrats are there?

87. Carol walked from her aunt's to her father's house. Halfway there she picked up her brother Arnold and one-third of the rest of the way there she picked up her sister Julie. The three of them walked a distance of $1\frac{1}{2}$ miles before arriving at their father's house. How far did Carol walk?

88. José takes the same amount of time to walk 3 miles as it takes Carlos to ride his bike 5 miles. If Carlos travels 5 miles per hour faster than José, what is Carlos' rate of speed?

89. It takes Charles $2\frac{1}{2}$ days to refinish a room and it takes Ellen $2\frac{1}{3}$ days to refinish the same room. How long would it take them to refinish the room if they worked together?

90. It takes Jerry twice as long as Sue to clean up the kitchen. If Sue can clean up the kitchen in $\frac{1}{3}$ hour, how long would it take to clean up the kitchen if she works together with Jerry?

91. How much of a 35% solution of alcohol should a chemist mix with 5 liters of a 70% alcohol solution in order to get a solution which is 60% alcohol?

92. How much pure water should be added to 3 liters of a 65% solution of sulfuric acid in order to dilute it to 30%?

93. Children's tickets at a movie theater sold for $1.50, while adult tickets sold for $4.25. On a single weekday the movie *Fred* took in $3,010. If 980 tickets were sold that day, how many of each type were sold?

94. Sam has eighty-two coins in pennies, nickels, and dimes, totaling $3.32. If she has 4 times as many nickels as dimes, how many of each coin does she have?

CHAPTER 4   Practice Test

1. Express the following in simplest form.

(a) $\dfrac{24x^2y^4}{64x^3y}$

(b) $\dfrac{x^2 - 9}{x^2 - 6x + 9}$

(c) $\dfrac{6x^3 - 9x^2 - 6x}{5x^3 - 10x^2}$

2. Perform the indicated operations and express your answer in simplest form.

(a) $\dfrac{4xy^3}{5ab^4} \cdot \dfrac{15}{16x^4y^5}$

(b) $\dfrac{3x}{18y^2} + \dfrac{5}{8x^2y}$

(c) $\dfrac{r^2 - rs - 2s^2}{2s^2 + 4rs} \div \dfrac{r - 2s}{4s^2 + 8rs}$

(d) $\dfrac{9x - 2}{4x - 3} + \dfrac{x + 4}{3 - 4x}$

(e) $\dfrac{3x}{x^2 - 4} + \dfrac{4}{x^2 - 5x + 6} - \dfrac{2x}{x^2 - x - 6}$

(f) $\left(\dfrac{3}{x} - \dfrac{2}{x + 1} \right) \div \dfrac{1}{x + 1}$

3. Express as a simple fraction reduced to lowest terms.

$$\dfrac{\dfrac{3}{x + 1} - 2}{\dfrac{5}{x} + 1}$$

4. Solve the following:

(a) $\dfrac{2x - 4}{6} - \dfrac{5x - 2}{3} < 5$

(b) $\dfrac{3}{x - 8} = 2 - \dfrac{5 - x}{x - 8}$

(c) $\dfrac{3}{2x + 1} + \dfrac{4}{2x - 1} = \dfrac{29}{4x^2 - 1}$

5. Solve explicitly for x:

$$y = \dfrac{x - 2}{2x + 1}$$

6. How much of a 30% alcohol solution must be mixed with 8 liters of a 45% solution in order to get a mixture which is 42% alcohol?

7. It takes Jackie $3\frac{1}{2}$ hours to complete a job and it takes 2 hours for Eleanor to do the same job. How long would it take for them to complete the job working together?

5

Exponents

Although we have dealt with exponents in previous chapters, in this chapter we will engage in a formal treatment of exponents. Beginning with the definition of *natural number* exponents, we will discuss the rules for multiplying and dividing simple expressions. Then we will examine what happens when we allow exponents to take on integer and rational values.

5.1

Natural Number Exponents

First, recall the definition of exponential form:

$$x^n = x \cdot x \cdot x \cdot \cdots \cdot x \qquad \text{where the factor } x \text{ occurs } n \text{ times in the product.}$$

Observe that for this definition to make sense, n must be a natural number. (How can you have -3 factors of x?) With this definition of exponents, and the suitable properties of the real number system, we will state the first three rules of natural number exponents:

Rules of Natural Number Exponents

1. $x^n \cdot x^m = x^{n+m}$
2. $(x^n)^m = x^{nm}$
3. $(xy)^n = x^n y^n$

As we stated in Chapter 1, rule 1 is a matter of counting x's:

$$x^n \cdot x^m = \underbrace{(x \cdot x \cdot x \cdot \cdots \cdot x)}_{n \text{ times}}\underbrace{(x \cdot x \cdot x \cdot \cdots \cdot x)}_{m \text{ times}} = \underbrace{x \cdot x \cdot x \cdot x \cdot \cdots \cdot x}_{n + m \text{ times}} = x^{n+m}$$

Rule 2 is derived mainly by applying rule 1:

$$(x^n)^m = \underbrace{x^n \cdot x^n \cdot \cdots \cdot x^n}_{x^n \text{ occurs } m \text{ times in the product.}}$$

$$= \overbrace{x^{n+n+\cdots+n}} \qquad \textit{By rule 1 the exponent n is added m times.}$$

$$= x^{nm}$$

Rule 1 and rule 2 are often confused (when do I add exponents and when do I multiply?). Keep the differences in mind: Rule 1 states that when *two powers of the same base are to be multiplied*, the exponents are *added*; rule 2 states that when *a power is raised to a power*, the exponents are *multiplied*.

EXAMPLE 1   Perform the indicated operations and simplify.

(a) $x^7 \cdot x^6$ (b) $3^8 \cdot 3^2$ (c) $(x^7)^6$

(d) $(4^6)^3$ (e) $(r^2)^5(r^3)^9$

Solution

(a) $x^7 \cdot x^6$ *Apply rule 1.*

$= x^{7+6} = \boxed{x^{13}}$

(b) $3^8 \cdot 3^2$ *Apply rule 1.*

$= 3^{8+2} = \boxed{3^{10}}$ *Note that the answer is **not** 9^{10}.*

(c) $(x^7)^6$ *Apply rule 2.*

$= x^{7 \cdot 6} = \boxed{x^{42}}$

(d) $(4^6)^3$ *Apply rule 2.*

$= 4^{6 \cdot 3} = \boxed{4^{18}}$

(e) $(r^2)^5(r^3)^9$ *Apply rule 2.*

$= r^{10} \cdot r^{27}$ *Apply rule 1.*

$= \boxed{r^{37}}$

■

Rule 3 is derived mainly from the associative and commutative properties of multiplication:

$$(xy)^n = \underbrace{(xy) \cdot (xy) \cdot \cdots \cdot (xy)}_{n \text{ times}}$$

$$= (\underbrace{x \cdot x \cdot \cdots \cdot x}_{n \text{ times}})(\underbrace{y \cdot y \cdot \cdots \cdot y}_{n \text{ times}})$$ *Reorder and regroup by the associative and commutative properties of multiplication.*

$$= x^n y^n$$

EXAMPLE 2

Perform the indicated operations and simplify.

(a) $(3xy)^4$ **(b)** $(a^4 b^5)^7$

(c) $(-3xy^2)^4(-2x^2y^3)^5$ **(d)** $(x^2 + y^3)^2$

Solution

(a) $(3xy)^4$ *Apply rule 3.*

$= 3^4 x^4 y^4$ *Note that each **factor** is raised to the 4th power.*

$= \boxed{81x^4 y^4}$

(b) $(a^4 b^5)^7$ *First apply rule 3.*

$= (a^4)^7 (b^5)^7$ *Since a^4 and b^5 are factors of $a^4 b^5$ we can raise each to the 7th power and then use rule 2.*

$= \boxed{a^{28} b^{35}}$

(c) $(-3xy^2)^4(-2x^2y^3)^5$ *First apply rule 3.*

$$= (-3)^4x^4(y^2)^4(-2)^5(x^2)^5(y^3)^5$$

Note that the sign is part of the coefficient and is raised to the given power. Then apply rule 2.

$$= 81x^4y^8(-32)x^{10}y^{15} \qquad \text{Apply rule 1.}$$

$$= \boxed{-2{,}592x^{14}y^{23}}$$

(d) $(x^2 + y^3)^2 = (x^2 + y^3)(x^2 + y^3)$

$$= \boxed{x^4 + 2x^2y^3 + y^6}$$

*We cannot apply rule 2 for exponents in this case since rule 2 applies only to **factors** and **not terms**. We use polynomial multiplication instead.* ∎

Up to now we have concentrated on developing rules for multiplication. Let's now examine division. We consider the expression $\dfrac{x^n}{x^m}$ (we assume $x \neq 0$):

CASE (i): $n > m$

$$\frac{x^n}{x^m} = \frac{\overbrace{x \cdots \cdot x}^{n\ times}}{\underbrace{x \cdots \cdot x}_{m\ times}} = \frac{\overbrace{\cancel{x} \cdots \cdot \cancel{x}}^{m\ factors\ reduced} \cdot \overbrace{x \cdots \cdot x}^{\substack{n-m\ factors \\ left}}}{\underbrace{\cancel{x} \cdots \cdot \cancel{x}}_{\substack{m\ factors \\ reduced}}} = \overbrace{x \cdot x \cdots \cdot x}^{n-m\ factors} = x^{n-m}$$

CASE (ii): $n < m$

$$\frac{x^n}{x^m} = \frac{\overbrace{x \cdots \cdot x}^{n\ times}}{\underbrace{x \cdots \cdot x}_{m\ times}} = \frac{\overbrace{\cancel{x} \cdots \cdot \cancel{x}}^{\substack{n\ factors \\ reduced}}}{\underbrace{\cancel{x} \cdots \cdot \cancel{x}}_{\substack{n\ factors \\ reduced}} \cdot \underbrace{x \cdots \cdot x}_{\substack{m-n\ factors \\ left}}} = \frac{1}{\underbrace{x \cdot x \cdots \cdot x}_{m-n\ factors}} = \frac{1}{x^{m-n}}$$

CASE (iii): $n = m$

Since $n = m$, we have $x^n = x^m$.

Hence, $\dfrac{x^n}{x^m}$ is a fraction with the numerator identical to the denominator.

Thus,

$$\frac{x^n}{x^m} = \frac{x^n}{x^n} = 1$$

We summarize this discussion as follows:

Rule 4 of Natural Number Exponents	If $x \neq 0$, $$\frac{x^n}{x^m} = \begin{cases} x^{n-m} & \text{if } n > m \\ \dfrac{1}{x^{m-n}} & \text{if } n < m \\ 1 & \text{if } n = m \end{cases}$$

We had to split rule 4 into three cases in order to ensure natural number values for the exponents. In the next section, when we allow integer values for exponents, we will consolidate these three cases into one.

The next and final rule for natural number exponents is a result of the associative and commutative properties of rational multiplication.

We consider the expression $\left(\dfrac{x}{y}\right)^n$.

$$\left(\frac{x}{y}\right)^n = \overbrace{\left(\frac{x}{y}\right)\left(\frac{x}{y}\right) \cdots \cdots \left(\frac{x}{y}\right)}^{n \text{ times}} = \frac{\overbrace{x \cdot x \cdots \cdots x}^{n \text{ times}}}{\underbrace{y \cdot y \cdots \cdots y}_{n \text{ times}}} = \frac{x^n}{y^n}$$

Thus, we have:

Rule 5 of Natural Number Exponents	If $y \neq 0$, $$\left(\frac{x}{y}\right)^n = \frac{x^n}{y^n}$$

For convenience, we assume that all values of the variable are nonzero real numbers.

EXAMPLE 3

Perform the indicated operations and simplify.

(a) $\dfrac{x^7}{x^4}$ (b) $\dfrac{x^4}{x^7}$

(c) $\dfrac{c^4}{c^4}$ (d) $\left(\dfrac{r^5}{s^2}\right)^4$

Solution | These problems are straightforward applications of rules 4 and 5.

(a) $\dfrac{x^7}{x^4}$ *Apply rule 4, case* (i).

$$= x^{7-4} = \boxed{x^3}$$

(b) $\dfrac{x^4}{x^7}$ *Apply rule 4, case* (ii).

$$= \dfrac{1}{x^{7-4}} = \boxed{\dfrac{1}{x^3}}$$

(c) $\dfrac{c^4}{c^4} = \boxed{1}$

(d) $\left(\dfrac{r^5}{s^2}\right)^4 = \dfrac{(r^5)^4}{(s^2)^4}$ *Rule 5. Then apply rule 2.*

$$= \boxed{\dfrac{r^{20}}{s^8}}$$

∎

Let's examine problems requiring us to combine the use of several of the rules of natural number exponents.

EXAMPLE 4 | Perform the indicated operations and simplify.

(a) $\left(\dfrac{a^2 a^5}{a^6}\right)^4$

(b) $\dfrac{(3a^2)^4(2a)^2}{(-2a^3)^5}$

(c) $\left(\dfrac{2r^2}{s^3}\right)^{10}\left(\dfrac{s^4}{4r}\right)^6$

Solution | **(a)** We will show two ways we can approach this problem.

First applying rule 1 inside parentheses:

$$\left(\dfrac{a^2 a^5}{a^6}\right)^4 = \left(\dfrac{a^7}{a^6}\right)^4 \quad \text{\textit{Then rule 4}}$$

$$= (a)^4$$

$$= \boxed{a^4}$$

Applying rule 5 first:

$$\left(\dfrac{a^2 a^5}{a^6}\right)^4 = \dfrac{(a^2 a^5)^4}{(a^6)^4} \quad \text{\textit{Then rule 3}}$$

$$= \dfrac{(a^2)^4(a^5)^4}{(a^6)^4} \quad \text{\textit{Then rule 2}}$$

$$= \dfrac{a^8 a^{20}}{a^{24}} \quad \text{\textit{Then rule 1}}$$

$$= \dfrac{a^{28}}{a^{24}} \quad \text{\textit{Then rule 4}}$$

$$= \boxed{a^4}$$

(b) $\dfrac{(3a^2)^4(2a)^2}{(-2a^3)^5} = \dfrac{3^4(a^2)^4 2^2 a^2}{(-2)^5(a^3)^5}$ *Rule 3. Then apply rule 2.*

$= \dfrac{3^4 a^8 2^2 a^2}{(-2)^5 a^{15}}$ *Then rule 1*

$= \dfrac{3^4 2^2 a^{10}}{(-2)^5 a^{15}}$ *Then rule 4*

$= \dfrac{(81)(4)}{-32 a^5}$ *Reduce.*

$= \boxed{-\dfrac{81}{8a^5}}$

(c) $\left(\dfrac{2r^2}{s^3}\right)^{10}\left(\dfrac{s^4}{4r}\right)^6 = \dfrac{2^{10} r^{20}}{s^{30}}\dfrac{s^{24}}{4^6 r^6}$

$= \dfrac{2^{10} r^{20} s^{24}}{s^{30} 4^6 r^6}$

$= \dfrac{2^{10} r^{14}}{s^6 4^6}$ *Rather than evaluate 2^{10} or 4^6, rewrite 4 as 2^2.*

$= \dfrac{2^{10} r^{14}}{s^6 (2^2)^6}$

$= \dfrac{2^{10} r^{14}}{s^6 2^{12}}$

$= \dfrac{r^{14}}{2^2 s^6}$

$= \boxed{\dfrac{r^{14}}{4s^6}}$

Exercises 5.1

Perform the indicated operations and simplify. Assume that all variables represent nonzero real numbers.

1. $(x^2 x^5)(x^3 x)$

2. $(a^2 b^5)(a^4 b^7)$

3. $(-2a^2 b^3)(3a^5 b^7)$

4. $(-4r^3 s)(-2rst^3)$

5. $(a^2)^5$

6. $(y^3)^8$

7. $\left(\dfrac{2}{3}\right)^2$

8. $\left(\dfrac{3}{2}\right)^3$

9. $(x + y^3)^2$

10. $(a^2 b)^2$

11. $(xy^3)^2$

12. $(a^2 + b)^2$

13. $(2^3 \cdot 3^2)^2$

14. $(5^2 \cdot 2^3)^2$

15. $(x^4 y^3)^5 (x^3 y^2)^2$

16. $(a^2 a^3)^5 (a^3 a)^6$

17. $(-2a^2)^3 (ab^2)^4$

18. $(r^2 s^3)^2 (-rs^4)^3$

19. $(r^2 st)^3 (-2rs^2 t)^4$

20. $(-2x^2 y)^2 (2xy^3)^3$

21. $(-2ab)^4 (-3a^2 b)^2$

22. $(-3ab^2 c)^3 (-3abc^3)^2$

23. $\dfrac{x^5}{x^2}$

24. $\dfrac{x^2}{x^5}$

25. $\dfrac{x^3 y^2}{xy^4}$

26. $\dfrac{a^7 b^4}{a^7 b^4}$

27. $\dfrac{5^4 \cdot 2^2}{25^2 \cdot 4^2}$

28. $\dfrac{2^2 \cdot 3^4}{9^2 \cdot 4^3}$

29. $\dfrac{a^5 b^9 c}{a^4 b c^5}$

30. $\dfrac{2^2 x^2 y^3}{2^4 xy^3}$

31. $\dfrac{(-3)^2 xy^4}{-3^2 xy^5}$

32. $\dfrac{(-2)^2 xy^5}{-2^2 xy^5}$

33. $\dfrac{3^2(-2)^3}{(-9^2)(-4)^2}$

34. $\dfrac{(-9)^2(-4)^3}{-3^3 \cdot 2^4}$

35. $\dfrac{(x^2 y)^2 (x^3 y^4)^3}{(x^2 y^2)^3}$

36. $\dfrac{(a^2 b)^2 (ab^2)^3}{a^2 b^4}$

37. $\left(\dfrac{y^5}{y^8}\right)^3$

38. $\left(\dfrac{x^5}{x^4}\right)^5$

39. $\left(\dfrac{y^2 y^7}{y^4}\right)^3$

40. $\left(\dfrac{x^5}{x^3 x^9}\right)^4$

41. $\dfrac{(3r^2 s)^3(-2rs^2)^4}{(-18rs)^2}$

42. $\dfrac{(2xy^2)^3(-2xy)^2}{(-2x)^3}$

43. $\left(\dfrac{xy^2 x^4}{x^2 y^4}\right)^5$

44. $\left(\dfrac{ab^2 c^3}{a^4 b^5}\right)^4$

45. $\left(\dfrac{2x^2 y^3}{xy^4}\right)^2 \left(\dfrac{3xy^2}{6}\right)^3$

46. $\left(\dfrac{x^2 x^3}{x}\right)^4 \left(\dfrac{x^2 x^5}{x^3}\right)^2$

47. $\left(\dfrac{(6ab^2)^2}{-3ab}\right)^3$

48. $\left(\dfrac{(-5b^2 c)^2}{-10b^2 c^3}\right)^3$

49. $\left(\dfrac{-3r^2 s^3}{4rs^2}\right)^3 (-2rs^2)^3$

50. $\left(\dfrac{-2a^2 b^3}{ab}\right)^3 (-3xy^2)^3$

51. $\left(\dfrac{(2x^2 y)^2(3xy)^3}{xy}\right)^2$

52. $\left(\dfrac{(3ab^2)^3(5a^2 b)}{9ab}\right)^3$

53. $(-4x^2 y)^3(-2xy^3)^2(-3)^2$

54. $(-3ab^2)^3(-2ab)^2(-2a^2 b)^3$

55. $[(r^3 s^2)^3(rs^2)^4]^2$

56. $[(-3rs^2)^2(-2st^2)^3]^3$

57. $\left(\dfrac{(-8bc^3)^2(4a^2 b)}{(-32ac^2)^3}\right)^2$

58. $\left(\dfrac{(-5xy^2 z)^2(5xyz^2)^3}{125xy^2 z}\right)^2$

? QUESTIONS FOR THOUGHT

59. Discuss what is **wrong** (if anything) in each of the following.
    (a) $x^2 x^3 \overset{?}{=} x^6$
    (b) $(x^2)^3 \overset{?}{=} x^5$
    (c) $(x^2 + y^2)^3 \overset{?}{=} x^6 + y^6$
    (d) $\dfrac{6x^6}{2x^2} \overset{?}{=} 3x^3$

60. Fill in the operation(s) that makes the following true for all n:

$$(x \; ? \; y)^n = x^n \; ? \; y^n$$

61. Why do we have to separate rule 4 of natural number exponents into cases?

62. Suppose rule 4 were stated simply as

$$\frac{x^n}{x^m} = x^{n-m} \quad (x \neq 0)$$

What would you then get as an answer for each of the following?

(a) $\dfrac{x^2}{x^5}$

(b) $\dfrac{x^3}{x^3}$

MINI-REVIEW

63. *Evaluate.* $-3 - (-4) + (-5) - (-8)$

64. *Evaluate.* $-3 - (-2)^2 - (4 - 5)$

65. *Perform the operations and simplify.*

$$\frac{4}{x - 2} - \frac{2x}{x - 2}$$

66. *Perform the operations and simplify.*

$$\frac{(x - 4)(x + 2)}{x^3 + 8} \cdot \frac{x + 4}{x^2 - 16}$$

67. The first side of a triangle is twice the length of the second side; the third side is 24 inches. Find the length of the first and second sides if the perimeter is 75 inches.

68. If a kilogram is 2.2 lb, how many kilograms are there in 42 lb?

5.2

Integer Exponents

We originally defined x^n as being the product of n x's. Based on that definition, it does not make sense for n to be negative or 0. Hence, our exponents were natural numbers.

In this section, we would like to allow our exponents to take on any *integer* values, and determine what it means when exponents have nonpositive integer values. This is called extending the definition of exponents to the integers.

In previous discussions, our understanding of the definition of a positive integer exponent led quite naturally to the five rules for exponents. In order to define negative and zero exponents, we will work the other way around. Since we want the exponent rules developed in the last section to continue to apply, we assume that the exponent rules developed for natural number exponents will still be valid for integer exponents. We then will use these rules to show us how to define integer exponents.

Let's assume x^0 exists and $x \neq 0$; see what happens when we find the product $x^0 \cdot x^n$ by applying the first rule of exponents:

$$x^0 \cdot x^n = x^{0+n} \qquad \textit{Rule 1}$$
$$= x^n$$

Hence, $x^0 \cdot x^n = x^n$. Verbally stated, multiplying an expression by x^0 ($x \neq 0$) does not change the expression. Thus, x^0 has the same property as 1, the multiplicative identity. Since 1 is the only number that has this property, it would be convenient to have $x^0 = 1$. We therefore make the following definition:

DEFINITION

A zero exponent is defined as follows:

$$x^0 = 1 \qquad (x \neq 0)$$

Note that 0^0 *is* **undefined**.

For example (assume all variables are nonzero real numbers):

$$5^0 = 1$$

$$(xy^3)^0 = 1$$

$$(-165)^0 = 1 \qquad \textit{Even if the base is negative, raising to the zero power still yields } +1.$$

$$\left(\frac{x^2y^5z^2}{xy^4z^9}\right)^0 = 1 \qquad \textit{Think before simplifying an expression.}$$

Now let's examine x^{-n} when n is a natural number and $x \neq 0$. What does it mean to have a negative exponent? We assume x^{-n} exists and will examine what happens when we find the product $x^n \cdot x^{-n}$ by applying rule 1 again:

$$x^{-n} \cdot x^n = x^{-n+n} \qquad \textit{Rule 1}$$

$$= x^0$$

$$= 1 \qquad \textit{Definition of zero exponent}$$

Hence, $x^{-n} \cdot x^n = 1$.

Dividing both sides of the equation above by x^n, we get

$$x^{-n} = \frac{1}{x^n}$$

We therefore make the following definition:

DEFINITION

Negative exponents are defined as follows:

$$x^{-n} = \frac{1}{x^n} \qquad (x \neq 0)$$

Verbally stated, an expression with a negative exponent is the reciprocal of the same expression with the exponent made positive. Thus,

$$x^{-3} = \frac{1}{x^3} \qquad 5^{-4} = \frac{1}{5^4} \qquad 10^{-5} = \frac{1}{10^5}$$

Let's examine what happens when there is a negative exponent in the denominator of a fraction:

$$\frac{1}{x^{-n}} = \frac{1}{\dfrac{1}{x^n}} \qquad \textit{Definition of negative exponents}$$

$$= 1 \cdot \frac{x^n}{1} \qquad \textit{Rule for dividing fractions}$$

$$= x^n$$

Hence, $\quad \dfrac{1}{x^{-n}} = x^n$

Thus, rewriting both rules as

$$\frac{x^{-n}}{1} = \frac{1}{x^n} \qquad \text{and} \qquad \frac{1}{x^{-n}} = \frac{x^n}{1}$$

we find that *changing the sign of the exponents of a power changes the power into its reciprocal.*

For example:

$$\frac{1}{x^{-4}} = \frac{x^4}{1} = x^4 \qquad \frac{1}{2^{-2}} = \frac{2^2}{1} = 4$$

$$x^{-2} = \frac{1}{x^2} \qquad \frac{1}{(-2)^{-3}} = (-2)^{+3} = -8$$

Do not confuse the sign of an exponent with the sign of the base. For example, -2 is a number 2 units to the left of 0 on the number line, whereas $2^{-1} = \frac{1}{2}$ is a positive number, one-half unit to the right of 0, as shown in the figure:

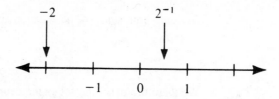

It is important to note that our definitions of zero and negative exponents are consistent with the five exponent rules we have been using, so we are free to use the exponent rules for expressions involving both positive and negative integer exponents.

Since we need not be concerned if the exponents are integers, we can now consolidate rule 4 for natural number exponents.

Rule 4 for Integer Exponents	$\dfrac{x^n}{x^m} = x^{n-m} \qquad (x \neq 0)$

This rule is consistent with our previous discussion regarding integer exponents. For example,

According to natural number exponent rule 4:

$$\frac{x^3}{x^3} = 1$$

$$\frac{x^2}{x^6} = \frac{1}{x^{6-2}} = \frac{1}{x^4}$$

According to integer exponent rule 4:

$$\frac{x^3}{x^3} = x^{3-3} = x^0 = 1$$

$$\frac{x^2}{x^6} = x^{2-6} = x^{-4}$$

$$= \frac{1}{x^4} \qquad \textit{By definition of negative exponents}$$

Definitions and Rules for Integer Exponents

Rule 1: $\quad x^n x^m = x^{n+m}$

Rule 2: $\quad (x^n)^m = x^{nm}$

Rule 3: $\quad (xy)^n = x^n y^n$

Rule 4: $\quad \dfrac{x^n}{x^m} = x^{n-m} \qquad (x \neq 0)$

Rule 5: $\quad \left(\dfrac{x}{y}\right)^n = \dfrac{x^n}{y^n} \qquad (y \neq 0)$

Definition: $\quad a^0 = 1 \qquad (a \neq 0)$

Definition: $\quad a^{-n} = \dfrac{1}{a^n} \qquad (a \neq 0)$

The only difference between the rules for integer and natural number exponents is the consolidation of rule 4.

Let's now examine some expressions involving integer exponents.

EXAMPLE 1

Perform the indicated operations and simplify. Express your answers with positive exponents only.

(a) $\quad y^{-4}y^8y^{-6}$ **(b)** $\quad \dfrac{x^{-6}}{x^{-8}}$ **(c)** $\quad (x^{-4})^{-6}$

(d) $\quad (a^{-2}b^3c^{-2})^2$ **(e)** $\quad \left(\dfrac{r^{-3}}{r^0}\right)^{-2}$

Solution

We *could* simplify all these expressions by using the definition of negative exponents and changing all negative exponents to positive exponents (substituting in reciprocals). For example:

(a) $\quad y^{-4}y^8y^{-6} = \left(\dfrac{1}{y^4}\right)(y^8)\left(\dfrac{1}{y^6}\right) \qquad$ *Definition of negative exponents*

$\qquad\qquad = \dfrac{1}{y^4} \cdot \dfrac{y^8}{1} \cdot \dfrac{1}{y^6}$

$\qquad\qquad = \dfrac{y^8}{y^4 y^6} \qquad$ *Apply rule 1 for **natural number exponents**.*

$\qquad\qquad = \dfrac{y^8}{y^{10}} \qquad$ *Apply rule 4 for **natural number exponents**.*

$\qquad\qquad = \boxed{\dfrac{1}{y^2}}$

A more efficient method would be to apply the rules for integer exponents first, and then use the definition, if needed, to express the final answer with positive exponents.

ALTERNATE SOLUTION FOR PART (a):

$y^{-4}y^8y^{-6}$ *Apply rule 1 for **integer exponents**.*

$\quad = y^{-4+8+(-6)}$

$\quad = y^{-2}$ *Apply the definition of negative exponents.*

$\quad = \boxed{\dfrac{1}{y^2}}$

When simplifying an expression with exponents it is easier, in general, to use the exponent rules rather than using the definition.

(b) $\dfrac{x^{-6}}{x^{-8}}$ *Apply rule 4 for **integer exponents**.*

$\quad = x^{-6-(-8)}$ *Note: Numerator exponent minus denominator exponent*

$\quad = x^{-6+8}$ *Remember that you are subtracting a negative exponent.*

$\quad = \boxed{x^2}$

(c) $(x^{-4})^{-6} = x^{(-6)(-4)} = \boxed{x^{24}}$ *Rule 2*

(d) $(a^{-2}b^3c^{-2})^2$ *Apply rule 3.*

$\quad = (a^{-2})^2(b^3)^2(c^{-2})^2$ *Rule 2*

$\quad = a^{-4}b^6c^{-4}$ *Apply the definition of negative exponents.*

$\quad = \left(\dfrac{1}{a^4}\right)(b^6)\left(\dfrac{1}{c^4}\right)$

$\quad = \boxed{\dfrac{b^6}{a^4c^4}}$

(e) $\left(\dfrac{r^{-3}}{r^0}\right)^{-2} = \left(\dfrac{r^{-3}}{1}\right)^{-2}$ *Definition of zero exponent*

$\quad = (r^{-3})^{-2}$ *Rule 2*

$\quad = \boxed{r^6}$ ∎

EXAMPLE 2

Perform the indicated operations and simplify. Express your answers with positive exponents only.

(a) $\dfrac{c^{-4}d^6}{c^5d^{-3}}$

(b) $[(x^{-3} \cdot x^{-2})^{-2}(y^{-2})^3]^{-4}$

(c) $\left(\dfrac{x^0x^{-3}x^2}{x^{-4}x}\right)^{-2}$

(d) $\dfrac{(2a^{-2}b)^{-3}(3ab)^{-2}}{(6a)^{-4}}$

(e) $\dfrac{9^{-3} \cdot 3^{-4}}{3^{-2} \cdot 9^2}$

Solution

(a)

$$\frac{c^{-4}d^6}{c^5 d^{-3}} = c^{-4-5}d^{6-(-3)} \qquad \textit{Rule 4}$$

$$= c^{-9}d^9$$

$$= \boxed{\frac{d^9}{c^9}}$$

(b) $[(x^{-3} \cdot x^{-2})^{-2}(y^{-2})^3]^{-4} \qquad \textit{Rule 1}$

$$= [(x^{-5})^{-2}(y^{-2})^3]^{-4} \qquad \textit{Rule 2}$$

$$= (x^{10}y^{-6})^{-4} \qquad \textit{Rule 3 and Rule 2}$$

$$= x^{-40}y^{24} \qquad \textit{Definition of negative exponents}$$

$$= \boxed{\frac{y^{24}}{x^{40}}}$$

(c) We will demonstrate two approaches for part **(c)**. First we simplify inside parentheses using rule 1:

$$\left(\frac{x^0 x^{-3} x^2}{x^{-4}x}\right)^{-2} = \left(\frac{x^{0-3+2}}{x^{-4+1}}\right)^{-2}$$

$$= \left(\frac{x^{-1}}{x^{-3}}\right)^{-2} \qquad \textit{Rule 4}$$

$$= (x^{-1-(-3)})^{-2}$$

$$= (x^2)^{-2} \qquad \textit{Rule 2}$$

$$= x^{-4}$$

$$= \boxed{\frac{1}{x^4}}$$

Or we could first apply rule 5:

$$\left(\frac{x^0 x^{-3} x^2}{x^{-4}x}\right)^{-2} \qquad \textit{Rule 5}$$

$$= \frac{(x^0 x^{-3} x^2)^{-2}}{(x^{-4}x)^{-2}} \qquad \textit{Rule 3 and rule 2}$$

$$= \frac{x^0 x^6 x^{-4}}{x^8 x^{-2}} \qquad \textit{Rule 1}$$

$$= \frac{x^2}{x^6} \qquad \textit{Rule 4}$$

$$= x^{-4}$$

$$= \boxed{\frac{1}{x^4}}$$

(d) $\dfrac{(2a^{-2}b)^{-3}(3ab)^{-2}}{(6a)^{-4}}$ *Rule 3 and rule 2*

$= \dfrac{2^{-3}a^6b^{-3}3^{-2}a^{-2}b^{-2}}{6^{-4}a^{-4}}$ *Factor 6 into* $3 \cdot 2$.

$= \dfrac{2^{-3}a^6b^{-3}3^{-2}a^{-2}b^{-2}}{(3 \cdot 2)^{-4}a^{-4}}$ *Rule 3*

$= \dfrac{2^{-3}a^6b^{-3}3^{-2}a^{-2}b^{-2}}{3^{-4}2^{-4}a^{-4}}$ *Rule 1 and rule 4*

$= 2^{-3-(-4)} \cdot 3^{-2-(-4)}a^{6+(-2)-(-4)}b^{-3+(-2)}$

$= 2 \cdot 3^2 a^8 b^{-5}$

$= \boxed{\dfrac{18a^8}{b^5}}$

(e) $\dfrac{9^{-3} \cdot 3^{-4}}{3^{-2} \cdot 9^2}$

If you notice that everything can be expressed as a power of 3, *you can proceed as follows.*

$= \dfrac{(3^2)^{-3} \cdot 3^{-4}}{3^{-2}(3^2)^2}$

$= \dfrac{3^{-6} \cdot 3^{-4}}{3^{-2} \cdot 3^4}$

$= \dfrac{3^{-10}}{3^2}$

$= 3^{-10-2}$

$= 3^{-12}$

$= \boxed{\dfrac{1}{3^{12}}}$ ■

EXAMPLE 3   Perform the indicated operations and simplify. Express your answer with positive exponents only.

(a) $(5x^{-2} + y^{-1})^{-2}$ **(b)** $\dfrac{a^{-2} - b^{-2}}{a^{-1} + b^{-1}}$

Solution   **(a)**   Do not try to apply rule 3 in this problem.

x^{-2} and y^{-1} are *terms*, not factors. Rule 3 applies to *factors only*.

$(5x^{-2} + y^{-1})^{-2} = \left(\dfrac{5}{x^2} + \dfrac{1}{y}\right)^{-2}$

By the definition of negative exponents. Note that 5 remains in the numerator of the first fraction. Now add fractions in parentheses; the LCD is x^2y.

$= \left(\dfrac{5y + x^2}{x^2y}\right)^{-2}$

Now we can change the sign of the exponent outside the parentheses and rewrite the entire fraction as its reciprocal.

$$= \left(\frac{x^2y}{5y + x^2}\right)^{+2}$$

$$= \frac{(x^2y)^2}{(5y + x^2)^2}$$

$$= \boxed{\frac{x^4y^2}{(5y + x^2)^2}}$$

(b) $\dfrac{a^{-2} - b^{-2}}{a^{-1} + b^{-1}}$ *We rewrite each **term** in the numerator and denominator using the definition of negative exponents.*

$$= \frac{\dfrac{1}{a^2} - \dfrac{1}{b^2}}{\dfrac{1}{a} + \dfrac{1}{b}}$$ *A complex fraction. Multiply the numerator and denominator of the complex fraction by a^2b^2. Why?*

$$= \frac{\left(\dfrac{1}{a^2} - \dfrac{1}{b^2}\right)a^2b^2}{\left(\dfrac{1}{a} + \dfrac{1}{b}\right)a^2b^2}$$

$$= \frac{b^2 - a^2}{ab^2 + a^2b}$$

$$= \frac{(b - a)(b + a)}{ab(b + a)}$$

$$= \boxed{\frac{b - a}{ab}}$$

Exercises 5.2

Perform the indicated operations and simplify. Express your answers with positive exponents only. (Assume that all variables represent nonzero real numbers.)

1. $x^{-2}x^4x^{-3}$

2. $a^2a^{-3}b^2$

3. $(x^5y^{-4})(x^{-3}y^2x^0)$

4. $(a^{-1}b^2)(a^3b^{-4})$

5. $(3^{-2})^{-3}$

6. $(2^{-3})^{-2}$

7. $(a^{-2}b^{-3})^2$

8. $(x^{-5}y^{-3}x^0)^{-2}$

9. $(r^{-3}s^2)^{-4}(r^2)^{-3}$

10. $(a^{-2}b^3)^{-5}(a^3)^{-2}$

11. -2^{-2}

12. $(-2)^{-2}$

13. $(-3)^{-2}$

14. -3^{-2}

15. $(2^{-2})^{-3}(3^{-3})^2$

16. $(3^{-2})^4(2^{-4})^{-3}$

17. $(3^{-2}s^3)^4(9s^{-3})^{-2}$

18. $(2r^2s)^{-2}(4r)^3$

19. $(-2r^{-3}s)^{-2}$

20. $(-3^2a^{-2}b^3)^{-3}$

21. $(-2x^2y^{-3})^{-2}(-6x^2y^{-1})^{-3}$

22. $(-3x^2y)^{-3}(-4x^2y^{-3})^{-2}$

23. $\dfrac{x^4}{x^{-2}}$

24. $\dfrac{x^{-2}}{x^4}$

25. $\dfrac{x^{-3}y^2}{x^{-5}y^0}$

26. $\dfrac{a^{-2}b^2}{a^{-3}b^4a^0}$

27. $\left(\dfrac{1}{2}\right)^{-1}$

28. $\left(-\dfrac{3}{4}\right)^{-1}$

29. $\left(-\dfrac{3}{5}\right)^{-3}$

30. $\left(\dfrac{5}{8}\right)^{-3}$

31. $\dfrac{x^{-1}xy^{-2}}{x^4y^{-3}y}$ **32.** $\dfrac{a^{-4}b^{-3}}{a^{-2}b^{-4}}$ **33.** $\dfrac{(a^{-2}b^2)^{-3}}{ab^{-2}}$

34. $\dfrac{(2r^{-4}s^{-2})^{-3}}{2r^2s^3}$ **35.** $\left(\dfrac{x^{-1}x^{-3}}{x^{-2}}\right)^{-3}$ **36.** $\left(\dfrac{a^{-2}a^3}{a^{-1}}\right)^{-2}$

37. $\dfrac{2^{-2}\cdot 3^2}{6^{-2}}$ **38.** $\dfrac{3^{-5}\cdot 9^{-1}}{27^2}$ **39.** $\dfrac{(-2)^{-3}(6)^2}{-3^2(-2)^{-1}}$

40. $\dfrac{(-3)^{-2}(9)^{-3}}{(-9)^{-4}(-3)^3}$ **41.** $\dfrac{(3x)^{-2}(2xy^{-1})^0}{(2x^{-2}y^3)^{-2}}$ **42.** $\dfrac{(2xy^2)^0(3x^2y^{-1})^{-1}}{(3x^{-1}y)^{-1}}$

43. $\left(\dfrac{x^{-3}y^{-4}}{x^{-5}y^{-7}}\right)^{-3}$ **44.** $\left(\dfrac{a^{-2}b^3}{a^3b^{-4}}\right)^{-2}$ **45.** $\dfrac{(-3a^{-4}b^{-2})(-4ab^{-3})^{-1}}{(12ab^2)^{-1}}$

46. $\dfrac{(-2a^2b^{-1})^{-2}(-5ab^2)^{-1}}{(50a^{-2}b)^{-1}}$ **47.** $\left(\dfrac{r^{-3}s^{-2}}{2r^4s^{-5}}\right)^{-2}(4r^3s^2)^{-1}$ **48.** $\left(\dfrac{x^{-2}y^{-3}}{3x^2y^{-4}}\right)^{-3}(9x^2y)^{-2}$

49. $\left(\dfrac{2r^2s^{-3}}{-4r^{-2}}\right)^{-2}\left(\dfrac{4^{-1}r^2s^{-3}}{2^{-1}r}\right)^{-3}$ **50.** $\left(\dfrac{5^{-1}x^2y^{-3}}{xy^{-4}}\right)^{-2}\left(\dfrac{10x^{-2}y}{x^2y^{-4}}\right)^2$ **51.** x^2y^{-3}

52. a^3b^{-2} **53.** x^2+y^{-3} **54.** a^3+b^{-2}

55. $(x^2y^{-3})^{-2}$ **56.** $(a^3b^{-2})^{-2}$ **57.** $(x^2+y^{-3})^{-2}$

58. $(a^3+b^{-2})^{-2}$ **59.** $(x^{-2}+y^{-2})^{-2}$ **60.** $(x^{-2}+y^{-2})(x^{-2}-y^{-2})$

61. $\left(\dfrac{x^{-4}y^{-7}z^{-6}}{x^{-24}y^{-16}}\right)^0$ **62.** $\left(\dfrac{x^{-2}+y^{-5}}{x^{-6}-y^{-4}}\right)^0$ **63.** $\dfrac{x^{-1}+y^{-1}}{xy^{-1}}$

64. $\dfrac{a^{-1}-b^{-1}}{(ab)^{-1}}$ **65.** $\dfrac{r^{-2}+s^{-1}}{r^{-1}+s^{-2}}$ **66.** $\dfrac{c^{-2}-d^{-3}}{c^{-3}-d^{-2}}$

67. $\dfrac{2a^{-1}+b^{-2}}{a^{-2}+b}$ **68.** $\dfrac{3x^{-2}-y^{-3}}{x+y^{-1}}$ **69.** $\dfrac{x^{-2}+y^{-2}}{(x+y)^{-2}}$

70. $\dfrac{a^{-1}+b^{-1}}{(a+b)^{-1}}$

? QUESTIONS FOR THOUGHT

71. Describe what is *wrong* (if anything) with the following:

(a) $8^{-1}\overset{?}{=}-8$

(b) $(-2)^{-3}\overset{?}{=}+8$

(c) $(-2)^{-3}\overset{?}{=}+6$

(d) $\dfrac{x^{16}}{x^{-5}}\overset{?}{=}x^{11}$

(e) $xy^{-1}\overset{?}{=}\dfrac{1}{xy}$

72. Suppose that we allow exponents to be fractions and that the rules of integer exponents still hold. Notice that if we square $9^{1/2}$, we get

$$(9^{1/2})^2=9^{2/2}=9^1=9$$

Hence,

$$(9^{1/2})^2=9$$

What does this imply about the values of $9^{1/2}$?

MINI-REVIEW

73. *Factor the following completely.* $162a^2y^2 - 32$

75. *Express as a simple fraction reduced to lowest terms.*

$$\frac{\dfrac{3}{x+1} + 2}{\dfrac{2}{x-1} + x}$$

77. How many ounces of a 30% solution of alcohol must be mixed with 6 ounces of a 70% solution of alcohol in order to get a mixture that is 60% alcohol?

74. *Factor the following completely.* $x^3 - xy^2 - x^2y + y^3$

76. A carpenter charges $30 per hour for his time and $18 per hour for his assistant's time. On a certain job the assistant works alone for 3 hours, and then the carpenter and his assistant complete the job together. If the total bill was $270, how many hours did the carpenter work?

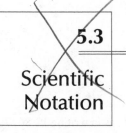

5.3

Scientific Notation

When using your calculator or computer, you may have noticed occasionally getting such answers as 6.43E12 or 2.38E−15. When an answer is given in this form, it is usually because the answer has more digits than the calculator or computer can display. The expression 6.43E12 is actually a shorthand form of scientific notation and means 6.43×10^{12}. Similarly, the expression 2.38E−15 means 2.38×10^{-15}. The E stands for exponent.

Scientific notation is a way of concisely expressing very large or very small numbers. Before we discuss how to write numbers using scientific notation, we will examine powers of 10 with integer exponents:

$$10^3 = 10 \cdot 10 \cdot 10 = 1{,}000$$

$$10^2 = 10 \cdot 10 = 100$$

$$10^1 = 10$$

$$10^0 = 1$$

$$10^{-1} = \frac{1}{10} = 0.1$$

$$10^{-2} = \frac{1}{10^2} = \frac{1}{100} = 0.01$$

$$10^{-3} = \frac{1}{10^3} = \frac{1}{1{,}000} = 0.001$$

You can see from these examples that the exponent of 10 tells the direction and number of places to move the decimal point from the 1. We start with $10^0 = 1.0$ and move the decimal point to the right if the exponent is positive, to the left if the exponent is negative. The number of places we move is equal to the absolute value of the exponent. For example,

$$10^{+4} = 1\ 0000. = 10,000$$

Start here, move 4 places right.

$$10^{-6} = 0.000001$$

Start here, move 6 places left.

We already know how to multiply and divide powers of 10 by applying the rules for integer exponents.

EXAMPLE 1

Evaluate each of the following:

(a) $10^{-3} \cdot 10^7 \cdot 10^{-6}$ **(b)** $\dfrac{10^{-4} \cdot 10^{-2}}{10^{-5}}$

Solution

(a) $10^{-3} \cdot 10^7 \cdot 10^{-6} = 10^{-3+7+(-6)} = 10^{-2} = \boxed{0.01}$

(b) $\dfrac{10^{-4} \cdot 10^{-2}}{10^{-5}} = \dfrac{10^{-6}}{10^{-5}} = 10^{-6+5} = 10^{-1} = \boxed{0.1}$

Scientific Notation

Standard notation is the decimal notation we normally use, as in writing 12.86, 7.954, or 0.0072. **Scientific notation** has the following form:

$$\boxed{a \times 10^i \quad \text{where } 1 \le a < 10 \text{ and } i \text{ is an integer}}$$

For example, 1.645×10^4, 6.8×10^{-3}, and 8.796×10^4 are numbers written in scientific notation.

The following are *not* in scientific notation:

62×10^1 *62 is not between 1 and 10. This should be written as* 6.2×10^2.

0.064 *0.064 is not between 1 and 10. This should be written as* 6.4×10^{-2}.

Converting from scientific to standard notation is straightforward; it is simply a matter of moving the decimal point right or left.

EXAMPLE 2

Express the following in standard notation:
(a) 1.642×10^5 **(b)** 7.3×10^{-4}

Solution

(a) $1.642 \times 10^5 = 1.642 \times 100,000 = \boxed{164,200}$

(b) $7.3 \times 10^{-4} = 7.3 \times 0.0001 = \boxed{0.00073}$

The middle step may be bypassed by observing that the number of places the decimal point is moved is given by the absolute value of the exponent of 10; the direction the decimal point is moved is determined by the sign of the exponent.

$$1.642 \times 10^5 = 164,200 \qquad \textit{Move 5 places right.}$$ ∎

Converting from standard to scientific notation takes a bit more thought, but again is simply a matter of shifting decimal points right or left. For example, suppose we want to convert 6,748 into scientific notation. We know it must take on the following form:

$$6,748 = 6.748 \times 10^?$$

All that remains is to fill in the question mark.

Since we moved the decimal point three places left to put 6,748 into scientific notation form, we must make up for this by multiplying the expression by 10^{+3}:

$$6,748 = 6.748 \times 10^? \qquad$$ *6.748 is the result of moving the decimal point in 6,748 **three places left**, which means it was multiplied by 10^{-3}. Hence, to make up for this, you must multiply 6.748 by 10^{+3}.*

Answer: $6,748 = 6.748 \times 10^{+3}$

A quicker way to state this is to say that the power of 10 restores the decimal point's original position.

EXAMPLE 3

Express the following in scientific notation:

(a) 78,964 (b) 0.00751 (c) 62

Solution

(a) $78,964 = 7.8964 \times 10^?$

$\qquad = \boxed{7.8964 \times 10^{+4}}$

Since we moved the decimal point 4 places left we must multiply by 10^{+4} to make up for it.

(b) $0.00751 = 7.51 \times 10^?$

$\qquad = \boxed{7.51 \times 10^{-3}}$

Since we moved the decimal point 3 places right we must multiply by 10^{-3}.

(c) $62 = \boxed{6.2 \times 10^1}$

CHECK: Always check your answers.

(a) $7.8964 \times 10^4 \overset{\checkmark}{=} 78,964$ (b) $7.51 \times 10^{-3} \overset{\checkmark}{=} 0.00751$

(c) $6.2 \times 10^1 \overset{\checkmark}{=} 62$ ∎

Computations can often be simplified using scientific notation. We would first convert all numbers from standard to scientific notation and then perform the computations.

EXAMPLE 4

Perform the operations using scientific notation:

$$\frac{(3,600)(40,000)}{(0.0012)(0.08)}$$

Solution

$\frac{(3,600)(40,000)}{(0.0012)(0.08)} = \frac{(3.6 \times 10^3)(4 \times 10^4)}{(1.2 \times 10^{-3})(8 \times 10^{-2})}$ *Separate out powers of 10.*

$= \frac{(3.6)(4)}{(1.2)(8)} \times \frac{10^3 \cdot 10^4}{10^{-3} \cdot 10^{-2}}$ *Reduce fraction and simplify powers of 10.*

$= \frac{\overset{3}{\cancel{(3.6)}}\overset{1}{\cancel{(4)}}}{\underset{1}{\cancel{(1.2)}}\underset{2}{\cancel{(8)}}} \times \frac{10^7}{10^{-5}}$

$= \frac{3}{2} \times 10^{12}$

$= \boxed{1.5 \times 10^{12} \quad \text{or} \quad 1,500,000,000,000}$ ∎

EXAMPLE 5

The sun is approximately 93 million miles from Earth. How long does it take light, which travels at approximately 186,000 miles per second, to reach us from the sun?

Solution

In order to compute the time it takes light to reach us from the sun, we will use the familiar relationship $d = rt$ in the form

$$t = \frac{d}{r}$$

$$t = \frac{93,000,000}{186,000}$$

This computation will be much easier to do with the numbers in scientific notation:

$$t = \frac{9.3 \times 10^7}{1.86 \times 10^5} = 5 \times 10^2 = 500 \text{ seconds} \quad \text{or} \quad \boxed{8 \text{ min. } 20 \text{ sec.}}$$ ∎

Exercises 5.3

In Exercises 1–10, perform the indicated operations. Express your answers in standard form.

1. $10^{-4} \cdot 10^7$

2. $10^{-6} \cdot 10^5 \cdot 10^{-3}$

3. $\frac{10^4}{10^{-5}}$

4. $(10^{-2})^{-3}$

5. $\frac{10^{-4} \cdot 10^2}{10^{-3}}$

6. $\frac{10^4 \cdot 10^{-5}}{10^{-7} \cdot 10^{-2}}$

7. $\frac{10^{-4} \cdot 10^2 \cdot 10^{-3}}{10^4 \cdot 10^{-3}}$

8. $\frac{10^{-5} \cdot 10^0 \cdot 10^2}{10^{-3} \cdot 10^{-2}}$

9. $\frac{10^{-4} \cdot 10^{-5} \cdot 10^7}{10^{-6} \cdot 10^{-2} \cdot 10^0}$

10. $\frac{10^5 \cdot 10^{-5} \cdot 10^4}{10^2 \cdot 10^{-2} \cdot 10^{-3}}$

In Exercises 11–18, convert into standard form.

11. 1.62×10^1

12. 8.654×10^{-2}

13. 7.6×10^8

14. 9.37×10^{-5}

15. 8.51×10^{-7}

16. 3.4×10^0

17. 6.0×10^3

18. 7.924×10^{-5}

In Exercises 19–30, convert into scientific notation.

19. 824   **20.** 72   **21.** 5   **22.** 0.06

23. 0.0093   **24.** 65,789   **25.** 827,546,000   **26.** 70,000,000,000

27. 0.00000072   **28.** 685.4   **29.** 79.32   **30.** 0.0000573

In Exercises 31–38, perform the computations by first converting the numbers into scientific notation. Express your answers in standard notation.

31. $\dfrac{(6,000)(0.007)}{(0.021)(12,000)}$ **32.** $\dfrac{(720,000)(0.005)}{(0.8)(0.0003)}$ **33.** $\dfrac{(120)(0.005)}{(10,000)(60)}$ **34.** $\dfrac{(200)(25)}{(0.00004)(0.005)}$

35. $\dfrac{(68,000)(3,000)}{(0.000034)}$ **36.** $\dfrac{(0.0087)(230)}{(0.00003)(4,600)}$ **37.** $\dfrac{(0.065)(120,000)}{(0.0003)(13,000)}$ **38.** $\dfrac{(64)(28)}{(8,000)(70,000)}$

39. Pluto is the planet in our solar system which is furthest from the sun, at a distance of 3,670,000,000 miles. How long does it take light from the sun to reach Pluto? (Light travels at approximately 186,000 miles per second.)

40. If it takes light 4 years to reach us from Proxima Centauri, the star closest to Earth, how far is Proxima Centauri from Earth?

41. Scientists estimate that the sun converts about 700 million tons of hydrogen to helium every second. How much hydrogen does the sun use in 1 day? In 1 year?

42. The sun contains about 1.49×10^{27} tons of hydrogen. Use the result of Exercise 41 to compute how many years it will take to use up the sun's supply of hydrogen.

43. Two common units of measurements in physics and chemistry are the *angstrom* unit (written Å), which is 10^{-8} cm, and the *micron* unit (written μ), which is 10^{-4} cm. How many angstroms are there in 1 micron?

44. The mass of the earth is approximately 5.98×10^{24} kg and the mass of the sun is approximately 1.99×10^{30} kg. How many times greater is the mass of the sun than that of the earth?

45. An atom of oxygen has an atomic radius of 0.66 Å. Use scientific notation to write the number in cm. (See Exercise 43.)

46. A colloidal suspension may contain particles which are 17 thousandths of a micron wide. Use scientific notation to write this number in cm. (See Exercise 43.)

47. One light-year (the distance that light travels in 1 year) is approximately equal to 5.86×10^{12} miles. If 1 mile is 1.6 kilometers, what is the distance of 1 light-year in kilometers?

48. Another unit of measurement used by astronomers is the *parsec*. One parsec is equal to 3.08×10^{13} km. How many light-years are there in 1 parsec? (See Exercise 47.)

49. The diameter of a typical cheek cell is 60 μ. Use scientific notation to write this number in cm. (See Exercise 43.)

CALCULATOR EXERCISES

50. If 40,000 cheek cells are lined up in a row, how far would they stretch in cm? (Use the result of Exercise 49.)

51. If one atom of hydrogen has a mass of 10^{-23} gm, what is the mass of 75,000 atoms of hydrogen?

? QUESTIONS FOR THOUGHT

52. If one atom of hydrogen has a mass of 10^{-23} gm, how many atoms are there in 1 gram of hydrogen? In 1 kilogram?

53. Explain how you would multiply 2.58×10^{-3} by 4.22×10^{4} without converting either number into standard notation.

54. In scientific notation, what is the sign of the exponent of 10 of a number that is greater than 1? Less than 1?

 MINI-REVIEW

55. *Simplify the following expression.* $3\{x - [2(x - 3y) - 2y] + x\}$

56. What does the multiplicative inverse of x yield when multiplied by x?

57. *Simplify.* $(3x^2y^3)^2$

58. *Simplify.*

$$\left(\frac{x^2y^3}{xy^4}\right)^4$$

59. *Perform the following operations and simplify.*

$$\frac{x^3 - 5x^2 + 6x}{x^2 - x - 6} \cdot \frac{x^2y + 2xy}{x^3y^2 - 8y^2}$$

5.4

Rational Exponents

Our original definition of exponents required natural number exponents. We then let the exponent rules tell us how to extend the definition of exponents to include integer exponents. Now we will extend the definition even further to define rational number exponents. We will approach this task in the same way we did in Section 5.2, except that we will primarily use rule 2 for exponents to extend our definition to include rational number exponents. Remember that a rational number is a number of the form $\frac{p}{q}$, where p and q are integers and $q \neq 0$.

We will start with the simplest case: $a^{1/2}$. How do we define $a^{1/2}$? First observe that if we apply rule 2 and square $9^{1/2}$ we get

$$(9^{1/2})^2 = 9^{2/2} = 9^1 = 9$$

So $9^{1/2}$ is a number which, *when squared*, yields 9. There are two possible answers:

$$3 \quad \text{since } 3^2 = 9 \quad \text{and} \quad -3 \quad \text{since } (-3)^2 = 9$$

Therefore, we create the following definition to avoid ambiguity:

DEFINITION

$a^{1/2}$ (called the **principal square root** of a) is the *nonnegative* quantity which, when squared, yields a.

Thus, $9^{1/2} = 3$.

EXAMPLE 1 | Evaluate the following:

(a) $16^{1/2}$ **(b)** $-16^{1/2}$ **(c)** $(-16)^{1/2}$

Solution | **(a)** $16^{1/2} = \boxed{4}$ *Since $4^2 = 16$*

(b) $-16^{1/2} = \boxed{-4}$ *Note: We are taking the negative of $16^{1/2}$.*

(c) $(-16)^{1/2}$ is $\boxed{\text{not a real number}}$ *Since **no** real number will yield a negative number when squared* ∎

We arrive at the definition of $a^{1/3}$ in the same way we did for $a^{1/2}$. For example, if we cube $8^{1/3}$, we get

$$(8^{1/3})^3 = 8^{3/3} = 8^1 = 8$$

Thus, $8^{1/3}$ is the number which, *when cubed*, yields 8.

Since $2^3 = 8$, we have $8^{1/3} = 2$.

DEFINITION

$a^{1/3}$ (called the ***cube root*** of a) is the quantity which, when cubed, yields a.

Hence,

$27^{1/3} = 3$ since $3^3 = 27$.
$(-125)^{1/3} = -5$ since $(-5)^3 = -125$.

Let's examine one more root before we generalize our findings. We will look at $a^{1/4}$. Let's raise $16^{1/4}$ to the fourth power:

$$(16^{1/4})^4 = 16^{4/4} = 16^1 = 16$$

Thus, $16^{1/4}$ is a number which, when raised to the fourth power, yields 16. As with the square root, we have two possible answers:

2   since $2^4 = 16$ and -2 since $(-2)^4 = 16$

Again, in order to avoid ambiguity, we define $a^{1/4}$ as follows:

DEFINITION

$a^{1/4}$ (called the ***principal fourth root*** of a) is the *nonnegative* quantity which, when raised to the fourth power, yields a.

Looking at the discussion preceding each of the definitions above, you will notice that there is a possibility for ambiguity when the root is even. That is, you will get two answers. To eliminate this ambiguity, we inserted the word *nonnegative* in our definitions of the square root and fourth root. On the other hand, for odd roots (such as cube roots), there is only one real-number answer, so we need not be concerned with ambiguity. This leads us to the following general definition.

DEFINITION

$a^{1/n}$ (called the ***principal nth root of a***) is the real number (positive when n is even) which, when raised to the nth power, yields a.

Earlier we pointed out that the square root of -16 is not a real number, since no real number squared will yield a negative number. In general, raising any real number to an even power will always yield a nonnegative number. Therefore, the even root of a negative number is not a real number.

Symbolically, we defined the ***nth root*** as follows:

DEFINITION

$$a^{1/n} = \begin{cases} b & \text{if } n \text{ is odd and } b^n = a \\ |b| & \text{if } n \text{ is even, } a \geq 0, \text{ and } b^n = a \end{cases}$$

We can summarize the various types of roots as indicated in the box.

	n is even	*n is odd*
$a > 0$	$a^{1/n}$ is the positive nth root of a	$a^{1/n}$ is the nth root of a
$a < 0$	$a^{1/n}$ is not a real number	$a^{1/n}$ is the nth root of a
$a = 0$	$0^{1/n} = 0$	$0^{1/n} = 0$

EXAMPLE 2

Evaluate the following:

(a) $(-32)^{1/5}$ (b) $(-64)^{1/6}$ (c) $(-8)^{1/3}$ (d) $\left(\dfrac{1}{81}\right)^{1/4}$

Solution

(a) $(-32)^{1/5} = \boxed{-2}$ *Since* $(-2)^5 = -32$

(b) $(-64)^{1/6}$ $\boxed{\text{is not a real number}}$ *What number raised to the sixth power will yield* -64?

(c) $(-8)^{1/3} = \boxed{-2}$ *Since* $(-2)^3 = -8$

(d) $\left(\dfrac{1}{81}\right)^{1/4} = \boxed{\dfrac{1}{3}}$ *Since* $\left(\dfrac{1}{3}\right)^4 = \dfrac{1}{81}$

We will often have occasion to use an alternate notation for fractional exponents, called *radical notation*:

$x^{1/n}$ is also written as $\sqrt[n]{x}$

For example:

$$x^{1/5} = \sqrt[5]{x}$$
$$x^{1/7} = \sqrt[7]{x}$$

In particular,

$$x^{1/2} \text{ is written as } \sqrt{x}$$

Thus far we have defined $a^{1/n}$ where n is a natural number. With some help from rule 2 for exponents, we can define the expression $a^{m/n}$ where n and m are natural numbers.

DEFINITION

If $a^{1/n}$ is a real number, then

$$a^{m/n} = (a^{1/n})^m$$

That is, $a^{m/n}$ is the nth root of a raised to the mth power.

We define $a^{-m/n}$ as follows:

DEFINITION

$$a^{-m/n} = \frac{1}{a^{m/n}} \qquad (a \neq 0)$$

Now that we have defined rational exponents, we assert that the *rules for integer exponents hold for rational exponents as well, provided the root is a real number* (that is, provided we avoid even roots of negative numbers). For example, in order to use the rule $(a^r)^s = a^{rs}$ where r and s are rational, it is necessary that both a^r *and* a^s be defined.

By the exponent rules, we find that $(a^{1/n})^m = (a^m)^{1/n}$. Hence, we can interpret $a^{m/n}$ in two ways, as indicated in the box.

If $a^{1/n}$ is a real number, then

$$a^{m/n} = (a^{1/n})^m = (a^m)^{1/n}$$

EXAMPLE 3

Evaluate each of the following:

(a) $27^{2/3}$ **(b)** $16^{3/4}$ **(c)** $(-64)^{2/3}$ **(d)** $-64^{2/3}$

(e) $36^{-1/2}$ **(f)** $(-125)^{-1/3}$ **(g)** $\left(\dfrac{1}{125}\right)^{-1/3}$ **(h)** $(-32)^{-3/5}$

Solution

In general, you will find it easier to find the root *before* raising to a power.

(a) $27^{2/3} = (27^{1/3})^2 = 3^2 = \boxed{9}$

(b) $16^{3/4} = (16^{1/4})^3 = 2^3 = \boxed{8}$

We could have done this same problem in the following way:

$$16^{3/4} = (16^3)^{1/4} = (4{,}096)^{1/4} = 8$$

But this approach requires more multiplication and finding a difficult root.

(c) $(-64)^{2/3} = [(-64)^{1/3}]^2 = [-4]^2 = \boxed{+16}$

(d) $-64^{2/3} = -[(64)^{1/3}]^2 = -[4]^2 = \boxed{-16}$ *Note the difference between this problem and part (c).*

(e) $36^{-1/2} = \dfrac{1}{36^{1/2}}$ *Begin by changing the negative exponent into a positive exponent by using the definition of a negative rational exponent.*

$= \boxed{\dfrac{1}{6}}$

(f) $(-125)^{-1/3} = \dfrac{1}{(-125)^{1/3}}$ *Definition of negative rational exponent*

$= \dfrac{1}{-5}$

$= \boxed{-\dfrac{1}{5}}$

Note that *the sign of the exponent has no effect on the sign of the base.*

(g)   We will approach this problem a bit differently, using the fact that an exponent of -1 turns an expression into its reciprocal.

$\left(\dfrac{1}{125}\right)^{-1/3} = \left[\left(\dfrac{1}{125}\right)^{-1}\right]^{1/3}$

$= \left[\dfrac{125}{1}\right]^{1/3}$ *Since the reciprocal of $\dfrac{1}{125}$ is $\dfrac{125}{1}$*

$= 125^{1/3}$

$= \boxed{5}$

(h) $(-32)^{-3/5} = \dfrac{1}{(-32)^{3/5}}$ *Change to positive exponents.*

$= \dfrac{1}{[(-32)^{1/5}]^3}$ *Again, the sign of the exponent has no effect on the sign of the base.*

$$= \frac{1}{(-2)^3}$$

$$= \boxed{-\frac{1}{8}}$$

Using radical notation, we can write $a^{m/n}$ as:

$$\sqrt[n]{a^m} \quad [= (a^m)^{1/n}] \qquad \text{or as} \qquad (\sqrt[n]{a})^m \quad [= (a^{1/n})^m]$$

(assuming $a^{1/n}$ is a real number and m/n is reduced to lowest terms). Hence,

$$a^{m/n} = \sqrt[n]{a^m} = (\sqrt[n]{a})^m$$

EXAMPLE 4   Rewrite the following using radical notation.

(a) $a^{1/6}$ **(b)** $b^{2/3}$ **(c)** $(8)^{4/9}$ **(d)** $5x^{1/3}$ **(e)** $x^{-3/4}$

Solution   **(a)** $a^{1/6} = \boxed{\sqrt[6]{a}}$

(b) $b^{2/3} = \boxed{\sqrt[3]{b^2} \quad \text{or} \quad (\sqrt[3]{b})^2}$

(c) $(8)^{4/9} = \boxed{\sqrt[9]{8^4} \quad \text{or} \quad (\sqrt[9]{8})^4}$

(d) $5x^{1/3} = \boxed{5\sqrt[3]{x}}$ *Note that the exponent applies only to x and not to 5.*

(e) $x^{-3/4}$ *Change to positive exponents first.*

$$= \frac{1}{x^{3/4}}$$

$$= \boxed{\frac{1}{\sqrt[4]{x^3}} \quad \text{or} \quad \frac{1}{(\sqrt[4]{x})^3}}$$

We deal with simplifying radicals in the next chapter.   ■

As we stated earlier, the rules for integer exponents are also valid for rational exponents. Thus, we simplify expressions involving rational exponents by following the same procedures we used for integer exponents. The main difficulty in simplifying rational exponent expressions is the fractional arithmetic involved.

One point of confusion that frequently arises involves the difference between negative exponents and fractional exponents. *Negative exponents involve* **reciprocals** *of the base*. They do *not* create fractional *exponents*. For example,

$$x^{-4} = \frac{1}{x^4} \qquad \text{or} \qquad 16^{-4} = \frac{1}{16^4} = \frac{1}{65{,}536}$$

On the other hand, *fractional exponents yield **roots** of the base*. For example,

$$x^{1/4} = \text{the fourth root of } x \qquad \text{or} \qquad 16^{1/4} = 2$$

As with integer exponents, when we are asked to simplify an expression, the bases and exponents should appear as few times as possible. Unless otherwise noted, we assume all variables represent positive real numbers.

EXAMPLE 5

Perform the operations and simplify. Express your answer with positive exponents only.

(a) $x^{1/2}x^{2/3}x^{3/4}$ **(b)** $\dfrac{x^{2/5}}{x^{3/4}}$ **(c)** $(y^{2/3}y^{-1/2})^2$

(d) $\dfrac{a^{1/2}a^{-2/3}}{a^{1/4}}$ **(e)** $\left(\dfrac{x^{-1/2}y^{-1/4}}{x^{1/4}}\right)^{-4}$ **(f)** $\dfrac{(x^{-1}y^{1/2})^{-1/3}}{x^{1/2}y^{-1/3}}$

Solution

We apply the rules of rational exponents.

(a) $x^{1/2}x^{2/3}x^{3/4}$ *Rule 1*

$= x^{1/2+2/3+3/4}$

$= x^{6/12+8/12+9/12}$

$= \boxed{x^{23/12}}$ *Leave the exponent as an improper fraction (reduced).*

(b) $\dfrac{x^{2/5}}{x^{3/4}}$ *Rule 4*

$= x^{2/5-3/4}$

$= x^{8/20-15/20}$

$= x^{-7/20}$ *Change negative exponents to positive and write the reciprocal.*

$= \boxed{\dfrac{1}{x^{7/20}}}$

(c) $(y^{2/3}y^{-1/2})^2$ *Make sure you can follow the arithmetic. Apply rule 1.*

$= (y^{2/3+(-1/2)})^2$

$= (y^{1/6})^2$ *Rule 2*

$= y^{2/6}$ *Reduce the exponent.*

$= \boxed{y^{1/3}}$

(d) $\dfrac{a^{1/2}a^{-2/3}}{a^{1/4}}$ *Rule 1*

$= \dfrac{a^{1/2-2/3}}{a^{1/4}}$

$= \dfrac{a^{-1/6}}{a^{1/4}}$ *Rule 4*

$$= a^{-1/6 - 1/4}$$

$$= a^{-5/12}$$

$$= \boxed{\frac{1}{a^{5/12}}}$$

(e) This example is a lot easier if we bring in the outside exponent by rules 5 and 3.

$$\left(\frac{x^{-1/2}y^{-1/4}}{x^{1/4}}\right)^{-4} \qquad \textit{First rule 5}$$

$$= \frac{(x^{-1/2}y^{-1/4})^{-4}}{(x^{1/4})^{-4}} \qquad \textit{Next rule 3}$$

$$= \frac{(x^{-1/2})^{-4}(y^{-1/4})^{-4}}{(x^{1/4})^{-4}} \qquad \textit{Now rule 2}$$

$$= \frac{x^2 y^1}{x^{-1}}$$

$$= x^{2-(-1)}y$$

$$= \boxed{x^3 y}$$

(f) $\dfrac{(x^{-1}y^{1/2})^{-1/3}}{x^{1/2}y^{-1/3}} \qquad \textit{Rules 3 and 2}$

$$= \frac{x^{1/3}y^{-1/6}}{x^{1/2}y^{-1/3}} \qquad \textit{Rule 4}$$

$$= x^{1/3 - 1/2}y^{-1/6-(-1/3)}$$

$$= x^{-1/6}y^{1/6}$$

$$= \boxed{\frac{y^{1/6}}{x^{1/6}}}$$

       ■

Again, keep in mind that rules 3 and 5 for exponents apply to *factors*, not terms.

EXAMPLE 6 Perform the indicated operations and simplify the following:

(a) $(x^{1/2} + 2x^{1/2})x^{-1/3}$ **(b)** $(5a^{1/2} + 3b^{1/2})^2$

Solution **(a)** $(x^{1/2} + 2x^{1/2})x^{-1/3}$ *Combine like terms in parentheses.*

$$= (3x^{1/2})x^{-1/3} \qquad \textit{Then apply rule 1.}$$

$$= 3x^{1/2 + (-1/3)}$$

$$= \boxed{3x^{1/6}}$$

(b) For squaring a binomial, we can use the perfect square special product.

$$(5a^{1/2} + 3b^{1/2})^2 = (5a^{1/2})^2 + 2(5a^{1/2})(3b^{1/2}) + (3b^{1/2})^2$$
$$= 5^2 a^{2/2} + 30a^{1/2}b^{1/2} + 3^2 b^{2/2}$$
$$= \boxed{25a + 30a^{1/2}b^{1/2} + 9b}$$

Exercises 5.4

In Exercises 1–32, evaluate the expression.

1. $8^{1/3}$

2. $16^{1/4}$

3. $(-32)^{1/5}$

4. $(-64)^{1/3}$

5. $(-81)^{1/2}$

6. $-100^{1/2}$

7. $-81^{1/2}$

8. $(-100)^{1/2}$

9. $(32)^{4/5}$

10. $32^{2/5}$

11. $(-32)^{4/5}$

12. $(-32)^{2/5}$

13. $(-32)^{3/5}$

14. $(-243)^{3/5}$

15. $32^{-1/5}$

16. $27^{-1/3}$

17. $(-32)^{-1/5}$

18. $(-27)^{-1/3}$

19. $-(81)^{-1/2}$

20. $-(16)^{-1/4}$

21. $(-64)^{-2/3}$

22. $(-8)^{-2/3}$

23. $(-32)^{-3/4}$

24. $(-81)^{-3/4}$

25. $-(-64)^{-2/3}$

26. $-(-125)^{-2/3}$

27. $\left(\dfrac{64}{27}\right)^3$

28. $\left(\dfrac{16}{81}\right)^4$

29. $\left(\dfrac{81}{16}\right)^{-1/4}$

30. $\left(\dfrac{27}{64}\right)^{-1/3}$

31. $\left(-\dfrac{1}{32}\right)^{-4/5}$

32. $\left(-\dfrac{1}{243}\right)^{-4/5}$

In Exercises 33–64, perform the operations and simplify. Express your answers with positive exponents only. Assume that all variables represent positive real numbers.

33. $x^{1/2}x^{2/3}$

34. $x^{1/3}x^{-2/5}$

35. $x^{1/5}x^{-2/3}x^{3/4}$

36. $a^{-1/2}a^{1/3}a^{2/5}$

37. $(a^{-1/2})^{-3/4}$

38. $(s^{-1/3})^{2/5}$

39. $(2^{-1} \cdot 4^{1/2})^{-2}$

40. $(3^2 \cdot 27^{-1/3})^{-2}$

41. $(r^{-1/3}r^{1/2})^{-6}$

42. $(r^{-2/3}s^{1/4})^{-15}$

43. $(r^{1/2}r^{-2/3}s^{1/2})^{-2}$

44. $(a^{1/2}b^{-1/3})^{-1/2}$

45. $(r^{-1}s^{1/2})^{-2}(r^{-1/2}s^{1/3})^2$

46. $(a^{1/3}b^{-2})^{-2}(a^{2/3}b^{-2})^3$

47. $\dfrac{x^{-1/2}}{x^{-1/3}}$

48. $\dfrac{x^{1/2}}{x^{-1/4}}$

49. $\dfrac{a^{-1/2}b^{1/3}}{a^{1/4}b^{1/5}}$

50. $\dfrac{x^{1/2}y^{-1/3}}{x^{1/3}y^{1/5}}$

51. $\left(\dfrac{x^{1/2}x^{-1}}{x^{1/3}}\right)^{-6}$

52. $\left(\dfrac{x^{-1/3}x^{1/5}}{x}\right)^{-15}$

53. $\left(\dfrac{x^{-1/2}y^{1/3}}{x^{1/3}y^{1/5}}\right)^{-1/2}$

54. $\dfrac{x^{1/3}y^{-2/5}}{x^{-1/2}y^{1/5}}$

55. $\dfrac{(4^{-1/2} \cdot 16^{3/4})^{-2}(64^{5/6})}{(-64)^{1/3}}$

56. $\left(\dfrac{9^{-1/2} \cdot 27^{2/3} \cdot 81^{1/4}}{(-3)^{-1} \cdot 81^{3/4}}\right)^{-2}$

57. $\dfrac{(x^{1/2}y^{1/3})^{-2}(x^{1/3}y^{1/4})^{-12}}{xy^{1/4}}$

58. $\dfrac{(a^{1/2}b^{1/3})^{-6}(a^{1/3}b^{-1/2})^2}{a^{-1/3}b}$

59. $(x^{1/2} + y)x^{1/2}$

60. $(x^{1/2} + y^{3/2})y^{1/3}$

61. $(a^{1/2} + b^{1/2})(a^{1/2} - b^{1/2})$

62. $(2r^{1/2} + s^{1/2})(3r^{1/2} - 2s^{1/2})$

63. $(x^{1/2} - 2x^{-1/2})^2$

64. $(a^{-1/2} + 3b^{1/2})^2$

In Exercises 65–72, change to radical notation.

65. $x^{2/3}$ **66.** $10^{3/5}$ **67.** $7^{2/3}$ **68.** $x^{3/2}$

69. $6x^{1/3}$ **70.** $(3x)^{1/3}$ **71.** $6x^{2/5}$ **72.** $(3x)^{2/5}$

? QUESTIONS FOR THOUGHT

73. Explain what is *wrong* (if anything) with each of the following:

 (a) $8^{1/3} \stackrel{?}{=} \dfrac{1}{8^3}$

 (b) $8^{-3} \stackrel{?}{=} -512$

 (c) $8^{-1/3} \stackrel{?}{=} -2$

74. What is the difference between a negative exponent and a fractional exponent?

75. What are the differences among the rules of exponents for rational, integer, and natural number exponents?

76. Discuss the similarities and differences among 9^{-2}, $9^{1/2}$, and $9^{-1/2}$.

MINI-REVIEW

Solve the following for a:

77. $5 - [3 - (a - 2)] = a - 2$ **78.** $|9 - 3a| \le 6$ **79.** $|9 - 3a| \ge 6$

80. $\dfrac{3}{1 - a} + 1 = 1$ **81.** $\dfrac{1}{a} - \dfrac{1}{b} = \dfrac{1}{c}$

CHAPTER 5   Summary

After having completed Chapter 5, you should:

1. Know the five rules of exponents and be able to use them to simplify expressions (Section 5.1).

 For example:

$$\frac{(2x^3y^2)^5}{8(x^2y^3)^4} = \frac{2^5(x^3)^5(y^2)^5}{8(x^2)^4(y^3)^4}$$
$$= \frac{32x^{15}y^{10}}{8x^8y^{12}}$$
$$= \frac{4x^7}{y^2}$$

2. Understand the definition of zero and negative exponents (Section 5.2).

 For example:

 (a) $18^0 = 1$

 (b) $4^{-3} = \dfrac{1}{4^3} = \dfrac{1}{4 \cdot 4 \cdot 4} = \dfrac{1}{64}$

3. Be able to use the exponent rules to simplify expressions with integer exponents (Section 5.2).

   For example: Simplify and express with positive exponents only.

   (a)
   $$\frac{(x^{-3}y^8)^{-2}}{(x^{-1}y^{-3})^4} = \frac{(x^{-3})^{-2}(y^8)^{-2}}{(x^{-1})^4(y^{-3})^4}$$
   $$= \frac{x^6 y^{-16}}{x^{-4} y^{-12}}$$
   $$= x^{6-(-4)} y^{-16-(-12)}$$
   $$= x^{10} y^{-4}$$
   $$= x^{10}\left(\frac{1}{y^4}\right)$$
   $$= \frac{x^{10}}{y^4}$$

   (b) $(x^{-2} + y^{-3})^{-1} = \left(\dfrac{1}{x^2} + \dfrac{1}{y^3}\right)^{-1}$ *Add fractions.*
   $$= \left(\frac{y^3 + x^2}{x^2 y^3}\right)^{-1}   \text{\textit{Take the reciprocal.}}$$
   $$= \frac{x^2 y^3}{y^3 + x^2}$$

4. Be able to write and compute with numbers in scientific notation (Section 5.3).

   For example: Compute.

   $$\frac{(4{,}000)(0.00015)}{0.005} = \frac{(4 \times 10^3)(1.5 \times 10^{-4})}{5 \times 10^{-3}}$$
   $$= \frac{(4)(1.5)}{5} \times \frac{10^3 10^{-4}}{10^{-3}}$$
   $$= \frac{6}{5} \times \frac{10^{-1}}{10^{-3}}$$
   $$= 1.2 \times 10^2$$

5. Understand and be able to evaluate expressions involving rational exponents (Section 5.4).

   For example:

   (a) $16^{1/2} = 4$ because $4^2 = 16$

   (b) $(-32)^{1/5} = -2$ because $(-2)^5 = -32$

   (c) $27^{-2/3} = \dfrac{1}{27^{2/3}} = \dfrac{1}{(27^{1/3})^2} = \dfrac{1}{(3)^2} = \dfrac{1}{9}$

6. Be able to simplify expressions involving rational exponents (Section 5.4).

   For example:

   (a)
   $$\frac{(4x^2)^{1/2}(x^{-2/3})}{(2x)^2(x^5)^{-1/2}} = \frac{4^{1/2}x \cdot x^{-2/3}}{2^2 x^2 \cdot x^{-5/2}}$$
   $$= \frac{2x^{1/3}}{4x^{-1/2}} = \frac{x^{5/6}}{2}$$

   (b)
   $$(x^{1/3} + x^{2/5})^2 = (x^{1/3} + x^{2/5})(x^{1/3} + x^{2/5})$$
   $$= x^{1/3}x^{1/3} + 2x^{1/3}x^{2/5} + x^{2/5}x^{2/5}$$
   $$= x^{2/3} + 2x^{1/3 + 2/5} + x^{4/5}$$
   $$= x^{2/3} + 2x^{11/15} + x^{4/5}$$

CHAPTER 5   Review Exercises

In Exercises 1–62, perform the operations and simplify; express your answers with positive exponents only. Assume all variables represent nonzero real numbers.

1. $(x^2x^5)(x^4x)$

2. $(x^5y^2)(x^6y^7)$

3. $(-3x^2y)(-2xy^4)(-x)$

4. $(-3xy^5)(-2xy)(-5x)$

5. $(a^3)^4$

6. $(b^5)^4$

7. $(a^2b^3)^7$

8. $(r^2s^3)^8$

9. $(a^2b^3)^2(a^2b)^3$

10. $(x^2y^5)^2(xy^4)^3$

11. $(2^2 \cdot 3^2)^5$

12. $(-2)^3(-2)^2$

13. $(a^2bc^2)^2(ab^2c)^3$

14. $(-2xy^2)(-3x^2y)^3(-3x)$

15. $\dfrac{a^5}{a^6}$

16. $\dfrac{x^7}{x^2}$

17. $\dfrac{x^2x^5}{x^4x^3}$

18. $\dfrac{y^5y^7}{y^8y^2}$

19. $\dfrac{(x^3y^2)^3}{(x^5y^4)^5}$

20. $\dfrac{(a^2b^3)^2}{(a^2b^4)^3}$

21. $\left(\dfrac{a^2b}{ab}\right)^4$

22. $\left(\dfrac{ab^2}{cb}\right)^4$

23. $\dfrac{(2ax^2)^2(3ax)^2}{(-2x)^2}$

24. $\dfrac{(-5xy)^2(-3x^2)^3}{(-15x)^2}$

25. $\left(\dfrac{-3xy}{x^2}\right)^2\left(\dfrac{-2xy^2}{x}\right)^3$

26. $\left(\dfrac{-3ab^2}{a^2b}\right)^2\left(\dfrac{-2ab}{5a^2}\right)^3$

27. $(2x^2y)\left(\dfrac{-3xy^2}{x^2y}\right)^2$

28. $[(-3ab)^2(-2a^2b)^3]^2$

29. $\left(\dfrac{(-3ab^2c)^2}{-6a^2b}\right)^2$

30. $\left(\dfrac{(-8rs^2)^3}{(-4rs^2)^5}\right)^2$

31. $a^{-3}a^{-4}a^5$

32. $x^{-5}x^{-4}x^0$

33. $(x^{-2}y^5)^{-4}$

34. $(x^{-2}x^3)^{-4}$

35. $(a^{-2}b^2)^{-3}(a^3b^{-4})^2$

36. $(-2a^2b)^{-2}$

37. $(-3)^{-4}(-2)^{-1}$

38. $(-2)^{-2}(-2)^2$

39. $(-3x^{-2}yx^{-3})^{-2}(9x^{-2}y)^{-3}$

40. $(-5a^2b^{-1})^{-2}(25a^2b^{-2})$

41. $\left(\dfrac{3x^{-5}y^2z^{-4}}{2x^{-7}y^{-4}}\right)^0$

42. $(-156,794)^0$

43. $\dfrac{x^{-3}x^{-6}}{x^{-5}x^0}$

44. $\dfrac{x^{-2}y^{-3}}{x^2y^{-4}}$

45. $\dfrac{x^{-3}y^{-5}x^{-2}}{x^4y^{-3}}$

46. $\dfrac{x^{-5}x^{-7}y^2}{y^{-2}y^4x}$

47. $\left(\dfrac{x^{-2}y^{-3}}{y^{-3}x^2}\right)^{-2}$

48. $\left(\dfrac{a^{-2}b^{-3}}{a^2b^3}\right)^{-2}$

49. $\left(\dfrac{r^{-2}s^{-3}r^{-2}}{s^{-4}}\right)^{-2}\left(\dfrac{r^{-1}}{s^{-1}}\right)^{-3}$

50. $\left(\dfrac{2r^{-1}s^{-2}}{r^{-3}s^2}\right)^{-2}\left(\dfrac{-3r^{-2}s^{-2}}{4r^{-3}s}\right)^{-1}$

51. $\left(\dfrac{2}{5}\right)^{-2}$

52. $\left(\dfrac{3}{4}\right)^{-4}$

53. $\dfrac{(2x^2y^{-1}z)^{-2}}{(3xy^2)^{-3}}$

54. $\dfrac{(3x^{-2}y^2z^4)^{-2}}{(2x^{-1}y)^{-3}}$

55. $\left(\dfrac{(2x^{-2}y)^{-2}}{(3x^{-1}y^{-1})^2}\right)^{-2}$

56. $\left(\dfrac{(-2xy^{-2})^{-2}}{(2x^{-1}y^{-3})^2}\right)^{-2}$

57. $(x^{-1} + y^{-1})(x - y)$

58. $(x^{-1} + 2y^{-2})^2$

59. $\dfrac{x^{-1} + y^{-3}}{x^{-1}y^2}$

60. $\dfrac{a^{-2} + b^{-1}}{a^{-1}b^{-2}}$

61. $\dfrac{x^{-2} - y^{-1}}{x^{-1} + y^{-2}}$

62. $\dfrac{a^{-1} + b^{-2}}{a^{-2} - b^{-3}}$

In Exercises 63–66, convert to standard form.

63. 2.83×10^4

64. 6.29×10^0

65. 7.96×10^{-5}

66. 8.264×10^{-7}

In Exercises 67–72, convert to scientific notation.

67. 7.936   **68.** 92.59   **69.** 0.00578   **70.** 8

71. 625,897   **72.** 0.0000073

In Exercises 73–74, perform the computations by first converting the numbers into scientific notation. Express your answers in standard notation.

73. $\dfrac{(0.0014)(9,000)}{(20,000)(63,000)}$ **74.** $\dfrac{(720)(1,000)}{(900,000)(0.008)}$

In Exercises 75–82, evaluate the expression.

75. $25^{1/2}$ **76.** $64^{2/3}$ **77.** $(-243)^{-3/5}$ **78.** $(-216)^{-1/3}$

79. $\left(\dfrac{64}{27}\right)^{1/3}$ **80.** $\left(\dfrac{16}{25}\right)^{1/2}$ **81.** $\left(-\dfrac{27}{125}\right)^{-1/3}$ **82.** $\left(-\dfrac{81}{16}\right)^{-3/4}$

In Exercises 83–100, perform the operations and simplify; express your answers with positive exponents only. Assume all variables represent positive real numbers only.

83. $x^{1/2}x^{1/3}$ **84.** $y^{1/3}y^{-1/2}$ **85.** $(x^{-1/2}x^{1/3})^{-6}$

86. $(y^{-1/3}y^{-1/2}y^{2/3})^{-1/2}$ **87.** $\dfrac{x^{1/2}x^{1/3}}{x^{2/5}}$ **88.** $\dfrac{a^{1/3}a^{-1/2}}{a^{-2/3}}$

89. $\dfrac{r^{-1/2}s^{-1/3}}{r^{1/3}s^{-2/3}}$ **90.** $\dfrac{r^{-1/5}s^{-1/2}}{s^{1/2}s^{1/3}}$ **91.** $\left(\dfrac{a^{1/2}a^{-1/3}}{a^{1/2}b^{1/5}}\right)^{-15}$

92. $\left(\dfrac{x^{-1/2}x^{1/3}}{x^{-1/3}y^{1/2}}\right)^{-6}$ **93.** $\dfrac{(x^{-1/2}y^{1/2})^{-2}}{(x^{-1/3}y^{-1/3})^{-1/2}}$ **94.** $\dfrac{(a^{1/2}a^{1/3}a^{-2/3})^{-2}}{a^{1/3}a^{-1/2}}$

95. $\left(\dfrac{4^{-1/2}\cdot 16^{-3/4}}{8^{1/3}}\right)^{-2}$ **96.** $\left(\dfrac{3^{-2}\cdot 9^{-1/2}}{81^{-3/4}}\right)^{-1}$ **97.** $(a^{1/2}-b^{1/2})(a^{-2/3})$

98. $(a^{1/3}+b^{1/3})(a^{1/3}-b^{1/3})$ **99.** $(a^{-1/2}+2b^{1/2})^2$ **100.** $(2a^{-1/2}+3b^{-1/2})^2$

In Exercises 101–104, change the expression to radical notation.

101. $x^{2/5}$ **102.** $x^{1/3}$ **103.** $3y^{2/5}$ **104.** $(3y)^{2/5}$

CHAPTER 5   Practice Test

Perform the indicated operations and express your answers in simplest form with positive exponents only. Assume that all variables represent positive real numbers.

1. $(a^2b^3)(ab^4)^2$ **2.** $(-2x^2y)^3(-5x)^2$ **3.** $(-3ab^{-2})^{-1}(-2x^{-1}y)^2$

4. $\dfrac{(-2x^2y)^3}{(-6xy^4)^2}$ **5.** $\left(\dfrac{5r^{-1}s^{-3}}{3rs^2}\right)^{-2}$ **6.** $\dfrac{(x^{-2}y^0)^{-3}}{(x^2y^{-1})^{-4}}$

7. $\dfrac{27^{2/3}\cdot 3^{-4}}{9^{-1/2}}$ **8.** $\left(\dfrac{x^{1/4}x^{-2/3}}{x^{-1}}\right)^4$ **9.** $\dfrac{x^{-3}+x^{-1}}{yx^{-2}}$

10. $x^{1/3}(x^{2/3}-x)$ **11.** $(x^{-1/2}+3x^{1/2})^2$

12. Evaluate the following:

    (a) $(-125)^{1/3}$ **(b)** $(-128)^{-3/7}$

13. Express $3a^{2/3}$ in radical form.

14. Light travels at approximately 186,000 miles per second. How far does light travel in 1 hour? (Use scientific notation to compute your answer.)

6

Radical Expressions

In Chapter 1, we stated that the real numbers consist of rational and irrational numbers. Recall that irrational numbers are defined as real numbers which cannot be expressed as a quotient of two integers. Some of the examples of irrational numbers given included expressions such as $\sqrt{2}$, $\sqrt{3}$, and $\sqrt{7}$, which are also examples of square root expressions.

In this chapter we will examine the more general radical expression, $\sqrt[n]{a}$. In addition, just as we found it necessary to extend whole numbers to integers, integers to rational numbers, and rational numbers to real numbers in order to solve various types of equations, we will eventually define a new system of numbers, called *complex numbers*, in order to find solutions to equations such as $x^2 = -5$.

6.1

Radicals and Rational Exponents

Recall from Chapter 5 that radicals are an alternative way of writing an expression with fractional exponents. That is, we have the following definition:

DEFINITION

$\sqrt[n]{a} = a^{1/n}$ where n is a positive integer.

$\sqrt[n]{a}$ is called the principal **nth root of a**.

In $\sqrt[n]{a}$, n is called the **index** of the radical, the symbol $\sqrt{}$ is called the **radical** or radical sign, and the expression, a, under the radical is called the **radicand**. Thus,

$\sqrt[4]{3} = 3^{1/4}$ is called the **fourth root** of 3.

$\sqrt[5]{9} = 9^{1/5}$ is called the **fifth root** of 9.

$\sqrt[3]{x} = x^{1/3}$ is usually called the **cube root** of x.

$\sqrt{y} = y^{1/2}$ is usually called the **square root** of y.

Note that we usually drop the index for square roots and write $\sqrt{a}$ rather than $\sqrt[2]{a}$.

Thus, $\sqrt[n]{a}$ is that quantity (nonnegative when n is even) which, when raised to the nth power, yields a.

Symbolically, we define the nth root for n a positive odd integer as follows:

For $a \in R$ and n a positive *odd* integer,

$$\sqrt[n]{a} = b \quad \text{if and only if} \quad b^n = a.$$

Thus,

$$\sqrt[3]{64} = 4 \quad \text{since } 4^3 = 64.$$

$$\sqrt[5]{-32} = -2 \quad \text{since } (-2)^5 = -32.$$

When the index of the radicand is even, we require the *root* to be positive:

For $a \in R$, and n a positive *even* integer,

$$\sqrt[n]{a} = b \quad \text{if } b^n = a \text{ and } b \geq 0$$

In either case, where n is odd or even, if $\sqrt[n]{a}$ is real, $\sqrt[n]{a}$ is called the ***principal nth root*** of a. Keep in mind that when n is even, the principal nth root cannot be negative.

$$\sqrt[4]{16} = 2 \quad \text{since } 2^4 = 16$$
$$\sqrt{9} = 3 \quad \text{since } 3^2 = 9 \qquad \text{\textit{Note that even though } } (-3)^2 = 9, \textit{ we are}$$
$$\textit{interested only in the \textbf{principal} or positive}$$
$$\textit{square root.}$$

$\sqrt[6]{-64}$ is not a real number since no real number when raised to the sixth power will yield a negative number.

As with rational exponents, we summarize the various types of roots as follows:

	n is even	*n is odd*
$a > 0$	$\sqrt[n]{a}$ is the positive nth root of a	$\sqrt[n]{a}$ is the nth root of a
$a < 0$	$\sqrt[n]{a}$ is not a real number	$\sqrt[n]{a}$ is the nth root of a
$a = 0$	$\sqrt[n]{0} = 0$	$\sqrt[n]{0} = 0$

EXAMPLE 1

Evaluate the following.

(a) $\sqrt[4]{81}$ (b) $\sqrt[5]{-243}$ (c) $\sqrt{-81}$ (d) $\sqrt[3]{0}$

Solution

(a) $\sqrt[4]{81} = \boxed{3}$ *Because $(3)^4 = 81$ (principal root only)*

(b) $\sqrt[5]{-243} = \boxed{-3}$ *Because $(-3)^5 = -243$*

(c) $\sqrt{-81}$ $\boxed{\text{is not a real number}}$ *Since the radicand is negative **and** the index is even*

(d) $\sqrt[3]{0} = \boxed{0}$ ∎

By defining $\sqrt[n]{a} = a^{1/n}$, we can define $a^{m/n}$ in terms of radicals assuming $a^{1/n}$ is a real number. Throughout the following discussion we will assume that m/n is reduced to lowest terms.

Using exponent rule 2, we have

$$a^{m/n} = (a^m)^{1/n} = \sqrt[n]{a^m}$$

Or, again by exponent rule 2, we can equivalently write

$$a^{m/n} = (a^{1/n})^m = (\sqrt[n]{a})^m$$

Hence, we have the property stated in the box.

If $\sqrt[n]{a}$ is a real number, m and n are positive integers, and m/n is reduced to lowest terms, then

$$\sqrt[n]{a^m} = (\sqrt[n]{a})^m$$

EXAMPLE 2

Express in radical form.

(a) $x^{5/8}$ **(b)** $(x^2 + y^2)^{3/2}$

Solution

(a) $x^{5/8} = \boxed{(\sqrt[8]{x})^5}$ or $\boxed{\sqrt[8]{x^5}}$

Note: It is usually standard practice to leave a radical in the form $\sqrt[8]{x^5}$ rather than the form $(\sqrt[8]{x})^5$.

(b) $(x^2 + y^2)^{3/2} = \boxed{\sqrt{(x^2 + y^2)^3}}$ ∎

As with rational exponents, depending on the problem, you may find one form more convenient to use than another. Using the form $(\sqrt[n]{a})^m$, or finding the root first, is most useful when evaluating a number. For example, to evaluate $16^{3/4}$, we *could* interpret this as $\sqrt[4]{16^3}$. Then

$$\sqrt[4]{16^3} = \sqrt[4]{4,096} = 8$$

which requires quite a bit of multiplication and being able to figure out that the fourth root of 4,096 is 8.

On the other hand, if we interpret $16^{3/4}$ as $(\sqrt[4]{16})^3$, then

$$16^{3/4} = (\sqrt[4]{16})^3 = (2)^3 = 8$$

which is obviously much less work.

When we evaluate numerical expressions it is preferable to use the second form and find the root first. The form $\sqrt[n]{a^m}$ will be most useful when we simplify radical expressions with variables, as we will discuss in the next section.

Let's examine $\sqrt[n]{a^n}$. It is tempting to say that $\sqrt[n]{a^n} = a$. However, if a is negative and n is even, this causes a problem. For example,

$$\sqrt[4]{(-2)^4} = \sqrt[4]{+16} = 2 \qquad \textit{Note that the answer is \textbf{not} } -2.$$

Regardless of the sign of a, if n is even, a^n is always nonnegative and therefore its root always exists. In addition, by definition, an even root is always nonnegative. We therefore have the conclusion given in the box.

For $a \in R$ and n an even positive integer,

$$\sqrt[n]{a^n} = |a|$$

On the other hand, if n is an odd positive integer, we *do* find the following:

For $a \in R$ and n an odd positive integer,

$$\sqrt[n]{a^n} = a$$

EXAMPLE 3

Evaluate the following:

(a) $\sqrt[3]{5^3}$ (b) $\sqrt{(-2)^2}$ (c) $\sqrt[7]{(-8)^7}$ (d) $\sqrt[4]{(-3)^4}$

Solution

(a) $\sqrt[3]{5^3} = \boxed{5}$

(b) $\sqrt{(-2)^2} = |-2| = \boxed{2}$ *Note that the answer is **not** -2.*

(c) $\sqrt[7]{(-8)^7} = \boxed{-8}$

(d) $\sqrt[4]{(-3)^4} = |-3| = \boxed{3}$ *Note:* $\sqrt[4]{(-3)^4} = \sqrt{81} = 3$

*Also note that $(\sqrt[4]{-3})^4$ is **not** equal to $\sqrt[4]{(-3)^4}$ since $\sqrt[4]{-3}$ is not a real number.*   ∎

Up until now we have restricted our discussion to numbers that are perfect nth powers (the nth root is a rational number). Consider the two numbers $\sqrt{9}$ and $\sqrt{21}$. They are numerically different, but conceptually, they are the same: $\sqrt{9}$ is the positive number which when squared yields 9; $\sqrt{21}$ is the positive number which when squared yields 21. The only difference is that $\sqrt{9}$ turns out to be a nice rational number but $\sqrt{21}$ does not.

Keep in mind that decimal representations of irrational numbers must be approximations since the decimal neither terminates nor repeats. When we write approximations, we usually use the symbol "≈." Thus, we can state

$\sqrt{21} \approx 4.58$ if we want an approximate value of $\sqrt{21}$ rounded to the nearest hundredth.

$\sqrt{21} \approx 4.583$ if we want the value rounded to the nearest thousandth.

On occasion we need to do computations involving square roots which are irrational numbers. In such situations we may use a calculator. In the absence of a calculator with the capacity to compute square roots, you may use Appendix C in the back of the book, which contains a table of square roots. For example, to find $\sqrt{41}$ rounded to the nearest thousandth, we look up the number 41 in the left column (under n) and find (under $\sqrt{n}$ column) that $\sqrt{41} \approx 6.403$ (remember that an irrational number can never be written as an exact decimal). To find $\sqrt{750}$ rounded to the nearest hundredth, we look up 75 (under n) and find our answer under the column marked $\sqrt{10n}$ (since $10n = 750$ if $n = 75$), and we obtain $\sqrt{750} \approx 27.386 = 27.39$ rounded to the nearest hundredth.

Exercises 6.1

In Exercises 1–36, evaluate (if possible).

1. $\sqrt[3]{64}$

2. $\sqrt[6]{64}$

3. $\sqrt[4]{81}$

4. $\sqrt{-81}$

5. $\sqrt[8]{-1}$

6. $\sqrt[3]{-125}$

7. $-\sqrt[9]{-1}$

8. $\sqrt[10]{0}$

9. $\sqrt[3]{-343}$

10. $\sqrt[6]{-78}$

11. $-\sqrt[4]{1{,}296}$

12. $-\sqrt[5]{-243}$

13. $\sqrt[8]{256}$

14. $\sqrt[3]{-729}$

15. $\sqrt[7]{78{,}125}$

16. $-\sqrt[3]{343}$

17. $9^{1/2}$

18. $81^{1/4}$

19. $(-125)^{1/3}$

20. $-(256)^{1/8}$

21. $(-256)^{1/4}$

22. $729^{2/3}$

23. $\sqrt[3]{8^2}$

24. $\sqrt[4]{81^3}$

25. $(\sqrt[4]{81})^3$

26. $(\sqrt[3]{125})^2$

27. $\sqrt[3]{(-27)^2}$

28. $\sqrt[4]{-(27)^2}$

29. $\sqrt{(-16)^2}$

30. $\sqrt{-16^2}$

31. $(\sqrt{-16})^2$

32. $\sqrt{(-3)^4}$

33. $-(\sqrt{16})^2$

34. $(\sqrt{-3})^4$

35. $\sqrt[n]{3^{2n}}$

36. $\sqrt[n]{2^n}$

In Exercises 37–50, change to exponential form.

37. $\sqrt[3]{xy}$

38. $\sqrt[4]{x^3y^2}$

39. $\sqrt{x^2 + y^2}$

40. $\sqrt[3]{x^3 - y^3}$

41. $\sqrt[5]{5a^2b^3}$

42. $3\sqrt[4]{2x^2y^3}$

43. $2\sqrt[3]{3xyz^4}$

44. $2\sqrt[3]{ab}$

45. $5\sqrt[3]{(x - y)^2}$

46. $3\sqrt[7]{(x - y)^2(x + y)^3}$

47. $\sqrt[n]{x^n - y^n}$

48. $\sqrt[n]{x^{2n}y^{3n}}$

49. $\sqrt[n]{x^{5n+1}y^{2n-1}}$

50. $\sqrt[n]{x^{2n-3}y^{4n}}$

In Exercises 51–64, change to radical form.

51. $x^{1/3}$

52. $a^{3/5}$

53. $mn^{1/3}$

54. $(mn)^{1/3}$

55. $(-a)^{2/3}$

56. $(-x)^{3/4}$

57. $-a^{2/3}$

58. $-x^{3/4}$

59. $(a^2b)^{1/3}$

60. $(x^3y^2)^{3/5}$

61. $(x^2 + y^2)^{1/2}$

62. $(x^3 - y^3)^{1/3}$

63. $(x^n - y^n)^{1/2}$

64. $(x^2 - y)^{1/n}$

In Exercises 65–70, use the table in Appendix C to find the following values to the nearest hundredth.

65. $\sqrt{38}$ **66.** $\sqrt{95}$ **67.** $\sqrt{120}$ **68.** $\sqrt{380}$

69. $-\sqrt{670}$ **70.** $-\sqrt{880}$

? QUESTIONS FOR THOUGHT

71. Prove the following using the rules of rational exponents, assuming all roots are real numbers:

$$\sqrt[m]{\sqrt[n]{a}} = \sqrt[mn]{a}$$

[*Hint:* Change the radicals to rational exponents.]

72. Given the result of Exercise 71, rewrite the following as a single radical:

   (a) $\sqrt[3]{\sqrt[4]{6}}$ **(b)** $\sqrt{\sqrt[3]{x^2 y}}$ **(c)** $\sqrt[4]{\sqrt[3]{\sqrt{x}}}$

73. Explain why the following is *wrong*:

$$\sqrt{(-4)^2} \overset{?}{=} (\sqrt{(-4)})^2$$

⟳ MINI-REVIEW

74. *Evaluate.* $(-32)^{-2/5}$

75. *Simplify the following and express your answer with positive exponents only.*

$$\frac{x^2 y^{-3}}{x^{-2} y^3}$$

76. *Simplify the following and express your answer with positive exponents only.*

$$\left(\frac{a^{1/3} b^{1/2}}{a^{-1/5}}\right)^4$$

77. *Solve explicitly for y:*

$$x = \frac{2 - y}{3y}$$

78. Maryann can shovel a walk in $\frac{1}{2}$ hour: Michael can shovel the same walk in $\frac{3}{4}$ hour. How long would it take them to shovel the same walk working together?

79. The speed of light is 186,000 miles per second. If 1 mile is approximately 1.6 kilometers, express the speed of light in kilometers per second using scientific notation.

6.2

Simplifying Radicals

Now that we have defined radicals, our next step is to determine how to put them in simplified form. Along with the definition of a radical, the following three properties of radicals will provide us with much of what we will need to simplify radicals:

Properties of Radicals

If $\sqrt[n]{a}$ and $\sqrt[n]{b}$ are real numbers, then

1. $\sqrt[n]{ab} = \sqrt[n]{a}\sqrt[n]{b}$

2. $\sqrt[n]{\dfrac{a}{b}} = \dfrac{\sqrt[n]{a}}{\sqrt[n]{b}}$ $(b \neq 0)$

3. $\sqrt[np]{a^{mp}} = \sqrt[n]{a^m}$ $(a \geq 0)$

Properties 1 and 2 are actually forms of rules 3 and 5 of rational exponents, while property 3 is a result of reducing a fractional exponent.

PROPERTY 1:
$$\sqrt[n]{ab} = (ab)^{1/n} = a^{1/n}b^{1/n} = \sqrt[n]{a}\sqrt[n]{b}$$

PROPERTY 2:
$$\sqrt[n]{\frac{a}{b}} = \left(\frac{a}{b}\right)^{1/n} = \frac{a^{1/n}}{b^{1/n}} = \frac{\sqrt[n]{a}}{\sqrt[n]{b}}$$

PROPERTY 3:
$$\sqrt[np]{a^{mp}} = a^{(mp)/(np)} = a^{m/n} = \sqrt[n]{a^m}$$

Our goal in this section is to write expressions in simplest radical form. We *define* simplest radical form in the accompanying box.

An expression is in ***simplest radical form*** if:

1. All factors of the radicand have exponents less than the index.

   For example, $\sqrt[3]{y^5}$ violates this condition.

2. There are no fractions under the radical.

   For example, $\sqrt{\dfrac{3}{5}}$ violates this condition.

3. There are no radicals in the denominator of a fraction.

   For example, $\dfrac{4}{\sqrt[3]{x}}$ violates this condition.

4. The greatest common factor of the index and the exponents of *all* the radicand factors is 1.

   For example, $\sqrt[8]{x^4}$ violates this condition.

To simplify matters, from this point on, all variables will represent positive real numbers.

Given the definition of a radical and the three properties of radicals, we can simplify many radical expressions according to the criteria given in the box. Here are a few examples of how we would use the properties to simplify radicals:

1. $\sqrt{18} = \sqrt{9 \cdot 2} = \sqrt{3^2 \cdot 2}$ *Factor out the greatest perfect square. Then apply property 1.*

$= \sqrt{3^2}\sqrt{2}$

$= 3\sqrt{2}$ *Since $\sqrt{3^2} = 3$*

2. $\sqrt[5]{64} = \sqrt[5]{2^6}$ *Factor out the greatest perfect fifth power. Then apply property 1.*

$= \sqrt[5]{2^5 \cdot 2}$

$= \sqrt[5]{2^5}\sqrt[5]{2}$

$= 2\sqrt[5]{2}$ *Since $\sqrt[5]{2^5} = 2$*

3. $\sqrt[4]{\dfrac{81}{16}} = \dfrac{\sqrt[4]{81}}{\sqrt[4]{16}}$ *Property 2*

$= \dfrac{\sqrt[4]{3^4}}{\sqrt[4]{2^4}}$

$= \dfrac{3}{2}$

4. $\sqrt[8]{x^6}$ *Factor the greatest common factor from the index and the exponent of the radicand.*

$= \sqrt[4 \cdot 2]{x^{3 \cdot 2}}$

$= \sqrt[4]{x^3}$

An alternative way to simplify this expression is to first convert the radical expression into rational exponents:

$$\sqrt[8]{x^6} = x^{6/8} = x^{3/4} = \sqrt[4]{x^3}$$

EXAMPLE 1

Express the following in simplest radical form. Assume all variables are positive numbers.

(a) $\sqrt{25x^8y^6}$ **(b)** $\sqrt[3]{\dfrac{27}{y^{18}}}$

Solution

(a) $\sqrt{25x^8y^6}$ *Apply property 1*

$= \sqrt{25}\sqrt{x^8}\sqrt{y^6}$

$= \boxed{5 \cdot x^4 \cdot y^3}$ *Since $5^2 = 25$, $(x^4)^2 = x^8$, and $(y^3)^2 = y^6$*

(b) $\sqrt[3]{\dfrac{27}{y^{18}}} = \dfrac{\sqrt[3]{27}}{\sqrt[3]{y^{18}}}$ *Property 2*

$= \boxed{\dfrac{3}{y^6}}$ *Since $3^3 = 27$ and $(y^6)^3 = y^{18}$*   ∎

For a power to be a perfect *n*th power, the exponent must be divisible by *n*. In all cases of Example 1, the factors were perfect *n*th powers, where *n* was the index of the radical. Our next step is to simplify radicals with radicand factors that are not perfect nth powers. In these cases we try to factor the greatest perfect nth power from the expression.

EXAMPLE 2

Express the following in simplest radical form:

(a) $\sqrt{x^7}$ (b) $\sqrt[6]{x^{17}}$ (c) $\sqrt[3]{24}$ (d) $\sqrt[4]{64a^{14}b^{17}c^{24}}$

Solution

(a) $\sqrt{x^7}$

$= \sqrt{x^6 \cdot x}$ *Factor the greatest perfect square factor of x^7, which is x^6,*
from the radicand.

$= \sqrt{x^6}\sqrt{x}$

$= \boxed{x^3\sqrt{x}}$ $\sqrt{x^6} = x^3$ because $(x^3)^2 = x^6$.

(b) $\sqrt[6]{x^{17}}$

$= \sqrt[6]{x^{12} \cdot x^5}$ *The greatest perfect sixth power factor of x^{17} is x^{12}. Thus, we*
factor x^{12} from x^{17}.

$= \sqrt[6]{x^{12}}\sqrt[6]{x^5}$

$= \boxed{x^2\sqrt[6]{x^5}}$ $\sqrt[6]{x^{12}} = x^2$ because $(x^2)^6 = x^{12}$.

Note that the exponent of x under the radical is less than the index, thereby satisfying criterion 1 for simplifying radicals.

(c) $\sqrt[3]{24} = \sqrt[3]{8 \cdot 3}$

$= \sqrt[3]{8}\sqrt[3]{3}$

$= 2\sqrt[3]{3}$

(d) $\sqrt[4]{64a^{14}b^{17}c^{24}}$

$= \sqrt[4]{(16 \cdot 4)(a^{12} \cdot a^2)(b^{16} \cdot b)(c^{24})}$ *Factor the greatest perfect fourth*
power from each factor.

$= \sqrt[4]{(16a^{12}b^{16}c^{24})(4a^2b)}$

$= \sqrt[4]{16}\sqrt[4]{a^{12}}\sqrt[4]{b^{16}}\sqrt[4]{c^{24}}\sqrt[4]{4a^2b}$ *By property 1*

$= \boxed{2a^3b^4c^6\sqrt[4]{4a^2b}}$

Again, note that all exponents of factors under the radical are less than the index. ■

EXAMPLE 3

Express in simplest radical form.

(a) $\sqrt[5]{(x^2 + 2)^8}$ (b) $\sqrt[3]{x^3 + y^3}$ (c) $3xy\sqrt{32x^7}$

Solution

(a) $\sqrt[5]{(x^2 + 2)^8} = \sqrt[5]{(x^2 + 2)^5(x^2 + 2)^3}$

$= \sqrt[5]{(x^2 + 2)^5}\sqrt[5]{(x^2 + 2)^3}$

$= \boxed{(x^2 + 2)\sqrt[5]{(x^2 + 2)^3}}$

(b) $\sqrt[3]{x^3 + y^3}$ $\boxed{\text{cannot be simplified}}$ *Remember, $x^3 + y^3$ is not the same as $(x + y)^3$.*

(c) $3xy\sqrt{32x^7}$ *Simplify the radical first.*

$= 3xy\sqrt{16 \cdot 2 \cdot x^6 x}$

$= 3xy\sqrt{16x^6}\sqrt{2x}$

$= 3xy(4x^3\sqrt{2x})$ *Then multiply.*

$= \boxed{12x^4y\sqrt{2x}}$ ∎

The properties of radicals can also be used to simplify products and quotients of radicals. When the first two properties are read from right to left, they state that *if the indices are the same*, the product (quotient) of radicands is the radicand of the product (quotient).

EXAMPLE 4

Perform the operations and simplify.

(a) $\sqrt{3a^2b}\sqrt{6ab^3}$ **(b)** $(ab\sqrt[3]{2a^2b})(2a\sqrt[3]{4ab})$

Solution

(a) We could simplify each radical first, but after we multiply it may be necessary to simplify again. It is usually better to multiply radicands first.

$\sqrt{3a^2b}\sqrt{6ab^3} = \sqrt{(3a^2b)(6ab^3)}$

$= \sqrt{18a^3b^4}$ *Then simplify.*

$= \sqrt{(9a^2b^4)(2a)}$

$= \sqrt{9a^2b^4}\sqrt{2a}$

$= \boxed{3ab^2\sqrt{2a}}$

(b) Since this is all multiplication, we would use the associative and commutative properties to reorder and regroup variables, and then use radical property 1 to multiply the radicals.

$(ab\sqrt[3]{2a^2b})(2a\sqrt[3]{4ab}) = (ab)(2a)\sqrt[3]{(2a^2b)(4ab)}$

$= 2a^2b\sqrt[3]{(8a^3b^2)}$ *Then simplify the radical.*

$= 2a^2b(\sqrt[3]{8a^3}\sqrt[3]{b^2})$

$= 2a^2b(2a\sqrt[3]{b^2})$ *Multiply the expressions outside the radical.*

$= \boxed{4a^3b\sqrt[3]{b^2}}$ ∎

Next we examine radicals of quotients.

EXAMPLE 5

Express in simplest radical form. $\sqrt{\dfrac{32s^9}{4s^{17}}}$

Solution

$\sqrt{\dfrac{32s^9}{4s^{17}}}$ *Do not apply property 2 without thinking—try to simplify fractions first.*

$= \sqrt{\dfrac{8}{s^8}}$ *Then apply property 2.*

$= \dfrac{\sqrt{8}}{\sqrt{s^8}}$ *We simplify the numerator and denominator by using property 1.*

$= \boxed{\dfrac{2\sqrt{2}}{s^4}}$

∎

Rationalizing the Denominator

If, after we reduce a fraction under a radical with index n and apply property 2, there is a factor in the denominator which is not a perfect nth power, a radical would remain in the denominator, violating criterion 3 for simplifying radicals. In such a case we would have to **rationalize**, or get rid of the radical, in the denominator. To do this, we apply the fundamental principle of fractions, as shown in Example 6.

EXAMPLE 6

Simplify. $\sqrt{\dfrac{2}{5}}$

Solution

$\sqrt{\dfrac{2}{5}}$ *This violates criterion 2 of simplified form for radicals. We apply property 2.*

$= \dfrac{\sqrt{2}}{\sqrt{5}}$ *This violates criterion 3 (a radical in the denominator).*

We apply the fundamental principle of fractions and multiply the numerator and denominator by $\sqrt{5}$. Why $\sqrt{5}$? We choose $\sqrt{5}$ because the product of $\sqrt{5}$ and the denominator will yield the square root of a perfect square, which yields a rational expression. Thus,

$$\dfrac{\sqrt{2}}{\sqrt{5}} = \dfrac{\sqrt{2} \cdot \sqrt{5}}{\sqrt{5} \cdot \sqrt{5}} = \dfrac{\sqrt{2 \cdot 5}}{\sqrt{5^2}} = \boxed{\dfrac{\sqrt{10}}{5}}$$

∎

In rationalizing a denominator, what we try to do is to multiply the numerator *and* the denominator of the fraction by the expression which will make the denominator the nth root of a perfect nth power.

EXAMPLE 7

Simplify. $\sqrt[3]{\dfrac{1}{y}}$

Solution

$$\sqrt[3]{\dfrac{1}{y}} = \dfrac{\sqrt[3]{1}}{\sqrt[3]{y}} = \dfrac{1}{\sqrt[3]{y}}$$

We multiply the numerator and denominator by $\sqrt[3]{y^2}$. Why $\sqrt[3]{y^2}$? Because multiplying $\sqrt[3]{y^2}$ by the original denominator, $\sqrt[3]{y}$, will yield $\sqrt[3]{y^3}$, the cube root of a perfect cube. Thus,

$$\sqrt[3]{\dfrac{1}{y}} = \dfrac{1}{\sqrt[3]{y}} = \dfrac{1 \cdot \sqrt[3]{y^2}}{\sqrt[3]{y}\sqrt[3]{y^2}}$$

$$= \dfrac{\sqrt[3]{y^2}}{\sqrt[3]{y \cdot y^2}}$$

$$= \dfrac{\sqrt[3]{y^2}}{\sqrt[3]{y^3}} \qquad\qquad \textit{Recall that } \sqrt[3]{y^3} = y.$$

$$= \boxed{\dfrac{\sqrt[3]{y^2}}{y}} \qquad\qquad \textit{Notice that we no longer have}$$
$$\textit{a radical in the denominator.}$$

EXAMPLE 8

Express the following in simplest radical form:

(a) $\sqrt{\dfrac{3}{2x^2}}$ **(b)** $\sqrt[3]{\dfrac{3}{2x^2}}$

Solution

(a) $\sqrt{\dfrac{3}{2x^2}}$ *Apply property 2.*

$$= \dfrac{\sqrt{3}}{\sqrt{2x^2}} \qquad \textit{Simplify the denominator by property 1: } \sqrt{2x^2} = x\sqrt{2}$$

$$= \dfrac{\sqrt{3}}{x\sqrt{2}} \qquad \textit{To make the denominator into the square root of a perfect}$$
$$\textit{square, mutiply the numerator and denominator by } \sqrt{2}.$$

$$= \dfrac{\sqrt{3} \cdot \sqrt{2}}{x\sqrt{2} \cdot \sqrt{2}}$$

$$= \dfrac{\sqrt{6}}{x\sqrt{2^2}} \qquad \textit{Now we have a perfect square under the radical in the}$$
$$\textit{denominator.}$$

$$= \boxed{\dfrac{\sqrt{6}}{2x}}$$

(b) $\sqrt[3]{\dfrac{3}{2x^2}}$ *Apply property 2.*

$$= \dfrac{\sqrt[3]{3}}{\sqrt[3]{2x^2}} \qquad \textit{To change } \sqrt[3]{2x^2} \textit{ into the cube root of a perfect cube, we}$$
$$\textit{must multiply the numerator and denominator by } \sqrt[3]{4x}.$$
$$\textit{Then } \sqrt[3]{2x^2}\sqrt[3]{4x} = \sqrt[3]{8x^3}.$$

$$= \dfrac{\sqrt[3]{3}\sqrt[3]{4x}}{\sqrt[3]{2x^2}\sqrt[3]{4x}}$$

$$= \dfrac{\sqrt[3]{12x}}{\sqrt[3]{8x^3}} \qquad \textit{Now the radicand in the denominator is a perfect cube.}$$

$$= \boxed{\dfrac{\sqrt[3]{12x}}{2x}}$$

EXAMPLE 9

Express the following in simplest radical form:

$$\frac{(5b\sqrt{2ab^2})(b\sqrt{24ab})}{4b\sqrt{18ab^4}}$$

Solution

We could simplify each radical first, but after simplifying, multiplying, and reducing, we may have to simplify again. Since this expression consists of products and quotients, we can use properties 1 and 2 to collect all radicands under one radical.

$$\frac{(5b\sqrt{2ab^2})(b\sqrt{24ab})}{4b\sqrt{18ab^4}}$$

$$= \frac{5b^2}{4b}\sqrt{\frac{(2ab^2)(24ab)}{18ab^4}}$$ *Then reduce the fractions.*

$$= \frac{5b^2}{4b}\sqrt{\frac{(2ab^2)(24ab)}{18ab^4}}$$

$$= \frac{5b}{4}\sqrt{\frac{8a}{3b}}$$ *At this point, we will take a slightly different approach. Instead of applying property 2 and then rationalizing, we can apply the fundamental principle to the fraction under the radical to change the denominator into a perfect square.*

Then apply property 2.

$$= \frac{5b}{4}\sqrt{\frac{8a \cdot 3b}{3b \cdot 3b}}$$

$$= \frac{5b}{4}\frac{\sqrt{(8)(3ab)}}{\sqrt{(3b)(3b)}}$$

$$= \frac{5b}{4}\frac{\sqrt{4 \cdot (2 \cdot 3ab)}}{\sqrt{(3b)^2}}$$

$$= \frac{5b}{4}\frac{2\sqrt{6ab}}{3b}$$ *Then (reduce and) multiply.*

$$= \boxed{\frac{5\sqrt{6ab}}{6}}$$

The last criterion for simplifying radical expressions requires us to factor the greatest common factor of the index and all the exponents of factors of the radicand; this requires property 3 for radicals:

$$\sqrt[np]{a^{mp}} = \sqrt[n]{a^m}$$ *We can factor out the greatest common factor of the index and the exponent of the radicand.*

For example, $\sqrt[3]{x}$ and $\sqrt[6]{x^2}$ are identical (assuming $x \geq 0$). If we convert to exponential notation, we can see that

$$\sqrt[6]{x^2} = x^{2/6} = x^{1/3} = \sqrt[3]{x}$$

Thus, we can find $\sqrt[6]{27^2}$ by the following:

$$\sqrt[6]{27^2} = \sqrt[6]{729} = 3$$

Or we can recognize that $\sqrt[6]{27^2} = \sqrt[3]{27} = 3$, which takes less effort to evaluate.

We can use property 3 to "reduce the index with the exponent of the radicand" or we can convert to fractional exponents, reduce the fractional exponent, and then rewrite the answer in radical form.

Property 3 is also used to find products and quotients of radicals with different indices. (Keep in mind that properties 1 and 2 require that the indices be the same.)

EXAMPLE 10

Solution

Simplify. $\sqrt[3]{x^2}\sqrt[5]{x^4}$

$\sqrt[3]{x^2}\sqrt[5]{x^4}$ *The indices are not the same; thus, we cannot use property 1.*

The least common multiple of the indices, 5 and 3, is 15. We change the radicals using property 3 as follows:

$$\sqrt[3]{x^2}\sqrt[5]{x^4} = \sqrt[3\cdot5]{x^{2\cdot5}}\,\sqrt[5\cdot3]{x^{4\cdot3}}$$
$$= \sqrt[15]{x^{10}}\,\sqrt[15]{x^{12}} \qquad \text{*Since the indices are the same*}$$
$$= \sqrt[15]{x^{22}} \qquad\qquad \text{*we can now apply property 1.*}$$
$$= \boxed{x\sqrt[15]{x^7}}$$

Actually, it is more instructive to change radicals into rational exponents and to simplify the problem as follows:

$$\sqrt[3]{x^2}\sqrt[5]{x^4} = x^{2/3}x^{4/5}$$
$$= x^{2/3 + 4/5}$$
$$= x^{10/15 + 12/15}$$
$$= x^{22/15} \qquad \text{*Change back to radical notation.*}$$
$$= \sqrt[15]{x^{22}} \qquad \text{*Simplify.*}$$
$$= \boxed{x\sqrt[15]{x^7}}$$ ∎

Exercises 6.2

Simplify the following. Assume that all variables represent positive real numbers.

1. $\sqrt{56}$

2. $\sqrt{45}$

3. $\sqrt{48}$

4. $\sqrt{54}$

5. $\sqrt[5]{64}$

6. $\sqrt[4]{243}$

7. $\sqrt{8}\sqrt{18}$

8. $\sqrt{12}\sqrt{75}$

9. $\sqrt{64x^8}$

10. $\sqrt[4]{64x^8}$

11. $\sqrt[4]{81x^{12}}$

12. $\sqrt{81x^{12}}$

13. $\sqrt{128x^{60}}$

14. $\sqrt[3]{128x^{60}}$

15. $\sqrt[4]{128x^{60}}$

16. $\sqrt[6]{128x^{60}}$

17. $\sqrt[5]{128x^{60}}$

18. $\sqrt[7]{128x^{60}}$

19. $\sqrt[3]{x^3y^6}$

20. $\sqrt{x^4y^8}$

21. $\sqrt{32a^2b^4}$

22. $\sqrt{54x^6y^{12}}$

23. $\sqrt[5]{a^{35}b^{75}}$

24. $\sqrt[6]{x^{36}y^{72}}$

25. $\sqrt{x^3y}\sqrt{xy^3}$

26. $\sqrt{6ab}\sqrt{6a^5b^5}$

27. $\sqrt{\dfrac{1}{2}}$

28. $\sqrt{\dfrac{2}{7}}$

29. $\dfrac{\sqrt{x}}{\sqrt{5}}$

30. $\dfrac{\sqrt{5}}{\sqrt{a}}$

31. $\sqrt{\dfrac{45}{4}}$

32. $\sqrt{\dfrac{8}{9}}$

33. $\dfrac{1}{\sqrt{75}}$

34. $\sqrt{\dfrac{3}{8}}$

35. $\sqrt{64x^5y^8}$

36. $\sqrt{12a^3b^9}$

37. $\sqrt[3]{81x^8y^7}$

38. $\sqrt[3]{24a^5b^2}$

39. $\sqrt[4]{54y^2}\sqrt[4]{48y^4}$

40. $\sqrt[3]{20ab^5}\sqrt[3]{50b^4}$

41. $\sqrt[6]{(x+y^2)^6}$

42. $\sqrt[6]{(a^2+b)^6}$

43. $\sqrt[4]{x^4-y^4}$

44. $\sqrt[3]{x^3-y^3}$

45. $(2s\sqrt{6t})(5t\sqrt{3s})$

46. $(4a\sqrt{10b})(3b\sqrt{2a})$

47. $(3a\sqrt[3]{2b^4})(2a^2\sqrt[3]{4b^2})$

48. $(3a\sqrt[3]{5b^2a})(2b\sqrt[3]{25a^2b})$

49. $\sqrt[3]{\dfrac{x^3y^6}{8}}$

50. $\sqrt[5]{\dfrac{r^{10}s^{15}}{32}}$

51. $\sqrt[4]{\dfrac{32x^9}{y^{12}}}$

52. $\sqrt[3]{\dfrac{32x^9}{y^{12}}}$

53. $\dfrac{\sqrt{54xy}}{\sqrt{2xy}}$

54. $\dfrac{\sqrt{6x^2y}}{\sqrt{3xy}}$

55. $\dfrac{\sqrt[4]{x^2y^{17}}}{\sqrt[4]{x^{14}y}}$

56. $\dfrac{\sqrt[3]{a^{13}b^2}}{\sqrt[3]{a^4b^5}}$

57. $\sqrt{\dfrac{3xy}{5x^2y}}$

58. $\sqrt{\dfrac{5ab}{3a^2b}}$

59. $\sqrt{\dfrac{3x^2y}{x^3y^4}}$

60. $\sqrt{\dfrac{2a^2b^3}{a^5b^{18}}}$

61. $\sqrt[3]{\dfrac{3}{2}}$

62. $\sqrt[3]{\dfrac{2}{3}}$

63. $\sqrt[3]{\dfrac{9}{4}}$

64. $\sqrt[3]{\dfrac{4}{9}}$

65. $\sqrt[4]{\dfrac{9}{4}}$

66. $\sqrt[4]{\dfrac{4}{9}}$

67. $\sqrt[3]{\dfrac{81x^2y^4}{2x^3y}}$

68. $\sqrt[3]{\dfrac{64a^2b^5}{9a^4b^2}}$

69. $\dfrac{3a^2\sqrt{a^2x^5}}{9a^5\sqrt{a^6x}}$

70. $\dfrac{5x^2\sqrt{a^3b^2}}{4y^2\sqrt{a^7b^3}}$

71. $\dfrac{-3r^2s\sqrt{32r^2s^5}}{2r\sqrt{2r^5}}$

72. $\dfrac{-7a^2b\sqrt{81a^2b}}{14a\sqrt{9a^7}}$

73. $\sqrt[12]{a^6}$

74. $\sqrt[16]{x^4}$

75. $(\sqrt[3]{x})(\sqrt[4]{x^3})$

76. $(\sqrt[3]{x^2})(\sqrt{x^3})$

77. $\dfrac{\sqrt[3]{a^2}}{\sqrt{a}}$

78. $\dfrac{\sqrt[3]{a}}{\sqrt[4]{a^2}}$

79. $\sqrt[n]{x^{5n}y^{3n}}$

80. $\sqrt[n]{x^{2n}y^{4n}}$

? QUESTION FOR THOUGHT

81. Discuss what is *wrong* (if anything) with each of the following:

(a) $(\sqrt{-2})^2 \overset{?}{=} \sqrt{(-2)^2} \overset{?}{=} \sqrt{4} \overset{?}{=} 2$

(b) $2 \overset{?}{=} \sqrt[6]{64} \overset{?}{=} \sqrt[6]{(-8)^2} \overset{?}{=} \sqrt[3]{-8} \overset{?}{=} -2$

(c) $\sqrt{x^2 - y^2} \overset{?}{=} x - y$

(d) $(a^5 + b^5)^{1/5} \overset{?}{=} a + b$

⟳ MINI-REVIEW

82. *Simplify and express your answer as a single fraction with positive exponents only.*

$$(x^{-1} + 2)^{-1}$$

83. Which property is illustrated by the following?

$$(x + 3)(x - 4) = x(x - 4) + 3(x - 4)$$

84. *Perform the operations and express your answer in simplest form:*

$$\dfrac{3x}{x^2 + x - 6} - \dfrac{2}{x^2 + 2x - 3}$$

85. If two-thirds of a number is 5 more than three-fifths of the number, what is the number?

6.3

Combining Radicals

We combine terms with radical factors in the same way we combined terms with variable factors—through the use of the distributive property. Just as we can combine

$$3x + 4x = (3 + 4)x \qquad \textit{Distributive property}$$
$$= 7x$$

we can also combine

$$3\sqrt{2} + 4\sqrt{2} = (3 + 4)\sqrt{2} \qquad \textit{Distributive property}$$
$$= 7\sqrt{2}$$

EXAMPLE 1

Simplify.

(a) $7\sqrt{3} - 4\sqrt{3} + 6\sqrt{3}$ **(b)** $5\sqrt{2} - 8\sqrt{3} - (\sqrt{2} - 7\sqrt{3})$

Solution

(a) $7\sqrt{3} - 4\sqrt{3} + 6\sqrt{3}$ *Apply the distributive property.*

$$= (7 - 4 + 6)\sqrt{3}$$

$$= \boxed{9\sqrt{3}}$$

(b) $5\sqrt{2} - 8\sqrt{3} - (\sqrt{2} - 7\sqrt{3})$ *First remove parentheses.*

$$= 5\sqrt{2} - 8\sqrt{3} - \sqrt{2} + 7\sqrt{3}$$

$$= (5 - 1)\sqrt{2} + (-8 + 7)\sqrt{3}$$

$$= \boxed{4\sqrt{2} - \sqrt{3}}$$

We may encounter a problem where, at first glance, it may seem that the radicals cannot be combined, as in the following case:

Simplify. $\sqrt{18x^3} - x\sqrt{32x}$

However, if we simplify each radical term first, we will find that we *can* combine the two radicals:

$$\sqrt{18x^3} - x\sqrt{32x} = \sqrt{9x^2}\sqrt{2x} - x\sqrt{16}\sqrt{2x}$$
$$= 3x\sqrt{2x} - 4x\sqrt{2x}$$
$$= (3 - 4)x\sqrt{2x}$$
$$= \boxed{-x\sqrt{2x}}$$

Thus, our first step should be to simplify each radical expression.

EXAMPLE 2

Simplify each of the following.

(a) $\sqrt{27} - \sqrt{81} + \sqrt{12}$ **(b)** $5\sqrt{4x^3} + 7x\sqrt{8x} - 2x\sqrt{9x} + \sqrt{2x}$

(c) $3x\sqrt[3]{24x^2} + 4x\sqrt[3]{54x^5} - 2\sqrt[3]{81x^5}$

Solution

(a) $\sqrt{27} - \sqrt{81} + \sqrt{12}$ *Simplify each term first.*

$= 3\sqrt{3} - 9 + 2\sqrt{3}$ *Then combine where possible.*

$= \boxed{5\sqrt{3} - 9}$

Note that we cannot combine $5\sqrt{3} - 9$, just as we cannot combine $5x - 9$.

(b) $5\sqrt{4x^3} + 7x\sqrt{8x} - 2x\sqrt{9x} + \sqrt{2x}$

$= 5(\sqrt{4x^2}\sqrt{x}) + 7x(\sqrt{4}\sqrt{2x}) - 2x(\sqrt{9}\sqrt{x}) + \sqrt{2x}$

$= 5(2x\sqrt{x}) + 7x(2\sqrt{2x}) - 2x(3\sqrt{x}) + \sqrt{2x}$

$= 10x\sqrt{x} + 14x\sqrt{2x} - 6x\sqrt{x} + \sqrt{2x}$

$= \boxed{4x\sqrt{x} + 14x\sqrt{2x} + \sqrt{2x} \quad \text{or} \quad 4x\sqrt{x} + (14x + 1)\sqrt{2x}}$

(c) $3x\sqrt[3]{24x^2} + 4x\sqrt[3]{54x^5} - 2\sqrt[3]{81x^5}$ *Simplify each radical.*

$= 3x(\sqrt[3]{8}\sqrt[3]{3x^2}) + 4x(\sqrt[3]{27x^3}\sqrt[3]{2x^2}) - 2(\sqrt[3]{27x^3}\sqrt[3]{3x^2})$

$= 3x(2\sqrt[3]{3x^2}) + 4x(3x\sqrt[3]{2x^2}) - 2(3x\sqrt[3]{3x^2})$

$= 6x\sqrt[3]{3x^2} + 12x^2\sqrt[3]{2x^2} - 6x\sqrt[3]{3x^2}$ *Then combine where possible.*

$= \boxed{12x^2\sqrt[3]{2x^2}}$ ∎

EXAMPLE 3

Perform the indicated operations and simplify.

(a) $\sqrt{3} + \dfrac{6}{\sqrt{3}}$ **(b)** $5\sqrt{\dfrac{x}{y}} - \dfrac{\sqrt{xy}}{y}$

Solution

(a) $\sqrt{3} + \dfrac{6}{\sqrt{3}}$ *Rationalize the denominator.*

$= \sqrt{3} + \dfrac{6\sqrt{3}}{\sqrt{3}\sqrt{3}}$

$= \sqrt{3} + \dfrac{6\sqrt{3}}{3}$ *Reduce.*

$= \sqrt{3} + 2\sqrt{3}$ *Combine.*

$= \boxed{3\sqrt{3}}$

(b) $5\sqrt{\dfrac{x}{y}} - \dfrac{\sqrt{xy}}{y}$ *Apply property 2.*

$= \dfrac{5}{1} \cdot \dfrac{\sqrt{x}}{\sqrt{y}} - \dfrac{\sqrt{xy}}{y}$

$= \dfrac{5\sqrt{x}}{\sqrt{y}} - \dfrac{\sqrt{xy}}{y}$ *Rationalize the first denominator.*

$= \dfrac{5\sqrt{x}\sqrt{y}}{\sqrt{y}\sqrt{y}} - \dfrac{\sqrt{xy}}{y}$

$= \dfrac{5\sqrt{xy}}{y} - \dfrac{\sqrt{xy}}{y}$ *Combine fractions.*

$= \dfrac{5\sqrt{xy} - \sqrt{xy}}{y}$

$= \boxed{\dfrac{4\sqrt{xy}}{y}}$ ∎

EXAMPLE 4   *Perform the operations and simplify.* $2\sqrt[3]{25} + 4\sqrt[3]{\dfrac{8}{5}}$

Solution $2\sqrt[3]{25} + 4\sqrt[3]{\dfrac{8}{5}}$ *Apply property 2.*

$= 2\sqrt[3]{25} + \dfrac{4\sqrt[3]{8}}{\sqrt[3]{5}}$ *Simplify $\sqrt[3]{8}$.*

$= 2\sqrt[3]{25} + \dfrac{4 \cdot 2}{\sqrt[3]{5}}$

$= 2\sqrt[3]{25} + \dfrac{8}{\sqrt[3]{5}}$ *Rationalize the denominator; multiply numerator and denominator by $\sqrt[3]{25}$. (Why $\sqrt[3]{25}$?)*

$= 2\sqrt[3]{25} + \dfrac{8\sqrt[3]{25}}{\sqrt[3]{5}\sqrt[3]{25}}$

$= 2\sqrt[3]{25} + \dfrac{8\sqrt[3]{25}}{\sqrt[3]{125}}$

$= 2\sqrt[3]{25} + \dfrac{8\sqrt[3]{25}}{5}$ *Now we add; find the LCD and change to equivalent fractions.*

$= \dfrac{5 \cdot 2\sqrt[3]{25}}{5} + \dfrac{8\sqrt[3]{25}}{5}$

$= \dfrac{10\sqrt[3]{25} + 8\sqrt[3]{25}}{5}$

$= \boxed{\dfrac{18\sqrt[3]{25}}{5}}$ ∎

Exercises 6.3

*Perform the indicated operations. Express your answers in simplest radical form. Assume
that all variables represent positive real numbers.*

1. $5\sqrt{3} - \sqrt{3}$

2. $8\sqrt{7} - 2\sqrt{7}$

3. $2\sqrt{5} - 4\sqrt{5} - \sqrt{5}$

4. $9\sqrt{2} - \sqrt{2} - 12\sqrt{2}$

5. $8\sqrt{3} - (4\sqrt{3} - 2\sqrt{6})$

6. $6\sqrt{5} - (3\sqrt{7} - \sqrt{5})$

7. $2\sqrt{3} - 2\sqrt{5} - (\sqrt{3} - \sqrt{5})$

8. $\sqrt{6} - 2\sqrt{2} - (\sqrt{2} - \sqrt{6})$

9. $2\sqrt{x} - 5\sqrt{x} + 3\sqrt{x}$

10. $5\sqrt{ab} - 7\sqrt{ab} + 3\sqrt{a}$

11. $5a\sqrt{b} - 3a^3\sqrt{b} + 2a\sqrt{b}$

12. $3xy - 2\sqrt{y} - 5x\sqrt{y}$

13. $3x\sqrt[3]{x^2} - 2\sqrt[3]{x^2} + 6\sqrt[3]{x^2}$

14. $6 - 4x\sqrt[3]{x^2} + 3\sqrt[3]{x^2} - 8$

15. $(7 - 3\sqrt[3]{a}) - (6 - \sqrt[3]{a})$

16. $(2 - 5\sqrt[3]{x}) - (6 + \sqrt[3]{x})$

17. $\sqrt{12} - \sqrt{27}$

18. $\sqrt{18} - \sqrt{8} + \sqrt{32}$

19. $\sqrt{24} - \sqrt{27} + \sqrt{54}$

20. $\sqrt{12} + \sqrt{18} + \sqrt{24}$

21. $6\sqrt{3} - 4\sqrt{81}$

22. $5\sqrt{2} - 6\sqrt{16}$

23. $3\sqrt{24} - 5\sqrt{48} - \sqrt{6}$

24. $3\sqrt{8} - 5\sqrt{32} + 2\sqrt{27}$

25. $3\sqrt[3]{24} - 5\sqrt[3]{48} - \sqrt[3]{6}$

26. $3\sqrt[3]{8} - 5\sqrt[3]{32} + 2\sqrt[3]{27}$

27. $2a\sqrt{ab^2} - 3b\sqrt{a^2b} - ab\sqrt{ab}$

28. $x\sqrt{x^3y} + y\sqrt{y^3}$

29. $\sqrt[3]{x^4} - x\sqrt[3]{x}$

30. $3\sqrt[3]{16x} - 5\sqrt[3]{2x} - 3x\sqrt[3]{2}$

31. $\sqrt{20x^9y^8} + 2xy\sqrt{5x^7y^6}$

32. $2b^2\sqrt{48a^7b^6} - 5a\sqrt{27a^5b^{10}}$

33. $5\sqrt[3]{9x^5} - 3x\sqrt[3]{x^2} + 2x\sqrt[3]{72x^2}$

34. $3a\sqrt[3]{ab^4} - 5b\sqrt[3]{8a^4b} - ab\sqrt[3]{2ab}$

35. $4\sqrt[4]{16x} - 7\sqrt[4]{x^5} + x\sqrt[4]{81x}$

36. $5\sqrt[4]{32x^5} - 3x\sqrt[4]{2x} + 7\sqrt[4]{x^5}$

37. $\dfrac{1}{\sqrt{5}} + 2$

38. $\dfrac{1}{\sqrt{3}} - 3$

39. $\dfrac{12}{\sqrt{6}} - 2\sqrt{6}$

40. $\dfrac{15}{\sqrt{3}} - 3\sqrt{3}$

41. $\sqrt{\dfrac{1}{2}} + \sqrt{2}$

42. $\sqrt{\dfrac{2}{7}} - \sqrt{7}$

43. $\sqrt{\dfrac{5}{2}} + \sqrt{\dfrac{2}{5}}$

44. $\sqrt{\dfrac{3}{5}} - \sqrt{\dfrac{5}{3}}$

45. $\sqrt{\dfrac{1}{7}} - 3\sqrt{\dfrac{1}{5}}$

46. $2\sqrt{\dfrac{1}{5}} + \sqrt{\dfrac{2}{3}}$

47. $\dfrac{1}{\sqrt[3]{2}} - 6\sqrt[3]{4}$

48. $\dfrac{1}{\sqrt[3]{5}} - 4\sqrt[3]{25}$

49. $\sqrt{\dfrac{1}{x}} + \sqrt{\dfrac{1}{y}}$

50. $\sqrt{\dfrac{1}{x}} - \sqrt{y}$

51. $\dfrac{1}{\sqrt[3]{9}} - \dfrac{3}{\sqrt[3]{3}}$

52. $\dfrac{5}{\sqrt[3]{25}} - \dfrac{15}{\sqrt[3]{5}}$

53. $3\sqrt{\dfrac{2}{49}} + 3\sqrt{7}$

54. $2\sqrt{2} - 3\sqrt{\dfrac{5}{4}}$

55. $3\sqrt{10} - \dfrac{4}{\sqrt{10}} + \dfrac{2}{\sqrt{10}}$

56. $2\sqrt{30} - \dfrac{5}{\sqrt{30}} + \dfrac{1}{\sqrt{30}}$

57. $6\sqrt[3]{25} - \dfrac{15}{\sqrt[3]{5}} + 5\sqrt[3]{\dfrac{1}{5}}$

58. $2\sqrt[3]{49} - \dfrac{14}{\sqrt[3]{7}} + 5\sqrt[3]{\dfrac{1}{7}}$

MINI-REVIEW

Simplify the following and express your answers with positive exponents only.

59. $(xy^{-1})^{-2}(x^{-2}y)^3$

60. $\left(\dfrac{x^{-1/3}y^{1/2}}{x^{1/3}}\right)^6$

Perform the following operations and express your answers in simplest form.

61. $\dfrac{5x + 1}{x^2 + x - 6} - \dfrac{2}{2 - x}$

62. $\dfrac{x^2 - 25}{(x - 5)^2} \cdot (x^3 - 125)$

63. How long will it take a car traveling at a rate of 50 mph to catch up with a car traveling at a rate of 40 mph with an hour head start?

64. Susan had 32 coins in nickels and dimes. If the total value of her coins was $2.00, how many of each coin did she have?

6.4

Multiplying and Dividing Radical Expressions

As with multiplying polynomials, we use the distributive property to multiply expressions with more than one radical term. For example,

$$\sqrt{2}(\sqrt{2} - \sqrt{3}) = \sqrt{2}\sqrt{2} - \sqrt{2}\sqrt{3} \qquad \textit{Distributive property}$$
$$= 2 - \sqrt{6}$$

EXAMPLE 1

Perform the indicated operations.

(a) $(2\sqrt{3} - 5\sqrt{2})(2\sqrt{3} - \sqrt{2})$ **(b)** $(2\sqrt{x} - 3\sqrt{y})(2\sqrt{x} + 3\sqrt{y})$

Solution

We apply what we have learned about multiplying polynomials. Each term in the first set of parentheses multiplies each term in the second set.

(a) $(2\sqrt{3} - 5\sqrt{2})(2\sqrt{3} - \sqrt{2})$

$$= (2\sqrt{3})(2\sqrt{3}) + (2\sqrt{3})(-\sqrt{2}) - (5\sqrt{2})(2\sqrt{3}) - (5\sqrt{2})(-\sqrt{2})$$
$$= 4 \cdot 3 - 2\sqrt{6} - 10\sqrt{6} + 5 \cdot 2$$
$$= 12 - 12\sqrt{6} + 10$$
$$= \boxed{22 - 12\sqrt{6}}$$

(b) $(2\sqrt{x} - 3\sqrt{y})(2\sqrt{x} + 3\sqrt{y})$ *This is a difference of squares.*

$$= (2\sqrt{x})^2 - (3\sqrt{y})^2 \qquad \textit{Square each term.}$$
$$= 2^2(\sqrt{x})^2 - 3^2(\sqrt{y})^2$$
$$= \boxed{4x - 9y}$$

∎

EXAMPLE 2

Perform the indicated operations.

(a) $(\sqrt{x + y})^2 - (\sqrt{x} + \sqrt{y})^2$ **(b)** $(\sqrt[3]{x} - \sqrt[3]{y})(\sqrt[3]{x^2} + \sqrt[3]{xy} + \sqrt[3]{y^2})$

Solution

(a) $(\sqrt{x + y})^2 - (\sqrt{x} + \sqrt{y})^2$

$$= x + y - [(\sqrt{x})^2 + 2\sqrt{x}\sqrt{y} + (\sqrt{y})^2]$$
$$= x + y - [x + 2\sqrt{xy} + y]$$
$$= x + y - x - 2\sqrt{xy} - y$$
$$= \boxed{-2\sqrt{xy}}$$

Note the difference in the way we handle $(\sqrt{x + y})^2$ and $(\sqrt{x} + \sqrt{y})^2$.

(b) $(\sqrt[3]{x} - \sqrt[3]{y})(\sqrt[3]{x^2} + \sqrt[3]{xy} + \sqrt[3]{y^2})$

$= \sqrt[3]{x}\sqrt[3]{x^2} + \sqrt[3]{x}\sqrt[3]{xy} + \sqrt[3]{x}\sqrt[3]{y^2} - \sqrt[3]{y}\sqrt[3]{x^2} - \sqrt[3]{y}\sqrt[3]{xy} - \sqrt[3]{y}\sqrt[3]{y^2}$

$= \sqrt[3]{x^3} + \sqrt[3]{x^2y} + \sqrt[3]{xy^2} - \sqrt[3]{x^2y} - \sqrt[3]{xy^2} - \sqrt[3]{y^3}$

$= \sqrt[3]{x^3} - \sqrt[3]{y^3}$

$= \boxed{x - y}$ ∎

Part **(b)** of Examples 1 and 2 illustrate that any expression that is a difference can be factored if we lift our restriction that requires factors to be polynomials with integer coefficients. For example, $3a - 2b$ can be factored into

$$(\sqrt{3a} - \sqrt{2b})(\sqrt{3a} + \sqrt{2b})   \text{ as a } \textit{difference of squares}$$

or

$$(\sqrt[3]{3a} - \sqrt[3]{2b})(\sqrt[3]{9a^2} + \sqrt[3]{6ab} + \sqrt[3]{4b^2})   \text{ as a } \textit{difference of cubes.}$$

Multiply these out and check to verify that they equal $3a - 2b$.

The last operation to cover is division of radical expressions. We begin with division by a single term.

EXAMPLE 3   *Simplify.* $\dfrac{\sqrt{24} - 8}{10}$

Solution

$\dfrac{\sqrt{24} - 8}{10}$ *Simplify the radical.*

$= \dfrac{\sqrt{4}\sqrt{6} - 8}{10}$

$= \dfrac{2\sqrt{6} - 8}{10}$ *Do not try to reduce yet.*

$= \dfrac{\overset{1}{\cancel{2}}(\sqrt{6} - 4)}{\underset{5}{\cancel{10}}}$ *Factor **first**. Then reduce.*

$= \boxed{\dfrac{\sqrt{6} - 4}{5}}$ ∎

In Section 6.3 we discussed rationalizing the denominator of a fraction with a single radical term in the denominator. In this section we will discuss how to rationalize the denominator of a fraction with more than one term in the denominator.

We cannot rationalize the denominator of $\dfrac{2}{3 + \sqrt{5}}$ as easily as we rationalized denominators consisting of only a single term. What shall we multiply the numerator by: $\sqrt{5}$ or $3 + \sqrt{5}$? Let's try $\sqrt{5}$ and see what happens:

$$\frac{2}{3 + \sqrt{5}} = \frac{2 \cdot \sqrt{5}}{(3 + \sqrt{5}) \cdot \sqrt{5}} = \frac{2\sqrt{5}}{3\sqrt{5} + \sqrt{5}\sqrt{5}} = \frac{2\sqrt{5}}{3\sqrt{5} + 5}$$

Notice that we still have a radical remaining in the denominator. If we were to try $3 + \sqrt{5}$, we would find that a radical would still remain in the denominator.

The denominator can be rationalized by another method, however. We exploit the difference of squares and multiply the numerator and denominator by a *conjugate* of the denominator, $3 - \sqrt{5}$ (recall that a conjugate of $a + b$ is $a - b$):

$$\frac{2}{3 + \sqrt{5}} = \frac{2(3 - \sqrt{5})}{(3 + \sqrt{5})(3 - \sqrt{5})} \qquad \text{\textit{Multiply numerator and denominator by } } 3 - \sqrt{5}.$$

$$= \frac{2(3 - \sqrt{5})}{(3)^2 - (\sqrt{5})^2} \qquad \text{\textit{The denominator is the difference of squares.}}$$

$$= \frac{2(3 - \sqrt{5})}{9 - 5}$$

$$= \frac{2(3 - \sqrt{5})}{4} \qquad \text{\textit{Then reduce.}}$$

$$= \frac{\overset{1}{2}(3 - \sqrt{5})}{\underset{2}{\cancel{4}}}$$

$$= \frac{3 - \sqrt{5}}{2}$$

EXAMPLE 4

Simplify.

(a) $\dfrac{\sqrt{a} + \sqrt{b}}{\sqrt{a} - \sqrt{b}}$ **(b)** $\dfrac{3\sqrt{5} - 2}{2\sqrt{7} - \sqrt{3}}$

Solution

(a) $\dfrac{\sqrt{a} + \sqrt{b}}{\sqrt{a} - \sqrt{b}}$ *Multiply the numerator and denominator by $\sqrt{a} + \sqrt{b}$, a conjugate of the denominator.*

$$= \frac{(\sqrt{a} + \sqrt{b})(\sqrt{a} + \sqrt{b})}{(\sqrt{a} - \sqrt{b})(\sqrt{a} + \sqrt{b})}$$

$$= \frac{a + \sqrt{ab} + \sqrt{ba} + b}{a - b}$$

$$= \boxed{\frac{a + 2\sqrt{ab} + b}{a - b}}$$

(b) $\dfrac{3\sqrt{5} - 2}{2\sqrt{7} - \sqrt{3}}$ *Multiply numerator and denominator by $2\sqrt{7} + \sqrt{3}$, a conjugate of the denominator.*

$$= \frac{(3\sqrt{5} - 2)(2\sqrt{7} + \sqrt{3})}{(2\sqrt{7} - \sqrt{3})(2\sqrt{7} + \sqrt{3})}$$

$$= \frac{(3\sqrt{5})(2\sqrt{7}) + (3\sqrt{5})(\sqrt{3}) - (2)(2\sqrt{7}) - 2(\sqrt{3})}{(2\sqrt{7})^2 - (\sqrt{3})^2}$$

$$= \frac{6\sqrt{35} + 3\sqrt{15} - 4\sqrt{7} - 2\sqrt{3}}{4 \cdot 7 - 3}$$

$$= \boxed{\frac{6\sqrt{35} + 3\sqrt{15} - 4\sqrt{7} - 2\sqrt{3}}{25}}$$

EXAMPLE 5

Perform the operations and simplify. $\dfrac{21}{3 - \sqrt{2}} - \dfrac{6}{\sqrt{2}}$

Solution

First we rationalize the denominator of each fraction:

$$\frac{21(3 + \sqrt{2})}{(3 - \sqrt{2})(3 + \sqrt{2})} = \frac{21(3 + \sqrt{2})}{9 - 2} = \frac{\overset{3}{\cancel{21}}(3 + \sqrt{2})}{\cancel{7}} = 3(3 + \sqrt{2}) = 9 + 3\sqrt{2}$$

$$\frac{6\sqrt{2}}{\sqrt{2}\sqrt{2}} = \frac{\overset{3}{\cancel{6}}\sqrt{2}}{\cancel{2}} = 3\sqrt{2}$$

Thus, the original expression

$$\frac{21}{3 - \sqrt{2}} - \frac{6}{\sqrt{2}}$$

becomes

$$9 + 3\sqrt{2} - 3\sqrt{2} = \boxed{9}$$

Try finding the solution to the original problem by using a calculator and see how close your answer is to ours. ∎

Exercises 6.4

Perform the operations. Express your answer in simplest radical form.

1. $5(\sqrt{5} - 3)$

2. $3(\sqrt{7} + 3)$

3. $2(\sqrt{3} - \sqrt{5}) - 4(\sqrt{3} + \sqrt{5})$

4. $4(\sqrt{6} - \sqrt{2}) - 3(\sqrt{2} - \sqrt{6})$

5. $\sqrt{a}(\sqrt{a} + \sqrt{b})$

6. $\sqrt{x}(\sqrt{x} - \sqrt{y})$

7. $\sqrt{2}(\sqrt{5} + \sqrt{2})$

8. $\sqrt{3}(\sqrt{2} - \sqrt{3})$

9. $3\sqrt{5}(2\sqrt{3} - 4\sqrt{5})$

10. $5\sqrt{2}(7\sqrt{3} - 6\sqrt{2})$

11. $\sqrt{2}(\sqrt{3} + \sqrt{2}) - 3(2\sqrt{6} - 4)$

12. $2\sqrt{5}(\sqrt{3} - \sqrt{5}) + 2(\sqrt{10} - 3)$

13. $(\sqrt{5} - 2)(\sqrt{3} + 1)$

14. $(\sqrt{7} - 2)(\sqrt{5} + 3)$

15. $(\sqrt{5} - \sqrt{3})(\sqrt{5} + \sqrt{3})$

16. $(\sqrt{2} + \sqrt{5})^2$

17. $(\sqrt{5} - \sqrt{3})^2$

18. $(\sqrt{2} + \sqrt{5})(\sqrt{2} - \sqrt{5})$

19. $(2\sqrt{7} - 5)(2\sqrt{7} + 5)$

20. $(2\sqrt{5} - 7)(2\sqrt{5} + 7)$

21. $(5\sqrt{2} - 3\sqrt{5})(5\sqrt{2} + 3\sqrt{5})$

22. $(3\sqrt{5} - 2\sqrt{3})(3\sqrt{5} + 2\sqrt{3})$

23. $(2\sqrt{a} - \sqrt{b})^2$

24. $(5\sqrt{x} - \sqrt{y})^2$

25. $(3\sqrt{x} - 2\sqrt{y})(3\sqrt{x} + 2\sqrt{y})$

26. $(5\sqrt{a} - 3\sqrt{b})(5\sqrt{a} + 3\sqrt{b})$

27. $(\sqrt{x} - 3)^2$

28. $(\sqrt{a} + 5)^2$

29. $(\sqrt{x - 3})^2$

30. $(\sqrt{a + 5})^2$

31. $(\sqrt{x + 1})^2 - (\sqrt{x} + 1)^2$

32. $(\sqrt{a - 2})^2 - (\sqrt{a} - 2)^2$

33. $(\sqrt[3]{2} - \sqrt[3]{3})(\sqrt[3]{4} + \sqrt[3]{6} + \sqrt[3]{9})$

34. $(\sqrt[3]{4} + \sqrt[3]{5})(\sqrt[3]{16} - \sqrt[3]{20} + \sqrt[3]{25})$

35. $\dfrac{4\sqrt{2} - 6\sqrt{3}}{2}$

36. $\dfrac{12\sqrt{3} - 8\sqrt{2}}{4}$

37. $\dfrac{5\sqrt{8} - 2\sqrt{7}}{8}$

38. $\dfrac{2\sqrt{27} + 6\sqrt{5}}{3}$

39. $\dfrac{3\sqrt{50} + 5\sqrt{5}}{5}$

40. $\dfrac{2\sqrt{12} + 5\sqrt{8}}{12}$

41. $\dfrac{4 + \sqrt{28}}{4}$

42. $\dfrac{20 + \sqrt{60}}{4}$

43. $\dfrac{1}{\sqrt{2} - 3}$ $(\sqrt{2}+3)$ $\sqrt{4}$ $=9$

44. $\dfrac{1}{2 - \sqrt{3}}$

45. $\dfrac{10}{\sqrt{5} + 1}$

46. $\dfrac{15}{\sqrt{6} - 1}$

47. $\dfrac{2}{\sqrt{3} - \sqrt{a}}$

48. $\dfrac{3}{\sqrt{x} - \sqrt{2}}$

49. $\dfrac{\sqrt{x}}{\sqrt{x} - \sqrt{y}}$

50. $\dfrac{\sqrt{x}}{\sqrt{x} + \sqrt{y}}$

51. $\dfrac{\sqrt{2}}{\sqrt{5} - \sqrt{2}}$

52. $\dfrac{\sqrt{5}}{\sqrt{7} + \sqrt{3}}$

53. $\dfrac{2\sqrt{2}}{2\sqrt{5} - \sqrt{2}}$

54. $\dfrac{3\sqrt{5}}{4\sqrt{3} + \sqrt{5}}$

55. $\dfrac{2\sqrt{5} - \sqrt{2}}{2\sqrt{2}}$

56. $\dfrac{4\sqrt{3} + \sqrt{5}}{3\sqrt{5}}$

57. $\dfrac{\sqrt{3} + \sqrt{2}}{\sqrt{3} - \sqrt{2}}$

58. $\dfrac{\sqrt{x} + \sqrt{y}}{\sqrt{x} - \sqrt{y}}$

59. $\dfrac{3\sqrt{5} - 2\sqrt{2}}{2\sqrt{5} - 3\sqrt{2}}$

60. $\dfrac{2\sqrt{7} + 3\sqrt{2}}{3\sqrt{7} + 2\sqrt{2}}$

61. $\dfrac{x - y}{\sqrt{x} - \sqrt{y}}$

62. $\dfrac{a^2 - b^2}{\sqrt{a} + \sqrt{b}}$

63. $\dfrac{x^2 - x - 2}{\sqrt{x} - \sqrt{2}}$

64. $\dfrac{x^2 - 3x - 4}{\sqrt{x} + 2}$

65. $\dfrac{12}{\sqrt{6} - 2} - \dfrac{36}{\sqrt{6}}$

66. $\dfrac{15}{4 + \sqrt{11}} + \dfrac{33}{\sqrt{11}}$

67. $\dfrac{20}{\sqrt{7} + \sqrt{3}} + \dfrac{28}{\sqrt{7}}$

68. $\dfrac{30}{\sqrt{10} - \sqrt{5}} - \dfrac{15}{\sqrt{5}}$

 CALCULATOR EXERCISE

69. Compute the value of the expressions given in Exercises 65 and 66 using a calculator. How do your answers compare to the answers found as a result of simplifying the expressions?

MINI-REVIEW

70. *Factor completely.* $16 - (x + y)^2$

71. *Solve for x and graph its solution set.* $-4 \le 3x - 2 < 5$

Simplify the following and express your answers with positive exponents only.

72. $\left(\dfrac{3^{-2}x^{-2}y^{-1}}{9^{-1}xy^{-3}}\right)^{-2}$

73. $\left(\dfrac{64x^{1/2}y^{1/5}}{x^{-1/2}}\right)^{-1/2}$

6.5

Radical Equations

In Chapter 4, when we solved equations containing fractions, our first step was to transform the equation into an equivalent one without fractions. Our strategy for treating radicals in equations will be similar. That is, our first step in solving an equation with radicals will be to transform the equation into one that does not contain radicals.

We utilize the following:

$$\boxed{\text{If } a = b \text{ then } a^2 = b^2.}$$

In words, this says that squares of equal quantities are equal to each other. We can apply this property to "get rid of the radical." For example, to solve the equation

$$\sqrt{x} = 5$$

we would square both sides of the equation to get

$$(\sqrt{x})^2 = 5^2$$

or

$$x = 25$$

Simply squaring both sides of an equation containing a radical does not necessarily eliminate the radical, however. For example, in solving

$$\sqrt{x} + 6 = 9$$

if we simply square both sides we would have

$$(\sqrt{x} + 6)^2 = 9^2$$
$$x + 12\sqrt{x} + 36 = 81 \qquad \textit{Remember that when squaring a}$$
$$\textit{binomial, we will have middle terms.}$$

Note that a radical still remains in the equation. Therefore, we should *isolate the radical before squaring both sides of the equation.* Thus, we would solve the equation as follows:

$$\sqrt{x} + 6 = 9 \qquad \textit{First isolate } \sqrt{x}.$$
$$\sqrt{x} = 3 \qquad \textbf{Then } \textit{square both sides of the equation.}$$
$$(\sqrt{x})^2 = 3^2$$
$$x = 9$$

EXAMPLE 1

Solve for x.

(a) $\sqrt{2x - 7} = 9$ **(b)** $\sqrt{2x} - 7 = 9$ **(c)** $2\sqrt{x - 7} = 9$

Solution

(a) $\sqrt{2x - 7} = 9$ *Since the radical is already isolated, we square both sides of*
$$(\sqrt{2x - 7})^2 = 9^2 \qquad \textit{the equation.}$$
$$2x - 7 = 81$$
$$2x = 88$$
$$\boxed{x = 44}$$

CHECK $x = 44$: $\sqrt{2x - 7} = 9$

$$\sqrt{2(44) - 7} \overset{?}{=} 9$$

$$\sqrt{88 - 7} \overset{?}{=} 9$$

$$\sqrt{81} \overset{\checkmark}{=} 9$$

(b) $\sqrt{2x} - 7 = 9$ *First isolate the radical.*

$\sqrt{2x} = 16$ *Then square both sides of the equation.*

$(\sqrt{2x})^2 = 16^2$

$2x = 256$

$$\boxed{x = 128}$$

CHECK $x = 128$: $\sqrt{2x} - 7 = 9$

$$\sqrt{2(128)} - 7 \overset{?}{=} 9$$

$$\sqrt{256} - 7 \overset{?}{=} 9$$

$$16 - 7 \overset{\checkmark}{=} 9$$

Note the similarities and differences between two equations in parts **(a)** and **(b)** and how they were solved.

(c) $2\sqrt{x - 7} = 9$ *Isolate the radical; divide both sides by 2.*

$\sqrt{x - 7} = \dfrac{9}{2}$ *Square both sides of the equation.*

$$(\sqrt{x - 7})^2 = \left(\frac{9}{2}\right)^2$$

$x - 7 = \dfrac{81}{4}$ *Add 7 to both sides of the equation.*

$$x = \frac{81}{4} + 7 = \boxed{\frac{109}{4}}$$

You should check to see that this is a valid solution. ∎

Based on our original properties of equality (Chapter 2), we were able to state that certain operations performed on both sides of an equation yield *equivalent equations*. By *equivalent* we mean that the transformed equation and the original equation have identical solutions.

We cannot make the same statement for the property of squaring used in Example 1. Note that it is written "if $a = b$ then $a^2 = b^2$." For equations, this means that solutions for the equation $a = b$ are also solutions for $a^2 = b^2$, but all solutions for $a^2 = b^2$ are not necessarily solutions for $a = b$.

For example, let's suppose we have the equation

$$x = -2$$

If we square both sides, we get

$$x^2 = (-2)^2$$
$$x^2 = 4$$

We had only one solution to the first equation, $x = -2$, but you can check to see that the transformed equation $x^2 = 4$ has two solutions: $x = -2$ and $x = 2$.

By squaring both sides of the equation, we picked up an extra number which is a solution to the transformed equation, but *not* a solution to the original equation. This "extra" solution is called an **extraneous solution**, or an **extraneous root**. Because of this problem, we *have to check* all solutions in the original equation to ensure that our solution is valid. (In the next chapter we will encounter some radical equations which seem to yield two solutions and we will have to check to see if both are valid.)

EXAMPLE 2

Solve for the variable.

(a) $\sqrt{3a + 1} + 13 = 8$ **(b)** $\sqrt{4a - 1} - \sqrt{2a + 3} = 0$

Solution

(a) $\sqrt{3a + 1} + 13 = 8$ *First isolate the radical $\sqrt{3a + 1}$. Subtract 13 from both sides of the equation.*

$\sqrt{3a + 1} = -5$ *Does this make sense?*

Let's continue and see what we get:

$$(\sqrt{3a + 1})^2 = (-5)^2$$
$$3a + 1 = 25$$
$$3a = 24$$
$$a = 8$$

CHECK $a = 8$: $\sqrt{3a + 1} + 13 = 8$
$\sqrt{3(8) + 1} + 13 \stackrel{?}{=} 8$
$\sqrt{24 + 1} + 13 \stackrel{?}{=} 8$
$\sqrt{25} + 13 \stackrel{?}{=} 8$
$5 + 13 \neq 8$

The only solution is extraneous.

Therefore, there is $\boxed{\text{no solution}}$

We could have stopped at the point where we had $\sqrt{3a + 1} = -5$ and concluded then that there is no solution to this problem. Why?

(b) $\sqrt{4a - 1} - \sqrt{2a + 3} = 0$ *Isolate one of the radicals.*

$\sqrt{4a - 1} = \sqrt{2a + 3}$ *Then square both sides of the equation.*

$(\sqrt{4a - 1})^2 = (\sqrt{2a + 3})^2$

$4a - 1 = 2a + 3$ *Isolate a.*

$2a = 4$

$a = 2$

CHECK $a = 2$: $\quad \sqrt{4a - 1} - \sqrt{2a + 3} = 0$

$$\sqrt{4(2) - 1} - \sqrt{2(2) + 3} \overset{?}{=} 0$$

$$\sqrt{8 - 1} - \sqrt{4 + 3} \overset{?}{=} 0$$

$$\sqrt{7} - \sqrt{7} \overset{\checkmark}{=} 0$$

The solution is $\boxed{a = 2}$ ∎

We can generalize the squaring property to other powers:

$$\boxed{\text{If } a = b \text{ then } a^n = b^n \text{ where } n \text{ is a rational number.}}$$

This property allows us to solve radical equations of a higher index, as shown in the next example.

EXAMPLE 3

Solve each of the following equations:

(a) $\sqrt[3]{x} = 5$ **(b)** $\sqrt[4]{x - 7} + 2 = 6$

Solution

(a) $\qquad \sqrt[3]{x} = 5 \qquad$ *Cube both sides of the equation.*

$$(\sqrt[3]{x})^3 = (5)^3$$

$$x = 125$$

CHECK $x = 125$: $\quad \sqrt[3]{x} = 5$

$$\sqrt[3]{125} \overset{\checkmark}{=} 5$$

The solution is $\boxed{x = 125}$

(b) $\qquad \sqrt[4]{x - 7} + 2 = 6 \qquad$ *Isolate the radical.*

$$\sqrt[4]{x - 7} = 4 \qquad \textit{Raise each side to the fourth power.}$$

$$(\sqrt[4]{x - 7})^4 = (4)^4$$

$$x - 7 = 256$$

$$x = 263$$

CHECK $x = 263$: $\quad \sqrt[4]{x - 7} + 2 = 6$

$$\sqrt[4]{263 - 7} + 2 \overset{?}{=} 6$$

$$\sqrt[4]{256} + 2 \overset{?}{=} 6$$

$$4 + 2 \overset{\checkmark}{=} 6$$

The solution is $\boxed{x = 263}$ ∎

Since we can rewrite radicals using rational exponents, the same principles can be applied to equations with rational exponents. Note that in raising an expression such as $a^{1/6}$ to the sixth power, we use the second rule of exponents as follows:

$$(a^{1/6})^6 = a^{(1/6)(6)} = a^1 = a$$

EXAMPLE 4

Solve each of the following equations:

(a) $x^{1/4} = 2$ (b) $x^{1/3} - 2 = 5$

(c) $(x - 2)^{1/3} = 5$ (d) $a^{-1/3} = 4$

Solution

(a) $x^{1/4} = 2$ *Raise each side to the fourth power.*

    $(x^{1/4})^4 = 2^4$

       $x = 2^4$ *Since $(x^{1/4})^4 = x^{(1/4)4} = x^1 = x$*

       $x = 16$

    CHECK $x = 16$: $x^{1/4} = 2$

                      $16^{1/4} \overset{\checkmark}{=} 2$

Thus, $\boxed{x = 16}$

(b) $x^{1/3} - 2 = 5$ *Isolate $x^{1/3}$.*

       $x^{1/3} = 7$ *Cube both sides.*

    $(x^{1/3})^3 = 7^3$

      $\boxed{x = 343}$

The check is left to the student.

(c) $(x - 2)^{1/3} = 5$ *Cube both sides.*

   $[(x - 2)^{1/3}]^3 = 5^3$

       $x - 2 = 125$

        $\boxed{x = 127}$

The check is left to the student.

Note the differences between parts (b) and (c).

(d) $a^{-1/3} = 4$ *We raise both sides to the -3 power in order to change $a^{-1/3}$ to $a^1 = a$.*

   $(a^{-1/3})^{-3} = 4^{-3}$

       $a^1 = 4^{-3}$ *Rewrite 4^{-3} using positive exponents.*

        $a = \dfrac{1}{4^3}$

        $a = \dfrac{1}{64}$

    CHECK $a = \dfrac{1}{64}$: $a^{-1/3} = 4$

                    $\left(\dfrac{1}{64}\right)^{-1/3} \overset{?}{=} 4$

                     $64^{1/3} \overset{\checkmark}{=} 4$

The solution is $\boxed{a = \dfrac{1}{64}}$

Exercises 6.5

In Exercises 1–62, solve for the given variable.

1. $\sqrt{a} = 7$ **2.** $\sqrt{b} = 14$ **3.** $\sqrt{x - 3} = 5$

4. $6 = \sqrt{a + 7}$ **5.** $\sqrt{x + 7} = 8$ **6.** $\sqrt{a} - 4 = 6$

7. $\sqrt{3a} + 4 = 5$ **8.** $\sqrt{5x} - 1 = 4$ **9.** $\sqrt{3a + 4} = 5$

10. $\sqrt{5x - 1} = 4$ **11.** $\sqrt{y + 3} - 5 = 4$ **12.** $\sqrt{s + 7} + 6 = 3$

13. $2 = 5 - \sqrt{3x - 1}$ **14.** $\sqrt{5q + 2} + 3 = 1$ **15.** $\sqrt{2 - 3y} + 5 = 8$

16. $-4 = 7 - \sqrt{5 - 2x}$ **17.** $3\sqrt{x} = 6$ **18.** $5\sqrt{y} = 20$

19. $4\sqrt{a} = 9$ **20.** $5\sqrt{y} = 8$ **21.** $4\sqrt{x} + 2 = 17$

22. $6\sqrt{y} + 3 = 21$ **23.** $5\sqrt{y} - 2 = 7$ **24.** $7\sqrt{s} + 3 = 8$

25. $42 = 7\sqrt{2x - 1}$ **26.** $27 = 9\sqrt{4x + 3}$ **27.** $3 + \sqrt{x - 1} = 7$

28. $5 + \sqrt{x + 2} = 10$ **29.** $7 + 2\sqrt{y + 1} = 12$ **30.** $6 + 3\sqrt{r + 2} = 14$

31. $\sqrt{2x - 3} - \sqrt{x + 5} = 0$ **32.** $\sqrt{3x - 2} - \sqrt{x + 6} = 0$ **33.** $\sqrt{2x + 1} + \sqrt{x + 1} = 0$

34. $\sqrt{3x - 5} + \sqrt{x - 3} = 0$ **35.** $-4 + 5\sqrt{2a + 1} = 6$ **36.** $2 + 6\sqrt{3x - 1} = 14$

37. $4 = \sqrt[3]{x}$ **38.** $\sqrt[3]{y} + 5 = 7$ **39.** $\sqrt[4]{y} + 2 = 5$

40. $2 = \sqrt[4]{a}$ **41.** $\sqrt[5]{s} = -3$ **42.** $\sqrt[5]{x} = -2$

43. $\sqrt[3]{x - 1} - 2 = 5$ **44.** $-3 = \sqrt[3]{y + 5}$ **45.** $-4 = \sqrt[3]{2x + 3}$

46. $\sqrt[3]{4 - 3y} + 2 = 1$ **47.** $x^{1/3} = 5$ **48.** $y^{1/3} = -3$

49. $x^{1/3} + 7 = 5$ **50.** $y^{1/3} - 6 = 2$ **51.** $(a + 7)^{1/3} = 4$

52. $(b - 7)^{1/3} = 5$ **53.** $x^{-1/4} = 4$ **54.** $x^{-1/3} = 8$

55. $(x + 3)^{1/3} + 4 = 2$ **56.** $(2x - 1)^{1/3} - 2 = 5$

 **QUESTION FOR THOUGHT**

57. What is the difference between squaring the expression $\sqrt{x + 4}$ and squaring the expression $\sqrt{x} + 4$?

MINI-REVIEW

58. *Perform the operations and simplify.*

$$\frac{3x}{x^2 - 1} + \frac{2}{x^3 + x^2 - x - 1}$$

59. *Simplify.*

$$\frac{\sqrt{3}}{\sqrt{3} - \sqrt{2}}$$

60. Jay has $50,000 invested in two bonds. One bond yields 6% annual interest and the other yields $7\frac{1}{2}$% annual interest. How much did he invest in each bond if the combined yearly interest from both bonds is $3,150?

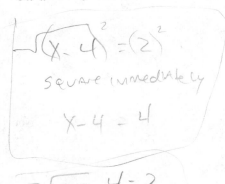

6.6

Complex Numbers

In Chapter 1 we pointed out the need to extend number systems in order to solve various types of equations. For example, the set of rational numbers contains the solution to the equation $3x + 7 = 2$. However, for the equation $x^2 = 3$, we have to "go beyond" the rationals and define the real numbers in order to locate the solutions $x = +\sqrt{3}$ and $x = -\sqrt{3}$.

It seems as though our system is complete, but we are still unable to solve all polynomial equations. For example, the equation $x^2 = -5$ has no real number solution because no real number will yield a negative number when squared. Thus, we again define a system even larger than the real numbers in order to have the solutions to such equations. We will begin by defining the quantity i.

DEFINITION

The *imaginary unit*, *i*, is defined by

$$i^2 = -1$$

Hence $i = \sqrt{-1}$.

Given this definition, we have:

$$i$$
$$i^2 = (\sqrt{-1})^2 = -1$$
$$i^3 = i^2 \cdot i = (-1)i = -i$$
$$i^4 = i^2 \cdot i^2 = (-1)(-1) = +1$$
$$i^5 = i^4 i = (1)i = i$$
$$i^6 = i^4 i^2 = (1)(-1) = -1$$

Notice that this cycle repeats itself after i^4, so that any power of i can be written as i, -1, $-i$, or 1.

EXAMPLE 1

Simplify each of the following:
(a) i^{39} **(b)** i^{26}

Solution

Because $i^4 = 1$, it would be most convenient to factor the largest perfect fourth power of i.

(a) $i^{39} = i^{36} \cdot i^3$ *36 is the largest multiple of 4 less than 39; hence i^{36} is the*
      $= (i^4)^9 i^3$ *greatest perfect fourth power factor of i^{39}. We rewrite i^{36} as*
               a perfect fourth power.
      $= (1)^9 i^3$ *Since $i^4 = 1$*
      $= i^3$
      $= -i$

Hence, $i^{39} = \boxed{-i}$

Actually, we find that $i^s = i^r$, where r is the remainder when s is divided by 4, and then rewrite the expression as i, $-i$, 1, or -1.

(b) $i^{26} = (i^4)^6 i^2$ *Note:* $4\overline{)\,26}$
$$\begin{array}{r} 6 \\ 4\,\overline{)\,26} \\ -24 \\ \hline 2 \end{array}$$

$= 1^6 i^2$ *We could have divided 4 into 26 to find a remainder of 2; hence,*
$= i^2$ $i^{26} = i^2$.

$= \boxed{-1}$ ∎

Using i, we can now translate square roots with negative radicands as follows:

$$\sqrt{-4} = \sqrt{4(-1)} = \sqrt{4i^2} = \sqrt{4}\sqrt{i^2} = 2i$$
$$\sqrt{-\frac{1}{16}} = \sqrt{\frac{1}{16}(-1)} = \sqrt{\frac{1}{16}i^2} = \sqrt{\frac{1}{16}}\sqrt{i^2} = \frac{1}{4}i$$

We will use i, the imaginary unit, to define a new type of number: the *complex number*.

DEFINITION

A ***complex number*** is a number that can be written in the form $a + bi$, where a and b are real numbers, and i is the imaginary unit. a is called the ***real part*** of $a + bi$ and bi is called the ***imaginary part***.

For example, in the complex number $3 + 4i$, 3 is the real part and $4i$ is the imaginary part.

If a is a real number, then we can rewrite it as a complex number in the following way:

$$a = a + 0i$$

Since we can put any real number into complex form (with b equal to 0), we conclude that all real numbers are complex numbers.

Hence, $R \subset C$.

If a nonzero complex number does not have a real part (real part is 0), then we say that the number is ***pure imaginary***. For example, $3i$, $-4i$, and $2i$ are pure imaginary numbers.

EXAMPLE 2   Put the following in complex number form:

(a) $5 + \sqrt{-16}$ **(b)** 0   **(c)** -5 **(d)** $\dfrac{6 - \sqrt{-3}}{2}$

Solution

(a) Since $\sqrt{-16} = \sqrt{16(-1)} = \sqrt{16i^2} = 4i$, we have

$$5 + \sqrt{-16} = \boxed{5 + 4i}$$

(b) $0 = \boxed{0 + 0i}$

(c) $-5 = \boxed{-5 + 0i}$

(d) $\dfrac{6 - \sqrt{-3}}{2} = \dfrac{6 - i\sqrt{3}}{2}$ *We place i before the radical so it is not confused with $\sqrt{3i}$.*

$$= \frac{6}{2} - \frac{\sqrt{3}}{2}i$$

$$= \boxed{3 - \frac{\sqrt{3}}{2}i}$$ *The real part is 3; the imaginary part is $(-\sqrt{3}/2)i$.*   ∎

Now that we have defined complex numbers, our next step will be to examine operations on complex numbers. First we will define the following:

DEFINITION

$$a + bi = c + di \quad \text{if and only if} \quad a = c \text{ and } b = d.$$

Verbally stated, two complex numbers are equal if and only if their real parts are identical *and* their imaginary parts are identical.

Addition and Subtraction of Complex Numbers

Addition and subtraction of complex numbers are relatively straightforward:

Addition and Subtraction of Complex Numbers	$(a + bi) + (c + di) = (a + c) + (b + d)i$ Addition $(a + bi) - (c + di) = (a - c) + (b - d)i$ Subtraction

The real part of the sum (difference) is the sum (difference) of the real parts and the imaginary part of the sum (difference) is the sum (difference) of the imaginary parts.

EXAMPLE 3

Perform the indicated operations.

(a) $(3 + 4i) + (5 - 6i)$ (b) $(6 + 9i) - (3 - 2i)$

Solution

We could either use the definition of addition and subtraction given in the box, or treat i as a variable and combine "like" terms.

(a) $(3 + 4i) + (5 - 6i) = 3 + 4i + 5 - 6i = \boxed{8 - 2i}$

By treating i as a variable and combining terms

Or, alternatively, by using the definition, we have:

$$(3 + 4i) + (5 - 6i) = (3 + 5) + (4 - 6)i = \boxed{8 - 2i}$$

(b) $(6 + 9i) - (3 - 2i)$ *Use the definition.*

$= (6 - 3) + [9 - (-2)]i$

$= \boxed{3 + 11i}$ ■

Products of Complex Numbers

We can treat a complex number as if it were a binomial and multiply two complex numbers using binomial multiplication:

$(a + bi)(c + di) = ac + adi + bci + bdi^2$ *Binomial multiplication*

$= ac + adi + bci + bd(-1)$ *Since $i^2 = -1$*

$= ac + adi + bci - bd$ *Rearrange terms such that the real parts are together and the imaginary parts are together, then factor i from $adi + bci$.*

$= ac - bd + (ad + bc)i$

Thus, we have the rule for multiplying complex numbers:

Multiplication of Complex Numbers	$(a + bi)(c + di) = (ac - bd) + (ad + bc)i$

It is probably not very useful to memorize this rule as a special product at this point. It is better to multiply two complex numbers as binomials, substitute -1 for i^2, and finally combine the real parts and imaginary parts to conform to complex number form.

EXAMPLE 4

Perform the indicated operations.

(a) $(3 + 2i)(4 - 5i)$ (b) $(3 - 7i)^2$

(c) $(5 - 2i)(5 + 2i)$ (d) $(3 - i)^2 - 6(3 - i)$

Solution

(a) $(3 + 2i)(4 - 5i) = 12 + 8i - 15i - 10i^2$ *Binomial multiplication*

$= 12 + 8i - 15i + 10$ *Since $i^2 = -1$,*
$-10i^2 = -10(-1) = +10$.

$= \boxed{22 - 7i}$

(b) $(3 - 7i)^2 = (3)^2 - 2(3)(7i) + (7i)^2$ *This is a special product: a perfect square.*

$= 9 - 42i + 49i^2$ *Since $i^2 = -1$, $49i^2 = 49(-1) = -49$.*

$= 9 - 42i - 49$

$= \boxed{-40 - 42i}$

(c) $(5 - 2i)(5 + 2i) = 5^2 - (2i)^2$ *This is a special product: the difference of squares.*

$= 25 - 4i^2$

$= 25 + 4$ *Since $i^2 = -1$*

$= \boxed{29}$ *Note that the result is a real number.*

(d) $(3 - i)^2 - 6(3 - i)$ *Powers first: square $(3 - i)$.*

$= 9 - 6i + i^2 - 6(3 - i)$ *Then multiply.*

$= 9 - 6i + i^2 - 18 + 6i$ *$i^2 = -1$*

$= 9 - 6i - 1 - 18 + 6i$

$= \boxed{-10}$ ∎

EXAMPLE 5

Show that $2 + 3i$ is a solution to the equation $x^2 - 4x + 13 = 0$.

Solution

In order to determine if $2 + 3i$ is a solution to the equation, we substitute the value $2 + 3i$ for x in the equation and check to see if, after simplifying, the right- and left-hand sides of the equation agree:

$$x^2 - 4x + 13 = 0$$
$$(2 + 3i)^2 - 4(2 + 3i) + 13 \overset{?}{=} 0$$
$$4 + 12i + 9i^2 - 8 - 12i + 13 \overset{?}{=} 0$$
$$4 + 12i - 9 - 8 - 12i + 13 \overset{\checkmark}{=} 0$$

$\boxed{\text{Thus, } 2 + 3i \text{ does satisfy the equation } x^2 - 4x + 13 = 0.}$ ∎

We now define the *conjugate* of a complex number.

DEFINITION

The **complex conjugate** of $a + bi$ is $a - bi$.

By this definition, the complex conjugate of $a - bi$ is $a + bi$.

Observe, as in part **(c)** of Example 4, that the product of complex conjugates will yield a real number. In general,

$$
\begin{aligned}
(a + bi)(a - bi) &= a^2 - (bi)^2 && \textit{Difference of squares} \\
&= a^2 - b^2 i^2 && \textit{Since } i^2 = -1,\ -b^2 i^2 = -b^2(-1) = +b^2 \\
&= a^2 + b^2 && \textit{Note that } a^2 + b^2 \textit{ is a real number.}
\end{aligned}
$$

We will use this result to help us find certain quotients of complex numbers.

Quotients of Complex Numbers

Our goal is to express a quotient of complex numbers in the form $a + bi$.

If we have a quotient of a complex number and a real number, where the real number is the divisor, then expressing the quotient in the form $a + bi$ is similar to dividing a polynomial by a monomial. For example,

$$
\frac{6 - 5i}{2} = \frac{6}{2} - \frac{5}{2}i = 3 - \frac{5}{2}i
$$

Keeping in mind that $i = \sqrt{-1}$ is a radical expression, we will find the rationalizing techniques we used with quotients of radicals useful when working with quotients of complex numbers when the divisor has an imaginary part.

EXAMPLE 6

Express the following in the form $a + bi$:

$$
\frac{4 + 3i}{2i}
$$

Solution

$$
\frac{4 + 3i}{2i}
$$

We will treat this expression in the same manner as we would if we were rationalizing a denominator. Multiply the numerator and denominator by i.

$$
= \frac{(4 + 3i) \cdot i}{2i \cdot i}
$$

$$
= \frac{4i + 3i^2}{2i^2}
$$

$$
= \frac{4i + 3(-1)}{2(-1)} \qquad \textit{Since } i^2 = -1
$$

$$
= \frac{-3 + 4i}{-2} = \boxed{\frac{3}{2} - 2i}
$$

∎

EXAMPLE 7

Express the following in the form $a + bi$:

$$
\frac{7 - 4i}{2 + i}
$$

Solution | Our goal is to get a real number in the denominator so that we may express the result in the form $a + bi$. We proceed as follows:

$$\frac{7 - 4i}{2 + i}$$

Multiply the numerator and the denominator by the conjugate of the denominator, which is $2 - i$.

$$= \frac{(7 - 4i)(2 - i)}{(2 + i)(2 - i)}$$

$$= \frac{14 - 7i - 8i + 4i^2}{2^2 - i^2}$$

Recall that $i^2 = -1$.

$$= \frac{14 - 7i - 8i - 4}{4 + 1}$$

Note that the denominator is now a real number. Simplify the numerator and denominator.

$$= \frac{10 - 15i}{5}$$

Then rewrite in complex number form.

$$= \frac{10}{5} - \frac{15}{5}i$$

Simplify.

$$= \boxed{2 - 3i}$$

∎

We have defined complex numbers and their operations. Complex numbers, together with the operations defined, obey the same properties as the real numbers (associative, commutative, distributive, inverses, etc.) and form a system called the **complex number system**.

We have extended the natural numbers to the whole numbers, the whole numbers to the integers, the integers to the reals, and finally, the reals to the complex numbers, C (see Figure 6.1):

$$N \subset W \subset Z \subset Q \subset R \subset C$$

Figure 6.1
The complex number system

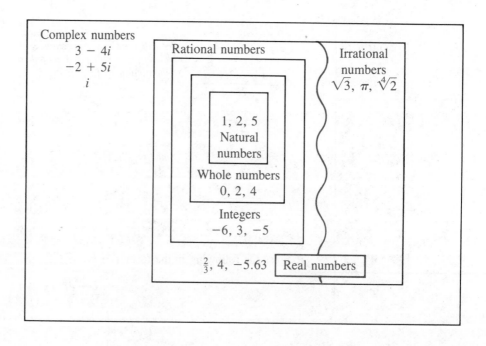

We assert that any polynomial equation has all its solutions in C. In the next chapter we will see that any second-degree equation has its solutions in C, the complex number system.

Exercises 6.6

In Exercises 1–12, express as i, $-i$, -1, or $+1$:

1. i^3

2. i^7

3. $-i^{21}$

4. $-i^{13}$

5. $(-i)^{29}$

6. $(-i)^{56}$

7. i^{72}

8. i^{35}

9. i^{67}

10. i^{103}

11. $-i^{16}$

12. $-i^{24}$

In Exercises 13–22, express in the form $a + bi$:

13. $\sqrt{-5}$

14. $\sqrt{-8}$

15. $3 - \sqrt{-2}$

16. $5 - \sqrt{-64}$

17. $2\sqrt{-4} - \sqrt{28}$

18. $\sqrt{-6} - \sqrt{64}$

19. $\dfrac{3 - \sqrt{-2}}{5}$

20. $\dfrac{5 - \sqrt{-3}}{2}$

21. $\dfrac{6 - \sqrt{-3}}{3}$

22. $\dfrac{8 - 4\sqrt{-2}}{2}$

In Exercises 23–70, perform the indicated operations and express your answers in the form $a + bi$.

23. $(3 + 2i) - (2 + 3i)$

24. $(5 - 4i) + (3 + 6i)$

25. $(2 - i) + (2 + i)$

26. $(5 - 3i) - (2 - 4i)$

27. $5(3 - 7i)$

28. $6(2 + 5i)$

29. $(2 - i)(-2)$

30. $(3 - 8i)(-4)$

31. $(5i)(3i)$

32. $(2i)(8i)$

33. $(-3i)^2$

34. $(-7i)^2$

35. $2i(3 + 2i)$

36. $3i(5 - 4i)$

37. $-2i(-3 + 5i)$

38. $-5i(-2 - 2i)$

39. $(3 + 2i)(0)$

40. $(0)(7 + 2i)$

41. $(7 - 4i)(3 + i)$

42. $(5 - 2i)(5 + 3i)$

43. $(3 + i)(3 - i)$

44. $(2 + i)(2 - i)$

45. $(2 - 7i)(2 + 7i)$

46. $(3 + 7i)(3 - 7i)$

47. $(i + 1)^2$

48. $(5 + i)^2$

49. $(2 - i)^2 - 4(2 - i)$

50. $(3 - 4i)^2 + (3 - 4i)$

51. $(1 + i)^2 - 2(1 + i) + 1$

52. $(5 - 2i)^2 + (5 - 2i) + 6$

53. $\dfrac{6 - 5i}{3}$

54. $\dfrac{-4 + 9i}{3}$

55. $\dfrac{3 - i}{i}$

56. $\dfrac{8 - i}{i}$

57. $\dfrac{5 - 2i}{2i}$

58. $\dfrac{3 - 5i}{3i}$

59. $\dfrac{2i}{5 - 2i}$

60. $\dfrac{3i}{3 - 5i}$

61. $\dfrac{2}{5 - 2i}$

62. $\dfrac{3}{3 - 5i}$

63. $\dfrac{2 + i}{2 - i}$

64. $\dfrac{3 + i}{3 - i}$

65. $\dfrac{2 - 5i}{2 + 5i}$

66. $\dfrac{2 - 3i}{2 + 3i}$

67. $\dfrac{3 - 7i}{5 + 2i}$

68. $\dfrac{4 + 5i}{3 + 7i}$

69. $\dfrac{5 - \sqrt{-4}}{3 + \sqrt{-25}}$

70. $\dfrac{3 - \sqrt{-9}}{3 + \sqrt{-16}}$

71. Show that $2i$ is a solution for $x^2 + 4 = 0$.

72. Show that $-3i$ is a solution for $x^2 + 9 = 0$.

73. Show that $2 - i$ is a solution for $x^2 - 4x + 5 = 0$.

74. Show that $2 + i$ is a solution for $x^2 - 4x + 5 = 0$.

? QUESTIONS FOR THOUGHT

75. Are the irrational numbers closed under multiplication? (Does the product of two irrational numbers always yield an irrational number?) If not, give an example.

76. The number i allows us to represent even roots of negative numbers other than square roots. Show how we can represent $\sqrt[6]{-1}$ using i. [*Hint:* $\sqrt[6]{x} = \sqrt{\sqrt[3]{x}}$.] How about $\sqrt[10]{-1}$? $\sqrt[14]{-1}$?

77. Express the following as i, $-i$, -1, or $+1$: i^{4n}, i^{4n+1}, i^{4n+2}, i^{4n+3}.

CHAPTER 6   Summary

After having completed this chapter, you should be able to:

1. Convert expressions from radical form to exponential form, and vice versa (Section 6.1).

For example:

(a) $2a^{5/4} = 2\sqrt[4]{a^5}$ or $2(\sqrt[4]{a})^5$

(b) $\sqrt[5]{x^3y} = (x^3y)^{1/5}$ or $x^{3/5}y^{1/5}$

2. Evaluate numerical expressions given in radical or exponential form (Section 6.1).

For example:

(a) $\sqrt{64} = 8$ because $8^2 = 64$

(b) $\sqrt[3]{-64} = -4$ because $(-4)^3 = -64$

(c) $\sqrt[6]{64} = 2$ because $2^6 = 64$

(d) $\sqrt{-64}$ is not a real number

(e) $64^{-2/3} = \dfrac{1}{64^{2/3}} = \dfrac{1}{(\sqrt[3]{64})^2} = \dfrac{1}{4^2} = \dfrac{1}{16}$

3. Write radical expressions in simplest radical form (Section 6.2).

For example:

(a) $\sqrt{24x^5y^6} = \sqrt{(4x^4y^6)(6x)} = \sqrt{4x^4y^6}\sqrt{6x} = 2x^2y^3\sqrt{6x}$

(b) $\sqrt[3]{24x^5y^6} = \sqrt[3]{(8x^3y^6)(3x^2)} = \sqrt[3]{8x^3y^6}\sqrt[3]{3x^2} = 2xy^2\sqrt[3]{3x^2}$

(c) $(a\sqrt{3a^2})(b\sqrt{9ab}) = ab(\sqrt{27a^3b})$

$\qquad\qquad\qquad\quad = ab(\sqrt{9a^2}\sqrt{3ab})$

$\qquad\qquad\qquad\quad = ab(3a\sqrt{3ab})$

$\qquad\qquad\qquad\quad = 3a^2b\sqrt{3ab}$

(d) $\sqrt{\dfrac{3}{x}} = \dfrac{\sqrt{3}}{\sqrt{x}} = \dfrac{\sqrt{3}\sqrt{x}}{\sqrt{x}\sqrt{x}} = \dfrac{\sqrt{3x}}{\sqrt{x^2}} = \dfrac{\sqrt{3x}}{x}$

(e) $\dfrac{\sqrt[3]{3}}{\sqrt[3]{x^2y}} = \dfrac{\sqrt[3]{3}\sqrt[3]{xy^2}}{\sqrt[3]{x^2y}\sqrt[3]{xy^2}} = \dfrac{\sqrt[3]{3xy^2}}{\sqrt[3]{x^3y^3}} = \dfrac{\sqrt[3]{3xy^2}}{xy}$

(f) $\sqrt[3]{3}\sqrt{3} = \sqrt[6]{3^2}\sqrt[6]{3^3} = \sqrt[6]{3^5} = \sqrt[6]{243}$

4. Combine radicals (Section 6.3).

For example:

(a) $2\sqrt{75} - \sqrt{12} = 2(\sqrt{25}\sqrt{3}) - \sqrt{4}\sqrt{3} = 2(5\sqrt{3}) - 2\sqrt{3} = 10\sqrt{3} - 2\sqrt{3}$
$$= 8\sqrt{3}$$

(b) $2\sqrt[3]{24} + 5\sqrt[3]{81} = 2(\sqrt[3]{8}\sqrt[3]{3}) + 5(\sqrt[3]{27}\sqrt[3]{3})$
$$= 2(2\sqrt[3]{3}) + 5(3\sqrt[3]{3})$$
$$= 4\sqrt[3]{3} + 15\sqrt[3]{3} = 19\sqrt[3]{3}$$

(c) $\sqrt{\dfrac{3}{2}} - 5\sqrt{6} = \dfrac{\sqrt{3}}{\sqrt{2}} - 5\sqrt{6} = \dfrac{\sqrt{3}\sqrt{2}}{\sqrt{2}\sqrt{2}} - 5\sqrt{6}$

$$= \dfrac{\sqrt{6}}{2} - 5\sqrt{6}$$

$$= \dfrac{\sqrt{6}}{2} - \dfrac{10\sqrt{6}}{2}$$

$$= \dfrac{-9\sqrt{6}}{2}$$

5. Find products and quotients of radicals (Section 6.4).

For example:

(a) $(2\sqrt{3} - 4)(3\sqrt{2} + 5) = (2\sqrt{3})(3\sqrt{2}) + 5(2\sqrt{3}) - 4(3\sqrt{2}) - 4(5)$
$$= 6\sqrt{6} + 10\sqrt{3} - 12\sqrt{2} - 20$$

(b) $\dfrac{18}{\sqrt{6} - \sqrt{3}} = \dfrac{18(\sqrt{6} + \sqrt{3})}{(\sqrt{6} - \sqrt{3})(\sqrt{6} + \sqrt{3})}$

$$= \dfrac{18(\sqrt{6} + \sqrt{3})}{6 - 3}$$

$$= \dfrac{\overset{6}{\cancel{18}}(\sqrt{6} + \sqrt{3})}{\cancel{3}}$$

$$= 6(\sqrt{6} + \sqrt{3})$$

6. Solve radical equations (Section 6.5).

For example:

(a) *Solve for x.* $\sqrt{3x - 2} = 5$

$$\sqrt{3x - 2} = 5$$
$$(\sqrt{3x - 2})^2 = 5^2$$
$$3x - 2 = 25$$
$$3x = 27$$
$$x = 9$$

CHECK: $\sqrt{3(9) - 2} \overset{?}{=} 5$
$$\sqrt{27 - 2} \overset{?}{=} 5$$
$$\sqrt{25} \overset{\checkmark}{=} 5$$

(b) *Solve for a.* $a^{1/3} + 2 = 7$

$$a^{1/3} = 5$$
$$(a^{1/3})^3 = 5^3$$
$$a = 125$$

7. Add, subtract, multiply, and divide complex numbers (Section 6.6).

For example:

(a) $(5 + 3i) + (2 - 6i) = 7 - 3i$

(b) $(5 + 3i)(2 - 6i) = 10 - 30i + 6i - 18i^2$
$$= 10 - 24i - 18(-1)$$
$$= 10 - 24i + 18$$
$$= 28 - 24i$$

(c) $(3 - 2i)^2 - 6(3 - 2i) = 9 - 12i + 4i^2 - 18 + 12i$
$$= 9 - 12i - 4 - 18 + 12i$$
$$= -13$$

(d) $\dfrac{5 + 3i}{2 - 6i} = \dfrac{(5 + 3i)(2 + 6i)}{(2 - 6i)(2 + 6i)}$

$$= \dfrac{10 + 36i - 18}{4 + 36}$$

$$= \dfrac{-8 + 36i}{40}$$

$$= \dfrac{-8}{40} + \dfrac{36}{40}i$$

$$= -\dfrac{1}{5} + \dfrac{9}{10}i$$

CHAPTER 6   Review Exercises

In Exercises 1–10, write the expression in radical form. Assume that all variables represent positive real numbers only.

1. $x^{1/2}$
2. $x^{1/3}$
3. $xy^{1/2}$
4. $(xy)^{1/2}$

5. $m^{2/3}$
6. $m^{3/2}$
7. $(5x)^{3/4}$
8. $5x^{3/4}$

9. $s^{-4/5}$
10. $s^{-5/4}$

In Exercises 11–18, write the expression in exponential form.

11. $\sqrt[3]{a}$
12. $\sqrt[5]{t}$
13. $-\sqrt[5]{n^4}$
14. $-\sqrt[4]{n^5}$

15. $(\sqrt[3]{t})^5$
16. $\sqrt[3]{t^5}$
17. $\dfrac{1}{\sqrt[5]{t^7}}$
18. $\dfrac{1}{\sqrt[7]{t^5}}$

In Exercises 19–48, express in simplest radical form.

19. $\sqrt{54}$
20. $\sqrt[3]{54}$
21. $\sqrt{x^{60}}$
22. $\sqrt[4]{x^{60}}$

23. $\sqrt[5]{x^{60}}$
24. $\sqrt[3]{x^{60}}$
25. $\sqrt[3]{48x^4y^8}$
26. $\sqrt{28x^9y^{13}}$

27. $\sqrt[4]{81x^9y^{10}}$ **28.** $\sqrt[5]{32y^4z^6}$ **29.** $\sqrt{75xy}\sqrt{3x}$ **30.** $\sqrt{48a^2b}\sqrt{12b}$

31. $(x\sqrt{xy})(2x^2y\sqrt{xy^2})$ **32.** $(2x^3\sqrt{x^2})(3x^2\sqrt{x^2y^4})$ **33.** $\dfrac{\sqrt{28}}{\sqrt{63}}$ **34.** $\dfrac{\sqrt{12}}{\sqrt{27}}$

35. $\dfrac{y}{x\sqrt{y}}$ **36.** $\dfrac{y}{\sqrt{xy}}$ **37.** $\sqrt{\dfrac{48a^2b}{3a^5b^2}}$ **38.** $\sqrt{\dfrac{81x^3y^2}{2x^4}}$

39. $\dfrac{\sqrt{x^3y^2}\sqrt{2xy}}{\sqrt{xy^5}}$ **40.** $\dfrac{2a\sqrt{18a^2}\sqrt{3a}}{\sqrt{a^5}}$ **41.** $\sqrt{\dfrac{5}{a}}$ **42.** $\sqrt[3]{\dfrac{5}{a}}$

43. $\dfrac{4}{\sqrt[3]{2a}}$ **44.** $\dfrac{4}{\sqrt[3]{2a^2}}$ **45.** $\sqrt[6]{x^4}$ **46.** $\sqrt[16]{x^6}$

47. $\sqrt{2}\sqrt[3]{2}$ **48.** $\dfrac{\sqrt{2}}{\sqrt[3]{2}}$

In Exercises 49–74, perform the indicated operations and simplify as completely as possible.

49. $8\sqrt{3} - 3\sqrt{3} - 6\sqrt{3}$ **50.** $4\sqrt{2} + \sqrt{2} - 5\sqrt{2}$ **51.** $7\sqrt{54} + 6\sqrt{24}$

52. $6\sqrt{12} - 4\sqrt{27}$ **53.** $b\sqrt{a^3b} + a\sqrt{ab^3}$ **54.** $2t\sqrt{s^5t^2} - 3s\sqrt{s^2t^5}$

55. $\sqrt{\dfrac{3}{2}} + \sqrt{\dfrac{5}{3}}$ **56.** $\sqrt{\dfrac{5}{7}} + \sqrt{\dfrac{2}{3}}$ **57.** $2\sqrt[3]{\dfrac{1}{9}} - 3\sqrt[3]{3}$

58. $5\sqrt[3]{\dfrac{5}{49}} + 2\sqrt[3]{7}$ **59.** $\sqrt{3}(\sqrt{6} - \sqrt{2}) + \sqrt{2}(\sqrt{3} - 3)$ **60.** $\sqrt{5}(\sqrt{2} - 2) + \sqrt{10}(\sqrt{2} - 2)$

61. $(3\sqrt{7} - \sqrt{3})(\sqrt{7} - 2\sqrt{3})$ **62.** $(2\sqrt{x} - \sqrt{5})(3\sqrt{x} + \sqrt{4})$ **63.** $(\sqrt{x} - 5)^2$

64. $(\sqrt{7} - \sqrt{3})^2$ **65.** $(\sqrt{a + 7})^2 - (\sqrt{a} + 7)^2$ **66.** $(\sqrt{b} - 3)^2 - (\sqrt{b - 3})^2$

67. $\dfrac{20}{\sqrt{3} - \sqrt{5}}$ **68.** $\dfrac{12}{\sqrt{5} + \sqrt{3}}$ **69.** $\dfrac{m - n^2}{\sqrt{m} - n}$

70. $\dfrac{8x - 20y}{\sqrt{2x} - \sqrt{5y}}$ **71.** $\dfrac{20}{3 - \sqrt{5}} - \dfrac{50}{\sqrt{5}}$ **72.** $\dfrac{15}{\sqrt{7} + \sqrt{2}} - \dfrac{21}{\sqrt{7}}$

73. $(\sqrt[3]{5} - \sqrt[3]{2})(\sqrt[3]{25} + \sqrt[3]{10} + \sqrt[3]{4})$ **74.** $(\sqrt[3]{s} + \sqrt[3]{t})(\sqrt[3]{s^2} - \sqrt[3]{st} + \sqrt[3]{t^2})$

In Exercises 75–82, solve the equations.

75. $\sqrt{2x} - 5 = 7$ **76.** $\sqrt{3x} + 2 = 4$ **77.** $\sqrt{2x - 5} = 7$ **78.** $\sqrt{3x + 2} = 4$

79. $\sqrt[4]{x - 4} = 3$ **80.** $\sqrt[3]{x + 3} = 2$ **81.** $x^{1/4} + 1 = 3$ **82.** $(x + 1)^{1/4} = 3$

In Exercises 83–86, express as i, $-i$, 1, or -1.

83. i^{11} **84.** i^{29} **85.** $-i^{14}$ **86.** $(-i)^{14}$

In Exercises 87–98, perform the indicated operations with complex numbers and express your answers in complex number form.

87. $(5 + i) + (4 - 2i)$ **88.** $(7 - 2i) - (6 + 3i)$ **89.** $(7 - 2i)(2 - 3i)$

90. $(5 - 4i)(5 + 4i)$ **91.** $2i(3i - 4)$ **92.** $5i(i - 2)$

93. $(3 - 2i)^2$ **94.** $(2 - 3i)^2$ **95.** $(6 - i)^2 - 12(6 - i)$

96. $(7 - i)^2 - 2(7 - 4i)$ **97.** $\dfrac{4 - 3i}{3 + i}$ **98.** $\dfrac{2 + 3i}{-3 + 2i}$

99. Show that $1 - 2i$ is a solution for $x^2 - 2x + 5 = 0$.

100. Show that $1 + 2i$ is a solution for $x^2 - 2x + 5 = 0$.

CHAPTER 6   Practice Test

Perform the indicated operations and express the following in simplest radical form. Assume that all variables represent positive real numbers.

1. $\sqrt[3]{27x^6y^9}$

2. $\sqrt[3]{4x^2y^2}\sqrt[3]{2x}$

3. $(2x^2\sqrt{x})(3x\sqrt{xy^2})$

4. $\dfrac{\sqrt{98}}{\sqrt{8}}$

5. $\sqrt{\dfrac{5}{7}}$

6. $\sqrt[3]{\dfrac{8}{9}}$

7. $\dfrac{(xy\sqrt{2xy})(3x\sqrt{y})}{\sqrt{4x^3}}$

8. $\sqrt[4]{5}\sqrt{5}$

9. $\sqrt{50} - 3\sqrt{8} + 2\sqrt{18}$

10. $(\sqrt{x} - 3)^2$

11. $\sqrt{24} - 4\sqrt{\dfrac{2}{3}}$

12. $\dfrac{\sqrt{6}}{\sqrt{6} - 2}$

Solve the following equations:

13. $\sqrt{x - 3} + 4 = 8$

14. $\sqrt{x} - 3 = 8$

15. $\sqrt[3]{x - 3} + 4 = 2$

16. $x^{1/4} = 5$

Express the following in the form $a + bi$:

17. i^{51}

18. $(3i + 1)(i - 2)$

19. $\dfrac{2i + 3}{i - 2}$

In Exercises 1–4, reduce to lowest terms.

1. $\dfrac{18a^3b^2}{16a^5b}$

2. $\dfrac{x^2 - 9y^2}{x^2 - 6xy + 9y^2}$

3. $\dfrac{2a^3 - 5a^2b - 3ab^2}{a^4 - 2a^3b - 3a^2b^2}$

4. $\dfrac{27x^3 - y^3}{18x^2 + 6xy + 2y^2}$

In Exercises 5–14, perform the operations. Express your answer in simplest terms.

5. $\dfrac{3xy^2}{5a^3b} \div \dfrac{21x^3y}{25ab^3}$

6. $\dfrac{2x^2 - xy - y^2}{x + y} \cdot \dfrac{x^2 - y^2}{x - y}$

7. $\dfrac{3x}{2y} + \dfrac{2y}{3x}$

8. $\dfrac{6}{x - 2} - \dfrac{3x}{x - 2}$

9. $\left(\dfrac{2x^3 - 2x^2 - 24x}{x + 2}\right)\left(\dfrac{x + 2}{4x^2 + 12x}\right)$

10. $\dfrac{9a^2 - 6ab + b^2}{3a + b} \div \dfrac{3a^2 - 4ab + b^2}{3a^2 - 2ab - b^2}$

11. $\dfrac{2}{x - 5} - \dfrac{3}{x - 2}$

12. $\dfrac{3}{2x^2 - 5xy - 3y^2} - \dfrac{5}{x^2 - 2xy - 3y^2}$

13. $\dfrac{x^3 + x^2y}{2x^2 + xy} \div \left[\left(\dfrac{x^2 + 2xy - 3y^2}{2x^2 - xy - y^2}\right)(x + y)\right]$

14. $\dfrac{x}{x - 5y} - \dfrac{2y}{x + 5y} - \dfrac{20y^2}{x^2 - 25y^2}$

In Exercises 15–16, write as a simple fraction reduced to lowest terms.

15. $\dfrac{x - \dfrac{2}{x}}{\dfrac{1}{2} - x}$

16. $\dfrac{\dfrac{x}{x + y} - \dfrac{4y}{x + 4y}}{\dfrac{x - 2y}{x + y} + 1}$

In Exercises 17–22, solve the equations or inequalities.

17. $\dfrac{2}{x} - \dfrac{1}{2} = 2 - \dfrac{1}{x}$

18. $\dfrac{4}{x + 4} - 2 = 0$

19. $\dfrac{x}{2} - \dfrac{x + 1}{3} > \dfrac{2}{3}$

20. $\dfrac{3}{x - 2} + 3 = \dfrac{x + 1}{x - 2}$

21. $\dfrac{3}{x - 2} + \dfrac{5}{x + 1} = \dfrac{1}{x^2 - x - 2}$

22. $\dfrac{x + 3}{4} - \dfrac{2x + 1}{3} \le \dfrac{1}{2}$

In Exercises 23–26, solve for the given variable.

23. $2a + 3b = 5b - 4a$ (for a)

24. $3xy - 2y = 5x + 3y$ (for y)

25. $\dfrac{x - y}{y} = x$ (for y)

26. $\dfrac{2x + 3}{x - 2} = y$ (for x)

In Exercises 27–30, solve algebraically.

27. The ratio of foreign-made cars to American-made cars in a town is 5 to 6. If there are 1,200 American-made cars in the town, how many cars are there all together?

28. How much of a 20% solution of alcohol should be mixed with 8 liters of a 35% solution of alcohol in order to get a solution that is 30% alcohol?

(continued)

29. Carmen can paint a room in 4 hours and Judy can paint the same room in $4\frac{1}{2}$ hours. How long would it take for them to paint the room working together?

30. General admission tickets at a theater sold for $3.50, while reserved seats sold for $4.25. If 505 tickets were sold for a performance which took in $1,861.25, how many of each type of ticket were sold?

In Exercises 31–42, perform the indicated operations and simplify. Express your answer with positive exponents only.

31. $(-2x^2y)(-3xy^2)^3$

32. $\dfrac{8x^4y^5}{16x^6y}$

33. $\left(\dfrac{2xy^2}{4x^2y^3}\right)^2$

34. $\dfrac{(-2ab^2)^2(-3a)^2}{-6a^2b}$

35. $(x^{-5}y^{-2})(x^{-7}y)$

36. $(2x^{-3}y)^{-2}$

37. $\dfrac{r^{-3}s^{-2}}{r^{-2}s^0}$

38. $\left(-\dfrac{2}{3}\right)^{-3}$

39. $(-2r^{-3}s)^{-1}(-4r^{-2}s^{-3})^2$

40. $\left(\dfrac{5x^{-1}y}{(-5x^{-2}y^2)^2}\right)^{-1}$

41. $(x^{-1}-y^{-1})(x+y)$

42. $\dfrac{a^{-1}+b^{-2}}{a^{-1}+1}$

In Exercises 43–44, express the following using scientific notation:

43. 56,429.32

44. 0.0000752

45. How far does light travel in 1 day if it travels 186,000 miles per second?

46. Perform the computations using scientific notation and express your answer in standard notation:

$$\dfrac{(0.00016)(400,000)}{(0.08)(5,000)}$$

In Exercises 47–48, evaluate.

47. $(-1,000)^{1/3}$

48. $\left(\dfrac{27}{64}\right)^{-2/3}$

In Exercises 49–52, perform the indicated operations and express your answers with positive exponents only.

49. $a^{2/3}a^{-1/2}$

50. $\dfrac{x^{2/5}}{x^{1/3}}$

51. $[(81)^{-1/3}3^2]^{-3}$

52. $\left(\dfrac{x^{-1/3}x^{2/3}}{x^{1/4}}\right)^{-6}$

In Exercises 53–54, change to radical notation.

53. $x^{3/4}$

54. $2x^{1/2}$

In Exercises 55–62, express in simplest radical form. Assume that all variables represent positive real numbers only.

55. $\sqrt{16a^4b^8}$

56. $\sqrt{8a^3b^{10}}$

57. $(3x\sqrt{2x^2y})(2x\sqrt{8xy^3})$

58. $\sqrt[3]{81x^4y^5}$

59. $\dfrac{\sqrt{48}}{\sqrt{3}}$

60. $\sqrt{\dfrac{3}{xy^2}}$

61. $\sqrt[8]{x^4}$

62. $\sqrt[3]{\dfrac{2}{5x}}$

In Exercises 63–72, perform the indicated operations and simplify as completely as possible.

63. $2\sqrt{5} - 3\sqrt{5} + 8\sqrt{5}$

64. $2\sqrt{27} - \sqrt{75} - \sqrt{3}$

65. $3a\sqrt[3]{a^4} - a^2\sqrt[3]{a}$

66. $\sqrt{\dfrac{2}{3}} - \sqrt{\dfrac{3}{2}}$

67. $\sqrt{2}(\sqrt{2} - 1) + 2\sqrt{2}$

68. $(\sqrt{5} - \sqrt{3})^2$

69. $(\sqrt{5} - \sqrt{3})(\sqrt{5} + \sqrt{3})$

70. $(\sqrt{x + 4})^2 - (\sqrt{x} + 4)^2$

71. $\dfrac{5}{\sqrt{3} + \sqrt{2}} - \dfrac{3}{\sqrt{3}}$

72. $\dfrac{2\sqrt{3}}{\sqrt{5} + \sqrt{3}}$

In Exercises 73–76, solve the equation.

73. $\sqrt{3x + 2} = 7$

74. $\sqrt{3x} + 2 = 7$

75. $\sqrt[3]{x - 1} = -2$

76. $x^{1/5} + 5 = 4$

In Exercises 77–78, express as i, -1, $-i$, or 1.

77. i^{35}

78. i^{-2}

In Exercises 79–80, express in the form $a + bi$.

79. $(3 - 2i)(5 + i)$

80. $\dfrac{1 + 3i}{1 + 2i}$

CHAPTERS 4–6   CUMULATIVE PRACTICE TEST

1. *Perform the indicated operations and express your answer in simplest form.*

(a) $\left(\dfrac{25x^2 - 9y^2}{x - y}\right)\left(\dfrac{2x^2 + xy - 3y^2}{10x^2 + 9xy - 9y^2}\right)$

(b) $\dfrac{2y}{2x - y} + \dfrac{4x}{y - 2x}$

(c) $\dfrac{3x}{x^2 - 10x + 21} - \dfrac{2}{x^2 - 8x + 15}$

2. *Write as a simple fraction reduced to lowest terms:*

$$\dfrac{\dfrac{1}{x} - 2}{3 + \dfrac{1}{x + 1}}$$

3. *Solve the following equations:*

(a) $\dfrac{2}{x + 1} + \dfrac{3}{2x} = \dfrac{6}{x^2 + x}$

(b) $\dfrac{x}{3} - \dfrac{x + 2}{7} < 4$

(c) $\dfrac{5}{x + 3} + 2 = \dfrac{x + 8}{x + 3}$

(continued)

4. *Solve for a.*

$$y = \frac{a}{a + 1}$$

5. Carol can process 80 forms in 6 hours, while Joe can process the same 80 forms in $5\frac{1}{2}$ hours. How long would it take them to process the same forms if they work together?

6. Perform the indicated operations and simplify. Express your answer with positive exponents.

(a) $(3x^2y)^2(-2xy^3)^3$

(b) $(-2x^{-1}y^3)(-3x^2y^{-3})^2$

(c) $\left(\dfrac{3x^{-2}y^{-1}}{9xy^{-3}}\right)^{-2}$

(d) $\dfrac{a^{-1}}{a^{-1} + b^{-1}}$

7. Express 0.000034 using scientific notation.

8. Perform the operations using scientific notation. Express your answer in standard form.

$$\frac{(150,000)(0.00028)}{(0.07)(0.0002)}$$

9. Evaluate $(-128)^{-3/7}$.

10. Perform the operations and simplify. Express your answers using positive exponents only. Assume that all variables represent positive real numbers.

(a) $(x^{1/3}x^{-1/5})^5$

(b) $\dfrac{x^{-2/3}y^{-3/4}}{x^{1/2}}$

11. Express the following in simplest form. Assume that all variables represent positive real numbers.

(a) $\sqrt{24x^2y^5}$

(b) $(a^2\sqrt{2ab^2})(b\sqrt{4a^2b^3})$

(c) $\sqrt[3]{\dfrac{5}{a}}$

12. Peform the operations and simplify as completely as possible.

(a) $5\sqrt{8} - 5\sqrt{2} - \sqrt{50}$

(b) $\sqrt{\dfrac{x}{y}} - \sqrt{\dfrac{x}{2}}$

(c) $(2 - \sqrt{3})(\sqrt{3} - 2)$

(d) $\dfrac{5\sqrt{2}}{\sqrt{5} - \sqrt{2}}$

13. *Solve for x.* $\sqrt{2x} + 3 = 4$

14. *Express the following in the form a + bi.* $\dfrac{2 - i}{3 - i}$

7

Second–Degree Equations and Inequalities

In Chapters 3 and 4 we solved first-degree equations and inequalities. In this chapter we will turn our attention to solving second-degree equations and inequalities.

A **second-degree equation** is a polynomial equation in which the highest degree of the variable is 2. In particular, a second-degree equation in one unknown is called a **quadratic equation**. We define the *standard form* of a quadratic equation as

$$Ax^2 + Bx + C = 0 \qquad (A > 0)$$

where A is the coefficient of the second-degree term, x^2; B is the coefficient of the first-degree term, x; and C is a numerical constant.

As with all other equations, the solutions of quadratic equations are values which, when substituted for the variable, will make the equation a true statement. The solutions to the equation $Ax^2 + Bx + C = 0$ are also called the **roots** of the polynomial $Ax^2 + Bx + C$.

In the next few sections we will discuss various methods for solving quadratic equations. Then we will apply these methods to more complex equations as well as to inequalities.

7.1

The Factoring and Square Root Methods

The Factoring Method

We begin by noting that if the product of two quantities is 0, then either one or both of the quantities must be 0. Symbolically written, if $a \cdot b = 0$, then $a = 0$, $b = 0$, or both $a = b = 0$. In mathematics, the word *or* includes the possibility of both. Thus, we can write

The Zero-Product Rule	If $a \cdot b = 0$, then $a = 0$ or $b = 0$.

For example, if we want to find the solution to the equation

$$(x - 2)(x + 3) = 0$$

then either the factor $x - 2$ must equal 0 or the factor $x + 3$ must equal 0. Now we have two first-degree equations:

$$x - 2 = 0 \quad \text{and} \quad x + 3 = 0$$

If $x - 2 = 0$, then $x = 2$; if $x + 3 = 0$, then $x = -3$.

Hence, our solutions to the equation $(x - 2)(x + 3) = 0$ are $x = 2$ and $x = -3$.

EXAMPLE 1

Solve each of the following equations.

(a) $(3x - 2)(x + 3) = 0$ **(b)** $y(y + 4) = 0$ **(c)** $5x(x - 1)(x + 8) = 0$

Solution

(a) If $(3x - 2)(x + 3) = 0$, then

$3x - 2 = 0 \quad \text{or} \quad x + 3 = 0$ *Solve each first-degree equation.*
$3x = 2$

Hence, $\boxed{x = \dfrac{2}{3} \quad \text{or} \quad x = -3}$

Let's check the solution $x = -3$:

$(3x - 2)(x + 3) = 0$ *Substitute -3 for x.*
$[3(-3) - 2](-3 + 3) \overset{?}{=} 0$

Note that this arithmetic is not necessary once we establish that one of the factors is 0.

$(-11)(0) \overset{\checkmark}{=} 0$

(b) If $y(y + 4) = 0$, then

$y = 0 \quad \text{or} \quad y + 4 = 0$ *If the product of factors is 0, then at least one of the factors must be 0.*

Hence, $\boxed{y = 0 \quad \text{or} \quad y = -4}$

(c) We can generalize the zero-product rule to more than two factors. If $5x(x - 1)(x + 8) = 0$, then

$$x = 0 \quad \text{or} \quad x - 1 = 0 \quad \text{or} \quad x + 8 = 0$$

If the product of factors is 0, then at least one of the factors must be 0. We solve each equation.

Note that we can ignore the constant factor 5 since $5 \neq 0$.

Hence, $\boxed{x = 0, \quad x = 1, \quad \text{or} \quad x = -8}$

We mentioned that we can ignore the *factor* 5 because $5 \neq 0$. We would arrive at the same conclusion if we were to first divide both sides of the equation by 5. ∎

If we multiply out the left side in part **(a)** of the previous example, we get $3x^2 + 7x - 6 = 0$, which is a quadratic equation in standard form. If the second-degree expression of a quadratic equation (in standard form) factors into two first-degree factors, we can apply the above principle to solve the quadratic equation. This is called the ***factoring method*** of solving quadratic equations.

For example, to solve the second-degree equation $5x^2 - 2 = 3x^2 - x + 4$, we must first put the equation in standard form:

$$5x^2 - 2 = 3x^2 - x + 4 \qquad \textit{Put in standard form.}$$

$$2x^2 + x - 6 = 0 \qquad \textit{Factor the left-hand side.}$$

$$(2x - 3)(x + 2) = 0 \qquad \textit{Since the product is 0, set each factor equal to 0.}$$

$$2x - 3 = 0 \quad \text{or} \quad x + 2 = 0 \qquad \textit{Then solve each first-degree equation.}$$
$$2x = 3 \qquad\qquad x + 2 = 0$$
$$x = \frac{3}{2} \quad \text{or} \quad x = -2$$

CHECK $x = \dfrac{3}{2}$:
$$5x^2 - 2 = 3x^2 - x + 4$$
$$5\left(\frac{3}{2}\right)^2 - 2 \overset{?}{=} 3\left(\frac{3}{2}\right)^2 - \frac{3}{2} + 4$$
$$5\left(\frac{9}{4}\right) - 2 \overset{?}{=} 3\left(\frac{9}{4}\right) - \frac{3}{2} + 4$$
$$\frac{45}{4} - 2 \overset{?}{=} \frac{27}{4} - \frac{3}{2} + 4$$
$$\frac{45}{4} - \frac{8}{4} \overset{?}{=} \frac{27}{4} - \frac{6}{4} + \frac{16}{4}$$
$$\frac{37}{4} \overset{\checkmark}{=} \frac{37}{4}$$

CHECK $x = -2$:
$$5x^2 - 2 = 3x - x + 4$$
$$5(-2)^2 - 2 \overset{?}{=} 3(-2)^2 - (-2) + 4$$
$$5(4) - 2 \overset{?}{=} 3(4) + 2 + 4$$
$$20 - 2 \overset{\checkmark}{=} 12 + 2 + 4$$

EXAMPLE 2

Solve each of the following equations.

(a) $2x^2 - 5x + 3 = 1$ **(b)** $3a^2 = 5a$

Solution

(a) Do not try to factor first; $2x^2 - 5x + 3 = 1$ is not in standard form. Remember, the principle upon which the factoring method is based requires the product be equal to 0.

$$2x^2 - 5x + 3 = 1$$ *First we put the equation in standard form (subtract 1 from both sides of the equation).*

$$2x^2 - 5x + 2 = 0$$ *Then factor.*

$$(2x - 1)(x - 2) = 0$$ *And solve each first-degree equation.*

$$2x - 1 = 0 \qquad x - 2 = 0$$
$$\qquad\quad \text{or}$$
$$2x = 1 \qquad\qquad x = 2$$

$$\boxed{x = \frac{1}{2} \quad \text{or} \quad x = 2}$$

We leave it to the student to check the solutions whenever the check is omitted.

(b) Do not make the mistake of dividing both sides of the equation by a; a is a variable which may be 0 and division by 0 is not allowed.

$$3a^2 = 5a$$ *First put in standard form.*

$$3a^2 - 5a = 0$$ *Then factor.*

$$a(3a - 5) = 0$$ *Set each factor equal to 0.*

$$a = 0 \qquad \text{or} \qquad 3a - 5 = 0$$
$$3a = 5$$

$$\boxed{a = 0 \quad \text{or} \quad a = \frac{5}{3}}$$ ■

EXAMPLE 3

Solve each of the following equations.

(a) $(3x + 1)(x - 1) = (5x - 3)(2x - 3)$

(b) $\dfrac{x}{x - 5} - \dfrac{3}{x + 1} = \dfrac{30}{x^2 - 4x - 5}$

Solution

(a)
$$(3x + 1)(x - 1) = (5x - 3)(2x - 3)$$

Before we put this equation in standard form, we must first simplify each side. First multiply out.

$$3x^2 - 2x - 1 = 10x^2 - 21x + 9$$

Then put in standard form.

$$0 = 7x^2 - 19x + 10$$

Factor the right-hand side of the equation.

$$0 = (7x - 5)(x - 2)$$

Then set each factor equal to 0.

$$7x - 5 = 0 \qquad \text{or} \qquad x - 2 = 0$$
$$7x = 5 \qquad\qquad\qquad x = 2$$

$$\boxed{x = \frac{5}{7} \quad \text{or} \quad x = 2}$$

(b) First we find the LCD of the fractions, which is $(x - 5)(x + 1)$.

$$\frac{x}{x - 5} - \frac{3}{x + 1} = \frac{30}{x^2 - 4x - 5}$$

$$(x - 5)(x + 1)\left(\frac{x}{x - 5} - \frac{3}{x + 1}\right) = \frac{30}{(x - 5)(x + 1)}(x - 5)(x + 1)$$

Multiply both sides of the equation by $(x - 5)(x + 1)$.

$$\cancel{(x - 5)}(x + 1)\frac{x}{\cancel{x - 5}} - (x - 5)\cancel{(x + 1)}\frac{3}{\cancel{x + 1}} = \frac{30}{\cancel{(x - 5)(x + 1)}}\cancel{(x - 5)(x + 1)}$$

Reduce.

$$(x + 1)x - (x - 5)(3) = 30$$

$$x^2 + x - 3x + 15 = 30 \qquad \text{\textit{Simplify.}}$$

$$x^2 - 2x + 15 = 30 \qquad \text{\textit{Put in standard form.}}$$

$$x^2 - 2x - 15 = 0 \qquad \text{\textit{Factor and solve each first-}}$$

$$(x - 5)(x + 3) = 0 \qquad \text{\textit{degree equation.}}$$

$$x - 5 = 0 \qquad \text{or} \qquad x + 3 = 0$$

$$x = 5 \qquad \text{or} \qquad x = -3$$

Always check the solutions to rational equations to determine if the solutions are valid.

As we pointed out in Chapter 4, we can observe at the outset that neither 5 nor -1 are admissible values since they make a denominator of one of the fractions 0 and therefore the fraction undefined. [Also, multiplying both sides of the equation by $(x - 5)(x + 1)$ is not valid if x is 5 or -1, since multiplying both sides by 0 does not yield an equivalent equation.] Since 5 is one of the solutions we arrived at, we must eliminate this value, and therefore -3 is the only solution.

Hence, the solution is $\boxed{x = -3}$ ∎

The Square Root Method

The solutions to the equation $x^2 = d$ are numbers which, when squared, yield d. We defined the square root of d, $\sqrt{d}$, in Chapter 6 as the *positive* expression which, when squared, yields d. However, since we have no information as to whether x is positive or negative, we must take into account the negative square root as well. For example, in solving $x^2 = 4$, since $(+2)^2 = 4$ and $(-2)^2 = (-2)(-2) = 4$, then both $+2$ *and* -2 are solutions. Hence, if $x^2 = d$, then $x = +\sqrt{d}$ or $x = -\sqrt{d}$.

A shorter way to write a quantity and its opposite would be to write "$\pm$" in front of the quantity to symbolize both answers. Thus, the answers to $x^2 = 4$ can be written as $x = \pm 2$ rather than $x = +2$ and $x = -2$.

THEOREM

If $x^2 = d$, then $x = \pm\sqrt{d}$.

The square root method of solving quadratic equations is used mainly when there is no x term in the standard form of the quadratic equation—that is, when $B = 0$ in $Ax^2 + Bx + C = 0$.

The method is to isolate x^2 on one side of the equation and then apply the theorem. When we use this theorem we will say that we are *taking square roots*. For example:

Solve for x if $x^2 + 5 = 8$.

We proceed as follows:

$$x^2 + 5 = 8 \qquad \text{\textit{Isolate }} x^2.$$
$$x^2 = 3 \qquad \text{\textit{Take square roots.}}$$
$$x = \pm\sqrt{3} \qquad \text{\textit{(which means }} x = +\sqrt{3} \text{ \textit{or} } x = -\sqrt{3}\text{)}$$

EXAMPLE 4

Solve each of the following equations.

(a) $x^2 + 2 = 5 - 2x^2$ **(b)** $3x^2 - 2 = 6$ **(c)** $2x^2 + 15 = 7$

Solution

(a) $x^2 + 2 = 5 - 2x^2$ *First isolate x^2 on the left side of the equation.*

$$3x^2 = 3$$
$$x^2 = 1 \qquad \text{\textit{Then take the square roots.}}$$
$$x = \pm\sqrt{1} \qquad \text{\textit{Do not forget the negative root.}}$$
$$\boxed{x = \pm 1}$$

(b) $3x^2 - 2 = 6$ *Isolate x^2 (add $+2$ to both sides of the equation).*

$$3x^2 = 8 \qquad \text{\textit{Divide both sides of the equation by 3.}}$$
$$x^2 = \frac{8}{3} \qquad \text{\textit{Take square roots.}}$$
$$x = \pm\sqrt{\frac{8}{3}} \qquad \text{\textit{Simplify your answers: rationalize the denominator.}}$$
$$x = \frac{\pm\sqrt{8}}{\sqrt{3}} = \frac{\pm\sqrt{8}\sqrt{3}}{\sqrt{3}\sqrt{3}} = \frac{\pm\sqrt{24}}{3} = \frac{\pm2\sqrt{6}}{3}$$

Recall that this means the solutions are $\boxed{\dfrac{-2\sqrt{6}}{3} \text{ and } \dfrac{+2\sqrt{6}}{3}}$

(c) $2x^2 + 15 = 7$ *Isolate x^2.*

$$2x^2 = -8$$
$$x^2 = \frac{-8}{2} = -4 \qquad \text{\textit{Take the square roots.}}$$
$$x = \pm\sqrt{-4}$$

This means there are no *real* solutions, but if we use complex numbers, then

$$x = \pm\sqrt{4(-1)} = \pm\sqrt{4i^2} = \pm\sqrt{4}\sqrt{i^2} = \boxed{\pm 2i}$$

Let's check the solutions:

CHECK $x = +2i$: $2x^2 + 15 = 7$

$$2(2i)^2 + 15 \stackrel{?}{=} 7$$

$$2(4i^2) + 15 \stackrel{?}{=} 7$$

$$8i^2 + 15 \stackrel{?}{=} 7$$

$$-8 + 15 \stackrel{\checkmark}{=} 7 \qquad \textit{Since } i^2 = -1$$

CHECK $x = -2i$: $2x^2 + 15 = 7$

$$2(-2i)^2 + 15 \stackrel{?}{=} 7$$

$$2(4i^2) + 15 \stackrel{?}{=} 7$$

$$8i^2 + 15 \stackrel{?}{=} 7$$

$$-8 + 15 \stackrel{\checkmark}{=} 7 \qquad \textit{Since } i^2 = -1$$ ∎

We can solve literal quadratic equations by the same methods.

EXAMPLE 5

Solve the following explicitly for x:

$$9x^2 + a = b$$

Solution

$9x^2 + a = b$ *Isolate x^2.*

$9x^2 = b - a$

$$\frac{9x^2}{9} = \frac{b - a}{9}$$

$$x^2 = \frac{b - a}{9} \qquad \textit{Take square roots.}$$

$$x = \pm\sqrt{\frac{b - a}{9}} \qquad \textit{Simplify.}$$

$$\boxed{x = \pm\frac{\sqrt{b - a}}{3}}$$ ∎

We mentioned that the square root method is used mainly when the standard form of a quadratic equation has no first-degree term. The square root method can also be used immediately if we have the square of a binomial equal to a constant. For example, consider solving the following equation for x:

$$(x - 5)^2 = 7$$

If we tried to solve this equation by first putting it in standard form, we would square $x - 5$ to get $x^2 - 10x + 25 = 7$, which becomes

$$x^2 - 10x + 18 = 0$$

Note that $x^2 - 10x + 18$ does not factor (with integer coefficients), so we do not use the factoring method. In addition, we cannot use the square root method on the equation in standard form since there is a first-degree term.

We can, however, take advantage of the form of the given equation and "take square roots" as our *first* step in solving the equation.

$(x - 5)^2 = 7$ *Take square roots.*

$x - 5 = \pm\sqrt{7}$ *Now isolate x by adding 5 to both sides of the equation.*

$x = 5 \pm \sqrt{7}$

The two solutions are $5 + \sqrt{7}$ and $5 - \sqrt{7}$.

EXAMPLE 6   *Solve for x.* $(x - 3)^2 = 4$

Solution

$(x - 3)^2 = 4$ *Take square roots.*

$(x - 3) = \pm\sqrt{4}$

$x - 3 = \pm 2$ *Then add 3 to both sides of the equation.*

$x = \pm 2 + 3$

Hence,

$$x = 2 + 3 = 5 \quad \text{or} \quad x = -2 + 3 = 1$$

Hence, $\boxed{x = 5 \quad \text{or} \quad x = 1}$ ■

Exercises 7.1

In Exercises 1–66, solve the equation using the factoring or square root method where appropriate.

1. $(x + 2)(x - 3) = 0$ **2.** $(a - 5)(a + 7) = 0$ **3.** $0 = (2y - 1)(y - 4)$ **4.** $0 = (2a + 3)(a - 4)$

5. $x^2 = 25$ **6.** $a^2 = 81$ **7.** $x(x - 4) = 0$ **8.** $y(y + 3) = 0$

9. $12 = x(x - 4)$ **10.** $15 = y(y + 2)$ **11.** $5y(y - 7) = 0$ **12.** $3x(x + 2) = 0$

13. $x^2 - 16 = 0$ **14.** $a^2 - 225 = 0$ **15.** $0 = 9c^2 - 16$ **16.** $0 = 25x^2 - 4$

17. $8a^2 - 18 = 0$ **18.** $2x^2 = 32$ **19.** $0 = x^2 - x - 6$ **20.** $0 = a^2 + 2a - 8$

21. $2y^2 - 3y + 1 = 0$ **22.** $5a^2 - 14a - 3 = 0$ **23.** $0 = 8x^2 + 4x - 112$ **24.** $0 = 6x^2 + 8x - 8$

25. $4x^2 = 24$ **26.** $5y^2 = 65$ **27.** $7a^2 - 19 = 0$ **28.** $6y^2 - 25 = 0$

29. $10 = x^2 - 3x$ **30.** $1 - 2x = 3x^2$ **31.** $6a^2 + 3a - 1 = 4a$ **32.** $9a^2 + 4 = 12a^2$

33. $3y^2 - 5y + 8 = 9y^2 - 10y + 2$ **34.** $5y^2 + 15y + 4 = 2y + 10$ **35.** $0 = 3x^2 + 5$

36. $0 = 2x^2 + 7$ **37.** $(2s - 3)(3s + 1) = 7$ **38.** $x(x - 2) = -1$

39. $(x + 2)^2 = 25$ **40.** $(y - 15)^2 = 36$ **41.** $(x - 2)(3x - 1) = (2x - 3)(x + 1)$

42. $(x + 5)(2x - 3) = (x - 1)(3x + 1)$ **43.** $(2a + 5)(a + 3) = 11(a + 2)$ **44.** $(5a - 2)(a - 3) = -17a + 10$

45. $(x + 3)^2 = x(x + 5)$ **46.** $(2x + 1)^2 = 4x(x - 3)$ **47.** $x^2 + 3 = 1$

48. $x^2 - 7 = -12$ **49.** $8 = (x - 8)^2$ **50.** $12 = (x + 12)^2$

51. $(y - 5)^2 = -16$ **52.** $(y + 3)^2 = -4$ **53.** $\dfrac{2}{x - 1} + x = 4$

54. $\dfrac{6}{x - 3} + x = 8$ **55.** $2a - 5 = \dfrac{3(a + 2)}{a + 4}$ **56.** $3a - 4 = \dfrac{a + 2}{a}$

57. $\dfrac{5x + 2}{x - 2} = \dfrac{2x + 1}{x - 2}$

58. $\dfrac{2y + 5}{12} = \dfrac{y + 1}{5y - 2}$

59. $\dfrac{x}{x + 2} - \dfrac{3}{x} = \dfrac{x + 1}{x}$

60. $\dfrac{x}{x - 1} + \dfrac{4}{x} = \dfrac{x + 2}{x}$

61. $\dfrac{1}{x} + x = 2$

62. $2x - \dfrac{5}{x} = 9$

63. $\dfrac{3}{x - 2} + \dfrac{7}{x + 2} = \dfrac{x + 1}{x - 2}$

64. $\dfrac{3}{a + 4} + \dfrac{2}{a - 2} = \dfrac{4a + 5}{a + 4}$

65. $\dfrac{2}{x - 1} + \dfrac{3x}{x + 2} = \dfrac{2(5x + 9)}{x^2 + x - 2}$

66. $\dfrac{3}{y - 1} + \dfrac{2}{y + 1} = \dfrac{2y + 3}{3(y - 1)}$

In Exercises 67–76, solve explicitly for the given variable using the factoring or square root method.

67. $8a^2 + 3b = 5b$ for a

68. $5r^2 - 8a = 2a - 3r^2$ for r

69. $5x^2 + 7y^2 = 9$ for y

70. $14a^2 + 3b^2 = c$ for b

71. $V = \dfrac{2}{3}\pi r^2$ for $r > 0$

72. $K = \dfrac{2gm}{s^2}$ for $s > 0$

73. $a^2 - 4b^2 = 0$ for a

74. $x^2 - 9y^2 = 0$ for x

75. $x^2 - xy - 6y^2 = 0$ for x

76. $a^2 + 3ab - 10b^2 = 0$ for a

? QUESTIONS FOR THOUGHT

77. On what property of real numbers is the factoring method based?

78. Discuss what is *wrong* (and why) with the solutions to the following equations:

    (a) Problem: Solve $(x - 3)(x - 4) = 7$.

        Solution: $x - 3 = 7$ $x - 4 = 7$

                     $x = 10$ $x = 11$

    (b) Problem: Solve $3x(x - 2) = 0$.

        Solution: $x = 3$, $x = 0$ or $x = 2$

79. Solve each of the following by the factoring method *and* the square root method.

    (a) $x^2 = 9$

    (b) $9x^2 - 16 = 0$

    (c) $7x^2 - 5 = 3x^2 + 4$

80. Look at the solution of the following problem. See if you can justify each step.

    Problem: $x^2 + 4x - 7 = 0$

    Solution: **1.** $x^2 + 4x - 7 = 0$

               2. $x^2 + 4x = 7$

               3. $x^2 + 4x + 4 = 7 + 4$

               4. $x^2 + 4x + 4 = 11$

               5. $(x + 2)^2 = 11$

               6. $x + 2 = \pm\sqrt{11}$

               7. $x = -2 \pm \sqrt{11}$

81. Check that $2 + \sqrt{3}$ is a solution to the equation $x^2 - 4x + 1 = 0$ by substituting $2 + \sqrt{3}$ for x in the equation. Check to see if $2 - \sqrt{3}$ is also a solution of the equation

$$x^2 - 4x + 1 = 0$$

7.2

Completing the Square

The factoring method is the easiest method for solving quadratic equations, but it cannot be applied to a problem such as $x^2 + 3x - 5 = 0$ because the expression $x^2 + 3x - 5$ is not factorable using integer coefficients.

The square root method is straightforward but, as we mentioned, we usually do not apply this method if there is a first-degree term in the standard form of the quadratic equation.

Our interest in this section is to find a method for solving quadratic equations which can be applied to all cases.

If we can take any equation and put it in the form $(x + p)^2 = d$, where p and d are constants, then all that remains is to apply the square root method—that is, take the square roots, solve for x, and simplify the answer. But can all quadratic equations be put in the form $(x + p)^2 = d$, where p and d are constants?

To answer this, we first examine the squares of binomials of the form $(x + p)$, where p is a constant. When the squares are multiplied out, we call them *perfect squares*.

First we will square binomials of the form $x + p$ and look at the relationship between the x coefficient and the numerical term. Observe the following:

	Coefficient of the x term	*The numerical term*
$(x - 3)^2 = x^2 - 6x + 9$	-6	$+9$
$(x + 5)^2 = x^2 + 10x + 25$	$+10$	$+25$
$(x + 4)^2 = x^2 + 8x + 16$	$+8$	$+16$

Let's examine what happens when we square $(x + p)$:

Square of second term in the binomial

$$(x + p)^2 = x^2 + 2px + p^2$$

Twice the product of the terms in the binomial
Square of first term in the binomial

Given that p is a constant and x is a variable, the middle (first-degree) term *coefficient* will be $2p$, and the numerical term will be p^2. What is the relationship between the middle term coefficient, $2p$, and the numerical term, p^2?

If you take half of $2p$ and square it, you will get p^2:

$$\left[\frac{1}{2}(2p)\right]^2 = \left(\frac{2p}{2}\right)^2 = p^2$$

The square of half of $2p$ is p^2

Thus, if you multiply out the binomial $(x + p)^2$, *the square of one-half the middle term coefficient will yield the numerical term.*

We now return to our examples:

$$(x - 3)^2 = x^2 - 6x + 9 \qquad\qquad (x + 5)^2 = x^2 + 10x + 25$$

1. Take half of −6 to get −3.   **1.** Take half of 10 to get 5.

2. Square −3 to get the numerical   **2.** Square 5 to get the numerical
term, +9.   term, 25.

Now we will demonstrate how we can take any equation and put it in the form $(x + p)^2 = d$, where p and d are constants. Suppose we have an equation such as

$$x^2 + 6x - 8 = 0$$

In order to make it clearer how to make the left-hand side into a perfect square, we first add 8 to both sides of the equation:

$x^2 + 6x - 8 = 0$ *Add 8 to both sides of the equation.*
$\quad x^2 + 6x \;\;= 8$

Now what is missing to make the left side a perfect square? Take half the middle term coefficient, 6, and square it:

$$\left[\frac{1}{2}(6)\right]^2 = (3)^2 = 9$$

Thus, 9 must be added to make the left side a perfect square. But since we are dealing with equations, we must add 9 to *both sides* of the equation:

$x^2 + 6x + 9 = 8 + 9$
$x^2 + 6x + 9 = 17$ *Write as a perfect square in factored form.*
$\quad (x + 3)^2 = 17$

The only difference between the equation $x^2 + 6x = 8$ and the perfect square version $x^2 + 6x + 9 = 17$ is that 9 was added to both sides. Why 9? To make the left-hand side a perfect square so that it could be written in factored form. This process of adding a number to make a perfect square is called ***completing the square.***

Similarly, *any* quadratic equation can be put in the form $(x + p)^2 = d$, where p and d are constants. Now we can solve this equation by the square root method, as demonstrated in Section 7.1.

$(x + 3)^2 = 17$ *Take square roots.*
$\quad x + 3 = \pm\sqrt{17}$ *Then isolate x by adding −3 to both sides of the equation.*
$\qquad x = -3 \pm \sqrt{17}$

EXAMPLE 1

Solve the following by completing the square:

$$x^2 + 8x - 4 = 0$$

Solution

1. Add $+ 4$ to both sides of the equation. $x^2 + 8x \quad = 4$

2. Take one-half the middle term coefficient and square it:
$[\frac{1}{2}(8)]^2 = 4^2 = 16$

3. Add result of step 2 to both sides of the equation. $x^2 + 8x \boxed{+ 16} = 4 \boxed{+ 16}$
$x^2 + 8x + 16 = 20$

4. Factor the left-hand side as a perfect square. $(x + 4)^2 = 20$

5. Take square roots. $x + 4 = \pm\sqrt{20}$

6. Isolate x by adding -4 to both sides of the equation. $x = -4 \pm \sqrt{20}$

7. Simplify the radical. $x = -4 \pm 2\sqrt{5}$

The solutions are $\boxed{-4 + 2\sqrt{5} \quad \text{and} \quad -4 - 2\sqrt{5}}$ ∎

EXAMPLE 2

Solve $2x^2 - 6x + 5 = 0$ by completing the square.

Solution

Keep in mind that the number we added to the quadratic expression to make it a perfect square was found by exploiting the relationship between coefficients in $(x + p)^2 = x^2 + 2px + p^2$, when x^2 has a coefficient of 1. However, in this example, the coefficient of x^2 is not 1. Thus, the relationship between the x term coefficient and the numerical coefficient (being the square of half the x-term coefficient) cannot be applied yet. We must first divide both sides of the equation by 2, so that the leading (highest-degree) term has a coefficient of 1.

1. Divide both sides of the equation by the leading term coefficient, 2. $\dfrac{2x^2}{2} - \dfrac{6x}{2} + \dfrac{5}{2} = \dfrac{0}{2}$

$x^2 - 3x + \dfrac{5}{2} = 0$

2. Add $-\dfrac{5}{2}$ to both sides of equation. $x^2 - 3x = -\dfrac{5}{2}$

3. Take half the middle term coefficient, square it, and add the result to both sides of the equation.
$\left[\dfrac{1}{2}(-3)\right]^2 = \left(\dfrac{-3}{2}\right)^2 = \dfrac{9}{4}$ $x^2 - 3x \boxed{+ \dfrac{9}{4}} = -\dfrac{5}{2} \boxed{+ \dfrac{9}{4}}$

4. Factor the left-hand side as a perfect square and simplify the right-hand side. $\left(x - \dfrac{3}{2}\right)^2 = -\dfrac{10}{4} + \dfrac{9}{4} = -\dfrac{1}{4}$

5. Take square roots. $x - \dfrac{3}{2} = \pm\sqrt{-\dfrac{1}{4}}$

This equation has no real solutions, but if we use complex numbers,

6. Isolate x by adding $\frac{3}{2}$ to both sides.

$$x = \frac{3}{2} \pm \sqrt{-\frac{1}{4}}$$

7. Simplify.

$$\boxed{x = \frac{3}{2} \pm \frac{i}{2}} \quad \blacksquare$$

In general, we solve quadratic equations by completing the square as indicated in the following example:

Solve:　$3x^2 - 5 = -12x$

1. Simplify to standard quadratic form.

$$3x^2 + 12x - 5 = 0$$

2. Divide both sides of the equation by the leading coefficient (if not 1).

$$\frac{3x^2}{3} + \frac{12x}{3} - \frac{5}{3} = \frac{0}{3}$$

$$x^2 + 4x - \frac{5}{3} = 0$$

3. Put the numerical term on the other side of the equation by adding its opposite to both sides of the equation.

$$x^2 + 4x = \frac{5}{3}$$

4. Take half the middle term coefficient, square it, and add this result to both sides of the equation.

$$\left[\frac{1}{2}(4)\right]^2 = 2^2 = 4$$

$$x^2 + 4x + 4 = \frac{5}{3} + 4$$

Add 4 to both sides of the equation.

5. Factor the left-hand side as a perfect square and simplify the right-hand side.

$$(x + 2)^2 = \frac{5}{3} + \frac{12}{3} = \frac{17}{3}$$

6. Take square roots.

$$x + 2 = \pm\sqrt{\frac{17}{3}}$$

7. Isolate x.

$$x = -2 \pm \sqrt{\frac{17}{3}}$$

8. Express your answer in simplest radical form.

$$x = -2 \pm \frac{\sqrt{17}}{\sqrt{3}} \qquad \textit{Rationalize the denominator.}$$

$$= -2 \pm \frac{\sqrt{17}\sqrt{3}}{\sqrt{3}\sqrt{3}} = \boxed{-2 \pm \frac{\sqrt{51}}{3}}$$

$$\text{or} \quad = \frac{-6}{3} \pm \frac{\sqrt{51}}{3} = \boxed{\frac{-6 \pm \sqrt{51}}{3}}$$

Exercises 7.2

Solve the following by completing the square only.

1. $x^2 - 6x - 1 = 0$

2. $s^2 - 2s - 15 = 0$

3. $0 = c^2 - 2c - 5$

4. $0 = a^2 - 2a - 4$

5. $y^2 + 5y - 2 = 0$

6. $x^2 + 3x - 2 = 0$

7. $2x^2 + 3x - 1 = x^2 - 2$

8. $y^2 - 3y - 2 = 2y - 3$

9. $(a - 2)(a + 1) = 2$

10. $(2b + 3)(b - 4) = 3b$

11. $10 = 5a^2 + 10a + 20$

12. $y + 1 = (y + 2)(y + 3)$

13. $2x^2 - 3x = x^2 - 5x + 2$

14. $3x^2 - 7x - 4 = 2x^2 - 3x + 1$

15. $(a - 2)(a + 1) = 6$

16. $(c - 3)(2c + 1) = c(c - 1)$

17. $2x - 7 = x^2 - 3x + 4$

18. $2t - 4 = t^2 - 3$

19. $5x^2 + 10x - 14 = 20$

20. $2a^2 - 3a - 5 = 0$

21. $2t^2 + 3t - 4 = 2t - 1$

22. $3s^2 + 12s + 4 = 0$

23. $(2x + 5)(x - 3) = (x + 4)(x - 1)$

24. $(3x - 2)(x - 1) = (2x - 1)(x - 4)$

25. $\dfrac{2x}{2x - 3} = \dfrac{3x - 1}{x + 1}$

26. $\dfrac{x + 1}{2x + 3} = \dfrac{3x + 2}{x + 2}$

27. $a^2 - a + 1 = 0$

28. $a^2 + a + 1 = 0$

29. $0 = 2y^2 + 2y + 5$

30. $0 = 2y^2 + 2y - 5$

31. $5n^2 - 3n = 2n^2 - 6$

32. $(3n - 1)(n - 2) = n + 8$

33. $(3t + 5)(t + 1) = (t + 4)(t + 2)$

34. $(2z - 1)(z + 3) = (3z - 2)(2z + 1)$

35. $\dfrac{3}{x + 2} - \dfrac{2}{x - 1} = 5$

36. $\dfrac{2y + 1}{y + 1} - \dfrac{2}{y + 4} = 7$

? QUESTIONS FOR THOUGHT

37. Verbally state the relationship between the middle term coefficient and the numerical term of a perfect square.

38. Solve the equation $3x^2 + 11x = 4$ by completing the square. What does the fact that the answers are rational tell you? Solve the same equation by factoring. Which method was easier?

39. Solve for x in the equation $x^2 + rx + q = 0$ by completing the square; that is, solve for x in terms of r and q.

40. Solve for x in the equation $Ax^2 + Bx + C = 0$ (where $A > 0$) by completing the square.

 MINI-REVIEW

41. *Evaluate:* $3\{6 - [(-4 - 5)^2 + 5] - 9\}$

Perform the following operations and express your answer in simplest form:

42. $\dfrac{2x + 1}{x^2 - 9} + \dfrac{3x}{3 - x}$

43. $\dfrac{2x + 1}{x^2 - 9} \div \dfrac{3x}{3 - x}$

44. Solve the following explicitly for q:

$$3q = \frac{2q - 1}{a + 4}$$

45. *Simplify:*

(a) $\dfrac{8 + \sqrt{28}}{10}$ (b) $\dfrac{2\sqrt{5}}{\sqrt{6} - \sqrt{2}}$

46. The width of a rectangle is three-fourths its length. If the perimeter is 100 feet, find its dimensions.

7.3

The Quadratic Formula

Completing the square is a useful algebraic technique which will be needed elsewhere in intermediate algebra, as well as in precalculus and calculus. It is the most powerful of the methods for solving quadratic equations covered so far because, unlike the previous methods, completing the square works for *all* quadratic equations. It is, however, usually the most tedious method for solving quadratic equations.

What we would like is a method which works for all quadratic equations without the effort required in completing the square. Algebraically, we can derive a formula which will allow us to produce the solutions to the "general" quadratic equation.

To derive the formula, we start with the general equation $Ax^2 + Bx + C = 0$ and solve it for x by the method of completing the square. (We assume $A > 0$.)

Start with the equation $Ax^2 + Bx + C = 0$

1. Divide both sides of the equation by A.

$$\frac{Ax^2}{A} + \frac{Bx}{A} + \frac{C}{A} = \frac{0}{A}$$

$$x^2 + \frac{Bx}{A} + \frac{C}{A} = 0$$

2. Subtract $\dfrac{C}{A}$ from both sides of the equation.

$$x^2 + \frac{B}{A}x = -\frac{C}{A}$$

3. Take half the middle term coefficient, square it,

$$\left[\frac{1}{2} \cdot \frac{B}{A}\right]^2 = \left(\frac{B}{2A}\right)^2 = \frac{B^2}{4A^2}$$

and add the result to both sides of the equation.

$$x^2 + \frac{B}{A}x + \frac{B^2}{4A^2} = \frac{B^2}{4A^2} - \frac{C}{A}$$

4. Factor the left-hand side and simplify the right-hand side.

$$\left(x + \frac{B}{2A}\right)^2 = \frac{B^2}{4A^2} - \frac{4AC}{4A^2}$$

$$\left(x + \frac{B}{2A}\right)^2 = \frac{B^2 - 4AC}{4A^2}$$

5. Take square roots.

$$x + \frac{B}{2A} = \pm \sqrt{\frac{B^2 - 4AC}{4A^2}}$$

6. Isolate x.

$$x = -\frac{B}{2A} \pm \sqrt{\frac{B^2 - 4AC}{4A^2}}$$

7. Simplify the solution.

$$x = \frac{-B}{2A} \pm \frac{\sqrt{B^2 - 4AC}}{\sqrt{4A^2}}$$

Note: Since $A > 0$, $\sqrt{4A^2} = 2A$.

$$= \frac{-B}{2A} \pm \frac{\sqrt{B^2 - 4AC}}{2A}$$

$$x = \frac{-B \pm \sqrt{B^2 - 4AC}}{2A}$$

This solution is known as the ***quadratic formula.***

The Quadratic Formula

The solutions to the equation $Ax^2 + Bx + C = 0$, $A > 0$, are given by

$$x = \frac{-B \pm \sqrt{B^2 - 4AC}}{2A}$$

As long as we can put an expression in standard quadratic form, we can identify A (the coefficient of x^2), B (the coefficient of x), and C (the constant). Once we have identified A, B, and C, we substitute the numbers into the quadratic formula and find the solutions to the equation.

EXAMPLE 1   Solve each of the following equations by using the quadratic formula.

(a) $y^2 - 4y - 3 = 0$ **(b)** $(x - 3)(x - 4) = 8$

(c) $\dfrac{2}{z + 2} + \dfrac{1}{z} = 1$ **(d)** $x^2 + 3x + 4 = 0$

Solution   **(a)**   Since the quadratic equation is already in standard form, we can identify A, B, and C:

$A = 1$ (the coefficient of y^2)

$B = -4$ (the coefficient of y)

$C = -3$ (the constant)

We now find y using the quadratic formula:

$$y = \frac{-B \pm \sqrt{B^2 - 4AC}}{2A} = \frac{-(-4) \pm \sqrt{(-4)^2 - 4(1)(-3)}}{2(1)}$$

$$= \frac{4 \pm \sqrt{16 + 12}}{2}$$

$$= \frac{4 \pm \sqrt{28}}{2} = \frac{4 \pm 2\sqrt{7}}{2} = \frac{2(2 \pm \sqrt{7})}{2}$$

$$= \frac{\cancel{2}(2 \pm \sqrt{7})}{\cancel{2}} = \boxed{2 \pm \sqrt{7}}$$

Notice that we must factor the numerator first before reducing.

We will check these solutions:

CHECK $y = 2 + \sqrt{7}$:

$$y^2 - 4y - 3 = 0$$
$$(2 + \sqrt{7})^2 - 4(2 + \sqrt{7}) - 3 \overset{?}{=} 0$$
$$4 + 4\sqrt{7} + 7 - 4(2 + \sqrt{7}) - 3 \overset{?}{=} 0$$
$$4 + 4\sqrt{7} + 7 - 8 - 4\sqrt{7} - 3 \overset{\checkmark}{=} 0$$

CHECK $y = 2 - \sqrt{7}$:

$$y^2 - 4y - 3 = 0$$
$$(2 - \sqrt{7})^2 - 4(2 - \sqrt{7}) - 3 \overset{?}{=} 0$$
$$4 - 4\sqrt{7} + 7 - 4(2 - \sqrt{7}) - 3 \overset{?}{=} 0$$
$$4 - 4\sqrt{7} + 7 - 8 + 4\sqrt{7} - 3 \overset{\checkmark}{=} 0$$

(b) $(x - 3)(x - 4) = 8$ *We put the equation in standard form.*

$$x^2 - 7x + 12 = 8$$
$$x^2 - 7x + 4 = 0$$

Therefore,

$$A = 1 \quad \text{(coefficient of } x^2)$$
$$B = -7 \quad \text{(coefficient of } x)$$
$$C = +4 \quad \text{(numerical term)}$$

Thus,

$$x = \frac{-B \pm \sqrt{B^2 - 4AC}}{2A} = \frac{-(-7) \pm \sqrt{(-7)^2 - 4(1)(4)}}{2(1)}$$
$$x = \frac{7 \pm \sqrt{49 - 16}}{2}$$
$$x = \boxed{\frac{7 \pm \sqrt{33}}{2}}$$

(c) $$\frac{2}{z + 2} + \frac{1}{z} = 1$$ *First put the equation in standard form. Multiply both sides by $z(z + 2)$.*

$$z(z + 2)\left(\frac{2}{z + 2} + \frac{1}{z}\right) = 1 \cdot z(z + 2)$$
$$2z + (z + 2) = z(z + 2)$$
$$3z + 2 = z^2 + 2z$$
$$0 = z^2 - z - 2$$

Now identify A, B, and C:

$$A = 1, \quad B = -1, \quad C = -2$$

Solve for z using the quadratic formula:

$$z = \frac{-B \pm \sqrt{B^2 - 4AC}}{2A} = \frac{-(-1) \pm \sqrt{(-1)^2 - 4(1)(-2)}}{2(1)}$$

$$= \frac{1 \pm \sqrt{1 + 8}}{2}$$

$$= \frac{1 \pm \sqrt{9}}{2} = \frac{1 \pm 3}{2}$$

Thus,

$$z = \frac{1 + 3}{2} = \frac{4}{2} = 2 \quad \text{or} \quad z = \frac{1 - 3}{2} = \frac{-2}{2} = -1$$

The solutions are $\boxed{2 \quad \text{and} \quad -1}$
Since the solutions are rational, we could have solved the equation by the factoring method. Try it for yourself.

You should check these solutions to see that both are valid.

(d) Identify A, B, and C:

$$A = 1 \qquad B = 3 \qquad C = 4$$

Solve for x by the quadratic formula:

$$x = \frac{-B \pm \sqrt{B^2 - 4AC}}{2A} = \frac{-3 \pm \sqrt{(3)^2 - 4(1)(4)}}{2(1)}$$

$$= \frac{-3 \pm \sqrt{9 - 16}}{2}$$

$$= \frac{-3 \pm \sqrt{-7}}{2}$$

Hence, there are no real solutions. The complex number solutions are

$$\boxed{\frac{-3 + i\sqrt{7}}{2} \quad \text{and} \quad \frac{-3 - i\sqrt{7}}{2}}$$ ∎

Here are a few things to be careful about when you use the quadratic formula:

1. If B is a negative number, remember that $-B$ will be positive (see parts **(a)**–**(c)** of Example 1).

2. If C is negative (and A is positive), then you will end up *adding* the expressions under the radical [see parts **(a)** and **(c)** of Example 1].

3. Do not forget that $2A$ is in the denominator of the *entire expression*:

$$\frac{-B \pm \sqrt{B^2 - 4AC}}{2A}$$

The Discriminant

Let us further examine the quadratic formula; in particular, let's look at the radical portion of the quadratic formula: $\sqrt{B^2 - 4AC}$. Note that if we were solving a quadratic equation and the quantity $B^2 - 4AC$ were negative [that is, if $B^2 - 4AC < 0$, as in part **(d)** of Example 1], then we would have the square root of a negative number. As we discussed in Chapter 6, the square root of a negative number is not a real number. Thus, if $B^2 - 4AC < 0$, then the solutions or roots

$$x = \frac{-B + \sqrt{B^2 - 4AC}}{2A} \quad \text{and} \quad x = \frac{-B - \sqrt{B^2 - 4AC}}{2A} \quad \text{are not real.}$$

On the other hand, if $B^2 - 4AC = 0$, then $\sqrt{B^2 - 4AC} = \sqrt{0} = 0$. Thus, the two roots

$$x = \frac{-B + \sqrt{0}}{2A} = \frac{-B}{2A} \quad \text{and} \quad x = \frac{-B - \sqrt{0}}{2A} = \frac{-B}{2A}$$

are equal. They are also real, since the quotient of two real numbers yields a real number.

Finally, if $B^2 - 4AC$ is positive (that is, $B^2 - 4AC > 0$), then the roots are real and distinct (unequal):

$$x = \frac{-B + \sqrt{B^2 - 4AC}}{2A} \quad \text{and} \quad x = \frac{-B - \sqrt{B^2 - 4AC}}{2A}$$

We call $B^2 - 4AC$ the ***discriminant*** of the equation $Ax^2 + Bx + C = 0$. The above discussion is summarized in the box.

For the equation $Ax^2 + Bx + C = 0$ $(A \neq 0)$:

If $B^2 - 4AC < 0$, the roots are not real.

If $B^2 - 4AC = 0$, the roots are real and equal.

If $B^2 - 4AC > 0$, the roots are real and distinct.

EXAMPLE 2

Without solving the given equation, determine the nature of the roots of

$$3x^2 - 2x + 5 = 0$$

Solution

Identify A, B, and C:

$$A = 3 \quad B = -2 \quad C = 5$$

The discriminant is $B^2 - 4AC = (-2)^2 - 4(3)(5) = 4 - 60 = -56 < 0$. Thus, the roots are not real

Which Method To Use in Solving Quadratic Equations

The method of completing the square and the quadratic formula will work for all quadratic equations, but completing the square can be cumbersome, and there are quite a few places to make errors using the quadratic formula. The factoring and square root methods are easiest, but do not always work. In general, we recommend the following:

Unless the equation is in the form $(x + a)^2 = b$ or $(x + a)(x + b) = 0$, get the equation in the standard form $Ax^2 + Bx + C = 0$.

If there is no x term (that is, if $B = 0$), then use the square root method.

If there is an x term, try the factoring method.

If the quadratic expression does not factor easily, then use either the quadratic formula or completing the square.

Exercises 7.3

In Exercises 1–14, solve using the quadratic formula only.

1. $x^2 - 4x - 5 = 0$
2. $a^2 + 4a - 5 = 0$
3. $2a^2 - 3a - 4 = 0$
4. $5x^2 + x - 2 = 0$
5. $(3y - 1)(2y - 3) = y$
6. $(5s + 2)(s - 1) = 2s + 1$
7. $y^2 - 3y + 4 = 2y^2 + 4y - 3$
8. $3x^2 + 2 = 5x^2 + 1$
9. $2x^2 - 3x - 7 = 0$
10. $2a^2 - 4a - 9 = 0$
11. $3a^2 + a + 2 = 0$
12. $2y^2 - 3y = -4$
13. $(s - 3)(s + 4) = (2s - 1)(s + 2)$
14. $\dfrac{2x}{x + 1} + \dfrac{1}{x} = 1$

In Exercises 15–26, use the discriminant to determine the nature of the roots.

15. $3a^2 - 2a + 5 = 0$
16. $(2z - 3)(z + 4) = z - 2$
17. $x^2 - 10x = -25$
18. $x^2 - 3x + 1 = 0$
19. $(3y + 5)(2y - 8) = (y - 4)(y + 1)$
20. $\dfrac{3a - 2}{2a + 1} = \dfrac{1}{a - 2}$
21. $2a^2 + 4a = 0$
22. $(5s + 3)(2s + 1) = s^2 - s - 1$
23. $(x - 1)^2 = 0$
24. $(x - 1)^2 = -2$
25. $(2y + 3)(y - 1) = y + 5$
26. $\dfrac{2}{2x + 1} + \dfrac{3x}{x - 2} = 4$

In Exercises 27–68, solve by any method.

27. $a^2 - 3a - 4 = 0$
28. $x^2 - 5x + 4 = 0$
29. $8y^2 = 3$
30. $12y^2 = 5$
31. $(5x - 4)(2x - 3) = 0$ factor
32. $(2z - 3)(3z + 1) = 0$
33. $(5x - 4)(2x - 3) = 17$
34. $(2z - 3)(3z + 1) = 17$
35. $(a - 1)(a + 2) = -2$
36. $(5x - 1)(x + 2) = 2x^2 - 3x + 10$
37. $(y - 3)(y + 3) = 12$
38. $(a - 2)(a + 1) = 2a^2 + 3a$
39. $(5x - 1)(2x + 3) = 3x - 3$
40. $(y - 1)(y + 1) = (2y - 1)(y - 1)$

41. $x^2 - 3x + 5 = 0$

42. $3z^2 - 5z - 4 = -4z - 3$

43. $2s^2 - 5s - 12 = -5s$

44. $5a^2 - 2a + 4 = 8 - 2a$

45. $3x^2 - 2x + 9 = 2x^2 - 3x - 1$

46. $(3z - 2)(2z + 1) = 2z - 3$

47. $x^2 + 3x - 8 = x^2 - x + 11$

48. $t^2 - 6t + 1 = t^2 - 5t + 3$

49. $(y - 4)(y + 4) = (y - 4)^2$

50. $(x - 1)^2 = (x - 1)(x + 1)$

51. $3a^2 - 4a + 2 = 0$

52. $x^2 - 2x + 4 = 0$

53. $(x - 4)(2x + 3) = x^2 - 4$

54. $(a - 5)(a + 2) = (2a - 3)(a - 1)$

55. $(a + 1)(a - 3) = (3a + 1)(a - 2)$

56. $(2y + 3)(y + 5) = (y - 1)(y + 4)$

57. $(z + 3)(z - 1) = (z - 2)^2$

58. $(w - 3)^2 = (w + 4)(w - 2)$

59. $(x - 3)(x + 4) + 2 = x - 3$

60. $(x - 2)(x - 5) = (2x - 7)x$

61. $\dfrac{y}{y - 2} = \dfrac{y - 3}{y}$

62. $\dfrac{x - 4}{3} = \dfrac{2}{x + 5}$

63. $\dfrac{5}{y - 2} + \dfrac{3}{y + 2} = \dfrac{y}{y^2 - 4}$

64. $\dfrac{2y}{y + 2} = \dfrac{y - 2}{y}$

65. $\dfrac{3a}{a + 1} + \dfrac{2}{a - 2} = 5$

66. $\dfrac{3x}{2} + \dfrac{2}{x + 1} = 4$

67. $\dfrac{3}{a + 2} - \dfrac{5}{a - 2} = 2$

68. $\dfrac{2}{a + 4} - \dfrac{3}{a + 1} = 4$

🖩 CALCULATOR EXERCISES

69. $2.4x^2 - 12.72x + 3.6 = 0$

70. $2.3x^2 - 13.11x + 8.05 = 0$

❓ QUESTIONS FOR THOUGHT

71. Compare and contrast the various methods for solving quadratic equations.

72. Discuss what is *wrong* (if anything) with the solutions to the following problems:

(a) Solve $x^2 - 3x + 1 = 0$.

Solution: $x \overset{?}{=} \dfrac{-3 \pm \sqrt{9 - 4(1)}}{2} \overset{?}{=} \dfrac{-3 \pm \sqrt{5}}{2}$

(b) Solve $x^2 - 5x - 3 = 0$.

Solution: $x \overset{?}{=} \dfrac{5 \pm \sqrt{25 - 12}}{2} \overset{?}{=} \dfrac{5 \pm \sqrt{13}}{2}$

(c) Solve $x^2 - 3x + 2 = 0$.

Solution: $x \overset{?}{=} -3 \pm \dfrac{\sqrt{9 - 8}}{2} \overset{?}{=} -3 \pm \dfrac{\sqrt{1}}{2} \overset{?}{=} -3 \pm \dfrac{1}{2}$

(d) Solve $x^2 - 6x - 3 = 0$.   (*Discuss what is **wrong** with this solution.*)

Solution: $x = \dfrac{6 \pm \sqrt{36 + 12}}{2} = \dfrac{6 \pm \sqrt{48}}{2} = \dfrac{6 \pm 4\sqrt{3}}{2}$

$$= \dfrac{6 \pm \overset{2}{\cancel{4}}\sqrt{3}}{\cancel{2}} = 6 \pm 2\sqrt{3}$$

73. Write an equation whose roots are:

(a) 3 and 4

(b) −4 and +5

(c) $\frac{3}{5}$ and −2

(d) $3i$ and $-3i$

7.4

Applications

In this section we will examine applications that require solving quadratic equations. We first discuss the straightforward problems which translate directly into algebraic sentences.

EXAMPLE 1 | The sum of a number and its reciprocal is $\frac{29}{10}$. Find the numbers.

Solution | Let $x =$ the number.

$$\underbrace{\text{The sum of}}_{} \underbrace{\text{a number}}_{x} \underbrace{\text{and}}_{+} \underbrace{\text{its reciprocal}}_{\frac{1}{x}} \underbrace{\text{is } \frac{29}{10}}_{= \frac{29}{10}}$$

The equation is

$$x + \frac{1}{x} = \frac{29}{10}$$ *Multiply both sides of the equation by the LCD, 10x.*

$$10x\left(x + \frac{1}{x}\right) = \left(\frac{29}{10}\right)10x$$ *Distribute 10x and reduce.*

$$10x(x) + 10x\left(\frac{1}{x}\right) = \frac{29}{10}10x$$

$$10x^2 + 10 = 29x$$ *Now we have a quadratic equation and we solve for x.*

$$10x^2 - 29x + 10 = 0$$

$$(5x - 2)(2x - 5) = 0$$

$$5x - 2 = 0 \quad \text{or} \quad 2x - 5 = 0$$

$$5x = 2 \qquad\qquad 2x = 5$$

$$x = \frac{2}{5} \quad \text{or} \quad x = \frac{5}{2}$$

Thus, the answers are $\boxed{\frac{2}{5} \quad \text{and} \quad \frac{5}{2}}$. Notice that the two answers are reciprocals of each other.   ∎

Often we find that the relationship between two variables can be expressed as a second-degree equation in two unknowns.

EXAMPLE 2

The profit in dollars, P, on each television set made daily by AAA Television Company is related to the number of television sets, x, produced at the AAA factory as follows:

$$P = -\frac{x^2}{4} + 45x - 1,625$$

(a) What is the company's profit for each TV set if it produces 90 TV sets?

(b) How many sets must be produced for the company to make a profit of $175?

Solution

(a) Since profit per set is

$$P = -\frac{x^2}{4} + 45x - 1,625$$

where x is the number of sets produced daily, we simply substitute the daily number of sets produced for x and find P. For $x = 90$:

$$P = -\frac{(90)^2}{4} + 45(90) - 1,625$$

$$= -\frac{8,100}{4} + 4,050 - 1,625$$

$$= -2,025 + 4,050 - 1,625$$

$$= \boxed{\$400 \text{ profit per TV set}}$$

(b) To find the number of sets to be produced in order to make a profit of $175, we set $P = 175$ and find x by solving the quadratic equation:

$175 = -\dfrac{x^2}{4} + 45x - 1,625$ *Multiply both sides of the equation by 4.*

$700 = -x^2 + 180x - 6,500$ *Put in standard form.*

$0 = x^2 - 180x + 7,200$ *Solve by factoring.*

$0 = (x - 60)(x - 120)$

$$x = 60 \quad \text{or} \quad x = 120$$

Thus, in order for the company to make a profit of $175 per set, it must produce either 60 TV sets or 120 TV sets daily ∎

EXAMPLE 3

A rectangle has length 4 inches greater than its width. If its area is 77 square inches, what are the dimensions of the rectangle?

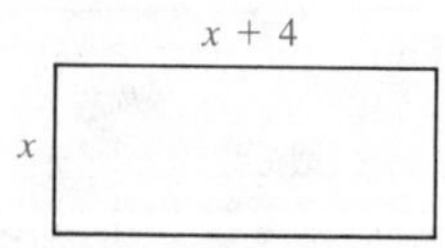

Figure 7.1
Rectangle for Example 3

Solution | Draw a diagram and label the sides, as indicated in Figure 7.1.

Let x = width.

Then $x + 4$ = length   (since the length is 4 more than the width).

The formula for the area of a rectangle is

$$A = (\text{Length})(\text{Width})$$

$$77 = (x + 4)x$$

Solve:

$$77 = x^2 + 4x$$
$$0 = x^2 + 4x - 77$$
$$(x - 7)(x + 11) = 0$$

$x = 7$ or $x = -11$ *We eliminate the negative answer since length is always positive.*

Thus, $x = 7$ inches (the width) and $x + 4 = 7 + 4 = 11$ inches (the length).

The dimensions are $7''$ by $11''$ ■

EXAMPLE 4 | A 20′ by 60′ rectangular pool is surrounded by a concrete walkway of uniform width. If the total area of the walkway is 516 square feet, how wide is the walkway?

Solution | Since the walkway is the same width at all points around the pool, we label this width x and our picture is as shown in Figure 7.2.

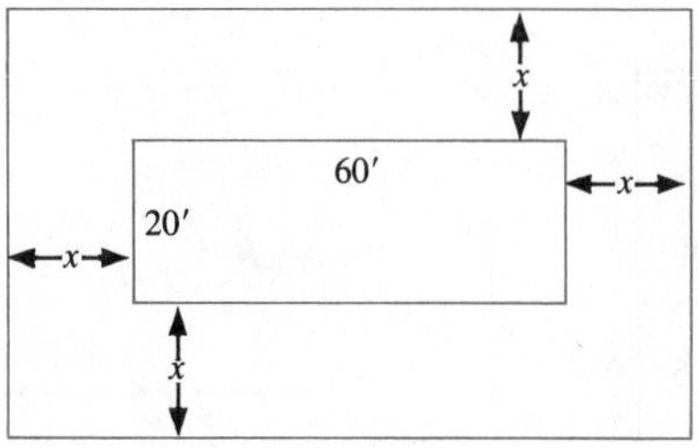

Figure 7.2
Diagram for Example 4

Notice that the length of the outer rectangle is $60 + x + x = 60 + 2x$, and the width of the outer rectangle is $20 + x + x = 20 + 2x$. In terms of x, the area of the outer rectangle (the pool *and* walkway) is

$$(20 + 2x)(60 + 2x)$$

The area of the inner rectangle (the pool) is $(60)(20) = 1,200$ square feet, and we are given that the area of the walkway itself is 516 square feet.

The diagram gives us the following relationship:

Area of outer rectangle = Area of inner rectangle + Area of walkway

$$(20 + 2x)(60 + 2x) = \qquad 1{,}200 \qquad + \qquad 516$$

We solve the quadratic equation:

$$(20 + 2x)(60 + 2x) = 1{,}200 + 516$$

$$1{,}200 + 160x + 4x^2 = 1{,}716 \qquad \textit{Put in standard form.}$$

$$4x^2 + 160x - 516 = 0 \qquad \textit{Divide both sides of the equation by 4.}$$

$$\frac{4(x^2 + 40x - 129)}{4} = \frac{0}{4}$$

$$x^2 + 40x - 129 = 0 \qquad \textit{Factor.}$$

$$(x - 3)(x + 43) = 0$$

$$x = 3 \quad \text{or} \quad x = -43 \qquad \textit{Eliminate the negative answer. (Why?)}$$

Hence, $x = 3$.

The width of the walkway is $\boxed{3 \text{ feet}}$ ∎

EXAMPLE 5

Samantha throws a ball up into the air. The equation

$$s = -16t^2 + 30t + 4$$

gives the distance, s (in feet), above the ground that the ball reaches t seconds after she throws it (see Figure 7.3).

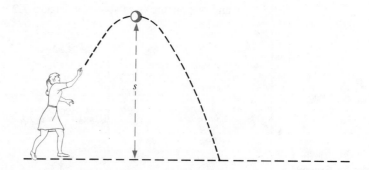

Figure 7.3
Diagram for Example 5

(a) How far above the ground is the ball exactly 1 second after she throws it? After $\frac{3}{4}$ of a second?

(b) How many seconds will it take for the ball to hit the ground from the time she throws it?

Solution

(a) Since the ball is s feet above the ground in t seconds, to find the height of the ball at 1 second, we substitute 1 for t in the given equation:

$$s = -16t^2 + 30t + 4$$
$$= -16(1)^2 + 30(1) + 4$$
$$= -16 + 30 + 4 = 18$$

The ball is $\boxed{\text{18 feet above the ground}}$ at 1 second.

At $\frac{3}{4}$ of a second:

$$s = -16t^2 + 30t + 4$$
$$= -16\left(\frac{3}{4}\right)^2 + 30\left(\frac{3}{4}\right) + 4$$
$$= -16\left(\frac{9}{16}\right) + 30\left(\frac{3}{4}\right) + 4$$
$$= -9 + \frac{90}{4} + 4$$
$$= -9 + \frac{45}{2} + 4$$
$$= -5 + \frac{45}{2} = \frac{-10}{2} + \frac{45}{2} = \frac{35}{2} = 17\frac{1}{2}$$

Hence, the ball is $\boxed{17\frac{1}{2} \text{ feet above the ground}}$ at $\frac{3}{4}$ of a second.

(b)   In this part of the problem, we need to find t, but in order to find t, we need to know s. Since s is the distance the ball is from the ground, when the ball hits the ground, s must be 0. Hence, we are being asked to find t when $s = 0$.

$$s = -16t^2 + 30t + 4 \qquad \textit{Let } s = 0.$$
$$0 = -16t^2 + 30t + 4 \qquad \textit{Find } t.$$
$$0 = -2(8t^2 - 15t - 2) \qquad \textit{Divide both sides by } -2.$$
$$\frac{0}{-2} = \frac{-2(8t^2 - 15t - 2)}{-2}$$
$$0 = 8t^2 - 15t - 2$$
$$0 = (8t + 1)(t - 2)$$

$$8t + 1 = 0 \quad \text{or} \quad t - 2 = 0$$
$$8t = -1 \qquad\qquad t = 2$$
$$t = -\frac{1}{8}$$

So $s = 0$ at $t = -\frac{1}{8}$ and at $t = 2$. But $-\frac{1}{8}$ second does not make any sense and we therefore eliminate this answer.

Thus, the ball reaches the ground in $\boxed{\text{2 seconds}}$

Note also that at $t = 0$ seconds, the distance, s, is 4 feet, indicating that the ball actually starts 4 feet above the ground. ∎

EXAMPLE 6

The rate of a stream is 4 mph. Meri rows her boat downstream a distance of 9 miles and then back upstream to her starting point. If the total trip took 10 hours, what was her rate in still water?

Solution

First we must realize that when Meri is traveling downstream, the net rate (the rate her boat is traveling relative to the land) is equal to the rate she can row in still water plus the rate of the stream. That is,

$$\text{Net rate downstream} = r_{\text{rowing}} + r_{\text{stream}}$$

On the other hand, the net rate upstream (the rate the boat travels upstream relative to the land) is the difference of the rate rowing in still water and the rate of the stream:

$$\text{Net rate upstream} = r_{\text{rowing}} - r_{\text{stream}}$$

Let's call Meri's rate rowing r. Then

Net rate downstream $= r + 4$

Net rate upstream $= r - 4$

Since distance $=$ (rate)(time), the *time* it takes for her to go downstream is

$$\text{Time downstream} = \frac{\text{Distance}}{\text{Net rate downstream}} = \frac{9}{r + 4}$$

The *time* it takes for her to go upstream is

$$\text{Time upstream} = \frac{\text{Distance}}{\text{Net rate upstream}} = \frac{9}{r - 4}$$

Since the total time for the round trip is 10 hours, we have

$$10 = \frac{9}{r + 4} + \frac{9}{r - 4}$$

We solve for r. First multiply both sides of the equation by $(r + 4)(r - 4)$:

$$(r + 4)(r - 4) \cdot 10 = \left(\frac{9}{r + 4} + \frac{9}{r - 4} \right)(r + 4)(r - 4)$$

$$10(r + 4)(r - 4) = 9(r - 4) + 9(r + 4)$$

$$10(r^2 - 16) = 9r - 36 + 9r + 36$$

$$10r^2 - 160 = 18r$$

$$10r^2 - 18r - 160 = 0$$

$$2(5r^2 - 9r - 80) = 0 \qquad \textit{Divide both sides by 2.}$$

$$5r^2 - 9r - 80 = 0$$

$$(5r + 16)(r - 5) = 0$$

$$r = -\frac{16}{5} \quad \text{or} \quad r = 5 \qquad \textit{We eliminate the negative value.}$$

Thus, her rate in still water is $\boxed{5 \text{ mph}}$

CHECK:

Going downstream takes her $\dfrac{9}{r+4} = \dfrac{9}{5+4} = \dfrac{9}{9} = 1$ hour.

Going upstream takes her $\dfrac{9}{r-4} = \dfrac{9}{5-4} = \dfrac{9}{1} = 9$ hours.

Total time $= 1 + 9 \overset{\checkmark}{=} 10$ hours. ∎

The Pythagorean Theorem

A **right triangle** is a triangle with a right (90°) angle. The sides forming the right angle in a right triangle are called the **legs** of the right triangle. The side opposite the right angle is called the **hypotenuse**. (See Figure 7.4.)

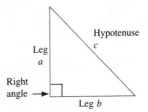

Figure 7.4
Right triangle

The Pythagorean Theorem	Given a right triangle with legs of length a and b and hypotenuse c, we have the following relationship: $$a^2 + b^2 = c^2$$

In other words, the sum of the squares of the legs of a right triangle is equal to the square of the hypotenuse. This is called the ***Pythagorean Theorem***. The converse is also true; that is, if the sum of the squares of two sides of a triangle is equal to the square of the third side, then the triangle is a right triangle.

EXAMPLE 7

Given a right triangle with one leg equal to 9″ and the hypotenuse equal to 15″, find the other leg.

Solution

Let $a = 9″$ and $c = 15″$ (the hypotenuse) and use the Pythagorean Theorem to find the other leg (see Figure 7.5):

$$a^2 + b^2 = c^2$$
$$9^2 + b^2 = 15^2$$
$$81 + b^2 = 225$$
$$b^2 = 144$$
$$b = \pm\sqrt{144} = \pm 12 \qquad \textit{Eliminate the negative answer.}$$

Figure 7.5
Diagram for Example 7

Thus, the other leg $(b) = \boxed{12″}$ ∎

EXAMPLE 8

Solution

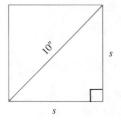

Figure 7.6
Diagram for Example 8

Find the side of a square with diagonal 10″.

We draw a diagram and label the diagonal as shown in Figure 7.6.
    A square has all right angles and so its diagonal will cut the square into two right triangles. Since a square has equal sides, the legs of the right triangles will be equal. The hypotenuse is 10″ since the hypotenuse of the right triangle is the diagonal of the square.
    Using the Pythagorean Theorem with each leg equal to s and hypotenuse equal to 10″, we have

$$s^2 + s^2 = 10^2$$
$$2s^2 = 100$$
$$s^2 = \frac{100}{2} = 50$$
$$s = \pm\sqrt{50} = \pm 5\sqrt{2}$$

The length of the sides of the square is $\boxed{5\sqrt{2} \text{ inches} \approx 7.07 \text{ inches}}$ ■

Exercises 7.4

Solve each of the following problems algebraically.

1. Five less than the square of a positive number is 1 more than 5 times the number. Find the number.

2. Seven more than the square of a positive number is 1 less than 6 times the number. Find the number(s).

3. The sum of the squares of two positive numbers is 68. If one of the numbers is 8, find the other number.

4. The sum of the squares of two negative numbers is 30. If one of the numbers is -5, find the other number.

5. Find the numbers such that the square of the sum of the number and 6 is 169.

6. Find the numbers such that the square of the sum of the number and 3 is 100.

7. The sum of a number and its reciprocal is $\frac{13}{6}$. Find the number(s).

8. The sum of a number and its reciprocal is $\frac{25}{12}$. Find the number(s).

9. The sum of a number and twice its reciprocal is 3. Find the number(s).

10. The sum of a number and three times its reciprocal is 4. Find the number(s).

11. The profit in dollars (P) made on a concert is related to the price in dollars (d) of a ticket in the following way:

$$P = 10,000(-d^2 + 12d - 35)$$

  (a) How much profit is made by selling tickets for $5?

  (b) What must the price of a ticket be to make a profit of $10,000?

12. The daily cost in dollars (C) of producing widgets in a factory is related to the number of widgets (x) produced in the following way:

$$C = -\frac{x^2}{10} + 100x - 24,000$$

  (a) How much does it cost to produce 450 widgets?

  (b) How many widgets must be produced so that the daily cost is $1,000?

13. A man jumps off a diving board into a pool. The equation

$$s = -16t^2 + 40$$

gives the distance s (in feet) the man is above the pool t seconds after he jumps.

   (a) How high above the pool is the diver after the first second?

   (b) How long does it take for him to hit the water?

   (c) How high is the diving board? [*Hint:* Let $t = 0$.]

14. Alex throws a ball straight up into the air. The equation

$$s = -16t^2 + 80t + 44$$

gives the distance (s) in feet the ball is above the ground t seconds after he throws it.

   (a) How high is the ball at $t = 2$ seconds?

   (b) How long does it take for the ball to hit the ground?

15. Find the dimensions of a rectangle whose area is 80 square feet and whose length is 2 feet more than its width.

16. Find the dimensions of a rectangle whose area is 108 square feet and whose length is 3 feet more than its width.

17. Find the dimensions of a square with area 60 square inches.

18. Find the dimensions of a square with area 75 square inches.

19. The product of two numbers is 120. What are the numbers if one number is 2 less than the other?

20. The sum of two numbers is 20. Find the numbers if their product is 96.

21. The product of two numbers is 85. What are the numbers if one number is 2 more than 3 times the other?

22. The sum of two numbers is 25. Find the numbers if their product is 144.

23. The product of two numbers is 40. Find the numbers if the difference between the two numbers is 3.

24. The product of two numbers is 72. Find the numbers if the difference between the two numbers is 1.

25. In a right triangle, the legs are 3″ and 8″. Find the hypotenuse.

26. In a right triangle, the legs are 5″ and 7″. Find the hypotenuse.

27. In a right triangle, a leg is 4″ and the hypotenuse is 7″. Find the length of the other leg.

28. A leg of a right triangle is 7″ and the hypotenuse is 15″. Find the length of the other leg.

29. Find the dimensions of the right triangle shown in the accompanying figure.

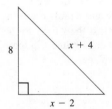

Diagram for Exercise 29

30. Find the dimensions of the right triangle shown in the figure.

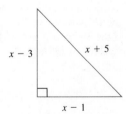

Diagram for Exercise 30

31. Find the length of the diagonals of a 7″ by 12″ rectangle.

32. Find the length of the diagonals of a 4″ by 5″ rectangle.

33. Find the length of the diagonals of a square with side equal to 5″.

34. Find the length of the diagonals of a square with side equal to 12″.

35. Find the length of the side of a square with diagonal equal to 8″.

36. Find the length of a rectangle if the width is 3″ and the diagonal is 18″.

37. Harold leans a 30′ ladder against a building. If the base of the ladder is 8′ from the building, how high up the building does the ladder reach? (See the accompanying figure.)

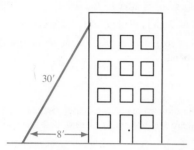

38. Harold leans a 30′ ladder against a building. If the base of the ladder is 4′ from the building, how high up the building does the ladder reach?

$$8^2 + x^2 = 30^2$$
$$64 + x^2 = 900$$
$$x^2 = 836$$
$$x = 2\sqrt{209}$$

39. Find the area of a square if its diagonal is 8″.

40. Find the area of a square if its diagonal is 9″.

41. Find the area of a rectangle with diagonal 12″ and width 5″.

42. Find the area of a rectangle with diagonal 15″ and width 3″.

43. A picture has a 1-inch (uniform width) frame as its border. If the length of the picture is twice its width and the total area of the picture and frame is 60 square inches, find the dimensions of the picture.

44. A 7″ by 10″ rectangular picture has a frame of uniform width. If the area of the frame is 60 square inches, how wide is the frame?

45. A 20′ by 55′ rectangular swimming pool is surrounded by a concrete walkway of uniform width. If the area of the concrete walkway is 400 sq. feet, find the width of the walkway.

46. A 21′ by 21′ square swimming pool is surrounded by a path of uniform width. If the area of the path is 184 sq. feet, find the width of the path.

47. A circular garden is surrounded by a path of uniform width. If the path has area 44π sq. feet and the radius of the garden is 10 feet, find the width of the path. (*Hint:* See Section 3.1, Example 7.)

48. A circular garden is surrounded by a path of uniform width. If the path has area 57π sq. feet and the radius of the garden is 8 feet, find the width of the path.

49. A boat travels upstream for 20 km against a current of 5 kph, then it travels downstream for 10 km with the same current. If the 30-km trip took $1\frac{1}{3}$ hours, how fast was the boat travelling in still water?

50. An airplane heads into the wind to get to its destination 200 miles away and returns travelling with the wind to its starting point. If the wind currents were at a constant 20 mph and the total round trip took $2\frac{1}{4}$ hours, find the speed of the airplane.

 MINI-REVIEW

51. *Factor the following completely.* $125y^6 - 1$

52. Perform the operation and express your answer in the form $a + bi$. $(3 + 5i)(3 - 5i)$

53. Perform the operation and express your answer in the form $a + bi$.

$$\frac{3}{2 - i}$$

54. *Solve for x.* $\sqrt{3x - 6} + 1 = 4$

55. *Solve for x.* $2x^{1/4} - 1 = 9$

7.5

Equations Reducible to Quadratic Form (and More Radical Equations)

Radical Equations

In Chapter 6, we solved radical equations by first isolating the radical and then squaring both sides of the equation to eliminate the radical where possible.

In this chapter we employ the same techniques to solve radical equations, except, unlike the radical equations we covered in Chapter 6, the radical equations we will solve in this section may give rise to quadratic equations.

EXAMPLE 1

Solve for x. $x - \sqrt{2x} = 0$

Solution

We remind you that simply squaring both sides of an equation containing a radical does not necessarily eliminate the radical. We should isolate the radical first.

$$x - \sqrt{2x} = 0 \qquad \text{\textit{Isolate the radical.}}$$
$$x = \sqrt{2x} \qquad \text{\textit{Square both sides of the equation.}}$$
$$(x)^2 = (\sqrt{2x})^2$$
$$x^2 = 2x \qquad \text{\textit{Now we have a quadratic equation which}}$$
$$x^2 - 2x = 0 \qquad \text{\textit{we can solve by factoring.}}$$
$$x(x - 2) = 0$$

$$x = 0 \qquad \text{or} \qquad x - 2 = 0$$
$$x = 2$$

As usual, we check the solutions to all radical equations to ensure that all our solutions are valid.

CHECK $x = 0$: $x - \sqrt{2x} = 0$ CHECK $x = 2$: $x - \sqrt{2x} = 0$
$$0 - \sqrt{2(0)} \overset{\checkmark}{=} 0$$
$$2 - \sqrt{2(2)} \overset{?}{=} 0$$
$$2 - \sqrt{4} \overset{\checkmark}{=} 0$$

Therefore, the solutions are $\boxed{x = 0 \quad \text{and} \quad x = 2}$ ∎

EXAMPLE 2

Solution

Solve for z. $\sqrt{z + 3} = 2z$

$$\sqrt{z + 3} = 2z$$ *First square both sides to eliminate the radical.*

$$(\sqrt{z + 3})^2 = (2z)^2$$

$$z + 3 = 4z^2$$ *Now we have a quadratic equation to solve. Put it in standard form.*

$$0 = 4z^2 - z - 3$$ *Solve for z.*

$$0 = (4z + 3)(z - 1)$$

$$0 = 4z + 3 \quad \text{or} \quad 0 = z - 1$$

$$z = \frac{-3}{4} \qquad\qquad z = 1$$

Now we must check both solutions in the original equation.

CHECK $z = \dfrac{-3}{4}$: $\sqrt{z + 3} = 2z$

$$\sqrt{\frac{-3}{4} + 3} \overset{?}{=} 2\left(\frac{-3}{4}\right)$$

$$\sqrt{\frac{9}{4}} \neq \frac{-3}{2}$$

*This is impossible since the square root must be positive. Remember, $\sqrt{a}$ means the **nonnegative** root of a.*

$z = \dfrac{-3}{4}$ is an extraneous root.

CHECK $z = 1$: $\sqrt{z + 3} = 2z$

$$\sqrt{1 + 3} \overset{?}{=} 2(1)$$

$$\sqrt{4} \overset{\checkmark}{=} 2$$

Hence, the *only* solution is $\boxed{z = 1}$ ∎

EXAMPLE 3

Solution

Solve for x. $x - \sqrt{x - 4} = 4$

$$x - \sqrt{x - 4} = 4$$ *First isolate the radical.*

$$-\sqrt{x - 4} = 4 - x$$ *Then square both sides of the equation.*

$$(-\sqrt{x - 4})^2 = (4 - x)^2$$

$$x - 4 = 16 - 8x + x^2$$

$$0 = x^2 - 9x + 20$$

$$0 = (x - 5)(x - 4)$$

$$\boxed{x = 5 \quad \text{or} \quad x = 4}$$

The student should check to see that both solutions are valid.   ∎

When we are confronted with two radicals in an equation, the algebra is more complicated. For example, the process of squaring $\sqrt{5a} - \sqrt{2a-1}$ can be messy. It is usually less trouble to *first isolate the more complicated radical and then square both sides of the equation* as demonstrated in Example 4.

EXAMPLE 4

Solve for a. $\sqrt{5a} - \sqrt{2a-1} = 2$

Solution

$$\sqrt{5a} - \sqrt{2a-1} = 2 \qquad \text{\textit{Isolate} } \sqrt{2a-1}.$$

$$\sqrt{5a} - 2 = \sqrt{2a-1} \qquad \text{\textit{Square both sides of the equation.}}$$

$$(\sqrt{5a} - 2)^2 = (\sqrt{2a-1})^2$$

$$5a - 4\sqrt{5a} + 4 = 2a - 1 \qquad \text{\textit{Note how each side is squared differently. Isolate} } 4\sqrt{5a}.$$

$$3a + 5 = 4\sqrt{5a} \qquad \text{\textit{Square both sides of the equation again.}}$$

$$(3a + 5)^2 = (4\sqrt{5a})^2$$

$$9a^2 + 30a + 25 = 16(5a) \qquad \text{\textit{Simplify.}}$$

$$9a^2 + 30a + 25 = 80a$$

$$9a^2 - 50a + 25 = 0 \qquad \text{\textit{Solve for a by factoring.}}$$

$$(9a - 5)(a - 5) = 0$$

$$a = \frac{5}{9} \qquad \text{or} \qquad a = 5$$

Check for extraneous solutions.

CHECK $a = \dfrac{5}{9}$: $\qquad \sqrt{5a} - \sqrt{2a-1} = 2$

$$\sqrt{5\left(\frac{5}{9}\right)} - \sqrt{2\left(\frac{5}{9}\right) - 1} \overset{?}{=} 2$$

$$\sqrt{\frac{25}{9}} - \sqrt{\frac{10}{9} - 1} \overset{?}{=} 2$$

$$\sqrt{\frac{25}{9}} - \sqrt{\frac{1}{9}} \overset{?}{=} 2$$

$$\frac{5}{3} - \frac{1}{3} \neq 2$$

$\dfrac{5}{9}$ is extraneous.

CHECK $a = 5$: $\qquad \sqrt{5a} - \sqrt{2a-1} = 2$

$$\sqrt{5 \cdot 5} - \sqrt{2 \cdot 5 - 1} \overset{?}{=} 2$$

$$\sqrt{25} - \sqrt{9} \overset{?}{=} 2$$

$$5 - 3 \overset{\checkmark}{=} 2$$

The solution is $\boxed{a = 5}$

∎

We follow the same principles in solving literal equations, as demonstrated in Example 5.

EXAMPLE 5 *Solve for a.* $\sqrt{ab + c} - b = d$

Solution

$\sqrt{ab + c} - b = d$	*Isolate the radical.*
$\sqrt{ab + c} = d + b$	*Square both sides of the equation.*
$(\sqrt{ab + c})^2 = (d + b)^2$	*Isolate a.*
$ab + c = d^2 + 2db + b^2$	*Subtract c from both sides of the equation.*
$ab = d^2 + 2db + b^2 - c$	*Divide both sides of the equation by b.*

$$a = \frac{d^2 + 2db + b^2 - c}{b}$$

∎

Miscellaneous Equations

In Section 7.1 we saw that we could solve equations of degree greater than 2 by the factoring method. For example,

$x^3 - x^2 - 12x = 0$	*Can be solved by factoring,*
$x(x - 4)(x + 3) = 0$	*setting each factor equal to 0,*

$x = 0$	or	$x - 4 = 0$	or	$x + 3 = 0$	*and solving each first-degree*
$x = 0$	or	$x = 4$	or	$x = -3$	*equation.*

The factoring method can be applied to any higher-degree equation which can be factored into first-degree factors. For example:

$x^4 - 13x^2 + 36 = 0$	*We factor the left-hand expression.*
$(x^2 - 4)(x^2 - 9) = 0$	*This can be factored further.*
$(x - 2)(x + 2)(x - 3)(x + 3) = 0$	*Then set each factor equal to 0.*

$x - 2 = 0,$	$x + 2 = 0,$	$x - 3 = 0,$	$x + 3 = 0$	*Solve each first-degree equation.*
$x = 2$ or	$x = -2$ or	$x = 3$ or	$x = -3$	*Check these solutions in the original equation.*

By applying our knowledge of solving quadratic equations, we can also solve equations which have second-degree factors that cannot be factored using integer coefficients. For example:

$x^3 - 5x = 0$	*We factor the left-hand side.*
$x(x^2 - 5) = 0$	*Set each factor equal to 0.*

$$x = 0 \quad \text{or} \quad x^2 - 5 = 0$$

The first equation is a simple linear equation, $x = 0$. The second equation is a quadratic equation which factors no further with integer coefficients, but we can solve it using the square root method as follows:

$$x^2 - 5 = 0$$
$$x^2 = 5$$
$$x = \pm\sqrt{5}$$

Therefore, the solutions are $x = 0$, $-\sqrt{5}$, and $+\sqrt{5}$.

EXAMPLE 6

Solve for y.
(a) $y^3 = 25y$ **(b)** $y^4 - 15y^2 = 16$

Solution

(a)

$$y^3 = 25y \qquad \textit{Rewrite the equation so one side of the equation is 0.}$$
$$y^3 - 25y = 0 \qquad \textit{Factor.}$$
$$y(y^2 - 25) = 0$$
$$y(y - 5)(y + 5) = 0 \qquad \textit{Set each factor equal to 0 and solve each equation.}$$

$$y = 0, \qquad y - 5 = 0, \qquad y + 5 = 0$$

$$\boxed{y = 0 \quad \text{or} \quad y = 5 \quad \text{or} \quad y = -5}$$

Check these solutions.

(b)

$$y^4 - 15y^2 = 16 \qquad \textit{Rewrite the equation so that one side of the equation is 0.}$$
$$y^4 - 15y^2 - 16 = 0 \qquad \textit{Factor.}$$
$$(y^2 - 16)(y^2 + 1) = 0$$
$$(y - 4)(y + 4)(y^2 + 1) = 0 \qquad \textit{Set each factor equal to 0, and solve each equation.}$$

$$y - 4 = 0, \qquad y + 4 = 0, \qquad y^2 + 1 = 0$$
$$y = 4 \quad \text{or} \qquad y = -4 \quad \text{or} \qquad y^2 = -1$$

Note that two of the solutions are real and two are not. If we use complex numbers, we simplify the imaginary solutions to get:

$$y = \pm\sqrt{-1} = \pm i$$

The solutions are $\boxed{-4, 4, -i, \text{ and } i}$

The student should check to see that the answers are correct. ∎

Let's reexamine how we solved the equation $x^4 - 13x^2 + 36 = 0$ on the previous page. To begin with, this equation is a fourth-degree equation, and at that point our only method of solving an equation of degree greater than 2 was by the factoring method. Although $x^4 - 13x^2 + 36$ is not a quadratic expression, we initially factored it as though it were. That is,

We factored $x^4 - 13x^2 + 36$ into $(x^2 - 9)(x^2 - 4)$ as though we were factoring the quadratic $u^2 - 13u + 36$ into $(u - 9)(u - 4)$ where u would equal x^2.

Thus, although $x^4 - 13x^2 + 36 = 0$ is not a quadratic equation, it does have the form of a quadratic equation, which is why we call it *an equation in quadratic form.*

Look back at page 328 and examine how we originally solved the equation. Now let's approach it in a slightly different way.

Suppose we let $u = x^2$. [Then $u^2 = (x^2)^2 = x^4$.] Now let's substitute u for x^2 and u^2 for x^4 in the original equation. Then

$$x^4 - 13x^2 + 36 = 0 \quad \text{becomes} \quad u^2 - 13u + 36 = 0$$

Now let's solve for u:

$$u^2 - 13u + 36 = 0$$
$$(u - 9)(u - 4) = 0$$
$$u = 9 \quad \text{or} \quad u = 4$$

Since we were asked originally to solve for x, we substitute x^2 back for u:

$u = 9$	or	$u = 4$	*Substitute x^2 for u.*
$x^2 = 9$	or	$x^2 = 4$	*Now we solve for x. Take square roots.*
$x = \pm\sqrt{9}$	or	$x = \pm\sqrt{4}$	
$x = \pm 3$	or	$x = \pm 2$	*Note that we get the same solutions.*

We used this example to illustrate the method of *substitution of variables.* We are substituting a single variable for a more complex expression in the equation in order to help us determine if the equation is in quadratic form.

You may be able to factor $x^4 - 13x^2 + 36$ by recognizing that it is in quadratic form, and therefore the substitution of variables method may seem to be an unnecessary complication. However, the method can help us in solving more complicated equations such as those in Example 7.

EXAMPLE 7

Solve the following. $2a^{1/2} + 3a^{1/4} - 2 = 0$

Solution

The given equation is not a quadratic equation (for that matter, it is not even a polynomial equation), but it is an equation in quadratic form. Let's examine

$$2a^{1/2} + 3a^{1/4} - 2 = 0$$

Is it obvious that we can solve this equation by factoring? It is probably not obvious that

$$2a^{1/2} + 3a^{1/4} - 2 \quad \text{factors into} \quad (2a^{1/4} - 1)(a^{1/4} + 2)$$

Let's see what happens when we make the appropriate substitution of variables. Let $u = a^{1/4}$. [Then $u^2 = (a^{1/4})^2 = a^{2/4} = a^{1/2}$.] Then

$$2a^{1/2} + 3a^{1/4} - 2 = 0 \quad \text{becomes} \quad 2u^2 + 3u - 2 = 0$$

We can solve for u by factoring:

$$(2u - 1)(u + 2) = 0$$
$$2u - 1 = 0 \quad \text{or} \quad u + 2 = 0$$

Hence, $u = \dfrac{1}{2}$ or $u = -2$.

We found the values of u, but we are looking for the values of a, so we substitute $a^{1/4}$ back in for u:

$$a^{1/4} = \dfrac{1}{2} \quad \text{or} \quad a^{1/4} = -2 \qquad \textit{Since } a^{1/4} = u$$

The equations above are similar to the equations we solved in Chapter 6. Raise both sides of each equation to the fourth power:

$$(a^{1/4})^4 = \left(\dfrac{1}{2}\right)^4 \qquad \text{or} \qquad (a^{1/4})^4 = (-2)^4$$

$$a = \dfrac{1}{16} \qquad \text{or} \qquad a = 16$$

Now we must check for extraneous roots:

CHECK $a = \dfrac{1}{16}$:
$$2a^{1/2} + 3a^{1/4} - 2 = 0$$
$$2\left(\dfrac{1}{16}\right)^{1/2} + 3\left(\dfrac{1}{16}\right)^{1/4} - 2 \overset{?}{=} 0$$
$$2\left(\dfrac{1}{4}\right) + 3\left(\dfrac{1}{2}\right) - 2 \overset{?}{=} 0$$
$$\dfrac{1}{2} + \dfrac{3}{2} - 2 \overset{?}{=} 0$$
$$2 - 2 \overset{\checkmark}{=} 0$$

CHECK $a = 16$:
$$2a^{1/2} + 3a^{1/4} - 2 = 0$$
$$2(16)^{1/2} + 3(16)^{1/4} - 2 \overset{?}{=} 0$$
$$2(4) + 3(2) - 2 \overset{?}{=} 0$$
$$8 + 6 - 2 \overset{?}{=} 0$$
$$12 \neq 0$$

Thus, $a = 16$ is an extraneous solution.

The solution is $\boxed{a = \dfrac{1}{16}}$

How do we know an equation is in quadratic form? How do we know what or where to substitute? To begin with, we observe that an equation in standard quadratic form should look like

$$Ax^2 + Bx + C = 0$$

But x can represent any expression. For example,

$$5(3a + 2)^2 + 7(3a + 2) + 8 = 0 \quad \text{and} \quad 5(y^{1/3})^2 + 7y^{1/3} + 8 = 0$$

are expressions in quadratic form. The important condition is that, ignoring the coefficients, in general, the higher-degree expression should be the square of the lower-degree expression. (The reverse is true if the exponents are negative.)

The following are more examples of equations in quadratic form:

$$a^6 - 7a^3 - 8 = 0 \qquad a^6 \text{ is the square of } a^3.$$
$$2x^{2/3} + x^{1/3} - 15 = 0 \qquad x^{2/3} \text{ is the square of } x^{1/3}.$$
$$a^{-4} - 6a^{-2} - 7 = 0 \qquad a^{-4} \text{ is the square of } a^{-2}.$$

We let u equal the middle term literal expression. Note that we always find u^2 to make sure it checks out. For example, suppose in $2x^{2/3} - 7x^{1/3} + 3 = 0$, we let $u = x^{2/3}$. Then $u^2 = (x^{2/3})^2 = x^{4/3}$. Since $x^{4/3}$ is not in the equation, we reason that we chose the wrong variable for substituting u. Therefore, if the equation *is* in quadratic form, then u *must* be $x^{1/3}$.

EXAMPLE 8

Solution

Solve for x. $3(x + 4)^2 - 2(x + 4) - 1 = 0$

We could multiply out the left-hand side of this equation, put it in standard form, and then solve it. Observe, however, that the equation is in quadratic form.

Let $u = x + 4$. [Then $u^2 = (x + 4)^2$.] Then

$$3(x + 4)^2 - 2(x + 4) - 1 = 0$$

becomes

$3u^2 - 2u - 1 = 0$ *Now solve for u by factoring.*
$(3u + 1)(u - 1) = 0$

$$u = -\frac{1}{3} \quad \text{or} \quad u = 1$$

Now we find x by substituting back $x + 4$ for u:

$$x + 4 = -\frac{1}{3} \quad \text{or} \quad x + 4 = 1 \qquad \text{\textit{Solve for x: add} } -4 \text{ \textit{to both sides of each equation.}}$$

$$\boxed{x = -\frac{13}{3} \quad \text{or} \quad x = -3}$$

■

Exercises 7.5

In Exercises 1–20, solve the equation.

1. $\sqrt{x + 3} = 2x$

2. $\sqrt{a - 2} = a - 4$

3. $\sqrt{x + 5} = 7 - x$

4. $\sqrt{y - 2} = 22 - y$

5. $\sqrt{5a - 1} + 5 = a$

6. $\sqrt{7a + 4} - 2 = a$

7. $\sqrt{a + 1} + a = 11$

8. $\sqrt{3a + 1} + a = 9$

9. $\sqrt{3x + 1} + 3 = x$

10. $5a - \sqrt{2a - 3} = 4a + 9$

11. $5a - 2\sqrt{a + 3} = 2a - 1$

12. $6a - 3\sqrt{a + 5} = 4a - 1$

13. $\sqrt{y + 3} = 1 + \sqrt{y}$

14. $\sqrt{3x + 4} = 2 + \sqrt{x}$

15. $\sqrt{a + 7} = 1 + \sqrt{2a}$

16. $\sqrt{2r - 1} = \sqrt{5r} - 2$

17. $\sqrt{7s + 1} - 2\sqrt{s} = 2$

18. $\sqrt{3s + 4} - \sqrt{s} = 2$

19. $\sqrt{7 - a} - \sqrt{3 + a} = 2$

20. $\sqrt{3 - 2a} - \sqrt{3 - 3a} = -1$

In Exercises 21–26, solve for the given variable.

21. $\sqrt{x} + a = b$ for x

22. $\sqrt{3a} - x = b$ for a

23. $\dfrac{\sqrt{\pi L}}{g} = T$ for L

24. $K = \sqrt{\dfrac{2gs}{l}}$ for s

25. $\sqrt{5x + b} = 6 + b$ for x

26. $\sqrt{3x + y} = 5 - y$ for x

In Exercises 27–68, solve the equation.

27. $x^3 - 2x^2 - 15x = 0$

28. $x^3 + x^2 - 20x = 0$

29. $6a^3 - a^2 - 2a = 0$

30. $2s^4 - 7s^3 + 3s^2 = 0$

31. $y^4 - 17y^2 + 16 = 0$

32. $2t^4 - 34t^2 = -32$

33. $3a^4 + 24 = 18a^2$

34. $a^4 + 45 = 14a^2$

35. $b^4 + 112 = 23b^2$

36. $b^4 + 75 = 28b^2$

37. $9 - \dfrac{8}{x^2} = x^2$

38. $3 - \dfrac{1}{x^2} = 2x^2$

39. $x^3 + x^2 - x - 1 = 0$

40. $a^3 - 2a^2 - 4a + 8 = 0$

41. $x^{2/3} - 4 = 0$

42. $a^{2/3} = 9$

43. $x + x^{1/2} - 6 = 0$ [*Hint:* Let $u = x^{1/2}$.]

44. $x - 5x^{1/2} + 6 = 0$

45. $y^{2/3} - 4y^{1/3} - 5 = 0$

46. $b^{2/3} - 4b^{1/3} = -3$

47. $x^{1/2} + 8x^{1/4} + 7 = 0$

48. $x^{1/2} + 3x^{1/4} - 10 = 0$

49. $x^{-2} - 5x^{-1} + 6 = 0$ [*Hint:* Let $u = x^{-1}$.]

50. $x^{-2} - 2x^{-1} - 15 = 0$

51. $6x^{-2} + x^{-1} - 1 = 0$

52. $15x^{-2} + 7x^{-1} - 2 = 0$

53. $x^{-4} - 13x^{-2} + 36 = 0$

54. $x^{-4} - 3x^{-2} = 4$ [*Hint:* Let $u = x^{-2}$.]

55. $\sqrt{a} - \sqrt[4]{a} - 6 = 0$ [*Hint:* Change radicals to rational exponents.]

56. $\sqrt{b} + 2\sqrt[4]{b} = 3$

57. $\sqrt{x} - 4\sqrt[4]{x} = 5$

58. $\sqrt{x} - 2\sqrt[4]{x} = 8$

59. $(a + 4)^2 + 6(a + 4) + 9 = 0$

60. $(a - 1)^2 - 3(a - 1) = 10$

61. $(x + 2)^2 + 4(x + 2) = 5$

62. $(2x - 3)^2 - 6(2x - 3) - 7 = 0$

63. $2(3x + 1)^2 - 5(3x + 1) - 3 = 0$

64. $3(2x - 1)^2 - 1 = 0$

65. $(x^2 + x)^2 - 4 = 0$

66. $(x^2 - 4x)^2 - 25 = 0$

67. $\left(a - \dfrac{10}{a}\right)^2 - 12\left(a - \dfrac{10}{a}\right) + 27 = 0$

68. $\left(y + \dfrac{12}{y}\right)^2 - 15\left(y + \dfrac{12}{y}\right) + 56 = 0$

? QUESTION FOR THOUGHT

69. Discuss what is *wrong* (if anything) with solving the equation $\sqrt{3x-2} - \sqrt{x} = 2$ in the following way:

$$\sqrt{3x-2} - \sqrt{x} = 2$$
$$(\sqrt{3x-2} - \sqrt{x})^2 = 2^2$$
$$3x - 2 - x = 4$$
$$2x - 2 = 4$$
$$2x = 6$$
$$x = 3$$

⟳ MINI-REVIEW

Solve the following for a:

70. $\dfrac{2a+1}{4} + a \le \dfrac{2}{3}$

71. $\dfrac{a+2}{a-1} + 5 = \dfrac{3}{a-1}$

72. $|5x+1| < 9$

73. $|5x+1| \ge 9$

7.6

Quadratic Inequalities

In this section we will solve quadratic inequalities such as $x^2 + x - 12 > 0$. Recall that in solving quadratic equations we found it convenient to work with them in standard form. We will find similarly that quadratic inequalities will be easier to work with if they are in standard form.

We say that a quadratic inequality is in **standard form** if it is in the form $Ax^2 + Bx + C > 0$ or $Ax^2 + Bx + C < 0$ with $A > 0$. (Note that $<$ and $>$ can be replaced respectively by $\ge$ and $\le$.)

In order to solve the inequality $x^2 + x - 12 > 0$ we must realize that we are looking for values of x which make $x^2 + x - 12$ *positive*. (Remember that $u > 0$ means u is positive.) On the other hand, to solve the inequality $x^2 + x - 12 < 0$ means that we are seeking the values of x which make the expression negative. Hence, we are concerned only with the *sign* of the expression $x^2 + x - 12$.

When is the expression $x^2 + x - 12$ positive and when is it negative? The easiest way to determine this is to factor the expression and *examine the sign of each factor* as x takes on various values on the number line. Then use the multiplication rules for signed numbers to determine the sign of the product in order to find the solution.

We begin by factoring $x^2 + x - 12$ into $(x + 4)(x - 3)$ and for convenience we will use the factored form.

Looking at $(x + 4)(x - 3)$, we can tell at a glance that at $x = -4$ and $x = +3$, $(x + 4)(x - 3)$ is 0. What happens when x takes on values around -4 and $+3$? Let's examine the signs of each factor of $(x + 4)(x - 3)$ using the number line.

First we examine the sign of $x + 4$ as x takes on various values on the number line. Notice that $x + 4$ is negative for $x < -4$ (if we let $x = -5$, then $x + 4 = -1$). Then $x + 4$ is positive for $x > -4$.

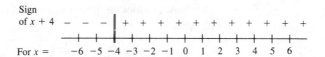

We do the same for the factor $x - 3$. Notice that $x - 3$ is negative for $x < 3$ and positive for $x > 3$.

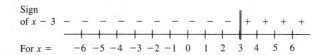

If we put these two pictures together, we can determine the sign of the product by looking at the signs of the factors above the number line.

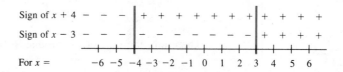

By using the multiplication rules, at a glance we arrive at the signs of the product. The sign of the product is positive when $x < -4$ or $x > 3$, negative when $-4 < x < 3$.

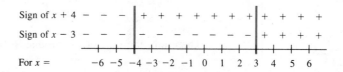

By the picture above, we find that $(x + 4)(x - 3) > 0$ or, equivalently, that $x^2 + x - 12 > 0$ (positive), when $x < -4$ or $x > 3$.

The x values where the quadratic expression $x^2 + x - 12$ is exactly 0 are the x values where the factors $x + 4$ and $x - 3$ are 0: at $x = -4$ and $x = +3$. These are called the **cutpoints** for the inequality $(x + 4)(x - 3) > 0$. The cutpoints serve as endpoints for the solution set(s) of the quadratic inequality.

Let's demonstrate with another example.

EXAMPLE 1

Solve for x. $2x^2 - 9x - 5 < 0$

Solution

Keep in mind that we are looking for the values of x which make the expression $2x^2 - 9x - 5$ negative.

$2x^2 - 9x - 5 < 0$ *This inequality is asking: When is $2x^2 - 9x - 5$ negative? Factor.*

$(2x + 1)(x - 5) < 0$ *Find the cutpoints.*

$$2x + 1 = 0 \rightarrow x = -\frac{1}{2} \quad \text{and} \quad x - 5 = 0 \rightarrow x = 5$$

Now we look at the sign of each factor as x takes on different values around the cutpoints.

$2x + 1$ is negative when $x < -\frac{1}{2}$, and positive when $x > -\frac{1}{2}$:

$x - 5$ is negative when $x < 5$, and positive when $x > 5$:

Therefore, the sign of the product is as shown here:

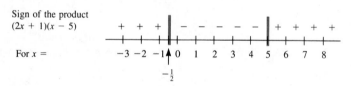

As we can see, $(2x + 1)(x - 5)$ is negative when $-\frac{1}{2} < x < 5$ or

$$(2x + 1)(x - 5) < 0 \quad \text{when} \quad -\frac{1}{2} < x < 5$$

Therefore, the solution to $2x^2 - 9x - 5 < 0$ is $\boxed{-\frac{1}{2} < x < 5}$

We can graph the solution set to the inequality $2x^2 - 9x - 5 < 0$ as shown in Figure 7.7.

Figure 7.7
Solution set for Example 1

Note we exclude the points $-\frac{1}{2}$ and 5. ∎

Looking at the number line representation of the signs of $(2x + 1)(x - 5)$ in Example 1, we can observe that the cutpoints break up the number line into three distinct intervals.

Most important, note that the sign of the product does not change within an interval (excluding the cutpoints). If the expression is positive (or negative) for one value of x within an interval, it is positive (or negative) for *all* values within that interval.

Hence, if we want to know the sign of the product in an interval, *we need only check one value within the interval*. For example, we found that the cutpoints for $(2x + 1)(x - 5)$ are $-\frac{1}{2}$ and 5. This breaks up the number line into three intervals, as shown in Figure 7.8.

Figure 7.8

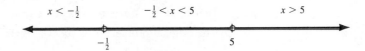

Choose any value in the leftmost interval; that is, choose any number less than $-\frac{1}{2}$, say $x = -6$.

When $x = -6$, the sign of $(2x + 1)(x - 5) = [2(-6) + 1][-6 - 5]$ is positive and therefore the sign of $(2x + 1)(x - 5)$ is positive for *all* values of x in the interval $x < -\frac{1}{2}$.

Choose any value in the middle interval, $-\frac{1}{2} < x < 5$, say $x = 0$.

When $x = 0$, the sign of $(2x + 1)(x - 5) = (2 \cdot 0 + 1)(0 - 5)$ is negative and therefore the sign of $(2x + 1)(x - 5)$ is negative for *all* values of x in the interval $-\frac{1}{2} < x < 5$.

Finally, choose any value in the rightmost interval, $x > 5$, say $x = 10$.

When $x = 10$, the sign of $(2x + 1)(x - 5) = (2 \cdot 10 + 1)(10 - 5)$ is positive and therefore the sign of $(2x + 1)(x - 5)$ is positive for *all* values of x in the interval $x > 5$.

The values we choose in the intervals to check the sign of the product are called the ***test values*** for the interval. We demonstrate the use of this method in the following example.

EXAMPLE 2

Solve for a and sketch a graph of the solution set. $a^2 - 7a + 6 > 0$

Solution

$a^2 - 7a + 6 > 0$ *This inequality is asking: When is $a^2 - 7a + 6$ positive? First factor the left-hand side.*

$(a - 1)(a - 6) > 0$ *Find the cutpoints.*

$$a - 1 = 0 \rightarrow a = 1 \quad \text{and} \quad a - 6 = 0 \rightarrow a = 6$$

Therefore, the cutpoints break up the number line into three intervals:

$$a < 1, \quad 1 < a < 6, \quad \text{and} \quad a > 6$$

To check the interval $a < 1$, let $a = 0$ be our test value. Then $(a - 1)(a - 6) = (0 - 1)(0 - 6)$, which is positive.

Hence, $(a - 1)(a - 6)$ is positive or $(a - 1)(a - 6) > 0$ for $a < 1$.

To check the interval $1 < a < 6$, let $a = 4$ be our test value. Then $(a - 1)(a - 6) = (4 - 1)(4 - 6)$, which is negative.

Hence, $(a - 1)(a - 6)$ is negative or $(a - 1)(a - 6) < 0$ for $1 < a < 6$.

To check the interval $a > 6$, let $a = 10$ be our test value. Then $(a - 1)(a - 6) = (10 - 1)(10 - 6)$, which is positive.

Hence, $(a - 1)(a - 6)$ is positive or $(a - 1)(a - 6) > 0$ for $a > 6$.

The original problem asks for what values of a is $a^2 - 7a + 6 > 0$.

The solution is $\boxed{a < 1 \quad \text{or} \quad a > 6}$

We can graph the solution to $a^2 - 7a + 6 > 0$ on the number line as shown in Figure 7.9.

Figure 7.9
Solution set for Example 2

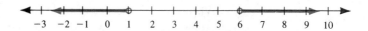

The general procedure for solving quadratic inequalities is summarized in the box.

Procedure for Solving Quadratic Inequalities	
	1. *Put the inequality into standard form* (the right side of the inequality is 0).
	2. *Factor* the quadratic expression.
	3. *Find the cutpoints* by setting each factor equal to 0.
	4. The cutpoints divide the number line into intervals. *Write the intervals* (excluding the cutpoints). (Draw a number line if helpful.)
	5. *Pick test values* within each interval determined by the cutpoints and check the sign of the product for each test value.
	6. *Choose the interval(s)* which satisfy the sign requirements of the inequality in step 1.

EXAMPLE 3

Solve for x and graph the solution on the number line:

$$x^2 - 2x \geq 15$$

Solution

$x^2 - 2x \geq 15$ *First rewrite the inequality so that one side of the inequality is 0.*

$x^2 - 2x - 15 \geq 0$ *Then factor.*

$(x - 5)(x + 3) \geq 0$

Find the cutpoints:

$$x - 5 = 0 \rightarrow x = 5 \qquad x + 3 = 0 \rightarrow x = -3 \qquad \textit{Cutpoints are 5 and } -3.$$

The intervals are $x < -3$, $-3 < x < 5$, and $x > 5$.

For $x < -3$, let $x = -10$ be the test value.
When $x = -10$, $(x - 5)(x + 3) = (-10 - 5)(-10 - 3)$ is positive.
For $-3 < x < 5$, let 0 be the test value.
When $x = 0$, $(x - 5)(x + 3) = (0 - 5)(0 + 3)$ is negative.
For $x > 5$, let 10 be the test value.
When $x = 10$, $(x - 5)(x + 3) = (10 - 5)(10 - 3)$ is positive.

Thus, $(x - 5)(x + 3)$ is positive for $x < -3$ or $x > 5$ and $(x - 5)(x + 3)$ is 0 for $x = -3$ and $x = 5$.

Thus, $(x - 5)(x + 3) \geq 0$ (positive or equal to 0) for

$$\boxed{x \leq -3 \quad \text{or} \quad x \geq 5}$$

We graph the *solution* in Figure 7.10.

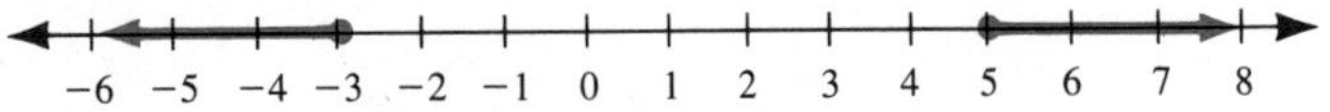

Figure 7.10
Solution set for Example 3

Notice the endpoints are included. (Why?) ∎

We can apply this same method and reasoning to solve rational inequalities such as

$$\frac{a - 7}{a + 2} < 0 \quad \text{or} \quad \frac{6}{a - 3} \leq 3$$

Notice that these inequalities are different from those covered in Chapter 4; the rational inequalities in Chapter 4 did not have variables in the denominator.

In attempting to solve $\dfrac{a - 7}{a + 2} < 0$, your first inclination may be to get rid of the denominator by multiplying both sides of the inequality by $a + 2$. With *equations* this approach is appropriate provided we keep in mind $a \neq -2$. With inequalities, however, we need to know whether this multiplier is positive or negative in order to determine whether the inequality symbol should be reversed. Is $a + 2$ positive or negative? We could reason out an answer on a case-by-case basis, but an easier approach is to use the same method we used for quadratic inequalities, as illustrated in the next example.

EXAMPLE 4

Solve for a.

(a) $\dfrac{a - 7}{a + 2} < 0$ (b) $\dfrac{6}{a - 3} \leq 3$

Solution

(a) We can analyze a quotient in the same way we analyzed a product: If the signs are the same, the quotient is positive; if the signs are opposite, the quotient is negative.

First we find the cutpoints. In the case of rational expressions, the cutpoints are the points where the rational expression is 0 or undefined. Note that *the rational expression is 0 when the numerator is 0 and undefined when the denominator is 0.* Thus, the cutpoints are:

$$\text{The value of } a \text{ where } a - 7 = 0 \rightarrow a = 7$$

$$\text{The value of } a \text{ where } a + 2 = 0 \rightarrow a = -2$$

The intervals are

$$a < -2, \quad -2 < a < 7, \quad \text{and} \quad a > 7$$

For $a < -2$, let our test value be -5.

When $a = -5$, $\dfrac{a - 7}{a + 2} = \dfrac{-5 - 7}{-5 + 2}$ is positive.

For $-2 < a < 7$, let our test value be 0.

When $a = 0$, $\dfrac{a - 7}{a + 2} = \dfrac{0 - 7}{0 + 2}$ is negative.

For $a > 7$, let our test value be 10.

When $a = 10$, $\dfrac{a - 7}{a + 2} = \dfrac{10 - 7}{10 + 2}$ is positive.

Since $\dfrac{a - 7}{a + 2}$ is negative (or less than 0) when $-2 < a < 7$, the solution is

$$\boxed{-2 < a < 7}$$

(b)   Our method of solving quadratic and rational inequalities is a matter of determining when the expression is positive or negative. Therefore, we must take the original inequality and transform it into the form $R \leq 0$, where R is a rational expression.

$$\frac{6}{a - 3} \leq 3 \qquad \textit{Subtract 3 from both sides of the inequality.}$$

$$\frac{6}{a - 3} - 3 \leq 0$$

Then we combine the left-hand side into one fraction.

$$\frac{6}{a - 3} - \frac{3(a - 3)}{a - 3} \leq 0 \qquad \textit{Find the LCD and change to equivalent fractions.}$$
$$\textit{Add numerators.}$$

$$\frac{6 - 3(a - 3)}{a - 3} \leq 0$$

$$\frac{6 - 3a + 9}{a - 3} \leq 0 \qquad \textit{Combine terms in the numerator.}$$

$$\frac{15 - 3a}{a - 3} \leq 0$$

The cutpoints are:

$$\text{The value of } a \text{ where } 15 - 3a = 0 \rightarrow a = 5$$
$$\text{The value of } a \text{ where } a - 3 = 0 \rightarrow a = 3$$

The intervals are

$$a < 3, \quad 3 < a < 5, \quad \text{and} \quad a > 5$$

Given the test value, you should be able to determine the sign of the expression mentally:

For $a < 3$, let our test value be 0. Then for $a = 0$, $\dfrac{15 - 3a}{a - 3}$ is negative.

For $3 < a < 5$, let $a = 4$. Then $\dfrac{15 - 3a}{a - 3}$ is positive.

For $a > 5$, let $a = 10$. Then $\dfrac{15 - 3a}{a - 3}$ is negative.

Thus, $\dfrac{15 - 3a}{a - 3} < 0$ when $a < 3$ or $a > 5$.

Our problem asks for a solution where the rational expression is less than *or equal to* 0. Since $\dfrac{15 - 3a}{a - 3} = 0$ when $a = 5$, we include the point 5 in the solution.

Therefore, the solution is $\boxed{a < 3 \quad \text{or} \quad a \geq 5}$

Notice that we cannot include the point $a = 3$ since this value makes the fraction undefined.

Exercises 7.6

In Exercises 1–8, solve the inequalities and sketch a graph of the solution set.

1. $(x + 4)(x - 2) < 0$

2. $(x - 7)(x + 2) > 0$

3. $(x + 4)(x - 2) > 0$

4. $(x - 7)(x + 2) < 0$

5. $(x + 2)(x - 5) \leq 0$

6. $(2x - 1)(x + 3) \geq 0$

7. $(x - 3)(2x - 1) \geq 0$

8. $(5x - 8)(x - 2) < 0$

In Exercises 9–36, solve the inequalities.

9. $a^2 - a - 20 < 0$

10. $x^2 - 3x - 18 > 0$

11. $x^2 + x - 12 \geq 0$

12. $y^2 + 7y - 8 \leq 0$

13. $2a^2 - 9a \leq 5$

14. $6x^2 + 7x > -2$

15. $6y^2 - y > 1$

16. $5a^2 < 7 - 34a$

17. $3x^2 \leq 10 - 13x$

18. $2y^2 \geq 5 - 9y$

19. $x^2 + 2x + 1 \geq 0$

20. $y^2 - 4y + 4 < 0$

21. $x^2 - 6x + 9 < 0$

22. $a^2 - 10a + 25 \geq 0$

23. $2x^2 - 13x > -15$

24. $10x^2 < -2 - 9x$

25. $3y^2 \geq 5y + 2$

26. $2x^2 \leq 12 - 5x$

27. $x^2 + 2x \leq -1$

28. $x^2 - 6x \leq -9$

29. $\dfrac{x - 2}{x + 1} < 0$

30. $\dfrac{x + 7}{x + 1} > 0$

31. $\dfrac{y + 3}{y - 5} \geq 0$

32. $\dfrac{y - 6}{y - 2} \leq 0$

33. $\dfrac{a - 6}{a + 4} < 0$

34. $\dfrac{z - 5}{z + 2} \geq 0$

35. $\dfrac{5}{y - 4} > 0$

36. $\dfrac{7}{y + 2} < 0$

In Exercises 37–50, solve the inequalities and sketch a graph of the solution set.

37. $\dfrac{3}{y - 1} < 1$

38. $\dfrac{4}{x + 5} > 1$

39. $\dfrac{y + 1}{y + 2} > 3$

40. $\dfrac{y + 4}{y - 1} < 2$

41. $\dfrac{2y + 3}{y - 1} \leq 2$

42. $\dfrac{x + 3}{2x + 1} \geq 4$

43. $\dfrac{y - 3}{y + 1} > -2$

44. $\dfrac{a + 3}{a - 5} < -3$

45. $\dfrac{x}{x - 1} \leq \dfrac{3}{x - 1}$

46. $\dfrac{a}{2a + 1} > \dfrac{3a + 1}{2a + 1}$

47. $\dfrac{x - 2}{x + 3} > \dfrac{x + 4}{x + 3}$

48. $\dfrac{y + 1}{y - 1} < \dfrac{y - 1}{y - 1}$

49. $\dfrac{1}{x - 2} + \dfrac{2}{x + 3} \leq \dfrac{3}{x + 3}$

50. $\dfrac{2}{x + 1} + \dfrac{3}{x + 2} \geq \dfrac{5}{x + 2}$

? QUESTION FOR THOUGHT

51. For what values of x would $\dfrac{3}{x - 2} = 0$?

 MINI-REVIEW

52. *Simplify:*

$$\dfrac{3 + \dfrac{2}{x - 1}}{5x + \dfrac{3x - 1}{x + 3}}$$

53. *Simplify with positive exponents only:*

$$\left(\dfrac{4x^{-4}y^{1/2}}{9x^{1/2}y^{-1}} \right)^{1/2}$$

54. *Perform the operations and simplify.*
$(3\sqrt{2} + \sqrt{3})(2\sqrt{2} - \sqrt{3})$

55. *Perform the operations and simplify:*

$$\dfrac{3\sqrt{6}}{\sqrt{5} - \sqrt{2}} - \dfrac{6}{\sqrt{6}}$$

CHAPTER 7   Summary

After having completed this chapter, you should be able to:

1. Solve quadratic equations by the square root method or factoring method (Section 7.1).

For example: Solve the following:

(a) $10x^2 + 13x = 3$ **(b)** $\dfrac{2x + 3}{x + 3} = \dfrac{x - 2}{x - 1}$

Solution:

(a) $10x^2 + 13x = 3$ *Put in standard form.*

$10x^2 + 13x - 3 = 0$ *Factor.*

$(5x - 1)(2x + 3) = 0$ *Set each factor equal to 0.*

$5x - 1 = 0$ or $2x + 3 = 0$ *Solve each equation.*

$x = \dfrac{1}{5}$ or $x = -\dfrac{3}{2}$

(b) $\dfrac{2x + 3}{x + 3} = \dfrac{x - 2}{x - 1}$ *Eliminate denominators [multiply both sides by $(x - 1)(x + 3)$].*

$(2x + 3)(x - 1) = (x + 3)(x - 2)$ *Simplify each side.*

$2x^2 + x - 3 = x^2 + x - 6$

$x^2 = -3$ *Take square roots.*

$x = \pm\sqrt{-3}$ *There are no real solutions, but we will use complex numbers.*

$x = \pm i\sqrt{3}$

The solutions are $-i\sqrt{3}$ and $+i\sqrt{3}$.

Check to ensure that denominators are not 0.

2. Solve a quadratic equation by completing the square (Section 7.2).

For example: Solve for y. $2y^2 + 2y - 3 = 0$

Solution:

$2y^2 + 2y - 3 = 0$ *Divide both sides of the equation by 2.*

$y^2 + y - \dfrac{3}{2} = 0$ *Next, add $\dfrac{3}{2}$ to both sides of the equation.*

$y^2 + y = \dfrac{3}{2}$ *Take $\dfrac{1}{2}$ of middle term coefficient and square it: $\left[\dfrac{1}{2}(1)\right]^2 = \dfrac{1}{4}$*

Add $\dfrac{1}{4}$ to both sides of the equation.

$y^2 + y + \boxed{\dfrac{1}{4}} = \dfrac{3}{2} + \boxed{\dfrac{1}{4}}$ *Factor the left-hand side and simplify the right-hand side.*

$\left(y + \dfrac{1}{2}\right)^2 = \dfrac{7}{4}$ *Take square roots.*

$y + \dfrac{1}{2} = \pm\sqrt{\dfrac{7}{4}}$ *Isolate y.*

$y = -\dfrac{1}{2} \pm \sqrt{\dfrac{7}{4}}$ *Simplify the answer.*

$y = \dfrac{-1 \pm \sqrt{7}}{2}$

3. Solve a quadratic equation by using the quadratic formula (Section 7.3).

For example: $3t^2 - 2t - 2 = 0$

Solution: $A = 3, \quad B = -2, \quad C = -2$

$$t = \frac{-(-2) \pm \sqrt{(-2)^2 - 4(3)(-2)}}{2(3)}$$

$$t = \frac{2 \pm \sqrt{4 + 24}}{6} = \frac{2 \pm \sqrt{28}}{6} = \frac{2 \pm \sqrt{4 \cdot 7}}{6}$$

$$= \frac{2 \pm 2\sqrt{7}}{6}$$

$$= \frac{\cancel{2}(1 \pm \sqrt{7})}{\cancel{6}_{3}}$$

The solutions are: $\dfrac{1 \pm \sqrt{7}}{3}$

4. Determine the nature of the roots of a quadratic equation without directly solving the equation (Section 7.3).

For example: Determine the nature of the roots of

$$3x^2 - 2x + 5 = 0$$

Solution: $A = 3, \quad B = -2, \quad C = 5$

The discriminant, $B^2 - 4AC = (-2)^2 - 4(3)(5) = 4 - 60 = -54 < 0$

Therefore, the roots are not real.

5. Solve verbal problems involving quadratic equations (Section 7.4).

6. Solve radical equations (Section 7.5).

For example: Solve for x. $\sqrt{3x + 1} - \sqrt{x - 1} = 2$

Solution:

$\sqrt{3x + 1} - \sqrt{x - 1} = 2$	*Isolate one radical.*
$\sqrt{3x + 1} = 2 + \sqrt{x - 1}$	*Square both sides of the equation.*
$(\sqrt{3x + 1})^2 = (2 + \sqrt{x - 1})^2$	*Then simplify.*
$3x + 1 = 4 + 4\sqrt{x - 1} + (x - 1)$	*Isolate the remaining radical.*
$2x - 2 = 4\sqrt{x - 1}$	*Divide both sides by 2.*
$x - 1 = 2\sqrt{x - 1}$	*Square both sides of the equation.*
$(x - 1)^2 = (2\sqrt{x - 1})^2$	*Simplify both sides of the equation.*
$x^2 - 2x + 1 = 4(x - 1)$	
$x^2 - 2x + 1 = 4x - 4$	*Put in standard form and solve by factoring.*
$x^2 - 6x + 5 = 0$	
$(x - 5)(x - 1) = 0$	
$x = 5 \quad \text{or} \quad x = 1$	

Check for extraneous roots.

CHECK $x = 5$:　$\sqrt{3x + 1} - \sqrt{x - 1} = 2$

$$\sqrt{3(5) + 1} - \sqrt{5 - 1} \overset{?}{=} 2$$
$$\sqrt{16} - \sqrt{4} \overset{?}{=} 2$$
$$4 - 2 \overset{\checkmark}{=} 2$$

CHECK $x = 1$:　$\sqrt{3x + 1} - \sqrt{x - 1} = 2$

$$\sqrt{3(1) + 1} - \sqrt{1 - 1} \overset{?}{=} 2$$
$$\sqrt{4} - \sqrt{0} \overset{?}{=} 2$$
$$2 - 0 \overset{\checkmark}{=} 2$$

Therefore, the solutions are $x = 1$ and $x = 5$.

7.　Solve equations in quadratic form (Section 7.5).

For example:　Solve for x.　$x^{2/3} + 2x^{1/3} = 15$

Solution:　Let $u = x^{1/3}$. [Then $u^2 = (x^{1/3})^2 = x^{2/3}$.] Substitute $x^{1/3}$ for u and

$$x^{2/3} + 2x^{1/3} = 15$$

becomes

$$u^2 + 2u = 15$$
$$u^2 + 2u - 15 = 0 \qquad \textit{Solve for u.}$$
$$(u + 5)(u - 3) = 0$$
$$u = -5 \quad \text{or} \quad u = 3$$

Since $x^{1/3} = u$, then

$$x^{1/3} = -5 \quad \text{or} \quad x^{1/3} = 3 \qquad \textit{Solve for x. Cube both sides of the equation.}$$
$$(x^{1/3})^3 = (-5)^3 \quad \text{or} \quad (x^{1/3})^3 = (3)^3$$
$$x = -125 \quad \text{or} \quad x = 27$$

8.　Solve quadratic and rational inequalities (Section 7.6).

For example:　Solve for x.　$x^2 - 9x - 10 \le 0$

Solution:

$$(x - 10)(x + 1) \le 0 \qquad \textit{Factor.}$$
$$x - 10 = 0 \quad x + 1 = 0 \qquad \textit{Find cutpoints.}$$
$$x = 10 \qquad x = -1$$

The intervals are

$$x < -1, \quad -1 < x < 10, \quad \text{and} \quad x > 10$$

For $x < -1$, let $x = -5$ be the test value.
Then $(x - 10)(x + 1) = (-5 - 10)(-5 + 1)$ is positive.
For $-1 < x < 10$, let $x = 0$ be the test value.
Then $(x - 10)(x + 1) = (0 - 10)(0 + 1)$ is negative.

For $x > 10$, let $x = 20$ be the test value.

Then $(x - 10)(x + 1) = (20 - 10)(20 + 1)$ is positive.

$(x + 10)(x + 1)$ is negative when $-1 < x < 10$.

$(x + 10)(x + 1)$ is 0 when $x = -1$ and $x = 10$.

Therefore, $x^2 - 9x - 10 \leq 0$ when $-1 \leq x \leq 10$ *Note that the cutpoints are included.*

CHAPTER 7   Review Exercises

Solve Exercises 1–18 by factoring or the square root method.

1. $(x + 7)(x - 4) = 0$

2. $(2x - 1)(x + 3) = 0$

3. $2y^2 - y - 1 = 0$

4. $3a^2 - 13a - 10 = 0$

5. $3x^2 - 17x = 28$

6. $10a^2 - 3a = 1$

7. $81 = a^2$

8. $0 = y^2 - 65$

9. $z^2 + 7 = 2$

10. $3r^2 + 5 = r^2$

11. $4x^2 + 36 = 24x$

12. $5a^2 - 5a = 10$

13. $(a + 7)(a + 3) = (3a + 1)(a + 1)$

14. $(y + 2)(y + 1) = (3y + 1)(y - 1)$

15. $x - 2 = \dfrac{1}{x + 2}$

16. $a - 5 = \dfrac{1}{a - 5}$

17. $\dfrac{2}{x - 2} - \dfrac{5}{x + 2} = 1$

18. $\dfrac{3}{x + 4} + \dfrac{5}{x + 2} = 6$

Solve Exercises 19–26 by completing the square.

19. $x^2 + 2x - 4 = 0$

20. $x^2 - 2x - 4 = 0$

21. $2y^2 + 4y - 3 = 0$

22. $2y^2 + 4y + 3 = 0$

23. $3a^2 + 6a - 5 = 0$

24. $3a^2 - 6a + 5 = 0$

25. $\dfrac{1}{a - 5} + \dfrac{3}{a + 2} = 4$

26. $\dfrac{3}{a - 1} - \dfrac{1}{a + 2} = 2$

Solve Exercises 27–48 by any method.

27. $6a^2 - 13a = 5$

28. $5x^2 - 18x = 8$

29. $3a^2 - 7a = 6$

30. $2y^2 + 5y = 7$

31. $5a^2 - 3a = 3 - 3a + 2a^2$

32. $5a^2 + 6a - 4 = 3a^2 + 10a + 26$

33. $(x - 4)(x + 1) = x - 2$

34. $(y + 5)(y - 7) = 3$

35. $(t + 3)(t - 4) = t(t + 2)$

36. $(u - 5)(u + 1) = (u - 3)(u + 5)$

37. $8x^2 = 12$

38. $3x^2 - 14 = 5$

39. $3x^2 - 2x + 5 = 7x^2 - 2x + 5$

40. $7y^2 - 36 = -1$

41. $(x + 2)(x - 4) = 2x - 10$

42. $(z - 2)(z - 3) = z - 5$

43. $\dfrac{1}{z + 2} = z - 4$

44. $\dfrac{3}{z - 2} = z$

45. $\dfrac{1}{x + 4} - \dfrac{3}{x + 2} = 5$

46. $\dfrac{2}{x - 1} - \dfrac{3}{x + 4} = 2$

47. $\dfrac{3}{x - 4} + \dfrac{2x}{x - 5} = \dfrac{3}{x - 5}$

48. $\dfrac{2}{x - 6} - \dfrac{x}{x + 6} = \dfrac{3}{x^2 - 36}$

In Exercises 49–52, solve for the given variable.

49. $A = \pi r^2 h$ for $r > 0$

50. $l = \dfrac{gt^2}{2}$ for $t > 0$

51. $2x^2 + xy - 3y^2 = 0$ for x

52. $5x^2 + 4xy - y^2 = 0$ for y

In Exercises 53–60, solve the equation.

53. $\sqrt{2a + 3} = a$

54. $\sqrt{5a - 4} = a$

55. $\sqrt{3a + 1} + 1 = a$

56. $\sqrt{7y + 4} - 2 = y$

57. $\sqrt{2x + 1} - \sqrt{x - 3} = 4$

58. $\sqrt{3x + 1} - \sqrt{x - 1} = 2$

59. $\sqrt{3x + 4} - \sqrt{x - 3} = 3$

60. $\sqrt{2x - 1} - \sqrt{9 - x} = 1$

In Exercises 61–64, solve for y.

61. $\sqrt{3y + z} = x$

62. $\sqrt{5y - z} = x$

63. $\sqrt{3y} + z = x$

64. $\sqrt{5y} - z = x$

In Exercises 65–82, solve the equation.

65. $x^3 - 2x^2 - 15x = 0$

66. $x^4 - 2x^3 = 35x^2$

67. $4x^3 - 10x^2 - 6x = 0$

68. $9x^3 + 3x^2 - 6x = 0$

69. $a^4 - 17a^2 = -16$

70. $y^4 - 18y^2 = -81$

71. $y^4 - 3y^2 = 4$

72. $a^4 - 5a^2 = 36$

73. $z^4 = 6z^2 - 5$

74. $z^4 = 11z^2 - 18$

75. $a^{1/2} - a^{1/4} - 6 = 0$

76. $a^{2/3} - a^{1/3} - 6 = 0$

77. $2x^{2/3} = 5x^{1/3} + 3$

78. $2x^{1/2} = 5x^{1/4} + 3$

79. $\sqrt{x} + 2\sqrt[4]{x} - 35 = 0$

80. $\sqrt{x} - 5\sqrt[4]{x} + 6 = 0$

81. $3x^{-2} + x^{-1} - 2 = 0$

82. $5x^{-2} - 2x^{-1} - 3 = 0$

In Exercises 83–86, solve the inequality.

83. $(x - 2)(x + 1) > 0$

84. $(a + 5)(a - 1) \geq 0$

85. $(3x + 1)(x - 2) \leq 0$

86. $(y - 6)(2y + 1) < 0$

In Exercises 87–96, solve the inequality and graph the solution set on the number line.

87. $y^2 - 5y + 4 > 0$

88. $a^2 - 13a + 36 \leq 0$

89. $a^2 < 81$

90. $a^2 > 81$

91. $5s^2 - 18s \geq 8$

92. $2a^2 \leq 4a + 30$

93. $\dfrac{x - 3}{x + 2} < 0$

94. $\dfrac{x + 4}{x - 2} > 0$

95. $\dfrac{x - 3}{x + 2} \geq 0$

96. $\dfrac{x + 4}{x - 2} \leq 0$

In Exercises 97–100, solve the inequality.

97. $\dfrac{2x + 1}{x - 3} < 2$

98. $\dfrac{3}{x - 2} \leq 4$

99. $\dfrac{5}{x + 4} \geq 4$

100. $\dfrac{2 - x}{5 + x} < 2$

101. The sum of the square of a number and 4 is 36. What are the numbers?

102. The square of the sum of a number and 4 is 36. What are the numbers?

103. The sum of a number and its reciprocal is $\frac{53}{14}$. Find the number(s).

104. The difference between a number and its reciprocal is $\frac{40}{21}$. Find the number(s).

105. Find the dimensions of a rectangle whose length is twice its width if its area is 50 square feet.

106. Find the dimensions of a rectangle whose length is 2 more than 3 times its width if its area is 85 square inches.

107. A 5″ by 8″ rectangular picture has a frame of uniform width. If the area of the frame is 114 square inches, what is the width of the frame?

108. A 20′ by 30′ rectangular pool is to be bordered by a cement walkway of uniform width. If the area of the walkway is 216 square feet, how wide is the walkway?

109. Find the hypotenuse of a right triangle with legs 5″ and 15″.

110. In a right triangle, the hypotenuse is 30 feet and one leg is 15 feet. Find the length of the other leg.

111. Find the length of the diagonals of a 5" by 4" rectangle.

112. Find the length of the side of a square with a 20" diagonal.

113. An airplane heads into the wind to get to its destination 300 miles away, and returns travelling with the wind to its starting point. If the airplane's speed is 200 mph, what was the rate of the wind if it took $3\frac{1}{8}$ hours for the round trip?

CHAPTER 7   Practice Test

1. Solve the following by either the factoring method or the square root method.

(a) $(3z - 1)(z - 4) = z^2 - 8z + 7$ (b) $3 + \dfrac{5}{x^2} = 4$

2. Solve the following by any method.

(a) $y^2 + 4y - 1 = 0$

(b) $(a - 2)(a + 1) = 3a^2 - 4$

(c) $\dfrac{3}{x - 2} + \dfrac{3x}{x + 2} = \dfrac{66}{x^2 - 4}$

3. Use the discriminant to determine the nature of the roots of $2x^2 - 3x + 5 = 0$.

4. Solve for $r > 0$. $V = 2\pi r^2 h$

5. Solve the following radical equation. $\sqrt{3x} = 2 + \sqrt{x + 4}$

6. Solve the following: $x^{1/2} - x^{1/4} - 20 = 0$

7. Solve the following and graph the solution set:

(a) $x^2 - 5x \geq 36$ (b) $\dfrac{3x - 2}{x - 6} < 0$

8. Find the length of the diagonal of a square with side equal to 9 feet.

9. A boat travels upstream for 15 miles against a current of 3 mph, then it travels downstream for 12 miles with the same current. If the 27-mile trip took $2\frac{1}{4}$ hours, how fast was the boat travelling in still water?

8

Graphing Linear Equations and Inequalities

1-55 odds

8.1

The Rectangular Coordinate System and Graphing Straight Lines

In Chapter 3 we discussed solving first-degree equations and inequalities in one variable. We now turn our attention to first-degree equations and inequalities in *two* variables. In order to "solve" such equations and inequalities we will need to expand our understanding of "solving an equation."

Let's begin by considering a first-degree equation in two variables, such as

$$3x + y = 6$$

What does it mean to have a solution to this equation? A moment's thought reveals that a solution to such an equation consists of a *pair* of numbers. We need an x value and y value in order to have *one* solution to this equation.

For example, the pair of numbers

$$x = 1 \text{ and } y = 3$$

makes the equation true:

$$3x + y = 6 \qquad \textit{Substitute } x = 1, y = 3.$$
$$3(1) + 3 \stackrel{?}{=} 6$$
$$6 \stackrel{\checkmark}{=} 6$$

The pair

$$x = 2 \text{ and } y = 0$$

also satisfies the equation:

$$3x + y = 6 \qquad \textit{Substitute } x = 2, y = 0.$$
$$3(2) + 0 \stackrel{?}{=} 6$$
$$6 \stackrel{\checkmark}{=} 6$$

However, the pair

$$x = 3 \text{ and } y = 1$$

does *not* satisfy the equation:

$$3x + y = 6 \qquad \textit{Substitute } x = 3, y = 1.$$
$$3(3) + 1 \overset{?}{=} 6$$
$$9 + 1 \neq 6$$

Thus, we see that the equation $3x + y = 6$ is neither always true nor always false. Nevertheless, this equation has infinitely many solutions. In fact, we can produce as many solutions to this equation as we like by simply picking *any* value for one of the variables and solving for the other variable. (Remember that a single solution to this equation consists of two numbers.)

For instance, in $3x + y = 6$ we can choose $x = 3$ and solve for y:

$$3x + y = 6 \qquad \textit{Substitute } x = 3.$$
$$3(3) + y = 6$$
$$9 + y = 6$$
$$y = -3 \qquad \textit{Thus, } x = 3 \textit{ and } y = -3 \textit{ is a solution.}$$

Or, we can choose $y = 6$ and solve for x:

$$3x + y = 6 \qquad \textit{Substitute } y = 6.$$
$$3x + 6 = 6$$
$$3x = 0$$
$$x = 0 \qquad \textit{Thus, } x = 0 \textit{ and } y = 6 \textit{ is a solution.}$$

One way to keep track of the solutions to $3x + y = 6$ that we have found thus far is to make a table:

x	y
0	6
1	3
2	0
3	-3

However, there is another way of listing the solutions which we will find more convenient and useful. We use **ordered pair** notation, which tells us to write the pair of numbers $x = 1$ and $y = 3$, for example, as $(1, 3)$. That is, an ordered pair of numbers is of the form (x, y), where the first number (sometimes called the first component or **abscissa**) is the x value and the second number or component (sometimes called the **ordinate**) is the y value. These are called *ordered pairs* for obvious reasons—the order of the numbers matters. As we saw earlier, the pair $(1, 3)$ satisfies the equation $3x + y = 6$, while the pair $(3, 1)$ does not.

As we have already pointed out, we can produce as many solutions as we want to a first-degree equation in two variables by simply choosing a value for one of the variables and solving for the other. How then can we exhibit all the solutions if there are infinitely many of them? Since we cannot list all of them, we need to develop an alternative method for displaying the solution set.

We introduce a two-dimensional coordinate system called a ***rectangular*** or ***Cartesian coordinate system*** (named after the French mathematician and philosopher René Descartes, 1596–1650). It is obtained by taking two number lines, one horizontal and one vertical, perpendicular to each other at their respective origins.

As usual, the horizontal number line (usually called the ***x-axis***) is labeled positive to the right of 0 and negative to the left of 0. The vertical number line (usually called the ***y-axis***) is, by convention, labeled positive above the 0 point and negative below the 0 point. The common 0 point of both axes is called the ***origin***. Our coordinate system is illustrated in Figure 8.1. Usually, but not necessarily, the units of length are the same on both axes.

The *x*- and *y*-axes divide the plane into four parts, called ***quadrants***. These are numbered in a conventional way, as indicated in the figure.

Figure 8.1

Rectangular coordinate system

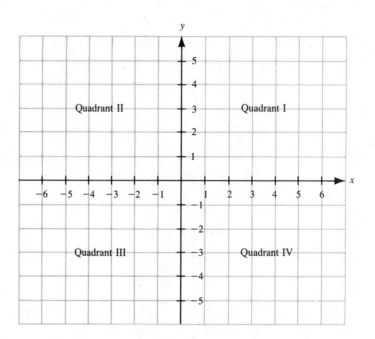

The *x*- and *y*-axes are referred to as the ***coordinate axes***. The first and second members of the ordered pair are often called the *x*-coordinate and *y*-coordinate, respectively. *Note that points **on the coordinate axes** are not considered as being **in any of** the quadrants*.

This coordinate system allows us to associate a point in the plane with each ordered pair (*x, y*).

To plot (graph) the point associated with an ordered pair (*x, y*), we start at the origin and move |*x*| units to the right if *x* is positive, to the left if *x* is negative, and then |*y*| units up if *y* is positive, down if *y* is negative.

The point at which we arrive is the graph of the ordered pair (x, y):

$$(x, \quad y)$$
$$\uparrow \qquad \uparrow$$

Tells you   Tells you
right/left   up/down

EXAMPLE 1

Plot (graph) the points with coordinates (3, 4), (−2, 1), (−3, −5), (2, −3), (5, 0), and (0, 4) on the rectangular coordinate system.

Solution

To graph the point (3, 4) we start at the origin and move 3 units to the right and then 4 units up. (Or, you could first move 4 units up and then 3 units to the right.) In a similar manner, we plot all the points as shown in Figure 8.2.

Figure 8.2
Solution for Example 1

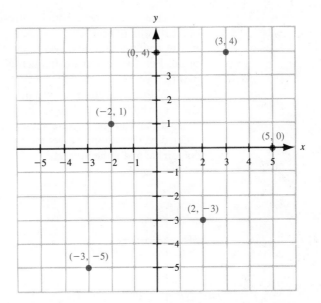

We call particular attention to the points (5, 0) and (0, 4). Notice that the point (5, 0) is on the *x*-axis because its *y-coordinate is* 0; the point (0, 4) is on the *y*-axis because its *x-coordinate is* 0.   ■

Just as every ordered pair is associated with a point, so too is every point associated with an ordered pair.

If we want to see which ordered pair is associated with a specific point *P,* we drop a perpendicular line from *P* to the *x*-axis. The point at which the perpendicular intersects the *x*-axis is the *x*-coordinate of the ordered pair associated with *P*. Similarly, we drop a perpendicular from *P* to the *y*-axis to find the *y*-coordinate of the ordered pair. Figure 8.3 (page 354) illustrates this process.

Figure 8.3

Determining the coordinates
of a point

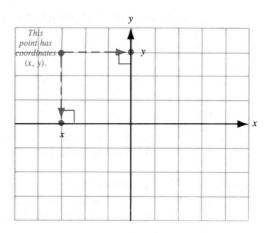

Thus, we have a 1-to-1 correspondence between the points in the plane and ordered pairs of real numbers; that is, every point on the plane is assigned a unique pair of real numbers, and every pair of real numbers is assigned a unique point on the plane. For this reason we frequently say "the point (x, y)" rather than "the ordered pair (x, y)."

With this coordinate system in hand we can now return to the question, "How can we exhibit the solution set of the equation $3x + y = 6$?" We have seen that the ordered pairs in the accompanying table satisfy $3x + y = 6$. If we write them in ordered pair notation, we get the following ordered pairs (remember that x is the first coordinate and y the second):

$$(0, 6) \qquad (1, 3) \qquad (2, 0) \qquad (3, -3)$$

x	y
0	6
1	3
2	0
3	−3

Now we can plot these points, as shown in Figure 8.4. The figure strongly suggests that we "connect the dots" and draw a straight line. In so doing, we are saying two things: first, that every ordered pair which satisfies the equation is a point on the line, and second, that every point on the line gives an ordered pair which satisfies the equation.

We will find the following definition useful.

DEFINITION

The *graph* of an equation is the set of all points whose coordinates satisfy the equation.

Figure 8.4

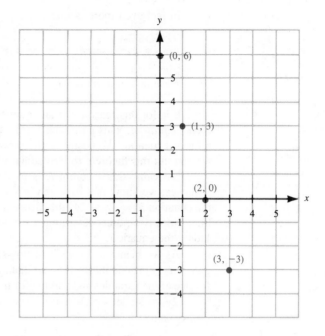

Thus, in drawing the straight line in Figure 8.5 we are saying that the graph of the equation $3x + y = 6$ is the straight line. This graph is now the "picture" of the solution set to the equation $3x + y = 6$.

Figure 8.5
The graph of the equation
$3x + y = 6$

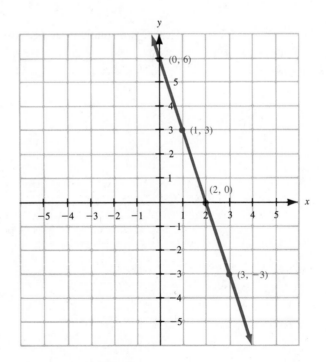

In fact even more is true:

THEOREM

The graph of an equation of the form $Ax + By = C$, where A and B are *not* both equal to 0, is a straight line.

It is for this reason that an equation of the form $Ax + By = C$ is often called a ***linear equation***. This particular form, $Ax + By = C$, is called the ***standard form*** of a linear equation. We postpone the proof of this theorem until Section 8.3; however, we can put this theorem to immediate use.

Basically, this theorem tells us that if we have a first-degree equation in two variables, we know that its graph is a straight line. For example, if we want to graph the equation $3x - 2y = 24$, the theorem tells us that the graph is going to be a straight line. Since two points determine a straight line, all we need find are two points which satisfy the equation.

As we mentioned earlier in this section, we can find as many points—that is, generate as many solutions to this equation—as we please, by simply picking a value for one of the variables and solving for the other variable. For example, for the equation $3x - 2y = 24$, let's pick two values for x and then find the y value associated with each x value.

LET $x = 4$: $3x - 2y = 24$ LET $x = 6$: $3x - 2y = 24$
 $3(4) - 2y = 24$ $3(6) - 2y = 24$
 $12 - 2y = 24$ $18 - 2y = 24$
 $-2y = 12$ $-2y = 6$
 $y = -6$ $y = -3$

Hence, two solutions are $(4, -6)$ and $(6, -3)$.

Figure 8.6
The graph of $3x - 2y = 24$

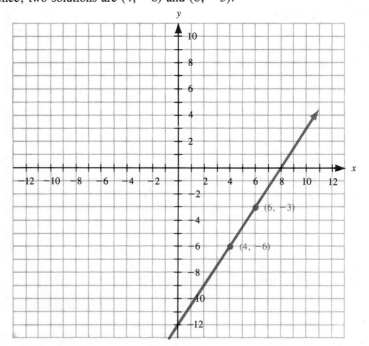

We would plot the two points and draw a line through them, as indicated in Figure 8.6. It is a good idea to check your graph by finding a third solution to the equation and checking to see that it lies on the line.

We see that we can graph a linear equation by arbitrarily finding two points. However, throughout our work in graphing there are certain points we want to pay particular attention to.

DEFINITION

The *x-intercepts* of a graph are the x values of the points where the graph crosses the x-axis. The *y-intercepts* of a graph are the y values of the points where the graph crosses the y-axis.

If we look back at the graph of $3x - 2y = 24$ (Figure 8.6), we can see that the x-intercept is 8 and the y-intercept is -12.

Since the graph crosses the x-axis when $y = 0$ (why?),

the x-intercept of a graph occurs when $y = 0$.

Similarly, since the graph crosses the y-axis when $x = 0$ (why?),

the y-intercept of a graph occurs when $x = 0$.

Whenever possible (and practical), we label the x- and y-intercepts of a graph.

EXAMPLE 2

Sketch the graphs of the following equations and label the intercepts.
(a) $5x - 2y = 20$ **(b)** $-2x + 3y = 15$

Solution

(a) Since the graph of $5x - 2y = 20$ is going to be a straight line (why?), it is sufficient for us to find two points which satisfy the equation. We will find the x- and y-intercepts.

To find the x-intercept, we set $y = 0$ and solve for x:

$$5x - 2y = 20$$
$$5x - 2(0) = 20$$
$$5x = 20$$
$$x = 4 \qquad \text{\textit{The x-intercept is 4.}}$$

Hence, the graph crosses the x-axis at (4, 0).

To find the y-intercept, we set $x = 0$ and solve for y:

$$5x - 2y = 20$$
$$5(0) - 2y = 20$$
$$-2y = 20$$
$$y = -10 \qquad \text{\textit{The y-intercept is} -10.}$$

Hence, the graph crosses the y-axis at (0, -10).

Now we can sketch the graph (see Figure 8.7 on page 358).

Figure 8.7

The graph of $5x - 2y = 20$

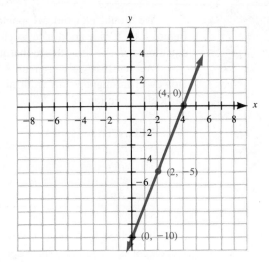

Again, since any two points determine a line, one wrong point will yield a wrong line. Hence, it is a good idea to find a check point just to make sure we have not made an error. We pick some value for x or y and solve for the other.

CHOOSE $x = 2$: $5x - 2y = 20$

$$5(2) - 2y = 20$$

$$10 - 2y = 20$$

$$-2y = 10$$

$$y = -5 \qquad \textit{Our check point is } (2, -5).$$

Looking at our graph (Figure 8.7), we can see that $(2, -5)$ is on the line.

(b)   The graph of $-2x + 3y = 15$ will be a straight line. Thus, we need find only two points—the intercepts:

$$-2x + 3y = 15 \qquad \textit{To find the x-intercept, set } y = 0 \textit{ and solve for x.}$$

$$-2x + 3(0) = 15$$

$$-2x = 15$$

$$x = -\frac{15}{2} \qquad \textit{The x-intercept is } -\frac{15}{2}. \textit{ Hence, the graph crosses the} \\ \textit{x-axis at } (-\frac{15}{2}, 0).$$

$$-2x + 3y = 15 \qquad \textit{To find the y-intercept, set } x = 0 \textit{ and solve for y.}$$

$$-2(0) + 3y = 15$$

$$3y = 15$$

$$y = 5 \qquad \textit{The y-intercept is 5. Hence, the graph crosses the} \\ \textit{y-axis at } (0, 5).$$

Now we can sketch the graph (see Figure 8.8).

Figure 8.8

The graph of
$-2x + 3y = 15$

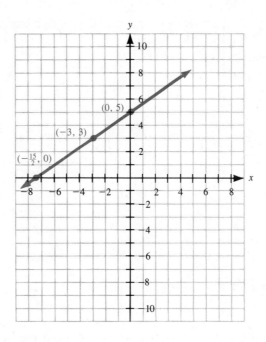

We check with a third point. Letting $y = 3$, we find that $x = -3$. From the graph we can see that $(-3, 3)$ is on the line. ∎

The method we have outlined in Example 2 is called the ***intercept method*** for graphing a straight line. It is usually the preferred method to use. However, there are occasions when it does not quite do the job.

EXAMPLE 3

Sketch the graph of $y = -2x$.

Solution

Again, since this is a first-degree equation in two variables, we know that the graph is going to be a straight line. We will find the intercepts.

To find the x-intercept, we set $y = 0$ and solve for x:

$y = -2x$

$0 = -2x$

$0 = x$ *The x-intercept is $(0, 0)$.*

But $(0, 0)$, as the origin, is on the y-axis as well, and so $(0, 0)$ is also the y-intercept. Since we get only one point from our search for the intercepts, we must find another point on the line.

Again, we simply choose a convenient value for x or y. We choose $x = 1$:

$y = -2x$

$y = -2(1)$

$y = -2$ *Our point on the line is $(1, -2)$.*

Now we can sketch our graph (see Figure 8.9). We leave the check of this graph to the student.

Figure 8.9
The graph of $y = -2x$

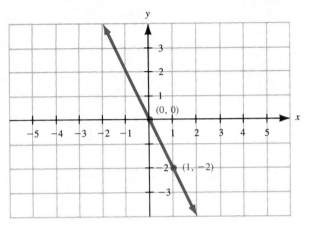

EXAMPLE 4

Sketch the graphs of:

(a) $x = -3$ **(b)** $y = 2$

Solution

First we must keep in mind that we are working in a two-dimensional coordinate system. Recall that we defined the standard form of a first-degree equation in two variables to be $Ax + By = C$, with A and B not both equal to 0.

(a) The equation $x = -3$ is of this form with $A = 1$, $B = 0$, and $C = -3$:

$$Ax + By = C$$
$$1 \cdot x + 0 \cdot y = -3$$

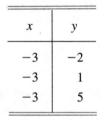

x	y
-3	-2
-3	1
-3	5

Try substituting various values for x and y in the above form and you will find that it does not matter what value we substitute for y, the x value must still be -3, as indicated in the accompanying table.

In order for an ordered pair to satisfy the equation $x = -3$, its x-coordinate must be -3. The equation $x = -3$ places no condition whatsoever on the y-coordinate—the y-coordinate can be anything. Hence, the graph is a vertical line 3 units to the left of the y-axis, as illustrated in Figure 8.10.

Figure 8.10
The graph of $x = -3$

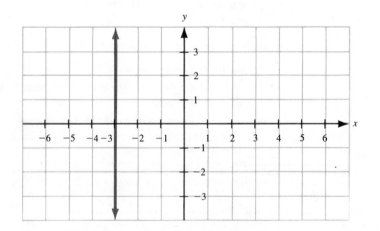

(b) Similarly, we can rewrite the equation $y = 2$ in standard form with $A = 0$, $B = 1$, and $C = 2$:

$$Ax + By = C$$
$$0 \cdot x + 1 \cdot y = 2$$

We can substitute numbers in for x and find that no matter what value is substituted for x, the y value will always be 2.

In order for an ordered pair to satisfy the equation $y = 2$, its y-coordinate must be 2, but the equation places no condition on the x-coordinate. Hence, the graph is a horizontal line 2 units above the x-axis (see Figure 8.11).

Figure 8.11
The graph of $y = 2$

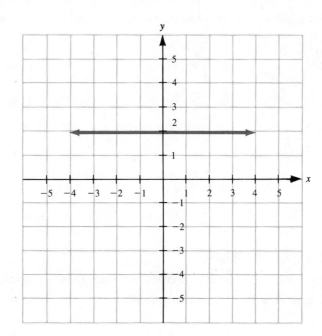

Exercises 8.1

$1-55$ odds

In Exercises 1–8, determine whether the given ordered pair satisfies the equation.

1. $3x - 5y = 17$; $(4, 1)$

2. $3x - 5y = 17$; $(-6, -7)$

3. $4y - 3x = 7$; $(1, -1)$

4. $4y - 3x = 20$; $(2, -4)$

5. $2x + 3y = 2$; $\left(\frac{3}{2}, -\frac{1}{3}\right)$

6. $5x - 4y = 0$; $\left(\frac{1}{10}, \frac{1}{8}\right)$

7. $\frac{2}{3}x - \frac{1}{4}y = 1$; $(6, 12)$

8. $\frac{3}{4}y - \frac{4}{5}x = 4$; $(20, 16)$

In Exercises 9–14, fill in the missing component of the given ordered pairs for the equations:

9. $x + y = 8$: $(-1, \ )$, $(0, \ )$, $(1, \ )$, $(\ , -2)$, $(\ , 0)$, $(\ , 4)$

10. $x - 2y = 5$: $(-3, \ )$, $(-1, \ )$, $(0, \ )$, $(\ , -3)$, $(\ , -1)$, $(\ , 0)$

11. $5x + 4y = 20$: $(-2, \ )$, $(0, \ )$, $(4, \ )$, $(\ , -5)$, $(\ , 0)$, $(\ , 4)$

12. $3x + 7y = 15$: $(-7, \ )$, $(0, \ )$, $(5, \ )$, $(\ , -3)$, $(\ , 0)$, $(\ , 2)$

13. $\dfrac{x}{3} + \dfrac{y}{4} = 1$: $(-3, \quad), (0, \quad), (3, \quad), (\quad, -4), (\quad, 0), (\quad, 4)$

14. $\dfrac{2x}{7} - \dfrac{y}{2} = 2$: $(-7, \quad), (0, \quad), (7, \quad), (\quad, -2), (\quad, 0), (\quad, 2)$

In Exercises 15–28, find the x- and y-intercepts of the graphs of the given equations.

15. $x + y = 6$ 16. $y + x = 5$ 17. $x - y = 6$ 18. $y - x = 5$

19. $y - x = 6$ 20. $x - y = 5$ 21. $2x + 4y = 12$ 22. $3x + 6y = 12$

23. $3y + 4x = 12$ 24. $4y + 6x = 12$ 25. $3x - 5y = 15$ 26. $3y - 5x = 15$

27. $2y - 3x = 7$ 28. $2x - 3y = 8$

In Exercises 29–56, sketch the graphs of the given equations. Label the x- and y-intercepts.

29. $4x + 3y = 0$ 30. $5x - 2y = 0$ 31. $y = x$ 32. $y = -x$

33. $\dfrac{x}{2} - \dfrac{y}{3} = 1$ 34. $\dfrac{y}{3} + \dfrac{x}{5} = 1$ 35. $y = 3x - 1$ 36. $y = -2x + 3$

37. $y = -\dfrac{2}{3}x + 4$ 38. $y = \dfrac{3}{5}x - 6$ 39. $5x - 4y = 20$ 40. $3x + 6y = 18$

41. $5x - 7y = 30$ 42. $3x + 8y = 15$ 43. $5x + 7y = 30$ 44. $3x - 8y = 16$

45. $x = 5$ 46. $y = -3$ 47. $-\dfrac{3}{4}x + y = 2$ 48. $\dfrac{5}{3}x - y = 6$

49. $\dfrac{3}{4}x - y = 2$ 50. $-\dfrac{5}{3}x + y = 6$ 51. $y + 5 = 0$ 52. $x - 7 = 0$

53. $5x - 4y = 0$ 54. $3x + 6y = 0$ 55. $5x - 4 = 0$ 56. $3 + 6y = 0$

? QUESTIONS FOR THOUGHT

57. Describe the x- and y-intercepts of a graph both geometrically and algebraically.

58. Sketch the graph of $2s + 3t = 6$ on a set of coordinate axes with t being the horizontal axis and s the vertical axis.

59. Repeat Exercise 58 with the axes reversed. Does reversing the labeling of the axes affect the graph? How?

60. Although we usually use the same scale on both the horizontal and vertical axes, sometimes it is necessary (or more convenient) to use different scales for each axis. Use appropriate scales on the x- and y-axes to sketch a graph of the following:

    (a) $y = 200x + 400$ **(b)** $y = 0.04x$

 MINI-REVIEW

61. *Factor completely.* $2x^3 + 2x^2 - 24x$

62. *Perform the operations and express your answer in simplest form:*
$$\dfrac{x^2 - 1}{x^2 - 4} + x - 1$$

63. *Express as a single fraction with positive exponents only.* $(x + y^{-2})^{-2}$

64. *Simplify:*
$$\sqrt[3]{\dfrac{27x^5y^8}{2x^7y^2}}$$

65. *Solve for x.* $2x^2 + 2x = 5$

66. The sum of two numbers is 23. Find the numbers if their product is 132.

8.2

The Slope

In the last section we discussed the question of "solving" a first-degree equation in two variables. We saw that a first-degree equation in two variables can be viewed as a condition on the two variables. If we let the variables x and y represent the first and second coordinates of points in a rectangular coordinate system, then certain "points" satisfy this condition and others do not. The set of all points which satisfy this condition—that is, the graph of a first-degree equation in two variables—is a straight line.

What if we now reverse the situation? Suppose we are given the graph of a straight line and we want to produce an equation for this straight line. Keeping in mind that an equation of a line is the condition that the points on the line must satisfy, we must naturally ask what condition must the points on the given line satisfy.

Look at Figure 8.12. From basic geometry we find that triangles ABC and DEF are similar triangles. ($\angle BAC \cong \angle EDF$ since AC and DF are parallel and similarly, $\angle ABC \cong \angle DEF$ since BC and EF are parallel. So corresponding angles are equal.) Therefore, their corresponding sides are in proportion. That is,

$$\frac{|BC|}{|AC|} = \frac{|EF|}{|DF|} \qquad [\textit{Note:} \quad |BC| \text{ means the length of line segment } BC.]$$

Figure 8.12

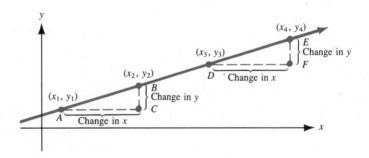

In other words, as we move from any point on a nonvertical line to any other point on the line, the ratio of the change in the y-coordinates of the points to the change in the x-coordinates of the points is constant for each line. This fact is what we are looking for—a condition which all points on a line must satisfy.

The remainder of this section is devoted to further development of this idea. In the next section we will return to answer the question raised earlier about how to obtain an equation for a line when we have its graph.

We define the following:

DEFINITION

Let $P_1(x_1, y_1)$ and $P_2(x_2, y_2)$ be any two points on a nonvertical line L. The **slope** of the line L, denoted by m, is given by

$$m = \frac{y_2 - y_1}{x_2 - x_1} \qquad (x_1 \neq x_2)$$

Note that this definition uses what we saw in Figure 8.12, that the ratio of the change in y to the change in x is *independent* of the points chosen; that is, for any two points on a particular line, the ratio will remain the same.

For example, let's see how we would find the slope of the line passing through the points $(2, -3)$ and $(8, 1)$. Although it is not always necessary, it is usually helpful to draw a diagram with the given information (see Figure 8.13).

Figure 8.13

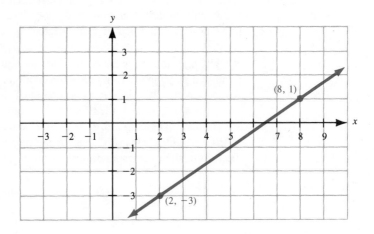

The formula for the slope of a line is

$$m = \frac{y_2 - y_1}{x_2 - x_1}$$

but which of our points is (x_1, y_1) and which is (x_2, y_2)? The fact of the matter is that it does not make any difference which we call the "first" point and which we call the "second" point, so long as we are consistent for both the x- and y-coordinates.

We can let $P_1(x_1, y_1) = (2, -3)$ and $P_2(x_2, y_2) = (8, 1)$ and we get

$$m = \frac{y_2 - y_1}{x_2 - x_1} = \frac{1 - (-3)}{8 - 2} = \frac{4}{6} = \frac{2}{3}$$

or we can let $P_1(x_1, y_1) = (8, 1)$ and $P_2(x_2, y_2) = (2, -3)$ and we get

$$m = \frac{y_2 - y_1}{x_2 - x_1} = \frac{-3 - 1}{2 - 8} = \frac{-4}{-6} = \frac{2}{3}$$

Thus, the slope is $m = \frac{2}{3}$.

EXAMPLE 1

Find the slope of the line passing through the given pair of points.

(a) $P(3, 8)$ and $Q(7, 2)$

(b) $P(3, -5)$ and $Q(-1, -9)$

Solution **(a)** Let $(x_1, y_1) = (3, 8)$ and $(x_2, y_2) = (7, 2)$ and we get

$$m = \frac{y_2 - y_1}{x_2 - x_1} = \frac{2 - 8}{7 - 3} = \frac{-6}{4} = -\frac{3}{2}$$

Therefore, the slope is $\boxed{-\dfrac{3}{2}}$

(b) Letting $(x_1, y_1) = (3, -5)$ and $(x_2, y_2) = (-1, -9)$, we get

$$m = \frac{y_2 - y_1}{x_2 - x_1} = \frac{-9 - (-5)}{-1 - 3} = \frac{-4}{-4} = 1$$

Therefore, the slope is $\boxed{1}$ ∎

EXAMPLE 2 Find the slope of the line with the equation $y = 3x - 4$.

Solution In the next section, we will find a quick way to arrive at the solution to this problem. But for now, in order to find the slope of a line, we need two points on the line. Since the slope is independent of the points chosen, we can arbitrarily choose any two points which satisfy the equation and therefore lie on the line.

Let $x = 0$. Then $y = 3(0) - 4 = -4$. Thus, $(0, -4)$ is one point on the line.

Let $x = 3$. Then $y = 3(3) - 4 = 5$. Thus, $(3, 5)$ is another point on the line.

Letting $(x_1, y_1) = (0, -4)$ and $(x_2, y_2) = (3, 5)$, we have

$$m = \frac{y_2 - y_1}{x_2 - x_1} = \frac{5 - (-4)}{3 - 0} = \frac{9}{3} = 3$$

Hence, the slope of the line $y = 3x - 4$ is $\boxed{3}$

If we choose two other points satisfying the equation, such as $(1, -1)$ and $(5, 11)$ (check that these points do satisfy the equation), we will still get the same slope since the slope of the line is independent of the points chosen:

$$m = \frac{11 - (-1)}{5 - 1} = \frac{12}{4} = \boxed{3}$$ ∎

It is important to note that in the definition of the slope of a line, we have specified that the line be nonvertical. The reason for this is that for a vertical line, all x-coordinates are the same; we are forced to divide by 0 in the computation of the slope.

For example, if we try to compute the slope of the vertical line passing through the points $(1, -4)$ and $(1, 2)$ (see Figure 8.14), we get

$$m = \frac{2 - (-4)}{1 - 1} = \frac{6}{0} \qquad \text{which is undefined}$$

Figure 8.14

The vertical line through $(1, -4)$ and $(1, 2)$

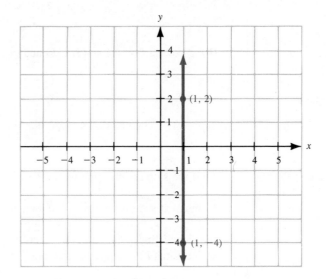

Thus, we find

Vertical lines have undefined slope.

Let us now examine what this number, the slope, tells us about a line.

Whenever we describe a graph, we describe it for increasing values of x; that is, moving from left to right.

The line in Figure 8.15**a** is rising (as we move from left to right), while the line in Figure 8.15**b** is falling (as we move from left to right).

Figure 8.15

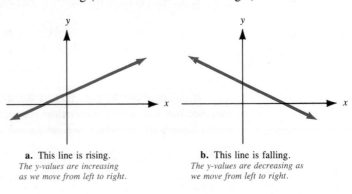

a. This line is rising.
The y-values are increasing as we move from left to right.

b. This line is falling.
The y-values are decreasing as we move from left to right.

The slope of a line is a number which tells us the *rate* at which its *y* values are increasing or decreasing. In other words, it is a measure of the steepness of a line. For example, if the slope of a line is $\frac{2}{5}$, this tells us that

$$m = \frac{2}{5} = \frac{\text{Change in } y}{\text{Change in } x}$$

This means that a 5-unit change in *x* gives a 2-unit change in *y*. Therefore, the line has the *steepness* shown in Figure 8.16.

Figure 8.16
A line whose slope is $\frac{2}{5}$

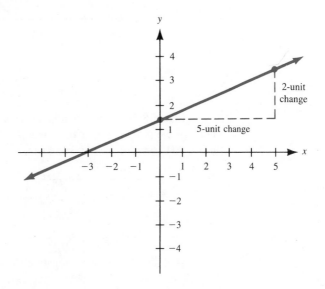

A line whose slope is $\frac{5}{2}$ has the steepness indicated in Figure 8.17.

Figure 8.17
A line with slope $\frac{5}{2}$

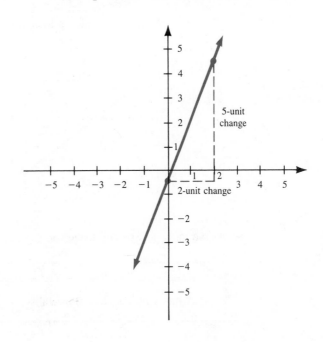

Figure 8.18

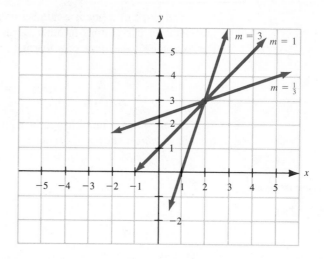

Figure 8.18 shows what happens when we vary the slope of a rising line passing through the point (2, 3). Notice that the greater the slope, the steeper the line. As we move **left to right** we can see that

A line with positive slope rises.

What about lines with negative slope? Let's look at a line with a slope of $-\frac{3}{4}$.

We can view $m = -\frac{3}{4}$ as $\frac{-3}{4}$ or $\frac{3}{-4}$.

Thus, we can draw the line as indicated in Figure 8.19**a** or Figure 8.19**b**. In both cases we get the same steepness.

Figure 8.19
Line passing through the point (3, 1) with slope $-\frac{3}{4}$

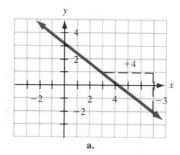

a.

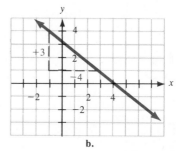

b.

As we move **left to right**, we can see that

A line with negative slope falls.

Figure 8.20 illustrates that the greater the *absolute value* of the slope, the steeper the line.

Figure 8.20

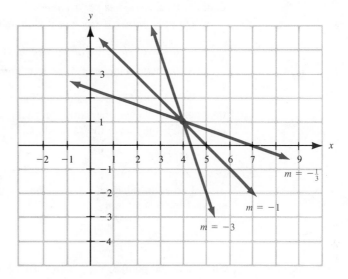

EXAMPLE 3

What is the slope of the line through the points (1, 4) and (3, 4)?

Solution

$$m = \frac{4 - 4}{3 - 1} = \frac{0}{2} = 0$$

This line is horizontal, as shown in Figure 8.21.

Figure 8.21

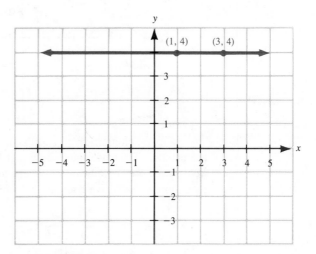

A horizontal line has zero slope.

Any horizontal line, for which the y-coordinates will all be equal, will have slope equal to 0. It is perfectly reasonable that a horizontal line should have a slope equal to 0, since it has *no steepness*.

Do not confuse a slope of 0 *with undefined slope.*

A line with slope 0 *has no steepness (it is horizontal).*

A line with undefined slope has "infinite" steepness (it is vertical).

Based on our previous discussion, we can conclude that since lines which have the same slope have the same steepness, they are parallel. Conversely, if two lines are parallel, they have the same slope.

THEOREM

Two distinct lines, L_1 and L_2, with slopes m_1 and m_2, respectively, are parallel if and only if $m_1 = m_2$.

EXAMPLE 4

Show that the points $P(2, 4)$, $Q(8, 0)$, $R(3, -3)$, and $S(-3, 1)$ form the vertices of a parallelogram.

Solution

We first plot the given points as shown in Figure 8.22.

Figure 8.22

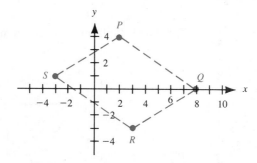

A parallelogram is a quadrilateral (a four-sided polygon) whose opposite sides are parallel. Thus, we must show that PQ is parallel to SR and that SP is parallel to RQ. By the previous theorem, we can find out if those sides are parallel by examining their slopes. Thus, we compute the slopes of each side:

$$m_{PQ} = \frac{4 - 0}{2 - 8} = \frac{4}{-6} = -\frac{2}{3} \qquad m_{SR} = \frac{1 - (-3)}{-3 - 3} = \frac{4}{-6} = -\frac{2}{3}$$

$$m_{SP} = \frac{1 - 4}{-3 - 2} = \frac{-3}{-5} = \frac{3}{5} \qquad m_{RQ} = \frac{-3 - 0}{3 - 8} = \frac{-3}{-5} = \frac{3}{5}$$

Since the slopes of the opposite sides are the same, the opposite sides are parallel and therefore the figure is a parallelogram. ∎

If two lines pass through a single point and have the same slope, then they must be the same line. We can restate this as follows:

A point and a slope determine a line.

Thus, just as we can draw a line by knowing two points, we should also be able to draw a line given one point and the slope of the line.

EXAMPLE 5

(a) Graph the line passing through the point (3, 1) with slope $= \frac{2}{5}$.

(b) Graph the line passing through the point $(-2, 1)$ with slope $= -4$.

Solution

(a) First we plot the point (3, 1). Since the line must have slope $m = \frac{2}{5}$, this means that

$$m = \frac{\text{Change in } y}{\text{Change in } x} = \frac{2}{5}$$

or that for every 5 units we move off the line to the right (change in x equal to $+5$), we must move 2 units up (change in y equal to $+2$) in order to return to the line. Hence, we start at (3, 1), count 5 units right and then 2 units up, and the point where we arrive is another point on the line. We draw a line through the two points, as illustrated in Figure 8.23.

Figure 8.23

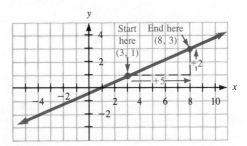

Notice that if we decided to move 10 units right (change in x equal to $+10$), then we would have to move 4 units up (change in y equal to $+4$) in order to make the ratio, $m = \frac{4}{10} = \frac{2}{5}$. On the other hand, if we moved 5 units *left* (change in x equal to -5), then we would have to move 2 units *down* (change in x equal to -2) in order to keep $m = \frac{-2}{-5} = \frac{2}{5}$.

(b) To graph the line passing through $(-2, 1)$ with slope $= -4$, we start by plotting the point $(-2, 1)$. To find the next point, we note that since the slope, m, is -4, we can rewrite -4 as:

$$\frac{-4}{1} = \frac{\text{Change in } y}{\text{Change in } x}$$

Therefore, for every 1 unit we travel right of the line (change in x equal to $+1$), we must travel *down* 4 units (change in y equal to -4) before we find another point on the line. Thus, we start at $(-2, 1)$, move 1 unit right and 4 units down to find another point, and then draw a line through the two points (see Figure 8.24).

Figure 8.24

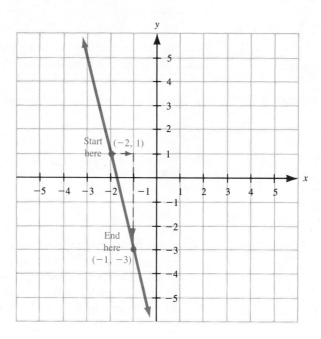

Example 5 demonstrated that given a point and a slope, we can geometrically determine the line. In the next section we will find that given the same information, we can (algebraically) determine the equation of the line.

We have already discussed parallel lines. What about the slopes of perpendicular lines?

EXAMPLE 6

Sketch the lines with slopes $\frac{6}{7}$ and $-\frac{7}{6}$ which pass through the point $(10, 4)$.

Solution

$m = \dfrac{6}{7}$ means a 7-unit change (increase) in x produces a 6-unit change (increase) in y.

$m = -\dfrac{7}{6}$ means a -6-unit change (decrease) in x produces a 7-unit change (increase) in y.

Let's sketch two lines with slopes $\frac{6}{7}$ and $-\frac{7}{6}$ (see Figure 8.25).

Figure 8.25

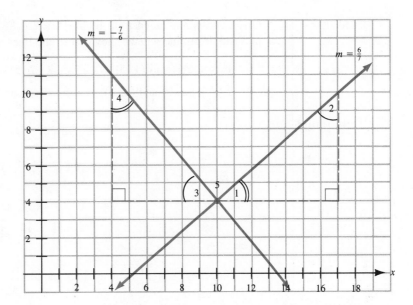

On the figure we have drawn in two triangles to help us visualize the following:

We know that $\angle 1 + \angle 2 = 90°$.

Since these two triangles are congruent (compare the lengths of their sides), we also know that $\angle 2 = \angle 3$ (they are both opposite the side of the triangle whose length is 7).

Therefore, $\angle 1 + \angle 3 = 90°$.

But then $\angle 5$ must also be 90° (because $\angle 1 + \angle 3 + \angle 5 = 180°$) and so the two lines are perpendicular.

Note that $\left(\dfrac{6}{7}\right)\left(-\dfrac{7}{6}\right) = -1$.

Example 6 is a particular case of the following theorem.

THEOREM

Two nonvertical lines L_1 and L_2 with slopes m_1 and m_2, respectively, are perpendicular if and only if $m_1 \cdot m_2 = -1$.

If $m_1 \cdot m_2 = -1$, we can write $m_2 = -\dfrac{1}{m_1}$ or $m_1 = -\dfrac{1}{m_2}$.
For this reason we often say that the slopes of nonvertical perpendicular lines are "negative reciprocals of each other."

We should note that throughout the previous discussion we have insisted on nonvertical lines. While it is true that a horizontal line and a vertical line are perpendicular, a vertical line has undefined slope and so is not covered by the last theorem.

EXAMPLE 7

Given two points on each of the lines L_1, L_2, L_3, and L_4, compute the slope of each line and determine whether any two of the lines are parallel or perpendicular.

$$L_1: \quad (1, 3) \text{ and } (4, 5) \qquad L_2: \quad (-2, 5) \text{ and } (0, 2)$$

$$L_3: \quad (-3, 0) \text{ and } (0, 2) \qquad L_4: \quad (2, 1) \text{ and } (1, 3)$$

Solution

$$m_1 = \frac{5 - 3}{4 - 1} = \frac{2}{3}$$

$$m_2 = \frac{2 - 5}{0 - (-2)} = \frac{-3}{2} = -\frac{3}{2}$$

$$m_3 = \frac{2 - 0}{0 - (-3)} = \frac{2}{3}$$

$$m_4 = \frac{3 - 1}{1 - 2} = \frac{2}{-1} = -2$$

Since $m_1 = m_3$, $\boxed{L_1 \text{ and } L_3 \text{ are parallel.}}$

Since $m_2 = -\dfrac{1}{m_1}$ and $m_2 = -\dfrac{1}{m_3}$, $\boxed{L_2 \text{ is perpendicular to } L_1 \text{ and } L_3.}$ ∎

EXAMPLE 8

Find a value for c so that the line passing through the points $(3, 4)$ and $(-1, c)$ has slope $\frac{1}{2}$.

Solution

Using the formula for the slope of a line, we determine that the slope is

$$\frac{c - 4}{-1 - 3} = \frac{c - 4}{-4}$$

and we want this to be equal to $\frac{1}{2}$. Our equation is

$$\frac{c - 4}{-4} = \frac{1}{2} \qquad \textit{Multiply both sides by } -4.$$

$$\frac{-4}{1} \cdot \frac{c - 4}{-4} = \frac{1}{2} \cdot \frac{-4}{1}$$

$$c - 4 = -2$$

$$\boxed{c = 2}$$

CHECK: The slope of the line passing through $(3, 4)$ and $(-1, 2)$ is

$$\frac{4 - 2}{3 - (-1)} = \frac{2}{4} \overset{\checkmark}{=} \frac{1}{2}$$ ∎

EXAMPLE 9

Find the value(s) of d so that the line L_1 passing through the points $(0, 2)$ and $(3, d)$ is perpendicular to the line L_2 passing through the points $(0, 3)$ and $(-2, d)$.

Solution

$$m_1 = \frac{d - 2}{3 - 0} = \frac{d - 2}{3} \qquad m_2 = \frac{d - 3}{-2 - 0} = \frac{d - 3}{-2}$$

If we want L_1 to be perpendicular to L_2, we require $m_1 = -\dfrac{1}{m_2}$. Thus, we want

$$\frac{d-2}{3} = \frac{2}{d-3} \qquad \frac{2}{d-3} \text{ is the negative reciprocal of } \frac{d-3}{-2}.$$

We clear the denominators by multiplying both sides of the equation by $3(d-3)$.

$$\frac{\cancel{3}(d-3)}{1} \cdot \frac{d-2}{\cancel{3}} = \frac{2}{\cancel{d-3}} \cdot \frac{3\cancel{(d-3)}}{1}$$

$$(d-3)(d-2) = 6$$

$$d^2 - 5d + 6 = 6$$

$$d^2 - 5d = 0 \qquad \textit{This is a quadratic equation which we can solve}$$

$$d(d-5) = 0 \qquad \textit{by factoring.}$$

$$\boxed{d=0} \quad \text{or} \quad d-5 = 0 \rightarrow \boxed{d=5}$$

Let's check the answers:

If $d = 0$, then $m_1 = \dfrac{0-2}{3} = -\dfrac{2}{3}$ and $m_2 = \dfrac{0-3}{-2} = \dfrac{-3}{-2} = \dfrac{3}{2}$.

These are negative reciprocals and so L_1 is perpendicular to L_2.

If $d = 5$, then $m_1 = \dfrac{5-2}{3} = \dfrac{3}{3} = 1$ and $m_2 = \dfrac{5-3}{-2} = \dfrac{2}{-2} = -1$.

And again L_1 and L_2 are perpendicular. ∎

Exercises 8.2

In Exercises 1–6, sketch the line through the given pair of points and compute its slope.

1. $(1, -2)$ and $(-3, 1)$

2. $(2, -3)$ and $(-1, 4)$

3. $(0, 2)$ and $(2, 0)$

4. $(5, 0)$ and $(0, 5)$

5. $(-3, -4)$ and $(-2, -5)$

6. $(-4, -3)$ and $(-2, -1)$

In Exercises 7–14, compute the slope of the line passing through the given pair of points.

7. $(2, 4)$ and $(-3, 4)$

8. $(1, -5)$ and $(1, 3)$

9. $(4, 2)$ and $(4, -3)$

10. $(-5, 1)$ and $(3, 1)$

11. (a, a) and (b, b) $(a \neq b)$

12. (a, b) and (b, a) $(a \neq b)$

13. (a, a^2) and (b, b^2) $(a \neq b)$

14. (a, b^2) and (b, a^2) $(a \neq b)$

15. Find the slope of the line which crosses the x-axis at $x = 5$ and the y-axis at $y = -3$.

16. Find the slope of the line which crosses the y-axis at $y = 7$ and the x-axis at $x = -4$.

In Exercises 17–28, sketch the graph of the line L which contains the given point and has the given slope m.

17. $(1, 3)$, $m = 2$

18. $(2, 5)$, $m = 1$

19. $(1, 3)$, $m = -2$

20. $(2, 5)$, $m = -1$

21. $(0, 3)$, $m = -\dfrac{1}{4}$

22. $(0, 3)$, $m = -4$

23. $(4, 0), \quad m = \dfrac{2}{5}$

24. $(-4, 0), \quad m = -\dfrac{2}{5}$

25. $(2, 5), \quad m = 0$

26. $(-1, 3), \quad$ undefined slope

27. $(2, 5), \quad$ undefined slope

28. $(-1, 3), \quad m = 0$

In Exercises 29–36, determine whether the line passing through the points P_1 and P_2 is parallel to or perpendicular to (or neither) the line passing through the points P_3 and P_4.

29. $P_1(1, 2), \quad P_2(3, 4), \quad P_3(-1, -2), \quad P_4(-3, -4)$

30. $P_1(5, 6), \quad P_2(7, 8), \quad P_3(-5, -6), \quad P_4(-7, -8)$

31. $P_1(0, 4), \quad P_2(-1, 2), \quad P_3(-3, 5), \quad P_4(1, 7)$

32. $P_1(2, 3), \quad P_2(3, 0), \quad P_3(-2, -5), \quad P_4(1, -6)$

33. $P_1(2, -3), \quad P_2(-3, 0), \quad P_3(0, 2), \quad P_4(3, 7)$

34. $P_1(-2, -1), \quad P_2(0, 11), \quad P_3(2, 3), \quad P_4(8, 2)$

35. $P_1(3, 5), \quad P_2(-2, 5), \quad P_3(1, 4), \quad P_4(1, -2)$

36. $P_1(a, b), \quad P_2(b, a), \quad P_3(c, d),$
$P_4(-d, -c) \quad (a \neq b, c \neq d)$

37. Find a number h so that the line passing through the points $(4, -2)$ and $(1, h)$ has slope -5.

38. Find a number k so that the line passing through the points $(k, 2)$ and $(-3, -1)$ has slope 4.

39. Find the value(s) of n so that the line passing through the points $(1, n)$ and $(3, n^2)$ is parallel to the line passing through the points $(-6, 0)$ and $(-5, 6)$.

40. Find the value(s) of a so that the line passing through the points $(-1, a)$ and $(2, a^2)$ is perpendicular to the line passing through the points $(0, 2)$ and $(2, 1)$.

41. Find the value(s) of c so that the line passing through the points $(0, 1)$ and (c, c) is perpendicular to the line passing through the points $(0, 2)$ and (c, c).

42. Find the value(s) of t so that the line passing through the points $(-2, t)$ and $(-t, 1)$ is parallel to the line passing through the points $(-4, t)$ and $(-t, 3)$.

43. Using slopes, show that the points $(0, 0)$, $(2, 1)$, $(-2, 5)$, and $(0, 6)$ are the vertices of a parallelogram.

44. Using slopes, show that the points $(-3, 1)$, $(-7, 4)$, $(0, 5)$, and $(-4, 8)$ are the vertices of a rectangle.

45. Using slopes, show that the points $(-3, 2)$, $(-1, 6)$, and $(3, 4)$ are the vertices of a right triangle.

46. Using slopes, show that the points $(1, 1)$, $(3, 5)$, $(-1, 6)$, and $(-5, -2)$ are the vertices of a trapezoid.

? QUESTIONS FOR THOUGHT

47. How could you use the idea of slope to determine whether the three points $(-2, -1)$, $(0, 4)$, and $(2, 9)$ all lie on the same line (are collinear)?

48. How would you describe a line with positive slope? Negative slope? Zero slope? Undefined slope?

49. Suppose we do not insist on the units along the x- and y-axes being the same. Sketch the graph of the line with slope 3 passing through the point $(1, 2)$ if the units on the x-axis are twice as large as the units on the y-axis.

50. Repeat Exercise 49 if the units on the y-axis are twice as large as the units on the x-axis.

 MINI-REVIEW

51. *Evaluate.* $(64)^{-5/6}$

52. *Perform operations and express your answer in simplest form:*
$$\sqrt{27} - 5\sqrt{3} + 4\sqrt{12}$$

53. *Perform the operations and express your answer in simplest form:*
$$\dfrac{3\sqrt{5}}{\sqrt{10} - 1}$$

54. *Solve for x.* $\quad x^2 + x - 6 = 24$

55. Find the length of the diagonals of a $12''$ by $10''$ rectangle.

8.3

Equations of a Line

In our previous discussions we have stressed the idea that a first-degree equation in two variables is a *condition* that an ordered pair must satisfy. We have seen that the set of all points which satisfy such an equation is a straight line.

We are now prepared to reverse the question. Suppose we "specify a line." How can we find its equation? In other words, what condition must all the points on the line satisfy?

To answer this question we must first understand what is meant by the phrase "specify a line." We know that a line is determined by two points, but we also saw in the last section that, for a nonvertical line, if we know one point on the line and its slope, we can find as many additional points on the line as we choose. Therefore, equivalently we can say that to specify a nonvertical line we specify a point on the line and its slope.

Let us suppose we have a line passing through a given point (x_1, y_1) with slope m (see Figure 8.26). In order to determine whether another point (x, y) is on this line, we must check whether the points (x_1, y_1) and (x, y) give us the required slope of m.

If $\dfrac{y - y_1}{x - x_1} = m$, then (x, y) is on the line.

If $\dfrac{y - y_1}{x - x_1} \neq m$, then (x, y) is not on the line.

Figure 8.26

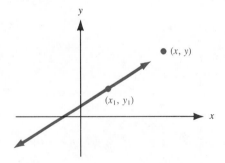

Thus, the equation

$$\frac{y - y_1}{x - x_1} = m$$

is the condition which a point (x, y) must satisfy in order for it to be on this line.

If we multiply both sides of the equation by $x - x_1$, we get the relationship given in the box.

Point–Slope Form of the Equation of a Straight Line

An equation of the line with slope m passing through (x_1, y_1) is

$$y - y_1 = m(x - x_1)$$

EXAMPLE 1

Write an equation of the line with slope $\frac{2}{3}$ which passes through $(-3, 4)$.

Solution

The given point $(-3, 4)$ corresponds to (x_1, y_1), and $m = \frac{2}{3}$.

$$y - y_1 = m(x - x_1)$$

$$y - 4 = \frac{2}{3}[x - (-3)]$$

$$\boxed{y - 4 = \frac{2}{3}(x + 3)}$$

This is the equation of the line shown in Figure 8.27.

Figure 8.27
Line with slope $\frac{2}{3}$ passing
through $(-3, 4)$

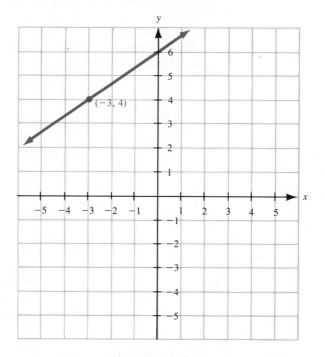

The point–slope form allows us to write an equation of a nonvertical line if we know *any* point on the line and its slope. What if the given point happens to be where the graph of the line crosses the *y*-axis? That is, suppose the line has slope *m* and *y*-intercept equal to *b*. Then the graph passes through the point $(0, b)$. Using the point–slope form, we can write its equation as

$$y - y_1 = m(x - x_1)$$
$$y - b = m(x - 0)$$
$$y - b = mx$$
$$y = mx + b$$

This last form is called the ***slope–intercept form*** of the equation of a straight line.

Slope–Intercept Form of the Equation of a Straight Line	An equation of the line with slope m and y-intercept b is $$y = mx + b$$

EXAMPLE 2

Write an equation of the line with slope 4 and y-intercept -3, and graph the equation.

Solution

Since we are given the slope and y-intercept, it seems appropriate to use the slope–intercept form.

$y = mx + b$ *We are given $m = 4$ and $b = -3$.*

$y = 4x + (-3)$

$\boxed{y = 4x - 3}$

We could also use the point–slope form with $m = 4$ and $(x_1, y_1) = (0, -3)$.

$$y - y_1 = m(x - x_1)$$
$$y - (-3) = 4(x - 0)$$
$$y + 3 = 4x$$
$$\boxed{y = 4x - 3}$$

We graph the equation as we graphed equations in the last section, given the slope and a point. Since the point is the y-intercept, we start at the y-axis at -3. Since the slope $= 4 = \frac{4}{1}$, we move right 1 unit and up 4 units to find another point on the line and then draw the line (see Figure 8.28).

Figure 8.28

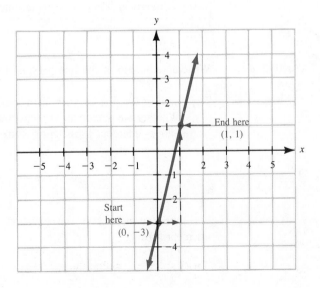

EXAMPLE 3

Find the slopes of the lines with the following equations:

(a) $y = -3x + 5$ **(b)** $3y + 5x = 8$

Solution

(a)   Perhaps the most useful feature of the slope–intercept form is the fact that when the equation of a straight line is written in this form, it is easy to "read off" the slope (as well as the y-intercept):

$$y = -3x + 5$$
$$y = \ \ mx + b$$

Therefore, the slope is $\boxed{-3}$

(b)   In order to "read off" the slope of a line from its equation, the equation must be *exactly* in slope–intercept form. That is, the equation must be in the form $y = mx + b$.

$$3y + 5x = 8 \qquad \textit{We solve for } y.$$
$$3y = -5x + 8$$
$$y = \frac{-5}{3}x + \frac{8}{3}$$

Therefore, the slope is $\boxed{-\dfrac{5}{3}}$ ■

In Section 8.1 we said that the graph of a first-degree equation in two variables is a straight line. On the other hand,

---

Any nonvertical line can be represented by an equation of the form $y = mx + b$.

EXAMPLE 4

Write an equation of the line passing through the points $(1, -2)$ and $(-3, -5)$.

Solution

The graph of the line appears in Figure 8.29.
    Whether we decide to write an equation for this line using the point–slope form or the slope–intercept form, either form requires us to know the slope of the line:

$$m = \frac{-2 - (-5)}{1 - (-3)} = \frac{3}{4}$$

We can now write an equation using the point–slope form *or* the slope–intercept form.

Figure 8.29

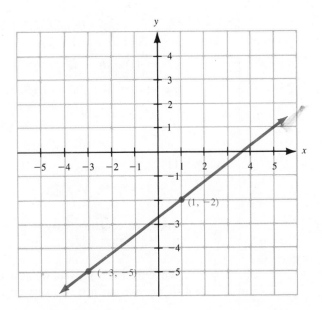

Using point–slope form: If we choose to use the point–slope form, we have another choice to make. We can choose to use either $(1, -2)$ or $(-3, -5)$ as our given point (x_1, y_1).

USING $(1, -2)$: $y - (-2) = \dfrac{3}{4}(x - 1)$

$$\boxed{y + 2 = \dfrac{3}{4}(x - 1)}$$

USING $(-3, -5)$: $y - (-5) = \dfrac{3}{4}[x - (-3)]$

$$\boxed{y + 5 = \dfrac{3}{4}(x + 3)}$$

As we shall see in a moment, these two equations are in fact equivalent, although they may look different.

Using slope–intercept form: The slope–intercept form is $y = mx + b$. Since we know $m = \dfrac{3}{4}$, we can write

$$y = \dfrac{3}{4}x + b$$

We do not yet know the value of b since we do not know the y-intercept. However, we do know that the points $(1, -2)$ and $(-3, -5)$ are on the line and so must satisfy the equation. Therefore, if we substitute one of these points, say $(1, -2)$, into the equation $y = \dfrac{3}{4}x + b$, we can solve for b.

$$y = \frac{3}{4}x + b \qquad \textit{Substitute } (1, -2).$$

$$-2 = \frac{3}{4}(1) + b$$

$$-2 = \frac{3}{4} + b$$

$$-2 - \frac{3}{4} = b$$

$$-\frac{11}{4} = b$$

Thus, our equation is $\boxed{y = \frac{3}{4}x - \frac{11}{4}}$

While the three answers we have obtained look different, they are in fact equivalent. If we take our first two answers and solve for y we get:

First answer	*Second answer*
$y + 2 = \frac{3}{4}(x - 1)$	$y + 5 = \frac{3}{4}(x + 3)$
$y + 2 = \frac{3}{4}x - \frac{3}{4}$	$y + 5 = \frac{3}{4}x + \frac{9}{4}$
$y = \frac{3}{4}x - \frac{3}{4} - 2$	$y = \frac{3}{4}x + \frac{9}{4} - 5$
$y = \frac{3}{4}x - \frac{11}{4}$	$y = \frac{3}{4}x - \frac{11}{4}$

and so all three answers are equivalent. ∎

The standard form for the equation obtained in Example 4 would be either $-3x + 4y = -11$ or $3x - 4y = 11$. Using the slope–intercept form in the example required the most work, but it had the advantage of giving us a uniform answer.

Unless there are instructions to the contrary, you may use whichever form *you* find most convenient. As the last example illustrates, anytime you are asked to write an equation of a nonvertical line, you may use either the point–slope or slope–intercept form. The given information in each problem will determine which form is the easier to use.

EXAMPLE 5

Write an equation of the line passing through the point $(2, 3)$ which is parallel to the line $3x - 6y = 12$.

Solution

Figure 8.30 illustrates what the example is asking. We are being asked to find an equation of the dashed line.

Figure 8.30

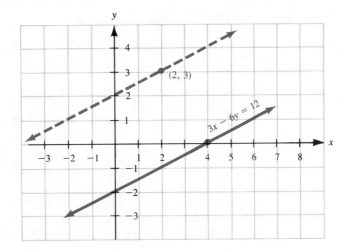

In order to write an equation of a line, it is sufficient to know a point on the line [we are given the point (2, 3)] and the slope of the line. (Then we can use the point–slope form.) Since our line is to be parallel to $3x - 6y = 12$, it must have the same slope. Thus, the entire problem boils down to finding the slope of the line $3x - 6y = 12$.

As we saw in Example 3, the easiest way to find the slope of a line whose equation we have is to put the equation in slope–intercept form ($y = mx + b$) and then "read off" the slope:

$$3x - 6y = 12$$
$$-6y = -3x + 12$$
$$y = \frac{-3}{-6}x + \frac{12}{-6}$$
$$y = \frac{1}{2}x - 2$$

Our slope must be $m = \frac{1}{2}$.

Now we can use the point–slope form to write the equation.

$$y - y_1 = m(x - x_1) \qquad \textit{Our given point is (2, 3); } m = \frac{1}{2}.$$

$$\boxed{y - 3 = \frac{1}{2}(x - 2)}$$

Notice that in this solution we used both the slope–intercept and point–slope forms. We used the slope–intercept form since it was the easiest way to find the slope of the parallel line, and then we used the point–slope form since it was the easiest method to write the equation. ■

EXAMPLE 6

Write an equation of the line passing through each pair of points:

(a)   (2, 5)   and   (−3, 5)   **(b)**   (1, 3)   and   (1, −4)

Solution

(a)   The slope of the line is

$$m = \frac{5 - 5}{2 - (-3)} = \frac{0}{5} = 0$$

Using the point–slope form with the point $(2, 5)$, we get

$$y - 5 = 0(x - 2)$$
$$y - 5 = 0$$
$$\boxed{y = 5}$$

Alternatively, we may recognize at the outset that the line passing through the points $(2, 5)$ and $(-3, 5)$ is a horizontal line 5 units above the x-axis (see Figure 8.31). As we saw in Section 8.1, the equation of this line is $y = 5$.

Figure 8.31

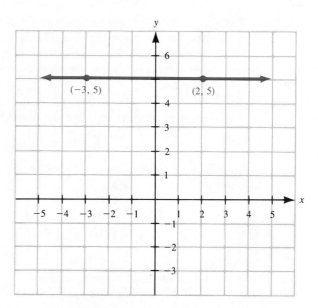

(b)   If we attempt to compute the slope of this line, we get

$$m = \frac{3 - (-4)}{1 - 1} = \frac{7}{0} \quad \text{which is undefined!}$$

Therefore, we cannot use the point–slope or slope–intercept forms to write an equation. (Both these forms require the line to have a slope. That is why in the discussion leading up to obtaining those forms we always specified a "non-vertical line.")

However, once we recognize that this is a vertical line 1 unit to the right of the y-axis (see Figure 8.32), we can write its equation as

$$\boxed{x = 1}$$

Figure 8.32

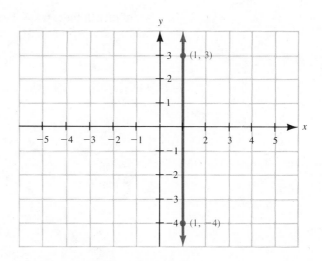

EXAMPLE 7

A manufacturer determines that the relationship between the profit earned, P, and the number of items produced, x, is linear. Suppose the profit is $1,500 on 45 items and $2,500 on 65 items.

(a) Write an equation relating P and x.

(b) What would the expected profit be if 100 items were produced?

Solution

(a) The fact that the relationship is linear means we can write a first-degree equation in the two variables, x and P. If we think of our points as (x, P), we have the two points $(45, 1,500)$ and $(65, 2,500)$. Thus, the slope of our line is

$$m = \frac{2,500 - 1,500}{65 - 45} = \frac{1,000}{20} = 50$$

We can now write the equation using the point–slope form with the point $(45, 1,500)$ and $m = 50$ (keep in mind that we are using P instead of y):

$$P - 1,500 = 50(x - 45)$$

Or, if we put this in slope–intercept form, we have

$$\boxed{P = 50x - 750}$$

(b) We want to know the expected profit if 100 items are produced. We want to find P when $x = 100$:

$P = 50x - 750$ *Substitute $x = 100$.*

$P = 50(100) - 750$

$P = 5,000 - 750$

$P = \$4,250$

Thus, the expected profit if 100 items are produced is $\boxed{\$4,250}$

It is often helpful to graph the equation to visualize the relationship between the variables (see Figure 8.33).

Figure 8.33

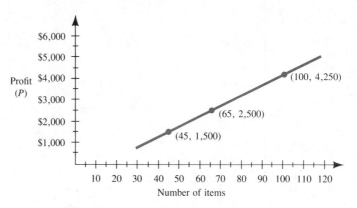

One final comment: In this section we have shown that the equation of a straight line is a first-degree equation in *x* and *y*. However, we have still not proven the theorem stated in Section 8.1 that the graph of a first-degree equation in two variables is a straight line. A proof of this fact is outlined in Exercise 63.

Exercises 8.3

In Exercises 1–48, write an equation of the line L satisfying the given conditions. Where possible, express your answer in slope–intercept form.

1. *L* has slope 5 and passes through the point $(1, -3)$.

2. *L* has slope -4 and passes through the point $(-2, 4)$.

3. *L* has slope -3 and passes through the point $(-5, 2)$.

4. *L* has slope 2 and passes through the point $(-1, 6)$.

5. *L* passes through $(6, 1)$ and has slope $\frac{2}{3}$.

6. *L* passes through $(10, 3)$ and has slope $\frac{3}{5}$.

7. *L* passes through $(4, 0)$ and has slope $-\frac{1}{2}$.

8. *L* passes through $(0, 4)$ and has slope $-\frac{1}{2}$.

9. *L* has slope $\frac{3}{4}$ and passes through $(0, 5)$.

10. *L* has slope $\frac{4}{3}$ and passes through $(5, 0)$.

11. *L* has slope 0 and passes through $(-3, -4)$.

12. *L* has undefined slope and passes through $(-3, -4)$.

13. *L* has undefined slope and passes through $(-4, 7)$.

14. *L* has slope 0 and passes through $(-4, 7)$.

15. *L* passes through the points $(2, 3)$ and $(5, 7)$.

16. *L* passes through the points $(1, 4)$ and $(3, 8)$.

17. *L* passes through the points $(-2, -1)$ and $(-3, -5)$.

18. *L* passes through the points $(-3, -2)$ and $(-1, -4)$.

19. *L* passes through the points $(2, 3)$ and $(0, 5)$.

20. *L* passes through the points $(0, -1)$ and $(3, 1)$.

21. *L* has slope 4 and crosses the *y*-axis at $y = 6$.

22. *L* has slope 4 and crosses the *x*-axis at $x = 6$.

23. *L* has slope -2 and crosses the *x*-axis at $x = -3$.

24. *L* has slope $-\frac{1}{5}$ and crosses the *y*-axis at $y = -2$.

25. *L* passes through the points $(2, 3)$ and $(6, 3)$.

26. *L* passes through the points $(3, 2)$ and $(3, 6)$.

27. *L* is vertical and passes through $(-2, -4)$.

28. *L* is horizontal and passes through $(-5, -3)$.

29. *L* crosses the *x*-axis at $x = -3$ and the *y*-axis at $y = 2$.

30. *L* crosses the *y*-axis at $y = -5$ and the *x*-axis at $x = -2$.

31. *L* passes through $(2, 2)$ and is parallel to $y = 3x + 7$.

32. *L* passes through $(2, 2)$ and is perpendicular to $y = 3x + 7$.

33. L passes through $(-3, -3)$ and is perpendicular to $y = -\frac{2}{3}x - 1$.

34. L passes through $(-3, -3)$ and is parallel to $y = -\frac{2}{3}x - 1$.

35. L passes through $(0, 0)$ and is perpendicular to $y = x$.

36. L passes through $(0, 0)$ and is parallel to $x + y = 6$.

37. L passes through $(-1, -2)$ and is parallel to $2y - 3x = 12$.

38. L passes through $(-1, -2)$ and is perpendicular to $2y - 3x = 12$.

39. L passes through $(1, -3)$ and is perpendicular to $4x - 3y = 9$.

40. L passes through $(1, -3)$ and is parallel to $4x - 3y = 9$.

41. L is perpendicular to $8x - 5y = 20$ and has the same y-intercept.

42. L is perpendicular to $8x - 5y = 20$ and has the same x-intercept.

43. L passes through $(0, 4)$ and is parallel to the line passing through the points $(3, -6)$ and $(-1, 2)$.

44. L passes through the point $(4, 0)$ and is perpendicular to the line passing through the points $(-3, 1)$ and $(2, 6)$.

45. L passes through $(4, 3)$ and is parallel to the x-axis.

46. L passes through $(4, 3)$ and is parallel to the y-axis.

47. L passes through $(-1, -2)$ and is perpendicular to the x-axis.

48. L passes through $(-1, -2)$ and is perpendicular to the y-axis.

In Exercises 49–56, *determine if the given pairs of lines are perpendicular, parallel, or neither.*

49. $3x - 2y = 5$ and $3x - 2y = 6$

50. $5x - 7y = 4$ and $5x + 7y = 4$

51. $2x = 3y - 4$ and $2x + 3y = 4$

52. $6x - 2y = 7$ and $3x - y = 8$

53. $5x + y = 2$ and $5y = x + 3$

54. $2x - y = 8$ and $4x = 2y + 9$

55. $3x - 7y = 1$ and $6x = 14y + 5$

56. $3x + 5y = 2$ and $4 + 10x = 6y$

57. A manufacturer determines that the relationship between the profit earned, P, and the number of items produced, x, is linear. If the profit is \$200 on 18 items and \$2,660 on 100 items, write an equation relating P to x and determine what the expected profit would be if 200 items were produced.

58. Joe found that the relationship between the profit, P, he made on his wood carvings and the number of wood carvings he produced, x, is linear. If he made a profit of \$10 on 5 carvings and \$90 on 15 carvings, write an equation relating P to x and determine his expected profit if he produced 35 carvings.

59. A psychologist found that the relationship between the scores on two types of personality tests, test A and test B, was perfectly linear. An individual who scored a 35 on test A would score a 75 on test B, and an individual who scored a 15 on test A would score a 35 on test B. What test B score would an individual get if she scored a 40 on test A?

60. A factory foreman found a perfectly linear relationship between the number, D, of defective widgets produced weekly and the total number of overtime hours per week, h, put in by the widget inspectors. When the inspectors put in 100 hours total overtime, 85 defective widgets were found that week; when the inspectors put in 40 hours total overtime, 30 defective widgets were found. How many defective widgets should be found during the week the inspectors put in 150 hours total overtime?

61. A math teacher found that the performance, E, of her students on their first math exam was related linearly to their performance, V, on a videogame located in the recreation room. A student who scored a 70 on his math exam scored 35,000 points on the videogame. On the other hand, a student who scored an 85 on her exam scored a 20,000 on the videogame. Write an equation relating E and V and predict what exam score a student would have received if he scored 15,000 points on the videogame.

62. A physiologist found that the relationship between the length of the right-hand thumb, T, and the length of the left little toe, t, was perfectly linear for a group of hospital residents. For one of the residents, the right-hand thumb was 5 cm and the left little toe was 2.5 cm; for another resident, the right-hand thumb was 7 cm and the left little toe was 2 cm. Write an equation relating T to t and predict the size of a resident's toe if his thumb is 8 cm.

? QUESTIONS FOR THOUGHT

63. We have already proven that if a graph is a straight line, then its equation can be put in the form $y = mx + b$ (or $x = k$ if the line is vertical). In either case, its equation is first-degree. To prove the converse we need to show that any first-degree equation in two variables has as its graph a straight line.

Suppose we have the equation $Ax + By = C$, where $B \neq 0$. Show that *any* two ordered pairs (x_1, y_1) and (x_2, y_2) which satisfy the equation will yield the same slope. [*Hint:* If (x_1, y_1) satisfies the equation $Ax + By = C$, then

$$Ax_1 + By_1 = C$$

Now solve for y_1 and we have

$$y_1 = \frac{C - Ax_1}{B}$$

Therefore, we can write our ordered pair as $\left(x_1, \dfrac{C - Ax_1}{B}\right)$.

Similarly, for (x_2, y_2), we get $\left(x_2, \dfrac{C - Ax_2}{B}\right)$.

Now show that the slope you get from these two points is independent of x_1 and x_2.]

64. Put the equation of the line $Ax + By = C$ in slope–intercept form. What is the slope of the line (in terms of A, B, and C)?

 MINI-REVIEW

65. *Solve for z.* $-2 \leq 3z - 2 < 5$

66. *Simplify and express with positive exponents only:*

$$\left(\frac{x^{-1/3}y^{1/2}}{x^{1/3}y^{1/4}}\right)^2$$

67. *Solve for x.* $\sqrt{3x - 2} + x = 2$

68. *Solve for x.* $|2x - 3| > 5$

69. *Solve for x and graph its solution set.* $x^2 - 5x \geq 14$

8.4

Graphing Linear Inequalities in Two Variables

Let's begin by reexamining the *number line* graph of a first-degree inequality in *one* variable. As illustrated in Figure 8.34, on the number line, the graph of a first-degree conditional inequality in one variable, $x \leq a$, forms a half-line. This half-line has, as its endpoint, the point that is the solution of the *equation* $x = a$. In other words, on the number line, the graph of the inequality $x \leq a$ is bounded by the graph of the equation $x = a$.

Figure 8.34

The graph of $x \leq a$ on the number line

In graphing linear inequalities in *two variables* on the *rectangular coordinate system*, we find an analogous situation. The graph of $Ax + By \leq C$ will be a *half-plane* bounded by the *equation* $Ax + By = C$ (see Figure 8.35).

Figure 8.35

A possible graph of $Ax + By \leq C$

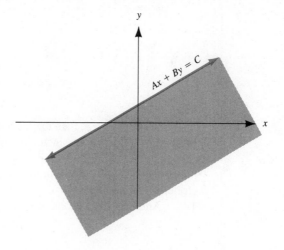

Using what we have learned thus far, we can graph linear inequalities in two variables.

EXAMPLE 1

Sketch the solution set to the inequality $x + y \leq 4$.

Solution

We begin by graphing the equation $x + y = 4$. We sketch the graph by the intercept method as outlined in Section 8.1 (see Figure 8.36).

Figure 8.36

Graph of $x + y = 4$

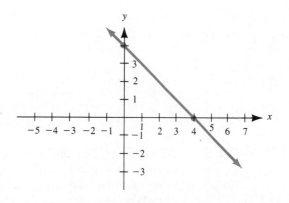

Suppose we let $x = 0$. Then the inequality would read $0 + y \leq 4$, or simply $y \leq 4$. Hence, $(0, 4)$, $(0, 3)$, $(0, 0)$, and $(0, -2)$ are a few of an infinite number of solutions for the inequality $x + y \leq 4$ (see Figure 8.37 on page 390).

Figure 8.37
Graph of $x + y \leq 4$

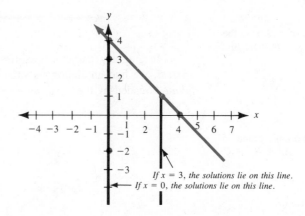

If $x = 3$, the solutions lie on this line.
If $x = 0$, the solutions lie on this line.

If we let $x = 3$, then the inequality becomes $y \leq 1$, and we can generate another infinite set of solutions, including $(3, 1)$, $(3, -2)$, and $(3, -5)$.

In continuing the process of picking values for x and finding values for y, we discover that the entire solution set consists of those points *below* the line $x + y = 4$ (refer to Figure 8.37).

Another way to look at it is to solve the inequality for y to get

$$x + y \leq 4$$
$$y \leq 4 - x$$

We want $y \leq 4 - x$. That is, we want every point whose y-coordinate is *less than or equal to* 4 minus its x-coordinate. Since the line in Figure 8.36 is the set of points with each y-coordinate *equal* to 4 minus its x-coordinate, we want those points *below* the line. (Remember that the y-coordinate of a point is its height. Asking for y to be less than something means we want the points to be *lower*.) Thus, our graph is as illustrated in Figure 8.38.

Figure 8.38
The graph of $x + y \leq 4$
(points on or below the line)

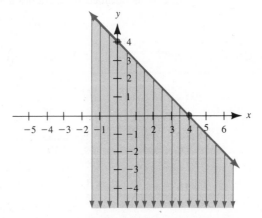

Note that we could also have proceeded as follows:

$$x + y \leq 4 \qquad \text{Solve for } x.$$
$$x \leq 4 - y$$

This means that we want the x-coordinate to be *less than* or equal to 4 minus the y-coordinate. Thus, we want the points to the *left* of the line, as indicated in Figure 8.39. Note that this gives us the same graph as in Figure 8.38.

Figure 8.39

The graph of $x + y \leq 4$ (points on or to the left of the line)

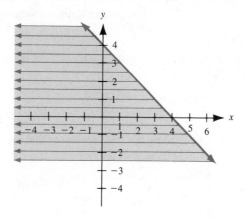

Let's look at another example.

EXAMPLE 2

Sketch the solution set of $3x - 2y > 10$.

Solution

Even though we have a strict inequality, we begin by sketching the graph of the equation $3x - 2y = 10$. The intercepts are $(0, -5)$ and $(\frac{10}{3}, 0)$ (see Figure 8.40). Note that we drew a *dashed* line instead of a solid line. This is because the inequality is strict and the points on the line are *not* included in the solution set.

Figure 8.40

The graph of $3x - 2y = 10$

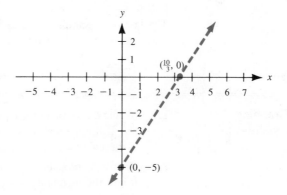

Rather than proceed as we did in the last example to solve the inequality for x or y, let's think a moment. We know that the graph of the inequality is going to be a half-plane lying either above or below (or to the left or right of, if you like) the graph of the line. We already know how to graph the line. The question that remains is, "To which side of the line will the solutions to the inequality lie?" Either *all* solutions will lie in one part or *all* solutions will lie in the other part.

Therefore, all we need do is pick a convenient point in one of the parts (above or below the line) and substitute it into the inequality. If that point satisfies the inequality, then *all* the points in that region do. If that point does not satisfy the inequality, then *all* the points in the *other* part do.

The point $(0, 0)$ is often a convenient point to choose (note that it lies *above* the line $3x - 2y = 10$):

$$3x - 2y > 10 \qquad \textit{Substitute } (0, 0) \textit{ to see if it satisfies the inequality.}$$
$$3(0) - 2(0) \overset{?}{>} 10$$
$$0 - 0 \not> 10$$

Since $(0, 0)$ does *not* satisfy the inequality and $(0, 0)$ is located in the region *above* the line, we know that the solution set consists of all points *below* the line (see Figure 8.41).

Figure 8.41
The graph of $3x - 2y > 10$

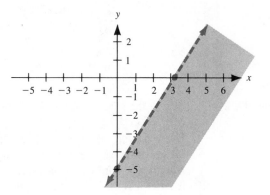

The last two examples lead us to the following outline, which can be used to graph linear inequalities in two variables.

Outline for Graphing Linear Inequalities	
	1. Sketch the graph of the equation which is the boundary of the solution set. If the inequality is *weak* (involves $\leq$ or $\geq$), draw a solid line. If the inequality is strict (involves $<$ or $>$), draw a dashed line.
	2. Choose a convenient test point (a point not on the line) and substitute it into the inequality.
	3. If the test point satisfies the inequality, then shade the region which contains the test point. Otherwise, shade the region on the other side of the line.

EXAMPLE 3 | Sketch the graph of $-5y - 2x < 20$.

Solution

Following our outline, we graph the equation $-5y - 2x = 20$, whose intercepts are $(-10, 0)$ and $(0, -4)$, as a dashed line (see Figure 8.42).

Figure 8.42
The graph of
$-5y - 2x = 20$

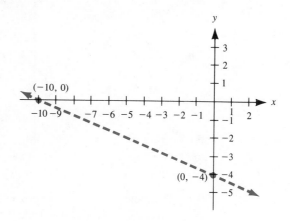

We choose the test point $(0, 0)$ (note that it lies *above* $-5y - 2x = 20$):

$$-5y - 2x < 20 \qquad \text{\textit{Substitute} } (0, 0).$$
$$-5(0) - 2(0) \overset{?}{<} 20$$
$$0 \overset{\checkmark}{<} 20$$

The point $(0, 0)$ *does* satisfy the inequality and we shade the region which contains $(0, 0)$, as shown in Figure 8.43.

Figure 8.43
The graph of
$-5y - 2x < 20$

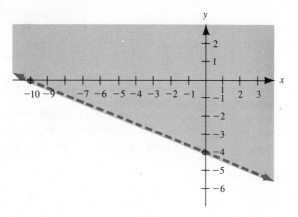

EXAMPLE 4

In a rectangular coordinate system, sketch the graphs of:
(a) $x \geq -2$ **(b)** $y < 5$

Solution

(a) The graph of $x = -2$ is a vertical line 2 units to the left of the y-axis. Since we want x to be *greater than or equal to* -2, we shade the region to the *right* of the line, as indicated in Figure 8.44 on page 394.

Figure 8.44
The graph of $x \geq -2$

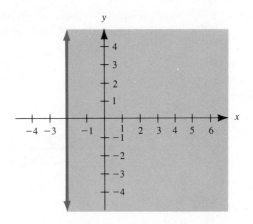

(b)   The graph of $y = 5$ is a horizontal line 5 units above the x-axis. Since we want y to be *less than* 5, we shade the region below the line (see Figure 8.45).

Figure 8.45
The graph of $y < 5$

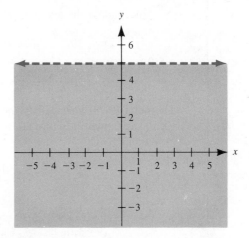

Notice that the example instructions said "In a rectangular coordinate system, ..." Without such instructions we might have just sketched the solution sets as inequalities in *one variable* on the number line.   ■

Exercises 8.4

Sketch the solution set to each of the following inequalities on a rectangular coordinate system.

1. $y \leq x + 3$ **2.** $x > y + 3$ **3.** $x > y - 2$ **4.** $y \leq x - 2$

5. $x + y < 3$ **6.** $x + y \geq 4$ **7.** $x + y \geq 3$ **8.** $x + y < 4$

9. $2x + y \le 6$ **10.** $x + 4y > 8$ **11.** $x + 2y > 6$ **12.** $4x + y \le 8$

13. $3x + 2y \ge 12$ **14.** $5x + 3y < 30$ **15.** $2x + 5y < 10$ **16.** $4y + 3x \ge 24$

17. $2x + 5y > 10$ **18.** $4y + 3x \le 24$ **19.** $2x + 5y \ge 10$ **20.** $4y + 3x < 24$

21. $2x - 5y \ge 10$ **22.** $4y - 3x < 24$ **23.** $y \le x$ **24.** $x > -y$

25. $2x - y < 4$ **26.** $x - 2y \ge 4$ **27.** $4x - y \ge 8$ **28.** $x - 4y < 8$

29. $3x - 4y > 12$ **30.** $4x - 3y \le 12$ **31.** $7x - 3y < 15$ **32.** $8x + 3y > 18$

33. $7x + 3y > 15$ **34.** $8x + 3y \le 18$ **35.** $\dfrac{x}{2} + \dfrac{y}{3} < 4$ **36.** $\dfrac{x}{3} - \dfrac{y}{2} > 5$

37. $\dfrac{x}{3} - \dfrac{y}{4} \ge 3$ **38.** $\dfrac{x}{5} + \dfrac{y}{2} \le 2$ **39.** $y < 3$ **40.** $y > -1$

41. $x \ge -2$ **42.** $x < 5$ **43.** $x < -2$ **44.** $x > 5$

45. $y < 0$ **46.** $y \ge 0$ **47.** $x \le 0$ **48.** $x < 0$

49. $x < \dfrac{1}{2}$ **50.** $y \ge \dfrac{2}{3}$ **51.** $\dfrac{y}{2} > 0$ **52.** $\dfrac{x}{3} < 0$

? | QUESTION FOR THOUGHT

53. Using set notation, we can rewrite $y + x \le 7$ as

$$\{(x, y) \mid y + x \le 7\}$$

meaning "the set of ordered pairs that satisfy the condition: $y + x \le 7$."

    Let $A = \{(x, y) \mid y + x \le 7\}$ and $B = \{(x, y) \mid y - x \le 7\}$.

(a) Graph the sets A and B on the same pair of coordinate axes.

(b) Where on your graph is $A \cap B$?

(c) Where on your graph is $A \cup B$?

 MINI-REVIEW

54. *Express as a simple fraction in simplest form:*

$$\dfrac{\dfrac{3x + 1}{x - 1} + 1}{2 - \dfrac{1}{x + 2}}$$

55. Convert 728 to scientific notation.

56. *Express in the form $a + bi$:*

$$\dfrac{2 + 3i}{2 - 3i}$$

57. The width of a rectangle is 12 cm. If the perimeter is to be at least 100 cm, how large must the length be?

CHAPTER 8   Summary

After completing this chapter, you should be able to:

1. Sketch the graph of a first-degree equation in two variables (Section 8.1).

For example: Sketch the graph of $6y - 3x = 12$. Label the intercepts.

To find the x-intercept: *To find the y-intercept:*
Set $y = 0$ and solve for x.   *Set $x = 0$ and solve for y.*

$$6y - 3x = 12 \qquad\qquad 6y - 3x = 12$$
$$6(0) - 3x = 12 \qquad\quad 6y - 3(0) = 12$$
$$-3x = 12 \qquad\qquad\quad 6y = 12$$
$$x = -4 \qquad\qquad\qquad\quad y = 2$$

The graph crosses the x-axis at $(-4, 0)$ and the y-axis at $(0, 2)$.

The graph of $6y - 3x = 12$ is shown in Figure 8.46. Finding a check point is left to the student.

Figure 8.46

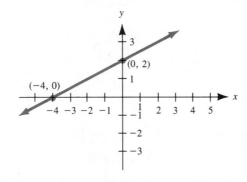

2. Use the definition of slope to compute the slope of a line when two points on the line are known (Section 8.2).

For example: Compute the slope of the line passing through the points $(-1, 3)$ and $(2, -5)$.

$$m = \frac{y_2 - y_1}{x_2 - x_1} = \frac{3 - (-5)}{-1 - 2} = \frac{8}{-3} = -\frac{8}{3}$$

3. Recognize that lines with equal slopes are parallel and that lines whose slopes are negative reciprocals of each other are perpendicular (Section 8.2).

For example:

If line L_1 has slope $\frac{2}{5}$, then any line parallel to L_1 must have slope $\frac{2}{5}$.

If line L_2 has slope $\frac{3}{7}$, then any line perpendicular to L_2 must have slope $-\frac{7}{3}$.

4. Write an equation of a line using either

the point–slope form $y - y_1 = m(x - x_1)$

or

the slope–intercept form $y = mx + b$

(Section 8.3).

For example: Write an equation of the line passing through the points $(-2, 1)$ and $(1, 2)$.

We first find the slope:

$$m = \frac{y_2 - y_1}{x_2 - x_1} = \frac{1 - 2}{-2 - 1} = \frac{-1}{-3} = \frac{1}{3}$$

The point–slope form is $y - y_1 = m(x - x_1)$. We can use either $(-2, 1)$ or $(1, 2)$ as our given point (x_1, y_1). Using $(1, 2)$ as the given point, we get

$$y - 2 = \frac{1}{3}(x - 1)$$

5. Find the slope of a line by putting its equation in slope–intercept form (Section 8.3).

For example: Find the slope of the line whose equation is $5x + 4y = 7$.

We can find the slope by putting the equation in slope–intercept form—that is, in the form $y = mx + b$:

$$5x + 4y = 7$$
$$4y = -5x + 7$$
$$y = \frac{-5}{4}x + \frac{7}{4}$$

Therefore, we can "read off" the slope, which is $-\frac{5}{4}$.

6. Write an equation of a line satisfying certain conditions (Section 8.3).

For example: Write an equation of the line which passes through the point $(3, -2)$ and is perpendicular to $2x - 7y = 11$.

We first find the slope of $2x - 7y = 11$ by putting it in slope–intercept form:

$$2x - 7y = 11$$
$$-7y = -2x + 11$$
$$y = \frac{2}{7}x - \frac{11}{7} \qquad \textit{The slope of this line is } \frac{2}{7}.$$

Since the required line is perpendicular to $2x - 7y = 11$, it will have slope $-\frac{7}{2}$. We can now write the equation using point–slope form:

$$y - y_1 = m(x - x_1)$$
$$y - (-2) = -\frac{7}{2}(x - 3)$$
$$y + 2 = -\frac{7}{2}(x - 3)$$

7. Sketch the graph of an inequality in two variables (Section 8.4).

For example: Sketch the graph of $x - 3y \leq 6$.

We first sketch the graph of $x - 3y = 6$, whose intercepts are $(6, 0)$ and $(0, -2)$, using a solid line.

Now we use (0, 0) as a check point:

$$x - 3y \leq 6$$
$$0 - 3(0) \overset{?}{\leq} 6$$
$$0 - 0 \overset{?}{\leq} 6$$
$$0 \overset{\checkmark}{\leq} 6$$

Since (0, 0) satisfies the inequality, we shade the region containing (0, 0), as indicated in Figure 8.47.

Figure 8.47

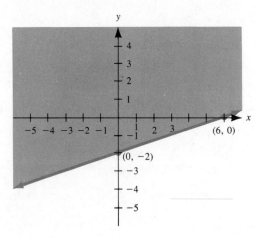

CHAPTER 8   Review Exercises

In Exercises 1–30, sketch the graph of the equation. Label the intercepts.

1. $2x + 6y = 6$ **2.** $2x + 6y = -6$ **3.** $2x - 6y = -6$ **4.** $6y - 2x = 6$

5. $5x - 3y = 10$ **6.** $3x - 5y = 10$ **7.** $2x + 5y = 7$ **8.** $3x + 4y = 5$

9. $3x - 8y = 11$ **10.** $3x + 8y = 11$ **11.** $5x + 7y = 21$ **12.** $4x - 5y = 20$

13. $y = x$ **14.** $y = -x$ **15.** $y = -2x$ **16.** $x = 3y$

17. $y = x + 1$ **18.** $y = x - 2$ **19.** $y = 3x - 4$ **20.** $y = -2x + 5$

21. $y = \frac{2}{3}x + 2$ **22.** $y = -\frac{1}{2}x - 3$ **23.** $\frac{x}{3} + \frac{y}{2} = 12$ **24.** $\frac{x}{4} - \frac{y}{2} = 16$

25. $x - 2y = 8$ **26.** $x - 2y = 0$ **27.** $x - 2 = 0$ **28.** $y + 2 = 0$

29. $2y = 5$ **30.** $3x = 4$

In Exercises 31–52, find the slope of the line satisfying the given condition(s).

31. Passing through the points $(-1, 0)$ and $(3, -2)$

32. Passing through the points $(6, -3)$ and $(-4, 3)$

33. Passing through the points $(8, 2)$ and $(3, -2)$

34. Passing through the points $(-5, 4)$ and $(5, -4)$

35. Its equation is $y = 3x - 5$.

36. Its equation is $3y = 2x + 1$.

37. Its equation is $4y - 3x = 1$.

38. Its equation is $x = 3y - 5$.

39. Parallel to the line passing through the points $(3, 5)$ and $(1, 4)$

40. Parallel to the line passing through the points $(2, 4)$ and $(5, 0)$

41. Perpendicular to the line passing through the points (4, 7) and (4, 9)

42. Perpendicular to the line passing through the points (3, 5) and (1, 4)

43. Passing through the points $(-7, 6)$ and $(2, 6)$

44. Passing through the points $(3, 5)$ and $(3, -2)$

45. Parallel to the line whose equation is $y = 3x - 7$

46. Perpendicular to the line whose equation is $y = 5x + 1$

47. Perpendicular to the line whose equation is $3y - 5x + 6 = 0$

48. Parallel to the line whose equation is $6x - 4y - 9 = 0$

49. Parallel to the line whose equation is $x = 3$

50. Perpendicular to the line whose equation is $x = 4$

51. Passing through the points $(a + b, a)$ and $(a, a + b)$ $(b \neq 0)$

52. Passing through the points $(1, a)$ and (a, a^2) $(a \neq 1)$

*In Exercises 53–58, find the value(s) of **a** which satisfy the given conditions.*

53. The line through the points $(4, a)$ and $(1, 2)$ has slope 4.

54. The line through the points $(a, 3)$ and $(2, 5)$ has slope 1.

55. The line through the points $(a, 4)$ and $(-2, a)$ has slope -2.

56. The line through the points $(-3, a)$ and $(1, a^2)$ has slope 5.

57. The line through the points $(-3, a)$ and $(0, 3)$ is parallel to the line through the points $(a, 7)$ and $(0, 0)$.

58. The line through the points $(2, a)$ and $(0, 5)$ is perpendicular to the line through the points $(0, -1)$ and $(4, a)$.

In Exercises 59–84, write an equation of the line satisfying the given conditions.

59. The line passes through the points $(-2, 3)$ and $(1, -4)$.

60. The line passes through the points $(-1, -4)$ and $(0, -2)$.

61. The line passes through the points $(3, 5)$ and $(-3, -5)$.

62. The line passes through the points $(2, -6)$ and $(-2, 6)$.

63. The line passes through the point $(2, 5)$ and has slope $\frac{2}{5}$.

64. The line passes through the point $(0, 4)$ and has slope -4.

65. The line passes through the point $(4, 7)$ and has slope 5.

66. The line passes through the point $(3, 8)$ and has slope $-\frac{3}{4}$.

67. The line has slope 5 and y-intercept 3.

68. The line has slope -3 and y-intercept -4.

69. The horizontal line passes through the point $(2, 3)$.

70. The vertical line passes through the point $(2, 3)$.

71. The vertical line passes through the point $(-3, -4)$.

72. The horizontal line passes through the point $(-3, -4)$.

73. The line passes through the point $(0, 0)$ and is parallel to the line $y = \frac{3}{2}x - 1$.

74. The line passes through the point $(-3, 0)$ and is perpendicular to the line $y = -5x + \frac{3}{7}$.

75. The line is perpendicular to $2y - 5x = 1$ and passes through the point $(0, 6)$.

76. The line is parallel to $6x - 7y + 3 = 0$ and passes through the point $(5, 5)$.

77. The line is perpendicular to $3x = -5y$ and passes through the point $(0, 0)$.

78. The line is parallel to $3x = -5y$ and passes through the point $(0, 0)$.

79. The line crosses the x-axis at $x = 3$ and the y-axis at $y = -5$.

80. The line has x-intercept 5 and y-intercept 8.

81. The line has x-intercept -2 and y-intercept 5.

82. The line passes through $(2, 2)$ and crosses the x-axis at $x = 1$.

83. The line is parallel to $3x - 2y = 5$ and has the same y-intercept as $5y = x + 3$.

84. The line is parallel to $2x + 5y = 4$ and has the same y-intercept as $2y = x - 4$.

In Exercises 85–100, graph the inequality on a rectangular coordinate system.

85. $y - 2x < 4$

86. $y + 2x < 4$

87. $2y - 3x > 6$

88. $2y + 3x < 12$

89. $5y - 8x \leq 20$

90. $2y - 7x \geq 14$

91. $3y \geq 2x - 8$

92. $2x \leq 3y - 2$

93. $4y + 8x < 16$ **94.** $2y + 4x < 8$ **95.** $\dfrac{x}{2} + \dfrac{y}{3} \geq 6$ **96.** $\dfrac{x}{5} - \dfrac{y}{3} < 15$

97. $x \leq -3$ **98.** $y > 2$ **99.** $y < 5$ **100.** $x \leq -1$

101. A manufacturer found that the relationship between his profit, P, and the number of gadgets produced, x, is linear. If he makes $12,000 by producing 250 gadgets and $20,000 by producing 300 gadgets, write an equation relating P to x and predict how much he would make if he produced 400 gadgets.

102. A psychologist found that the relationship between scores on two tests, test A and test B, was perfectly linear. Joe scored 32 on test A and 70 on test B, Sue scored 45 on test A and 96 on test B. If Jake scored 40 on test A, what would he score on test B? If Charles scored 80 on test B, what would he score on test A?

CHAPTER 8   Practice Test

1. Graph the following using the intercept method:

    (a) $3x - 5y = 30$ **(b)** $x - 7 = 0$

2. Find the slope of the line satisfying the given conditions:

    (a) Passing through $(3, 9)$ and $(-2, 4)$ **(b)** Equation is $3x - 2y = 8$

3. Find the value of a if a line with slope 2 passes through the points $(a, 2)$ and $(2, 5)$.

4. Write the equation of the line satisfying the given conditions.

    (a) The line passes through points $(2, -3)$ and $(3, 5)$.

    (b) The line passes through $(1, 0)$ with slope -4.

    (c) The line passes through $(2, 5)$ with y-intercept 3.

    (d) The line passes through $(2, -3)$ and is parallel to $x + 3y = 8$.

    (e) The line passes through $(2, -3)$ and is perpendicular to $x + 3y = 8$.

    (f) The horizontal line passes through $(3, -1)$.

5. Graph the inequality $3x - 8y > 12$.

6. A psychologist finds that the relationship between scores on two tests, test A and test B, is perfectly linear. A person scoring 60 on test A scores 90 on test B; someone scoring 80 on test A scores 150 on test B. What should a person receive on test B if he or she scores 85 on test A?

9

Systems of Linear Equations

It is frequently the case that we are considering a problem which requires us to satisfy simultaneously two or more different conditions. For example, given its particular circumstances, a business may want to choose a method of advertising which minimizes cost but which also maximizes exposure.

Each condition can sometimes be represented by an equation in two or more variables. We then seek the numbers (if any) which satisfy all the equations (conditions) simultaneously.

We begin by considering a system of two linear equations in two variables and then proceed to more complex situations.

9.1

Linear Systems: Two Variables

Let us begin by considering the following situation.

EXAMPLE 1

A homeowner wants to build a family room as an extension to her house. She receives bids from two competing contractors. The ILF (It Lasts Forever) Construction Company charges a flat fee of $2,250 for architectural plans and obtaining permits, plus $225 per square meter for the actual construction. The WBB (We Build Better) Construction Company charges a flat fee of $3,500 plus $175 per square meter. For what size room will the two companies charge the same price?

Solution

We can analyze the question as follows. The cost, C, of building the room can be expressed in terms of the area, A, of the room.

$$\text{Cost} = \text{Flat fee} + \text{Actual building costs}$$
$$= \text{Flat fee} + (\text{Cost per square meter})(\#\text{ of square meters})$$

For ILF this would be: $C = 2{,}250 + 225A$
For WBB this would be: $C = 3{,}500 + 175A$

Each of these equations is a first-degree equation in two variables (a linear equation) and hence has as its graph a straight line. Asking for what area, A, the costs will be the same is equivalent to asking where the lines intersect.

We sketch the graphs of each line using the methods discussed in Chapter 8, and then we estimate the point of intersection (see Figure 9.1). (Naturally, we draw the graph only for $A \geq 0$, to the right of the C-axis, since a negative area makes no sense.) It appears that the graphs intersect when $A = 25$.

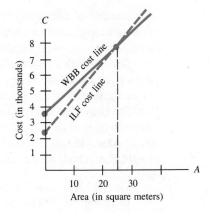

Figure 9.1

Graphs of
$C = 2,250 + 225A$ and
$C = 3,500 + 175A$

CHECK:

For ILF: The cost of building a 25-square-meter room is

$$C = 2,250 + 25(225) = 2,250 + 5,625 = \$7,875$$

For WBB: The cost of building a 25-square-meter room is

$$C = 3,500 + 25(175) = 3,500 + 4,375 = \$7,875$$

Thus, the answer to the example is $\boxed{25 \text{ square meters}}$ ■

A few comments are in order here about the "graphical solution" obtained in Example 1. What can be said in favor of this graphical solution is that not only did we answer the question which was asked, but in addition, by looking at the graphs in Figure 9.1, we can see that for a room whose area is less than 25 square meters the ILF Company is cheaper, while for a room whose area is greater than 25 square meters the WBB Company is cheaper.

What can be said against this graphical solution is that "reading" off the point of intersection from the graph is rather risky. What if the answer had actually been $A = 24.6$ square meters? We cannot possibly read the graph that precisely.

Since the graphical method can be imprecise, we would like to develop an algebraic method for finding the solution to such a problem. Let's first introduce some terminology.

DEFINITION

Two or more equations considered together are called a *system of equations*. In particular, if the equations are of the first degree it is called a *linear system*.

Thus,

$$\begin{cases} 2x - 3y = 6 \\ x - y = 1 \end{cases}$$

is an example of a linear system in two variables. This is called a **2 × 2** (read "2 by 2") **system**, since there are two equations and two variables. The "{" indicates that the two equations are to be considered together.

Solving such a *system of equations* means finding all the ordered pairs that satisfy *all* the equations in the system. Keep in mind that *one* solution to the system consists of two numbers—an *x* value and a *y* value.

We know that a single linear equation has an infinite number of solutions. How many solutions can a 2 × 2 linear system have? Or how many ordered pairs can satisfy two equations in two unknowns? Each of the equations in a 2 × 2 linear system is a straight line. Thus, we have the following three possibilities:

Case 1 The lines intersect in exactly one point. The coordinates of the point are the solution to the system.

Such a system is called **consistent** and **independent** (see Figure 9.2**a**).

Case 2 The lines are parallel and therefore never intersect. There are no solutions to the system.

Such a system is called **inconsistent** (see Figure 9.2**b**).

Case 3 The lines coincide. All the points which satisfy one of the equations also satisfy the other. Thus, there are infinitely many solutions.

Such a system is called **dependent** (see Figure 9.2**c**).

Figure 9.2

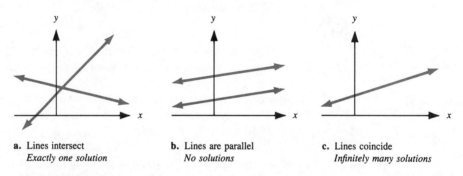

a. Lines intersect
Exactly one solution

b. Lines are parallel
No solutions

c. Lines coincide
Infinitely many solutions

The algebraic methods we are about to discuss will allow us to determine which situation we have, and in case 1, to find the unique solution.

The Elimination Method

The **elimination** (or **addition**) **method** is based on the following idea. We already know how to solve a variety of equations involving *one* variable. If we can manipulate our system of equations so that one of the variables is eliminated, we can then solve the resulting equation in one variable.

We illustrate the elimination method with several examples.

EXAMPLE 2

Solve the following system of equations:

$$\begin{cases} x + y = -5 \\ x - y = 9 \end{cases}$$

Solution

In solving an equation we are used to adding the same number or quantity to both sides of an equation. However, the addition property of equality allows us to add equal *quantities* to both sides of an equation. According to the second equation, $x - y$ and 9 are equal quantities; thus, we can just "add" that quantity to both sides of the equation $x + y = -5$, but we add $x - y$ to the left-hand side of $x + y = -5$ and 9 to the right-hand side. Thus, we "add" the two equations together:

$$\begin{array}{rl} x + y = & -5 \\ x - y = & 9 \quad \textit{Add.} \\ \hline 2x = & 4 \end{array}$$

Notice that we have *eliminated* one of the variables and now we have an equation in one variable. We solve for that variable:

$$x = 2$$

To find the other variable, y, we substitute $x = 2$ in either one of the equations and solve for y:

$$\begin{array}{ll} x - y = 9 & \textit{Substitute } x = 2 \textit{ in the second equation and solve for } y. \\ 2 - y = 9 & \\ y = -7 & \end{array}$$

The solution is $\boxed{(2, -7)}$

CHECK: We substitute $x = 2$ and $y = -7$ into both equations:

$$\begin{array}{ll} x + y = -5 & x - y = 9 \\ 2 + (-7) \overset{\checkmark}{=} -5 & 2 - (-7) \overset{\checkmark}{=} 9 \end{array}$$

EXAMPLE 3

Solve the following system of equations:

$$\begin{cases} 2x - 3y = 6 \\ x - y = 1 \end{cases}$$

Solution

We would like to "add" the two equations in such a way that one of the variables is eliminated. To do this we must change the coefficients of either the x or y variable so that they are exact opposites. Thus, when we add the two equations, the variables with opposite coefficients will be eliminated.

For example, to eliminate the y variable, we multiply the second equation by -3. This produces a y coefficient of $+3$ in the second equation, which is the opposite of -3, the y coefficient in the first equation.

We proceed as follows:

$$\begin{array}{lll} 2x - 3y = 6 & \textit{As is} \rightarrow & \\ x - y = 1 & \textit{Multiply by } -3 \rightarrow & \end{array} \qquad \begin{array}{rl} 2x - 3y = & 6 \\ -3x + 3y = & -3 \\ \hline -x = & 3 \\ x = & -3 \end{array} \qquad \begin{array}{l} \textit{Add the resulting} \\ \textit{equations. Solve} \\ \textit{for } x. \end{array}$$

Now we can substitute $x = -3$ into one of the original equations (we will use the first one) and solve for y:

$$2x - 3y = 6$$
$$2(-3) - 3y = 6$$
$$-6 - 3y = 6$$
$$-3y = 12$$
$$y = -4$$

Thus, our solution is $\boxed{(-3, -4)}$

CHECK: We substitute $x = -3$ and $y = -4$ into both equations:

$$2x - 3y = 6 \qquad\qquad x - y = 1$$
$$2(-3) - 3(-4) \stackrel{?}{=} 6 \qquad -3 - (-4) \stackrel{?}{=} 1$$
$$-6 + 12 \stackrel{\checkmark}{=} 6 \qquad\qquad -3 + 4 \stackrel{\checkmark}{=} 1$$

$\blacksquare$

Note that we could have chosen to eliminate x in Example 3 by multiplying the second equation by -2 and then adding the two equations. The elimination method is also called the *addition method*.

EXAMPLE 4

Solve the following system of equations:

$$\begin{cases} 3x = 4y + 6 \\ 5y = 2x - 4 \end{cases}$$

Solution

In order to make the elimination process easier to perform, we should first line up "like" variables to make it easier to see how to eliminate one of them.

$$3x = 4y + 6 \quad \rightarrow \quad 3x - 4y = 6$$
$$5y = 2x - 4 \quad \rightarrow \quad -2x + 5y = -4$$

A system in this form with the variables lined up is said to be in **standard form**.

We choose to eliminate x. To keep the arithmetic as simple as possible we convert the coefficients of x to $+6$ and -6. Note that 6 is the least common multiple (LCM) of 3 and 2.

$$\begin{array}{lll} 3x - 4y = 6 & \textit{Multiply by } 2 \rightarrow & 6x - 8y = 12 \\ -2x + 5y = -4 & \textit{Multiply by } 3 \rightarrow & \underline{-6x + 15y = -12} \\ & & 7y = 0 \\ & & y = 0 \end{array}$$

Add the resulting equations. Solve for y.

Substitute $y = 0$ into one of the original equations and solve for x:

$3x = 4y + 6$ *We substitute $y = 0$ into the first equation.*

$3x = 4(0) + 6$

$3x = 6$

$x = 2$

Thus, our solution is $\boxed{(2, 0)}$

You should check to see that $(2, 0)$ does indeed satisfy both equations.   ■

We can summarize the elimination method as follows.

The Elimination (Addition) Method	
	1. Put the system of equations in standard form—that is, make sure the variables and constants line up vertically.
	2. Decide which variable you want to eliminate.
	3. Multiply one or both equations by appropriate constants so that the variable you have chosen to eliminate appears with opposite coefficients.
	4. Add the resulting equations.
	5. Solve the resulting equation in *one* variable.
	6. Substitute the value of the variable obtained into one of the original equations, and solve for the other variable.
	7. Check your solution in both of the original equations.

EXAMPLE 5

Solve the following system of equations:

$$\begin{cases} 3a - \dfrac{b}{2} = 7 \\ \dfrac{a}{5} - \dfrac{2b}{3} = 3 \end{cases}$$

Solution

We begin by clearing the system of fractions.

$3a - \dfrac{b}{2} = 7$ *Multiply by 2 →* $6a - b = 14$

$\dfrac{a}{5} - \dfrac{2b}{3} = 3$ *Multiply by 15 →* $3a - 10b = 45$

We choose to eliminate a:

$$6a - b = 14 \qquad \textit{As is} \rightarrow \qquad\qquad 6a - b = 14$$
$$3a - 10b = 45 \qquad \textit{Multiply by } -2 \rightarrow \qquad \underline{-6a + 20b = -90} \qquad \textit{Add.}$$
$$19b = -76 \qquad \textit{Solve for } b.$$
$$b = -4$$

In order to obtain the value for a we may substitute $b = -4$ into any of the equations containing a and b. We substitute $b = -4$ into the equation $6a - b = 14$ (the first of our equations without fractional coefficients):

$$6a - b = 14$$
$$6a - (-4) = 14$$
$$6a + 4 = 14$$
$$6a = 10$$
$$a = \frac{10}{6} = \frac{5}{3}$$

Thus, our solution is $\boxed{a = \dfrac{5}{3},\ b = -4}$

CHECK: (in the original equations)

$$3a - \frac{b}{2} = 7 \qquad\qquad \frac{a}{5} - \frac{2b}{3} = 3$$

$$3\left(\frac{5}{3}\right) - \frac{(-4)}{2} \overset{?}{=} 7 \qquad\qquad \frac{\frac{5}{3}}{5} - \frac{2(-4)}{3} \overset{?}{=} 3$$

$$5 + 2 \overset{\checkmark}{=} 7 \qquad\qquad \frac{5}{3} \cdot \frac{1}{5} - \left(\frac{-8}{3}\right) \overset{?}{=} 3$$

$$\frac{1}{3} + \frac{8}{3} \overset{\checkmark}{=} 3 \qquad\blacksquare$$

The Substitution Method

There is another method we can use to solve a system of equations. While our goal remains the same, to obtain an equation in one variable, our approach will be slightly different. We illustrate the substitution method with several examples.

EXAMPLE 6

Solve the following system of equations:

$$\begin{cases} 5x - 3y = 7 \\ x = 6y - 4 \end{cases}$$

Solution | We notice that the second equation is solved explicitly for x. Since $x = 6y - 4$, we can substitute $6y - 4$ into the first equation in place of x:

$$5x - 3y = 7 \qquad \textit{Replace each occurrence of x by } 6y - 4.$$
$$5(6y - 4) - 3y = 7 \qquad \textit{Now we have an equation in one variable. Solve for y.}$$
$$30y - 20 - 3y = 7$$
$$27y - 20 = 7$$
$$27y = 27$$
$$y = 1$$

Now we substitute $y = 1$ into the second of our original equations (since it is already solved explicitly for x):

$$x = 6y - 4$$
$$x = 6(1) - 4$$
$$x = 2$$

Thus, our solution is $\boxed{(2, 1)}$

CHECK:

$$5x - 3y = 7 \qquad x = 6y - 4$$
$$5(2) - 3(1) \overset{?}{=} 7 \qquad 2 \overset{?}{=} 6(1) - 4$$
$$10 - 3 \overset{\checkmark}{=} 7 \qquad 2 \overset{\checkmark}{=} 6 - 4$$

■

We can summarize the substitution method as follows.

The Substitution Method

1. Solve one of the equations explicitly for one of the variables.
2. Substitute the expression obtained in step 1 into the other equation.
3. Solve the resulting equation in one variable.
4. Substitute the value obtained into one of the original equations (usually the one solved explicitly in step 1) and solve for the other variable.
5. Check the solution.

EXAMPLE 7 | Solve the following system of equations:

$$\begin{cases} \dfrac{5}{2}x + y = 4 \\ 2y + 5x = 10 \end{cases}$$

Solution | We solve the first equation explicitly for y:

$$\frac{5}{2}x + y = 4 \quad \rightarrow \quad y = -\frac{5}{2}x + 4$$

Substitute for y in the second equation:

$$2y + 5x = 10 \qquad \textit{Substitute } -\frac{5}{2}x + 4 \textit{ for } y.$$

$$2\left(-\frac{5}{2}x + 4\right) + 5x = 10$$

$$-5x + 8 + 5x = 10$$

$$8 = 10 \qquad \textit{This is a contradiction.}$$

Thus, there are no solutions common to both equations.

If we solve the *second* equation in our system for y, we get

$$y = -\frac{5}{2}x + 5 \qquad \textit{This equation is in the form } y = mx + b.$$

We can see that the two lines never meet (they both have the same slope of $-\frac{5}{2}$ but have different y-intercepts).

This system of equations has no solutions and is therefore inconsistent ∎

EXAMPLE 8 | Solve the following system of equations:

$$\begin{cases} 6x - 4y = 10 \\ 2y + 5 = 3x \end{cases}$$

Solution | If we choose to use the substitution method, then we must solve one of the equations explicitly for one of the variables. Whichever equation and whichever variable we choose, we are forced to work with fractional expressions. (Try it!) In this case the elimination method seems to be easier.

We begin by getting the system in standard form, and then eliminate y:

$$
\begin{array}{lllll}
6x - 4y = 10 \rightarrow & 6x - 4y = 10 & \textit{As is} \rightarrow & 6x - 4y = & 10 \\
2y + 5 = 3x \rightarrow & -3x + 2y = -5 & \textit{Multiply by 2} \rightarrow & -6x + 4y = & -10 \\
& & & \hline \\
& & & 0 = & 0 \\
& & & \textit{This is an identity.}
\end{array}
$$

If we solve both equations for y, we obtain

$$y = \frac{3}{2}x - \frac{5}{2}$$

and we see that the lines are identical (they have the same slope *and* the same *y*-intercept). Thus, every ordered pair which satisfies one of the equations also satisfies the other. There are infinitely many solutions. The equations are dependent.

The solution set is all the points on the line, that is,

$$\{(x, y) \mid 6x - 4y = 10\}$$

Having the ability to solve a system of equations gives us a great deal of flexibility in how we set up our solutions to verbal problems. In many cases we may be able to solve a verbal problem either by using a one-variable approach as we did in Chapters 3 and 4, or by writing a system of equations.

EXAMPLE 9

A stationery store ordered 50 cases of envelopes costing a total of $551.50. Among the 50 cases were some that contained legal-size envelopes and cost $11.95 each, while the remaining cases contained letter-size envelopes and cost $9.95 each. How many of each type were there?

Solution

If we want to use more than one variable, then we must have as many independent equations as we have variables.

Let x = # of cases of legal-size envelopes.

Let y = # of cases of letter-size envelopes.

We must create two equations—the first relating the *number* of cases, the second relating the *cost* of the cases.

Our equations are

$$\begin{cases} x + y = 50 \\ 11.95x + 9.95y = 551.50 \end{cases}$$

There are 50 cases all together.

The total cost is $551.50.

$$\begin{array}{lll} x + y = 50 & \text{\textit{Multiply by} } -995 \rightarrow & -995x - 995y = -49{,}750 \\ 11.95x + 9.95y = 551.50 & \text{\textit{Multiply by} } 100 \rightarrow & \underline{1{,}195x + 995y = \ \ 55{,}150} \\ & & \ \ \ \ \ 200x \ \ \ \ \ \ \ = \ \ \ \ 5{,}400 \\ & & \ \ \ \ \ \ \ \ \ x \ \ \ \ \ \ \ = \ \ \ \ \ \ \ \ \ 27 \end{array}$$

Substitute $x = 27$ into the first equation:

$$x + y = 50$$
$$27 + y = 50$$
$$y = 23$$

Hence, there are 27 cases of legal-size envelopes and 23 cases of letter-size envelopes.

EXAMPLE 10

A pharmacist must produce a 12% alcohol solution. She has 80 ml of a 7% alcohol solution and a large supply of 20% alcohol solution. If she wants to use all of the 7% solution, how much 20% solution must she mix with it to produce the required 12% solution?

Solution

We offer both a one- and a two-variable approach.

ONE-VARIABLE APPROACH: We can visualize the problem as shown in Figure 9.3.

Let x = amount of 20% alcohol solution to be added.

Figure 9.3
Diagram for one-variable
approach

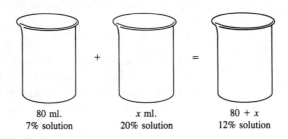

80 ml. x ml. $80 + x$
7% solution   20% solution   12% solution

Our equation relates the amount of actual alcohol contained in each part of the mixture.

$$0.07(80) + 0.20x = 0.12(80 + x)$$ *Clear the decimals.*

$$7(80) + 20x = 12(80 + x)$$
$$560 + 20x = 960 + 12x$$
$$8x = 400$$
$$x = 50$$

She must use   50 ml of the 20% solution

CHECK: $0.07(80) + 0.20(50) \overset{?}{=} 0.12(80 + 50)$
$$5.6 + 10 \overset{?}{=} 0.12(130)$$
$$15.6 \overset{\checkmark}{=} 15.6$$

TWO-VARIABLE APPROACH: We can also visualize the problem as indicated in Figure 9.4.

Let x = amount of 20% alcohol *solution* to be added.
Let y = amount of resulting 12% alcohol *solution*.

Figure 9.4

Diagram for two-variable approach

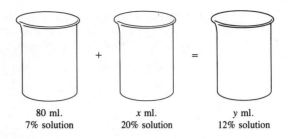

80 ml.	x ml.	y ml.
7% solution	20% solution	12% solution

We create two equations—one relating the amounts of *solution*, the other relating the amounts of *pure alcohol*. (Notice the similarities between this problem and Example 9.) We get the following system of equations:

$$\begin{cases} 80 + x = y & \textit{Amounts of solution} \\ 0.07(80) + 0.20x = 0.12y & \textit{Amounts of pure alcohol} \end{cases}$$

Using the first equation, we substitute into the second equation and get

$$0.07(80) + 0.20x = 0.12(80 + x)$$

This is exactly the same equation we obtained in using the one-variable approach. The rest of the solution is as it appears above. ∎

In Example 10, neither the one-variable nor the two-variable approach offers any particular advantages. Sometimes, however, as the following example illustrates, the way a problem is stated makes the two-variable approach significantly easier.

EXAMPLE 11

George and Ruth both go into a store to buy blank audio cassettes. George buys eight 60-minute cassettes and five 90-minute cassettes for a total of $39.45, while Ruth buys six 60-minutes cassettes and ten 90-minute cassettes for a total of $51.40. What are the prices of a single 60-minute and a single 90-minute cassette?

Solution

If we try the one-variable approach here we will find that it is difficult to represent both prices in terms of one variable. (Try it!) However, the statement of the problem allows us to use the two-variable approach quite naturally.

Let s = price of a single 60-minute cassette.

Let n = price of a single 90-minute cassette.

From the statement of the problem, we obtain the following system:

$8s + 5n = 39.45$	*This equation represents George's purchase.*
$6s + 10n = 51.40$	*This equation represents Ruth's purchase.*

We choose to eliminate n.

$8s + 5n = 39.45$	*Multiply by* $-2 \rightarrow$	$-16s - 10n = -78.90$
$6s + 10n = 51.40$	*As is* $\rightarrow$	$6s + 10n = 51.40$

$$-10s \qquad = -27.50$$
$$s = 2.75$$

Substitute $s = 2.75$ into the first equation:

$$8s + 5n = 39.45$$
$$8(2.75) + 5n = 39.45$$
$$22 + 5n = 39.45$$
$$5n = 17.45$$
$$n = 3.49$$

A 60-minute cassette costs $2.75 and a 90-minute cassette costs $3.49.

The check is left to the student.

Exercises 9.1

In Exercises 1–38, solve the system of equations. State whether the system is independent, inconsistent, or dependent. Use whichever method you prefer.

1. $\begin{cases} 2x + y = 12 \\ 3x - y = 13 \end{cases}$

2. $\begin{cases} x + 4y = 6 \\ -x + 3y = 8 \end{cases}$

3. $\begin{cases} -x + 5y = 11 \\ x - 2y = -2 \end{cases}$

4. $\begin{cases} 4x + y = 16 \\ 3x - y = 5 \end{cases}$

5. $\begin{cases} 3x - y = 0 \\ 2x + 3y = 11 \end{cases}$

6. $\begin{cases} 5x - y = 13 \\ 3x - 2y = 5 \end{cases}$

7. $\begin{cases} x + 7y = 20 \\ 5x + 2y = 34 \end{cases}$

8. $\begin{cases} -x + 5y = 12 \\ -3x + 4y = 3 \end{cases}$

9. $\begin{cases} 4x + 5y = 0 \\ 2x + 3y = -2 \end{cases}$

10. $\begin{cases} 5x - 3y = 18 \\ 4x - 6y = 0 \end{cases}$

11. $\begin{cases} 2x + 3y = 7 \\ 4x + 6y = 14 \end{cases}$

12. $\begin{cases} 3x - 5y = 4 \\ 6x - 10y = 9 \end{cases}$

13. $\begin{cases} 5x - 6y = 3 \\ 10x - 12y = 5 \end{cases}$

14. $\begin{cases} -2x + 14y = 8 \\ x - 7y = -4 \end{cases}$

15. $\begin{cases} 2x - 3y = 10 \\ 3x - 2y = 15 \end{cases}$

16. $\begin{cases} 2x + 3y = 18 \\ 3x + 2y = 12 \end{cases}$

17. $\begin{cases} y = 2x + 3 \\ 2x + y = -1 \end{cases}$

18. $\begin{cases} x = 3y - 4 \\ 3x + 2y = 10 \end{cases}$

19. $\begin{cases} 6a - 3b = 1 \\ 8a + 5b = 7 \end{cases}$

20. $\begin{cases} 2a - 6b = -4 \\ 5a - 7b = -4 \end{cases}$

21. $\begin{cases} s = 3t - 5 \\ t = 3s - 5 \end{cases}$

22. $\begin{cases} s = 5t - 8 \\ t = 5s - 8 \end{cases}$

23. $\begin{cases} 3m - 2n = 8 \\ 3n = m - 8 \end{cases}$

24. $\begin{cases} 5m - 3n = 2 \\ m - 4 = 2n \end{cases}$

25. $\begin{cases} 3p - 4q = 5 \\ 3q - 4p = -9 \end{cases}$

26. $\begin{cases} 5p + 6q = 1 \\ 5q + 6p = -1 \end{cases}$

27. $\begin{cases} \dfrac{u}{3} - v = 1 \\ u - \dfrac{v}{2} = 5 \end{cases}$

28. $\begin{cases} u - \dfrac{v}{4} = 4 \\ \dfrac{u}{5} - v = -3 \end{cases}$

29. $\begin{cases} \dfrac{w}{4} + \dfrac{z}{6} = 4 \\ \dfrac{w}{2} - \dfrac{z}{3} = 4 \end{cases}$

30. $\begin{cases} \dfrac{w}{6} - \dfrac{z}{5} = 2 \\ \dfrac{w}{2} - \dfrac{z}{10} = 1 \end{cases}$

31. $\begin{cases} \dfrac{x}{6} + \dfrac{y}{8} = \dfrac{3}{4} \\ \dfrac{x}{4} + \dfrac{y}{3} = \dfrac{17}{12} \end{cases}$

32. $\begin{cases} \dfrac{x}{5} + \dfrac{y}{3} = \dfrac{2}{3} \\ \dfrac{x}{10} - \dfrac{y}{4} = \dfrac{3}{4} \end{cases}$

33. $\begin{cases} \dfrac{x+3}{2} + \dfrac{y-4}{3} = \dfrac{19}{6} \\[2mm] \dfrac{x-2}{3} + \dfrac{y-2}{2} = 2 \end{cases}$

34. $\begin{cases} \dfrac{a-2}{4} - \dfrac{b+1}{2} = \dfrac{3}{2} \\[2mm] \dfrac{a-3}{3} + \dfrac{b+1}{4} = \dfrac{25}{4} \end{cases}$

35. $\begin{cases} 0.1x + 0.01y = .37 \\ .02x + 0.05y = .41 \end{cases}$

36. $\begin{cases} 0.3x - 0.7y = 2.93 \\ 0.06x - 0.2y = 0.58 \end{cases}$

37. $\begin{cases} \dfrac{x}{2} + 0.05y = .35 \\[2mm] 0.3x + \dfrac{y}{4} = .65 \end{cases}$

38. $\begin{cases} 0.02x + \dfrac{y}{2} = .3 \\[2mm] \dfrac{x}{2} - 0.4y = 2.34 \end{cases}$

Solve the following problems algebraically by writing an equation or a system of equations. Clearly label what each variable represents.

39. Susan wants to invest a total of $14,000 so that her yearly interest is $1,350. If she chooses to invest part in a certificate of deposit paying 9% and the remainder in a corporate bond paying 11%, how much should she invest at each rate?

40. Harry makes two investments. He invests $6,000 more at 8.5% than he invests at 7.5%. If the total yearly interest from both investments is $1,022, how much is invested at each rate?

41. To the nearest cent, how can $10,000 be split into two investments, one paying 10% interest and the other paying 8% interest, so that the yearly interest from the two investments are equal?

42. A person invests money at 12% and at 9%, earning a total yearly interest of $540. Had the amounts invested been reversed, the yearly interest would have been $510. How much was invested all together?

43. The perimeter of a rectangle is 36 cm. If the length is 2 cm more than the width, find the dimensions of the rectangle.

44. The side of a square is 2 less than 3 times the side of an equilateral triangle. If the perimeter of the square is 12 more than twice the perimeter of the triangle, find the lengths of the sides of both figures.

45. How much of each of 30% and 70% iodine solutions must be mixed together to produce 100 ml of a 54% iodine solution?

46. How much of an 18% saline solution must be mixed with 40 ml of a 30% saline solution to produce a 25% saline solution?

47. Albert and Audrey both go into a camera store to buy film. Albert spends $35.60 on 5 rolls of 35-mm film and 3 rolls of movie film. Audrey spends $43.60 on 3 rolls of 35-mm film and 5 rolls of movie film. What are the costs of a single roll of each type of film?

48. Jim has a part-time job selling newspaper and magazine subscriptions. One week he earns $62.20 by selling 10 newspaper and 6 magazine subscriptions. The following week he earns $79 by selling 12 newspaper and 8 magazine subscriptions. How much does he earn for each newspaper and each magazine subscription he sells?

49. A donut shop sells a box containing 7 cream-filled and 5 jelly donuts for $3.16, and a box containing 4 cream-filled and 8 jelly donuts for $3.04. Find the costs of a single cream-filled and a single jelly donut.

50. A candy shop sells a mixture containing 1 pound of hard candy and 2 pounds of chocolates for $15.39, and a mixture containing 2 pounds of hard candy and 1 pound of chocolates for $12.93. What are the costs for 1 pound of hard candy and 1 pound of chocolates?

51. An electronics manufacturer produces two types of transmitters. The more expensive model requires 6 hours to manufacture and 3 hours to assemble. The less expensive model requires 5 hours to manufacture and 2 hours to assemble. If the company can allocate 730 hours for manufacture and 340 hours for assembly, how many of each type can be produced?

52. The mathematics department hires tutors and graders. For the month of October, the department budgets $830 for 80 hours of tutoring and 45 hours of grading. In November, the department budgets $600 for 60 hours of tutoring and 30 hours of grading. How much does the department pay for each hour of tutoring and for each hour of grading?

53. A bank teller receives a deposit of 43 bills totaling $340. If the bills are all $5 and $10 bills, how many of each are there?

54. Dorothy purchases a total of 80 stamps for $15.85. If she bought 22-cent and 15-cent stamps, how many of each did she buy?

55. A car rental agency charges a flat fee plus a mileage rate for a 1-day rental. If the charge for a 1-day rental with 85 miles is $44.30 and the charge for a 1-day rental with 125 miles is $51.50, find the flat fee and the charge per mile.

56. A discount airline has a fixed charge for processing tickets plus a mileage fee. A ticket for a 200-mile trip costs $48, while a ticket for a 300-mile trip costs $61. Find the fixed charge and the charge per mile.

57. A plane can cover a distance of 2,310 km in 6 hours with a tailwind (with the wind) and a distance of 1,530 km in the same time with a headwind (against the wind). Find the speed of the plane and the speed of the wind.

58. An express train travels 35 kph faster than a freight train. After 3 hours they have travelled a total of 465 kilometers. Find the rate of each train.

59. Find the point (x, y) so that the line passing through the points (x, y) and $(2, 3)$ with a slope of -1 intersects the line passing through (x, y) and $(1, -2)$ with a slope of 2.

60. Find the values of A and B so that the line whose equation is $Ax + By = 8$ passes through the points $(1, -8)$ and $\left(5, \frac{8}{3}\right)$.

? QUESTIONS FOR THOUGHT

61. When asked to solve a system of equations, describe what you look for in deciding whether to use the elimination or substitution method.

62. Solve the following system by the elimination method:

$$\begin{cases} 4x - 9y = 3 \\ 10x - 6y = 7 \end{cases}$$

In using the elimination method, we eliminated a variable and then used *substitution* to find the value of the variable eliminated. A variation of the elimination method is to use the elimination process twice—once for each variable—to solve for each variable. Would this be easier for the system you just solved? Why?

63. Solve the following systems for u and v by first substituting x for $\frac{1}{u}$ and y for $\frac{1}{v}$:

(a) $$\begin{cases} \dfrac{1}{u} + \dfrac{1}{v} = 5 \\ \dfrac{1}{u} - \dfrac{1}{v} = 1 \end{cases}$$ **(b)** $$\begin{cases} \dfrac{2}{u} + \dfrac{1}{v} = 3 \\ \dfrac{6}{u} + \dfrac{1}{v} = 5 \end{cases}$$

↻ MINI-REVIEW

64. *Factor completely.* $(x - y)^2 - 25$

65. *Perform the operations and simplify:*

$$\frac{3}{x + 4} - \frac{3}{x - 5}$$

66. *Solve the following equation:*

$$\frac{3}{x + 4} - \frac{3}{x - 5} = \frac{3}{2}$$

67. *Perform the following operations and simplify:*

$$\sqrt{32} - 5\sqrt{18} + \frac{1}{\sqrt{2}}$$

68. If it takes Charlie 3 days to refinish a desk and it takes Janet 1 day to refinish the same desk, how long would it take them to refinish the desk together?

9.2

Linear Systems: Three Variables

The method of elimination which we described in Section 9.1 generalizes quite naturally to systems of equations with more than two variables.

A **3 × 3** (3 by 3) *system* is a system of three equations in three variables. Whereas the graph of a first-degree equation in two variables is a straight line in a two-dimensional coordinate system, a first-degree equation in three variables is a *plane* in a three-dimensional coordinate system. Instead of solutions being ordered pairs, they are *ordered triplets* (x, y, z) for a 3×3 system.

For example, the triplet $(1, -3, -2)$, which means $x = 1$, $y = -3$, $z = -2$, satisfies the system

$$\begin{cases} x + y + z = -4 \\ 4x - y + z = 5 \\ 3x + y - z = 2 \end{cases}$$

As in the case with a 2×2 system, a 3×3 system also has three possibilities. Geometric examples are illustrated in Figures 9.5**a–c**.

1. There is one unique solution to the system (see Figure 9.5**a**).

2. There are no solutions to the system (see Figure 9.5**b**).

3. There are infinitely many solutions to the system (see Figure 9.5**c**).

Figure 9.5

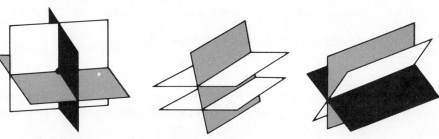

a. Three planes intersect at a point
One unique solution

b. Two planes are parallel
No solutions

c. Three planes intersect in a line
Infinitely many solutions

Let's illustrate how we can use the elimination method to solve a 3×3 system.

EXAMPLE 1

Solve the following system of equations:

$$\begin{cases} x + y + 2z = 11 & (1) \\ 2x - y - z = 0 & (2) \\ 3x + y + 4z = 25 & (3) \end{cases}$$

Note: We number the equations for ease of reference.

Solution

The elimination method for a 3×3 system requires us to take the 3×3 system, eliminate one of the variables to produce a 2×2 system, and then use the elimination process again to solve this 2×2 system.

We must first decide which variable we want to eliminate, and for a 3×3 system, making the "right" choice can significantly simplify the procedure. If we take a moment to look at the system, we can see that y can be eliminated most easily. We proceed as follows:

$$
\begin{array}{l}
x + y + 2z = 11 \qquad \textit{Add equations (1) and (2).} \\
\underline{2x - y - z = 0} \\
3x \quad + z = 11 \qquad \text{Let's call this equation (4).}
\end{array}
$$

If we can now get another equation involving x and z only, we will have a 2×2 system.

$$
\begin{array}{l}
2x - y - \; z = 0 \qquad \textit{Add equations (2) and (3).} \\
\underline{3x + y + 4z = 25} \\
5x \quad + 3z = 25 \qquad \text{Let's call this equation (5).}
\end{array}
$$

Equations (4) and (5) together are a 2×2 system which we can solve:

$$
\begin{array}{lll}
3x + \; z = 11 & \textit{Multiply by } -3 \rightarrow & -9x - 3z = -33 \\
5x + 3z = 25 & \textit{As is} \rightarrow & \underline{5x + 3z = \quad 25} \qquad \textit{Add the two equations.} \\
& & -4x \qquad = \; -8 \\
& & \qquad x = 2
\end{array}
$$

Now we can substitute $x = 2$ into either equation (4) or (5) to solve for z. We will substitute $x = 2$ into equation (4):

$$
\begin{array}{c}
3x + z = 11 \\
3(2) + z = 11 \\
6 + z = 11 \\
z = 5
\end{array}
$$

Having the values for x and z, we can go back to any of the original equations and solve for y. We will substitute $x = 2$, $z = 5$ into equation (1):

$$
\begin{array}{c}
x + y + 2z = 11 \\
2 + y + 2(5) = 11 \\
y + 12 = 11 \\
y = -1
\end{array}
$$

Thus, our solution is $\boxed{x = 2, \; y = -1, \; z = 5, \quad \text{or} \quad (2, -1, 5)}$

CHECK:

$$
\begin{array}{lll}
x + y + 2z = 11 & 2x - y - z = 0 & 3x + y + 4z = 25 \\
2 + (-1) + 2(5) \overset{?}{=} 11 & 2(2) - (-1) - 5 \overset{?}{=} 0 & 3(2) + (-1) + 4(5) \overset{?}{=} 25 \\
2 - 1 + 10 \overset{\checkmark}{=} 11 & 4 + 1 - 5 \overset{\checkmark}{=} 0 & 6 - 1 + 20 \overset{\checkmark}{=} 25
\end{array}
$$

The following outline summarizes the elimination method for a 3 × 3 system.

Elimination Method for a 3 × 3 System

1. Look over the system and choose the most convenient variable to eliminate (it may be that there is no preference).

2. Use any two of the equations to eliminate one of the variables. This gives an equation in at most two variables; call it equation (4).

3. Using a *different* pair of equations, eliminate the same variable as in step 2. This gives another equation in the same two variables; call it equation (5).

4. Use the elimination process to solve the system of equations (4) and (5).

5. Substitute the two values obtained in step 4 into one of the original equations to solve for the remaining variable.

6. Check the solution in all three original equations.

EXAMPLE 2

Solve the following system of equations:

$$\begin{cases} 6x + 2y - 3z = 6 & (1) \\ 12x - y = 2 & (2) \\ 3x + 4y + 2z = 9 & (3) \end{cases}$$

Solution

Noticing that equation (2) already involves only x and y, we use equations (1) and (3) to eliminate z as well.

$$\begin{array}{ll} 2 \text{ times equation (1)} \rightarrow & 12x + 4y - 6z = 12 \\ 3 \text{ times equation (3)} \rightarrow & \underline{9x + 12y + 6z = 27} \qquad \textit{Add the two equations.} \\ & 21x + 16y = 39 \qquad (4) \end{array}$$

Our 2 × 2 system consists of equations (2) and (4):

$$\begin{array}{lll} 12x - y = 2 & \textit{Multiply by 16} \rightarrow & 192x - 16y = 32 \\ 21x + 16y = 39 & \textit{As is} \rightarrow & \underline{21x + 16y = 39} \\ & & 213x = 71 \\ & & x = \dfrac{71}{213} = \dfrac{1}{3} \end{array}$$

Substitute $x = \frac{1}{3}$ into (2):

$$12x - y = 2$$

$$12\left(\frac{1}{3}\right) - y = 2$$

$$4 - y = 2$$

$$-y = -2$$

$$y = 2$$

Now substitute $x = \frac{1}{3}$ and $y = 2$ into equation (1):

$$6x + 2y - 3z = 6$$

$$6\left(\frac{1}{3}\right) + 2(2) - 3z = 6$$

$$2 + 4 - 3z = 6$$

$$6 - 3z = 6$$

$$-3z = 0$$

$$z = 0$$

Thus, our solution is $\boxed{x = \frac{1}{3},\ y = 2,\ z = 0, \quad \text{or} \quad \left(\frac{1}{3}, 2, 0\right)}$

The check is left to the student.   ■

EXAMPLE 3   Solve the following system of equations:

$$\begin{cases} 4x - 5y + 8z = 12 & (1) \\ -4x + 7y + z = 9 & (2) \\ x - \frac{7}{4}y - \frac{1}{4}z = 1 & (3) \end{cases}$$

Solution   We choose to eliminate x. Adding equations (1) and (2) we get

$$2y + 9z = 21 \qquad (4)$$

Next we eliminate x again:

Equation (2) →	$-4x + 7y + z = \ \ 9$	
4 times equation (3) →	$\underline{4x - 7y - z = \ \ 4}$	*Add the two equations.*
	$0 = 13$	*This is always false.*

Therefore, the system of equations has $\boxed{\text{no solutions}}$ ■

EXAMPLE 4   Solve the following system of equations:

$$\begin{cases} x + 2y + 3z = 5 & (1) \\ \dfrac{x}{6} + \dfrac{y}{3} + \dfrac{z}{2} = \dfrac{5}{6} & (2) \\ 4x + 8y + 12z = 20 & (3) \end{cases}$$

Solution   Whichever variable we choose to eliminate, any pair of equations will always yield an identity. (Try it.) In fact, if we look carefully at the system, we can see that if we multiply equation (2) by 6 we get equation (1), and if we multiply equation (1) by 4 we get equation (3). Thus, all three equations are equivalent. All three equations represent the same plane.

The solution set to this system is the set of all points which satisfy any one of the equations.

Thus, the solution set is $\boxed{\{(x,\, y,\, z) \mid x + 2y + 3z = 5\}}$ ■

EXAMPLE 5   A manufacturer produces three types of thread. Each type uses a different amount of three raw materials: cotton, rayon, and polyester. Table 9.1 contains the relevant information on the amount of each raw material used to produce one spool of each type of thread. If the manufacturer has 27 kilograms (kg) of cotton, 10.8 kg of rayon, and 8.2 kg of polyester in stock and wants to use it all up, how many spools of each type should be produced?

TABLE 9.1

Type of Thread	Amount of Cotton	Amount of Rayon	Amount of Polyester
Type A	15 gm	3 gm	2 gm
Type B	10 gm	6 gm	4 gm
Type C	8 gm	6 gm	6 gm

Solution   Let a = # of spools of type A thread that should be produced.

Let b = # of spools of type B thread that should be produced.

Let c = # of spools of type C thread that should be produced.

Each equation in our system represents the total number of grams of each material used to manufacture all the spools of thread. (Note that all units must be the same, so we have written 27 kg as 27,000 grams, etc.)

$$\begin{cases} 15a + 10b + 8c = 27{,}000 & (1) \\ 3a + 6b + 6c = 10{,}800 & (2) \\ 2a + 4b + 6c = 8{,}200 & (3) \end{cases}$$

This equation represents the amount of cotton used.

This equation represents the amount of rayon used.

This equation represents the amount of polyester used.

We choose to eliminate c:

$$
\begin{array}{rl}
\textit{Equation (2)} \rightarrow & 3a + 6b + 6c = 10{,}800 \\
-1 \textit{ times equation (3)} \rightarrow & \underline{-2a - 4b - 6c = -8{,}200} \\
& a + 2b = 2{,}600 \quad (4)
\end{array}
$$

$$
\begin{array}{rl}
3 \textit{ times equation (1)} \rightarrow & 45a + 30b + 24c = 81{,}000 \\
-4 \textit{ times equation (2)} \rightarrow & \underline{-12a - 24b - 24c = -43{,}200} \\
& 33a + 6b = 37{,}800 \quad (5)
\end{array}
$$

Thus, our 2 × 2 system is

$$
\begin{array}{lll}
a + 2b = 2{,}600 & (4) & \textit{Multiply by } -3 \rightarrow \\
33a + 6b = 37{,}800 & (5) & \textit{As is} \rightarrow
\end{array}
\qquad
\begin{array}{rl}
-3a - 6b = -7{,}800 \\
\underline{33a + 6b = 37{,}800} \\
30a = 30{,}000 \\
a = 1{,}000
\end{array}
$$

Substitute $a = 1{,}000$ into equation (4):

$$
\begin{aligned}
a + 2b &= 2{,}600 \\
1{,}000 + 2b &= 2{,}600 \\
2b &= 1{,}600 \\
b &= 800
\end{aligned}
$$

Substitute $a = 1{,}000$, $b = 800$ into equation (3):

$$
\begin{aligned}
2a + 4b + 6c &= 8{,}200 \\
2(1{,}000) + 4(800) + 6c &= 8{,}200 \\
2{,}000 + 3{,}200 + 6c &= 8{,}200 \\
5{,}200 + 6c &= 8{,}200 \\
6c &= 3{,}000 \\
c &= 500
\end{aligned}
$$

> Thus, our solution is 1,000 spools of type A, 800 spools of type B, and 500 spools of type C thread.

The check is left to the student. ∎

Exercises 9.2

In Exercises 1–26, *solve the system of equations.*

1. $\begin{cases} x + y + z = 9 \\ 2x - y + z = 9 \\ x - y + z = 3 \end{cases}$

2. $\begin{cases} x + y + z = 6 \\ 3x + y - z = 6 \\ 2x + y - z = 4 \end{cases}$

3. $\begin{cases} -x + 2y + z = 0 \\ x - y + 2z = 1 \\ x + 3y + z = 5 \end{cases}$

4. $\begin{cases} 2x + y - z = 8 \\ 3x - y - z = 11 \\ 2x - y + 2z = 2 \end{cases}$

5. $\begin{cases} x + y - z = 1 \\ 2x + 2y + 2z = 0 \\ x - y + z = 3 \end{cases}$

6. $\begin{cases} 2x - 3y + z = 1 \\ -2x + y - 2z = 6 \\ 2x - y - z = 3 \end{cases}$

7. $\begin{cases} x + 2y + 3z = 1 \\ 3x + 6y + 9z = 3 \\ 4x + 8y + 12z = 4 \end{cases}$

8. $\begin{cases} 3x + 2y + z = 2 \\ 6x + 4y + 2z = 4 \\ 15x + 10y + 5z = 10 \end{cases}$

9. $\begin{cases} 3x - 2y + 5z = 2 \\ 4x - 7y - z = 19 \\ 5x - 6y + 4z = 13 \end{cases}$

10. $\begin{cases} 5x + 4y - 3z = 3 \\ 6x + 3y - 2z = 7 \\ x - 2y + 4z = 8 \end{cases}$

11. $\begin{cases} 2a + b - 3c = -6 \\ 4a - 4b + 2c = 10 \\ 6a - 7b + c = 12 \end{cases}$

12. $\begin{cases} 2a + 3b - c = 2 \\ 5a - 6b + 4c = 3 \\ 6a - 9b + 5c = 4 \end{cases}$

13. $\begin{cases} x + 3y + 2z = 3 \\ x + 3z = 4 \\ x - 4y - z = 0 \end{cases}$

14. $\begin{cases} 2x - 3y + 4z = 1 \\ 5y - 2z = 5 \\ 3x + 2y - 5z = 8 \end{cases}$

15. $\begin{cases} \frac{1}{2}s + \frac{1}{3}t + u = 3 \\ \frac{1}{3}s - \frac{1}{2}t - 2u = 1 \\ \frac{2}{3}s - \frac{1}{6}t + \frac{1}{2}u = 6 \end{cases}$

16. $\begin{cases} \frac{s}{4} + \frac{t}{6} - \frac{u}{3} = 1 \\ \frac{s}{2} + \frac{t}{3} + u = -3 \\ \frac{s}{8} + \frac{t}{4} - u = 5 \end{cases}$

17. $\begin{cases} p + q + r = 6 \\ 2q + r - p = 6 \\ r - p + q = 4 \end{cases}$

18. $\begin{cases} 2r - s + t = 6 \\ s - 2r + t = 0 \\ 3t - r + s = 20 \end{cases}$

19. $\begin{cases} x + y = 0 \\ y + z = 0 \\ x + z = 2 \end{cases}$

20. $\begin{cases} 2x - y = 0 \\ y - z = 0 \\ 3x + 2z = 0 \end{cases}$

21. $\begin{cases} a + b = 2b + c \\ a - 2b = c + 3 \\ 2a - b = 3c - 9 \end{cases}$

22. $\begin{cases} m + 3n = p + 8 \\ 2m - 4n = 2p + 6 \\ 3m - n = -p + 2 \end{cases}$

23. $\begin{cases} 12a + 5b + 3c = 24,000 \\ 10a + 6b + 4c = 13,300 \\ 8a + 7b + 5c = 8,700 \end{cases}$

24. $\begin{cases} 12a + 10b + 8c = 24,000 \\ 5a + 6b + 7c = 13,300 \\ 3a + 4b + 5c = 8,700 \end{cases}$

25. $\begin{cases} 0.06x + 0.07y + 0.08z = 440 \\ 0.05x + 0.06y + 0.08z = 410 \\ 0.04x + 0.05y + 0.06z = 320 \end{cases}$

26. $\begin{cases} 0.08x + 0.10y + 0.12z = 17,000 \\ 0.06x + 0.05y + 0.08z = 12,300 \\ 0.04x + 0.10y + 0.06z = 11,000 \end{cases}$

In Exercises 27–34, solve the problems algebraically. Clearly label what each variable represents.

27. A collection of 48 coins consists of dimes, quarters, and half-dollars, and has a total value of $10.55. If there are 2 fewer dimes than quarters and half-dollars combined, how many of each type of coin are there?

28. A bank teller gave a customer change for a $500 bill in $5, $10, and $20 bills. The number of $5 and $20 bills combined was 5 less than twice the number of $10 bills. If there were 40 bills in all, how many of each type of bill were there?

29. Martha splits up a total of $12,000 into three investments. She has a bank account paying 8.7%, a bond paying 9.3%, and a stock paying 12.66%. Her annual interest from the three investments is $1,266. If the interest from the stock is equal to the interest from the bank account and the bond combined, how much is invested at each rate?

30. George wants to divide $17,000 into three investments so that the interests from the three investments are equal. If he is going to invest part in a certificate of deposit paying 8.5%, part in a corporate bond paying 13.6%, and the rest in a high-risk real-estate deal paying 17%, how much should he invest at each rate?

31. A theater group plans to sell 750 tickets for a play. They are charging $12 for orchestra seats, $8 for mezzanine seats, and $6 for balcony seats, and they plan to collect $7,290. If there are 100 more orchestra tickets than mezzanine and balcony tickets combined, how many tickets of each type are there?

32. An opera house is planning to put on a performance and wants to determine its ticket prices. The theater has an orchestra section which holds 550 people, a mezzanine which holds 140 people, and a balcony which holds 275 people. If they want a balcony seat to cost 20% less than a mezzanine seat and an orchestra seat to cost $5 more than a mezzanine seat, and they want a full house to bring in $16,400, how much should they charge for each type of seat?

33. A computer manufacturer produces three models of personal computers—model A, model B, and model C. The company knows how much production, assembly, and testing time is needed for each model. This information is found in the accompanying table. If the company has allocated 721 hours for production, 974 hours for assembly, and 168 hours for testing, how many of each model can the company produce if it wants to use up all the time allocated for each phase of the process?

34. A nutritionist wants to create a food supplement out of three substances: A, B, and C. She wants the food supplement to have the following characteristics: 5 grams of the supplement should supply 1.54 gm of iron and cost 48¢. The iron content and cost of substances A, B, and C are entered in the accompanying table. How many grams of each substance should be used to make such a food supplement?

Substance	Iron Content per Gram	Cost per Gram
Substance A	0.3 gm	10¢
Substance B	0.28 gm	8¢
Substance C	0.4 gm	12¢

Model	Hours Required for Production	Hours Required for Assembly	Hours Required for Testing
Model A	2.1	3.2	0.5
Model B	2.8	3.6	0.6
Model C	3.2	4.0	0.8

 MINI-REVIEW

35. Sketch the graph of $3x - 2y = 12$.

36. *Simplify:*

$$\sqrt[3]{\frac{7}{3}}$$

37. *Perform the following operations, simplify, and express your answer using positive exponents only:*

$$\left(\frac{2x^{1/4}y^{1/5}}{3x^2}\right)^3$$

38. Find the slope of the line passing through $(-2, 4)$ and $(3, -5)$.

39. How much pure water must be added to 5 quarts of a 60% solution of alcohol in order to dilute it to 40%?

9.3

Solving Linear Systems Using Matrices

In the process of using the elimination method for solving systems of equations, we concentrated mainly on the constants and the coefficients of the variables. We were concerned only with the actual variables when we wanted to make sure we were "matching up" the correct variables to be eliminated. Hence, as long as we have some way of keeping track of the variables, we should be able to solve systems focusing primarily on the coefficients and constants of the system. In this and the next section we discuss methods of solving systems using the constants and the coefficients of the variables of the system.

A rectangular array of numbers is called a *matrix*. The numbers in a matrix are called the *elements* or *entries* of the matrix. The horizontal arrays of entries are called the *rows*, and the vertical arrays of entries are called the *columns*. The following are two examples of matrices:

$$\begin{bmatrix} 1 & -2 & 3 & 4 \\ 0 & 5 & 0 & -3 \end{bmatrix} \qquad \begin{matrix} \text{Row 1} \to \\ \text{Row 2} \to \\ \text{Row 3} \to \end{matrix} \begin{bmatrix} 2 & 3 & -1 \\ 0 & -5 & 2 \\ 1 & 3 & 5 \end{bmatrix}$$

where the columns are labeled Column 1, Column 2, Column 3.

We usually specify the size of a matrix by first giving the number of rows and then the number of columns. The first example is a 2×4 (2 by 4) matrix; the second is a 3×3 matrix. A *square matrix* is a matrix that contains an equal number of rows and columns; the 3×3 matrix above is a square matrix.

We will now examine how we can use matrices to solve systems of equations. We begin by taking a system and rewriting it as an *augmented matrix* as follows:

$$\begin{cases} 2x - 3y = 4 \\ 5x + y = -1 \end{cases}$$

is written as the augmented matrix

$$\begin{bmatrix} 2 & -3 & \bigm| & 4 \\ 5 & 1 & \bigm| & -1 \end{bmatrix}$$

The matrix $\begin{bmatrix} 2 & -3 \\ 5 & 1 \end{bmatrix}$, made up of the coefficients of the variables of the system, is called the *coefficient matrix*. The *augmented matrix* of a system of equations consists of the coefficient matrix of the system with the constants of the system adjoined to the right.

Consider another example:

$$\begin{cases} 3x + 5y - z = 7 \\ 2x + 3z = -2 \\ x - 2y - 2z = 4 \end{cases} \quad \text{is written as} \quad \begin{matrix} x & y & z & \\ \end{matrix} \begin{bmatrix} 3 & 5 & -1 & \bigm| & 7 \\ 2 & 0 & 3 & \bigm| & -2 \\ 1 & -2 & -2 & \bigm| & 4 \end{bmatrix}$$

Notice that a 0 is entered in the second row of the augmented matrix for the missing y variable in the second equation. Also observe that 1 is entered in row 3 as the coefficient of x in the third equation of the system.

The relative positions of the column entries in the coefficient matrix indicate the variable: the first column consists of the coefficients of the x variable; the second column, the coefficients of the y variable; and the third column, the coefficients of the z variable. The constants are on the augmented side. Thus it is important that the variables be lined up columnwise before we change a system into its augmented matrix.

As we do in solving single equations, we define **equivalent systems of equations** as systems with the same solution(s). For example, the three following systems on the left-hand side are equivalent systems. When we solve each system we find that each system has the unique solution $(2, -1, 3)$. Look carefully at each system's augmented matrix on the right-hand side.

$$
\begin{array}{cc}
\textit{System} & \textit{Augmented matrix} \\
\end{array}
$$

$$
1 \begin{cases} 3x - y + z = 10 \\ x + y + z = 4 \\ x + 2y + 3z = 9 \end{cases}
\qquad
\begin{bmatrix} 3 & -1 & 1 & | & 10 \\ 1 & 1 & 1 & | & 4 \\ 1 & 2 & 3 & | & 9 \end{bmatrix}
$$

$$
2 \begin{cases} 3x - y + z = 10 \\ -x - y - z = -4 \\ y + 2z = 5 \end{cases}
\qquad
\begin{bmatrix} 3 & -1 & 1 & | & 10 \\ -1 & -1 & -1 & | & -4 \\ 0 & 1 & 2 & | & 5 \end{bmatrix}
$$

$$
3 \begin{cases} 3x - y + z = 10 \\ -4y - 2z = -2 \\ 6z = 18 \end{cases}
\qquad
\begin{bmatrix} 3 & -1 & 1 & | & 10 \\ 0 & -4 & -2 & | & -2 \\ 0 & 0 & 6 & | & 18 \end{bmatrix}
$$

The main diagonal

Notice that the last system, system 3, is easiest to solve. We simply start at the bottom equation and solve for z. Then we substitute the value we found for z in the next equation up and solve for y. Finally, we substitute the values we found for y and z into the first equation to find the value for x. This process of substitution is called **back substitution**.

System 3 and its augmented matrix are said to be in *triangular* (or *echelon*) *form*.* The elements 3, -4, 6 in this augmented matrix are called the **main diagonal**. Note that the main diagonal of an augmented matrix is the diagonal of the coefficient matrix. A matrix is in **triangular** (or **echelon**) **form** if it has all zero entries below the main diagonal.

This will be our goal: Change an augmented matrix into triangular form, convert the matrix to its associated system, and then solve the simpler system by back substitution. We begin with the following definition:

> Two matrices are **row-equivalent** if their associated systems of equations are equivalent.

Thus, all of the matrices shown on the right in the previous display are row-equivalent since their corresponding systems are all equivalent.

---

* Sometimes the term *triangular form* is restricted to the system of equations rather than its augmented matrix; however, we will use the term *triangular form* to describe both the system and its augmented matrix.

With equations, after we defined equivalence we discussed what we were allowed to do to an equation in order to transform it into a simpler equivalent equation. Now that we have defined row-equivalence for matrices, we will take the same approach and list the transformations we are allowed to perform on matrices in order to arrive at simpler, row-equivalent matrices (or those closer to triangular form). We call these transformations the ***elementary row operations***.

Elementary Row Operations	
	1. Multiply each entry in a given row by any nonzero constant.
	2. Interchange any two rows.
	3. Add a multiple of one row to another row.

For example,

1. $\begin{bmatrix} 2 & 3 & -2 \\ 3 & 2 & 5 \end{bmatrix}$ $\quad -3R_1 \rightarrow R_1$ $\quad \begin{bmatrix} -6 & -9 & 6 \\ 3 & 2 & 5 \end{bmatrix}$

This notation means multiply each entry in row 1 by -3 to get the new row 1.

2. $\begin{bmatrix} 2 & 5 & 0 \\ 3 & 1 & 2 \\ 1 & 0 & 4 \end{bmatrix}$ $\quad R_1 \longleftrightarrow R_2$ $\quad \begin{bmatrix} 3 & 1 & 2 \\ 2 & 5 & 0 \\ 1 & 0 & 4 \end{bmatrix}$

This notation means interchange row 1 and row 2. (Row 3 remains unchanged.)

3. $\begin{bmatrix} 2 & -3 & 3 \\ 3 & 0 & 2 \\ 1 & 2 & -1 \end{bmatrix}$ $\quad 2R_1 + R_3 \rightarrow R_3$ $\quad \begin{bmatrix} 2 & -3 & 3 \\ 3 & 0 & 2 \\ 5 & -4 & 5 \end{bmatrix}$

This notation means multiply each entry in row 1 by 2 and add the resulting entries to each entry in row 3 to get a new row 3. (Rows 1 and 2 remain unchanged.)

Since this last example is a bit more complex than the first two, we will demonstrate how we found the new row 3.

Multiply each entry in row 1 by 2:

$$ 2 \quad -3 \quad 3 \quad \longrightarrow \quad 4 \quad -6 \quad 6 \qquad \textit{This is } 2R_1. $$

Then add each entry in this multiple of row 1 to each entry in row 3:

$$ + \quad \underline{1 \quad 2 \quad -1} \qquad \textit{This is } R_3. $$

to get the new row 3:

$$ 5 \quad -4 \quad 5 \qquad \textit{This is the new } R_3. $$

Notice that the elementary row operations are equivalent to the transformations performed on *systems of equations*: Row operation 1 is equivalent to multiplying both sides of any equation by a nonzero constant; row operation 2 is equivalent to interchanging any two equations in a system; and row operation 3 is equivalent to adding a multiple of one equation to another.

Since elementary row operations produce associated systems that are equivalent to each other, we can conclude:

> A matrix can be transformed into a row-equivalent matrix by performing any elementary row operation on the matrix.

The method we will use in solving systems of equations by matrices, called *Gaussian elimination*, is described in the next box.

Method for Solving Linear Systems Using Matrices (Gaussian Elimination)	1. Set up the augmented matrix of the system.
	2. Use the elementary row operations to transform the augmented matrix into a row-equivalent matrix in triangular form.
	3. For the augmented matrix in triangular form, write the corresponding system of equations.
	4. Solve this system by back substitution.
	5. Check your solution in the original equations.

EXAMPLE 1

Solve the following system by Gaussian elimination:

$$\begin{cases} x + 2y = 8 \\ 4x - 3y = 21 \end{cases}$$

Solution

The first step is to set up the augmented matrix of the system:

$$\begin{bmatrix} 1 & 2 & | & 8 \\ 4 & -3 & | & 21 \end{bmatrix}$$ *Now we use the elementary row operations to transform this matrix into a row-equivalent matrix in triangular form.*

For a 2×2 matrix to be in triangular form, all we need is one 0 in the bottom left-hand corner of the matrix (below the main diagonal: $1, -3$):

$$\begin{bmatrix} 1 & 2 & | & 8 \\ 4 & -3 & | & 21 \end{bmatrix} \qquad -4R_1 + R_2 \rightarrow R_2 \qquad \begin{bmatrix} 1 & 2 & | & 8 \\ 0 & -11 & | & -11 \end{bmatrix}$$

Multiply the entries in row 1 by -4 and add the result to the entries in row 2 to get the new row 2.   *This matrix is now in triangular form.*

Now that the matrix is in triangular form, we can convert the new (row-equivalent) matrix into its associated system:

$$\begin{cases} x + 2y = 8 \\ -11y = -11 \end{cases}$$ *Now we can solve this system by back substitution.*

First solve for y:

$$-11y = -11 \qquad \text{*Divide both sides of the equation by -11.*}$$
$$y = 1$$

Then substitute 1 for y in the first equation and solve for x:

$x + 2y = 8$ *Substitute $y = 1$*

$x + 2(1) = 8$ *and solve for x to get:*

$x = 6$

The solution is $\boxed{x = 6, y = 1.}$ *You should check that this is the solution to the original system of equations.* ∎

EXAMPLE 2

Solve the following system by Gaussian elimination:

$$\begin{cases} x + 2y + 2z = 10 \\ 2x + 3y - z = 6 \\ 3x + y + 5z = 8 \end{cases}$$

Solution

First set up the augmented matrix of the system:

$$\begin{bmatrix} 1 & 2 & 2 & | & 10 \\ 2 & 3 & -1 & | & 6 \\ 3 & 1 & 5 & | & 8 \end{bmatrix}$$
Now use the elementary row operations to transform this matrix into a row-equivalent matrix in triangular form.

Begin by getting 0 below 1 in the first column:

$$\begin{bmatrix} 1 & 2 & 2 & | & 10 \\ 2 & 3 & -1 & | & 6 \\ 3 & 1 & 5 & | & 8 \end{bmatrix} \quad -2R_1 + R_2 \rightarrow R_2 \quad \begin{bmatrix} 1 & 2 & 2 & | & 10 \\ 0 & -1 & -5 & | & -14 \\ 3 & 1 & 5 & | & 8 \end{bmatrix}$$

Multiply row 1 by −2 and add it to row 2 to get the new row 2.

Next, we need to get another 0 in the bottom left-hand corner:

$$\begin{bmatrix} 1 & 2 & 2 & | & 10 \\ 0 & -1 & -5 & | & -14 \\ 3 & 1 & 5 & | & 8 \end{bmatrix} \quad -3R_1 + R_3 \rightarrow R_3 \quad \begin{bmatrix} 1 & 2 & 2 & | & 10 \\ 0 & -1 & -5 & | & -14 \\ 0 & -5 & -1 & | & -22 \end{bmatrix}$$

Multiply row 1 by −3 and add it to row 3 to get the new row 3.

The last step is to get 0 under -1 in the second column. Note that if we try to add a multiple of row 1 to row 3, we will lose the 0 in row 3, column 1. Because there is a 0 in row 2, column 1, adding a multiple of row 2 will not affect the 0 in row 3, column 1.

$$\begin{bmatrix} 1 & 2 & 2 & | & 10 \\ 0 & -1 & -5 & | & -14 \\ 0 & -5 & -1 & | & -22 \end{bmatrix} \quad -5R_2 + R_3 \rightarrow R_3 \quad \begin{bmatrix} 1 & 2 & 2 & | & 10 \\ 0 & -1 & -5 & | & -14 \\ 0 & 0 & 24 & | & 48 \end{bmatrix}$$

Multiply row 2 by −5 and add it to row 3 to get the new row 3.

This matrix is in triangular form.

Now that the matrix is in triangular form, we can convert the new (row-equivalent) matrix into its associated system:

$$\begin{cases} x + 2y + 2z = 10 \\ \quad -y - 5z = -14 \\ \qquad\qquad 24z = 48 \end{cases}$$ *Now we can solve this system by back substitution.*

First solve for z:

$24z = 48$ *Divide both sides of the equation by 24.*

$z = 2$

Then substitute 2 for z in the second equation and find y:

$-y - 5z\ = -14$ *Substitute $z = 2$*

$-y - 5(2) = -14$ *and solve for y to get:*

$-y - 10\ = -14$

$y = 4$

Then substitute 2 for z, and 4 for y in the first equation, and solve for x:

$x + 2y\ + 2z\ = 10$ *Substitute $z = 2$ and $y = 4$*

$x + 2(4) + 2(2) = 10$ *and solve for x to get:*

$x = -2$

Hence, the solution is $\boxed{(-2, 4, 2).}$ *Check this solution in the original system of equations.* ∎

EXAMPLE 3   Solve the following system by Gaussian elimination:

$$\begin{cases} 6x - 4y = 7 \\ 3x - 2y = 4 \end{cases}$$

Solution   First identify its augmented matrix:

$$\begin{bmatrix} 6 & -4 & | & 7 \\ 3 & -2 & | & 4 \end{bmatrix}$$

Again, for a 2×2 matrix to be in triangular form, all we need is one 0 in the bottom left-hand corner of the matrix. In order to get 0 in the bottom left-hand corner we would have to multiply row 1 by $-\frac{1}{2}$ and add it to row 2. In this example we can avoid computations with fractions by first applying elementary row operation 2—interchanging rows 1 and 2:

$$\begin{bmatrix} 6 & -4 & | & 7 \\ 3 & -2 & | & 4 \end{bmatrix} \qquad R_1 \longleftrightarrow R_2 \qquad \begin{bmatrix} 3 & -2 & | & 4 \\ 6 & -4 & | & 7 \end{bmatrix}$$

Now we can get 0 in the bottom left-hand corner:

$$\begin{bmatrix} 3 & -2 & | & 4 \\ 6 & -4 & | & 7 \end{bmatrix} \qquad -2R_1 + R_2 \rightarrow R_2 \qquad \begin{bmatrix} 3 & -2 & | & 4 \\ 0 & 0 & | & -1 \end{bmatrix}$$

Multiply the entries in row 1 by −2 and add this to the entries in row 2 to get the new row 2.

This matrix is now in triangular form. We write the associated system of the new matrix:

$$\begin{cases} 3x - 2y = 4 \\ \qquad\quad 0 = -1 \end{cases}$$

Since the second equation is a contradiction, the system is $\boxed{\text{inconsistent.}}$ We conclude that there is $\boxed{\text{no solution.}}$ ∎

EXAMPLE 4

Solve the following system by Gaussian elimination:

$$\begin{cases} 2x + 6y + 4z = 8 \\ 4x + 12y + 10z = 20 \\ 3x + 9y + 6z = 12 \end{cases}$$

Solution

First identify its augmented matrix:

$$\begin{bmatrix} 2 & 6 & 4 & | & 8 \\ 4 & 12 & 10 & | & 20 \\ 3 & 9 & 6 & | & 12 \end{bmatrix}$$

Now we get 0 below 2 in the first column:

$$\begin{bmatrix} 2 & 6 & 4 & | & 8 \\ 4 & 12 & 10 & | & 20 \\ 3 & 9 & 6 & | & 12 \end{bmatrix} \qquad -2R_1 + R_2 \rightarrow R_2 \qquad \begin{bmatrix} 2 & 6 & 4 & | & 8 \\ 0 & 0 & 2 & | & 4 \\ 3 & 9 & 6 & | & 12 \end{bmatrix}$$

Multiply the entries in row 1 by −2 and add the result to the entries in row 2 to get a new row 2.

Then we get 0 in the bottom left-hand corner. We could multiply the entries in row 1 by $-\frac{3}{2}$ and add the result to the entries in row 3 to get 0 in this spot. However, we can avoid fractions by first noting that the LCM of 2 and 3 is 6, and multiply row 1 by 3 and row 3 by 2 in order to get 6's in the first column. We will do both operations in this step:

$$\begin{bmatrix} 2 & 6 & 4 & | & 8 \\ 0 & 0 & 2 & | & 4 \\ 3 & 9 & 6 & | & 12 \end{bmatrix} \qquad \begin{array}{c} 3R_1 \rightarrow R_1 \\ \\ 2R_3 \rightarrow R_3 \end{array} \qquad \begin{bmatrix} 6 & 18 & 12 & | & 24 \\ 0 & 0 & 2 & | & 4 \\ 6 & 18 & 12 & | & 24 \end{bmatrix}$$

Multiply row 1 by 3 to get the new row 1 and multiply row 3 by 2 to get the new row 3.

Now we can easily get 0 in the bottom left-hand corner:

$$\begin{bmatrix} 6 & 18 & 12 & | & 24 \\ 0 & 0 & 2 & | & 4 \\ 6 & 18 & 12 & | & 24 \end{bmatrix} \qquad -R_1 + R_3 \rightarrow R_3 \qquad \begin{bmatrix} 6 & 18 & 12 & | & 24 \\ 0 & 0 & 2 & | & 4 \\ 0 & 0 & 0 & | & 0 \end{bmatrix}$$

We convert the matrix into its associated system:

$$\begin{cases} 6x + 18y + 12z = 24 \\ 2z = 4 \end{cases}$$ *Note that the third equation, which would be $0 = 0$, need not appear in the system.*

We solve the system by first solving for z in the last equation to get

$$2z = 4$$
$$z = 2$$

Then we substitute 2 for z in the first equation:

$$6x + 18y + 12z = 24 \qquad \textit{Substitute } z = 2 \textit{ to get:}$$
$$6x + 18y + 12(2) = 24 \qquad \textit{Which is equivalent to:}$$
$$6x + 18y = 0 \qquad \textit{or equivalently (dividing both sides of the equation by 6):}$$
$$x + 3y = 0$$

There is no unique solution for this system. The first and third equations of the original system are equivalent and we end up with two planes intersecting in a line similar to Figure 9.5c in Section 9.2. The best we can do is list the set of all points that satisfy this system in the following way:

$$\boxed{\{(x, y, 2) \mid x + 3y = 0\}} \qquad \textit{Notice } z = 2. \qquad \blacksquare$$

Gauss–Jordan Elimination

In the method described above, when we change a matrix into triangular form and identify its associated system of equations, we still had to back substitute in order to arrive at the system's solution. Also, keep in mind that triangular matrices for a system are not unique, that is, a system can have different (but equivalent) triangular matrices.

If we continued to apply the elementary row operations to a matrix in triangular form in an attempt to get all 1's on the main diagonal and 0's in the rest of the coefficient matrix as pictured in the 3×3 matrix below (where a, b, and c are constants), then we can read off the solution directly from the matrix. In the matrix below, $x = a$, $y = b$, and $z = c$:

$$\begin{bmatrix} 1 & 0 & 0 & | & a \\ 0 & 1 & 0 & | & b \\ 0 & 0 & 1 & | & c \end{bmatrix}$$

The form of this 3×3 matrix is called *reduced echelon form*.

A matrix is in ***reduced echelon form*** when:

1. Rows consisting entirely of 0 entries are on the bottom of the matrix.

2. The first nonzero entry in any row is 1, called a *pivot element*.

3. Each nonzero row is arranged so that the pivot element occurs farther to the right than the pivot element in the preceding row.

4. In each column with a pivot element, the remaining entries are all 0.

We can put a matrix into reduced echelon form using the elementary row operations. The method of solving systems by putting a system's augmented matrix into reduced echelon form is called ***Gauss–Jordan elimination***. The next example illustrates an efficient approach for putting a matrix into reduced echelon form.

EXAMPLE 5

Solve the following system by Gauss–Jordan elimination:

$$\begin{cases} 2x - y + z = 3 \\ x + y + z = 4 \\ x + 2y + 3z = 4 \end{cases}$$

Solution

We identify the system's augmented matrix:

$$\left[\begin{array}{ccc|c} 2 & -1 & 1 & 3 \\ 1 & 1 & 1 & 4 \\ 1 & 2 & 3 & 4 \end{array}\right]$$

1. First, we want to get a 1 in the upper left-hand corner. Rather than divide row 1 by 2 and end up with fractions in the first row, we can interchange row 1 with row 2:

$$\left[\begin{array}{ccc|c} 2 & -1 & 1 & 3 \\ 1 & 1 & 1 & 4 \\ 1 & 2 & 3 & 4 \end{array}\right] \qquad R_1 \longleftrightarrow R_2 \qquad \left[\begin{array}{ccc|c} 1 & 1 & 1 & 4 \\ 2 & -1 & 1 & 3 \\ 1 & 2 & 3 & 4 \end{array}\right]$$

2. Now we want to get 0's below the 1 in the first column. (This is called sweeping out the column.) We will take two steps at once; observe the transformations.

$$\left[\begin{array}{ccc|c} 1 & 1 & 1 & 4 \\ 2 & -1 & 1 & 3 \\ 1 & 2 & 3 & 4 \end{array}\right] \qquad \begin{array}{l} -2R_1 + R_2 \to R_2 \\ -R_1 + R_3 \to R_3 \end{array} \qquad \left[\begin{array}{ccc|c} 1 & 1 & 1 & 4 \\ 0 & -3 & -1 & -5 \\ 0 & 1 & 2 & 0 \end{array}\right]$$

3. Next we want to get 1 where -3 is (in the middle of the main diagonal). Again, rather than divide row 2 by -3 and get fractions, we can interchange row 2 and row 3:

$$\begin{bmatrix} 1 & 1 & 1 & | & 4 \\ 0 & -3 & -1 & | & -5 \\ 0 & 1 & 2 & | & 0 \end{bmatrix} \quad R_2 \longleftrightarrow R_3 \quad \begin{bmatrix} 1 & 1 & 1 & | & 4 \\ 0 & 1 & 2 & | & 0 \\ 0 & -3 & -1 & | & -5 \end{bmatrix}$$

4. Now we want to get 0's in the first and third rows of the second column (sweep out the rest of the second column). Again, we will take two steps:

$$\begin{bmatrix} 1 & 1 & 1 & | & 4 \\ 0 & 1 & 2 & | & 0 \\ 0 & -3 & -1 & | & -5 \end{bmatrix} \quad \begin{matrix} -R_2 + R_1 \to R_1 \\ 3R_2 + R_3 \to R_3 \end{matrix} \quad \begin{bmatrix} 1 & 0 & -1 & | & 4 \\ 0 & 1 & 2 & | & 0 \\ 0 & 0 & 5 & | & -5 \end{bmatrix}$$

5. Next we want to get 1 in the lower right-hand corner of the coefficient matrix:

$$\begin{bmatrix} 1 & 0 & -1 & | & 4 \\ 0 & 1 & 2 & | & 0 \\ 0 & 0 & 5 & | & -5 \end{bmatrix} \quad \tfrac{1}{5}R_3 \to R_3 \quad \begin{bmatrix} 1 & 0 & -1 & | & 4 \\ 0 & 1 & 2 & | & 0 \\ 0 & 0 & 1 & | & -1 \end{bmatrix}$$

6. Finally, we want to sweep out the rest of the third column:

$$\begin{bmatrix} 1 & 0 & -1 & | & 4 \\ 0 & 1 & 2 & | & 0 \\ 0 & 0 & 1 & | & -1 \end{bmatrix} \quad \begin{matrix} R_3 + R_1 \to R_1 \\ -2R_3 + R_2 \to R_2 \end{matrix} \quad \begin{bmatrix} 1 & 0 & 0 & | & 3 \\ 0 & 1 & 0 & | & 2 \\ 0 & 0 & 1 & | & -1 \end{bmatrix}$$

The matrix is in reduced echelon form and we can directly read off the matrix by columns to get the solution to the original system:

$$\boxed{x = 3, y = 2, \text{ and } z = -1.} \qquad \blacksquare$$

A few comments should be made about trying to put a matrix into triangular form versus reduced echelon form. Unlike triangular form, one advantage to reduced echelon form is that it is unique for a system. The key advantage to Gauss–Jordan elimination is that it lends itself well to programming. However, Gauss–Jordan elimination requires more elementary row operations than does Gaussian elimination and you may feel that it would be quicker to stop and begin back substitution when you have a matrix in triangular form. This is especially true when you cannot avoid extensive computations with fractions.

Exercises 9.3

In Exercises 1–4, set up the augmented matrices of the systems of equations.

1. $\begin{cases} 3x - 2y = 5 \\ x - y = 8 \end{cases}$

2. $\begin{cases} 2x + y = -6 \\ 7x + 2y = 0 \end{cases}$

3. $\begin{cases} x - 2y + 3z = 4 \\ y - z = -3 \\ 2x + 3y = 8 \end{cases}$

4. $\begin{cases} x - 2y + z = -1 \\ 3y - z = 2 \\ x - 2z = 0 \end{cases}$

In Exercises 5–26, solve the systems of equations by Gaussian elimination.

5. $\begin{cases} x - 2y = 7 \\ 2x - 3y = 12 \end{cases}$

6. $\begin{cases} x + 2y = 3 \\ 5x - 3y = -11 \end{cases}$

7. $\begin{cases} x - 3y = 6 \\ 3x + 5y = -10 \end{cases}$

8. $\begin{cases} 5x - 3y = 4 \\ 10x - 6y = 2 \end{cases}$

9. $\begin{cases} 2x - 5y = -8 \\ 7x + 3y = -28 \end{cases}$

10. $\begin{cases} 5x - 2y = 18 \\ 3x + 4y = -10 \end{cases}$

11. $\begin{cases} 6x + 2y = 9 \\ 4x - y = -1 \end{cases}$

12. $\begin{cases} 2x - 3y = 7 \\ 3x + 6y = 14 \end{cases}$

13. $\begin{cases} 6x + 2y = 5 \\ 3x - 4y = 0 \end{cases}$

14. $\begin{cases} 4x + 6y = 0 \\ 8x - 2y = 7 \end{cases}$

15. $\begin{cases} x + 3y + z = 8 \\ x + 2y + z = 7 \\ x - 2y + 2z = 6 \end{cases}$

16. $\begin{cases} x + y + z = 4 \\ 3x + 2y - z = 13 \\ 2x - y + 2z = -1 \end{cases}$

17. $\begin{cases} x + y - z = 2 \\ 3x + y - z = -2 \\ 4x - 2y + z = -13 \end{cases}$

18. $\begin{cases} x + 3y - z = 0 \\ 2x + y - z = 4 \\ 3x - y + 2z = 5 \end{cases}$

19. $\begin{cases} x - 2y + 3z = 7 \\ 2x + 3y - z = 0 \\ x + y + z = 1 \end{cases}$

20. $\begin{cases} x + 2y - 2z = 4 \\ 3x + 3y - z = 7 \\ 5x - 2y + 2z = 8 \end{cases}$

21. $\begin{cases} 4x - y + 2z = 6 \\ 2x + 3y - z = 4 \\ 2x - 2y + z = 0 \end{cases}$

22. $\begin{cases} 2x - 3y + 2z = 11 \\ x - 6y - z = 1 \\ 3x - 3y + z = 10 \end{cases}$

23. $\begin{cases} 2x + 3y + 2z = 4 \\ 4x + 6y + 4z = 8 \\ 2x + 3y + 5z = 13 \end{cases}$

24. $\begin{cases} x + y - z = 6 \\ 2x + 2y - 2z = 12 \\ 3x + 3y - 3z = 18 \end{cases}$

25. $\begin{cases} w - 2x + y - z = 2 \\ w + 2x + 2y + z = 0 \\ 2w - 2x + y - z = 3 \\ 2w - 2y + z = 5 \end{cases}$

26. $\begin{cases} w - x + y = -2 \\ x - z = -3 \\ -2x + 2y - z = -7 \\ w + 2y + z = -1 \end{cases}$

In Exercises 27–30, put the augmented matrix in reduced echelon form.

27. $\left[\begin{array}{cc|c} 2 & 3 & 5 \\ 1 & 4 & 0 \end{array}\right]$

28. $\left[\begin{array}{cc|c} 3 & 2 & 0 \\ 2 & 1 & 1 \end{array}\right]$

29. $\left[\begin{array}{ccc|c} 1 & 1 & 2 & 1 \\ 2 & 4 & 2 & 6 \\ 3 & 1 & 2 & 5 \end{array}\right]$

30. $\left[\begin{array}{ccc|c} 2 & 1 & 3 & 13 \\ 1 & 2 & 1 & 2 \\ 2 & 3 & 2 & 6 \end{array}\right]$

In Exercises 31–38, solve by the Gauss–Jordan method.

31. $\begin{cases} x - 3y = 5 \\ 3x + 5y = 1 \end{cases}$

32. $\begin{cases} x - 2y = 0 \\ 6x - 4y = 8 \end{cases}$

33. $\begin{cases} x + 2y + z = 8 \\ x + 4y - z = 12 \\ x - 2y + z = -4 \end{cases}$

34. $\begin{cases} x + y - 4z = -3 \\ 2x - y + z = 9 \\ 2x + 2y - 2z = 0 \end{cases}$

35. $\begin{cases} x - 2y = -4 \\ 2y + 2z = 4 \\ x + z = 1 \end{cases}$

36. $\begin{cases} x + z = 2 \\ 2x - 2y = -6 \\ 2y - z = 4 \end{cases}$

37. $\begin{cases} w + x + y + z = 7 \\ 2w + x - y + 2z = 6 \\ 3w + 2x + y - z = 3 \\ w - x - y - 2z = -10 \end{cases}$

38. $\begin{cases} w - 2x + y - z = 3 \\ w + y + z = -1 \\ 2w + x - y = 5 \\ 3w - x + z = 4 \end{cases}$

9.4

Determinants and Cramer's Rule

In the first two sections we solved many systems of equations. It is quite likely that you noticed the repetitive nature of the procedure. We kept repeating the same basic process—it was just the numbers that changed each time.

This type of situation leads us quite naturally to ask whether we can apply the procedure to the general case and produce a "formula" for the solutions of linear systems. You may recall that this is exactly the same type of question we asked in Chapter 7 regarding the method of completing the square, where the answer turned out to be the quadratic formula. The answer to the question here, regarding systems of linear equations, is called **Cramer's rule**.

EXAMPLE 1

Solve the following system of equations for x and y:

$$\begin{cases} a_1x + b_1y = k_1 \\ a_2x + b_2y = k_2 \end{cases}$$

Solution

This is the "general" 2×2 linear system. We solve the system by the elimination method. We first eliminate y.

$a_1x + b_1y = k_1$ *Multiply by b_2 →* $a_1b_2x + b_1b_2y = \quad k_1b_2$

$a_2x + b_2y = k_2$ *Multiply by $-b_1$ →* $\underline{-a_2b_1x - b_1b_2y = -k_2b_1} \qquad$ *Add.*

$$a_1b_2x - a_2b_1x = k_1b_2 - k_2b_1$$

Solve for x.

$$(a_1b_2 - a_2b_1)x = k_1b_2 - k_2b_1$$

$$x = \frac{k_1b_2 - k_2b_1}{a_1b_2 - a_2b_1}$$

This solution is valid provided $a_1b_2 - a_2b_1 \neq 0$.

Similarly, we can solve for y by eliminating x:

$a_1x + b_1y = k_1$ *Multiply by $-a_2$ →* $-a_1a_2x - a_2b_1y = -a_2k_1$

$a_2x + b_2y = k_2$ *Multiply by a_1 →* $\underline{a_1a_2x + a_1b_2y = \quad a_1k_2} \qquad$ *Add.*

$$a_1b_2y - a_2b_1y = a_1k_2 - a_2k_1$$

Solve for y.

$$(a_1b_2 - a_2b_1)y = a_1k_2 - a_2k_1$$

$$y = \frac{a_1k_2 - a_2k_1}{a_1b_2 - a_2b_1}$$

Again, this solution for y is valid provided $a_1b_2 - a_2b_1 \neq 0$.

Thus, the solution to this general 2×2 system is

$$\boxed{x = \frac{k_1b_2 - k_2b_1}{a_1b_2 - a_2b_1}, \quad y = \frac{a_1k_2 - a_2k_1}{a_1b_2 - a_2b_1}}$$ ∎

While the solution obtained in Example 1 is "general," it can be very confusing to remember. To help us keep this solution clear, we interrupt our discussion to introduce the idea of a *determinant*. We will return to the general solution in a moment.

DEFINITION

The symbol $\begin{vmatrix} a & c \\ b & d \end{vmatrix}$ is called a **2 × 2 *determinant***. Its value is defined to be

$\begin{vmatrix} a & c \\ b & d \end{vmatrix} = ad - bc$, which we may indicate as follows:

$$\begin{vmatrix} a & c \\ b & d \end{vmatrix} = ad - bc$$

EXAMPLE 2

Evaluate each of the following:

(a) $\begin{vmatrix} 7 & 3 \\ 5 & 4 \end{vmatrix}$ **(b)** $\begin{vmatrix} 8 & -2 \\ 3 & -4 \end{vmatrix}$

Solution

Using this definition is sometimes called *expanding the determinant*.

(a) $\begin{vmatrix} 7 & 3 \\ 5 & 4 \end{vmatrix} = 7(4) - 5(3) = 28 - 15 = \boxed{13}$

(b) $\begin{vmatrix} 8 & -2 \\ 3 & -4 \end{vmatrix} = 8(-4) - (3)(-2) = -32 + 6 = \boxed{-26}$ ∎

EXAMPLE 3

Solve for x: $\begin{vmatrix} x^2 & 3 \\ x & 1 \end{vmatrix} = 10$

Solution

$\begin{vmatrix} x^2 & 3 \\ x & 1 \end{vmatrix} = 10$ *Expand the determinant.*

$x^2 - 3x = 10$ *This is a quadratic equation. Solve by factoring.*

$x^2 - 3x - 10 = 0$

$(x - 5)(x + 2) = 0$

$x - 5 = 0$ or $x + 2 = 0$

$\boxed{x = 5 \quad \text{or} \quad x = -2}$

CHECK:

$\begin{vmatrix} 5^2 & 3 \\ 5 & 1 \end{vmatrix} = 25 \cdot 1 - 5 \cdot 3 \overset{\checkmark}{=} 10$ $\begin{vmatrix} (-2)^2 & 3 \\ -2 & 1 \end{vmatrix} = 4 \cdot 1 - (-2) \cdot (3) \overset{\checkmark}{=} 10$ ∎

We now return to the result of Example 1. We found the solution to the general system

$$a_1 x + b_1 y = k_1$$
$$a_2 x + b_2 y = k_2$$

to be

$$x = \frac{k_1 b_2 - k_2 b_1}{a_1 b_2 - a_2 b_1} \quad \text{and} \quad y = \frac{a_1 k_2 - a_2 k_1}{a_1 b_2 - a_2 b_1}$$

which we can write using determinant notation as follows:

$$x = \frac{\begin{vmatrix} k_1 & b_1 \\ k_2 & b_2 \end{vmatrix}}{\begin{vmatrix} a_1 & b_1 \\ a_2 & b_2 \end{vmatrix}} \quad \text{and} \quad y = \frac{\begin{vmatrix} a_1 & k_1 \\ a_2 & k_2 \end{vmatrix}}{\begin{vmatrix} a_1 & b_1 \\ a_2 & b_2 \end{vmatrix}}$$

It is useful to note that the denominator in each case is the determinant of the coefficients of x and y in our general system. This determinant is usually denoted by D:

$$D = \begin{vmatrix} a_1 & b_1 \\ a_2 & b_2 \end{vmatrix}$$

The numerator of the solution for each variable is obtained by taking D and replacing the column of coefficients of that variable by the column of constant terms. That is, the numerators are denoted as

$$D_x = \begin{vmatrix} k_1 & b_1 \\ k_2 & b_2 \end{vmatrix} \quad \text{and} \quad D_y = \begin{vmatrix} a_1 & k_1 \\ a_2 & k_2 \end{vmatrix}$$

We can summarize all that we have said thus far in the following theorem.

THEOREM: CRAMER'S RULE (part 1)

The solution to the system

$$a_1 x + b_1 y = k_1$$
$$a_2 x + b_2 y = k_2$$

is given by

$$x = \frac{D_x}{D} \quad \text{and} \quad y = \frac{D_y}{D}$$

where

$$D = \begin{vmatrix} a_1 & b_1 \\ a_2 & b_2 \end{vmatrix} \neq 0, \qquad D_x = \begin{vmatrix} k_1 & b_1 \\ k_2 & b_2 \end{vmatrix}, \qquad D_y = \begin{vmatrix} a_1 & k_1 \\ a_2 & k_2 \end{vmatrix}$$

If $D = 0$ there is no unique solution: the system is either inconsistent and has no solutions (if $D_x \neq 0$ or $D_y \neq 0$) or dependent and has infinitely many solutions (if $D_x = 0$ and $D_y = 0$).

The next example illustrates the use of Cramer's rule.

EXAMPLE 4

Solve the following system of equations:

$$\begin{cases} 5x - 3y = 7 \\ 7x - 8y = 4 \end{cases}$$

Solution

We first evaluate D, D_x, and D_y.

$$D = \begin{vmatrix} 5 & -3 \\ 7 & -8 \end{vmatrix} = 5(-8) - 7(-3) = -40 + 21 = -19$$

$$D_x = \begin{vmatrix} 7 & -3 \\ 4 & -8 \end{vmatrix} = 7(-8) - 4(-3) = -56 + 12 = -44$$

$$D_y = \begin{vmatrix} 5 & 7 \\ 7 & 4 \end{vmatrix} = 5(4) - 7(7) = 20 - 49 = -29$$

According to Cramer's rule, our solution is

$$x = \frac{D_x}{D} = \frac{-44}{-19} = \frac{44}{19}, \quad y = \frac{D_y}{D} = \frac{-29}{-19} = \frac{29}{19}$$

$$\boxed{x = \frac{44}{19}, \quad y = \frac{29}{19}}$$

The check is left to the student. ∎

In a similar manner, Cramer's rule can be extended to 3×3 linear systems. In order to do this we must define what we mean by a 3×3 determinant.

DEFINITION

A **3 × 3 *determinant*** is defined as follows:

$$\begin{vmatrix} a_1 & b_1 & c_1 \\ a_2 & b_2 & c_2 \\ a_3 & b_3 & c_3 \end{vmatrix} = a_1 \begin{vmatrix} b_2 & c_2 \\ b_3 & c_3 \end{vmatrix} - a_2 \begin{vmatrix} b_1 & c_1 \\ b_3 & c_3 \end{vmatrix} + a_3 \begin{vmatrix} b_1 & c_1 \\ b_2 & c_2 \end{vmatrix}$$

The 2×2 determinants in this definition are called the ***minors*** of their coefficients.

A minor is obtained by choosing an element of the determinant and crossing off the row and column which contain it. Thus, the minor of a_1 is obtained as follows:

$$\begin{vmatrix} a_1 & b_1 & c_1 \\ a_2 & b_2 & c_2 \\ a_3 & b_3 & c_3 \end{vmatrix} \quad \text{which gives} \quad \begin{vmatrix} b_2 & c_2 \\ b_3 & c_3 \end{vmatrix}$$

The other minors are found in a similar manner.

The procedure specified in the definition is called *expanding* the 3×3 determinant *down its first column.*

EXAMPLE 5

Evaluate the following determinant.

$$\begin{vmatrix} 3 & 2 & 2 \\ 4 & 1 & -1 \\ -2 & -3 & 5 \end{vmatrix}$$

Solution

Using the definition, we get

$$\begin{vmatrix} 3 & 2 & 2 \\ 4 & 1 & -1 \\ -2 & -3 & 5 \end{vmatrix} = 3 \begin{vmatrix} 3 & 2 & 2 \\ 4 & 1 & -1 \\ -2 & -3 & 5 \end{vmatrix} - 4 \begin{vmatrix} 3 & 2 & 2 \\ 4 & 1 & -1 \\ -2 & -3 & 5 \end{vmatrix} + (-2) \begin{vmatrix} 3 & 2 & 2 \\ 4 & 1 & -1 \\ -2 & -3 & 5 \end{vmatrix}$$

$$\begin{vmatrix} 3 & 2 & 2 \\ 4 & 1 & -1 \\ -2 & -3 & 5 \end{vmatrix} = 3\begin{vmatrix} 1 & -1 \\ -3 & 5 \end{vmatrix} - 4\begin{vmatrix} 2 & 2 \\ -3 & 5 \end{vmatrix} + (-2)\begin{vmatrix} 2 & 2 \\ 1 & -1 \end{vmatrix}$$

$$= 3(5 - 3) - 4(10 + 6) - 2(-2 - 2)$$

$$= 6 - 64 + 8 = \boxed{-50}$$

∎

In fact, a 3×3 determinant may be expanded across any row or down any column provided we *prefix* each entry with its proper sign, which is determined by its position. The **sign array**, as it is called, is

$$\begin{vmatrix} + & - & + \\ - & + & - \\ + & - & + \end{vmatrix}$$

which is obtained by putting a + sign in the upper-left-hand position and then alternating signs along each row and column.

EXAMPLE 6

Evaluate $\begin{vmatrix} 3 & 2 & 2 \\ 4 & 1 & -1 \\ -2 & 3 & 5 \end{vmatrix}$ by expanding across the second row.

Solution

$$\begin{vmatrix} 3 & 2 & 2 \\ 4 & 1 & -1 \\ -2 & 3 & 5 \end{vmatrix} = -4\begin{vmatrix} 2 & 2 \\ 3 & 5 \end{vmatrix} + 1\begin{vmatrix} 3 & 2 \\ -2 & 5 \end{vmatrix} - (-1)\begin{vmatrix} 3 & 2 \\ -2 & 3 \end{vmatrix}$$

Note the signs from the second row of the sign array.

$$= -4(10 - 6) + 1(15 + 4) - (-1)(9 + 4)$$

$$= -4(4) + 19 + 13 = \boxed{16}$$

∎

EXAMPLE 7

Evaluate. $\begin{vmatrix} 1 & 0 & 3 \\ 2 & 0 & 5 \\ -1 & 4 & 6 \end{vmatrix}$

Solution

We can expand the determinant using any row or column, so we might as well choose a row or column which contains zeros (if there is one) to make the computation easier. We will expand down the second column.

$$\begin{vmatrix} 1 & 0 & 3 \\ 2 & 0 & 5 \\ -1 & 4 & 6 \end{vmatrix} = -0\begin{vmatrix} 2 & 5 \\ -1 & 6 \end{vmatrix} + 0\begin{vmatrix} 1 & 3 \\ -1 & 6 \end{vmatrix} - 4\begin{vmatrix} 1 & 3 \\ 2 & 5 \end{vmatrix}$$

Signs from the second column of the sign array

$$= 0 + 0 - 4(-1) = \boxed{4}$$

∎

We can now state Cramer's rule for 3×3 linear systems.

THEOREM: CRAMER'S RULE (part 2)

The solution to the system

$$a_1x + b_1y + c_1z = k_1$$
$$a_2x + b_2y + c_2z = k_2$$
$$a_3x + b_3y + c_3z = k_3$$

is given by

$$x = \frac{D_x}{D}, \quad y = \frac{D_y}{D}, \quad z = \frac{D_z}{D}$$

where

$$D = \begin{vmatrix} a_1 & b_1 & c_1 \\ a_2 & b_2 & c_2 \\ a_3 & b_3 & c_3 \end{vmatrix} \neq 0 \qquad D_x = \begin{vmatrix} k_1 & b_1 & c_1 \\ k_2 & b_2 & c_2 \\ k_3 & b_3 & c_3 \end{vmatrix}$$

$$D_y = \begin{vmatrix} a_1 & k_1 & c_1 \\ a_2 & k_2 & c_2 \\ a_3 & k_3 & c_3 \end{vmatrix} \qquad D_z = \begin{vmatrix} a_1 & b_1 & k_1 \\ a_2 & b_2 & k_2 \\ a_3 & b_3 & k_3 \end{vmatrix}$$

EXAMPLE 8

Solve the following system of equations by using Cramer's rule.

$$\begin{cases} x + y + z = 1 \\ 3x + 2y - 6z = 1 \\ 9x - 4y + 12z = 3 \end{cases}$$

Solution

We begin by computing D, since if $D = 0$ there is no unique solution. We compute D by expanding across the first row:

$$D = \begin{vmatrix} 1 & 1 & 1 \\ 3 & 2 & -6 \\ 9 & -4 & 12 \end{vmatrix} = 1 \begin{vmatrix} 2 & -6 \\ -4 & 12 \end{vmatrix} - 1 \begin{vmatrix} 3 & -6 \\ 9 & 12 \end{vmatrix} + 1 \begin{vmatrix} 3 & 2 \\ 9 & -4 \end{vmatrix}$$

$$= (24 - 24) - (36 + 54) + (-12 - 18)$$
$$= 0 - 90 - 30$$
$$= -120$$

We compute D_x by expanding across the first row:

$$D_x = \begin{vmatrix} 1 & 1 & 1 \\ 1 & 2 & -6 \\ 3 & -4 & 12 \end{vmatrix} = 1 \begin{vmatrix} 2 & -6 \\ -4 & 12 \end{vmatrix} - 1 \begin{vmatrix} 1 & -6 \\ 3 & 12 \end{vmatrix} + 1 \begin{vmatrix} 1 & 2 \\ 3 & -4 \end{vmatrix}$$

$$= (24 - 24) - (12 + 18) + (-4 - 6)$$
$$= 0 - 30 - 10$$
$$= -40$$

Just for variety, we compute D_y by expanding down the third column:

$$D_y = \begin{vmatrix} 1 & 1 & 1 \\ 3 & 1 & -6 \\ 9 & 3 & 12 \end{vmatrix} = 1\begin{vmatrix} 3 & 1 \\ 9 & 3 \end{vmatrix} - (-6)\begin{vmatrix} 1 & 1 \\ 9 & 3 \end{vmatrix} + 12\begin{vmatrix} 1 & 1 \\ 3 & 1 \end{vmatrix}$$

$$= (9 - 9) + 6(3 - 9) + 12(1 - 3)$$
$$= 0 - 36 - 24$$
$$= -60$$

We expand D_z across the second row:

$$D_z = \begin{vmatrix} 1 & 1 & 1 \\ 3 & 2 & 1 \\ 9 & -4 & 3 \end{vmatrix} = -3\begin{vmatrix} 1 & 1 \\ -4 & 3 \end{vmatrix} + 2\begin{vmatrix} 1 & 1 \\ 9 & 3 \end{vmatrix} - 1\begin{vmatrix} 1 & 1 \\ 9 & -4 \end{vmatrix}$$

Note the signs from the second row of the sign array.

$$= -3(3 + 4) + 2(3 - 9) - (-4 - 9)$$
$$= -21 - 12 + 13$$
$$= -20$$

Therefore, the solution to the system is

$$x = \frac{D_x}{D} = \frac{-40}{-120} = \frac{1}{3}, \quad y = \frac{D_y}{D} = \frac{-60}{-120} = \frac{1}{2}, \quad z = \frac{D_z}{D} = \frac{-20}{-120} = \frac{1}{6}$$

$$\boxed{x = \frac{1}{3}, \quad y = \frac{1}{2}, \quad z = \frac{1}{6}}$$

The check is left to the student. ■

Keep in mind that in order to use Cramer's rule, the system of equations must be in standard form.

Exercises 9.4

In Exercises 1–18, compute the value of the determinant.

1. $\begin{vmatrix} 1 & 2 \\ 3 & 4 \end{vmatrix}$

2. $\begin{vmatrix} 3 & 1 \\ 4 & 2 \end{vmatrix}$

3. $\begin{vmatrix} 5 & 1 \\ -2 & 4 \end{vmatrix}$

4. $\begin{vmatrix} 4 & -3 \\ 2 & 6 \end{vmatrix}$

5. $\begin{vmatrix} -3 & -1 \\ 2 & -2 \end{vmatrix}$

6. $\begin{vmatrix} -5 & -4 \\ -3 & 2 \end{vmatrix}$

7. $\begin{vmatrix} 1 & 0 \\ 5 & 4 \end{vmatrix}$

8. $\begin{vmatrix} -1 & 7 \\ 8 & 0 \end{vmatrix}$

9. $\begin{vmatrix} 4 & 6 \\ 6 & 9 \end{vmatrix}$

10. $\begin{vmatrix} 2 & -3 \\ -6 & -9 \end{vmatrix}$

11. $\begin{vmatrix} 1 & 2 & -2 \\ 3 & -3 & 1 \\ -4 & 2 & -1 \end{vmatrix}$

12. $\begin{vmatrix} 5 & 2 & -1 \\ -3 & 4 & 1 \\ -2 & 2 & -2 \end{vmatrix}$

13. $\begin{vmatrix} 1 & 2 & 3 \\ 2 & 4 & 6 \\ 1 & 1 & 1 \end{vmatrix}$

14. $\begin{vmatrix} 3 & -1 & 2 \\ 4 & 5 & 6 \\ -3 & 1 & -2 \end{vmatrix}$

15. $\begin{vmatrix} -3 & 0 & 4 \\ 5 & 2 & -3 \\ 7 & 0 & 6 \end{vmatrix}$

16. $\begin{vmatrix} 1 & -2 & 6 \\ 0 & 8 & 0 \\ 5 & -3 & 4 \end{vmatrix}$

17. $\begin{vmatrix} 1 & 0 & 0 \\ 0 & 1 & 0 \\ 0 & 0 & 1 \end{vmatrix}$

18. $\begin{vmatrix} 2 & 0 & 0 \\ 0 & 2 & 0 \\ 0 & 0 & 2 \end{vmatrix}$

In Exercises 19–24, expand the determinant and solve the resulting equation for x.

19. $\begin{vmatrix} 2x & 4 \\ 3 & 5 \end{vmatrix} = 18$

20. $\begin{vmatrix} x+3 & -2 \\ 5 & 4 \end{vmatrix} = 30$

21. $\begin{vmatrix} x^2 & 5 \\ x & 1 \end{vmatrix} = 14$

22. $\begin{vmatrix} 2 & -x \\ -1 & x^2 \end{vmatrix} = 15$

23. $\begin{vmatrix} x & 3 \\ 2 & x+2 \end{vmatrix} = 9$

24. $\begin{vmatrix} x-5 & 3 \\ 2 & 3x \end{vmatrix} = 2x$

In Exercises 25–54, solve the system of equations by using Cramer's rule.

25. $\begin{cases} 9x + 2y = 1 \\ 5x + y = 0 \end{cases}$

26. $\begin{cases} 3x - 5y = 2 \\ 2x - 3y = 2 \end{cases}$

27. $\begin{cases} 2x - 4y = 5 \\ -3x + 6y = 7 \end{cases}$

28. $\begin{cases} -4x + 6y = 11 \\ 6x - 9y = 1 \end{cases}$

29. $\begin{cases} 3x - 5y - 2 = 0 \\ 4x - 3y - 10 = 0 \end{cases}$

30. $\begin{cases} 5x - 4y + 3 = 0 \\ 6x + 7y - 20 = 0 \end{cases}$

31. $\begin{cases} 2x - 7y = -4 \\ 3y - 4x = 8 \end{cases}$

32. $\begin{cases} 6y - 11x = -6 \\ 5x - 3y = 3 \end{cases}$

33. $\begin{cases} 6x = 5y - 7 \\ 4y = 3x - 5 \end{cases}$

34. $\begin{cases} 9a = 5b + 2 \\ 3b = 5a - 2 \end{cases}$

35. $\begin{cases} 2s - 9t = 4 \\ 3s + 5t = 6 \end{cases}$

36. $\begin{cases} 4m + 7n = 3 \\ 5m - 6n = 1 \end{cases}$

37. $\begin{cases} 6u + 7v = 3 \\ 5u + 8v = 9 \end{cases}$

38. $\begin{cases} 9w - 5z = 10 \\ 8w - 2z = 5 \end{cases}$

39. $\begin{cases} \dfrac{1}{2}x - 12y = 6 \\ \dfrac{1}{3}x - 8y = 6 \end{cases}$

40. $\begin{cases} \dfrac{2}{3}x + \dfrac{3}{4}y = 1 \\ 6x + 9y = 7 \end{cases}$

41. $\begin{cases} x + y + z = 3 \\ x - y + z = 2 \\ -x + y + z = 4 \end{cases}$

42. $\begin{cases} x + y + 2z = 4 \\ x - y + 2z = 2 \\ x - y - z = 2 \end{cases}$

43. $\begin{cases} 3x + 4y + 2z = 1 \\ 2x + 3y + z = 1 \\ 6x + y + 5z = 1 \end{cases}$

44. $\begin{cases} 4x + y + z = 0 \\ 2x - 3y + 4z = -7 \\ 3x + 4y - 2z = 6 \end{cases}$

45. $\begin{cases} r + 2s + 3t = 10 \\ 6r + 5s + 4t = 20 \\ 7r + 8s + 9t = 30 \end{cases}$

46. $\begin{cases} 2r - 3s + t = 5 \\ 3r - 2s + 4t = 5 \\ 5r - 3s + 2t = 5 \end{cases}$

47. $\begin{cases} 3x - y = 4 \\ 2y + z = 6 \\ 3x + 4z = 14 \end{cases}$

48. $\begin{cases} 5x - 3y = -1 \\ 3x + 4z = 11 \\ 5y + 2z = 16 \end{cases}$

49. $\begin{cases} 6x - 2y + 4z = 5 \\ 12x - 4y + 8z = 2 \\ -3x + y - 2z = 2 \end{cases}$

50. $\begin{cases} r - 2s + 3z = 3 \\ -3r + 6s - 9z = 1 \\ 2r - 4s + 6z = 9 \end{cases}$

51. $\begin{cases} x = y + z + 2 \\ y = x - z + 3 \\ z = x + y + 4 \end{cases}$

52. $\begin{cases} r = 2s - t + 1 \\ s = 3r + t + 4 \\ t = 5r - s - 1 \end{cases}$

53. $\begin{cases} 2a + b = 8 \\ -3a + c = -13 \\ 2b + 5c = 0 \end{cases}$

54. $\begin{cases} 3u + 2v = 0 \\ 5v + 6w = 5 \\ 4u - 3w = 0 \end{cases}$

? QUESTION FOR THOUGHT

55. Expand the determinant

$$\begin{vmatrix} a_1 & b_1 & c_1 \\ a_2 & b_2 & c_2 \\ a_3 & b_3 & c_3 \end{vmatrix}$$

across the first row and also down the first column. Show by regrouping terms and factoring that the results are the same.

MINI-REVIEW

56. *Solve for a:*

$$\frac{3}{a} = \frac{r}{1-a}$$

57. *Solve for x:*

$$\frac{2}{x-4} = \frac{1}{x+2} + 1$$

58. Find the equation of the line passing through $(-4, 5)$ and parallel to $3x - 2y = 5$.

59. Sketch the graph of the inequality $2x - 3y > 12$.

60. Find the length of the side of a square with diagonal 4 inches.

9.5

Systems of Linear Inequalities

In Section 8.4 we discussed solving linear inequalities in two variables by graphing the solution set. We will now apply the same procedure to solving systems of linear inequalities.

EXAMPLE 1

Solve the following system of inequalities:

$$\begin{cases} x + y \le 4 \\ 2x + y \ge 2 \end{cases}$$

Solution

As we outlined in Section 8.4, we graph the *equations* $x + y = 4$ and $2x + y = 2$, choose a test point, and shade the appropriate region.

The equation $x + y = 4$ is graphed in Figure 9.6**a**. Choose $(0, 0)$ as the test point. Since $(0, 0)$ satisfies the inequality $x + y \le 4$, we shade the region containing $(0, 0)$.

The equation $2x + y = 2$ is graphed in Figure 9.6**b**. Choose $(0, 0)$ as the test point. Since $(0, 0)$ does not satisfy the inequality $2x + y \ge 2$, we shade the region *not* containing $(0, 0)$.

Figure 9.6

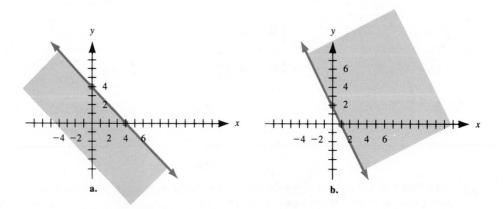

Combining the two graphs, we obtain the solution set to the system of inequalities—the *intersection* of the two shaded regions (see Figure 9.7).

Figure 9.7
Graph of the system
$$\begin{cases} x + y \leq 4 \\ 2x + y \geq 2 \end{cases}$$

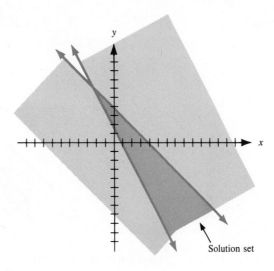

Solution set

We may choose a check point in this region, such as (3, 0), to see if it satisfies the system.

$$x + y \leq 4 \qquad \leftarrow Substitute\ (3, 0) \rightarrow \qquad 2x + y \geq 2$$
$$3 + 0 \overset{?}{\leq} 4 \qquad\qquad\qquad\qquad 2(3) + 0 \overset{?}{\geq} 2$$
$$3 \overset{\checkmark}{\leq} 4 \qquad\qquad\qquad\qquad 6 + 0 \overset{\checkmark}{\geq} 2 \qquad\qquad ∎$$

EXAMPLE 2

Solve the following system of inequalities:

$$\begin{cases} y - x > 4 \\ x - y > 3 \end{cases}$$

Solution

We proceed as in the last example, except that since the inequalities are strict, we draw dashed rather than solid lines (see Figures 9.8**a** and 9.8**b**).

Figure 9.8

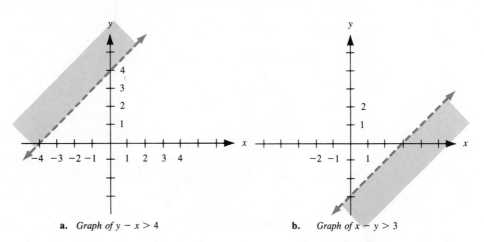

a. *Graph of* $y - x > 4$ **b.** *Graph of* $x - y > 3$

Putting the graphs together, we obtain Figure 9.9.

Figure 9.9
Graph of the system
$$\begin{cases} y - x > 4 \\ x - y > 3 \end{cases}$$

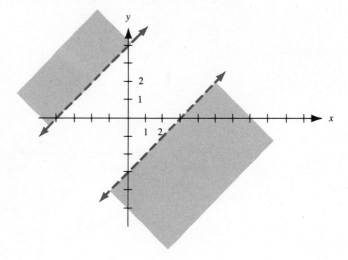

Looking at the equations

$$y - x = 4 \rightarrow y = x + 4$$
$$x - y = 3 \rightarrow y = x - 3$$

we can see that the lines are parallel (they both have slope equal to 1). Thus, the two lines never cross and the regions do not intersect.

There are ⎣ no solutions ⎦ to this system.   ■

We may use the same procedure to solve systems involving more than two inequalities.

EXAMPLE 3   Solve the following system of inequalities:

$$\begin{cases} 2x + 3y \le 12 \\ \quad\quad x \ge 0 \\ \quad\quad y \le 2 \end{cases}$$

Solution   We graph each inequality, as shown in Figure 9.10.

Figure 9.10

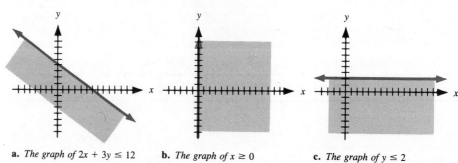

a. *The graph of* $2x + 3y \le 12$ **b.** *The graph of* $x \ge 0$ **c.** *The graph of* $y \le 2$

The solution set, which is the intersection of these three regions, is indicated in Figure 9.11.

Figure 9.11

Graph of the system
$$\begin{cases} 2x + 3y \leq 12 \\ x \geq 0 \\ y \leq 2 \end{cases}$$

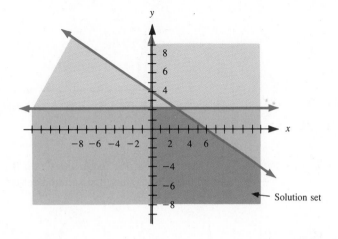

We may choose (1, 1) as a check point and verify that it satisfies the system. ∎

Systems of linear inequalities are often very useful in describing real-life situations.

EXAMPLE 4

The Cal-Q-Late Electronics Co. makes two types of calculators—a printing model and a programmable scientific model. The printing model requires $5 in materials and 1 hour to assemble and package. The scientific model requires $1 in material and 5 hours to assemble and package. The company decides to spend a maximum of $1,000 on material and to allot a maximum of 2,000 hours for packing and assembly. How many of each type of calculator can the company produce under these restrictions? Write a system of inequalities to describe this situation and sketch the solution set of the system.

Solution

Let $x =$ # of printing calculators the company can produce.

Let $y =$ # of scientific calculators the company can produce.

We can translate the information given in the problem into the following system of inequalities:

$$\begin{cases} 5x + y \leq 1{,}000 \\ x + 5y \leq 2{,}000 \\ x \geq 0 \\ y \geq 0 \end{cases}$$

Since the total cost for materials is $\leq$ $1,000

Since the total number of hours for packaging and assembly is $\leq$ 2,000

Because the number of calculators produced cannot be negative

We can now sketch the solution set of the first two inequalities (see Figures 9.12**a** and 9.12**b** on page 448).

Figure 9.12

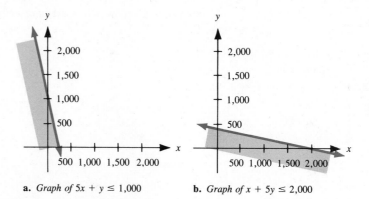

a. *Graph of $5x + y \leq 1,000$*   **b.** *Graph of $x + 5y \leq 2,000$*

We note that the inequalities $x \geq 0$ and $y \geq 0$ restrict us to the first quadrant and its boundary. Thus, the solution set of this system of inequalities is as shown in Figure 9.13.

Figure 9.13

Graph of the system
$$\begin{cases} 5x + y \leq 1,000 \\ x + 5y \leq 2,000 \\ x \geq 0 \\ y \geq 0 \end{cases}$$

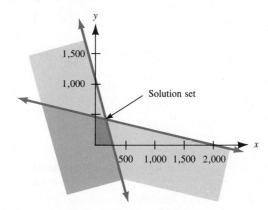

Solution set

This graph indicates that any point (x, y) in the darkest shaded region satisfies the system. However, since we cannot produce fractional parts of a calculator, only the ordered pairs in the darkest region with whole-number coordinates are actual choices for the company. ∎

Exercises 9.5

In Exercises 1–24, sketch the solution set of the system of inequalities.

1. $\begin{cases} x + y \leq 6 \\ x + 2y \geq 3 \end{cases}$

2. $\begin{cases} x + y \geq 4 \\ 3x + y < 6 \end{cases}$

3. $\begin{cases} 2y + x \geq 6 \\ 3y + x \geq 9 \end{cases}$

4. $\begin{cases} x - 2y \leq 8 \\ x - 3y \leq 12 \end{cases}$

5. $\begin{cases} x - y \geq 2 \\ y - x > -1 \end{cases}$

6. $\begin{cases} x - 2y \leq 2 \\ 4y - 2x \leq 0 \end{cases}$

7. $\begin{cases} x + y \leq 5 \\ 2x + y \leq 8 \end{cases}$

8. $\begin{cases} y - x \leq 4 \\ y + x \leq 4 \end{cases}$

9. $\begin{cases} 3x + 2y \leq 12 \\ y \leq x \end{cases}$

10. $\begin{cases} 3y - x > 6 \\ y \leq 5 \end{cases}$

11. $\begin{cases} 3x + 2y \geq 12 \\ y \geq x \end{cases}$

12. $\begin{cases} 3y - x < 6 \\ y \geq 5 \end{cases}$

13. $\begin{cases} 5x - 3y \leq 15 \\ x < 3 \end{cases}$

14. $\begin{cases} 2x + 7y > 14 \\ x \geq 7 \end{cases}$

15. $\begin{cases} 5x - 3y \geq 15 \\ x > 3 \end{cases}$

16. $\begin{cases} 2x + 7y < 14 \\ x \leq 7 \end{cases}$

17. $\begin{cases} x - 2y < 10 \\ x \geq 2 \\ y \leq 2 \end{cases}$ **18.** $\begin{cases} 3y - x > 6 \\ y \leq 5 \\ x \geq -3 \end{cases}$ **19.** $\begin{cases} 4x + 3y \leq 12 \\ x \geq 0 \\ y \geq 0 \end{cases}$ **20.** $\begin{cases} 2x - 5y \leq 10 \\ x \geq 0 \\ y \geq 0 \end{cases}$

21. $\begin{cases} x + 3y \geq 6 \\ 3x + 2y \leq 18 \\ y \leq 3 \\ x \geq 0 \end{cases}$ **22.** $\begin{cases} 2x + y \leq 8 \\ 3x - 2y \leq 12 \\ x \geq 0 \\ y \geq -4 \end{cases}$ **23.** $\begin{cases} x + 2y \geq 4 \\ 2x - 4y \leq -8 \\ x \geq 0 \\ y \geq 0 \end{cases}$ **24.** $\begin{cases} x > y \\ x + y < 8 \\ 3x - 4y < 24 \end{cases}$

In Exercises 25–28, write a system of inequalities to describe the given situation and sketch the graph of its solution set.

25. A jogger wants to establish a more balanced diet. She reads the nutritional information on cereal boxes and finds out that 1 ounce of Brand X cereal contains 10 gm of carbohydrates and 0.33 gm of sodium, while 1 ounce of Brand Y cereal contains 13 gm of carbohydrates and .37 gm of sodium. If she wants to create a 1-ounce mixture of Brands X and Y which will contain at least 21 gm of carbohydrates and no more than 1 gm of sodium, how much of each cereal can she use?

26. A discount appliance store wants to order a shipment of refrigerators and televisions. A refrigerator costs $400 and uses 30 cubic feet of storage space, while a TV costs $275 and uses 12 cubic feet of storage space. If the store wants to order at least $10,000 worth of merchandise but has at most 10,000 cubic feet of storage space in its warehouse, how many of each can it order?

27. A shoe manufacturer uses 1 square meter of leather and $\frac{1}{3}$ square meter of crepe rubber to make each pair of men's shoes, and $\frac{3}{4}$ square meter of leather and $\frac{1}{4}$ square meter of crepe rubber to make each pair of ladies' shoes. He has 500 square meters of leather and 150 square meters of crepe rubber on hand. If he cannot get any more leather or crepe rubber, how many pairs of each type of shoe can he produce?

28. A carpenter makes two kinds of bookcases—one type is custom-made to specifications and the other type is made from prefabricated parts. On the average, it takes 5 hours to make a custom-made bookcase and 3 hours to make one from prefabricated parts. He can make a profit of $175 on each custom-made bookcase and $100 on each prefabricated bookcase. If he can spend no more than 20 hours per week on bookcases and wants to earn at least $500 per week from making bookcases, how many of each type can he produce?

 MINI-REVIEW

29. *Solve for x:*

$$\frac{x + 1}{3} - \frac{x + 2}{5} = \frac{2}{3}$$

31. *Solve explicitly for y.* $x^2 - xy - 6y^2 = 0$

30. *Factor completely.* $4x^3 - 4$

32. Find the equation of the line perpendicular to $2x - 3y = 4$ and passing through $(2, -3)$.

CHAPTER 9 Summary

After having completed this chapter, you should be able to:

1. Solve a 2×2 system of linear equations by using either the elimination or the substitution method (Section 9.1).

For example: Solve for x and y:

$$\begin{cases} 3x - 4y = 10 \\ 4x - 5y = 13 \end{cases}$$

Solution: We use the elimination method. We choose to eliminate y.

$$3x - 4y = 10 \qquad \textit{Multiply by } -5 \rightarrow \qquad -15x + 20y = -50$$
$$4x - 5y = 13 \qquad \textit{Multiply by } 4 \rightarrow \qquad \underline{16x - 20y = 52}$$
$$x = 2$$

Substitute $x = 2$ into the first equation:

$$3x - 4y = 10$$
$$3(2) - 4y = 10$$
$$6 - 4y = 10$$
$$-4y = 4$$
$$y = -1$$

The solution is $(2, -1)$. Check in both equations.

2.   Solve a 3×3 system of linear equations by using the elimination method (Section 9.2).

For example: Solve for x, y, and z:

$$\begin{cases} x - y + 2z = 4 & (1) \\ 3x + 2y - z = 3 & (2) \\ 5x - 3y + 3z = 8 & (3) \end{cases}$$

Solution: We choose to eliminate z.

$$\textit{Equation (1)} \qquad x - y + 2z = 4$$
$$\textit{2 times equation (2)} \qquad \underline{6x + 4y - 2z = 6} \qquad \textit{Add.}$$
$$7x + 3y \phantom{{}- 2z} = 10 \qquad (4)$$

$$\textit{3 times equation (2)} \qquad 9x + 6y - 3z = 9$$
$$\textit{Equation (3)} \qquad \underline{5x - 3y + 3z = 8} \qquad \textit{Add.}$$
$$14x + 3y \phantom{{}+ 3z} = 17 \qquad (5)$$

Our 2×2 system is

$$7x + 3y = 10 \qquad \textit{Multiply by } -1 \rightarrow \qquad -7x - 3y = -10$$
$$14x + 3y = 17 \qquad \textit{As is} \rightarrow \qquad \underline{14x + 3y = 17}$$
$$7x \phantom{{}+ 3y} = 7$$
$$x \phantom{{}+ 3y} = 1$$

Substitute $x = 1$ into equation (4):

$$7x + 3y = 10$$
$$7(1) + 3y = 10$$
$$y = 1$$

Substitute $x = 1$, $y = 1$ into equation (1):

$$x - y + 2z = 4$$
$$1 - 1 + 2z = 4$$
$$z = 2$$

Thus, our solution is $x = 1$, $y = 1$, $z = 2$.

Check in all three original equations.

3. Solve a linear system by Gaussian elimination (Section 9.3).

For example: Solve the following system using Gaussian elimination:

$$\begin{cases} 2x + y - z = 4 \\ x - 2y + z = 7 \\ 3x - y + 2z = 11 \end{cases}$$

Solution:

Set up the augmented matrix, and transform the augmented matrix into a (row-equivalent) matrix in triangular form:

$$\begin{bmatrix} 2 & 1 & -1 & | & 4 \\ 1 & -2 & 1 & | & 7 \\ 3 & -1 & 2 & | & 11 \end{bmatrix} \quad R_1 \longleftrightarrow R_2 \quad \begin{bmatrix} 1 & -2 & 1 & | & 7 \\ 2 & 1 & -1 & | & 4 \\ 3 & -1 & 2 & | & 11 \end{bmatrix}$$

$$\begin{bmatrix} 1 & -2 & 1 & | & 7 \\ 2 & 1 & -1 & | & 4 \\ 3 & -1 & 2 & | & 11 \end{bmatrix} \quad \begin{matrix} -2R_1 + R_2 \to R_2 \\ -3R_1 + R_3 \to R_3 \end{matrix} \quad \begin{bmatrix} 1 & -2 & 1 & | & 7 \\ 0 & 5 & -3 & | & -10 \\ 0 & 5 & -1 & | & -10 \end{bmatrix}$$

$$\begin{bmatrix} 1 & -2 & 1 & | & 7 \\ 0 & 5 & -3 & | & -10 \\ 0 & 5 & -1 & | & -10 \end{bmatrix} \quad -R_2 + R_3 \to R_3 \quad \begin{bmatrix} 1 & -2 & 1 & | & 7 \\ 0 & 5 & -3 & | & -10 \\ 0 & 0 & 2 & | & 0 \end{bmatrix} \quad \begin{matrix} \textit{This matrix} \\ \textit{is now in} \\ \textit{triangular} \\ \textit{form.} \end{matrix}$$

Set up the associated system of the matrix in triangular form:

$$\begin{cases} x - 2y + z = 7 \\ 5y - 3z = -10 \\ 2z = 0 \end{cases}$$

Solve by back substitution:

First solve for z in the last equation:

$$2z = 0$$
$$z = 0$$

Substitute $z = 0$ in the second equation and solve for y:

$$5y - 3z = -10$$
$$5y - 3(0) = -10$$
$$y = -2$$

Substitute $z = 0$ and $y = -2$ in the first equation and solve for x:

$$x - 2y + z = 7$$
$$x - 2(-2) + 0 = 7$$
$$x = 3$$

The solution is $(3, -2, 0)$.

4. Solve a linear system by Gauss–Jordan elimination (Section 9.3).

For example: Solve the following system by Gauss–Jordan elimination:

$$\begin{cases} 2x + 4y = 24 \\ 3x + y = 11 \end{cases}$$

Set up the augmented matrix and transform the augmented matrix into reduced echelon form:

$$\begin{bmatrix} 2 & 4 & | & 24 \\ 3 & 1 & | & 11 \end{bmatrix} \qquad \tfrac{1}{2}R_1 \to R_1 \qquad \begin{bmatrix} 1 & 2 & | & 12 \\ 3 & 1 & | & 11 \end{bmatrix}$$

$$\begin{bmatrix} 1 & 2 & | & 12 \\ 3 & 1 & | & 11 \end{bmatrix} \qquad -3R_1 + R_2 \to R_2 \qquad \begin{bmatrix} 1 & 2 & | & 12 \\ 0 & -5 & | & -25 \end{bmatrix}$$

$$\begin{bmatrix} 1 & 2 & | & 12 \\ 0 & -5 & | & -25 \end{bmatrix} \qquad -\tfrac{1}{5}R_2 \to R_2 \qquad \begin{bmatrix} 1 & 2 & | & 12 \\ 0 & 1 & | & 5 \end{bmatrix}$$

$$\begin{bmatrix} 1 & 2 & | & 12 \\ 0 & 1 & | & 5 \end{bmatrix} \qquad -2R_2 + R_1 \to R_1 \qquad \begin{bmatrix} 1 & 0 & | & 2 \\ 0 & 1 & | & 5 \end{bmatrix}$$ *This matrix is in reduced echelon form.*

We can see from the matrix in reduced echelon form that the solution is (2, 5).

5. Evaluate 2×2 and 3×3 determinants (Section 9.4).

For example: Evaluate each of the following:

(a) $\begin{vmatrix} 3 & -2 \\ 4 & 5 \end{vmatrix} = 3(5) - 4(-2) = 15 + 8 = 23$

(b) $\begin{vmatrix} 2 & -1 & 3 \\ 1 & 4 & 2 \\ 3 & 1 & 2 \end{vmatrix} = 2\begin{vmatrix} 4 & 2 \\ 1 & 2 \end{vmatrix} - 1\begin{vmatrix} -1 & 3 \\ 1 & 2 \end{vmatrix} + 3\begin{vmatrix} -1 & 3 \\ 4 & 2 \end{vmatrix}$ *Expanding down the first column*

$$= 2(8 - 2) - (-2 - 3) + 3(-2 - 12)$$

$$= 12 + 5 - 42 = -25$$

6. Use Cramer's rule to solve 2×2 and 3×3 linear systems (Section 9.4).

For example: Solve the following system using Cramer's rule:

$$\begin{cases} 2x - 7y = 4 \\ 5x + 3y = 2 \end{cases}$$

Solution:

$$D = \begin{vmatrix} 2 & -7 \\ 5 & 3 \end{vmatrix} = 6 + 35 = 41$$

$$D_x = \begin{vmatrix} 4 & -7 \\ 2 & 3 \end{vmatrix} = 12 + 14 = 26$$

$$D_y = \begin{vmatrix} 2 & 4 \\ 5 & 2 \end{vmatrix} = 4 - 20 = -16$$

According to Cramer's rule,

$$x = \frac{D_x}{D} = \frac{26}{41} \quad \text{and} \quad y = \frac{D_y}{D} = \frac{-16}{41}$$

Thus, the solution is $\left(\dfrac{26}{41}, -\dfrac{16}{41}\right)$.

7. Solve a system of linear inequalities by the graphical method (Section 9.5).

For example: Solve the following system of inequalities:

$$\begin{cases} x + 2y \le 6 \\ y < x - 6 \end{cases}$$

Solution: We graph the lines $x + 2y = 6$ (with a solid line) and $y = x - 6$ (with a dashed line), as shown in Figures 9.14**a** and 9.14**b**.

Figure 9.14

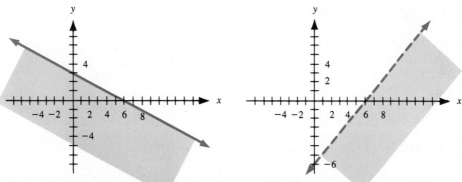

a. *Choose* $(0, 0)$ *as a test point. It satisfies* $x + 2y \le 6$, *so we take the region containing* $(0, 0)$.

b. *Choose* $(0, 0)$ *as a test point. It does not satisfy* $y < x - 6$, *so we take the region not containing* $(0, 0)$.

The solution set is the intersection of the two regions, as indicated in Figure 9.15.

Figure 9.15
Graph of the system
$$\begin{cases} x + 2y \le 6 \\ y < x - 6 \end{cases}$$

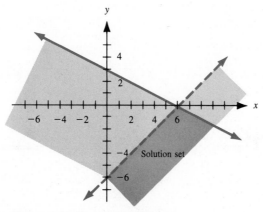

Solution set

We may choose $(6, -3)$ *as a check point in the shaded region and verify that it satisfies the system.*

8. Solve a variety of verbal problems which give rise to systems of equations (Sections 9.1, 9.2, 9.3, 9.4, and 9.5).

For example: A candy shop sells a mixture containing 1 lb of regular jelly beans and $\frac{1}{2}$ lb of gourmet jelly beans for $2.75, and a mixture containing $\frac{1}{2}$ lb of regular jelly beans and 1 lb of gourmet jelly beans for $3.25. What is the cost per pound of each type of jelly bean?

Solution:

Let $r =$ the cost per pound of regular jelly beans.

Let $g =$ the cost per pound of gourmet jelly beans.

Then

$$1r + \frac{1}{2}g = 2.75 \qquad \textit{Represents the cost of the first mixture}$$

$$\frac{1}{2}r + 1g = 3.25 \qquad \textit{Represents the cost of the second mixture}$$

Multiplying both sides of each equation by 2 yields the following system:

$$
\begin{cases}
2r + g = 5.5 \\
r + 2g = 6.5
\end{cases}
\begin{array}{l}
\textit{As is} \rightarrow \\
\textit{Multiply by } -2 \rightarrow
\end{array}
\begin{array}{l}
2r + g = 5.5 \\
\underline{-2r - 4g = -13} \qquad \textit{Add.} \\
\quad -3g = -7.5 \\
\qquad g = 2.5
\end{array}
$$

To find r, substitute 2.5 for g in $2r + g = 5.5$:

$$2r + g = 5.5$$
$$2r + 2.5 = 5.5$$
$$2r = 3$$
$$r = 1.5$$

The cost of regular jelly beans (r) is $1.50 per pound and the cost of gourmet jelly beans (g) is $2.50 per pound.

CHAPTER 9   Review Exercises

In Exercises 1–12, solve the systems of equations.

1. $\begin{cases} x - y = 4 \\ 2x - 3y = 7 \end{cases}$

2. $\begin{cases} 4x + 5y = 2 \\ 7x + 6y = 9 \end{cases}$

3. $\begin{cases} \dfrac{x}{6} - \dfrac{y}{4} = \dfrac{4}{3} \\ \dfrac{x}{5} - \dfrac{y}{2} = \dfrac{8}{5} \end{cases}$

4. $\begin{cases} \dfrac{2x}{3} + \dfrac{3y}{4} = \dfrac{7}{12} \\ \dfrac{6x}{5} - \dfrac{3y}{2} = \dfrac{1}{10} \end{cases}$

5. $\begin{cases} 3x - \dfrac{y}{4} = 2 \\ 6x - \dfrac{y}{2} = 4 \end{cases}$

6. $\begin{cases} \dfrac{x}{6} + y = 1 \\ \dfrac{x}{3} + 2y = 4 \end{cases}$

7. $\begin{cases} x = 2y - 3 \\ y = 3x + 2 \end{cases}$

8. $\begin{cases} 2s - 5t = 6 \\ 4t + 3s = 8 \end{cases}$

9. $\begin{cases} x + 2y - z = 2 \\ 2x + 3y + 4z = 9 \\ 3x + y - 2z = 2 \end{cases}$

10. $\begin{cases} 3x - 2y + 5z = 1 \\ 4x + 3y - 6z = 14 \\ 2x + 4y + 4z = 0 \end{cases}$

11. $\begin{cases} 2x - 3z = 7 \\ 3x + 4y = -11 \\ 3y - 2z = 0 \end{cases}$

12. $\begin{cases} a + 2b = 3c - 3 \\ 2a - 3b = c - 1 \\ 3b - 4c = a - 4 \end{cases}$

In Exercises 13–16, put the matrices into reduced echelon form.

13. $\begin{bmatrix} 1 & 2 & | & 6 \\ 2 & 6 & | & 8 \end{bmatrix}$

14. $\begin{bmatrix} 3 & 4 & | & 2 \\ 1 & 2 & | & 6 \end{bmatrix}$

15. $\begin{bmatrix} 2 & 2 & 6 & | & 4 \\ 2 & 5 & 9 & | & -2 \\ 1 & 2 & 3 & | & -1 \end{bmatrix}$

16. $\begin{bmatrix} 1 & 3 & -1 & | & -3 \\ 2 & 7 & -3 & | & -9 \\ 3 & 11 & -1 & | & -3 \end{bmatrix}$

In Exercises 17–20, solve by Gaussian elimination.

17. $\begin{cases} x - 2y = 1 \\ 2x + 3y = 9 \end{cases}$

18. $\begin{cases} 3x - y = 3 \\ x + 2y = 1 \end{cases}$

19. $\begin{cases} x + 2y - z = 5 \\ x - y + z = 0 \\ x + y + 2z = 1 \end{cases}$

20. $\begin{cases} x - 2y - z = -4 \\ 2x + y - z = 1 \\ 3x - y + 2z = -7 \end{cases}$

In Exercises 21–22, solve by Gauss–Jordan elimination.

21. $\begin{cases} x - 3y = -1 \\ 2x + y = 5 \end{cases}$

22. $\begin{cases} x + y + 9z = -6 \\ -x - 6z = 7 \\ 2x + y + 12z = -10 \end{cases}$

In Exercises 23–26, evaluate the given determinant.

23. $\begin{vmatrix} 2 & 3 \\ 4 & -1 \end{vmatrix}$

24. $\begin{vmatrix} 8 & 5 \\ 2 & 4 \end{vmatrix}$

25. $\begin{vmatrix} 1 & 3 & -2 \\ 2 & 4 & 3 \\ 5 & -1 & -3 \end{vmatrix}$

26. $\begin{vmatrix} 2 & -4 & 1 \\ 0 & 6 & 0 \\ 5 & 3 & 7 \end{vmatrix}$

In Exercises 27–32, solve the system using Cramer's rule.

27. $\begin{cases} 3x + 7y = 4 \\ 5x + 4y = 2 \end{cases}$

28. $\begin{cases} 5x - 2y = 3 \\ 6x - 7y = 1 \end{cases}$

29. $\begin{cases} 4a + 2b = 7 \\ 6a + 3b = 9 \end{cases}$

30. $\begin{cases} 9m + 6n = 12 \\ 3m + 2n = 4 \end{cases}$

31. $\begin{cases} 3s + 4t + u = 3 \\ 5s - 3t + 6u = 2 \\ 4s - 5t - 5u = 1 \end{cases}$

32. $\begin{cases} -4u + 3v + 6w = 4 \\ 3u - 5v - 7w = -1 \\ 5u - 4v + 2w = 3 \end{cases}$

In Exercises 33–36, graph the system of inequalities.

33. $\begin{cases} x + y \le 2 \\ 2x - 3y \le 6 \end{cases}$

34. $\begin{cases} y - 2x \le 8 \\ y > 8 - x \end{cases}$

35. $\begin{cases} 2x + 3y \le 12 \\ y < x \\ x \ge 0 \\ y \ge 0 \end{cases}$

36. $\begin{cases} y - x > 3 \\ x - y > 2 \end{cases}$

In Exercises 37–40, solve the problem algebraically.

37. A total of $8,500 is split into two investments. Part is invested in a certificate of deposit paying 8.75% per year and the rest is invested in a bond paying 9.65% per year. If the annual interest from the two investments is $768.95, how much is invested at each rate?

38. A total of $20,000 is split into three investments. Part is invested in a bank account paying 7.7% interest per year, part is invested in a corporate bond paying 8.6% interest per year, and the rest is invested in a municipal bond paying 9.8% interest per year. The sum of the amounts invested in the two bonds is $3,000 more than the amount invested in the bank account. If the annual interest from all three investments is $1,719.10, how much is invested at each rate?

39. Tom goes into a bakery and orders 3 pounds of bread and 5 pounds of cookies for a total of $22.02. Sarah buys 2 pounds of bread and 3 pounds of cookies in the same bakery for a total of $13.43. What are the prices per pound for bread and for cookies?

40. Jane, June, and Jean go shopping at a flea market. Jane buys one sweater, two blouses, and one pair of jeans. June buys two sweaters, one blouse, and one pair of jeans. Jean buys two sweaters, three blouses, and two pairs of jeans. If Jane spends $28.50, June spends $30, and Jean spends $52, what are the prices of an individual sweater, blouse, and pair of jeans?

CHAPTER 9   Practice Test

1. Solve the following systems of simultaneous equations:

(a) $\begin{cases} 2x + 3y = 1 \\ 3x + 4y = 4 \end{cases}$

(b) $\begin{cases} \dfrac{a}{3} + \dfrac{b}{2} = 2 \\ a = \dfrac{b}{3} - 5 \end{cases}$

(c) $\begin{cases} 2x - 3y + 4z = 2 \\ 3x + 2y - z = 10 \\ 2x - 4y + 3z = 3 \end{cases}$

2. Solve the following using matrices:

$\begin{cases} x - y = 2 \\ x + 2z = 7 \\ -2x + 3y + 4z = 5 \end{cases}$

3. Evaluate the following determinants:

(a) $\begin{vmatrix} 2 & 3 \\ -1 & 4 \end{vmatrix}$ (b) $\begin{vmatrix} 5 & 0 & 2 \\ 2 & 3 & 1 \\ 1 & 1 & 2 \end{vmatrix}$

4. Solve the following using Cramer's rule:

(a) $\begin{cases} 2x - 6y = -1 \\ 4x = 8y + 5 \end{cases}$ (b) $\begin{cases} 5x + y = 9 \\ 3x - z = 3 \\ y + 3z = 8 \end{cases}$

5. A bank teller gave a customer change for a $500 bill in $5, $10, and $20 bills. There were twice as many $10 bills as $20 bills and all together there were 40 bills. How many of each did the customer receive for his $500 bill?

6. Jerry invested $3,500 in two savings certificates; one yields $8\frac{1}{2}\%$ annual interest and the other yields 9% annual interest. How much was invested in each certificate if Jerry receives $309 in annual interest?

7. Sketch the solution set of the following system of inequalities: $\begin{cases} y < 4 \\ x \leq 2 \\ x + y > 3 \end{cases}$

In Exercises 1–4, solve by the factoring or square root method.

1. $a^2 - 2a - 15 = 0$

2. $a^2 - 6a + 9 = a + 3$

3. $2x^2 - 3x - 4 = 9 - 3(x - 2)$

4. $(x - 5)^2 = 28$

In Exercises 5–6, solve by completing the square.

5. $y^2 + 6y - 1 = 0$

6. $3x^2 + 6x = 2$

In Exercises 7–14, solve by any method.

7. $3a^2 - 2a - 2 = 0$

8. $(x - 2)(x + 3) = 2x^2 - 3x + 1$

9. $\dfrac{1}{x + 2} = x + 2$

10. $(y - 2)(y + 3) = 6$

11. $\dfrac{2}{x + 3} - \dfrac{3}{x} = -\dfrac{2}{3}$

12. $(b - 2)(b - 3) = (b - 4)(b + 1)$

13. $\dfrac{3}{x - 3} + \dfrac{2x}{x + 3} = \dfrac{5}{x - 3}$

14. $\dfrac{3}{x + 1} + \dfrac{2}{x - 5} = 5$

In Exercises 15–16, solve for the given variable.

15. $s = \dfrac{xy}{z^2}$ (for z)

16. $a^2 - 4ab - 5b^2 = 0$ (for b)

In Exercises 17–18, solve the equation.

17. $\sqrt{5x - 1} = x + 1$

18. $\sqrt{3x} - \sqrt{x + 1} = 1$

In Exercises 19–20, solve for the given variable.

19. $\sqrt{5x - 1} = 2y$ (for x)

20. $\sqrt{5x - 1} = 2y$ (for x)

In Exercises 21–22, solve the equation.

21. $x^4 - 81 = 0$

22. $a^{1/2} - 3a^{1/4} - 10 = 0$

In Exercises 23–26, solve the inequality and graph the solution set.

23. $(x - 5)(x - 8) < 0$

24. $2x^2 - 9x \geq 5$

25. $\dfrac{x - 2}{x + 3} > 0$

26. $\dfrac{x}{2x + 1} \leq 0$

In Exercises 27–30, solve algebraically.

27. The difference between a number and its reciprocal is $\frac{5}{6}$. Find the number.

28. Find the dimensions of a rectangle whose area is 21 square feet and whose length is 4 more than its width.

29. Find the length of the side of a square with diagonal 28 inches.

30. A 15′ by 20′ garden is to be bordered by a brick walkway of uniform width. If the area of the walkway is 294 square feet, how wide is the walkway?

In Exercises 31–36, sketch a graph of the equation using the intercept method.

31. $5x - 4y = 10$

32. $3x - 2y = 12$

33. $5y = 3x - 10$

34. $5x = 3y - 10$

35. $3x - 2 = 8$

36. $2y + 4 = 6$

In Exercises 37–42, find the slope of the line satisfying the given condition(s).

37. Passing through points $(2, -3)$ and $(3, 5)$

38. Passing through points $(5, -4)$ and $(-2, 3)$

39. Equation is $y = 5x - 8$

40. Equation is $3x - 2y = 11$

41. Parallel to the line passing through the points $(2, -1)$ and $(6, 4)$

42. Perpendicular to the line passing through the points $(2, -3)$ and $(4, 2)$

*In Exercises 43–44, find the values of **a** satisfying the given condition(s).*

43. The line through the points $(2, a)$ and $(a, -2)$ has slope 3.

44. The line through the points $(a, 2)$ and $(1, 3)$ is parallel to the line through the points $(4, -1)$ and $(5, 1)$.

In Exercises 45–52, find the equation of the line satisfying the given conditions.

45. The line passes through the point $(-2, 7)$ and has slope 3.

46. The line passes through the point $(2, 5)$ and has slope -2.

47. The line passes through the points $(2, 7)$ and $(3, 1)$.

48. The line passes through the points $(3, 5)$ and $(4, 5)$.

49. The line has y-intercept 2 and slope 4.

50. The line has x-intercept 3 and slope -1.

51. The line passes through the point $(2, -3)$ and is parallel to the line $3x + 5y = 4$.

52. The line passes through the point $(2, -3)$ and is perpendicular to the line $2y = 4x + 1$.

In Exercises 53–57, graph the inequality on the rectangular coordinate system.

53. $2x + 3y \geq 18$

54. $7x - y < 14$

55. $2y < 5x - 20$

56. $2y \leq 4$

57. $x - 2 > 5$

58. The owner of a clothing store found that the relationship between his daily profit on bathing suits (P) and the daily temperature (T) during the summer months is linear. If he makes \$450 when the temperature is 86°F and \$325 when the temperature is 80°F, write an equation relating P to T and predict his daily profit on a 90°F day.

In Exercises 59–72, solve the given system of equations.

59. $\begin{cases} 3x - 2y = 8 \\ 5x + y = 9 \end{cases}$

60. $\begin{cases} 4x - 3y = 2 \\ 6x - 5y = 3 \end{cases}$

61. $\begin{cases} 7u + 5v = 23 \\ 8u + 9v = 23 \end{cases}$

62. $\begin{cases} -10s + 7t = -6 \\ -4s + 6t = -4 \end{cases}$

63. $\begin{cases} 2m = 3n - 5 \\ 3n = 2m - 5 \end{cases}$

64. $\begin{cases} 5w = 4v + 7 \\ 4v = 5w - 7 \end{cases}$

65. $\begin{cases} \dfrac{2}{3}y - \dfrac{1}{2}x = 6 \\ \dfrac{4}{5}y - \dfrac{3}{4}x = 6 \end{cases}$

66. $\begin{cases} \dfrac{x}{4} + \dfrac{5y}{6} = -\dfrac{11}{12} \\ \dfrac{5x}{3} + \dfrac{y}{2} = 4 \end{cases}$

67. $\begin{cases} x + y + z = 6 \\ 2x + y - 2z = 6 \\ 3x - y + 3z = 10 \end{cases}$

68. $\begin{cases} 2x - 3y + 4z = 22 \\ 3x - y + 2z = 10 \\ 4x + 2y - z = -8 \end{cases}$

(continued)

69.
$$\begin{cases} 3x - 4y + 5z = 1 \\ 2x - y + 3z = 2 \\ x - 2y + z = 3 \end{cases}$$

70.
$$\begin{cases} \dfrac{x}{2} + \dfrac{y}{3} + \dfrac{z}{4} = 13 \\ x - \dfrac{y}{2} + \dfrac{z}{3} = 10 \\ \dfrac{x}{3} + \dfrac{y}{4} - \dfrac{z}{2} = 1 \end{cases}$$

71.
$$\begin{cases} x - 2y + 3z = 4 \\ \dfrac{3}{2}x - 3y + \dfrac{9}{2}z = 6 \\ -3x + 6y - 9z = -12 \end{cases}$$

72.
$$\begin{cases} 3x - y = 0 \\ 4x + z = 1 \\ 6y - 3z = 2 \end{cases}$$

In Exercises 73–80, evaluate the given determinant.

73. $\begin{vmatrix} 3 & 2 \\ 4 & 5 \end{vmatrix}$

74. $\begin{vmatrix} 1 & -2 \\ 3 & -1 \end{vmatrix}$

75. $\begin{vmatrix} 1 & 2 \\ 2 & 4 \end{vmatrix}$

76. $\begin{vmatrix} 3 & 5 \\ 3 & -2 \end{vmatrix}$

77. $\begin{vmatrix} 4 & -1 & 2 \\ 2 & 1 & 0 \\ -1 & 2 & -3 \end{vmatrix}$

78. $\begin{vmatrix} 3 & 2 & -2 \\ 6 & 4 & -4 \\ 5 & 1 & -1 \end{vmatrix}$

79. $\begin{vmatrix} 5 & 4 & 3 \\ 2 & 0 & 0 \\ 3 & 1 & 1 \end{vmatrix}$

80. $\begin{vmatrix} 2 & 2 & 1 \\ 2 & 2 & 1 \\ 3 & 4 & 1 \end{vmatrix}$

In Exercises 81–82, solve using matrices:

81.
$$\begin{cases} x + 2y = 0 \\ 2x - 3y = 7 \end{cases}$$

82.
$$\begin{cases} x + y - z = -2 \\ x + 2y + 2z = 9 \\ 2x + z = 1 \end{cases}$$

In Exercises 83–86, solve the given system of equations using Cramer's rule.

83.
$$\begin{cases} 3x + 7y = 2 \\ 10x + 5y = 11 \end{cases}$$

84.
$$\begin{cases} 5x - 6y = 4 \\ 11x + 8y = 5 \end{cases}$$

85.
$$\begin{cases} 2x + 3y + 4z = 3 \\ 5x + 2y - 3z = 2 \\ 3x - 7y + 5z = -7 \end{cases}$$

86.
$$\begin{cases} 4x - 6y - z = 8 \\ 2x + 3y + 3z = 9 \\ 6x + 9y + 5z = 10 \end{cases}$$

In Exercises 87–92, sketch the solution set to each system of inequalities.

87.
$$\begin{cases} x + y \le 6 \\ 2x + y \ge 4 \end{cases}$$

88.
$$\begin{cases} 3x + y > 6 \\ 2x - y < 4 \end{cases}$$

89.
$$\begin{cases} 2x + 3y < 12 \\ x < y \end{cases}$$

90.
$$\begin{cases} 4x - y \le 6 \\ x \le -2 \end{cases}$$

91.
$$\begin{cases} x + y \le 4 \\ x - y \le 4 \\ x \ge 0 \end{cases}$$

92.
$$\begin{cases} 3x + y \le 9 \\ x + 3y \le 9 \\ x \ge 0 \\ y \ge 0 \end{cases}$$

1. *Solve the following equations by any method.*

   (a) $2x^2 + x = 3$ **(b)** $\dfrac{x}{3} = \dfrac{4}{x}$

   (c) $2x^2 + 4x - 3 = 0$ **(d)** $\dfrac{x}{x + 1} + \dfrac{2}{x - 3} = 4$

2. *Solve for a.* $\sqrt{3a^2 + 2} = b$

3. *Solve for x:*

   (a) $\sqrt{2x + 1} = 2\sqrt{x} - 1$ **(b)** $x^3 + x^2 = 6x$

4. *Solve for x and graph the solution on the real number line:*

$$2x^2 - 7x - 4 \geq 0$$

5. Michael rows a boat downstream a distance of 4 miles, then back upstream to the starting point. If his rate in still water is 7 mph, what is the rate of the stream if he makes the total trip in $1\frac{2}{5}$ hours?

6. Sketch a graph of the equation $3x - 4y = 18$ using the intercept method.

7. Find the slope of a line passing through the points $(2, -3)$ and $(3, -4)$.

8. Find the equations of the line passing through the point $(2, -3)$ and

   (a) Parallel to the line $3y - 2x = 4$.   **(b)** Perpendicular to the line $3y - 2x = 4$.

9. Sketch a graph of $3x + 6y > 18$ on a rectangular coordinate system.

10. There is a perfect linear relationship between the scores on two tests, test A and test B. Anyone scoring 40 on test A will score 30 on test B; anyone scoring 60 on test A will score 25 on test B. What score on test B should a person get if he or she scored 48 on test A?

11. *Solve the following systems of equations:*

   (a) $\begin{cases} 5x + 2y = 4 \\ 2x - 3y = 13 \end{cases}$ **(b)** $\begin{cases} x + y + z = 6 \\ 3x + 2y - z = 11 \\ 2x - 4y - z = 12 \end{cases}$

12. *Sketch the solution set of the following system of inequalities:*

$$\begin{cases} x - y \leq 5 \\ x - 2y \leq 0 \\ \quad\ y \geq 0 \end{cases}$$

13. *Solve the following system using matrices:*

$$\begin{cases} x + 2y \quad\ = \quad 4 \\ x + \ y + \ z = \quad 0 \\ x - \ y + 2z = -6 \end{cases}$$

14. *Solve the following systems of equations by using Cramer's rule:*

   (a) $\begin{cases} 6x + 5y = 13 \\ 7x + 8y = 26 \end{cases}$ **(b)** $\begin{cases} 4x - 5y + 2z = 17 \\ 3x + 7y - 5z = 2 \\ 5x - 6y + 3z = 21 \end{cases}$

10

Conic Sections and Nonlinear Systems

In Chapter 8 we graphed first-degree equations in two variables, that is, equations of the form $Ax + By = C$, on the rectangular coordinate system. Graphs of equations of this form are straight lines and hence we called them *linear equations*.

In this chapter we will be concerned with graphs of second-degree equations—in particular, equations of the form

$$Ax^2 + By^2 + Cx + Dy + E = 0$$

where A and B are not both zero.

If the equation can be graphed, it can be shown that (with a few exceptions) the graph will be one of four figures: the circle, the parabola, the ellipse, or the hyperbola. These figures are called **conic sections** since they describe the intersection of a plane and a double cone, as illustrated in Figure 10.1.

Figure 10.1

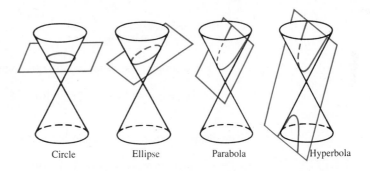

Circle   Ellipse   Parabola   Hyperbola

We will begin by deriving a formula for finding the distance between two points on a coordinate plane. This formula will be useful in deriving the general equations for conic sections.

10.1

The Distance Formula: Circles

The Distance Formula

In Section 7.4 we discussed the Pythagorean Theorem, which gives us the relationship between the sides of a right triangle. We will exploit this relationship to derive a method, or formula, for finding the distance between two points in a Cartesian plane.

Let's begin by first noting that *on the x-axis*, the distance between two points with x-coordinates x_1 and x_2 is $|x_2 - x_1|$ (see Figure 10.2 on page 464). This is also true for any two points on a line *parallel to the x-axis*.

On the y-axis, the distance between two points with *y*-coordinates y_1 and y_2 is $|y_2 - y_1|$ (see Figure 10.3). Again, the same is true for any two points on a line *parallel to the y-axis.*

Figure 10.2

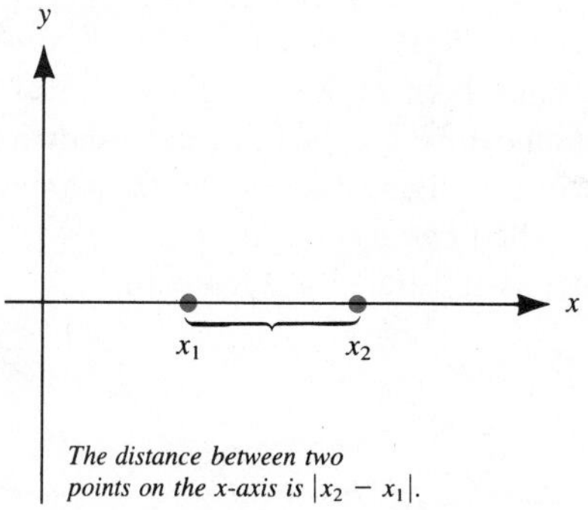

Figure 10.3

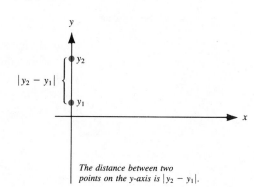

The distance between two points on the x-axis is $|x_2 - x_1|$.

The distance between two points on the y-axis is $|y_2 - y_1|$.

For example, the distance between points (4, 0) and (−2, 0) is

$$|x_2 - x_1| = |4 - (-2)| = |6| = 6 \quad \text{or} \quad |-2 - (4)| = |-6| = 6$$

(see Figure 10.4)

The distance between (4, 3) and (4, −5) is

$$|y_2 - y_1| = |3 - (-5)| = |8| = 8 \quad \text{or} \quad |-5 - 3| = |-8| = 8$$

(see Figure 10.5)

Figure 10.4
Distance between (4, 0) and
(−2, 0)

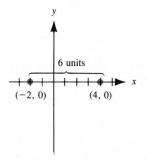

6 units

(−2, 0)　(4, 0)

Figure 10.5
Distance between (4, 3) and
(4, −5)

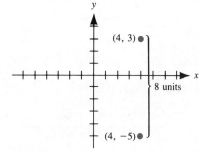

(4, 3)

8 units

(4, −5)

Now let's take any two points, P_1 with coordinates (x_1, y_1) and P_2 with coordinates (x_2, y_2), and see if we can find the distance, d, between the two points (see Figure 10.6).

Figure 10.6

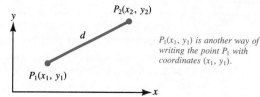

$P_2(x_2, y_2)$

d

$P_1(x_1, y_1)$

$P_1(x_1, y_1)$ is another way of writing the point P_1 with coordinates (x_1, y_1).

First we draw a line through P_2 which is perpendicular to the x-axis (and therefore parallel to the y-axis). Then we draw another line through P_1 which is perpendicular to the y-axis (and therefore parallel to the x-axis). The point where the lines intersect we will label Q (see Figure 10.7).

Figure 10.7

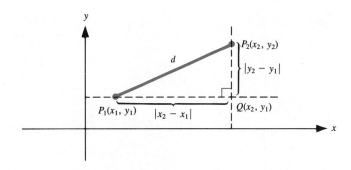

We note the following:

Q has coordinates (x_2, y_1). Why?

Triangle P_1QP_2 is a right triangle with right angle P_1QP_2.

Because our newly drawn lines are parallel to the x- and y-axes,

The distance between points P_1 and Q is $|x_2 - x_1|$.

The distance between points P_2 and Q is $|y_2 - y_1|$.

Applying the Pythagorean Theorem to triangle P_1QP_2, we have

$$(|x_2 - x_1|)^2 + (|y_2 - y_1|)^2 = d^2$$

We can drop the absolute values, since the squares of expressions are automatically nonnegative, to get

$(x_2 - x_1)^2 + (y_2 - y_1)^2 = d^2$ *Then take square roots.*

$\sqrt{(x_2 - x_1)^2 + (y_2 - y_1)^2} = d$ *We are interested only in the positive root since distances are always positive.*

We thus obtain the result stated in the box.

The Distance Formula

The distance, d, between points $P_1(x_1, y_1)$ and $P_2(x_2, y_2)$ is

$$d = \sqrt{(x_2 - x_1)^2 + (y_2 - y_1)^2}$$

EXAMPLE 1

Find the distance between the following pairs of points:

(a) $(2, 3)$ and $(5, 7)$ **(b)** $(4, 2)$ and $(-2, 4)$

Solution

Since the differences are squared it does not matter which point you designate (x_1, y_1) and (x_2, y_2), as long as you remember to subtract x-coordinates from x-coordinates and y-coordinates from y-coordinates.

(a)   Let $(x_1, y_1) = (2, 3)$ and $(x_2, y_2) = (5, 7)$. Then

$$
\begin{aligned}
d &= \sqrt{(5 - 2)^2 + (7 - 3)^2} \\
&= \sqrt{3^2 + 4^2} \\
&= \sqrt{9 + 16} \\
&= \sqrt{25} \\
&= \boxed{5}
\end{aligned}
$$

(b)   Let $(x_1, y_1) = (-2, 4)$ and $(x_2, y_2) = (4, 2)$. Then

$$
\begin{aligned}
d &= \sqrt{[4 - (-2)]^2 + (2 - 4)^2} \\
&= \sqrt{6^2 + (-2)^2} \\
&= \sqrt{36 + 4} \\
&= \sqrt{40} \\
&= \boxed{2\sqrt{10}} \qquad \textit{Always simplify radicals in answers.} \qquad ■
\end{aligned}
$$

EXAMPLE 2

Show that the points $P(1, 1)$, $Q(3, 4)$, and $R(6, 2)$ form the vertices of a right triangle.

Solution

If we join together the three points, we have a triangle. To determine if it is a right triangle, we refer to the converse of the Pythagorean Theorem in Section 7.4, which states that if the sum of the squares of two sides of a triangle is equal to the square of the third side, then the triangle is a right triangle. Thus, we must show that the sum of the squares of two sides of our triangle is equal to the square of the third side (see Figure 10.8).

Figure 10.8
Triangle for Example 2

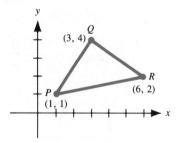

The lengths of the sides are computed as follows:

The distance from P to Q is

$$\sqrt{(1 - 3)^2 + (1 - 4)^2} = \sqrt{(-2)^2 + (-3)^2} = \sqrt{4 + 9} = \sqrt{13}$$

The distance from Q to R is

$$\sqrt{(3 - 6)^2 + (4 - 2)^2} = \sqrt{(-3)^2 + (2)^2} = \sqrt{9 + 4} = \sqrt{13}$$

The distance from P to R is

$$\sqrt{(1-6)^2 + (1-2)^2} = \sqrt{(-5)^2 + (-1)^2} = \sqrt{25+1} = \sqrt{26}$$

If we square the distance from P to Q, we get $(\sqrt{13})^2 = 13$.

If we square the distance from Q to R, we get $(\sqrt{13})^2 = 13$.

As we can see, they sum to the square of the distance from P to R: $(\sqrt{26})^2 = 26$.

Hence, the sum of the squares of two sides is equal to the square of the third side and therefore we do have a right triangle. ∎

Alternatively, we can solve this problem by showing that the slope of PQ is the negative reciprocal of the slope of QR and hence PQ is perpendicular to QR. Either method can be used.

Another formula which we will find useful is the midpoint formula, given in the box.

The Midpoint Formula	The ***midpoint*** of the line segment joining the two points $P(x_1, y_1)$ and $Q(x_2, y_2)$ has coordinates $$\left(\frac{x_1 + x_2}{2}, \frac{y_1 + y_2}{2}\right)$$

Hence, the point exactly midway between $(2, 5)$ and $(6, 1)$ has

$$x\text{-coordinate } \frac{2+6}{2} = \frac{8}{2} = 4 \quad \text{and} \quad y\text{-coordinate } \frac{5+1}{2} = \frac{6}{2} = 3$$

Thus, the midpoint is $(4, 3)$ (see Figure 10.9). Note that the x-coordinate of the midpoint, $\frac{x_1 + x_2}{2}$, is the *average* of the x-coordinates of P and Q, and the y-coordinate of the midpoint, $\frac{y_1 + y_2}{2}$, is the *average* of the y-coordinates of P and Q. Do not confuse the midpoint formula with the distance formula, which uses *differences*.

Figure 10.9

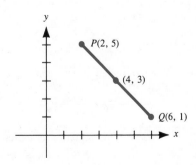

EXAMPLE 3

Find the midpoint of the following pair of points:

$$(6, -5) \quad \text{and} \quad (3, 7)$$

Solution

By the midpoint formula, the midpoint is

$$\left(\frac{6 + 3}{2}, \frac{-5 + 7}{2} \right) = \boxed{\left(\frac{9}{2}, 1 \right) \quad \text{or} \quad \left(4\frac{1}{2}, 1 \right)} \qquad \blacksquare$$

The Circle

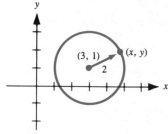

Figure 10.10
Circle with radius 2
and center (3, 1)

A *circle* is defined to be the set of all points whose distance from a fixed point is constant. The fixed point is called the center, C, and the constant distance from the center to the circle is called the radius, r.

Let's put a circle of radius 2 on a Cartesian plane and center it at the point (3, 1). Pick any point on the circle and call it (x, y) (see Figure 10.10).

Note that by definition of the circle, 2 is the distance from the center, (3, 1), to any point, (x, y), on the circle. Hence, by the distance formula, we have

$$\sqrt{(x - 3)^2 + (y - 1)^2} = 2$$

For convenience, we eliminate the radical by squaring both sides of the equation to get

$$(x - 3)^2 + (y - 1)^2 = 2^2 \qquad \text{\textit{This is the equation of the circle with center (3, 1) and radius 2.}}$$

In general, we have the result given in the box.

Standard Form of the Equation of a Circle

$$(x - h)^2 + (y - k)^2 = r^2$$

is the equation of a circle with center (h, k) and radius r.

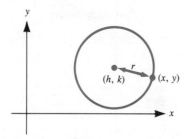

This equation is called the ***standard form of the equation of a circle.***

EXAMPLE 4 | Find the equation of the circle with:

(a)   Center (3, 5) and radius 6

(b)   Center (−2, 4) and radius 3

(c)   Center (0, 0) and radius 5

Solution | (a)   Looking at the standard form, we note that since the center is (3, 5), we have $h = 3$ and $k = 5$. The radius is 6; thus, $r = 6$. Therefore, the equation of the circle is

$$(x - h)^2 + (y - k)^2 = r^2$$
$$(x - 3)^2 + (y - 5)^2 = 6^2$$
$$\boxed{(x - 3)^2 + (y - 5)^2 = 36}$$ *Which, when multiplied out, is*

$$x^2 - 6x + 9 + y^2 - 10y + 25 = 36 \quad \text{or} \quad \boxed{x^2 + y^2 - 6x - 10y - 2 = 0}$$

(b)   The circle with center (−2, 4) and radius 3 has $h = -2$, $k = 4$, and $r = 3$. The equation is

$$(x - h)^2 + (y - k)^2 = r^2$$
$$[x - (-2)]^2 + (y - 4)^2 = 3^2 \quad \text{or} \quad \boxed{(x + 2)^2 + (y - 4)^2 = 9}$$

Note that the x-coordinate of the center is negative, which yields the expression $(x + 2)^2$.

(c)   The circle with center (0, 0) and radius 5 has $h = 0$, $k = 0$, and $r = 5$. Hence, its equation is

$$(x - 0)^2 + (y - 0)^2 = (5)^2 \quad \text{or} \quad \boxed{x^2 + y^2 = 25}$$ ∎

Given an equation of a circle in standard form, we can easily identify the center and radius, as indicated in the next example.

EXAMPLE 5 | Find the center and radius of the following circles and sketch their graphs:

(a) $(x - 5)^2 + (y - 3)^2 = 16$ (b) $(x + 4)^2 + (y - 2)^2 = 9$

Solution | (a)   We compare the given equation with the standard form of the equation of a circle:

$$(x - h)^2 + (y - k)^2 = r^2$$

We can identify $h = 5$, $k = 3$, and $r^2 = 16$, which means $r = 4$.

$$\boxed{\text{The center is (5, 3) and the radius is } r = 4.}$$ (see Figure 10.11 at the top of page 470)

Figure 10.11
Circle with radius 4
and center (5, 3)

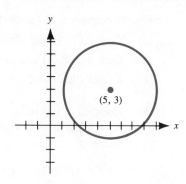

(b)   Note that the left-hand portion of the equation is not quite in standard form. In the standard form, the squared expressions are written as *differences*.

$$(x - h)^2 + (y - k)^2 = r^2$$
$$(x + 4)^2 + (y - 2)^2 = 9$$

Hence, $-h = 4$ or $h = -4$, $k = 2$, and $r^2 = 9$, or $r = 3$.

Thus, the center is $(-4, 2)$ and the radius is 3   (see Figure 10.12)

Figure 10.12
Circle with center (−4, 2)
and radius 3

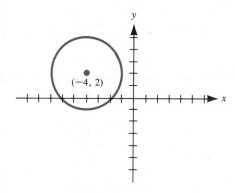

We can see that if we are given an equation of a circle in standard form, then finding the center and radius of the circle is quite straightforward. However, what if the equation is not in standard form, such as $x^2 + y^2 - 14x - 6y - 42 = 0$? How do we find the radius and center?

We want to change the equation into the form $(x - h)^2 + (y - k)^2 = r^2$. Note that the left-hand side of the equation consists of perfect squares. This suggests that we use the technique of completing the square. (You may want to review completing the square in Section 7.2.)

Let's look at an example.

EXAMPLE 6   Find the center and radius of the following circle:

$$x^2 + y^2 - 14x - 6y - 42 = 0$$

Solution

We want to change the equation into standard form. Since standard form consists of perfect squares, we complete the square *for each variable.*

$$x^2 + y^2 - 14x - 6y - 42 = 0 \qquad \textit{Add 42 to both sides of the equation.}$$
$$x^2 + y^2 - 14x - 6y = 42 \qquad \textit{Collect x and y terms.}$$
$$(x^2 - 14x \quad) + (y^2 - 6y \quad) = 42 \qquad$$

Complete the square for each quadratic expression: take half the middle term coefficient and square it.

$$\left[\frac{1}{2}(-14)\right]^2 = (-7)^2 = 49$$
$$\left[\frac{1}{2}(-6)\right]^2 = (-3)^2 = 9$$

*Add **both numbers,** 9 and 49, to both sides of the equation.*

$$(x^2 - 14x + 49) + (y^2 - 6y + 9) = 42 + 49 + 9 \qquad$$
$$(x - 7)^2 + (y - 3)^2 = 100$$

Rewrite quadratic expressions in factored form and simplify the right-hand side.

Thus, we have a circle with $\boxed{\text{center } (7, 3) \text{ and radius } \sqrt{100} = 10}$ ∎

EXAMPLE 7

Find the center and radius of the following circles:

(a) $x^2 + y^2 + 6x - 4y = 12$ **(b)** $2x^2 + 2y^2 + 20y = 10$

Solution

(a)
$$x^2 + y^2 + 6x - 4y = 12 \qquad \textit{Collect x and y terms.}$$
$$(x^2 + 6x \quad) + (y^2 - 4y \quad) = 12$$

Complete the square for each quadratic expression:

$$\left[\frac{1}{2}(6)\right]^2 = 3^2 = 9$$
$$\left[\frac{1}{2}(-4)\right]^2 = (-2)^2 = 4$$

Add both numbers to both sides of the equation.

$$(x^2 + 6x + 9) + (y^2 - 4y + 4) = 12 + 9 + 4$$

Rewrite quadratic expressions in factored form.

$$(x + 3)^2 + (y - 2)^2 = 25$$
$$\uparrow$$

Note that this sign is positive, so the x-coordinate of the center is negative.

Thus, we have a circle with $\boxed{\text{center } (-3, 2) \text{ and radius} = \sqrt{25} = 5}$

(b)   Since the standard form of the equation of a circle has a coefficient of 1 in the square terms, divide both sides of the equation by 2.

$$2x^2 + 2y^2 + 20y = 10 \qquad \textit{Divide both sides of the equation by 2.}$$
$$\frac{2(x^2 + y^2 + 10y)}{2} = \frac{10}{2}$$
$$x^2 + y^2 + 10y = 5 \qquad$$

We now put the equation in standard form. Collect x and y terms.

$$x^2 + (y^2 + 10y \qquad) = 5$$

We have only to complete the square for the quadratic expression in y:

$$\left[\frac{1}{2}(10)\right]^2 = 5^2 = 25$$

Add 25 to both sides.

$$x^2 + (y^2 + 10y + 25) = 5 + 25$$

$$(x - 0)^2 + (y + 5)^2 = 30$$

Rewrite quadratic expressions in factored form.

Thus, we have a circle with | center $(0, -5)$ and radius $\sqrt{30}$ | ■

Observe that with completing the square, we can take any equation of the form $Ax^2 + By^2 + Cx + Dy + E = 0$ and put it in the standard form of a circle *if the coefficients of the squared terms, A and B, are 1 (or can be made to equal 1).*

Exercises 10.1

In Exercises 1–12, find the distance between the two given points, P and Q.

1. $P(3, 5)$ and $Q(6, 9)$ **2.** $P(0, 2)$ and $Q(12, 7)$ **3.** $P(6, 3)$ and $Q(3, 6)$

4. $P(5, 7)$ and $Q(7, 5)$ **5.** $P(6, -9)$ and $Q(-6, 9)$ **6.** $P(-5, 4)$ and $Q(5, -4)$

7. $P(-8, -3)$ and $Q(-7, -3)$ **8.** $P(-6, 2)$ and $Q(-6, -4)$ **9.** $P\left(\frac{1}{2}, 0\right)$ and $Q\left(\frac{1}{3}, 2\right)$

10. $P\left(\frac{1}{4}, 2\right)$ and $Q\left(\frac{1}{3}, 0\right)$ **11.** $P(1.7, 1.2)$ and $Q(1.4, 0.8)$ **12.** $P(1.4, 3.8)$ and $Q(0.9, 2.6)$

In Exercises 13–20, find the midpoint between the two given points, P and Q.

13. $P(0, 5)$ and $Q(0, 7)$ **14.** $P(7, 0)$ and $Q(9, 0)$ **15.** $P(-3, 1)$ and $Q(3, 1)$

16. $P(6, -2)$ and $Q(-6, -2)$ **17.** $P(-3, 4)$ and $Q(3, -4)$ **18.** $P(5, -2)$ and $Q(-5, 2)$

19. $P\left(\frac{2}{5}, \frac{3}{4}\right)$ and $Q\left(\frac{1}{3}, 2\right)$ **20.** $P\left(\frac{2}{3}, \frac{3}{5}\right)$ and $Q\left(3, \frac{1}{2}\right)$

In Exercises 21–24, determine whether the given points are vertices of right triangles.

21. $P(5, 2)$, $Q(8, 2)$, $R(8, 6)$ **22.** $P(3, 5)$, $Q(7, 5)$, $R(6, 9)$

23. $P(4, 2)$, $Q(-1, 5)$, $R(3, 9)$ **24.** $P(0, 8)$, $Q(2, 11)$, $R(5, 9)$

In Exercises 25–28, plot the points P, Q, R, and S. If the diagonals of a quadrilateral bisect each other, then the quadrilateral is a parallelogram. Knowing this, determine if the quadrilateral formed by P, Q, R, and S is a parallelogram.

25. $P(5, 3)$, $Q(7, 4)$, $R(9, 7)$, $S(7, 6)$ **26.** $P(3, 9)$, $Q(0, 7)$, $R(7, 1)$, $S(10, 3)$

27. $P(-2, 3)$, $Q(5, -4)$, $R(-6, 5)$, $S(3, -4)$ **28.** $P(3, -5)$, $Q(6, -2)$, $R(9, -3)$, $S(6, -6)$

In Exercises 29–38, give the equation of the circle, given the center C and radius r.

29. $C(0, 0)$, $r = 1$ **30.** $C(0, 0)$, $r = 5$ **31.** $C(1, 0)$, $r = 7$ **32.** $C(0, 1)$, $r = 1$

33. $C(2, 5)$, $r = 6$ **34.** $C(5, 2)$, $r = 1$ **35.** $C(6, -2)$, $r = 5$ **36.** $C(-6, 2)$, $r = 4$

37. $C(-3, -2)$, $r = 1$ **38.** $C(-2, -3)$, $r = 1$

In Exercises 39–48, find the center and radius of the circle.

39. $x^2 + y^2 = 16$
40. $x^2 + y^2 = 36$
41. $x^2 + y^2 = 24$

42. $x^2 + y^2 = 98$
43. $(x - 3)^2 + y^2 = 16$
44. $(x - 4)^2 + y^2 = 25$

45. $(x - 2)^2 + (y - 1)^2 = 1$
46. $(x - 5)^2 + (y - 3)^2 = 36$
47. $(x + 1)^2 + (y - 3)^2 = 25$

48. $(x - 2)^2 + (y + 7)^2 = 49$

In Exercises 49–58, find the center and radius, and graph the given equation of the circle.

49. $(x + 2)^2 + (y + 3)^2 = 32$
50. $(x + 5)^2 + (y + 3)^2 = 27$
51. $(x + 7)^2 + (y + 1)^2 = 2$

52. $(y + 5)^2 + (x + 2)^2 = 8$
53. $x^2 + y^2 - 2x = 15$
54. $x^2 + y^2 - 6y = 7$

55. $x^2 + y^2 - 4x - 2y = 20$
56. $x^2 + y^2 - 2y = 6 + 6x$
57. $x^2 + y^2 - 2x = 20 + 4y$

58. $x^2 + y^2 - 2x - 6y = 6$

In Exercises 59–66, find the center and radius of the circle.

59. $x^2 + 10y = 71 - y^2 + 4x$
60. $x^2 + y^2 = 6x - 14y - 32$
61. $x^2 + y^2 = 2y - 6x - 2$

62. $2x - 6y = 2 + x^2 + y^2$
63. $2x^2 + 2y^2 - 4x + 4y = 22$
64. $3x^2 + 3y^2 + 18x - 6y = 45$

65. $x^2 + y^2 - x + 2y = \dfrac{59}{4}$
66. $x^2 + y^2 - 2x - 3y = \dfrac{75}{4}$

In Exercises 67–70, graph the given equation.

67. $x^2 + y^2 = 16$
68. $x^2 = 36 - y^2$
69. $x + y = 4$
70. $y = 6 - x$

? QUESTIONS FOR THOUGHT

71. Explain why we can drop the absolute value symbols in the equation

$$(|x_2 - x_1|)^2 + (|y_2 - y_1|)^2 = d^2$$

72. Show that the point with coordinates $\left(\dfrac{x_1 + x_2}{2}, \dfrac{y_1 + y_2}{2}\right)$ is the same distance from (x_1, y_1) as it is from (x_2, y_2).

73. Describe the graph of:
  (a) $x^2 + y^2 = 0$ **(b)** $x^2 + y^2 = -4$

 MINI-REVIEW

74. Perform the operations and simplify:

$$\frac{3x^2 + 3x - 6}{9x - 18} \cdot \frac{x^2 + 2x - 8}{x^2 + 9x + 20}$$

75. Perform the operations and express the answer in simplest form using positive exponents only:

$$\left(\frac{3x^2 y^{-2}}{x^5 y^{-8}}\right)^{-3}$$

76. *Simplify:*

$$\frac{1}{\sqrt{5} - \sqrt{4}}$$

77. Find the slope of the line $2x + 5y = 8$.

78. Find the distance between the points $(-3, 4)$ and $(5, 3)$.

10.2

The Parabola

A **parabola** is defined to be the set of points whose distance from a fixed point is equal to its distance from a fixed line. That is, for any point (x, y) on the parabola, the distance from the point (x, y) to a fixed point (a, b) is equal to the distance from the point (x, y) to a fixed line, $y = c$ (see Figure 10.13).

Figure 10.13

A parabola

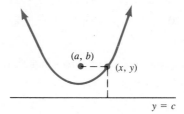

As with the circle, we can derive the general equation of the parabola using the distance formula in terms of a, b, c, x, and y, and arrive at an equation of the form $y = Ax^2 + Bx + C$, where A, B, and C are constants in terms of a, b, and c (see Exercise 65 at the end of this section). For our purposes, however, it would be more useful to examine the parabola $y = Ax^2 + Bx + C$ to see what effect the constants A, B, and C have on its shape. (We note that if $A = 0$, then we no longer have a parabola.) We begin by studying the basic parabola $y = Ax^2$.

Parabolas of the Form $y = Ax^2$

We are considering the case $y = Ax^2 + Bx + C$, where both B and C are 0. We will begin by examining the graph of $y = x^2$. First we let x take on various values and then find y:

x	-3	-2	-1	0	1	2	3
y	9	4	1	0	1	4	9

Sometimes it is more convenient to write tables horizontally.

We then plot the points (see Figure 10.14). This shape is called a *parabola*. Know this picture well.

The parabola shown in Figure 10.14 is the simplest parabola and it is important that we examine and understand its basic properties. The lowest point on this parabola is called the **vertex**. Notice that for each point on the parabola to the right of the vertex, there is a corresponding point to the left of the vertex. The right side, as a matter of fact, is the mirror image of the left side (put a mirror on the y-axis). We say that "the parabola is symmetric about the y-axis."

Figure 10.14
Graph of $y = x^2$

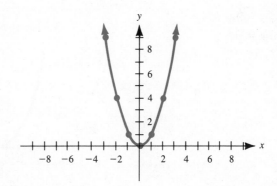

The **axis of symmetry** is the line which divides the parabola into two identical (mirror-image) parts: It is where we would place the mirror to get a full picture of the parabola. In this case, the axis of symmetry is the y-axis, which has the equation $x = 0$.

Now let's look at the graphs of $y = 2x^2$ and $y = \frac{1}{3}x^2$. In Figures 10.15 and 10.16, we plot the points given in the accompanying tables.

$y = 2x^2$

x	-3	-2	-1	0	1	2	3
y	18	8	2	0	2	8	18

Figure 10.15
Graph of $y = 2x^2$

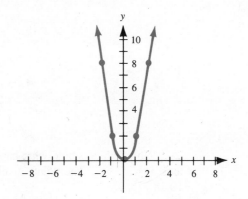

$y = \frac{1}{3}x^2$

x	-3	-2	-1	0	1	2	3
y	3	$\frac{4}{3}$	$\frac{1}{3}$	0	$\frac{1}{3}$	$\frac{4}{3}$	3

Figure 10.16

Graph of $y = \frac{1}{3}x^2$

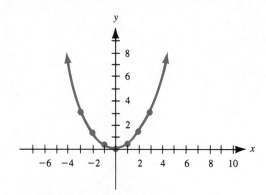

Notice that the graphs of $y = 2x^2$ (Figure 10.15) and $y = \frac{1}{3}x^2$ (Figure 10.16) have the same general shape, the same axis of symmetry ($x = 0$) and vertex (0, 0), and they both open upward. However,

$y = 2x^2$ is narrower than $y = x^2$ and $y = \frac{1}{3}x^2$ is wider than $y = x^2$.

We find that A, the coefficient of x^2, "stretches" the parabola. As Figure 10.17 illustrates, the larger the positive coefficient, the more stretched, or narrower, the parabola becomes. The smaller the positive coefficient, the wider the parabola becomes. Notice that the parabola stays at or above the x-axis, that is, y is always nonnegative.

Figure 10.17

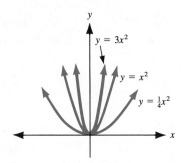

Now let's examine what happens when the coefficient of x^2 is negative. Let's look at $y = -x^2$. The points given in the accompanying table are plotted in Figure 10.18.

$y = -x^2$

x	-3	-2	-1	0	1	2	3
y	-9	-4	-1	0	-1	-4	-9

Figure 10.18

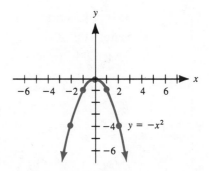

The highest point of this parabola is also called the *vertex*. Notice that the y values for $y = -x^2$ are opposite the y values for $y = x^2$. All y values [except at the point $(0, 0)$] are negative. Thus, we have the same parabola as $y = x^2$, but now it is upside down. The vertex (now the highest point) is still $(0, 0)$ and the axis of symmetry is still the y-axis. In this case, we say the parabola *opens down*.

The negative coefficient of x^2 still stretches the parabola, but now it is upside down. Figure 10.19 illustrates what happens to the parabola as $A < 0$ varies. Compare this figure with Figure 10.17, in which $A > 0$ varies.

Figure 10.19

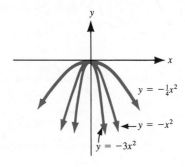

Hence, we have the results summarized in the box.

$y = Ax^2$ is a parabola with vertex $(0, 0)$ and the equation of the axis of symmetry is $x = 0$ (the y-axis).

   If $A > 0$, the parabola opens up.

   If $A < 0$, the parabola opens down.

EXAMPLE 1

Solution

Sketch the graph of $y = -5x^2$.

For this equation, we have $A = -5$. Thus, the parabola opens down (since $A < 0$). Plot a few points (see Figure 10.20 on page 498).

Axis of symmetry: $x = 0$ (the y-axis)

Vertex at $(0, 0)$

Stretched by a factor of 5

Figure 10.20

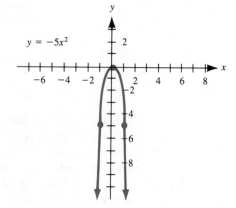

$y = -5x^2$

Parabolas of the Form $y = Ax^2 + C$

The next case is the form $y = Ax^2 + C$. Basically, we are adding a constant to the parabola $y = Ax^2$. For example, let's examine the table and graph of $y = x^2 + 3$. In this example, $A = 1$, $B = 0$, and $C = 3$.

$$y = x^2 + 3$$

x	-3	-2	-1	0	1	2	3
y	12	7	4	3	4	7	12

As we may expect, comparing this table with the table of $y = x^2$ on page 474, we find that for each x, the y value of $y = x^2 + 3$ is 3 more than the y value of $y = x^2$. Since we are simply adding 3 to each y value of the original parabola, we end up with the same shape, but the parabola is shifted up 3 units (see Figure 10.21).

Figure 10.21

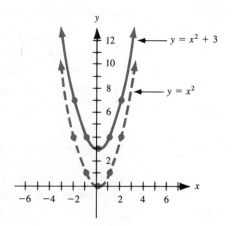

$y = x^2 + 3$

$y = x^2$

Notice that the axis of symmetry is still $x = 0$, but now the vertex is shifted up 3 units to $(0, 3)$. We can generalize this as follows:

The graph of $y = Ax^2 + C$ is a parabola with the same shape as the parabola $y = Ax^2$, but shifted $|C|$ units up if $C > 0$ or down if $C < 0$. The vertex is $(0, C)$ and the axis of symmetry is $x = 0$ (the y-axis).

EXAMPLE 2

Sketch a graph of $y = 9x^2 - 4$.

Solution

For this parabola, we have $A = 9$ and $C = -4$. Plot a few points (see Figure 10.22).

Graph has shape of $y = 9x^2$, but moved *down* 4 units ($C < 0$)

Axis of symmetry is $x = 0$

Vertex is $(0, -4)$

Parabola opens up ($A > 0$)

Figure 10.22

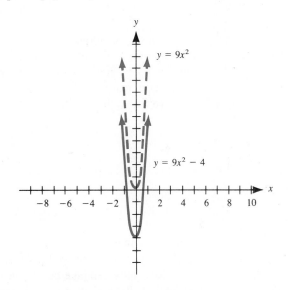

Thus far, the important information about a parabola includes its direction (opening up or down), its vertex, its axis of symmetry, and its stretching factor. We should also note where the parabola intersects the coordinate axes, that is, the x- and y-intercepts of the parabola.

The x- and y-Intercepts of a Parabola

From graphing linear or first-degree equations, we already know that the x- (or y-) intercept is the x (or y) value where the graph of the figure crosses the x- (or y-) axis. Thus, the x-intercept is the value of x when $y = 0$ (why?), and the y-intercept is the value of y when $x = 0$ (why?).

Locating the y-intercept is a straightforward process. For example, to find the y-intercept of $y = 9x^2 - 4$, we set $x = 0$ and solve for y:

$$y = 9x^2 - 4 = 9(0)^2 - 4 = -4$$

Thus, the y-intercept is -4, and the graph crosses the y-axis at $y = -4$, as illustrated in Figure 10.23.

Figure 10.23
Graph of $y = 9x^2 - 4$

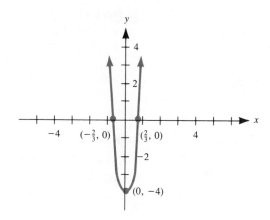

For the equations we have had thus far, those of the form $y = Ax^2 + C$, the y-intercept is $y = C$ (set $x = 0$ in $y = Ax^2 + C$). This corresponds to the vertex since the vertex lies on the axis of symmetry, which is the y-axis. (This will *not* be the case in the more general parabolas, as we shall soon see.)

To locate the x-intercepts of $y = 9x^2 - 4$, we set $y = 0$ and solve for x:

$y = 9x^2 - 4$ *Set $y = 0$.*

$0 = 9x^2 - 4$ *Solve for x.*

$0 = (3x - 2)(3x + 2)$

$$x = -\frac{2}{3} \quad \text{or} \quad x = +\frac{2}{3}$$

Thus, the x-intercepts are $-\frac{2}{3}$ and $+\frac{2}{3}$, and the graph crosses the x-axis at $x = +\frac{2}{3}$ and $x = -\frac{2}{3}$, as can be seen in Figure 10.23.

Now let's continue on to consider the case where $C = 0$ in the general form $y = Ax^2 + Bx + C$, that is, $y = Ax^2 + Bx$.

Parabolas of the Form $y = Ax^2 + Bx$

The case $y = Ax^2 + Bx$ will still be a parabola with a vertical axis of symmetry, but now the axis of symmetry and vertex will not lie on the y-axis. We will use the fact that the parabola will be symmetric (to a vertical line) to show how to find the vertex and axis of symmetry for $y = Ax^2 + Bx$.

First recall what it means for a parabola to be symmetric. There is a line which will cut the parabola into two parts so that one part of the parabola will be the mirror image of the other part (see Figure 10.24). In particular, let's look where the graph of the parabola crosses the x-axis (see Figure 10.23). The axis of symmetry lies exactly midway between these two points on the x-axis. Since the vertex lies on the axis of symmetry, *the x-coordinate of the vertex lies midway between the two points determined by the x-intercepts.*

Figure 10.24
Symmetry of parabola

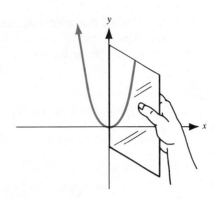

Let's see how we can use the x-intercepts to find the vertex of the graph of $y = x^2 - 4x$ and sketch its graph.

EXAMPLE 3

Find the axis of symmetry and the vertex of the parabola $y = x^2 - 4x$, and sketch its graph.

Solution

First we find the x-intercepts of $y = x^2 - 4x$:

$0 = x^2 - 4x$ *Set $y = 0$ and find x.*

$0 = x(x - 4)$

$x = 0$ or $x = 4$ *The x-intercepts are 0 and 4.*

We plot the points containing the x-intercepts (see Figure 10.25**a**).

Figure 10.25
Graph of $y = x^2 - 4x$

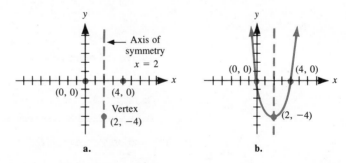

Now find the midpoint between the two points. Using the midpoint formula, we find the coordinates of the midpoint:

$$x = \frac{0 + 4}{2} = 2 \qquad y = \frac{0 + 0}{2} = 0$$

Remember, the x-intercepts have y-coordinates equal to 0.

The midpoint is (2, 0).

Therefore, since the axis of symmetry is a vertical line which passes through the midpoint (2, 0), our axis of symmetry must have equation $x = 2$. (Recall that the general equation of a vertical line is $x = k$.)

Since the vertex lies on the axis of symmetry, the vertex must have x-coordinate equal to 2 (why?). To find the y-coordinate of the vertex, we substitute 2 for x in $y = x^2 - 4x$; hence, the y-coordinate is

$$y = 2^2 - 4 \cdot 2 = 4 - 8 = -4$$

Thus, the vertex is $(2, -4)$.   (see Figure 10.25**a**)

Now we can sketch a graph of the parabola (see Figure 10.25**b**).   ■

Our next step is to apply the method used in Example 3 to our general form, $y = Ax^2 + Bx$, to derive a formula for finding the vertex and the equation of the axis of symmetry.

First we find the x-intercepts of $y = Ax^2 + Bx$:

$0 = Ax^2 + Bx$ *Set $y = 0$ and solve for x.*

$0 = x(Ax + B)$

$$x = 0 \quad \text{or} \quad Ax + B = 0$$
$$x = 0 \quad \text{or} \quad x = \frac{-B}{A}$$

Hence, the equation $y = Ax^2 + Bx$ has x-intercepts of 0 and $-\dfrac{B}{A}$.

Next, we find the x-coordinate of the midpoint between the two x-intercepts:

$$\frac{0 + \left(-\dfrac{B}{A}\right)}{2} = \frac{-\dfrac{B}{A}}{2} = -\frac{B}{2A}$$

Thus, we have the more general result given in the box.

For the equation $y = Ax^2 + Bx$:

The axis of symmetry is $x = -\dfrac{B}{2A}$.

The x-coordinate of the vertex is $-\dfrac{B}{2A}$.

We could substitute $-\dfrac{B}{2A}$ for x in the equation $y = Ax^2 + Bx$ and derive a formula for the y-coordinate of the vertex. However, instead of having another formula to memorize, it is much easier to substitute the x *value* in the equation, and then to find the y value of the vertex.

EXAMPLE 4

Sketch a graph of $y = -3x^2 + 6x$. Identify the vertex and axis of symmetry, and x- and y-intercepts.

Solution

Since $A = -3$ and $B = 6$, $-\dfrac{B}{2A} = -\dfrac{6}{2(-3)} = 1$.

Thus, the equation of the axis of symmetry is $x = 1$.

The x-coordinate of the vertex is 1; therefore, the y-coordinate is

$$y = -3(1)^2 + 6(1) = -3 \cdot 1 + 6 = 3$$

Thus, the vertex is (1, 3).

The y-intercept is 0 [since $-3(0)^2 + 6(0) = 0$].

The x-intercepts are 0 and 2 since

$$0 = -3x^2 + 6x \rightarrow 0 = -3x(x - 2) \rightarrow x = 0, x = 2.$$

The graph of the parabola is shown in Figure 10.26.

Figure 10.26
Graph of $y = -3x^2 + 6x$

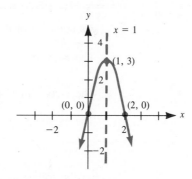

Our last step is to examine the most general case, $y = Ax^2 + Bx + C$.

Parabolas of the Form $y = Ax^2 + Bx + C$

When we discussed the case $y = Ax^2 + C$, we found that in adding the constant C to Ax^2 all we had to do was to shift the graph of $y = Ax^2$ vertically by C units. The same thing happens when we compare the graph of $y = Ax^2 + Bx$ and the graph of $y = Ax^2 + Bx + C$. That is, we shift the graph of $y = Ax^2 + Bx$ vertically by C units. Compare the graphs of $y = x^2 + 4x$ and $y = x^2 + 4x + 5$ shown in Figure 10.27 on page 484.

Notice that the graph of $y = x^2 + 4x$ is shifted up 5 units to make the graph of $y = x^2 + 4x + 5$.

Thus, the graph of $y = Ax^2 + Bx + C$ has the same shape as the graph of $y = Ax^2 + Bx$, but shifted $|C|$ units up if $C > 0$ or down if $C < 0$.

As we can see by the graphs in Figure 10.27, since we are shifting vertically, we no longer have the same x- or y-intercepts for the equations $y = Ax^2 + Bx$ and $y = Ax^2 + Bx + C$. We do find, however, that both equations will have the same axis of symmetry, and therefore the x-coordinate of the vertex for both equations will be the same.

Figure 10.27
Graphs of $y = x^2 + 4x$ and
$y = x^2 + 4x + 5$

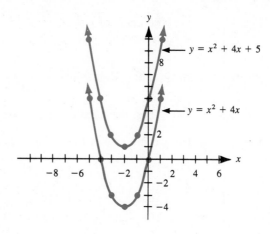

The general results are summarized in the box.

An equation of the form $y = Ax^2 + Bx + C$ $(A \neq 0)$ is a parabola.

The equation of the axis of symmetry is $x = -\dfrac{B}{2A}$.

The x-coordinate of the vertex is $-\dfrac{B}{2A}$.

The parabola opens up if $A > 0$ and down if $A < 0$.

We offer the following as a guide in sketching graphs of parabolas:

Outline for Graphing Parabolas

1. Find the *vertex* (and the axis of symmetry).
2. Find the *y-intercept*.
3. Find the *x-intercepts* (if any).
4. Find *additional points* if necessary.
5. *Sketch* the graph.

EXAMPLE 5

Sketch a graph of each of the following:

(a) $y = x^2 - 10x + 24$ **(b)** $y = -2x^2 + 6x + 2$ **(c)** $y = x^2 + 2x + 5$

Label the intercepts, vertex, and axis of symmetry.

Solution | **(a)** First find the vertex (and axis of symmetry). Since $A = 1$ and $B = -10$, we have $-\dfrac{B}{2A} = -\dfrac{-10}{2(1)} = 5$.

The equation of the axis of symmetry is $x = 5$.

The x-coordinate of the vertex is 5. The y-coordinate of the vertex is

$$y = 5^2 - 10(5) + 24 = -1$$

Thus, the vertex is $(5, -1)$.

Find the y-intercept:

$y = 0^2 - 10 \cdot 0 + 24$ *Set $x = 0$ and find y.*

$\quad = 24$

The y-intercept is 24.

Find the x-intercepts:

$0 = x^2 - 10x + 24$ *Set $y = 0$ and find x.*

$0 = (x - 6)(x - 4)$

$$x = 6, \quad x = 4$$

The x-intercepts are 6 and 4.

Sketch the graph (see Figure 10.28).

Figure 10.28

Graph of
$y = x^2 - 10x + 24$

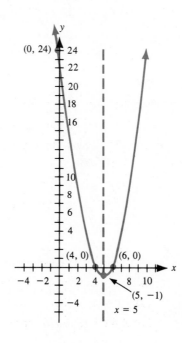

(b)   We find the vertex (and axis of symmetry) of $y = -2x^2 + 6x + 2$:

$$x = -\frac{B}{2A} = -\frac{6}{2(-2)} = \frac{3}{2}$$

The equation of the axis of symmetry is $x = \frac{3}{2}$.

The x-coordinate of the vertex is $\frac{3}{2}$. The y-coordinate of the vertex is

$$y = -2\left(\tfrac{3}{2}\right)^2 + 6\left(\tfrac{3}{2}\right) + 2 = \tfrac{13}{2} \text{ (you should check this)}$$

The vertex is $\left(\tfrac{3}{2}, \tfrac{13}{2}\right)$.

Find the y-intercept:

$$y = -2(0)^2 + 6(0) + 2 = 2 \qquad \textit{Set } x = 0 \textit{ and find } y.$$

The y-intercept is 2.

Find the x-intercepts:

$$0 = -2x^2 + 6x + 2 \qquad \textit{Set } y = 0 \textit{ and find } x.$$

Since $-2x^2 + 6x + 2$ does not factor, we will use the quadratic formula, with $A = -2$, $B = 6$, and $C = 2$, to find x:

$$x = \frac{-6 \pm \sqrt{6^2 - 4(-2)(2)}}{2(-2)} = \frac{-6 \pm \sqrt{36 + 16}}{-4} = \frac{-6 \pm \sqrt{52}}{-4}$$

$$= \frac{-6 \pm 2\sqrt{13}}{-4} = \frac{-2(+3 \pm \sqrt{13})}{-4}$$

$$= \frac{+3 \pm \sqrt{13}}{2}$$

Since $\sqrt{13} \approx 3.6$, the x-intercepts are:

$$\frac{+3 - \sqrt{13}}{2} \approx -0.3 \quad \text{and} \quad \frac{+3 + \sqrt{13}}{2} \approx 3.3$$

Sketch the graph (see Figure 10.29).

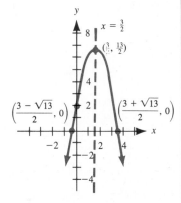

Figure 10.29
Graph of
$y = -2x^2 + 6x + 2$

(c)   We find the vertex and axis of symmetry of $y = x^2 + 2x + 5$:

$$x = -\frac{B}{2A} = -\frac{2}{2(1)} = -1$$

The equation of the axis of symmetry is $x = -1$.

The x-coordinate of the vertex is -1. The y-coordinate of the vertex is

$$y = (-1)^2 + 2(-1) + 5 = 4$$

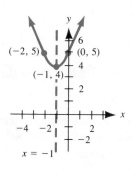

Figure 10.30

Graph of $y = x^2 + 2x + 5$

Thus, the vertex is $(-1, 4)$.

Find the y-intercept: $y = 0^2 + 2 \cdot 0 + 5 = 5$

The y-intercept is 5.

At this point you may notice that the vertex is above the y-axis and, since $A > 0$, the parabola opens up. You can therefore conclude that

There are *no x-intercepts*.

(You may check this for yourself by using the discriminant.)
We still should find another point: Set $x = -2$. Then

$$y = (-2)^2 + 2(-2) + 5 = 5$$

Plot the point $(-2, 5)$.
   Sketch the graph (see Figure 10.30).   ∎

Notice that in finding the x-intercepts of a parabola, we must solve a quadratic equation. As we discussed in Chapter 8, in solving quadratic equations, we can have three types of solutions: the roots may be real and unequal, the roots may be real and equal, or the roots may be imaginary. This translates into three possibilities for x-intercepts, as pictured in Figure 10.31.

Figure 10.31

Finding x-intercepts

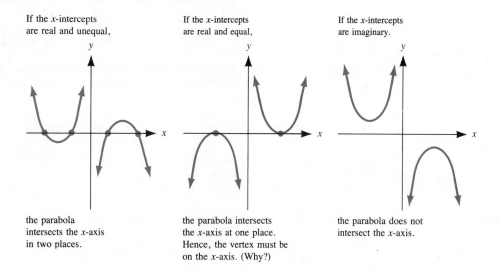

If the x-intercepts are real and unequal, the parabola intersects the x-axis in two places.

If the x-intercepts are real and equal, the parabola intersects the x-axis at one place. Hence, the vertex must be on the x-axis. (Why?)

If the x-intercepts are imaginary. the parabola does not intersect the x-axis.

Notice that in our outline for graphing parabolas, we found the y-intercepts first. This is because if the parabola opens up or down, it must have a y-intercept, but need not have x-intercepts.

EXAMPLE 6

The Popovics Furniture Company found that the relationship between its profit and the number of couches produced by the company could be expressed in the equation

$$P = -x^2 + 46x - 360$$

where x is the number of couches produced daily by the company and P is the daily profit in hundreds of dollars.

How many couches must the company produce daily in order to achieve the maximum daily profit? What is the maximum daily profit?

Solution

We notice that the relationship is quadratic and therefore, if we let the horizontal axis be the number of couches produced daily (x) and let the vertical axis be the daily profit in hundreds of dollars (P), we would have a parabola which opens down (why?). Our ordered pairs are of the form (x, P).

Since the parabola opens down, the vertex is the highest point. This means that the vertex is the point which yields the highest value of P, the profit. Thus, to find the highest profit (highest P), we find the vertex. Since $A = -1$ and $B = 46$, the x-coordinate of the vertex is

$$x = -\frac{B}{2A} = -\frac{46}{2(-1)} = 23$$

Thus, the Popovics Furniture Co. must make 23 couches daily in order to maximize the profit.

Since the maximum value of P occurs when $x = 23$, we substitute 23 for x in the equation and solve for P:

$$P = -(23)^2 + 46(23) - 360 = -529 + 1,058 - 360 = 169$$

Hence, the highest P (profit) is 169, which occurs when $x = 23$. This means that the maximum profit is \$16,900 (remember that P is in hundreds of dollars).

Figure 10.32 is the graph of the given equation.

Figure 10.32

Graph of
$P = -x^2 + 46x - 360$

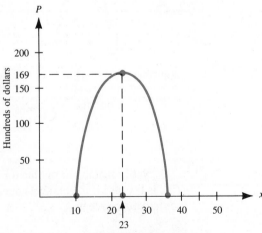

Daily number of couches

The Graph of $x = Ay^2 + By + C$

If the second-degree term involves y rather than x, then the roles of x and y are switched. That is,

---

An equation of the form $x = Ay^2 + By + C$ $(A \neq 0)$ is a parabola.

The equation of the axis of symmetry is $y = -\dfrac{B}{2A}$.

The y-coordinate of the vertex is $-\dfrac{B}{2A}$.

The parabola opens to the *right* if $A > 0$, and to the *left* if $A < 0$.

EXAMPLE 7

Graph $x = y^2 - 10y + 9$. Label the vertex, x-intercept, y-intercepts, and the axis of symmetry.

Solution

First find the vertex and axis of symmetry. Since $A = 1$ and $B = -10$, then

$$-\frac{B}{2A} = -\frac{-10}{2(1)} = 5$$

Hence, the equation of the axis of symmetry is $y = 5$. (Note that this is a *horizontal* line.)

The y-coordinate of the vertex is $y = 5$. Therefore, the x-coordinate of the vertex is $x = (5)^2 - 10(5) + 9 = -16$.

The vertex is $(-16, 5)$.

Find the x-intercept:

$x = 0^2 - 10 \cdot 0 + 9 = 9$ *Set $y = 0$, find x.*

The x-intercept is 9.

Then find the y-intercepts:

$$0 = y^2 - 10y + 9$$
$$0 = (y - 9)(y - 1)$$
$$y = 9, \quad y = 1$$

The y-intercepts are 9 and 1.

Sketch the graph (see Figure 10.33 on page 490).

Figure 10.33
Graph of $x = y^2 - 10y + 9$

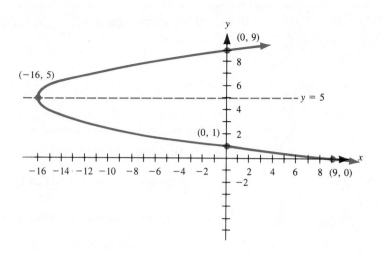

Observe that if an equation is in the form $Ax^2 + By^2 + Cx + Dy + E = 0$, then if either A or B but not both are 0, the equation will usually describe a parabola. (We will discuss the exceptions in Section 10.5.)

Exercises 10.2

Graph the following parabolas. Indicate the axis of symmetry, the vertex, and x- and y-intercepts, if they exist.

1. $y = 2x^2$

2. $y = 4x^2$

3. $y = -2x^2$

4. $y = -4x^2$

5. $y = \frac{1}{2}x^2$

6. $y = \frac{1}{4}x^2$

7. $y = -5x^2$

8. $y = 6x^2$

9. $y = x^2 - 4$

10. $y = x^2 - 9$

11. $y = 2x^2 + 8$

12. $y = 3x^2 + 12$

13. $y = 2x^2 - 8$

14. $y = 3x^2 - 12$

15. $y = -2x^2 + 8$

16. $y = -3x^2 + 12$

17. $y = -2x^2 - 8$

18. $y = -3x^2 - 12$

19. $y = 3x^2 - 9$

20. $y = 4x^2 - 8$

21. $y = 3x^2 + 9$

22. $y = 4x^2 + 8$

23. $y = 2x^2 - 3x$

24. $y = 3x^2 + 5x$

25. $y = 2x^2 + 5x$

26. $y = 4x^2 + 3x$

27. $y = x^2 - 10x + 25$

28. $y = x^2 - 8x + 16$

29. $y = x^2 + 10x + 25$

30. $y = x^2 + 8x + 16$

31. $y + 3 = -3x^2 + 6x$

32. $y - 12 = -3x^2 - 12x$

33. $y = \frac{1}{3}x^2 - \frac{2}{3}x + \frac{2}{3}$

34. $y = \frac{x^2}{2} - 2x + 2$

35. $y + 8x = x^2 + 15$

36. $y - x^2 = 4x - 3$

37. $y = -x^2 + 4x + 12$

38. $y = -x^2 + 6x + 16$

39. $x^2 + 2x = y - 3$

40. $y - x^2 = 6x - 8$

41. $x = y^2 - 2y - 35$

42. $x = y^2 - 2y - 15$

43. $y = -x^2 + 3x + 10$

44. $y = -x^2 + 3x + 28$

45. $y + 4x = 2x^2 + 1$

46. $y = 3x^2 - 12x + 3$

47. $x = 2y^2 + 4y + 1$

48. $x - 3y^2 = 12y + 3$

49. $y = -x^2 - 3x - 4$

50. $y = -x^2 + 5x - 7$

51. $y = 2x^2 - 3x + 2$

52. $y = 2x^2 + 3x - 4$

53. $x = 2y^2 + y + 4$

54. $x - 2 = 3y^2 - y$

55. The widget factory finds that its profit is related to the number of widgets produced as follows:

$$P = -x^2 + 70x$$

where x is the number of widgets produced weekly and P is the weekly profit in dollars. How many widgets must be produced weekly in order to maximize the profit? What would the maximum profit be?

56. The profit (P) made on a concert is related to the price (p) of a ticket in the following way:

$$P = 10,000(-p^2 + 12p - 35)$$

where P is the profit in dollars and p is the price of a ticket. What ticket price would produce the maximum profit?

57. The daily profit earned by the Barrie factory is related to the number of cases of candy canes produced in the following way:

$$P = -x^2 + 112x - 535$$

where P is the daily profit in dollars and x is the number of cases of candy canes made daily. Find the number of cases of candy canes to be made daily in order to maximize the daily profit. What is the maximum profit?

58. The number of portable widgets produced weekly by Widgets, Inc., is related to the weekly profit in the following way:

$$P = -2x^2 + 88x - 384$$

where P is the weekly profit in hundreds of dollars and x is the number of widgets produced weekly. How many widgets must be produced weekly for the maximum weekly profit? What is the maximum weekly profit?

59. Jerry fires a gun upward and the bullet travels according to the equation

$$y = -16t^2 + 400t$$

where y is the height of the bullet off the ground (in feet) at t seconds after he fires the gun. How many seconds does it take for the bullet to reach maximum height? What is the maximum height of the bullet?

60. Carol stands on the roof of a building and fires a gun upward. The bullet travels according to the equation

$$y = -16t^2 + 400t + 50$$

where y is the height of the bullet off the *ground* in feet at t seconds after it is fired.

(a) How far is Carol above the ground when she fires the bullet?

(b) How high does the bullet travel relative to the ground?

(c) After how many seconds does the bullet hit the ground?

61. Susan wanted to fence in a rectangular vegetable garden against her house. She needed to fence in only three sides since the house protected the fourth side. If she used 50 linear feet of fencing, what are the dimensions of the garden that would give her the maximum area possible? (See the accompanying figure.)

   [*Hint:* Let $x =$ one of the two equal sides being fenced. Then the side opposite the house must be $50 - 2x$ (why?). Write an equation for the area (A) of the rectangle in terms of x and then use what you know about parabolas to find the value of x that would produce the *maximum* area (A).]

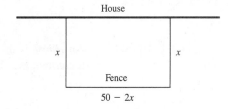

? QUESTIONS FOR THOUGHT

62. (a) In the equation $y = Ax^2 + Bx + C$, what can you say in general about the parabola if $A > 0$? If $A < 0$?

   (b) In the equation $x = Ay^2 + By + C$, what can you say in general about the parabola if $A < 0$? If $A > 0$?

63. Given the equation $y = Ax^2 + Bx + C$, what can be said of a parabola if $B^2 - 4AC > 0$? If $B^2 - 4AC < 0$? If $B^2 - 4AC = 0$?

64. Find the formula for the y-coordinate of the vertex of the parabola $y = Ax^2 + Bx + C$ by substituting the x value $-\dfrac{B}{2A}$ in the equation. Simplify your answer.

65. A parabola is defined to be the set of all points whose distance from a fixed point is equal to its distance from a fixed line. Suppose we let the fixed point be the point $F(0, a)$ and the fixed line be the line $y = -a$ (see the accompanying figure). Pick any point on the parabola and call it $P(x, y)$. Drop a perpendicular line down from the point (x, y) to the line $y = -a$. Then the perpendicular line intersects the line $y = -a$ at point D, which has coordinates $(x, -a)$ (why?).

Call d_1 the distance from P to F and d_2 the distance from P to D. By definition of the parabola, d_1 must equal d_2. Using the distance formula, write an equation in terms of x, y, and a. What is the difference between this equation and the equation of the parabola $y = Ax^2 + Bx + C$?

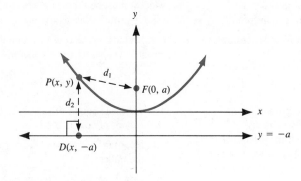

10.3

The Ellipse: Centered at the Origin

The Ellipse

An *ellipse* is the set of all points whose sum of the distances from two fixed points is a constant (see Figure 10.34).

Figure 10.34

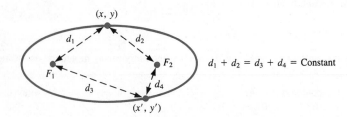

$$d_1 + d_2 = d_3 + d_4 = \text{Constant}$$

Instead of deriving the equation of the ellipse from the distance formula (see Exercise 42 at the end of this section), we will start with the standard form of the equation of an ellipse centered at the origin.

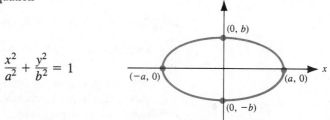

The equation

$$\frac{x^2}{a^2} + \frac{y^2}{b^2} = 1$$

is the standard form of the equation of an *ellipse* centered at the origin, with vertices $(-a, 0)$, $(a, 0)$, $(0, b)$, and $(0, -b)$, where a and b are *positive* constants.

Note that if we set $x = 0$ in $\frac{x^2}{a^2} + \frac{y^2}{b^2} = 1$ and solve for y, we obtain the y-intercepts:

$$\frac{(0)^2}{a^2} + \frac{y^2}{b^2} = 1 \qquad \textit{Set } x = 0.$$

$$\frac{y^2}{b^2} = 1 \qquad \textit{Multiply both sides by } b^2.$$

$$y^2 = b^2$$

$$y = \pm b \qquad \textit{The y-intercepts}$$

If we set $y = 0$, then we can find the x-intercepts, $\pm a$, as we would expect. The intercepts are also the vertices for the ellipse centered at the origin.

The ellipse has two axes of symmetry, which we will refer to simply as the *axes of the ellipse*.

The *axes of the ellipse* are the lines passing through the opposite vertices.

For the ellipse centered at the origin $(0, 0)$, the axes are the x- and y-coordinate axes (which have equations $y = 0$ and $x = 0$, respectively).

EXAMPLE 1

Graph the following. Label the vertices and identify the axes.

(a) $\dfrac{x^2}{9} + \dfrac{y^2}{16} = 1$ (b) $\dfrac{x^2}{25} + \dfrac{y^2}{20} = 1$

Solution

(a) This is the standard form of the equation of an ellipse. Therefore, we have an ellipse centered at the origin with the x- and y-coordinate axes as the axes of the ellipse.

Since the vertices of the equation in standard form are $(a, 0)$, $(-a, 0)$, $(0, b)$, and $(0, -b)$, all we need to do now is to identify a and b by simply observing the denominator of the squared terms:

$a^2 = 9$ (the denominator of x^2)   and $b^2 = 16$ (the denominator of y^2)

Thus, $a = 3$ and $b = 4$. (Remember that a and b are *positive* constants.)

Therefore, the vertices of the ellipse in this example are $(3, 0)$, $(-3, 0)$, $(0, 4)$, and $(0, -4)$.

The ellipse is graphed in Figure 10.35.

Alternatively, we could have found the x- and y-intercepts by substituting $y = 0$ and $x = 0$, respectively.

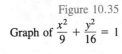

Figure 10.35

Graph of $\dfrac{x^2}{9} + \dfrac{y^2}{16} = 1$

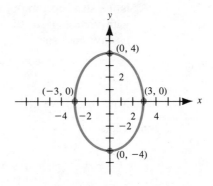

(b) This is the standard form of the equation of an ellipse. Thus, we have an ellipse centered at the origin. The axes of the ellipse are x- and y-coordinate axes.

Since the denominator of x^2 is 25, we have $a^2 = 25$. Therefore, $a = 5$. The denominator of y^2 is 20; therefore, $b^2 = 20$ and $b = \sqrt{20} = 2\sqrt{5}$ (≈ 4.47).

Thus, the vertices are $(5, 0)$, $(-5, 0)$, $(0, 2\sqrt{5})$, and $(0, -2\sqrt{5})$.

The ellipse is graphed in Figure 10.36.

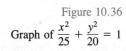

Figure 10.36

Graph of $\dfrac{x^2}{25} + \dfrac{y^2}{20} = 1$

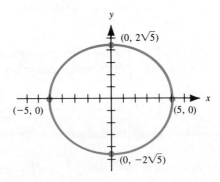

By algebraically manipulating the equation, we can identify the vertices and sketch a graph of the equation of an ellipse not in standard form, as demonstrated in the next example.

EXAMPLE 2 | Graph the following. Label the vertices and identify the axes.

(a) $4x^2 + 25y^2 = 100$ **(b)** $25x^2 + 25y^2 = 100$ **(c)** $100x^2 + y^2 = 25$

Solution | **(a)** This equation is not in standard form. To put the equation in standard form, we observe that the right-hand side of the ellipse in standard form is 1. Hence, we must divide both sides of $4x^2 + 25y^2 = 100$ by 100 in order to get 1 on the right-hand side.

$4x^2 + 25y^2 = 100$ *To get 1 on the right-hand side, divide both sides of the equation by 100.*

$\dfrac{4x^2 + 25y^2}{100} = \dfrac{100}{100}$ *Separate fractions.*

$\dfrac{4x^2}{100} + \dfrac{25y^2}{100} = 1$ *Reduce each fraction.*

$\dfrac{x^2}{25} + \dfrac{y^2}{4} = 1$

Now the equation is in standard form.

a^2 (the denominator of x^2) is 25; thus, $a = 5$.

b^2 (the denominator of y^2) is 4; thus, $b = 2$.

The vertices of the ellipse are $(-5, 0)$, $(5, 0)$, $(0, -2)$, and $(0, 2)$.
The ellipse is graphed in Figure 10.37.

Figure 10.37
Graph of $4x^2 + 25y^2 = 100$

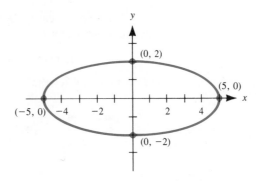

(b) This equation is not in standard form. We could divide both sides of the equation by 100 to get

$\dfrac{x^2}{4} + \dfrac{y^2}{4} = 1$ *Note that both vertices are the same.*

Now we sketch the graph, as shown in Figure 10.38 on page 496. We find that the figure is a circle with radius 2, centered at the origin.

Figure 10.38

Graph of
$25x^2 + 25y^2 = 100$

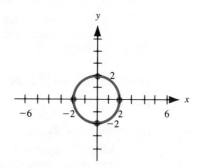

Notice that the coefficients of the squared terms of the original equation, $25x^2 + 25y^2 = 100$, are the same. As we discussed in Section 10.1, we should be able to put this equation in the standard form of a circle. Thus, if we divide both sides of the equation by 25 (rather than 100), we get

$25x^2 + 25y^2 = 100$ *Divide both sides by 25.*

$x^2 + y^2 = 4$ *This is the equation of a circle with radius 2, centered at the origin.*

(c)   This equation is not in standard form.

$100x^2 + y^2 = 25$ *Divide both sides of the equation by 25 and the right-hand side will be 1.*

$\dfrac{100x^2 + y^2}{25} = \dfrac{25}{25}$ *Separate fractions.*

$\dfrac{100x^2}{25} + \dfrac{y^2}{25} = 1$ *Reduce.*

$4x^2 + \dfrac{y^2}{25} = 1$

Now we can easily identify b^2 (the denominator of y^2) as 25, but what is a^2, the denominator of x^2? It is surely not 4, since 4 is not in the denominator of x^2. We have to rewrite $4x^2$ as $\dfrac{x^2}{a^2}$. For this we use our knowledge of quotients of fractions and rewrite

$$4x^2 \quad \text{as} \quad \dfrac{x^2}{\dfrac{1}{4}} \qquad \text{since} \qquad \dfrac{x^2}{\dfrac{1}{4}} = x^2\left(\dfrac{4}{1}\right) = 4x^2$$

Hence, $4x^2 + \dfrac{y^2}{25} = 1$ becomes $\dfrac{x^2}{\dfrac{1}{4}} + \dfrac{y^2}{25} = 1.$

We have

$$a^2 = \dfrac{1}{4} \rightarrow a = \sqrt{\dfrac{1}{4}} = \dfrac{1}{2} \quad \text{and} \quad b^2 = 25 \rightarrow b = 5$$

The equation is that of an ellipse centered at $(0, 0)$ with vertices $\left(\dfrac{1}{2}, 0\right)$, $\left(-\dfrac{1}{2}, 0\right)$, $(0, 5)$, and $(0, -5)$, as illustrated in Figure 10.39.

Figure 10.39
Graph of $100x^2 + y^2 = 25$

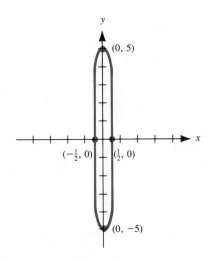

Again, for an ellipse centered at the origin, the vertices are the intercepts. Thus, an alternative way to find the vertices of this ellipse is to find the intercepts by substituting 0 for each variable in the original equation. (Try it.) ∎

Exercises 10.3

In Exercises 1–12, identify the vertices of the ellipse described by the given equation.

1. $\dfrac{x^2}{16} + \dfrac{y^2}{9} = 1$ **2.** $\dfrac{x^2}{4} + \dfrac{y^2}{9} = 1$ **3.** $\dfrac{x^2}{9} + \dfrac{y^2}{16} = 1$ **4.** $\dfrac{x^2}{9} + \dfrac{y^2}{4} = 1$

5. $\dfrac{y^2}{36} + \dfrac{x^2}{25} = 1$ **6.** $\dfrac{y^2}{49} + \dfrac{x^2}{81} = 1$ **7.** $\dfrac{x^2}{12} + \dfrac{y^2}{9} = 1$ **8.** $\dfrac{x^2}{16} + \dfrac{y^2}{8} = 1$

9. $\dfrac{x^2}{24} + \dfrac{y^2}{20} = 1$ **10.** $\dfrac{x^2}{20} + \dfrac{y^2}{18} = 1$ **11.** $\dfrac{x^2}{12} + \dfrac{y^2}{16} = 1$ **12.** $\dfrac{x^2}{8} + \dfrac{y^2}{9} = 1$

In Exercises 13–28, sketch a graph of the ellipse. Label the vertices.

13. $x^2 + \dfrac{y^2}{16} = 1$ **14.** $\dfrac{x^2}{9} + y^2 = 1$ **15.** $\dfrac{x^2}{8} + \dfrac{y^2}{4} = 1$ **16.** $\dfrac{x^2}{24} + \dfrac{y^2}{25} = 1$

17. $4x^2 + 25y^2 = 100$ **18.** $4y^2 + 25x^2 = 100$ **19.** $x^2 + 9y^2 = 9$ **20.** $16x^2 + y^2 = 1$

21. $8x^2 + 7y^2 = 56$ **22.** $7x^2 + 8y^2 = 56$ **23.** $12x^2 + 8y^2 = 72$ **24.** $12x^2 + 8y^2 = 96$

25. $4x^2 + y^2 = 1$ **26.** $x^2 + 9y^2 = 1$ **27.** $25x^2 + 16y^2 = 1$ **28.** $36x^2 + 16y^2 = 1$

In Exercises 29–34, identify the figure described by the equation.

29. $\dfrac{x^2}{4} + \dfrac{y^2}{16} = 1$ **30.** $\dfrac{x^2}{9} + \dfrac{y^2}{18} = 1$ **31.** $\dfrac{x}{4} + \dfrac{y}{16} = 1$

32. $\dfrac{x}{9} + \dfrac{y}{18} = 1$ **33.** $\dfrac{x^2}{16} + \dfrac{y^2}{16} = 1$ **34.** $\dfrac{x^2}{18} + \dfrac{y^2}{18} = 1$

In Exercises 35–40, identify and sketch a graph of the figure.

35. $2x^2 + 4y^2 = 8$ **36.** $3x^2 + 9y^2 = 18$ **37.** $2x + 4y = 8$

38. $3x + 9y = 18$ **39.** $2x^2 + 4y = 18$ **40.** $3x^2 + 9y = 18$

? QUESTIONS FOR THOUGHT

41. What type of figure is created by graphing $\dfrac{x^2}{a^2} + \dfrac{y^2}{a^2} = 1$?

42. An ellipse is defined to be the set of all points whose sum of the distances from two fixed points is a constant. The accompanying figure shows two fixed points, $F_1(-s, 0)$ and $F_2(s, 0)$. Pick any point $P(x, y)$ on the ellipse. We will call d_1 the distance from P to F_1 and d_2 the distance from P to F_2.

By definition of the ellipse, the sum of the distances, $d_1 + d_2$, must be constant. Let's call this constant k. Then we have $d_1 + d_2 = k$. Using the distance formula, write this as an equation in x, y, s, and k.

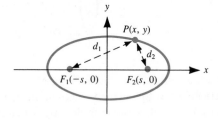

43. Look at the standard form of the equation of a circle with its center at the origin, and compare it with the standard form of a circle with center (h, k). Then generalize the standard form of an ellipse centered at the origin to the case where the center is at (h, k).

⟳ MINI-REVIEW

44. *Solve for x.* $(x + 1)^{1/3} = 4$

45. *Solve for x.* $(x - 2)^2 = (x + 1)^2 - 15$

46. Sketch the graph of $5x + 3y = 30$.

47. A person invests money at 8% and at 10%, earning a total yearly interest of $730. Had the amounts invested been reversed, the yearly interest would have been $710. How much was invested all together?

10.4

The Hyperbola: Centered at the Origin

A *hyperbola* is the set of all points such that the absolute value of the difference of their distances from two fixed points is a constant (see Figure 10.40).

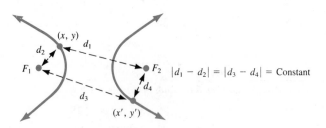

Figure 10.40
A hyperbola

Again, instead of deriving the equation of a hyperbola from the distance formula (see Exercise 47 at the end of this section), we will start with the standard form of the equation of a hyperbola.

The equation

$$\frac{x^2}{a^2} - \frac{y^2}{b^2} = 1$$

is the standard form of the equation of the *hyperbola* centered at the origin, with vertices $(-a, 0)$ and $(a, 0)$, where a and b are positive constants.

The axes of a hyperbola centered at the origin are the x- and y-axes.

Note the differences and similarities between the standard forms of the equations of the hyperbola and the ellipse. For example, in graphing the hyperbola $\frac{x^2}{9} - \frac{y^2}{16} = 1$, we note that

a^2 (the denominator of x^2) = 9 $\rightarrow a = 3$

b^2 (the denominator of y^2) = 16 $\rightarrow b = 4$

Hence, by the standard form given in the box, the vertices are $(-3, 0)$ and $(3, 0)$. Notice that if we set $y = 0$ in $\frac{x^2}{9} - \frac{y^2}{16} = 1$, we obtain $\frac{x^2}{9} = 1$, and hence

$$x^2 = 9 \rightarrow x = \pm 3$$

Thus, the x-intercepts and vertices are identical for hyperbolas of the form $\frac{x^2}{a^2} - \frac{y^2}{b^2} = 1$: $(-a, 0)$ and $(a, 0)$.

On the other hand, if we set $x = 0$ in $\frac{x^2}{9} - \frac{y^2}{16} = 1$, then we have $-\frac{y^2}{16} = 1$. This yields $y^2 = -16$, which has no real solutions (since y^2 must be positive).

Hence, x cannot be 0, which means that *the graph of the hyperbola*

$$\frac{x^2}{a^2} - \frac{y^2}{b^2} = 1$$

does not intersect the y-axis.

In graphing a hyperbola, we are guided by two intersecting lines which the hyperbola approaches but never touches. These lines are called **asymptotes**. For example, in graphing $\frac{x^2}{9} - \frac{y^2}{16} = 1$, we are guided by the asymptotes

$$y = \frac{4}{3}x \quad \text{and} \quad y = -\frac{4}{3}x$$

The hyperbola gets very close to, but never touches, the lines $y = \frac{4}{3}x$ and $y = -\frac{4}{3}x$ (see Figure 10.41).

Figure 10.41

Graph of $\frac{x^2}{9} - \frac{y^2}{16} = 1$

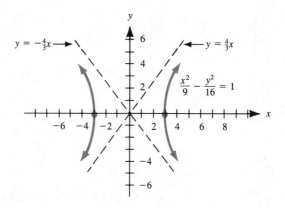

The asymptotes guide us in determining the shape of the hyperbola. We can manipulate the equation of the hyperbola (see Exercise 48 at the end of this section) to determine the asymptotes, as indicated in the box.

The asymptotes of the hyperbola $\frac{x^2}{a^2} - \frac{y^2}{b^2} = 1$ are

$$y = \pm\frac{b}{a}x$$

EXAMPLE 1

Sketch a graph of the hyperbola $\frac{x^2}{4} - \frac{y^2}{9} = 1$.

Solution

Since the hyperbola is in standard form, we can identify a and b in order to find the vertices and the asymptotes.

$a^2 = 4$ and $b^2 = 9$; therefore, $a = 2$ and $b = 3$.

Since $a = 2$, the vertices are $(+2, 0)$ and $(-2, 0)$.

We can sketch the asymptotes using a and b by the following procedure:

First *plot* the vertices—the points $(2, 0)$ and $(-2, 0)$—and *locate* the points $(0, 3)$ and $(0, -3)$.

Then draw a rectangle such that the points just found are the midpoints of the sides of the rectangle.

The lines containing the diagonals of the rectangle are the asymptotes of the hyperbola (see Figure 10.42).

Figure 10.42

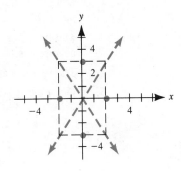

Now we sketch the hyperbola, using the asymptotes as a guide (see Figure 10.43).

Figure 10.43

Graph of $\dfrac{x^2}{4} - \dfrac{y^2}{9} = 1$

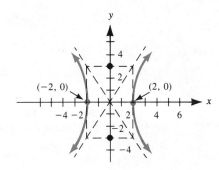

Why do the lines containing the diagonals of the rectangles have the same equation as the asymptotes? First, notice in Figure 10.42 that the slopes of the diagonals are $\pm\frac{3}{2}$. Then note that the diagonals pass through the origin, and hence their y-intercepts are 0. Therefore, by the slope–intercept form (Section 8.3), the equations of the lines containing the diagonals are

$$y = \pm\frac{3}{2}x$$

which are the asymptotes.

In general, we suggest the outline given in the box.

To graph the hyperbola $\dfrac{x^2}{a^2} - \dfrac{y^2}{b^2} = 1$:

1. Plot the points $(a, 0)$ and $(-a, 0)$, and locate the points $(0, b)$ and $(0, -b)$.
2. Draw a rectangle such that the points found in step 1 are the midpoints of the sides of the rectangle.
3. The lines containing the diagonals of the rectangle are the asymptotes.
4. The vertices are $(a, 0)$ and $(-a, 0)$.
5. Draw the hyperbola using the asymptotes and vertices as a guide.

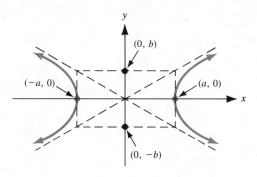

It is important for us to realize that the rectangle and the asymptotes are only guides to help us draw the hyperbola; they are not actually part of the graph of the hyperbola.

EXAMPLE 2

Sketch a graph of the following hyperbolas:

(a) $\dfrac{x^2}{25} - \dfrac{y^2}{9} = 1$ **(b)** $x^2 - 100y^2 = 100$

Solution

(a)   The equation $\dfrac{x^2}{25} - \dfrac{y^2}{9} = 1$ is the standard form of the equation of a hyperbola centered at the origin; hence,

$$a^2 = 25 \quad \text{and} \quad b^2 = 9; \quad \text{therefore, } a = 5 \text{ and } b = 3.$$

Therefore, we plot the points $(5, 0)$ and $(-5, 0)$, locate the points $(0, 3)$ and $(0, -3)$, and draw a rectangle such that the points we just plotted are the midpoints of the sides of the rectangle.

The lines containing the diagonals of the rectangle are the asymptotes $y = \pm\frac{b}{a}x$, which, in this example, are $y = \pm\frac{3}{5}x$. The vertices are (5, 0) and (−5, 0).

The graph of the hyperbola is shown in Figure 10.44.

Figure 10.44

Graph of $\dfrac{x^2}{25} - \dfrac{y^2}{9} = 1$

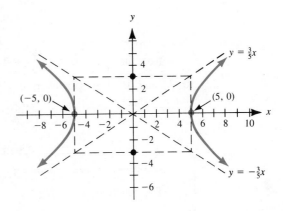

(b) The equation $x^2 - 100y^2 = 100$ is not in standard form. We divide both sides of the equation by 100 to get

$$\frac{x^2 - 100y^2}{100} = \frac{100}{100}$$

$$\frac{x^2}{100} - \frac{y^2}{1} = 1$$

Therefore, $a^2 = 100 \rightarrow a = 10$ and $b^2 = 1 \rightarrow b = 1$.

We plot the points (10, 0) and (−10, 0) and locate points (0, 1) and (0, −1), as shown in Figure 10.45. We then draw the rectangle and the diagonals. Finally, we graph the hyperbola centered at the origin with vertices (−10, 0) and (10, 0) and asymptotes $y = \pm\frac{1}{10}x$.

Figure 10.45

Graph of $x^2 - 100y^2 = 100$

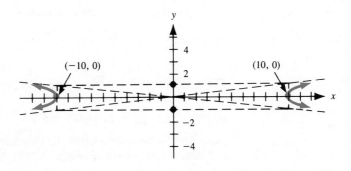

We have just discussed equations of the form $\dfrac{x^2}{a^2} - \dfrac{y^2}{b^2} = 1$. Note that the term $\dfrac{y^2}{b^2}$ is being subtracted from the term $\dfrac{x^2}{a^2}$. What if the situation were reversed, that is, what if $\dfrac{x^2}{a^2}$ were being subtracted from $\dfrac{y^2}{b^2}$? Then the form would be

$$\frac{y^2}{b^2} - \frac{x^2}{a^2} = 1$$

Note that if $x = 0$, then $\dfrac{y^2}{b^2} = 1$ and $y = \pm b$. Hence, we would have y-intercepts $(0, b)$ and $(0, -b)$. However, if $y = 0$, then we arrive at the impossible situation $-\dfrac{x^2}{a^2} = 1$. Therefore, the graph does not cross the x-axis.

The equation

$$\frac{y^2}{b^2} - \frac{x^2}{a^2} = 1$$

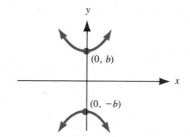

is the standard form of the equation of the hyperbola centered at the origin, with vertices $(0, -b)$ and $(0, b)$, where a and b are positive constants.

Fortunately, the asymptotes still remain the same:

The asymptotes of the hyperbola $\dfrac{y^2}{b^2} - \dfrac{x^2}{a^2} = 1$ are $y = \pm\dfrac{b}{a}x$.

Thus, we simply follow the same guidelines in graphing $\dfrac{y^2}{b^2} - \dfrac{x^2}{a^2} = 1$ as we did for graphing $\dfrac{x^2}{a^2} - \dfrac{y^2}{b^2} = 1$, except that the vertices, $(0, b)$ and $(0, -b)$, now lie on the y-axis and the hyperbola opens up and down rather than side to side. *Note that a^2 is still the denominator of x^2 and b^2 is still the denominator of y^2.*

EXAMPLE 3

Graph $9y^2 - 4x^2 = 144$.

Solution

We can put the equation in standard form by dividing both sides of the equation by 144 to get

$$\frac{y^2}{16} - \frac{x^2}{36} = 1$$

Since a^2 (the denominator of x^2) is 36 and b^2 (the denominator of y^2) is 16, we have

$$a = 6 \quad \text{and} \quad b = 4$$

We plot the points $(0, 4)$ and $(0, -4)$, locate the points $(6, 0)$ and $(-6, 0)$, and draw our rectangle with these points as midpoints of its sides (see Figure 10.46).

Figure 10.46
Graph of $9y^2 - 4x^2 = 144$

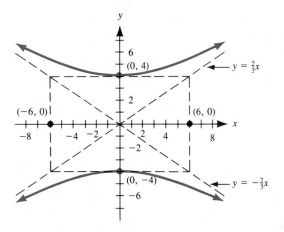

The vertices for this hyperbola are on the y-axis. Note that the asymptotes are still $y = \pm\frac{b}{a}x$, which are

$$y = \pm\frac{2}{3}x$$

Also, the hyperbola now opens up and down, as shown in the figure.   ■

We can recognize the equation of an ellipse or hyperbola by the signs of the coefficients of the squared terms: The coefficients of the squared terms of a hyperbola have opposite signs, whereas the coefficients of the squared terms of an ellipse have the same sign.

Once we recognize a hyperbola, we can also determine what type of hyperbola we have by locating its x- and y-intercepts. A hyperbola centered at the origin will have only one set of intercepts. By knowing which axis it crosses we know what type of hyperbola we have.

Exercises 10.4

In Exercises 1–9, identify the vertices and write the equations of the asymptotes for the hyperbolas described by the equation.

1. $\dfrac{x^2}{9} - \dfrac{y^2}{16} = 1$

2. $\dfrac{x^2}{25} - \dfrac{y^2}{36} = 1$

3. $\dfrac{x^2}{16} - \dfrac{y^2}{9} = 1$

4. $\dfrac{x^2}{36} - \dfrac{y^2}{25} = 1$

5. $\dfrac{y^2}{9} - \dfrac{x^2}{16} = 1$

6. $\dfrac{y^2}{25} - \dfrac{x^2}{36} = 1$

7. $x^2 - \dfrac{y^2}{16} = 1$

8. $\dfrac{y^2}{16} - x^2 = 1$

9. $\dfrac{x^2}{12} - \dfrac{y^2}{4} = 1$

In Exercises 10–26, sketch a graph of the hyperbola. Label the vertices and asymptotes.

10. $\dfrac{x^2}{8} - \dfrac{y^2}{12} = 1$

11. $\dfrac{y^2}{4} - \dfrac{x^2}{12} = 1$

12. $\dfrac{y^2}{8} - \dfrac{x^2}{12} = 1$

13. $9x^2 - 16y^2 = 144$

14. $25x^2 - 4y^2 = 100$

15. $x^2 - 25y^2 = 25$

16. $16x^2 - y^2 = 16$

17. $x^2 - 2y^2 = 2$

18. $2x^2 - y^2 = 4$

19. $2y^2 - x^2 = 4$

20. $y^2 - 2x^2 = 4$

21. $12y^2 - 5x^2 = 60$

22. $8y^2 - 9x^2 = 72$

23. $100x^2 - y^2 = 25$

24. $225x^2 - y^2 = 25$

25. $16x^2 - 4y^2 = 1$

26. $25x^2 - 36y^2 = 1$

In Exercises 27–35, identify the figure described by the equation.

27. $\dfrac{x^2}{25} - \dfrac{y^2}{100} = 1$

28. $\dfrac{x^2}{81} + \dfrac{y^2}{100} = 1$

29. $\dfrac{x^2}{100} + \dfrac{y^2}{25} = 1$

30. $\dfrac{y^2}{100} - \dfrac{x^2}{81} = 1$

31. $9y^2 + 16x^2 = 144$

32. $16y^2 - 9x^2 = 144$

33. $8y^2 - 8x^2 = 16$

34. $10x^2 - 10y^2 = 20$

35. $x + y = 4$

In Exercises 36–46, identify and sketch a graph of the figure.

36. $x - y = 9$

37. $x^2 + y = 4$

38. $x^2 - y = 9$

39. $x + y^2 = 4$

40. $x - y^2 = 9$

41. $4x^2 + y^2 = 16$

42. $9y^2 - x^2 = 81$

43. $4x^2 - y^2 = 16$

44. $9y^2 - x = 81$

45. $4x - y = 16$

46. $9y^2 + x^2 = 81$

? QUESTIONS FOR THOUGHT

47. A hyperbola is defined to be the set of all points such that the absolute value of the difference of their distances from two fixed points is a constant. The accompanying figure shows two fixed points, $F_1(-s, 0)$ and $F_2(s, 0)$, on the x-axis. Pick any point $P(x, y)$ on the hyperbola. We will call d_1 the distance from P to F_1 and d_2 the distance from P to F_2.

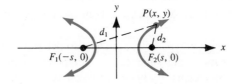

By definition of the hyperbola, the difference of the distances $|d_1 - d_2|$ must be constant. Let's call it k. Then we have $|d_1 - d_2| = k$. Using the distance formula, write this as an equation in x, y, s, and k.

48. **(a)** Show that if we start with the equation $\dfrac{x^2}{4} - \dfrac{y^2}{9} = 1$ and solve for y, we obtain

$$y = \pm\frac{3}{2}\sqrt{x^2 - 4}$$

(b) Compute y from the following two equations for these values of x: $x = 4, 10, 20, 100, 200, 1{,}000, 2{,}000$:

$$y = \pm\frac{3}{2}\sqrt{x^2 - 4} \quad \text{and} \quad y = \pm\frac{3}{2}x$$

Do you see why the graph of the hyperbola "gets closer" to $y = \pm\frac{3}{2}x$?

49. The graph of the equation $xy = k$, where k is a positive constant, is given here for $k = 1$. This figure is known as a *rectangular hyperbola*. What are the asymptotes?

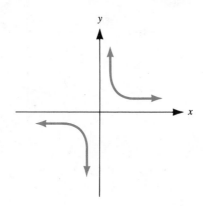

Graph of $xy = k$

50. Look at the standard form of the equation of a circle with its center at the origin, and compare it to the standard form of a circle with center (h, k). Then generalize the standard form of a hyperbola centered at the origin to the case where the center is at (h, k).

⟳ MINI-REVIEW

51. *Solve for y.* $(2y - 3)(3y + 1) = 17$

52. Sketch the graph of $\dfrac{x^2}{9} + \dfrac{y^2}{20} = 1$.

53. Pat leans a 40′ ladder against a building. If the base of the ladder is 6′ from the building, how high up the building does the ladder reach?

54. *Solve for x.* $\dfrac{x - 3}{x + 2} \leq 0$

10.5

Identifying Conic Sections

We can summarize the conic sections we have covered as follows:

Conic Sections

1. The parabola

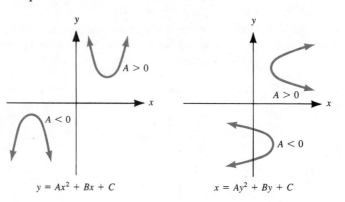

$$y = Ax^2 + Bx + C \qquad x = Ay^2 + By + C$$

2. The circle

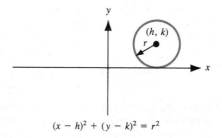

$$(x - h)^2 + (y - k)^2 = r^2$$

3. The ellipse

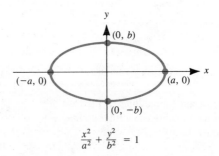

$$\frac{x^2}{a^2} + \frac{y^2}{b^2} = 1$$

4. The hyperbola

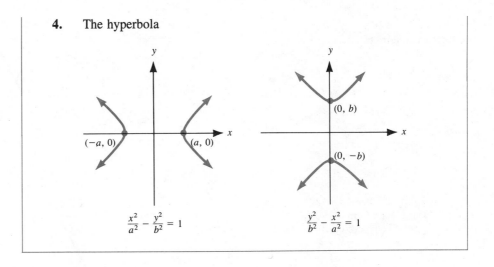

$$\frac{x^2}{a^2} - \frac{y^2}{b^2} = 1$$

$$\frac{y^2}{b^2} - \frac{x^2}{a^2} = 1$$

By now it may have occurred to you that (with the help of completing the square) we may be able to convert any equation of the form

$$Ax^2 + By^2 + Cx + Dy + E = 0 \quad \text{(where } A \text{ and } B \text{ are not both zero)}$$

into one of the conic section forms discussed in this chapter.

However, that is not necessarily possible, since the coefficients (A, B, and C) and h, k, and r must be real. For example, consider the equation $x^2 + y^2 = -4$. Since x and y must be real numbers, their squares must be nonnegative and can never sum to the negative number, -4. Since we require all constants and variables to be real, we can never put the equation in the form $(x - h)^2 + (y - k)^2 = r^2$, where h, k, and r are real numbers. Therefore, the equation $x^2 + y^2 = -4$ has no graph.

If the equation can be graphed, the next question to ask is, will we get a conic section? Here we run into another problem. The equation $x^2 + y^2 = 0$ is in the form of the equation of a circle with center $(0, 0)$ and radius $= 0$. Therefore its graph is not a circle, it is the point $(0, 0)$. As another example, the equation $\dfrac{x^2}{16} + \dfrac{y^2}{25} = 0$ *cannot* be put into one of the standard forms. It *can* be graphed, but it yields only the point $(0, 0)$. We call such forms ***degenerate forms***. The degenerate form of a circle and ellipse is a point.

Refer back to the pictures of the double cones (Figure 10.1) at the beginning of this chapter. Notice that if you hold the plane horizontally, you cut the cone so that its intersection with the cone yields a circle. If you move the plane to where the double cones meet, the radius of the circle is reduced until you get a point. The same is true of the ellipse: Its degenerate form is a point. (See Figure 10.47 on page 510.)

The degenerate form of the hyperbola is two intersecting lines. Referring back to Figure 10.1, we note that the hyperbola is formed by the plane intersecting both parts of the double cone. If you held the plane vertically and sliced the double cone exactly in half (where the two cones meet) the intersection of the double cone and the plane would yield two intersecting lines (see Figure 10.48).

Figure 10.47

Degenerate form of circle

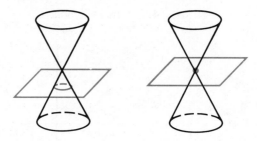

Figure 10.48

Degenerate form of
hyperbola

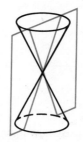

This degenerate form of the hyperbola can be represented by the equation

$$\frac{x^2}{a^2} - \frac{y^2}{b^2} = 0$$

If we solve for y, we get

$$\frac{x^2}{a^2} = \frac{y^2}{b^2}$$

$$\frac{b^2}{a^2}x^2 = y^2 \qquad \textit{Take square roots.}$$

$$\pm\frac{b}{a}x = y \qquad \textit{The equations of two intersecting lines}$$

The equation $x^2 - 4 = 0$ looks like it may be a parabola. However, this equation is equivalent to the equations $x = -2$ and $x = +2$, which, when graphed on a rectangular coordinate system, yields two vertical lines. This is a degenerate form of a parabola. The graph of $x^2 - 6x + 9 = 0$ is a single vertical line, which is also a degenerate form of a parabola.

Thus, we *can* say the following:

If the equation $Ax^2 + By^2 + Cx + Dy + E = 0$ (where A and B are not both 0) can be graphed, the graph will yield a conic section or one of its degenerate forms.

Actually, we can say more. If the figure can be graphed, we can identify what the figure may be by simply looking at the coefficients of the squared terms.

Assume that the equation $Ax^2 + By^2 + Cx + Dy + E = 0$, where A and B are not both 0, can be graphed.

1. If either $A = 0$ or $B = 0$, but not both, the equation will yield a parabola or one of its degenerate forms—a line or two parallel lines.

2. If the signs of A and B are the same, the equation will yield a circle if $A = B$, an ellipse if $A \neq B$ or, their degenerate form—a point.

3. If the signs of A and B are different, the equation will yield a hyperbola, or its degenerate form—two intersecting lines.

EXAMPLE 1

Identify the type of conic section by its coefficients, assuming the equation can be graphed.

(a) $3x^2 + 4y^2 - 2x - 3y = 0$ **(b)** $5x^2 - 3x + 2y = 4$

(c) $7y^2 - 2x^2 + 3x = 7$ **(d)** $-6x^2 - 6y^2 - x + y = 8$

(e) $3x + 2y = 5$

Solution

(a) Since the coefficients of the squared terms have the same sign but are not identical, the equation will yield an ellipse or a point if it can be graphed.

(b) Since there is no y^2 term ($B = 0$), the equation will yield a parabola or one of its degenerate forms if it can be graphed.

(c) Since the signs of the coefficients of the squared terms are different (7 and -2), the equation will yield a hyperbola or its degenerate form, if it can be graphed.

(d) Since the coefficients of the squared terms are identical, the equation will yield a circle or a point, if it can be graphed.

(e) Because the coefficients of both squared terms are 0, it is not the equation of a conic section. It is a linear equation (which will yield a straight line).   ■

In Appendix B we continue our discussion of conic sections with ellipses and hyperbolas centered at (h, k).

Exercises 10.5

In Exercises 1–20, assume the following equations can be graphed. Identify what figure the equation will yield without putting the equation in standard form.

1. $x^2 + y^2 = 12$

2. $9x^2 + 25y^2 = 220$

3. $6x^2 + 7y^2 = 42$

4. $4x^2 - 5y^2 = 1$

5. $3x^2 + 2x + 3y = 2$

6. $-7y^2 + 6x - 3y = -9$

7. $\dfrac{x}{3} + \dfrac{y}{2} = 1$

8. $5x^2 + 5y^2 = 12$

9. $x^2 - y^2 = 9$

10. $5y - 3x = 2$

11. $5y^2 - 9x^2 - 30y - 36x = 36$

12. $y^2 - 3x^2 - 4y + 24x = 45$

13. $x^2 + y^2 - 2x + 6y = -2$

14. $\dfrac{(x - 1)^2}{3} + \dfrac{(y - 1)^2}{3} = 1$

15. $3x^2 + 3y^2 + 18x - 12y = -24$

16. $5x^2 - 3x + 4y = -7$

17. $6x^2 - 7y^2 = 12$

18. $-x^2 - y^2 - 12x - 8y = 37$

19. $-6x^2 = 3y$

20. $14x^2 + 7y^2 - 84x - 28y = -56$

In Exercises 21–36, identify and sketch a graph of the equations, if possible. Label the important aspects of the figure.

21. $x^2 + y^2 = 16$

22. $16x^2 + 9y^2 = 144$

23. $x^2 + 2x - y = 9$

24. $16x^2 + 5y^2 = -80$

25. $x^2 - 100y^2 = 25$

26. $2y^2 + 16y - x = -24$

27. $16x^2 - 9y^2 = 0$

28. $100x^2 - 16y^2 = 1,600$

29. $3x^2 + 3y^2 = 24$

30. $2x^2 + 2y^2 = 0$

31. $2x^2 + 4x - y = -6$

32. $y^2 - 8y - x = -7$

33. $2x^2 + y^2 = 8$

34. $16x^2 + 5y^2 = 80$

35. $81y^2 - 36x^2 = 2,916$

36. $81y^2 + 36x^2 = 2,916$

? QUESTION FOR THOUGHT

37. Look over the problems in Exercises 1–20. By simply looking at the equations, can you tell the following?

(a) If it is a parabola, whether it opens up, down, left, or right.

(b) If it is a hyperbola, whether it opens up and down, or left and right.

10.6

Nonlinear Systems of Equations

In Chapter 9 we solved systems of first-degree equations. In this section we will discuss systems of equations involving second-degree equations.

In order to help us visualize the solutions, in the course of our discussion we will refer back to our work in this chapter on graphing second-degree equations (the conic sections).

As was the case with a linear system, when we solve a nonlinear system we often interpret the solution(s) as the point(s) of intersection of the graphs. Since we cannot graph complex-number solutions to systems of equations, we are going to restrict our attention to *real-number* solutions.

The next two examples illustrate that the two methods we used to solve linear systems—the elimination and substitution methods—can be applied to nonlinear systems as well.

EXAMPLE 1

Solve the following system of equations:

$$\begin{cases} x^2 + y^2 = 5 \\ 2x^2 + y = 0 \end{cases}$$

Solution

Although it is not necessary, it is often helpful to sketch the graphs of the equations in the system, so we know what to expect. Using the methods developed earlier in this chapter, we get the graphs in Figure 10.49. The graph of $x^2 + y^2 = 5$ is a circle with center $(0, 0)$ and radius $\sqrt{5}$. The graph of $2x^2 + y = 0$ or $y = -2x^2$ is a parabola with vertex $(0, 0)$ and opening downward.

Figure 10.49

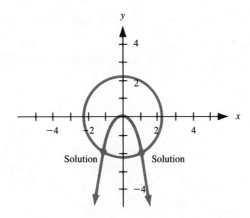

As these graphs clearly show, we expect to find *two* solutions to this sytem of equations (one solution for each point of intersection).

Since both equations contain an "x^2" term, we can use the elimination method to eliminate x^2 and get an equation involving the variable y only.

$x^2 + y^2 = 5$	*Multiply by* $-2 \rightarrow$	$-2x^2 - 2y^2 = -10$	
$2x^2 + y = 0$	*As is* $\rightarrow$	$\underline{2x^2 + y = 0}$	*Add.*
		$-2y^2 + y = -10$	

We now have a quadratic equation, which we put in standard form and solve:

$-2y^2 + y + 10 = 0$ *Multiply both sides of the equation by* -1.

$2y^2 - y - 10 = 0$ *We can factor or use the quadratic formula.*

$(2y - 5)(y + 2) = 0$

$$2y - 5 = 0, \qquad y + 2 = 0$$
$$y = \frac{5}{2}, \qquad y = -2$$

Now we substitute each y value into the second equation and solve for x:

Substitute $y = \dfrac{5}{2}$

$2x^2 + y = 0$

$2x^2 + \dfrac{5}{2} = 0$

$2x^2 = -\dfrac{5}{2}$

$x^2 = -\dfrac{5}{4}$

$x = \pm \sqrt{\dfrac{-5}{4}}$

No real solutions.

Substitute $y = -2$

$2x^2 + y = 0$

$2x^2 - 2 = 0$

$2x^2 = 2$

$x^2 = 1$

$x = \pm 1$

There are *two* x values for this *one* y value.

Thus, we have two solutions to this system, $\boxed{(1, -2) \text{ and } (-1, -2)}$

This result agrees quite well with what we saw in Figure 10.49. We check each solution in *both* equations.

CHECK $(1, -2)$:

$$x^2 + y^2 = 5 \qquad\qquad 2x^2 + y = 0$$
$$(1)^2 + (-2)^2 \overset{?}{=} 5 \qquad 2(1)^2 + (-2) \overset{?}{=} 0$$
$$1 + 4 \overset{\checkmark}{=} 5 \qquad\qquad 2 - 2 \overset{\checkmark}{=} 0$$

CHECK $(-1, -2)$:

$$x^2 + y^2 = 5 \qquad\qquad 2x^2 + y = 0$$
$$(-1)^2 + (-2)^2 \overset{?}{=} 5 \qquad 2(-1)^2 + (-2) \overset{?}{=} 0$$
$$1 + 4 \overset{\checkmark}{=} 5 \qquad\qquad 2 - 2 \overset{\checkmark}{=} 0$$

∎

If we reflect a moment on second-degree systems of equations and analyze the possibilities, we can see that such a system can have as many as four solutions or as few as none. That is, the graphs can have as many as four points of intersection or as few as none. For instance, if we consider the case of a circle and a parabola, we saw in Example 1 the situation in which they intersect in two points. Figure 10.50 illustrates five possibilities.

Figure 10.50

Possibilities for solutions of second-degree systems of equations

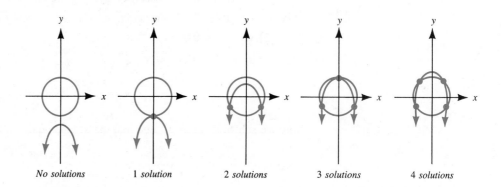

No solutions   *1 solution*   *2 solutions*   *3 solutions*   *4 solutions*

EXAMPLE 2

Solve the following system of equations:

$$\begin{cases} 4x^2 + y^2 = 16 \\ \quad\; x - y = 2 \end{cases}$$

Solution

Figure 10.51 illustrates this system of equations. We can see that we expect two solutions to this system—one of which, $(2, 0)$—we can identify from the graphs. The other must be found algebraically.

Figure 10.51

Graphs of the ellipse $4x^2 + y^2 = 16$ and the straight line $x - y = 2$

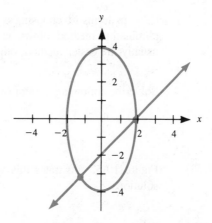

Unlike Example 1, we cannot use the elimination method here because we do not have "like" terms to eliminate. In such a case we can use the substitution method. We begin by solving the second equation for x.

From the second equation we get $x = y + 2$. Now substitute into the first equation:

$$4x^2 + y^2 = 16 \qquad \textit{Replace x with y + 2 and solve for y.}$$
$$4(y + 2)^2 + y^2 = 16$$
$$4(y^2 + 4y + 4) + y^2 = 16$$
$$4y^2 + 16y + 16 + y^2 = 16$$
$$5y^2 + 16y = 0$$
$$y(5y + 16) = 0$$

$$y = 0, \quad 5y + 16 = 0$$
$$y = -\frac{16}{5}$$

Now we substitute the y values into the equation $x = y + 2$ to get the x values and our solutions:

Substitute $y = 0$	*Substitute $y = -\dfrac{16}{5}$*
$x = y + 2$	$x = y + 2$
$x = 0 + 2$	$x = -\dfrac{16}{5} + 2$
$x = 2$	
$\boxed{(2, 0)}$	$x = -\dfrac{6}{5}$
	$\boxed{\left(-\dfrac{6}{5}, -\dfrac{16}{5}\right)}$

The check is left to the student. ∎

In terms of choosing a method of solution, we can say the following. If the elimination method allows us to eliminate one of the variables completely, then it is usually the easier method; otherwise, use the substitution method.

EXAMPLE 3

Solve the following system of equations:

$$\begin{cases} x^2 + 2y^2 = 8 & (1) \\ 2x^2 - y^2 = 6 & (2) \end{cases}$$

Solution

Figure 10.52 illustrates this system of equations and shows us that we expect four solutions.

Figure 10.52

Graphs of the ellipse $x^2 + 2y^2 = 8$ and the hyperbola $2x^2 - y^2 = 6$

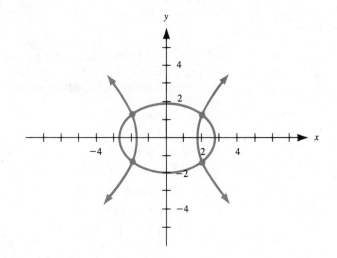

We proceed using the elimination method:

$$\begin{array}{lll} x^2 + 2y^2 = 8 & \textit{As is} \rightarrow & x^2 + 2y^2 = 8 \\ 2x^2 - y^2 = 6 & \textit{Multiply by 2} \rightarrow & \underline{4x^2 - 2y^2 = 12} \quad \textit{Add.} \\ & & 5x^2 = 20 \\ & & x^2 = 4 \\ & & x = \pm 2 \end{array}$$

Substitute $x = \pm 2$ into equation (1):

Substitute $x = 2$	*Substitute $x = -2$*
$x^2 + 2y^2 = 8$	$x^2 + 2y^2 = 8$
$(2)^2 + 2y^2 = 8$	$(-2)^2 + 2y^2 = 8$
$4 + 2y^2 = 8$	$4 + 2y^2 = 8$
$2y^2 = 4$	$2y^2 = 4$
$y^2 = 2$	$y^2 = 2$
$y = \pm\sqrt{2}$	$y = \pm\sqrt{2}$

Thus, we have four solutions to this system:

$$(2, \sqrt{2}), (2, -\sqrt{2}), (-2, \sqrt{2}), (-2, -\sqrt{2})$$

The check is left to the student.   ∎

EXAMPLE 4

Solve the following system of equations:

$$\begin{cases} x^2 + y^2 = 9 & (1) \\ y = 3 - x^2 & (2) \end{cases}$$

Solution

We offer two methods of solution.

USING THE SUBSTITUTION METHOD:
We use equation (2) to substitute for y in equation (1):

$$x^2 + y^2 = 9 \qquad \textit{Replace } y \textit{ with } 3 - x^2.$$
$$x^2 + (3 - x^2)^2 = 9$$
$$x^2 + 9 - 6x^2 + x^4 = 9$$
$$x^4 - 5x^2 = 0$$
$$x^2(x^2 - 5) = 0$$

$$x^2 = 0 \qquad x^2 - 5 = 0$$
$$x = 0 \qquad x = \pm\sqrt{5}$$

We can now substitute these x values into equation (2) to get their associated y values.

Substitute $x = 0$	*Substitute $x = \sqrt{5}$*	*Substitute $x = -\sqrt{5}$*
$y = 3 - x^2$	$y = 3 - (\sqrt{5})^2$	$y = 3 - (-\sqrt{5})^2$
$y = 3 - 0^2$	$y = 3 - 5$	$y = 3 - 5$
$y = 3$	$y = -2$	$y = -2$
$(0, 3)$	$(\sqrt{5}, -2)$	$(-\sqrt{5}, -2)$

There is an important point to be made here. Suppose we had decided to substitute our x values into equation (1) instead of equation (2). We would have obtained another "solution." That is, if we substitute $x = 0$ into equation (1) we get

$$x^2 + y^2 = 9$$
$$0^2 + y^2 = 9$$
$$y^2 = 9$$
$$y = \pm 3$$

which gives another "solution," $(0, -3)$. ***However***, $(0, -3)$ does *not* satisfy equation (2):

$$y = 3 - x^2 \qquad \textit{Substitute } (0, -3).$$
$$-3 \stackrel{?}{=} 3 - (0)^2$$
$$-3 \ne 3$$

Remember that a solution to a system of equations must satisfy *every* equation in the system. Thus, we should always check our solutions in *all* the equations in the system.

USING THE ELIMINATION METHOD:

We can rewrite equation (2) and eliminate x^2:

(1) $\quad x^2 + y^2 = 9 \qquad \to x^2 + y^2 = 9 \quad \textit{As is} \to \qquad\qquad x^2 + y^2 = 9$

(2) $\qquad\quad y = 3 - x^2 \to \ x^2 + y = 3 \quad \textit{Multiply by } -1 \to \quad \underline{-x^2 - y = -3} \quad \textit{Add.}$

$$y^2 - y = 6$$
$$y^2 - y - 6 = 0$$
$$(y - 3)(y + 2) = 0$$

$$y - 3 = 0 \qquad y + 2 = 0$$
$$y = 3 \qquad\qquad y = -2$$

Substitute these y values into equation (1) and solve for x:

Substitute $y = 3$	*Substitute* $y = -2$
$x^2 + y^2 = 9$	$x^2 + y^2 = 9$
$x^2 + 3^2 = 9$	$x^2 + (-2)^2 = 9$
$x^2 + 9 = 9$	$x^2 + 4 = 9$
$x^2 = 0$	$x^2 = 5$
$x = 0$	$x = \pm\sqrt{5}$
$\boxed{(0, 3)}$	$\boxed{(\sqrt{5}, -2)\ (-\sqrt{5}, -2)}$

■

EXAMPLE 5

Find the dimensions of a rectangle whose perimeter is 33 cm and whose area is 65 sq. cm.

Solution

We represent our rectangle as shown in Figure 10.53.

Let W = Width.

Let L = Length.

The information given in the problem allows us to write the following system of equations:

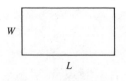

W

L

Figure 10.53
Rectangle for Example 5

$$\begin{cases} 2W + 2L = 33 & (1) \qquad \textit{Perimeter of the rectangle is 33 cm.} \\ WL = 65 & (2) \qquad \textit{Area of the rectangle is 65 sq. cm.} \end{cases}$$

Since we do not have "like" terms to eliminate, we will use the substitution method.

We can choose to solve either equation for one of the variables. We choose to solve the second equation for L.

$$WL = 65 \rightarrow L = \frac{65}{W} \qquad \textit{Substitute into equation (1).}$$

$$2W + 2\left(\frac{65}{W}\right) = 33$$

$$2W + \frac{130}{W} = 33 \qquad \textit{Clear fractions; multiply each side by W.}$$

$$2W^2 + 130 = 33W$$

$$2W^2 - 33W + 130 = 0$$

$$(2W - 13)(W - 10) = 0$$

$$2W - 13 = 0 \qquad W - 10 = 0$$
$$W = \frac{13}{2} \qquad\qquad W = 10$$

If $W = \frac{13}{2}$, then $L = \frac{65}{W} = \frac{65}{\frac{13}{2}} = \frac{65}{1} \cdot \frac{2}{13} = 10$.

Thus, the rectangle is $\frac{13}{2}$ by 10 or $\boxed{6.5 \text{ cm by } 10 \text{ cm}}$

Note that if we substitute $W = 10$, then $L = \frac{65}{W} = \frac{65}{10} = 6.5$, giving us the same dimensions.

CHECK:

The perimeter is $2(6.5) + 2(10) = 13 + 20 \overset{\checkmark}{=} 33$ cm.

The area is $(6.5)(10) \overset{\checkmark}{=} 65$ sq. cm. ∎

Exercises 10.6

In Exercises 1–36, solve the systems of equations.

1. $\begin{cases} x^2 + y^2 = 10 \\ x^2 - y^2 = 8 \end{cases}$
2. $\begin{cases} 2x^2 + y^2 = 6 \\ -x^2 + y^2 = 3 \end{cases}$
3. $\begin{cases} x^2 + y^2 = 25 \\ x - y = 5 \end{cases}$
4. $\begin{cases} x^2 - y^2 = 9 \\ x + y = 3 \end{cases}$

5. $\begin{cases} 16x^2 - 4y^2 = 64 \\ x^2 + y^2 = 9 \end{cases}$
6. $\begin{cases} 4x^2 - 16y^2 = 64 \\ x^2 - 5y^2 = 8 \end{cases}$
7. $\begin{cases} x^2 + y^2 = 9 \\ 2x - y = 3 \end{cases}$
8. $\begin{cases} x^2 + y^2 = 9 \\ 2x^2 - y^2 = -6 \end{cases}$

9. $\begin{cases} x^2 + y^2 = 13 \\ 3x^2 - y^2 = 3 \end{cases}$
10. $\begin{cases} x^2 + y^2 = 13 \\ 3x - y = 3 \end{cases}$
11. $\begin{cases} x^2 - y = 0 \\ x^2 - 2x + y = 6 \end{cases}$
12. $\begin{cases} y - 2x^2 = 0 \\ x^2 + 6x - y = 6 \end{cases}$

13. $\begin{cases} y = 1 - x^2 \\ x + y = 2 \end{cases}$
14. $\begin{cases} y = x^2 - 4 \\ x - y = 5 \end{cases}$
15. $\begin{cases} y = x^2 - 6x \\ y = x - 12 \end{cases}$
16. $\begin{cases} y = x^2 + 8x - 10 \\ y = 3x + 4 \end{cases}$

17. $\begin{cases} x^2 + y^2 = 16 \\ \quad\quad x = y^2 - 16 \end{cases}$ **18.** $\begin{cases} 4x^2 + y^2 = 16 \\ \quad\quad y = x - 4 \end{cases}$ **19.** $\begin{cases} x^2 + y^2 - 25 = 0 \\ \quad x + y - 7 = 0 \end{cases}$ **20.** $\begin{cases} \quad x^2 = y \\ 2x - y = 1 \end{cases}$

21. $\begin{cases} \quad x^2 - y^2 = 4 \\ 2x^2 + y^2 = 16 \end{cases}$ **22.** $\begin{cases} \quad x^2 + y^2 = 25 \\ 4x^2 + 3y^2 = 36 \end{cases}$ **23.** $\begin{cases} 9x^2 + y^2 = 9 \\ 3x + y = 3 \end{cases}$ **24.** $\begin{cases} x^2 + 4y^2 = 16 \\ \quad x + 4y = 4 \end{cases}$

25. $\begin{cases} \quad x^2 + y = 9 \\ 3x + 2y = 16 \end{cases}$ **26.** $\begin{cases} \quad x^2 - y = 10 \\ 2x - 3y = -10 \end{cases}$ **27.** $\begin{cases} \quad x = y^2 - 3 \\ x + y = 9 \end{cases}$ **28.** $\begin{cases} x - y^2 = 3 \\ x - y = 3 \end{cases}$

29. $\begin{cases} x^2 + 4x + y^2 - 6y = 7 \\ \quad\quad\quad\quad 2x + y = 5 \end{cases}$ **30.** $\begin{cases} x^2 - 2x + y^2 + 4y = 15 \\ \quad\quad\quad\quad x - 2y = -5 \end{cases}$ **31.** $\begin{cases} x^2 + y^2 = 10 \\ \quad\quad xy = 4 \end{cases}$

32. $\begin{cases} 2x^2 + 6 = y^2 \\ \quad\quad xy = 6 \end{cases}$ **33.** $\begin{cases} x^2 - y^2 = 4 \\ \quad xy = 2\sqrt{3} \end{cases}$ **34.** $\begin{cases} x^2 - 2y^2 = 0 \\ \quad\quad xy = 3\sqrt{2} \end{cases}$

35. $\begin{cases} \quad\quad x^2 + y^2 = 25 \\ x^2 - xy + y^2 = 13 \end{cases}$ **36.** $\begin{cases} \quad\quad x^2 - y^2 = 21 \\ x^2 + xy - y^2 = 31 \end{cases}$

[*Hint:* First eliminate x^2 or y^2, then solve for x and y and use the substitution method.]

In Exercises 37–46, solve the system of equations and sketch the graphs.

37. $\begin{cases} x^2 + y^2 = 29 \\ \quad x - y = 3 \end{cases}$ **38.** $\begin{cases} y = x^2 - 4x + 3 \\ y = 3x - 3 \end{cases}$ **39.** $\begin{cases} x^2 + y^2 = 29 \\ x^2 - y^2 = 3 \end{cases}$ **40.** $\begin{cases} y = x^2 + 4 \\ y = 16 - x^2 \end{cases}$

41. $\begin{cases} \quad x - y = 0 \\ x^2 - y^2 = 4 \end{cases}$ **42.** $\begin{cases} x^2 + y^2 = 18 \\ \quad x - y = 6 \end{cases}$ **43.** $\begin{cases} 2x^2 + y^2 = 8 \\ \quad x^2 - y^2 = 4 \end{cases}$ **44.** $\begin{cases} x = y^2 \\ y = x^2 \end{cases}$

45. $\begin{cases} 2x^2 + y^2 = 50 \\ \quad x^2 + 2y^2 = 25 \end{cases}$ **46.** $\begin{cases} \quad\quad x^2 + y^2 = 9 \\ x^2 + 2x + y^2 = 9 \end{cases}$

Solve each of the following problems algebraically.

47. Find the dimensions of a rectangle if the area is 210 square inches and the perimeter is 59 inches.

48. Find the dimensions of a rectangle if the perimeter is 57 cm and the area is 189 sq. cm.

49. Maria makes a round trip totaling 120 miles by car. Her return trip took 15 minutes less than her trip going, because she drove 12 mph faster returning than she did going. Find her rate and time in each direction.

50. An electronic mail sorter sorts two batches of 500 letters. It sorts the second batch, set at high speed, in 10 minutes less than it took to sort the first batch, at medium speed. If high speed is 150 letters per hour faster than medium speed, find the rates at medium and high speed and the time needed to sort each batch.

? QUESTION FOR THOUGHT

51. Discuss what is *wrong* (if anything) with the following "solution."

$\begin{cases} 4x^2 + y^2 = 25 \\ \quad x + y = 4 \end{cases}$ *Square both sides of the second equation.*

$\begin{array}{r} 4x^2 + y^2 = 25 \\ \underline{x^2 + y^2 = 16} \\ 3x^2 \quad\quad = 9 \\ x^2 \quad\quad = 3 \\ x \quad = \pm\sqrt{3} \end{array}$ *Subtract.*

⟲ MINI-REVIEW

52. *Express as a simple fraction reduced to lowest terms:*

$$\frac{\dfrac{3}{x} - \dfrac{2}{x-1}}{1 + \dfrac{3}{x}}$$

53. *Solve for x.* $\left|3 - 5x\right| \le -2.$

54. Sketch the graph of $9x^2 - 20y^2 = 180.$

55. Sketch the graph of $5x - 3y \ge 15.$

CHAPTER 10   Summary

After having completed this chapter, you should be able to:

1. Find the distance and midpoint between two points (Section 10.1).

For example: Find the distance between the points (5, 3) and (−4, 7), and find the midpoint between the two points.

Solution: Let $(x_1, y_1) = (5, 3)$ and $(x_2, y_2) = (-4, 7)$. Then the distance between the two points is

$$d = \sqrt{(x_1 - x_2)^2 + (y_1 - y_2)^2} = \sqrt{[5 - (-4)]^2 + (3 - 7)^2}$$
$$= \sqrt{(9)^2 + (-4)^2}$$
$$= \sqrt{81 + 16}$$
$$= \sqrt{97}$$

The midpoint between the two points is found by

$$M = \left(\frac{x_1 + x_2}{2}, \frac{y_1 + y_2}{2}\right) = \left(\frac{5 + (-4)}{2}, \frac{3 + 7}{2}\right)$$
$$= \left(\frac{1}{2}, 5\right)$$

2. Determine the equation of a circle given its radius and center (Section 10.1).

For example: Determine the equation of the circle with center (5, −2) and radius 3.

Solution:

$$(x - 5)^2 + [y - (-2)]^2 = 3^2$$
$$(x - 5)^2 + (y + 2)^2 = 9 \quad \text{(or } x^2 + y^2 - 10x + 4y + 20 = 0)$$

3. Graph the equation of a circle and identify its center and radius (Section 10.1).

For example: Graph the following circles and identify the center and radius.

(a) $(x - 3)^2 + (y + 2)^2 = 36$ **(b)** $x^2 + y^2 - 10x + 2y = -16$

Solution:

(a) Since the standard form of a circle is

$$(x - h)^2 + (y - k)^2 = r^2$$

then $h = 3$, $k = -2$, and $r^2 = 36 \rightarrow r = 6.$

Hence, the center is $(3, -2)$ and the radius is 6.

The circle is graphed in Figure 10.54.

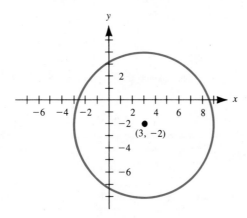

Figure 10.54   Graph of $(x - 3)^2 + (y + 2)^2 = 36$

(b)   We use the method of completing the square to put the equation in standard form:

$$x^2 + y^2 - 10x + 2y = -16$$

$$x^2 - 10x \quad + y^2 + 2y \quad = -16 \qquad \textit{Complete the squares:}$$

$$\left[\frac{1}{2}(-10)\right]^2 = 25 \qquad \left[\frac{1}{2}(2)\right]^2 = 1^2 = 1$$

Hence, we add $+25$ and $+1$ to both sides of the equation:

$$x^2 - 10x + 25 + y^2 + 2y + 1 = -16 + 25 + 1$$

$$(x - 5)^2 + (y + 1)^2 = 10 \qquad h = 5, \, k = -1, \textit{ and } r^2 = 10$$

Therefore, the center is $(5, -1)$ and the radius is $\sqrt{10}$.

A graph of the circle is given in Figure 10.55.

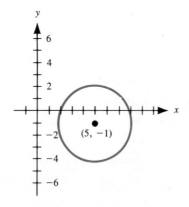

Figure 10.55   Graph of $x^2 + y^2 - 10x + 2y = -16$

4. Graph a parabola, identifying its vertex, axis of symmetry, and intercepts (if they exist) (Section 10.2).

   For example: Graph the following parabola and identify its vertex, axis of symmetry, and intercepts (if they exist):

   $$y = 3x^2 + x - 2$$

   Solution: The parabola is in the form $y = Ax^2 + Bx + C$, with $A = 3$, $B = 1$, and $C = -2$.

   The axis of symmetry is $x = -\dfrac{B}{2A} = -\dfrac{1}{2(3)} = -\dfrac{1}{6}$.

   The x component of the vertex is $-\frac{1}{6}$. The y component of the vertex is

   $$y = 3\left(-\tfrac{1}{6}\right)^2 + \left(-\tfrac{1}{6}\right) - 2 = -\tfrac{25}{12}$$

   The vertex is $\left(-\frac{1}{6}, -\frac{25}{12}\right)$.

   The y-intercept occurs at $x = 0$: $\quad y = 3(0)^2 + (0) - 2 = -2$

   The x-intercepts occur at $y = 0$: $\quad 0 = 3x^2 + x - 2$
   $$0 = (3x - 2)(x + 1)$$

   The x-intercepts are $\frac{2}{3}$ and -1.

   The graph of the parabola is shown in Figure 10.56.

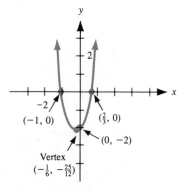

   Figure 10.56   Graph of $y = 3x^2 + x - 2$

5. Graph an ellipse centered at the origin and identify its vertices (Section 10.3).

   For example: Graph $x^2 + 2y^2 = 18$.

   Solution: Put in standard form:

   $$x^2 + 2y^2 = 18$$
   $$\frac{x^2}{18} + \frac{y^2}{9} = 1$$

   $$a^2 = 18 \rightarrow a = \sqrt{18} = 3\sqrt{2} \quad \text{and} \quad b^2 = 9 \rightarrow b = 3$$

The equation is that of an ellipse centered at $(0, 0)$ with vertices $(3\sqrt{2}, 0)$, $(-3\sqrt{2}, 0)$, $(0, 3)$, and $(0, -3)$. The axes are the x- and y-coordinate axes.

The graph is shown in Figure 10.57.

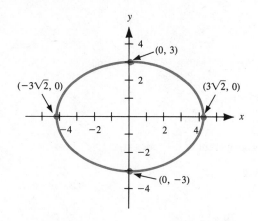

Figure 10.57   Graph of $x^2 + 2y^2 = 18$

6. Graph a hyperbola at the origin, identifying the vertices and the asymptotes (Section 10.4).

For example: Graph $9x^2 - 25y^2 = 225$.

Solution: Put in standard form:

$9x^2 - 25y^2 = 225$ *Divide both sides by 225.*

$$\frac{9x^2}{225} - \frac{25y^2}{225} = \frac{225}{225}$$

$$\frac{x^2}{25} - \frac{y^2}{9} = 1$$

Hence, $a^2 = 25 \rightarrow a = 5$, and $b^2 = 9 \rightarrow b = 3$.

The figure is a hyperbola with vertices at $(+5, 0)$ and $(-5, 0)$. The asymptotes are $y = \pm\frac{3}{5}x$. The graph is illustrated in Figure 10.58.

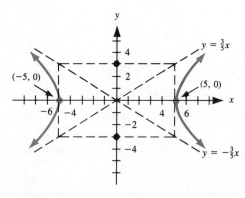

Figure 10.58   Graph of $9x^2 - 25y^2 = 225$

7. Identify conic sections (or their degenerate forms) by their equations in the general form $Ax^2 + By^2 + Cx + Dy + E = 0$ (Section 10.5).

For example: Assuming the equation $3x^2 - 4y^2 - 18x - 32y - 61 = 0$ can be graphed, identify what figure the equation will yield without putting the equation in standard form.

Solution: Since the signs of the coefficients of the squared terms are different, the equation will yield a hyperbola or its degenerate form, if it can be graphed.

8. Solve 2×2 nonlinear systems of equations and sketch their graphs (Section 10.6).

For example: Solve for x and y:

$$\begin{cases} x^2 + y^2 = 13 & (1) \\ x + y = 5 & (2) \end{cases}$$

Solution: Figure 10.59 illustrates the graphs of the equations in the system.

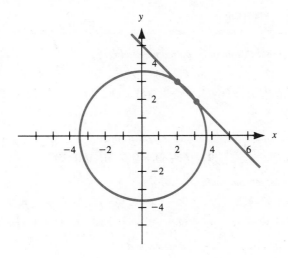

Figure 10.59 **Graphs of $x^2 + y^2 = 13$ and $x + y = 5$**

Since we do not have "like" terms to eliminate, we use the substitution method. We solve equation (2) for y:

$$y = 5 - x \qquad \text{Call this equation (3).}$$

Substitute $y = 5 - x$ into equation (1):

$$x^2 + y^2 = 13$$
$$x^2 + (5 - x)^2 = 13$$
$$x^2 + 25 - 10x + x^2 = 13$$
$$2x^2 - 10x + 12 = 0$$
$$2(x^2 - 5x + 6) = 0$$
$$2(x - 3)(x - 2) = 0$$

$$x - 3 = 0 \qquad x - 2 = 0$$
$$x = 3 \qquad\quad x = 2$$

Substitute x values into equation (3):

Substitute $x = 3$ *Substitute $x = 2$*

$y = 5 - x$ $y = 5 - x$

$y = 5 - 3$ $y = 5 - 2$

$y = 2$ $y = 3$

The solutions are $(3, 2)$ and $(2, 3)$.

CHAPTER 10   Review Exercises

In Exercises 1–10, find the distance between P and Q and the midpoint between P and Q.

1. $P(0, 0)$, $Q(2, 6)$ **2.** $P(4, 8)$, $Q(2, 12)$ **3.** $P(2, 5)$, $Q(-2, 5)$ **4.** $P(3, 8)$, $Q(3, -8)$

5. $P(6, -4)$, $Q(4, -6)$ **6.** $P(-7, -3)$, $Q(7, 3)$ **7.** $P(-2, -5)$, $Q(2, 5)$ **8.** $P(-2, 5)$, $Q(2, -5)$

9. $P(3, -4)$, $Q(8, -3)$ **10.** $P(3, -4)$, $Q(-4, 3)$

In Exercises 11–12, determine if the points P, Q, and R form the vertices of a right triangle.

11. $P(3, 6)$, $Q(5, 9)$, $R(8, 7)$ **12.** $P(4, 2)$, $Q(6, 5)$, $R(8, 3)$

In Exercises 13–18, graph the circle and indicate its center and radius.

13. $x^2 + y^2 = 100$ **14.** $x^2 + y^2 = 28$ **15.** $x^2 + y^2 - 4x - 14y = -52$

16. $x^2 + y^2 - 2x = 15$ **17.** $x^2 + y^2 - 6x + 4y = 68$ **18.** $x^2 + y^2 + 4x - 6y = 14$

In Exercises 19–36, graph the parabola and identify its vertex, axis of symmetry, and intercepts (if they exist).

19. $y = 7x^2$ **20.** $y = -6x^2$ **21.** $y = -7x^2 + 3$ **22.** $y = \frac{1}{6}x^2 - 3$

23. $y = x^2 - 6x$ **24.** $y = x^2 - 9x$ **25.** $y = x^2 - 2x - 8$ **26.** $y = x^2 - 12x + 35$

27. $x = y^2 - 2y - 8$ **28.** $x = y^2 - 12y + 35$ **29.** $y = x^2 - x - 12$ **30.** $y = x^2 - 2x - 3$

31. $y = x^2 - 2x - 2$ **32.** $y = x^2 - x - 10$ **33.** $y = x^2 - 2x + 5$ **34.** $y = x^2 - 4x + 7$

35. $y = -x^2 + 2x - 5$ **36.** $y = -x^2 + 4x - 7$

In Exercises 37–42, graph the ellipse and indicate the vertices.

37. $144x^2 + 9y^2 = 1,296$ **38.** $16x^2 + 27y^2 = 432$ **39.** $144x^2 + y^2 = 144$

40. $12x^2 + y^2 = 4$ **41.** $3x^2 + y^2 = 24$ **42.** $5x^2 + 9y^2 = 25$

In Exercises 43–48, graph the hyperbola and indicate the vertices and asymptotes.

43. $144x^2 - 9y^2 = 1,296$ **44.** $16x^2 - 27y^2 = 432$ **45.** $y^2 - 144x^2 = 144$

46. $y^2 - 12x^2 = 4$ **47.** $3x^2 - y^2 = 24$ **48.** $5x^2 - 9y^2 = 25$

In Exercises 49–72, identify the figure described by the equation.

49. $4x^2 + y^2 = 16$

50. $4x^2 - y^2 = 4$

51. $x^2 - 10x - y = -25$

52. $2x^2 + 3y^2 = 12$

53. $x^2 + y^2 = 108$

54. $25x^2 - 32y^2 = 800$

55. $x + y = 9$

56. $25x - 32y = 800$

57. $25x^2 + 32y^2 = 800$

58. $x^2 + 2x - y = 25$

59. $2x^2 - 3y^2 = 12$

60. $x^2 - y^2 = 10$

61. $y^2 + 2y - x = 25$

62. $x^2 + y^2 + 14x - 6y = -50$

63. $x^2 + y^2 - 14x - 6y = -50$

64. $x^2 - 6x - y = 17$

65. $x^2 + y^2 + 4x + 4y = -4$

66. $x^2 + y^2 + 4x + 4y = 4$

67. $x^2 - y^2 = 16$

68. $3x^2 + 2y^2 = 1$

69. $x - y = 16$

70. $3x + 2y = 1$

71. $x^2 + y^2 = 0$

72. $-3x^2 + y = 0$

In Exercises 73–92, identify and, if possible, sketch a graph of the equation. Label the important aspects of the figure.

73. $x^2 + y^2 + 6x - 4y = -15$

74. $x^2 + y^2 + 2x - 6y = 14$

75. $-3x^2 - y = -2$

76. $x^2 + y^2 = 32$

77. $y^2 + 2y - x = 8$

78. $5x^2 + 5y^2 - 10x = 15$

79. $-x^2 + 4x - y = 11$

80. $x^2 + 4x - y = 16$

81. $x + y = 36$

82. $4x - y = 36$

83. $x^2 + y^2 = 36$

84. $4x^2 - y^2 = 36$

85. $x^2 - y^2 = 36$

86. $4y^2 - x = 36$

87. $x^2 - y = 36$

88. $4y^2 + x^2 = 36$

89. $x^2 - y^2 = 32$

90. $x^2 + y^2 - 6x + 4y = -5$

91. $x^2 + 6x + y^2 - 8y = -9$

92. $x^2 - 2x - y = -6$

93. The Tabak Company found that the daily cost of producing widgets in a factory is related to the number of widgets produced in the following way:

$$C = \frac{-x^2}{10} + 100x - 24{,}000$$

where C is the cost in dollars and x is the number of widgets. What number of widgets would produce the maximum cost? What would this cost be?

95. Jake jumps off a diving board into a pool. The equation $S = -2t^2 + 5t + 5$ gives the distance (in feet) Jake is above the pool at t seconds after he jumps.

  (a) What is the maximum height Jake reaches when he dives?

  (b) How high is the diving board?

94. The profit (P) made on a concert is related to the price (p) of a ticket in the following way:

$$P = 10{,}000(-p^2 + 12p - 35)$$

where P is the profit in dollars and p is the price of a ticket. What ticket price would produce the maximum profit?

In Exercises 96–101, solve the nonlinear system of equations.

96. $\begin{cases} x^2 + y^2 = 8 \\ x + y = 4 \end{cases}$

97. $\begin{cases} x^2 - 3y^2 = 4 \\ 2x^2 + 5y^2 = 12 \end{cases}$

98. $\begin{cases} 2x^2 - y^2 = 14 \\ 2x - 3y = -8 \end{cases}$

99. $\begin{cases} x^2 + y^2 = 9 \\ x^2 + y = 3 \end{cases}$

100. $\begin{cases} x^2 + y^2 = 4 \\ 2x^2 + 3y^2 = 18 \end{cases}$

101. $\begin{cases} 2x^2 + 3y^2 = 4 \\ x^2 - 4y^2 = 9 \end{cases}$

CHAPTER 10   Practice Test

1. Find the distance between the points $(3, -2)$ and $(-2, 3)$.

2. Find the midpoint between the points $(3, 8)$ and $(5, 3)$.

3. Identify the center and radius of the following circles:

   (a) $(x - 3)^2 + (y + 2)^2 = 12$

   (b) $x^2 + y^2 - 4x - 6y - 3 = 0$

4. Sketch a graph of the following parabolas. Label the x- and y-intercepts, vertex, and axis of symmetry.

   (a) $y = 2x^2 + 3x + 1$

   (b) $y = -2x^2 + x - 3$

5. Sketch a graph of the following ellipse and indicate the vertices:

   $$8x^2 + 9y^2 = 72$$

6. Sketch a graph of the following hyperbola:

   $$\frac{x^2}{16} - \frac{y^2}{24} = 1$$

   Label the vertices and the asymptotes.

7. Sketch a graph of the following figures and label the important aspects of each figure.

   (a) $x^2 + 4y^2 = 64$

   (b) $y^2 + 6y = x$

   (c) $x^2 + y^2 - 2x + 4y - 3 = 0$

8. Solve the following nonlinear system of equations:

   $$\begin{cases} x^2 + 2y^2 = 6 \\ x - y = 3 \end{cases}$$

11

Functions

The concept of a function is one of the most important in mathematics. We often hear statements such as "Insurance rates are a function of your age" or "Crop yields are a function of the weather." We understand these statements to mean that one item is dependent on the other. In this chapter we will make the idea of one quantity being dependent on another more precise.

11.1

Relations and Functions: Basic Concepts

A *relation* is a correspondence between two sets where to each element of the first set there is associated or assigned one or more elements of the second set. The first set is called the *domain* and the second set is called the *range*.

A relation is simply a rule by which we decide how to match up or associate elements from one set with elements from another. For example, students are usually assigned an identification number, as illustrated below:

Names of students (Domain)		Student numbers (Range)
John Jones	⟶	#17345
Sam Klass	⟶	#65734
Carol Kane	⟶	#75664

The arrow leads from the domain to the range.

The relation shown above describes the assignment of numbers to names. The domain is the set of names, {John Jones, Sam Klass, Carol Kane}, and the range is the set of numbers assigned to the three names, {17345, 65734, 75664}. The relation *is* the assignment.

A relation can have many elements of the domain assigned to many elements of the range. For example, we can describe the following relation, by which we associate with each person his or her telephone number(s):

Domain		Range
Sue	⟶	555-2342
Carol	↘	
Harry	⟶	555-9235
Jake	⟶	555-3197
	↘	555-2816

Note that Jake has two telephone numbers, while the number 555-9235 is associated with two people, Carol and Harry.

Let's consider another example. Let the domain be the set $A = \{a, b, c\}$ and the range be the set $B = \{1, 2, 3\}$. Four of the many possible relations between the two sets are shown at the top of the next page:

Relation R

Domain		Range
a	$\longrightarrow$	1
b	$\longrightarrow$	2
c	$\longrightarrow$	3

Relation S

Domain		Range
a	$\longrightarrow$	2
b	$\longrightarrow$	3
c	$\longrightarrow$	1

Relation T

Domain		Range
a		1
b		2
c		3

Relation U

Domain		Range
a		1
b		2
c		3

In relation *R*, *a* is assigned 1, *b* is assigned 2, and *c* is assigned 3.

In relation *S*, *a* is assigned 2, *b* is assigned 3, and *c* is assigned 1.

In relation *T*, *a* is assigned 1, *b* is assigned 1, *c* is assigned 2, and *c* is also assigned 3.

In relation *U*, *a* is assigned 1, *a* is also assigned 2, *b* is assigned 3, and *c* is assigned 3.

Thus, we can describe any type of correspondence we want as long as we assign to each element of the domain an element or elements of the range.

An alternative way to describe a relation is to use ordered pairs. For example, instead of writing the correspondence of relation *R*:

a	$\longrightarrow$	1
b	$\longrightarrow$	2
c	$\longrightarrow$	3

we can write the correspondence without arrows as $(a, 1)$, $(b, 2)$, $(c, 3)$. Thus, the set $\{(a, 1), (b, 2), (c, 3)\}$ describes the relation, *R*, between the set *A* and the set *B*.

When using ordered pair notation, we will always assume that the first coordinate is an element of the domain and the second coordinate is the associated element of the range. Thus, the first component, *x*, of the ordered pair (x, y) is an element of the domain. The second component, *y*, is an element of the range such that *y* is assigned to *x*.

We usually call the variable representing possible values of the domain the **independent variable** and the variable representing possible values of the range the **dependent variable**.

We can rewrite the other previous examples as follows:

Relation *S*: $\{(a, 2), (b, 3), (c, 1)\}$

Relation *T*: $\{(a, 1), (b, 1), (c, 2), (c, 3)\}$

Relation *U*: $\{(a, 1), (a, 2), (b, 3), (c, 3)\}$

Note that this is all you need to describe the relation.

We define a relation using ordered pair notation as follows:

DEFINITION

A *relation* is a set of ordered pairs (x, y). The set of x values is called the *domain* and the set of y values is called the *range*.

For example:

The set of ordered pairs $R = \{(8, 2), (6, -3), (5, 7), (5, -3)\}$ is the relation between the sets $\{5, 6, 8\}$ and $\{2, -3, 7\}$ where $\{5, 6, 8\}$ is the domain and $\{2, -3, 7\}$ is the range; 8 is assigned 2, 6 is assigned -3, 5 is assigned 7, and 5 is also assigned -3.

Another example:

$S = \{(a, b), (a, c), (b, c), (c, a)\}$ is a set of ordered pairs describing the relation between set $\{a, b, c\}$ and itself. Thus, $\{a, b, c\}$ is both the domain and the range: a is assigned b and c, b is assigned c, and c is assigned a.

If the domain and range are infinite, we cannot write out each element assignment, so instead we can use set builder notation to describe the relation(ship) between the variables. For example:

$$S = \{(x, y) \mid y = x + 5, \quad x \text{ and } y \text{ are real numbers}\}$$

is a relation between the set of real numbers and itself (both the domain and range are the set of real numbers). Each value of x is assigned a value, y, which is 5 more than x. Thus,

3 is assigned value 8 since $8 = 3 + 5$ *or* $(3, 8) \in S$.
9 is assigned value 14 since $14 = 9 + 5$ *or* $(9, 14) \in S$.
-10 is assigned value -5 since $-5 = -10 + 5$ *or* $(-10, -5) \in S$.

Another example:

$$R = \{(x, y) \mid x < y \quad \text{and} \quad x \text{ and } y \text{ are real numbers}\}$$

In this relation, both the domain *and* range are the set of real numbers. If $x = 2$, then y must be greater than 2. Hence, $(2, 2\frac{1}{2})$, $(2, 3)$, $(2, 100)$, $(2, 50)$, and $(2, 2\frac{1}{10})$ satisfy the relation. Other ordered pairs which satisfy this relation are $(3, 8)$, $(7, 9)$, $(0, 4)$, $(-1, -\frac{1}{2})$, and $(-10, 16)$. (Why?)

We can drop the set notation and write $y = x + 5$ rather than

$$\{(x, y) \mid y = x + 5, \quad x \text{ and } y \text{ are real numbers}\}$$

When we write a relation as an equation in x and y, we assume that x represents the independent variable and y represents the dependent variable. *We also assume that the domain is the set of real numbers that will yield real-number values for y.* (The range is usually more difficult to identify.)

For example, suppose a relation is defined by

$$y = \frac{1}{x}$$

x can be any real number except 0 since 0 will produce an undefined y value. Therefore, the domain (allowable values of x) is all real numbers except 0. We can also write this as $\{x \mid x \neq 0\}$.

EXAMPLE 1

Find the domains of the relations defined by each of the following:

(a) $y = \dfrac{3x}{2x + 1}$ **(b)** $y = -2x + 7$ **(c)** $y = \sqrt{2x - 6}$

Solution

(a) We are looking for the domain of the relation defined by the given equation, that is, the set of allowable values of x (values of x which will yield real-number values of y).

By our previous experiences with rational expressions, we know that a sum, difference, product, or quotient of real numbers is a real number *except* when the divisor of a quotient is 0 (in which case the quotient is undefined). Therefore, the denominator, $2x + 1$, cannot be 0, or x cannot be $-\frac{1}{2}$.

Thus, the domain is $\boxed{\{x \mid x \neq -\frac{1}{2}\}}$

(b) Any real value substituted for x will produce a real value for y.

Therefore, the domain is $\boxed{\text{all real numbers}}$

(c) Since y must be a real number, we must make sure that the radicand, $2x - 6$, is nonnegative. (Why?) Therefore, $2x - 6$ must be greater than or equal to 0. We solve the inequality:

$$2x - 6 \geq 0$$
$$x \geq 3$$

Thus, the domain is $\boxed{\{x \mid x \geq 3\}}$

Notice that any x less than 3 will produce an imaginary y value. ∎

Since we can express a relation as a set of ordered pairs, we can get a picture of a relation by graphing the ordered pairs satisfying the relation on the rectangular coordinate system, and we can also identify its domain and range. We now demonstrate by example.

EXAMPLE 2

Graph the following relation. Identify its domain and range.

$$x^2 + 4y^2 = 16$$

Solution

We use what we know about graphing quadratic equations in two unknowns to arrive at the following:

$x^2 + 4y^2 = 16$ *Put into the standard form of an ellipse.*

$$\frac{x^2}{16} + \frac{y^2}{4} = 1$$

The graph is shown in Figure 11.1.

Figure 11.1
Graph of $x^2 + 4y^2 = 16$

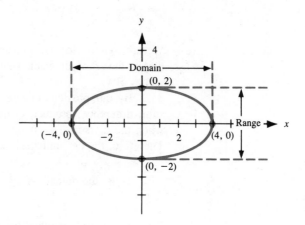

The domain is the set of possible values of x, which we can identify visually from the figure; the range is the set of possible values of y, which can also be identified in the same way.

Thus, the   domain is $\{x \mid -4 \le x \le 4\}$

and the   range is $\{y \mid -2 \le y \le 2\}$

EXAMPLE 3

Graph the following relation. Identify its domain and range.

$$y = 2x^2 + 2$$

Solution

This is a parabola, with the y-axis being the axis of symmetry (see Figure 11.2).
Notice that x can take any real value;

thus, the   domain is the set of all real numbers

Figure 11.2
Graph of $y = 2x^2 + 2$

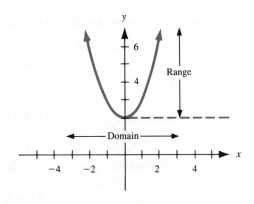

Since y is always greater than or equal to 2,

the   range is $\{y \mid y \geq 2\}$ ∎

Functions

In a relation it is possible to assign many elements of one set to many elements of the other set. For example, both of the following assignments describe relations:

Domain	Range		Domain	Range
a	1		a	1
b			b	2
c	3		c	3

Note that in the left example, both elements a and b of the domain are assigned the same element of the range, 1. In the example on the right, a is assigned two elements from the range, 1 and 2. A *function* is a type of relation which does *not* allow assignments such as that shown in the right-hand example.

DEFINITION

A *function* is a correspondence between two sets such that to each element of the first set (the domain) there is assigned *exactly* one element of the second set (the range).

Thus, for a relation to be a function, we cannot assign more than one element of the range to an element of the domain. However, in a function we can still assign an element of the range to more than one element of the domain.

For example, the relation

$$3 \longrightarrow 6$$
$$2 \longrightarrow 5$$
$$7 \longrightarrow 4$$

which is the set $\{(3, 6), (2, 5), (7, 4)\}$, is a function since each element in the domain $\{3, 2, 7\}$ is assigned only one element in the range $\{6, 5, 4\}$.

Another example is the relation

$$
\begin{array}{rcl}
3 & \longrightarrow & 6 \\
2 & \nearrow & \\
7 & \longrightarrow & 4
\end{array}
$$

which is the set {(3, 6), (2, 6), (7, 4)}. Even though the range element 6 is assigned to two elements of the domain, 3 and 2, this relation is still a function since each element in the *domain*, {3, 2, 7}, is assigned only *one element of the range*, {4, 6}: 3 is assigned only one value, 6; 2 is assigned only one value, 6; and 7 is assigned only one value, 4.

On the other hand, the relation

$$
\begin{array}{rcl}
3 & \longrightarrow & 6 \\
7 & \longrightarrow & 4
\end{array}
$$

which is the set {(3, 6), (3, 4), (7, 4)}, is *not* a function, since the domain element 3 is assigned two elements of the range, 6 and 4.

Refer back to our earlier example of telephone numbers, given at the beginning of this section. The relation that assigns telephone number(s) to each person is often not a function since a person can have more than one number.

The following is a useful metaphor:

> We can describe a function as a machine into which you throw elements of the domain, and out of which come elements of the range (see Figure 11.3).

Figure 11.3
Function machine

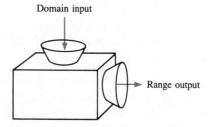

Domain input

Range output

Consider the function {(3, 5), (2, 5), (7, 4)}. Note that if we put 3 into the machine, out comes 5; if we put 2 into the machine, out comes 5; and if we put 7 into the machine, out comes 4.

If we tried to use the function machine for the relation {(3, 5), (3, 4), (2, 6)}, what would happen when we throw in 3? If we throw in 3, we cannot be sure if we will get 5 or 4. This is precisely why the relation is *not* a function.

Remember

All functions are relations, but not all relations are functions. Functions are a special kind of relation such that every value in the *domain* is assigned *exactly one* value in the *range*.

EXAMPLE 4

Indicate the domain of the following relations and determine whether each relation is a function.

(a)   {(3, 5), (4, 2), (3, 6), (5, 7)}   **(b)**   {(7, −2), (6, −4), (5, 8), (−4, 8)}

Solution

(a)   The domain is {3, 4, 5}.

The relation is │ not a function │

since 3 is assigned more than one element of the range (3 is assigned both 5 and 6).

(b)   The domain is {−4, 5, 6, 7}.

The relation │ is a function │

since each element in the domain is assigned no more than one element of the range.   ∎

As with relations, we can describe functions using two variables, and we can graph the function. Given the graph of a relation, we can visually determine whether the relation is a function by a simple test called the *vertical line test*.

The Vertical Line Test

If any vertical line intersects the graph of a relation at more than one point, the relation is *not* a function.

For example, suppose we have the graph of $(x − 5)^2 + (y − 3)^2 = 16$: a circle with radius 4 centered at (5, 3), as shown in Figure 11.4. If we place the vertical line $x = 4$ on this graph, we notice that the line intersects the circle at two points. Thus, there are two distinct points on the circle which have the same x component of 4—that is, two y values are assigned to the x value 4. This violates the definition of a function and therefore the circle is not a function.

Figure 11.4

Graph of
$(x − 5)^2 + (y − 3)^2 = 16$
intersecting the vertical line
$x = 4$

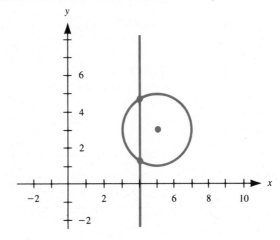

On the other hand, notice that the graph in Figure 11.5 is a function, since any vertical line placed on the graph will intersect the graph at no more than one point.

Figure 11.5

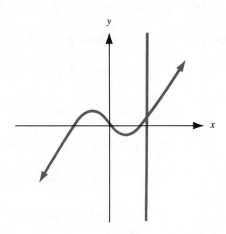

EXAMPLE 5

Visually determine which of the following relations are functions, by the vertical line test.

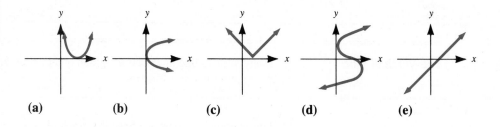

(a)   (b)   (c)   (d)   (e)

Solution

The relations in parts **(a)**, **(c)**, and **(e)** are functions, since any vertical line will intersect the graphs at only one point.

The relations in parts **(b)** and **(d)** are not functions, since some vertical lines will intersect the graphs at more than one point, as shown in the accompanying figures.

(b)

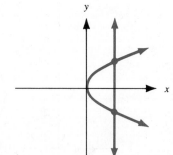

This vertical line intersects at two points; therefore, this relation is not a function.

(d)

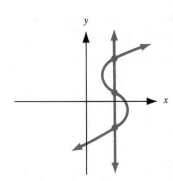

This vertical line intersects at three points; therefore, this relation is not a function.

Exercises 11.1

In Exercises 1–4, write the diagram using ordered pair notation.

1. $3 \longrightarrow 9$
 $8 \searrow 7$
 $7 \longrightarrow 2$

2. $4 \longrightarrow 8$
 $2 \nearrow 7$
 $6 \nearrow$

3. $3 \longrightarrow a$
 $8 \rightrightarrows b$
 $-1 \longrightarrow c$

4. $2 \longrightarrow 1$
 $3 \leftrightarrows 2$
 $0 \longrightarrow 5$

In Exercises 5–8, determine the domain and range.

5. $\{(3, 2), (4, 2), (5, 3)\}$

6. $\{(-1, 5), (-2, 6), (-3, 7)\}$

7. $\{(3, -2), (-2, 3), (3, -1), (4, 3)\}$

8. $\{(5, -1), (-1, 5)\}$

In Exercises 9–26, determine the domain.

9. $y = \dfrac{3}{x}$

10. $y = \dfrac{5}{x + 1}$

11. $y = 3x - 7$

12. $y = 2x - 3$

13. $y = \dfrac{4x}{2x + 3}$

14. $y = \dfrac{5x}{3x - 2}$

15. $y = x^2 - 3x + 4$

16. $y = 3x^2 - 5x + 2$

17. $y = \sqrt{x - 4}$

18. $y = \sqrt{x + 3}$

19. $y = \sqrt[3]{x - 3}$

20. $y = \sqrt[3]{x + 5}$

21. $y = \sqrt{5 - 4x}$

22. $y = \sqrt{6 - 5x}$

23. $y = 5 - \sqrt{4x}$

24. $y = 6 - \sqrt{5x}$

25. $y = \dfrac{x}{\sqrt{x - 3}}$

26. $y = \dfrac{2x}{\sqrt{3 - x}}$

In Exercises 27–38, determine the domain and range (sketch a graph if it would be helpful).

27. $x^2 + y^2 = 9$

28. $x^2 + y^2 = 25$

29. $y = x^2 - 4$

30. $y = 3x^2 + 2$

31. $x = y^2 - 4$

32. $x = 3y^2 + 2$

33. $4x + 2y = 12$

34. $3x + 9y = 9$

35. $\dfrac{x^2}{9} + \dfrac{y^2}{16} = 1$

36. $\dfrac{x^2}{81} + \dfrac{y^2}{9} = 1$

37. $\dfrac{x^2}{4} - \dfrac{y^2}{25} = 1$

38. $\dfrac{x^2}{25} - \dfrac{y^2}{16} = 1$

In Exercises 39–58, determine which of the relations are functions.

39. $2 \longrightarrow 6$
 $3 \nearrow$
 $4 \longrightarrow 1$

40. $5 \longrightarrow 7$
 $4 \nearrow$
 $2 \longrightarrow 1$

41. $6 \longrightarrow 3$
 $\searrow 1$
 $5 \longrightarrow 4$

42. $3 \longrightarrow 9$
 $5 \longrightarrow 4$
 $\searrow 6$

43. $\{(6, 3), (5, 3)\}$

44. $\{(8, -2), (7, -2)\}$

45. $\{(3, 1), (3, 2)\}$

46. $\{(5, 1), (5, 7)\}$

47. $\{(6, 5), (5, 6)\}$

48. $\{(7, -2), (-2, 7)\}$

49. $\{(3, 1), (5, 2), (6, 2)\}$

50. $\{(6, 2), (5, 8), (6, 3)\}$

51. $\{(9, -1), (6, -2), (9, 3)\}$

52. $\{(7, 4), (5, -2), (3, -2)\}$

53. $y = x + 3$

54. $2y = x + 4$

55. $x^2 + y^2 = 81$

56. $x^2 + 3y^2 = 9$

57. $x = y^2 - 4$

58. $y = x^2 + 2$

In Exercises 59–66, determine which of the relations are functions.

59.

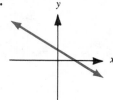

60.

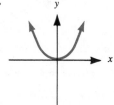

61.

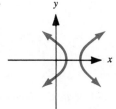

62.

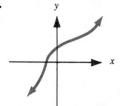

63.

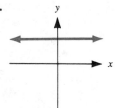

64.

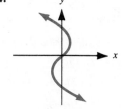

65.

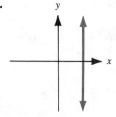

66.

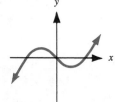

67. The relationship between the monthly profit (P) generated from all two-bedroom units and the rent (r) charged for each unit in an apartment complex is described in the accompanying graph.

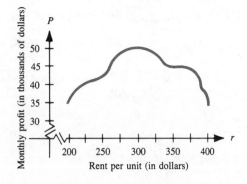

(a) Is this the graph of a function?

(b) What are its domain and range?

(c) What rent per unit generates the maximum profit? What is the maximum profit?

? QUESTION FOR THOUGHT

68. State in words what makes a relation a function.

11.2

Function Notation and the Algebra of Functions

In the previous section we discussed real-valued functions such as $y = x + 2$ and $y = 3x^2 - 2$. Because these equations are explicitly solved for y, we can say that y is a function of x, or that y is dependent on the values of x (hence, y is the dependent variable). Another way to state that y is a function of x is to write

$$y = f(x) \qquad \textit{f(x) is read "f of x."}$$

We can bypass the y variable completely and describe an expression such as $3x^2 - 2$ as being a function of x by writing

$$f(x) = 3x^2 - 2 \qquad \textit{The expression } 3x^2 - 2 \textit{ is a function of x.}$$

Most times we use the letters f, g, and h for functions. Other examples of functions are as follows:

$$f(x) = x^2 - 2x + 5$$
$$g(x) = 2 - 5x$$
$$h(a) = \frac{a}{a + 1}$$

Note that the parentheses in this notation are *not* used as grouping symbols—that is, $f(x)$ is *not* the product of f and x. The parentheses are being used to specify the independent variable.

The notation $f(x)$ is a useful shorthand for evaluating expressions or substituting variables. For example, if we define $f(x) = 7x - 4$, then if we want to know the value of the function $f(x)$ when $x = 3$, we write $f(3)$.

If $f(x) = 7x - 4$, and we want to find the function value when $x = 3$, then

$$f(3) = 7(3) - 4 = 21 - 4 = 17 \qquad \textit{Thus, f(3) = 17.}$$

In the same way, $f(-5)$ is the value of $f(x)$ when $x = -5$:

$$f(-5) = 7(-5) - 4 = -35 - 4 = -39 \qquad \textit{Thus, f(-5) = -39.}$$

Thus,

$f(a)$ is the value of $f(x)$ when a is substituted for x in $f(x)$.

We can also view $f(x)$ as the function, f, evaluated at x. For example, if we have $g(x) = 2x^2 - 4x + 5$, then $g(3)$ is the value of $g(x) = 2x^2 - 4x + 5$ when x is replaced by 3. Hence,

$$g(3) = 2(3)^2 - 4(3) + 5 \qquad \text{$g(3)$ is $g(x)$ evaluated when $x = 3$.}$$
$$= 2(9) - 4(3) + 5$$
$$= 11$$

We are not necessarily restricted to numbers as possible replacements for x. For example, if $f(x) = 7x - 4$, then

$$f(a) = 7a - 4 \qquad \text{Replace x with a in $f(x)$.}$$
$$f(z) = 7z - 4 \qquad \text{Replace x with z in $f(x)$.}$$

EXAMPLE 1

If $f(x) = 3x - 5$ and $g(x) = 2x^2 - 3x + 1$, find each of the following:

(a) $f(-2)$ **(b)** $g(4)$ **(c)** $g(-3)$ **(d)** $g(r)$

Solution

(a)   Evaluate $f(x)$ for $x = -2$:

$$f(-2) = 3(-2) - 5 = -11$$

Thus, $f(-2) = -11$

(b)   Evaluate $g(x)$ for $x = 4$:

$$g(4) = 2(4)^2 - 3(4) + 1 = 2 \cdot 16 - 3 \cdot 4 + 1 = 21$$

Thus, $g(4) = 21$

(c)   Evaluate $g(x)$ for $x = -3$:

$$g(-3) = 2(-3)^2 - 3(-3) + 1 = 2 \cdot 9 + 9 + 1 = 28$$

Thus, $g(-3) = 28$

(d)   Substitute r for x in $g(x)$:

$$g(r) = 2r^2 - 3r + 1$$

Thus, $g(r) = 2r^2 - 3r + 1$ ∎

EXAMPLE 2

Given $f(x) = 3x - 1$, find each of the following:

(a) $f(a + 1)$ **(b)** $f(2x - 1)$

Solution

(a) $f(a + 1)$ is the expression we get when we substitute $a + 1$ for x in $f(x)$. Since

$$f(x) = 3x - 1, \text{ we have:}$$

$$f(a + 1) = 3(a + 1) - 1 \qquad \textit{Simplify.}$$
$$= 3a + 3 - 1$$
$$= 3a + 2$$

$$\boxed{\text{Thus, } f(a + 1) = 3a + 2}$$

(b) $f(2x - 1)$ is the expression we get when we substitute $2x - 1$ for x in $f(x)$. Since

$$f(x) = 3x - 1,$$
$$f(2x - 1) = 3(2x - 1) - 1 \qquad \textit{Then simplify.}$$
$$= 6x - 3 - 1$$
$$= 6x - 4$$

$$\boxed{\text{Thus, } f(2x - 1) = 6x - 4}$$ ∎

EXAMPLE 3

Given $g(x) = x^2 - 3x + 1$, find each of the following:

(a) $g(a + 2)$ (b) $g(2x)$

Solution

(a) $g(a + 2)$ is the expression we get when we substitute $a + 2$ for x in $g(x)$. Since

$$g(x) = x^2 - 3x + 1,$$
$$g(a + 2) = (a + 2)^2 - 3(a + 2) + 1 \qquad \textit{Then simplify.}$$
$$= a^2 + 4a + 4 - 3a - 6 + 1$$
$$= a^2 + a - 1$$

$$\boxed{\text{Thus, } g(a + 2) = a^2 + a - 1}$$

(b) $g(2x)$ is the expression we get when we substitute $2x$ for x in $g(x)$. Since

$$g(x) = x^2 - 3x + 1,$$
$$g(2x) = (2x)^2 - 3(2x) + 1 \qquad \textit{Simplify.}$$
$$= 4x^2 - 6x + 1$$

$$\boxed{\text{Thus, } g(2x) = 4x^2 - 6x + 1}$$ ∎

EXAMPLE 4

Given $f(x) = 4x - 1$, find each of the following:

(a) $f(x) + 2$ (b) $f(x + 2)$ (c) $f(x) + f(2)$ (d) $f(x + 2) - f(x)$

Solution

(a) $f(x) + 2$ *Means add 2 to f(x); since f(x) = 4x − 1,*

$$= 4x - 1 + 2$$
$$= 4x + 1$$

Hence, $\boxed{f(x) + 2 = 4x + 1}$

(b) $f(x + 2)$ means to substitute $x + 2$ for x in $f(x)$. Since $f(x) = 4x - 1$, we have

$$f(x) = 4x - 1$$
$$f(x + 2) = 4(x + 2) - 1$$
$$= 4x + 8 - 1$$
$$= 4x + 7$$

Hence, $\boxed{f(x + 2) = 4x + 7}$

(c) $f(x) + f(2) = 4x - 1 + 4(2) - 1$ *Since f(x) = 4x − 1 and*
$$\qquad\qquad\qquad\qquad\qquad f(2) = 4(2) - 1$$
$$= 4x - 1 + 8 - 1$$
$$= 4x + 6$$

Hence, $\boxed{f(x) + f(2) = 4x + 6}$

Note the differences between part **(c)** and parts **(a)** and **(b)**. You cannot simply add $f(2)$ to $f(x)$ to get $f(x + 2)$. In general, $f(a + b) \neq f(a) + f(b)$.

(d) First find $f(x + 2)$; we found in part **(b)** that $f(x + 2) = 4x + 7$. Then

$$f(x + 2) - f(x) = 4x + 7 - (4x - 1)$$
$$= 4x + 7 - 4x + 1$$
$$= 8$$

Hence, $\boxed{f(x + 2) - f(x) = 8}$ ■

EXAMPLE 5

Given $g(x) = x^2 - 2x + 1$, find $\dfrac{g(x + 3) - g(x)}{3}$.

Solution

First find $g(x + 3)$:

$$g(x + 3) = (x + 3)^2 - 2(x + 3) + 1$$
$$= x^2 + 6x + 9 - 2x - 6 + 1$$
$$= x^2 + 4x + 4$$

Hence,

$$\frac{g(x + 3) - g(x)}{3} = \frac{x^2 + 4x + 4 - (x^2 - 2x + 1)}{3}$$

$$= \frac{x^2 + 4x + 4 - x^2 + 2x - 1}{3}$$

$$= \frac{6x + 3}{3} = \frac{\cancel{3}(2x + 1)}{\cancel{3}}$$

$$= \boxed{2x + 1}$$ ∎

EXAMPLE 6

If $f(x) = 2x^2 + x - 1$ and $g(x) = x - 2$, find each of the following:
(a) $f[g(3)]$ **(b)** $g[f(3)]$ **(c)** $f[g(x)]$ **(d)** $g[f(x)]$

Solution

(a) $f[g(3)]$ means to evaluate $f(x)$ when $x = g(3)$. Since $f(x) = 2x^2 + x - 1$,

$f[g(3)] = 2[g(3)]^2 + g(3) - 1$ *Let's find $g(3)$ by substituting 3 for x in $g(x)$:*
$\quad\quad\quad = 2(1)^2 + 1 - 1$ *$g(3) = 3 - 2 = 1$. Now substitute $g(3) = 1$.*
$\quad\quad\quad = 2 + 1 - 1$
$\quad\quad\quad = 2$

$$\boxed{\text{Thus, } f[g(3)] = 2}$$

(b) $g[f(3)]$ means to evaluate $g(x)$ when $x = f(3)$. Since $g(x) = x - 2$,

$g[f(3)] = f(3) - 2$ *We find $f(3)$: $f(3) = 2(3)^2 + 3 - 1 = 20$. Now*
$\quad\quad\quad = 20 - 2 = 18$ *substitute $f(3) = 20$.*

$$\boxed{\text{Thus, } g[f(3)] = 18}$$

(c) $f[g(x)]$ is the expression we get when we substitute $g(x)$ for x in $f(x)$. Since $f(x) = 2x^2 + x - 1$,

$f[g(x)] = 2[g(x)]^2 + [g(x)] - 1$ *And since $g(x) = x - 2$, we get*
$\quad\quad\quad = 2(x - 2)^2 + (x - 2) - 1$ *Simplify.*
$\quad\quad\quad = 2(x^2 - 4x + 4) + (x - 2) - 1$
$\quad\quad\quad = 2x^2 - 8x + 8 + x - 2 - 1$
$\quad\quad\quad = 2x^2 - 7x + 5$

$$\boxed{\text{Thus, } f[g(x)] = 2x^2 - 7x + 5}$$

(d) $g[f(x)]$ is the expression we get when we substitute $f(x)$ for x in $g(x)$. Since $g(x) = x - 2$,

$$g[f(x)] = f(x) - 2 \qquad \textit{Since } f(x) = 2x^2 + x - 1, \textit{ we get}$$
$$= (2x^2 + x - 1) - 2$$
$$= 2x^2 + x - 3$$

Thus, $g[f(x)] = 2x^2 + x - 3$ ■

Example 6 illustrates an operation on functions called ***composition of functions***. While $f[g(x)]$ is the composition of f with g, $g[f(x)]$ is the composition of g with f. Note that, in general, $g[f(x)] \neq f[g(x)]$.

The Algebra of Functions

We can define operations with functions as long as the domains have elements in common. If $f(x)$ and $g(x)$ are functions with the same domain, then we have the definitions listed in the box.

$(f + g)(x) = f(x) + g(x)$	Sum of functions
$(f - g)(x) = f(x) - g(x)$	Difference of functions
$(fg)(x) = f(x) \cdot g(x)$	Product of functions
$\left(\dfrac{f}{g}\right)(x) = \dfrac{f(x)}{g(x)}; \quad g(x) \neq 0$	Quotient of functions

Thus, if $f(x)$ and $g(x)$ are functions, their sum, product, difference, and quotient are functions, provided $f(x)$ and $g(x)$ exist. In the case of the quotient, we must eliminate any value of x which makes $g(x) = 0$ and therefore the quotient $\dfrac{f(x)}{g(x)}$ undefined.

EXAMPLE 7

Given $f(x) = x^2 - 16$ and $g(x) = x - 3$, express each of the following as a function of x:

(a) $(f + g)(x)$ **(b)** $\left(\dfrac{f}{g}\right)(x)$

Solution

(a) $(f + g)(x) = f(x) + g(x) \qquad \textit{By definition of } (f + g)$
$$= (x^2 - 16) + (x - 3)$$
$$= x^2 + x - 19$$

Thus, $(f + g)(x)$ is the function $x^2 + x - 19$

(b) $\left(\dfrac{f}{g}\right)(x) = \dfrac{f(x)}{g(x)}$ *By definition of* $\left(\dfrac{f}{g}\right)(x)$

$$= \dfrac{x^2 - 16}{x - 3}; \quad x \neq 3$$

Thus, $\left(\dfrac{f}{g}\right)(x)$ is the function $\dfrac{x^2 - 16}{x - 3}$ provided $x \neq 3$ (Why?)

Exercises 11.2

In Exercises 1–12, given $f(x) = 2x - 3$, $g(x) = 3x^2 - x + 1$, *and* $h(x) = \sqrt{x + 5}$, *find:*

1. $f(0)$ **2.** $g(0)$ **3.** $g(2)$ **4.** $f(2)$

5. $g(-2)$ **6.** $f(-2)$ **7.** $h(3)$ **8.** $h(4)$

9. $h(-3)$ **10.** $h(-4)$ **11.** $h(a)$ **12.** $g(a)$

In Exercises 13–22, given $f(x) = x^2 + 2$ *and* $g(x) = 2x - 3$, *find (and simplify):*

13. $f(x + 1)$ **14.** $g(x + 1)$ **15.** $g(x - 2)$ **16.** $f(x - 2)$

17. $g(x + 2)$ **18.** $f(x + 2)$ **19.** $f(2x)$ **20.** $g(3x)$

21. $g(3x + 2)$ **22.** $f(2x + 3)$

In Exercises 23–32, given $f(x) = x^2 + 2x - 3$, *find (and simplify):*

23. $f(x + 1)$ **24.** $f(x + 3)$ **25.** $f(x) + 1$ **26.** $f(x) + 3$

27. $f(x) + f(1)$ **28.** $f(x) + f(3)$ **29.** $f(3x)$ **30.** $2f(x)$

31. $3f(x)$ **32.** $f(2x)$

In Exercises 33–46, given $g(x) = \sqrt{3x + 2}$, *find (and simplify):*

33. $g(3x + 1)$ **34.** $2g(x) + 3$ **35.** $3g(x) + 1$ **36.** $g(2x + 3)$

37. $g(x + h)$ **38.** $g(x - h)$ **39.** $g(x) + g(h)$ **40.** $g(x) - g(h)$

41. $g(x) + h$ **42.** $g(x) - h$ **43.** $g(ax)$ **44.** $g(kx)$

45. $ag(x)$ **46.** $kg(x)$

In Exercises 47–50, given $f(x) = x^2 + 3x - 4$ *and* $h \neq 0$, *find (and simplify):*

47. $\dfrac{f(x + 2) - f(x)}{2}$ **48.** $\dfrac{f(x + 3) - f(x)}{3}$ **49.** $\dfrac{f(x + h) - f(x)}{h}$ **50.** $\dfrac{f(x - h) - f(x)}{h}$

In Exercises 51–64, given $f(x) = x^2 - 4$, $g(x) = \sqrt{x + 1}$, *and* $h(x) = \dfrac{1}{x}$, *find (and simplify):*

51. $f[g(3)]$ **52.** $f[g(0)]$ **53.** $g[f(3)]$ **54.** $g[f(2)]$

55. $f[g(x)]$ **56.** $f[g(a)]$ **57.** $g[f(x)]$ **58.** $g[f(a)]$

59. $g[h(3)]$ **60.** $h[g(3)]$ **61.** $f\left[g\left(\dfrac{1}{2}\right)\right]$ **62.** $h\left[f\left(\dfrac{1}{2}\right)\right]$

63. $g[h(x)]$ **64.** $h[g(x)]$

In Exercises 65–74, given $f(x) = x^2 - 2x - 3$ and $g(x) = x - 1$, find:

65. $(f + g)(0)$ **66.** $(f - g)(0)$ **67.** $(fg)(0)$ **68.** $(fg)(2)$

69. $\left(\dfrac{f}{g}\right)(3)$ **70.** $\left(\dfrac{g}{f}\right)(1)$ **71.** $\left(\dfrac{g}{f}\right)(3)$ **72.** $\left(\dfrac{f}{g}\right)(1)$

73. $\left(\dfrac{f}{g}\right)(x)$ **74.** $\left(\dfrac{g}{f}\right)(x)$

? QUESTION FOR THOUGHT

75. Given $f(x) = 5x + 1$, discuss what is **wrong** (if anything) with the following:

(a) $f(x + 3) \stackrel{?}{=} 5x + 1 + 3 \stackrel{?}{=} 5x + 4$

(b) $f(3x) \stackrel{?}{=} 3(5x + 1) \stackrel{?}{=} 15x + 3$

(c) $f(x + 3) \stackrel{?}{=} (5x + 1)(x + 3) \stackrel{?}{=} 5x^2 + 16x + 3$

⟳ MINI-REVIEW

76. *Express the following in simplest radical form:*

$$\frac{3}{\sqrt{6} - \sqrt{3}} + \frac{9}{\sqrt{3}}$$

77. Find the equation of the line passing through the points $(-2, 3)$ and $(6, 7)$.

78. Sketch the graph of $y = x^2 - 4x - 5$.

79. *Solve the following system of equations:*

$$\begin{cases} 2x + 3y - z = 4 \\ 3x - y + z = -2 \\ x + y + 2z = -1 \end{cases}$$

80. A bank teller receives a deposit of 50 bills totalling $430. If the money was all in $5 and $20 bills, how many of each are there?

11.3

Types of Functions

When we graph a function the horizontal axis represents values of the independent variable and the vertical axis represents the values of the dependent variable. Thus, $f(x) = x - 2$ and $C(p) = p^2 + 5$, where $p > 0$, can be graphed as shown in Figure 11.6.

Much of our previous work with graphs in this text has involved functions, although we have used y rather than $f(x)$ as the dependent variable. In this section we will present a few of the basic types of functions.

Linear Functions

DEFINITION

A *linear function* is a function of the form

$$f(x) = mx + b$$

Figure 11.6

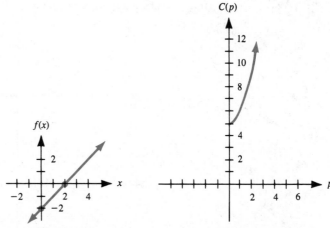

a. x is the independent variable and is therefore represented on the horizontal axis; $f(x)$ is the dependent variable and is therefore represented on the vertical axis.

b. p is the independent variable and is therefore represented on the horizontal axis; $C(p)$ is the dependent variable and is represented on the vertical axis.

We have worked with linear functions of the form $y = mx + b$ in Chapter 8. As the name suggests, the graph of a linear function is a straight line. For the function $f(x) = mx + b$, m is the slope and b is the y-intercept. Thus, all nonvertical lines can be represented by linear functions.

EXAMPLE 1

Graph the linear functions:

(a) $f(x) = 3x - 4$ **(b)** $f(x) = 5$

Solution

(a) It is helpful to write $y = f(x) = 3x - 4$ to remind us that $f(x)$ is just another name for the dependent variable, y. The function is graphed in Figure 11.7.

Figure 11.7

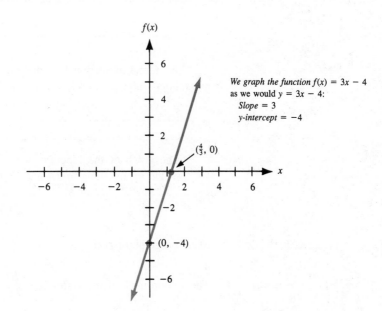

We graph the function $f(x) = 3x - 4$ as we would $y = 3x - 4$:
Slope = 3
y-intercept = −4

(b)   We are graphing $y = f(x) = 5$. The graph of the function is shown in Figure 11.8.

Figure 11.8

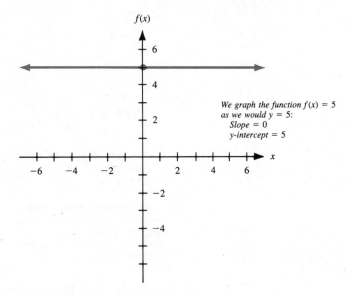

We graph the function $f(x) = 5$
as we would $y = 5$:
    Slope = 0
    y-intercept = 5

The function $f(x)$ in part **(b)** of Example 1 is a special case of a linear function. In general, the function $f(x) = k$, where k is a constant, is called a **constant function**.

Quadratic Functions

DEFINITION

A **quadratic function** is a function of the form

$$f(x) = ax^2 + bx + c \quad \text{where } a \neq 0$$

We found in Section 10.2 that the equation $y = ax^2 + bx + c$ $(a \neq 0)$ describes a parabola. Hence,

The graph of the function $f(x) = ax^2 + bx + c$ $(a \neq 0)$ is a parabola which opens either up (if $a > 0$) or down (if $a < 0$).

EXAMPLE 2   Sketch a graph of $f(x) = x^2 - 4x$.

Solution   The x-coordinate of the vertex is $-\dfrac{b}{2a} = -\dfrac{-4}{2(1)} = 2.$

Next we need to find the second coordinate when $x = 2$. Instead of saying "the y-coordinate when $x = 2$" or the $f(x)$ coordinate when $x = 2$, in the language of functions we would write $f(2)$—that is, $f(2)$ is the value of $f(x)$ when $x = 2$. Hence,

$$f(2) = 2^2 - 4(2) = 4 - 8 = -4 \qquad \text{"}f(2) = -4\text{" means } f(x) = -4 \text{ when } x = 2.$$

The vertex is at $(2, -4)$.

The y-intercept is $f(0) = 0^2 - 4(0) = 0$.

The x-intercept is the value of x when $f(x) = 0$:

$$0 = f(x) = x^2 - 4x$$
$$0 = x(x - 4) \rightarrow x = 0 \quad \text{and} \quad x = 4$$

The x-intercepts are 0 and 4.

The graph of the parabola is shown in Figure 11.9.

Figure 11.9
Graph of $f(x) = x^2 - 4x$

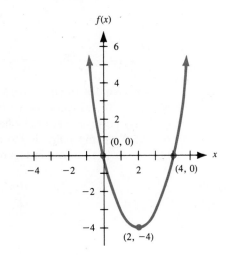

Polynomial Functions

The linear and quadratic functions are special cases of the polynomial function:

DEFINITION

Polynomial functions are functions of the form

$$f(x) = a_n x^n + a_{n-1} x^{n-1} + \cdots + a_2 x^2 + a_1 x + a_0 \qquad (a_n \neq 0)$$

where n is a nonnegative integer.

For example, the polynomial function $f(x) = x^3$ is graphed in Figure 11.10.

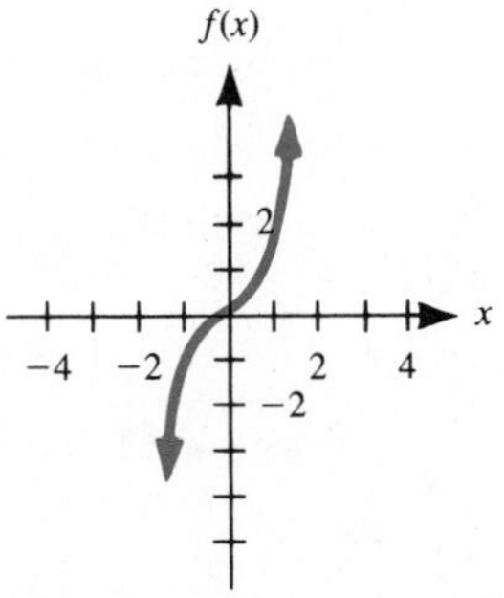

$f(x)$

$f(x) = x^3$

Figure 11.10

The Square Root Function

DEFINITION

The *square root function* is a function of the form

$$f(x) = \sqrt{x}$$

For example, if we plot the table of values for $\sqrt{x}$, we find:

x	$f(x) = \sqrt{x}$
-2	$\sqrt{-2}$ (not a real number)
-1	$\sqrt{-1}$ (not a real number)
0	0
1	1
2	$\sqrt{2}$
4	2
9	3

Notice that when x is negative, $f(x)$ is not a real number. Therefore, the domain is $\{x \mid x \geq 0\}$, which is the set of nonnegative real numbers. Notice as well that $f(x)$ is always positive. Recall that the square root is the principal or positive root. Hence, $f(x) \geq 0$, or the range is the set of all nonnegative real numbers. The graph of $f(x) = \sqrt{x}$ is as shown in Figure 11.11.

Figure 11.11
Graph of $f(x) = \sqrt{x}$

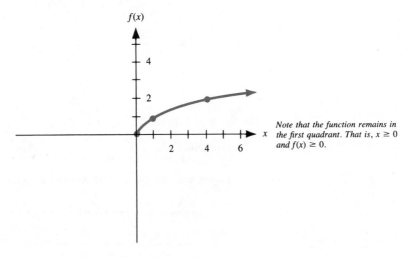

Note that the function remains in the first quadrant. That is, $x \geq 0$ and $f(x) \geq 0$.

EXAMPLE 3

Sketch a graph of the following functions:

(a) $f(x) = \sqrt{x - 3}$ **(b)** $f(x) = \sqrt{x} - 3$

Solution

(a) The domain of $f(x)$ is $\{x \mid x \geq 3\}$ (so that $\sqrt{x - 3}$ will be a real number). We use the values of x and $f(x)$ shown in the accompanying table to sketch a graph of $f(x) = \sqrt{x - 3}$ (see Figure 11.12).

Figure 11.12
Graph of $f(x) = \sqrt{x-3}$

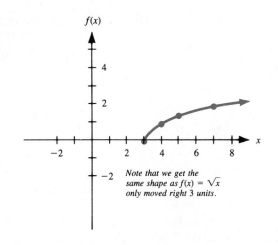

x	$f(x) = \sqrt{x-3}$
3	0
4	1
5	$\sqrt{2}$
7	2
12	3

Note that we get the same shape as $f(x) = \sqrt{x}$ only moved right 3 units.

From the figure we see that the domain must be $\{x \mid x \geq 3\}$; the range is $\{y \mid y \geq 0\}$.

(b) From the accompanying table of values, we see that the graph of $f(x)$ can take on negative values. We sketch a graph of $f(x)$ as shown in Figure 11.13.

x	$f(x) = \sqrt{x} - 3$
0	-3
1	-2
2	$\sqrt{2} - 3 \approx -1.59$
4	-1
9	0
16	1

Figure 11.13
Graph of $f(x) = \sqrt{x} - 3$

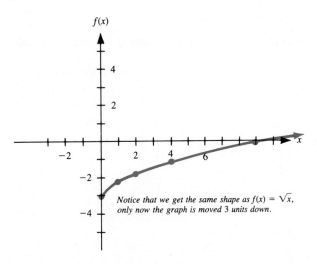

Notice that we get the same shape as $f(x) = \sqrt{x}$, only now the graph is moved 3 units down.

The domain is $\{x \mid x \geq 0\}$ and the range is $\{y \mid y \geq -3\}$. ∎

The Absolute Value Function

DEFINITION

The *absolute value function* is a function of the form

$$f(x) = |x|$$

We defined the absolute value of x algebraically as:

$$|x| = \begin{cases} x & \text{if } x \geq 0 \\ -x & \text{if } x < 0 \end{cases}$$

If we make a table of values, we obtain the following:

Figure 11.14
Graph of $f(x) = |x|$

x	$f(x) = \|x\|$
-2	2
-1	1
0	0
1	1
2	2

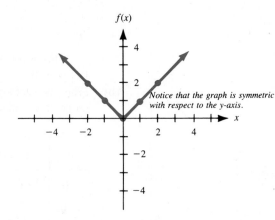

Notice that the graph is symmetric with respect to the y-axis.

We graph the function as shown in Figure 11.14. The domain is the set of all real numbers. Since $f(x)$ is always nonnegative, the range is $\{y \mid y \geq 0\}$.

EXAMPLE 4

Sketch a graph of each of the following:

(a) $f(x) = |x + 2|$ **(b)** $f(x) = |x| + 2$

Solution

(a) We construct the following table:

Figure 11.15
Graph of $f(x) = |x + 2|$

x	$f(x) = \|x + 2\|$
-6	4
-4	2
-2	0
0	2
2	4

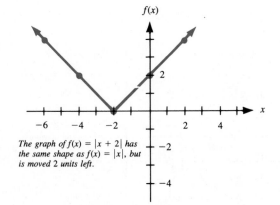

The graph of $f(x) = |x + 2|$ has the same shape as $f(x) = |x|$, but is moved 2 units left.

The graph is sketched in Figure 11.15. From the figure we see that the domain is the set of all real numbers, and the range is $\{y \mid y \geq 0\}$.

(b) We make a table:

Figure 11.16
Graph of $f(x) = |x| + 2$

x	$f(x) = |x| + 2$
-2	4
-1	3
0	2
1	3
2	4

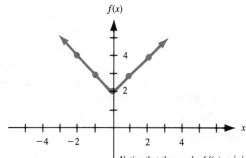

Notice that the graph of $f(x) = |x| + 2$ has the same shape as $f(x) = |x|$, but is moved 2 units up.

We sketch the graph in Figure 11.16. The domain is still the set of all real numbers, but the range is now $\{y \mid y \geq 2\}$. ∎

Exercises 11.3

In Exercises 1–16, determine the type of function.

1. $f(x) = 5x + 2$ **2.** $f(x) = 3x^2 + 2$ **3.** $f(x) = x^2 - 4x + 1$ **4.** $f(x) = 5x - 4$

5. $f(x) = 3x^3 - 2x + 4$ **6.** $f(x) = \sqrt{3x + 5}$ **7.** $f(x) = 2x^2 + 1$ **8.** $f(x) = 4x^4 - 2x^3 + 3x$

9. $f(x) = \sqrt{2x - 5}$ **10.** $f(x) = 3x^2 - 2$ **11.** $f(x) = |x| - 1$ **12.** $f(x) = |x - 2|$

13. $f(x) = 2x - 1$ **14.** $f(x) = 3x^2 - 2$ **15.** $f(x) = 2x^2 - 1$ **16.** $f(x) = \sqrt{2x + 5}$

In Exercises 17–24, evaluate $f(-2)$, $f(-1)$, $f(0)$, $f(1)$, and $f(2)$ for the functions.

17. $f(x) = |4x - 1|$ **18.** $f(x) = |3x - 2|$ **19.** $f(x) = |4x| - 1$ **20.** $f(x) = |3x| - 2$

21. $f(x) = |4x| + 1$ **22.** $f(x) = |3x| + 2$ **23.** $f(x) = |4x + 1|$ **24.** $f(x) = |3x + 2|$

In Exercises 25–58, determine the type of function and graph it.

25. $f(x) = 4 - 3x$ **26.** $f(x) = 5 - 4x$ **27.** $f(x) = x^2 - 9$ **28.** $f(x) = 4 - x^2$

29. $f(x) = x^3$ **30.** $f(x) = x^4$ **31.** $f(x) = 2x^3$ **32.** $f(x) = 2x^4$

33. $f(x) = \sqrt{x - 5}$ **34.** $f(x) = \sqrt{x + 4}$ **35.** $f(x) = \sqrt{x + 5}$ **36.** $f(x) = \sqrt{x - 4}$

37. $f(x) = \sqrt{x} + 5$ **38.** $f(x) = \sqrt{x} - 4$ **39.** $f(x) = \sqrt{x} - 5$ **40.** $f(x) = \sqrt{x} + 4$

41. $f(x) = x^2 - 4x + 1$ **42.** $f(x) = x^2 + 4x - 1$ **43.** $f(x) = \sqrt{3x - 2}$ **44.** $f(x) = \sqrt{2x - 5}$

45. $f(x) = 8 - 2x - x^2$ **46.** $f(x) = 4 + 3x - x^2$ **47.** $f(x) = \sqrt{8 - 2x}$ **48.** $f(x) = \sqrt{9 - 3x}$

49. $f(x) = \sqrt{6 - 4x}$ **50.** $f(x) = \sqrt{8 - 6x}$ **51.** $f(x) = |x + 5|$ **52.** $f(x) = |x + 4|$

53. $f(x) = |x| + 5$ **54.** $f(x) = |x| + 4$ **55.** $f(x) = x + 5$ **56.** $f(x) = x + 4$

57. $f(x) = |5 - x|$ **58.** $f(x) = |4 - x|$

? QUESTIONS FOR THOUGHT

59. **(a)** Sketch a graph of the ellipse $4x^2 + 9y^2 = 36$.

(b) Sketch a graph of the following two functions on the same set of coordinate axes:

$$y = \sqrt{4 - \frac{4x^2}{9}} \qquad y = -\sqrt{4 - \frac{4x^2}{9}}$$

(c) Compare the graph of part **(a)** with the graphs of part **(b)**.

(d) Demonstrate algebraically that the equation in part **(a)** can be solved explicitly for y to obtain the two equations in part **(b)**.

60. **(a)** Graph the circle $x^2 + y^2 = 25$.

(b) Solve $x^2 + y^2 = 25$ explicitly for y.

(c) Graph the results of part **(b)** and compare the graphs with that obtained in part **(a)**.

61. **(a)** Graph the hyperbola $x^2 - 4y^2 = 16$.

(b) Solve $x^2 - 4y^2 = 16$ explicitly for y.

(c) Graph the results of part **(b)** and compare the graphs with the one obtained in part **(a)**.

 MINI-REVIEW

62. Perform the operations and express your answer in simplest form using positive exponents only:

$$\left(\frac{16x^{1/2} y^{1/3}}{x^{-1/2}} \right)^{1/4}$$

63. *Solve the following for x:*

$$\frac{3}{2x - 4} = \frac{x - 2}{x + 1}$$

64. Find the midpoint between $(-2, 5)$ and $(3, 4)$.

65. Find the equation of the circle with center $(3, -4)$ and radius 5.

66. *Compute:*

$$\begin{vmatrix} 3 & -2 & 0 \\ 1 & 5 & 1 \\ 5 & 0 & 4 \end{vmatrix}$$

11.4

Inverse Relations and Functions

We defined a relation, R, as a correspondence between two sets, called the domain and range, such that to each element of the domain there is assigned one or more elements of the range.

If we interchange the domain and range of R and reverse the assignment, we have what is called the *inverse of R*.

For example, let's define the relation R as the following correspondence between sets $A = \{1, 2, 3\}$ and $B = \{5, 8\}$:

Domain   Range

1 $\longrightarrow$ 5

2

3 $\longrightarrow$ 8

which is the set $\{(1, 5), (2, 5), (3, 8)\}$, where A is the domain and B is the range.

The inverse of R, designated R^{-1}, is the following correspondence between the same sets, B and A:

Domain   Range

5 $\longrightarrow$ 1

2

8 $\longrightarrow$ 3

which is the set $\{(5, 1), (5, 2), (8, 3)\}$, where the domain is now B and the range is now A.

Using ordered pair notation, we define the following:

DEFINITION

If R is a relation, then the *inverse of R*, designated R^{-1}, is the relation consisting of all ordered pairs (y, x) such that (x, y) belongs to R.

Thus, the inverse of a relation interchanges the dependent variable with the independent variable.

A word about notation: Although the notation we use is the same, R^{-1} is *not* the reciprocal of R. With functions and relations, the exponent "-1" symbolizes the inverse of the relation or function. The use of the notation R^{-1} to mean the inverse of the relation R rather than the reciprocal of R is usually clear in the context of the given example or discussion.

EXAMPLE 1

Find the inverse of each of the following relations:

(a) $R = \{(1, 3), (5, 2), (2, 4)\}$ **(b)** $S = \{(3, 5), (6, 8), (6, 2), (4, 5)\}$

Solution

(a) R^{-1} is found by interchanging the independent variable with the dependent variable. Thus,

$$R^{-1} = \{(3, 1), (2, 5), (4, 2)\}$$

(b) S^{-1} is found by interchanging the independent variable with the dependent variable. Thus,

$$S^{-1} = \{(5, 3), (8, 6), (2, 6), (5, 4)\}$$

■

EXAMPLE 2

Find the inverse of the relation defined by $y = 2x + 3$.

Solution

The relation $y = 2x + 3$ is defined by all ordered pairs (x, y) such that y is 3 more than twice x. For example, the ordered pairs $(1, 5)$, $(0, 3)$, and $(-1, 1)$ are members of this relation.

To find the inverse we again interchange the independent variable with the dependent variable. We can do this with the equation.

If R is represented by $y = 2x + 3$,
then R^{-1} is represented by $x = 2y + 3$.

Since we prefer to have our function expressed in a form which is solved explicitly for the dependent variable, we now solve for y:

$$x = 2y + 3$$
$$x - 3 = 2y$$
$$\frac{x - 3}{2} = y$$

Hence, R^{-1} is represented by $\boxed{y = \dfrac{x - 3}{2}}$ ■

In Example 2, note that $(5, 1)$, $(3, 0)$, and $(1, -1)$ are members of R^{-1}, as we would expect. Let's graph this last example, $y = 2x + 3$, and graph its inverse. See Figure 11.17.

Figure 11.17

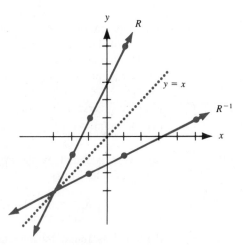

We will find that by interchanging the x and y variables of the relation R, we arrive at a picture which is the mirror image of the graph of R, with the mirror placed on the line $y = x$; this is R^{-1}. We call such a graph the **reflection** of R about the line $y = x$. We offer another example in Figure 11.18.

Figure 11.18

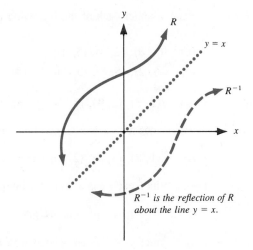

R^{-1} *is the reflection of R about the line y = x.*

Our next task, naturally, is to examine the inverses of functions. Let's examine the function

$$\{(1,\ 3),\ (2,\ 3),\ (4,\ 5)\}$$

This relation is a function since each x-coordinate has only one y-coordinate. If we define its inverse in the same way we defined the inverse of a relation, we would get as an inverse

$$\{(3,\ 1),\ (3,\ 2),\ (5,\ 4)\}$$

Note, however, that this relation is no longer a function because the x-component 3 is assigned two y-component values, 1 and 2.

We need an additional restriction on functions in order to guarantee that interchanging the roles of x and y will produce another *function*.

DEFINITION

A *one-to-one function* is a function in which at most one x value is associated with a y value.

For a relation to be a function, each x must be assigned a unique y, and for a function to be one-to-one, each y must be assigned to a unique x. This means that a one-to-one function is a one-to-one correspondence between elements of the domain and elements of the range. The relationship among relations, functions, and one-to-one functions is depicted in Figure 11.19.

Figure 11.19

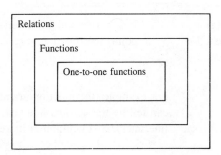

EXAMPLE 3

State whether each of the following relations is a function, a one-to-one function, or neither.

(a) $\{(2, 3), (2, 4), (5, 1)\}$ (b) $\{(1, 3), (5, 7), (2, 7)\}$

(c) $\{(6, 2), (5, 8), (3, 1)\}$ (d) $x + 2y = 4$ (e) $y = x^2$

Solution

(a) $\{(2, 3), (2, 4), (5, 1)\}$ is $\boxed{\text{not a function}}$

since $x = 2$ is assigned two values of y: $y = 3$ and $y = 4$.

(b) $\{(1, 3), (5, 7), (2, 7)\}$ $\boxed{\text{is a function}}$

since each x is assigned a unique y.

It is $\boxed{\text{not one-to-one}}$

since $y = 7$ is associated with two values of x: $x = 5$ and $x = 2$.

(c) $\{(6, 2), (5, 8), (3, 1)\}$ $\boxed{\text{is a one-to-one function}}$

Each x component is assigned a unique y-component *and* each y-component is associated with only one x-component.

(d) $x + 2y = 4$ $\boxed{\text{is a one-to-one function}}$

If you substitute any value for x you will get one value for y, and if you substitute any value for y you will get one value for x.

(e) $y = x^2$ $\boxed{\text{is a (quadratic) function}}$

as we discussed in Section 11.3. If we let $x = 3$, then $y = 3^2 = 9$; if we let $x = -3$, then $y = (-3)^2 = 9$. Hence, $(3, 9)$ *and* $(-3, 9)$ are members of this function. Since we have one y value, $y = 9$, assigned to two x values, $x = -3$ and $x = 3$,

the function is $\boxed{\textit{not} \text{ a one-to-one function}}$ ■

Just as we used a vertical line to tell if a relation is a function, we can use a horizontal line to tell whether a function is one-to-one.

The Horizontal Line Test	If a horizontal line intersects the graph of a function at more than one point, the function is *not* a one-to-one function.

For example, the function graphed in Figure 11.20, $y = x^2 + 2$, is a function, as we can tell by the vertical line test, but it is not a one-to-one function since a horizontal line intersects the graph at more than one point.

Figure 11.20
Graph of $y = x^2 + 2$

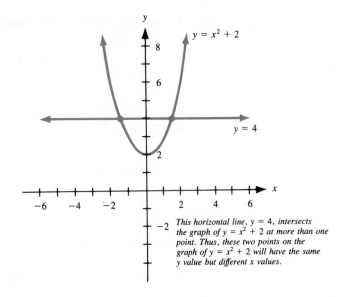

This horizontal line, $y = 4$, intersects
the graph of $y = x^2 + 2$ at more than one
point. Thus, these two points on the
graph of $y = x^2 + 2$ will have the same
y value but different x values.

Now we are in the position to determine whether the inverse of a function is still a function.

THEOREM

If $f(x)$ is a one-to-one function, then the inverse of $f(x)$ is also a function.

If a function $f(x)$ is one-to-one, we can define its inverse, designated $f^{-1}(x)$, by interchanging the roles of the dependent and independent variables. However, keep in mind that we generally write our function solved explicitly for the dependent variable; so after we interchange the roles of the variables, we solve the equation explicitly for the dependent variable.

EXAMPLE 4

Find the inverse of each of the following functions:

(a) $\{(3, 4), (5, 8), (2, 9)\}$ **(b)** $f(x) = 5x - 4$

(c)

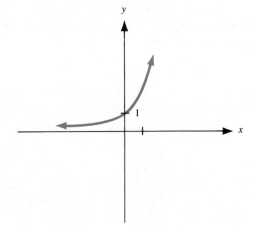

Solution

(a) The inverse is found by interchanging the x-components with the y-components. Therefore, the inverse is

$$\{(4, 3), (8, 5), (9, 2)\}$$

(b) The inverse is found by interchanging the dependent variable with the independent variable. For convenience, we let $f(x) = y$. Then the function $f(x)$ is defined by:

$$y = 5x - 4 \qquad \textit{Interchange x with y.}$$
$$x = 5y - 4 \qquad \textit{Then solve explicitly for y.}$$
$$x + 4 = 5y$$
$$\frac{x + 4}{5} = y$$

The inverse $f^{-1}(x)$ is defined by $y = \dfrac{x + 4}{5}$ or simply $\boxed{f^{-1}(x) = \dfrac{x + 4}{5}}$

(c)

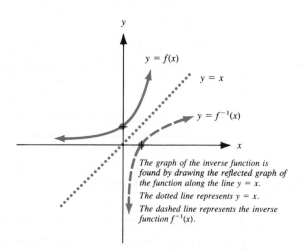

The graph of the inverse function is found by drawing the reflected graph of the function along the line $y = x$.
The dotted line represents $y = x$.
The dashed line represents the inverse function $f^{-1}(x)$.

EXAMPLE 5

Find the inverse of each of the following. Indicate the domain of the function and the domain of its inverse.

(a) $g(x) = \dfrac{x + 1}{x - 2}$ **(b)** $h(x) = x^3 - 3$

Solution

(a) First note that $\boxed{\text{the domain of } g(x) \text{ is } \{x \mid x \neq 2\}}$
Let $g(x) = y$. Then

$$y = \frac{x + 1}{x - 2} \qquad \textit{Then interchange x with y.}$$

$$x = \frac{y + 1}{y - 2} \qquad \textit{Solve explicitly for y: multiply both sides of the equation by } y - 2.$$

$$x(y - 2) = y + 1 \qquad \textit{Use the distributive property.}$$

$$xy - 2x = y + 1$$ *Collect all y terms on one side and non-y terms on the other side.*

$$xy - y = 2x + 1$$ *Factor y from the left-hand side.*

$$y(x - 1) = 2x + 1$$ *Divide both sides of the equation by x − 1.*

$$y = \frac{2x + 1}{x - 1}$$

The inverse is defined by $y = \dfrac{2x + 1}{x - 1}$ or $\boxed{g^{-1}(x) = \dfrac{2x + 1}{x - 1}}$

$$\boxed{\text{The domain of } g^{-1}(x) \text{ is } \{x \mid x \neq 1\}}$$

(b)   First note that $\boxed{\text{the domain of } h(x) \text{ is the set of all real numbers}}$

Let $h(x) = y$; then

$$y = x^3 - 3$$ *Defines the function h(x)*

The inverse is found by interchanging the x and y variables:

$$x = y^3 - 3$$ *Then solve for y. Isolate y^3.*

$$x + 3 = y^3$$ *Take cube roots.*

$$\sqrt[3]{x + 3} = y$$

Therefore, the inverse function is defined by

$$y = \sqrt[3]{x + 3} \text{ or, simply, } \boxed{h^{-1}(x) = \sqrt[3]{x + 3}}$$

$$\boxed{\text{The domain of } h^{-1}(x) \text{ is the set of all real numbers}}$$ ■

Exercises 11.4

In Exercises 1–14, find the inverses of the given relations.

1. $\{(1, 3), (2, 5)\}$

2. $\{(6, 3), (2, 4)\}$

3. $\{(3, 2), (3, -1), (-1, 3)\}$

4. $\{(5, -2), (-2, 5), (6, 5)\}$

5. $\{(3, 5), (2, -5), (2, 5), (2, 6)\}$

6. $\{(2, 5), (3, 2), (4, 2), (4, 3)\}$

7. $y = 3x - 4$

8. $y = 2x + 5$

9. $y = x^2$

10. $y = x^3$

11. $y = 2x^3$

12. $y = 3x^2$

13. $y = x^2 + 4$

14. $y = x^2 - 5$

In Exercises 15–32, determine if the relations are functions, one-to-one functions, or neither.

15. $\{(3, 1), (4, 1)\}$

16. $\{(5, 2), (6, 2)\}$

17. $\{(3, 2), (2, 3)\}$

18. $\{(5, -6), (-6, 5)\}$

19. $\{(3, 7), (3, 8)\}$

20. $\{(4, 9), (4, 1)\}$

21. $\{(-2, 3), (3, -2), (5, -2)\}$

22. $\{(6, 4), (6, -3), (5, 2)\}$

23. $\{(6, -1), (-1, 6), (3, 4)\}$

24. $\{(5, -2), (3, -1), (-2, 5)\}$ **25.** $y = 3x - 4$ **26.** $y = 2x + 5$

27. $y = 2x^2$ **28.** $y = x^2 - 4$ **29.** $y^2 = x$

30. $y^2 = x + 2$ **31.** $y = \sqrt{x - 3}$ **32.** $y = \sqrt{x + 5}$

In Exercises 33–56, find the domain of the function, the inverse of the function (if the inverse is a function), and the domain of the inverse. If the inverse is not a function, state so.

33. $\{(3, -2), (2, -3)\}$ **34.** $\{(5, 4), (4, 5)\}$ **35.** $\{(6, -3), (2, -4), (-3, 6)\}$

36. $\{(5, 1), (2, -3), (-3, 2)\}$ **37.** $\{(2, 3), (3, 3), (4, 2)\}$ **38.** $\{(3, 4), (2, 5), (7, 4)\}$

39. $f(x) = 3x + 4$ **40.** $f(x) = 5x - 3$ **41.** $g(x) = 2x - 3$ **42.** $g(x) = 3x - 2$

43. $h(x) = 4 - 5x$ **44.** $h(x) = 5 - 4x$ **45.** $f(x) = x^2 + 2$ **46.** $f(x) = x^2 - 4$

47. $f(x) = x^3 + 4$ **48.** $f(x) = x^3 - 6$ **49.** $g(x) = \dfrac{1}{x}$ **50.** $g(x) = \dfrac{5}{x}$

51. $g(x) = \dfrac{2}{x + 3}$ **52.** $g(x) = \dfrac{4}{x - 1}$ **53.** $g(x) = \dfrac{x - 1}{x}$ **54.** $g(x) = \dfrac{x + 2}{x}$

55. $h(x) = \dfrac{x + 2}{x - 1}$ **56.** $h(x) = \dfrac{x - 2}{x + 1}$

In Exercises 57–60, sketch a graph of the inverse of the graph of the function:

57.

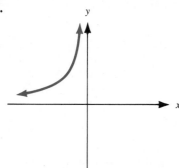

58.

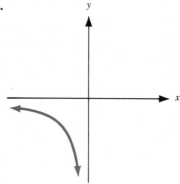

59.

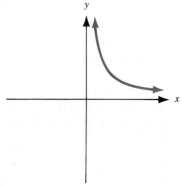

60.

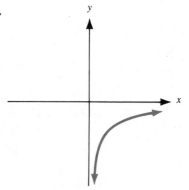

? QUESTIONS FOR THOUGHT

61. (a) Given $f(x) = 2x - 3$, find $f^{-1}(x)$.

    (b) Find $f[f^{-1}(2)]$ and $f^{-1}[f(2)]$.

    (c) Find $f[f^{-1}(x)]$ and $f^{-1}[f(x)]$.

62. **(a)** Graph the relation $x^2 + y^2 = 25$.
    (b) Graph its inverse.
    (c) Compare the graphs of parts **(a)** and **(b)**, and discuss the relationship between them.

 MINI-REVIEW

63. *Express in the form a + bi.* $(5 - 3i)(5 - 2i)$

64. *Solve for x.* $|2x + 3| < 7$

65. *Solve for x.* $x^2 - 4x < 5$

66. Identify what type of figure the equation will yield:

$$x^2 - y^2 - 4x + 20y = 15$$

67. Sketch the solution set of the following system of inequalities:

$$\begin{cases} 2x - 3y < 6 \\ x + 3y > 3 \end{cases}$$

11.5

Variation

The expressions "varies directly" and "varies inversely" have precise meanings in mathematics. When we say that the cost, C, of an object varies directly as the price, P, of materials, we are stating that C is a particular function of P, specifically,

$$C = kP$$

where k is a nonzero constant which does not change as either C or P changes.

In this section we will look at functions which are constructed in this manner. We begin with the following definition.

DEFINITION

y varies directly as x means that there is a constant, $k \neq 0$, such that $y = kx$.

We can also say that *y is directly proportional to x*. The quantity k is called the **constant of proportionality**.

For example, the relationship between the circumference of a circle and its radius is

$$C = 2\pi r$$

We could say that the circumference, C, varies directly as r. The constant of proportionality is 2π.

Suppose we know that y varies directly as x and $y = 15$ when $x = 3$. We can write

$y = kx$ *Since y varies directly as x*

Then we can find k by substituting $y = 15$ and $x = 3$. Hence,

$$y = kx$$
$$15 = k(3)$$
$$5 = k$$

Thus, we have the exact relationship between x and y:

$$y = 5x$$

Now if we want to find y given another value for x, say $x = 7$, we can substitute 7 for x in the equation $y = 5x$ to get

$$y = 5x$$
$$y = 5 \cdot 7$$
$$y = 35$$

EXAMPLE 1

If r is directly proportional to s and $r = 8$ when $s = 3$, find r when $s = 12$.

Solution

Since r varies directly as s, we have

$r = ks$ *Now we find k by substituting in the given values of r and s.*

$8 = k(3)$

$\dfrac{8}{3} = k$

Thus, the equation is $r = \dfrac{8}{3}s$.

To find r when $s = 12$, we substitute in 12 for s to get

$$r = \frac{8}{3}(12)$$
$$r = 32$$

Hence, $r = 32$ when $s = 12$

An alternative approach is to realize what we mean when we say that r is directly proportional to s, or that $r = ks$. First, we see that if $r = ks$, then

$$\frac{r}{s} = k$$

and since k is a constant, the proportion $\dfrac{r}{s}$ will never change. Thus, given $r = 8$ when $s = 3$, we can find r when $s = 12$ by setting up the following proportion:

$$\frac{r_1}{s_1} = \frac{r_2}{s_2} \qquad \textit{Let } r_1 = 8, \ s_1 = 3, \textit{ and } s_2 = 12.$$

$$\frac{8}{3} = \frac{r_2}{12}$$

Solving for r_2, we get

$$\boxed{r_2 = 32} \qquad \textit{Note that this is the same answer obtained previously.}$$

Although both approaches result in the same answer, the first has the advantage that it yields the formula $r = \frac{8}{3}s$, which can be used for additional values of r or s. ∎

DEFINITION

y varies directly as the nth power of x means there is a constant, $k \neq 0$, such that $y = kx^n$.

EXAMPLE 2

The cost of constructing a carton in the shape of a cube is directly proportional to the volume. If it costs $10.80 to construct a box which is 3 ft by 3 ft by 3 ft, how much would it cost to construct a box which is 2 ft by 2 ft by 2 ft?

Solution

The cost is directly proportional to the volume:

$$C = kV$$

Recall that the volume of a cube is the cube of the length of an edge, that is,

$$V = s^3 \quad \text{where } s \text{ is the length of an edge.}$$

Then $\qquad C = ks^3 \qquad$ *Since* $C = 10.80$ *when* $s = 3$*, we have*

$$10.80 = k(3)^3 \qquad \textit{Solve for k.}$$

$$10.80 = 27k$$

$$\frac{10.80}{27} = k$$

$$0.4 = k \qquad \textit{Hence, our equation is}$$

$$\boxed{C = 0.4s^3}$$

The cost of constructing a box which is 2 by 2 by 2 is

$$C = 0.4(2)^3$$
$$= 0.4(8)$$
$$= \boxed{\$3.20}$$

∎

DEFINITION

y varies inversely as x means that there is a constant, $k \neq 0$, such that $y = \dfrac{k}{x}$.

We also say that *y is inversely proportional to x.*

EXAMPLE 3

If *y* varies inversely as *x* and $y = 5$ when $x = 4$, find *x* when $y = 7$.

Solution

We are given that *y* varies inversely as *x*; hence,

$y = \dfrac{k}{x}$ *Since $y = 5$ when $x = 4$, we have*

$5 = \dfrac{k}{4}$

$20 = k$

Thus, our equation becomes

$$y = \frac{20}{x}$$

and we find *x* when $y = 7$ by substituting 7 for *y* to get

$$7 = \frac{20}{x}$$

Hence, $\boxed{x = \dfrac{20}{7}}$ ■

The dependent variable may be related to more than one independent variable.

DEFINITION

z varies jointly as x and y means that there is a constant, $k \neq 0$, such that $z = kxy$.

EXAMPLE 4

If *z* varies jointly as *x* and *y* and $z = 24$ when $x = 2$ and $y = 4$, find *z* when $x = 2$ and $y = 5$.

Solution

Since *z* varies jointly as *x* and *y*, we have

$z = kxy$ *Since $z = 24$ when $x = 2$ and $y = 4$, we have*

$24 = k(2)(4)$

$3 = k$

Thus, our equation is

$$z = 3xy \qquad \textit{Now find z when x = 2 and y = 5.}$$

$$z = 3(2)(5)$$

$$z = 30$$

∎

EXAMPLE 5

The volume, V, of a gas varies directly as its temperature, T, and inversely as its pressure, P. If 40 m³ of a gas yields a pressure of 20 kg/m² at a temperature of 200°K (°K stands for degrees Kelvin, another scale for measuring temperature), what will be the volume of the same gas if the pressure is decreased to 10 kg/m² and the temperature is increased to 250°K?

Solution

We translate the information that V varies directly as T and inversely as P:

$$V = \frac{kT}{P} \qquad \textit{Note that we need only one constant—k.}$$

To find k, we let $V = 40$, $P = 20$, and $T = 200$:

$$40 = \frac{k(200)}{20} \qquad \textit{Solve for k.}$$

$$\frac{(20)(40)}{200} = k$$

$$4 = k$$

Our equation is

$$V = \frac{4T}{P}$$

Now we find V when $P = 10$ and $T = 250$:

$$V = \frac{4(250)}{10} = 100$$

Thus, $V = 100$ m³

Alternatively, we could have solved for k in $V = \dfrac{kT}{P}$ to get

$$\frac{PV}{T} = k$$

and set up the proportion

$$\frac{P_1V_1}{T_1} = \frac{P_2V_2}{T_2}$$

Letting $P_1 = 20$, $V_1 = 40$, $T_1 = 200$, $P_2 = 10$, and $T_2 = 250$, we get

$$\frac{(20)(40)}{200} = \frac{(10)V_2}{250} \qquad \textit{Solving for } V_2, \textit{ we get}$$

$$\frac{(20)(40)(250)}{(10)(200)} = V_2$$

$$100 = V_2$$

This method gives the same answer: $\boxed{100 \text{ m}^3}$ ∎

Exercises 11.5

Solve each of the following problems algebraically.

1. If y varies directly as x, and $y = 8$ when $x = 4$, find y when $x = 3$.

2. If y is directly proportional to x, and $y = 36$ when $x = 12$, find y when $x = 14$.

3. If y varies directly as x, and $y = 25$ when $x = 15$, find y when $x = 8$.

4. If y varies directly as x, and $y = 14$ when $x = 21$, find y when $x = 9$.

5. If y is directly proportional to x, and $y = 22$ when $x = 3$, find x when $y = 5$.

6. If y is directly proportional to x, and $y = 30$ when $x = 7$, find x when $y = 3$.

7. If a varies directly as the square of b, and $a = 4$ when $b = 3$, find a when $b = 9$.

8. If r varies directly as the cube of s, and $r = 8$ when $s = 2$, find r when $s = 5$.

9. If r varies directly as the fourth power of s, and $r = 12$ when $s = 2$, find r when $s = 3$.

10. If r varies directly as the fifth power of s, and $r = 16$ when $s = 2$, find r when $s = 3$.

11. If y varies inversely as x, and $y = 20$ when $x = 4$, find y when $x = 8$.

12. If y is inversely proportional to x, and $y = 24$ when $x = 6$, find y when $x = 8$.

13. If y is inversely proportional to x, and $y = 21$ when $x = 12$, find x when $y = 9$.

14. If y varies inversely as x, and $y = 24$ when $x = 8$, find x when $y = 10$.

15. If y is inversely proportional to x, and $y = 25$ when $x = 10$, find x when $y = 12$.

16. If y varies inversely as x, and $y = 35$ when $x = 14$, find x when $y = 15$.

17. If a is inversely proportional to the cube of b, and $a = 6$ when $b = 2$, find a when $b = 16$.

18. If r varies inversely as the square of s, and $r = 16$ when $s = 2$, find r when $s = 12$.

19. If a varies inversely as the square root of b, and $a = 16$ when $b = 4$, find a when $b = 9$.

20. If x varies inversely as the cube root of y, and $x = 27$ when $y = 27$, find x when $y = 8$.

21. If z varies jointly as x and y, and $z = 12$ when $x = 2$ and $y = 4$, find z when $x = 5$ and $y = 2$.

22. If z varies jointly as x and y, and $z = 24$ when $x = 3$ and $y = 4$, find z when $x = 3$ and $y = 2$.

23. If z varies jointly as x and y, and $z = 20$ when $x = 3$ and $y = 4$, find z when $x = 2$ and $y = 5$.

24. If z varies jointly as x and y, and $z = 32$ when $x = 3$ and $y = 4$, find z when $x = 3$ and $y = 5$.

25. If a varies jointly as c and d, and $a = 20$ when $c = 2$ and $d = 4$, find d when $a = 25$ and $c = 8$.

26. If a varies jointly as c and d, and $a = 15$ when $c = 4$ and $d = 5$, find c when $a = 25$ and $d = 2$.

27. If z varies jointly as x and the square of y, and $z = 20$ when $x = 4$ and $y = 2$, find z when $x = 2$ and $y = 4$.

28. If z varies jointly as x and the square of y, and $z = 40$ when $x = 5$ and $y = 4$, find z when $x = 4$ and $y = 5$.

29. If z varies directly as x and inversely as y, and $z = 16$ when $x = 3$ and $y = 2$, find z when $x = 5$ and $y = 3$.

30. If z varies directly as x and inversely as y, and $z = 12$ when $x = 2$ and $y = 5$, find z when $x = 3$ and $y = 4$.

31. If z varies directly as the square of x and inversely as y, and $z = 20$ when $x = 2$ and $y = 4$, find z when $x = 4$ and $y = 2$.

32. If z varies directly as the square of x and inversely as y, and $z = 50$ when $x = 5$ and $y = 2$, find z when $x = 4$ and $y = 5$.

33. If z varies directly as x and inversely as the square of y, and $z = 32$ when $x = 4$ and $y = 2$, find z when $x = 3$ and $y = 3$.

34. If z varies directly as x and inversely as the square of y, and $z = 45$ when $x = 5$ and $y = 3$, find z when $x = 3$ and $y = 5$.

35. At a constant pressure, the volume of a gas is directly proportional to the temperature. If the volume of a gas is 250 m³ when the temperature is $30°$K, what is the volume of the gas when the temperature is $40°$K?

36. At a constant temperature, the volume of a gas is inversely proportional to the pressure. If the volume of a gas is 300 cm³ when the pressure is 5 kg/m², find the volume when the pressure is 12 kg/m².

37. The volume of a sphere is directly proportional to the cube of its radius. If the volume of a sphere is 36π cm³ when its radius is 3 cm, find its volume when the radius is 4 cm.

38. The volume of a sphere is directly proportional to the cube of its radius. If the volume of a sphere is 36π cm³ when its radius is 3 cm, find its radius when the volume is $\dfrac{32\pi}{3}$ cm³.

39. The distance an object falls is directly proportional to the square of the length of time it falls. If an object falls 256 feet in 4 seconds, how long does it take for it to fall 800 feet?

40. The distance an object falls is directly proportional to the square of the length of time it falls. If an object falls 144 feet in 3 seconds, how far does it fall in 8 seconds?

41. The intensity of illumination on a surface, E, in footcandles, varies inversely as the square of the distance, d, in feet, of the light source from the surface. If the illumination from a source is 25 footcandles when d is 4 feet, find the illumination when the distance is 8 feet.

42. The intensity of illumination on a surface, E, in footcandles, varies inversely as the square of the distance, d, in feet, of the light source from the surface. If the illumination from a source is 25 footcandles when d is 8 feet, find the distance when the illumination is 4 footcandles.

43. The volume of a cone varies jointly as its height and the square of its radius. If its volume is 4π m³ when its height is 3 m and its radius is 2 m, find its volume when its height is 2 m and its radius is 3 m.

44. The volume of a cylinder varies jointly as the height and the square of its radius. If the volume of a cylinder is 36π cm³ when the height is 4 cm and its radius is 3 cm, find its volume when its height is 5 cm and its radius is 4 cm.

45. The resistance (R) of a wire is directly proportional to the length (l) of the wire and inversely proportional to the square of the diameter (d) of the wire. If $R = 12$ ohms when the length is 80 feet and the diameter is 0.01 in., find R when the length is 100 feet and the diameter is 0.02 in.

46. The volume (V) of a gas is directly proportional to its temperature (T) and inversely proportional to the pressure (P). The volume of the gas is 20 m³ when the temperature is $100°$K and the pressure is 15 kg/m². What is the volume when the temperature is $150°$K and the pressure is 20 kg/m²?

⟳ MINI-REVIEW

47. *Write the following as a simple fraction reduced to lowest terms:*

$$\dfrac{\dfrac{2x}{y} + 7 + \dfrac{5y}{x}}{\dfrac{3x}{y} + 2 - \dfrac{y}{x}}$$

48. Find the equation of the line passing through $(2, 0)$ and parallel to the line $7x - 3y = 5$.

49. *Perform the operations and express your answer in simplest form:*

$$\sqrt[3]{5x^2y^5} \ \sqrt[3]{50x^5y^4}$$

50. *Solve the following system of equations:*

$$\begin{cases} 2x - 3y = 5 \\ 5x + 3y = -19 \end{cases}$$

CHAPTER 11   Summary

After having completed this chapter, you should be able to:

1. Understand the meaning of a relation and identify its domain and range (Section 11.1).

    For example:

    (a) The relation $\{(2, -3), (3, 4), (-2, -3)\}$ has domain $\{-2, 2, 3\}$ and range $\{-3, 4\}$. The relation assigns -3 to 2, 4 to 3, and -3 to -2.

    (b) The relation $4x^2 + y^2 = 4$ has the graph shown in Figure 11.21. By looking at the graph, we see that the domain is $\{x \mid -1 \le x \le 1\}$ and the range is $\{y \mid -2 \le y \le 2\}$.

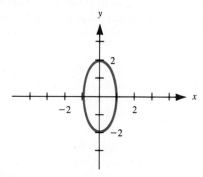

Figure 11.21   Graph of $4x^2 + y^2 = 4$

    (c) The relation described by $y = \dfrac{2}{x - 1}$ has domain $\{x \mid x \ne 1\}$ since $x = 1$ is the only value of x which produces either an undefined or a nonreal value for y.

2. Understand the meaning of a function (Section 11.1).

    For example:

    (a) The relation $\{(2, -3), (2, 4)\}$ is not a function since the x value, 2, is assigned two y values, $y = -3$ and $y = 4$.

    (b) $\{(3, 5), (2, 5)\}$ is a function since no x value is assigned more than one y value.

    (c) Consider the relation described by the following graph:

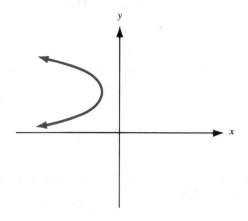

In the next figure we apply the vertical line test:

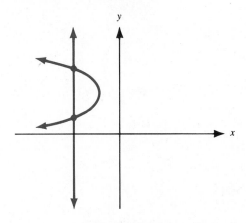

Since the graph of the relation intersects a vertical line at more than one point, this relation is not a function.

3. Use function notation (Section 11.2).

   For example:

   (a) If $f(x) = 3x^2 - 4$, then

   $$f(2) = 3(2)^2 - 4 = 3 \cdot 4 - 4 = 8$$
   $$f(-1) = 3(-1)^2 - 4 = 3 \cdot 1 - 4 = -1$$
   $$f(s) = 3s^2 - 4$$

   (b) If $f(x) = 2x - 1$ and $g(x) = x^2 + 3$, then

   $$f(x + 2) = 2(x + 2) - 1 = 2x + 4 - 1 = 2x + 3$$
   $$f(x) + 2 = (2x - 1) + 2 = 2x + 1$$
   $$f(x) + f(2) = 2x - 1 + 2 \cdot 2 - 1$$
   $$= 2x - 1 + 4 - 1 = 2x + 2$$
   $$f[g(x)] = 2[g(x)] - 1 = 2(x^2 + 3) - 1$$
   $$= 2x^2 + 6 - 1 = 2x^2 + 5$$

4. Find sums, products, differences, and quotients of functions (Section 11.2).

   For example: If $f(x) = 2x^2 - 3$ and $g(x) = x - 4$, then

   (a) $(f + g)(3) = f(3) + g(3) = [2(3)^2 - 3] + [3 - 4]$
   $$= 2 \cdot 9 - 3 + 3 - 4 = 14$$

   (b) $\left(\dfrac{f}{g}\right)(x) = \dfrac{f(x)}{g(x)} = \dfrac{2x^2 - 3}{x - 4} \qquad (x \neq 4)$

5. Graph and classify the graph as one of the basic types of functions (Section 11.3).

   For example:

   (a) $f(x) = \sqrt{3x - 1}$ is a square root function which has the graph shown in Figure 11.22 on page 574.

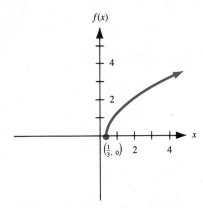

Figure 11.22   Graph of $f(x) = \sqrt{3x - 1}$

(b) $f(x) = 3x^2 - 4x + 1$ is a quadratic function which has the graph shown in Figure 11.23.

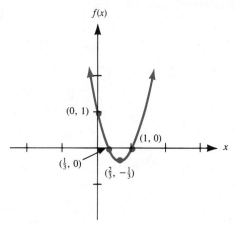

Figure 11.23   Graph of $f(x) = 3x^2 - 4x + 1$

6.   Find the inverse of a relation (Section 11.4).

For example:

(a)   The inverse of the relation $R = \{(6, 4), (2, 4)\}$ is the relation

$$R^{-1} = \{(4, 6), (4, 2)\}$$

(b)   The inverse of the relation $y = 3x^2$ is

$x = 3y^2$ *Interchange x and y variables.*

$\dfrac{x}{3} = y^2$ *Solve for y.*

$\pm\sqrt{\dfrac{x}{3}} = y$

7. Identify a one-to-one function (Section 11.4).

   For example:

   (a) $\{(3, 2), (5, 2)\}$ is a function but it is not a one-to-one function, since the y value, 2, is assigned to more than one x value, $x = 3$ and $x = 5$.

   (b) $\{(3, 2), (4, 5)\}$ is a one-to-one function since each x value is assigned a unique y value and each y value is assigned a unique x value.

   (c) $y = 3x^2 - 1$ is a function for which

   $$x = -2 \rightarrow y = 3(-2)^2 - 1 = 3 \cdot 4 - 1 = 11$$
   $$x = 2 \rightarrow y = 3(2)^2 - 1 = 3 \cdot 4 - 1 = 11$$

   Since $y = 11$ is assigned to two different values of x, $x = 2$ and $x = -2$, the function is not one-to-one.

8. Find the inverse of a function (Section 11.4).

   For example:

   (a) Given $f(x) = 3x - 4$. To find $f^{-1}(x)$, let $y = f(x)$. Then

   $y = 3x - 4$ *Interchange x and y variables.*

   $x = 3y - 4$ *Solve explicitly for y.*

   $$\frac{x + 4}{3} = y$$

   Hence, $f^{-1}(x) = \dfrac{x + 4}{3}$.

   (b) Sketch the graph of the inverse of a function given the graph of the function $f(x)$. See Figure 11.24.

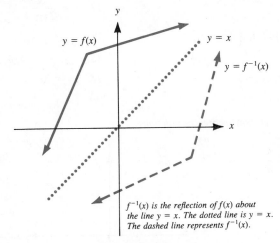

$f^{-1}(x)$ is the reflection of $f(x)$ about the line $y = x$. The dotted line is $y = x$. The dashed line represents $f^{-1}(x)$.

Figure 11.24

9. Solve variation problems (Section 11.5).

   For example: Suppose z varies directly as x and inversely as y. If $z = 15$ when $x = 3$ and $y = 2$, find z when $x = 3$ and $y = 4$.

Let $z = \dfrac{kx}{y}$. To find k we use the information that $z = 15$ when $x = 3$ and $y = 2$:

$$15 = \frac{k(3)}{2}$$

$$30 = 3k$$

$$10 = k$$

Thus,

$$z = \frac{10x}{y}$$

Now we find z when $x = 3$ and $y = 4$:

$$z = \frac{10(3)}{4} = \frac{30}{4} = \frac{15}{2}$$

CHAPTER 11   Review Exercises

In Exercises 1–6, identify the domain and range of the relation (sketch a graph if necessary).

1. $\{(3, 4), (2, 5), (3, 5)\}$ **2.** $\{(-2, 6), (-7, -3), (-3, -7)\}$ **3.** $y = 2x^2 - 1$

4. $x = 2y^2 - 1$ **5.** $x^2 + y^2 = 36$ **6.** $4x^2 + 25y^2 = 100$

In Exercises 7–12, identify the domain of the relation.

7. $y = 2x + 1$ **8.** $2y = 3x$ **9.** $y = \sqrt{4 - x}$

10. $y = \sqrt{x + 3}$ **11.** $y = \dfrac{3x}{x + 2}$ **12.** $y = \dfrac{x}{2x - 1}$

In Exercises 13–20, determine whether the given relation is a function.

13. $\{(2, -5), (3, 8), (4, -5)\}$ **14.** $\{(6, 2), (5, 1), (6, 8)\}$ **15.** $\{(3, -1), (4, 2), (4, 7)\}$

16. $\{(-3, 1), (1, -3)\}$ **17.** $y = 2x + 3$ **18.** $2y = 3x - 1$

19. $y = x^2 - 4$ **20.** $x = y^2 - 4$

In Exercises 21–24, determine which of the graphs of relations represent functions.

21.

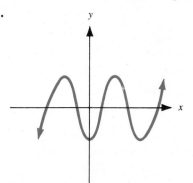

22.

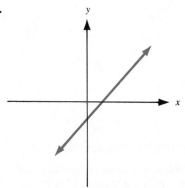

23.

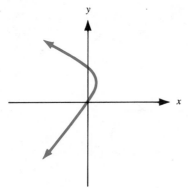

24.

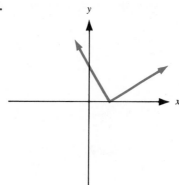

In Exercises 25–34, evaluate the functions at the given values. If a value is not in the domain of the function, state so.

25. $f(x) = 3x + 5$; $f(-1), f(0), f(1), f(2)$

26. $g(x) = 5 - 4x$; $g(-1), g(0), g(1), g(2)$

27. $f(x) = 2x^2 - 3x + 2$; $f(-1), f(0), f(1), f(2)$

28. $g(x) = 3x^3 + 2x - 3$; $g(-1), g(0), g(1), g(2)$

29. $h(x) = \sqrt{x - 5}$; $h(6), h(5), h(4)$

30. $g(x) = \sqrt{5 - 3x}$; $g(1), g(2), g(-1)$

31. $h(x) = \dfrac{x - 1}{x + 3}$; $h(1), h(3), h(-3)$

32. $h(x) = \dfrac{x + 1}{x}$; $h(-1), h(0), h(4)$

33. $f(x) = 2x^2 + 4x - 1$; $f(a), f(z)$

34. $f(a) = 2a^2 - 4a + 2$; $f(x), f(z)$

In Exercises 35–50, $f(x) = 5x + 2$ and $g(x) = \sqrt{3 - 2x}$. Find:

35. $f(x + 2)$

36. $f(x + 3)$

37. $f(x) + 2$

38. $f(x) + 3$

39. $f(x) + f(2)$

40. $f(x) + f(3)$

41. $g(x + 2)$

42. $g(x + 3)$

43. $g(2x)$

44. $g(3x)$

45. $2g(x)$

46. $3g(x)$

47. $f(x + h) - f(x)$

48. $g(x + h) - g(x)$

49. $f[g(x)]$

50. $g[f(x)]$

In Exercises 51–58, graph the function and identify its type.

51. $f(x) = 2x - 3$

52. $f(x) = 3x - 1$

53. $f(x) = \sqrt{2x - 3}$

54. $f(x) = |3x - 1|$

55. $f(x) = 4x - 2x^2$

56. $f(x) = \sqrt{3x - 1}$

57. $f(x) = |2x| - 3$

58. $f(x) = 6x - x^2$

In Exercises 59–62, find the inverse of the relation.

59. $\{(0, -1), (2, 0), (3, -1)\}$

60. $\{(2, 0), (-1, 3), (-1, 2)\}$

61. $y = x^2 - 3$

62. $y = x^3 - 3$

In Exercises 63–66, determine whether the relation is a function, a one-to-one function, or neither.

63. $\{(2, 0), (3, 0)\}$

64. $\{(5, 2), (5, 4)\}$

65. $y = 3x$

66. $2y = x + 1$

In Exercises 67–70, determine whether the graph of the relation is a function, a one-to-one function, or neither.

67.

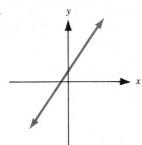

68.

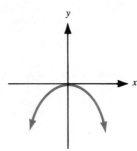

69.

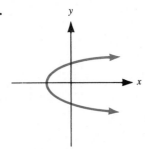

70.

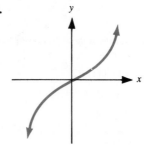

In Exercises 71–78, find the inverse of the function.

71. $\{(2, 3), (3, 4)\}$

72. $\{(5, -8), (3, 7), (2, 6)\}$

73. $y = 3x + 8$

74. $y = 5x - 1$

75. $y = x^3$

76. $y = x^5$

77. $y = \dfrac{3}{x + 1}$

78. $y = \dfrac{2x}{x + 3}$

In Exercises 79–80, sketch the inverse of the function given on the graph.

79.

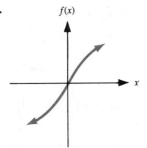

80.

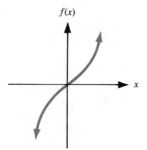

81. If x varies directly as y, and $x = 8$ when $y = 3$, find x when $y = 2$.

82. If x varies inversely as y, and $x = 8$ when $y = 3$, find x when $y = 2$.

83. If r is inversely proportional to the square of s, and $s = 2$ when $r = 8$, find s when $r = 4$.

84. If r is directly proportional to the cube of s, and $r = 16$ when $s = 2$, find s when $r = 4$.

85. If z varies jointly as x and y, and $z = 16$ when $y = 2$ and $x = 7$, find z when $x = 3$ and $y = 4$.

86. If z is directly proportional to x and inversely proportional to y, and $z = 12$ when $x = 2$ and $y = 4$, find z when $x = 3$ and $y = 2$.

87. The volume (V) of a sphere is directly proportional to the cube of its radius. If the volume is $\dfrac{500\pi}{3}$ in.³ when its radius is 5 in., find its volume when the radius is 6 in.

88. The weight of an object on or above the surface of the earth is inversely proportional to the square of its distance to the center of the earth. A man weighs 175 lb on the surface of the earth. Assuming the diameter of the earth is 4,000 miles, how much would the man weigh 200 miles above the earth's surface?

89. The volume (V) of a gas is directly proportional to its temperature (T) and inversely proportional to its pressure (P). If $V = 80$ m³ when $T = 20°$K and $P = 30$ kg/m², find V when $T = 10°$K and $P = 20$ kg/m².

90. The production cost of manufacturing widgets varies jointly as the number of widget machines running and the number of hours the machines are in operation. When 5 widget machines are running 8 hours, the production cost is \$3,800. What is the production cost for 6 widget machines running 20 hours?

CHAPTER 11   Practice Test

1. Identify the domain and range of each of the following relations (sketch a graph if necessary):

   (a) $\{(2, -3), (2, 5), (3, 5), (4, 6)\}$

   (b) $x^2 + y^2 = 1$

2. Identify the domain of each of the following relations:

   (a) $y = \sqrt{x - 4}$

   (b) $y = \dfrac{x}{3x - 4}$

3. Identify which of the following relations are functions:

   (a) $\{(2, 5), (2, 4)\}$

   (b) $\{(3, 2), (4, 3), (5, 2)\}$

   (c) $x = 3y^2$

   (d)

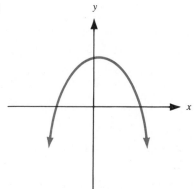

4. Given $f(x) = 3x^2 - 4$ and $g(x) = \sqrt{2x + 1}$, find:

   (a) $g(2)$ **(b)** $f(-3)$ **(c)** $f(x - 2)$ **(d)** $f[g(4)]$

5. Sketch a graph of $f(x) = |2x - 1|$.

6. Find the inverse of the relation $\{(2, -3), (5, 8), (3, 8)\}$.

7. Determine if the following function is a one-to-one function:

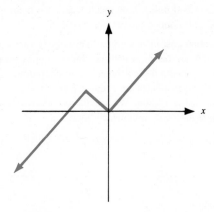

8. Find the inverse of each of the following functions:

   (a) $f(x) = 2x - 4$

   (b) $f(x) = \dfrac{x}{x + 3}$

9. If y varies inversely as the square of x, and $y = 24$ when $x = 2$, find y when $x = 6$.

10. If z varies directly as the square of x and inversely as y, and $z = 120$ when $x = 2$ and $y = 3$, find z when $x = 3$ and $y = 2$.

12

Exponential and Logarithmic Functions

12.1

Exponential Functions

In Chapter 11 we discussed a variety of equations and saw how they defined various relations and functions. We now turn our attention to a new type of equation.

DEFINITION

An equation of the form

$$y = b^x \quad \text{where } b > 0 \text{ and } b \neq 1$$

is called an *exponential equation*.

A question which should immediately come to mind is: For what values of x does this equation make sense? For example, for what values of x is $y = 3^x$ defined?

Based on our work in Chapter 5, we know what 3^x means when x is a rational number. In other words, we know what 3^x means when x takes on values such as $x = 5$, $x = -2$, and $x = \frac{2}{7}$.

$$3^5 = 3 \cdot 3 \cdot 3 \cdot 3 \cdot 3 = 243$$

$$3^{-2} = \frac{1}{3^2} = \frac{1}{9}$$

$$3^{2/7} = (\sqrt[7]{3})^2$$

But what about irrational values of x? What do we mean by $3^{\sqrt{2}}$?

While a formal definition of b^x for x irrational is beyond the scope of this book, we can make the following remarks. We know that we can get arbitrarily accurate approximations to $\sqrt{2}$. For example, we find that

$$1.414 < \sqrt{2} < 1.415$$

and so it seems reasonable that

$$3^{1.414} < 3^{\sqrt{2}} < 3^{1.415}$$

Since 1.414 and 1.415 are rational numbers, $3^{1.414}$ and $3^{1.415}$ are well defined (not easy to compute, perhaps, but well defined nonetheless).

In this way we can get better and better approximations to what we expect $3^{\sqrt{2}}$ to be. Consequently, we will make the following assumption.

The exponential function is defined for all real numbers x, and all the properties of exponents developed in Chapter 5 extend to real exponents as well.

EXAMPLE 1

Sketch the graph of $y = 3^x$.

Solution

Since we have not worked with this equation before, let's begin by constructing a brief table of values:

x	-3	-2	-1	0	1	2	3
$y = 3^x$	$\dfrac{1}{27}$	$\dfrac{1}{9}$	$\dfrac{1}{3}$	1	3	9	27

If we plot these points and connect them with a smooth curve, we obtain the graph shown in Figure 12.1.

Figure 12.1
Graph of $y = 3^x$

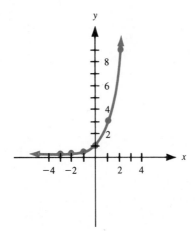

Notice that as x takes on smaller and smaller values (that is, negative numbers with larger and larger absolute values), the y values are getting closer and closer to 0. The graph is getting closer and closer to the x-axis but never touches it. The graph of $y = 3^x$ never intersects the x-axis because there is no value of x for which $3^x = 0$.

You will recall from our discussion of hyperbolas in Chapter 9 that such a line is called an *asymptote*. Thus, the x-axis is an asymptote for the graph of $y = 3^x$. ∎

The equation $y = 5^x$ will have a graph very similar to that constructed in Example 1 for $y = 3^x$. (*Try it!*) However, because the base is larger, it will *rise* more sharply.

Looking at the graph of $y = 3^x$, we notice that as x increases, y also increases, but very rapidly when compared to x. In general, the graph of the equation $y = b^x$ for $b > 1$ exhibits what is called **exponential growth**.

EXAMPLE 2

Sketch the graph of $y = \left(\dfrac{1}{3}\right)^x$.

Solution | Again, we begin by constructing a table of values.

x	-3	-2	-1	0	1	2	3
$y = \left(\dfrac{1}{3}\right)^x$	27	9	3	1	$\dfrac{1}{3}$	$\dfrac{1}{9}$	$\dfrac{1}{27}$

We plot these points and connect them with a smooth curve to obtain the graph shown in Figure 12.2.

Figure 12.2

Graph of $y = \left(\dfrac{1}{3}\right)^x$

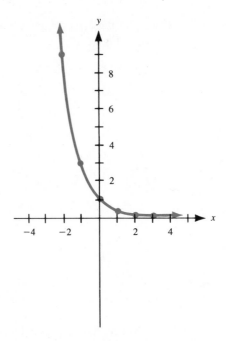

The graph again approaches, but does not touch, the x-axis (this time on the right side). ■

The equation $y = \left(\frac{1}{5}\right)^x$ will have a graph very similar to the one obtained in Example 2. (*Try it!*) However, because the base is smaller, it will *fall* more sharply.

Notice that as x increases, y decreases very rapidly when compared to x. In general, the graph of the equation $y = b^x$ when $0 < b < 1$ exhibits what is called **exponential decay**.

By looking at the graphs in Examples 1 and 2, we can see that the graphs satisfy the vertical line test for functions discussed in Section 11.1. In general, the equation $y = b^x$ defines y as a function of x.

DEFINITION

A function of the form $y = f(x) = b^x$, where $b > 0$ and $b \neq 1$, is called an *exponential function*.

Based on Examples 1 and 2 and their accompanying graphs, we make the following observations about the exponential function $f(x) = b^x$:

1.   The domain (the set of possible x values) for an exponential function is R, the set of all real numbers.

2.   The range (the set of possible y values) for an exponential function is $\{y \mid y > 0\}$. That is, the range of an exponential function is the set of all positive real numbers.

3.   The graph of an exponential function always passes through the point $(0, 1)$ (because $b^0 = 1$ for all $b \neq 0$).

4.   The graph of an exponential function rises if $b > 1$ and falls if $0 < b < 1$ (see Figure 12.3). Remember that we always describe a graph as we move from left to right, that is, for increasing values of x.

Figure 12.3
Graph of $y = b^x$

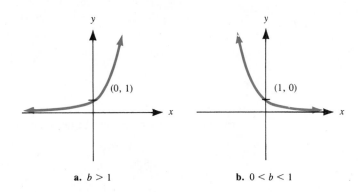

a. $b > 1$ **b.** $0 < b < 1$

5.   The graph of $y = b^x$ also satisfies the *horizontal line test*. That is, a horizontal line crosses the graph of an exponential function at most once. Based on our discussion in Section 11.4, this means that the exponential function is one-to-one and so has an inverse.

EXAMPLE 3   Solve each of the following exponential equations.

(a) $2^{3x-1} = 2^x$ (b) $3^{5x+1} = 9^{2x}$ (c) $\dfrac{4^{x^2}}{4^{3x}} = \dfrac{1}{16}$

Solution   The fact that the exponential function is one-to-one means that if $b^r = b^t$, then $r = t$.

(a) $2^{3x-1} = 2^x$ *Since the bases are equal, the exponents must be equal.*

           $3x - 1 = x$

           $2x = 1$

           $\boxed{x = \dfrac{1}{2}}$

CHECK $x = \dfrac{1}{2}$:

$$2^{3x-1} = 2^x$$
$$2^{3(1/2)-1} \overset{?}{=} 2^{1/2}$$
$$2^{(3/2)-1} \overset{?}{=} 2^{1/2}$$
$$2^{1/2} \overset{\checkmark}{=} 2^{1/2}$$

(b)   We try to express both sides of the equation in terms of the same base.

$$3^{5x+1} = 9^{2x} \qquad \textit{Write 9 as } 3^2.$$
$$3^{5x+1} = (3^2)^{2x} \qquad \textit{Use exponent rule 2.}$$
$$3^{5x+1} = 3^{4x} \qquad \textit{Since the bases are equal, the exponents must be equal.}$$
$$5x + 1 = 4x$$

$$\boxed{x = -1}$$

CHECK $x = -1$:

$$3^{5x+1} = 9^{2x}$$
$$3^{5(-1)+1} \overset{?}{=} 9^{2(-1)}$$
$$3^{-4} \overset{?}{=} 9^{-2}$$
$$\frac{1}{81} \overset{\checkmark}{=} \frac{1}{81}$$

(c)   We express both sides in terms of the base 4.

$$\frac{4^{x^2}}{4^{3x}} = \frac{1}{16}$$
$$4^{x^2-3x} = \frac{1}{4^2}$$
$$4^{x^2-3x} = 4^{-2}$$
$$x^2 - 3x = -2 \qquad \textit{This is a quadratic equation.}$$
$$x^2 - 3x + 2 = 0$$
$$(x - 2)(x - 1) = 0$$

$$\boxed{x = 2} \quad \text{or} \quad \boxed{x = 1}$$

The check is left to the reader.   ∎

Many "real-world" situations can be described in terms of exponential functions.

EXAMPLE 4

A bacteria culture initially has 1,000 bacteria of a type whose number doubles every 12 hours.

(a)   How many bacteria are there after 12 hours?

(b)   How many bacteria are there after 24 hours?

(c)   How many bacteria are there after 36 hours?

(d)   How many bacteria are there after t hours?

(e)   How many bacteria are there after 15 hours?

Solution

(a) Since the number of bacteria doubles every 12 hours, after 12 hours there will be twice as many as there were initially:

$$1{,}000(2) = \boxed{2{,}000 \text{ bacteria}}$$

(b) After 24 hours the number of bacteria doubles *twice* $\left(\frac{24}{12} = 2\right)$:

$$[1{,}000(2)](2) = 1{,}000(2)^2 = 1{,}000(4) = \boxed{4{,}000 \text{ bacteria}}$$

(c) After 36 hours the number of bacteria has doubled *three* times $\left(\frac{36}{12} = 3\right)$:

$$[1{,}000(2)(2)](2) = 1{,}000(2)^3 = 1{,}000(8) = \boxed{8{,}000 \text{ bacteria}}$$

(d) Since the number of bacteria doubles every 12 hours, in t hours the number will double $\dfrac{t}{12}$ times.

If we let N = the number of bacteria present after t hours, then

$$\boxed{N = 1{,}000 \cdot 2^{t/12}}$$

(e) We use the answer obtained in part (d) with $t = 15$:

$$N = 1{,}000 \cdot 2^{t/12} \qquad \textit{Substitute } t = 15.$$
$$= 1{,}000 \cdot 2^{15/12}$$
$$= 1{,}000 \cdot 2^{1.25}$$

In order to complete this example we need to be able to calculate $2^{1.25}$. With the aid of a calculator, we can compute this value as $2^{1.25} = 2.378$ correct to three decimal places.

Thus, $N = 1{,}000(2.378) = \boxed{2{,}378 \text{ bacteria}}$

In the coming sections we will develop an alternative method for computing $2^{1.25}$ without a calculator. ∎

Exercises 12.1

In Exercises 1–10, sketch the graph of the given equation.

1. $y = 2^x$

2. $y = \left(\frac{1}{2}\right)^x$

3. $y = \left(\frac{1}{5}\right)^x$

4. $y = 5^x$

5. $y = 3^{-x}$

6. $y = \left(\frac{1}{3}\right)^x$

7. $y = 2^{x+1}$

8. $y = 3^x - 1$

9. $y = 2^x + 1$

10. $y = 3^{x-1}$

In Exercises 11–24, solve the given exponential equation.

11. $2^x = 2^{3x-2}$ **12.** $3^{5x-3} = 3^{x+5}$ **13.** $5^x = 25^{x-1}$ **14.** $9^x = 3^{x+3}$

15. $4^{1-x} = 16$ **16.** $4^{1-x} = 8$ **17.** $8^x = 4^{x+1}$ **18.** $27^{x-1} = 9^{3x}$

19. $\dfrac{9^{x^2}}{9^x} = 81$ **20.** $\dfrac{4^{x^2}}{4^{2x}} = 64$ **21.** $4^{\sqrt{x}} = 2^{x-3}$ **22.** $25^{\sqrt{x}} = 5^x$

23. $16^{x^2-1} = 8^{x-1}$ **24.** $16^{x-1} = 8^{x^2-1}$

25. The number of bacteria in a culture which initially contains 2,500 bacteria doubles every 10 hours. How many bacteria are there after 10 hours? After 20 hours? After 50 hours? After t hours?

26. The population of Centerville is 8,000 and triples every 25 years. What is the population in 25 years? In 75 years? In 200 years? In Y years?

27. In 1985 the population of Capitol City is 20,000. Due to industrial development, the population will double every 14 years. What will the population be in 1999? In 2013? In year Y?

28. In 1985 the rabbit population in a certain area is estimated at 12,000, and is assumed to quadruple (grow by a factor of 4) every 3 years unless controlled by some outside agent. Estimate what the uncontrolled population will be in 1988? In 1997? In year Y?

29. An antibacterial substance is introduced into a bacterial culture which contains 10,000 bacteria. The substance destroys bacteria so that there are half as many bacteria as there were 1 hour ago. What is the number of bacteria present 2 hours after the substance is introduced? After 3 hours? After t hours?

30. The oil reserves of a certain oil field which contains 1 million barrels of oil are being depleted at the rate of $\frac{1}{10}$ every 5 years. How many barrels are left after 5 years? After 10 years? After 30 years?

 MINI-REVIEW

31. *Perform the operations, simplify, and express your answer with positive exponents only:*

$$\left(\frac{2x^{-5}y^{-3}}{x^{-4}y^2}\right)^{-1}$$

32. *Perform the operations and simplify:*

$$\frac{\sqrt{15}}{\sqrt{5} + \sqrt{2}}$$

33. *Solve for x.* $(x - 3)^2 = 4$

34. Given $f(x) = 3x^2 - 2x + 5$, find $f(-2)$ and $f(x - 2)$.

35. Given $f(x) = 3x^3 - 2$, find $f^{-1}(x)$.

12.2

Logarithms

In the last section we discussed exponential functions, that is, functions of the form

$$y = f(x) = b^x \qquad (b > 0, \quad b \neq 1)$$

and pointed out that the graphs of exponential functions satisfy the horizontal line test and so an exponential function has an inverse. For example, if we sketch the graph of $y = f(x) = 3^x$ (refer to Figure 12.1), we can see that each y value in the range is associated with exactly one x value in the domain. That is, each horizontal line crosses the graph at most once, and so the graph of $y = 3^x$ satisfies the horizontal line test. Thus, the function $y = f(x) = 3^x$ has an inverse function.

Let's look at the functions $y = x^3$ and $y = 3^x$ and their inverse functions. Keep in mind that we let x represent the *independent* variable and y the *dependent* variable, so that whenever possible we want to write y explicitly as a function of x.

Recall that in order to obtain the inverse of a given function, we interchange x and y and then solve for y.

Function	*Inverse function*
$y = x^3$	$y^3 = x$ or $y = \sqrt[3]{x}$
$y = 3^x$	$x = 3^y$ or $y = ?$

Notice that for the inverse function of the exponential function $y = 3^x$, we have no way of writing y explicitly as a function of x. (Actually, if you think about it, had we learned about functions before radicals and fractional exponents, then in order to take the function $x = y^3$ and solve for y, we would have had to "invent" radical notation so that we could write $y = \sqrt[3]{x}$.) In order to write the function $x = 3^y$, which is the inverse of the exponential function, explicitly as a function of x we must invent a new notation.

DEFINITION

We write $y = \log_b x$ to mean $x = b^y$, where $b > 0$, $x > 0$, and $b \neq 1$.
$y = \log_b x$ is read "$y = \log$ base b of x."

In words, $y = \log_b x$ means that "y is the exponent of b that gives x."

Remember	$y = \log_b x$ and $x = b^y$ are alternative ways of expressing the same relationship.
	A logarithm is just an exponent.

EXAMPLE 1

On the same set of coordinate axes, sketch the graph of each pair of equations.

(a) $y = 3^x$ and $y = \log_3 x$ **(b)** $y = \left(\dfrac{1}{2}\right)^x$ and $y = \log_{1/2} x$

Solution

Using the properties of inverse functions which we developed in Chapter 11, we note the following:

1. Since $y = b^x$ has as its domain R, the set of all real numbers, and as its range $\{y \mid y > 0\}$, the domain of the inverse function $y = \log_b x$ is $\{x \mid x > 0\}$ and the range of the inverse function is R, the set of all real numbers.

2. Since the graphs of all exponential functions pass through the point $(0, 1)$, the graphs of all logarithmic functions pass through the point $(1, 0)$.

3.   We can draw the graph of an inverse function by reflecting the graph of the original function about the line $y = x$.

Based on these observations, we obtain the graphs appearing in Figure 12.4(a) and (b).

Figure 12.4

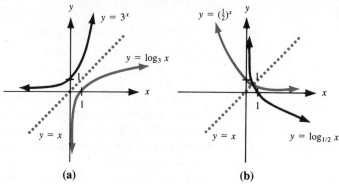

(a)   (b)

Note that since $y = 3^x$ and $y = \log_3 x$ are inverses of each other, we interchange the domain and range: the domain of $y = 3^x$ (all real numbers) is the range of $y = \log_3 x$, and the range of $y = 3^x$ ($\{y \mid y > 0\}$) is the domain of $y = \log_3 x$ ($\{x \mid x > 0\}$).

We have established the fact that

$$y = \log_b x \quad \text{is equivalent to} \quad x = b^y$$

We will often use the following terminology:

When an expression is written in the form $x = b^y$, it is said to be in *exponential form*.

When an expression is written in the form $y = \log_b x$, it is said to be in *logarithmic form*.

Thus, $5^2 = 25$ and $\log_5 25 = 2$ are the exponential and logarithmic forms, respectively, of the same relationship.

EXAMPLE 2   Write each of the following in exponential form.

(a) $\log_{10} 100 = 2$ (b) $\log_4 \dfrac{1}{64} = -3$ (c) $\log_7 \sqrt{7} = \dfrac{1}{2}$

(d) $\log_9 27 = \dfrac{3}{2}$

Solution   We use the fact that $\log_b x = y$ means that $b^y = x$.

(a) $\log_{10} 100 = 2$ means $\boxed{10^2 = 100}$

(b) $\log_4 \dfrac{1}{64} = -3$ means $\boxed{4^{-3} = \dfrac{1}{64}}$

(c) $\log_7 \sqrt{7} = \dfrac{1}{2}$ means $\boxed{7^{1/2} = \sqrt{7}}$

(d) $\log_9 27 = \dfrac{3}{2}$ means $\boxed{9^{3/2} = 27}$

Keep in mind that all we have been doing in Example 2 is *translating* from logarithmic form to exponential form. You should verify for yourself that we are translating *true* statements. ∎

EXAMPLE 3

Write each of the following in logarithmic form.

(a) $2^4 = 16$ **(b)** $4^2 = 16$ **(c)** $3^{-4} = \dfrac{1}{81}$ **(d)** $100^{1/2} = 10$

Solution

We use the fact that $b^y = x$ means that $\log_b x = y$.

(a) $2^4 = 16$ means $\boxed{\log_2 16 = 4}$

(b) $4^2 = 16$ means $\boxed{\log_4 16 = 2}$ *Notice the difference between parts* **(a)** *and* **(b)**.

(c) $3^{-4} = \dfrac{1}{81}$ means $\boxed{\log_3 \dfrac{1}{81} = -4}$

(d) $100^{1/2} = 10$ means $\boxed{\log_{100} 10 = \dfrac{1}{2}}$ ∎

Being able to translate from logarithmic form to exponential form often makes it easier to evaluate logarithmic expressions. For example, evaluating

$$\log_5 5^3 = ?$$

is the same as asking (in exponential form)

$$5^? = 5^3$$

and so the answer is clearly 3. Remember that since the exponential function is one-to-one, $b^t = b^s$ implies that $t = s$.

Thus, we have $\log_5 5^3 = 3$.

Similarly, $\log_6 6^{1/4} = ?$ is asking $6^? = 6^{1/4}$ and so the answer is $\frac{1}{4}$.

Thus, $\log_6 6^{1/4} = \dfrac{1}{4}$.

In general, we have

$$\boxed{\log_b b^r = r}$$

In words, this says that "the exponent of b which gives b^r is r."

EXAMPLE 4

Find each of the following:

(a) $\log_2 32$ **(b)** $\log_5 \dfrac{1}{25}$ **(c)** $\log_6 \sqrt[3]{6}$ **(d)** $\log_8 16$

(e) $\log_5 5$ **(f)** $\log_5 1$ **(g)** $\log_3(-8)$

Solution

We will try to make use of the fact just stated that $\log_b b^r = r$. Consequently, we try to express the number whose logarithm we are trying to find as a power of the base b.

(a) $\log_2 32 = \log_2 2^5 = \boxed{5}$

(b) $\log_5 \dfrac{1}{25} = \log_5 \dfrac{1}{5^2} = \log_5 5^{-2} = \boxed{-2}$

(c) $\log_6 \sqrt[3]{6} = \log_6 6^{1/3} = \boxed{\dfrac{1}{3}}$

Notice that since a logarithm is an exponent, when we are working with logarithms we generally prefer to write radical expressions in exponential form.

(d) In looking for $\log_8 16$, it is not quite so obvious how to express 16 in terms of the base 8. Let's call the answer t, and translate the required logarithm into exponential form.

$\log_8 16 = t$ means $8^t = 16$ *Let's write both 8 and 16 as powers of the same base, 2.*

$(2^3)^t = 2^4$

$2^{3t} = 2^4$ *Since the exponential function is one-to-one, if $b^r = b^t$ then $r = t$.*

$3t = 4$

$t = \dfrac{4}{3}$

Thus, $\log_8 16 = \boxed{\dfrac{4}{3}}$

(e) $\log_5 5 = \log_5 5^1 = \boxed{1}$

(f) $\log_5 1 = \log_5 5^0 = \boxed{0}$

(g) $\log_3(-8)$ $\boxed{\text{does not exist}}$ If we translate this expression into exponential form, we get $3^? = -8$. To what power do we raise 3 in order to get a negative number? We cannot take the log of 0, either. (For what value of x does $b^x = 0$?) Since log bases are always positive, *we cannot take the log of a nonpositive number.* ∎

Note that since $b^0 = 1$ for all $b \neq 0$, we have

$$\boxed{\log_b 1 = 0}$$

If we substitute $y = \log_b x$ into the equivalent equation $x = b^y$, we obtain the relationship

$$\boxed{b^{\log_b x} = x}$$

Thus, for example, $5^{\log_5 8} = 8$.

EXAMPLE 5

Evaluate $\log_8(\log_4 16)$.

Solution

We first find $\log_4 16 = \log_4 4^2 = 2$. Therefore, our original expression becomes

$\log_8(\log_4 16) = \log_8 2 = t$ *Let's call the answer t and translate into exponential form.*

$8^t = 2$

$(2^3)^t = 2$

$2^{3t} = 2$ *Recall that $2 = 2^1$.*

$3t = 1$

$$\boxed{t = \frac{1}{3}}$$

        ■

The same idea of translating a logarithmic statement into exponential form allows us to solve some logarithmic equations.

EXAMPLE 6

Solve each of the following equations for t.

(a) $\log_7 t = 3$ **(b)** $\log_9 \dfrac{1}{3} = t$ **(c)** $\log_t 125 = 3$

Solution

(a) $\log_7 t = 3$ means $7^3 = t$

$$\boxed{343 = t}$$

(b) $\log_9 \dfrac{1}{3} = t$ means $9^t = \dfrac{1}{3}$

$(3^2)^t = 3^{-1}$

$3^{2t} = 3^{-1}$

$2t = -1$

$$\boxed{t = -\frac{1}{2}}$$

(c) $\log_t 125 = 3$ means $t^3 = 125$

$$t = \sqrt[3]{125}$$

$$\boxed{t = 5}$$

The checks are left to the student.   ■

EXAMPLE 7

Sketch a graph of $y = \log_4 x$.

Solution

As in Example 1, we could first sketch the inverse of $y = \log_4 x$, which is $y = 4^x$, and then sketch the graph of $y = \log_4 x$ by reflecting the graph of $y = 4^x$ about the line $y = x$. Instead, we will graph the log equation by plotting points.

It may be more convenient to translate the log equation into its exponential equivalent, $x = 4^y$, first. In the accompanying table, we have chosen a few values for y and found the corresponding x values. The (x, y) ordered pairs are plotted and the curve is sketched in Figure 12.5.

y	-2	-1	0	1	2
$x = 4^y$	$\dfrac{1}{16}$	$\dfrac{1}{4}$	1	4	16

Figure 12.5
Graph of $y = \log_4 x$

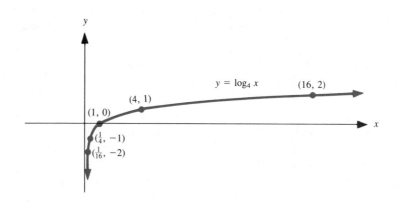

Let's end this section by summarizing the basic facts about logarithms we have developed in this section.

Summary

1.   The functions $y = \log_b x$ and $y = b^x$ are inverse functions. In other words, $y = \log_b x$ is equivalent to $x = b^y$ where $b > 0$ and $b \neq 1$.

2. The domain of the function $y = \log_b x$ is $\{x \mid x > 0\}$ and the range is R.

3. $\log_b b^r = r$

4. $\log_b 1 = 0$

5. $b^{\log_b x} = x$

In the next section we will derive some important properties of logarithms.

Exercises 12.2

In Exercises 1–38, write each logarithmic statement in exponential form and each exponential statement in logarithmic form.

1. $\log_7 49 = 2$ 2. $\log_4 64 = 3$ 3. $3^4 = 81$ 4. $2^6 = 64$

5. $\log_{10} 10{,}000 = 4$ 6. $\log_2 32 = 5$ 7. $10^3 = 1{,}000$ 8. $4^5 = 1{,}024$

9. $\log_9 81 = 2$ 10. $\log_5 125 = 3$ 11. $\log_{81} 9 = \frac{1}{2}$ 12. $\log_{125} 5 = \frac{1}{3}$

13. $6^{-2} = \frac{1}{36}$ 14. $10^{-3} = .001$ 15. $\log_3 \frac{1}{3} = -1$ 16. $8^{1/3} = 2$

17. $25^{1/2} = 5$ 18. $\log_4 \frac{1}{16} = -2$ 19. $\log_8 8 = 1$ 20. $11^1 = 11$

21. $8^0 = 1$ 22. $\log_{11} 11 = 1$ 23. $\log_{16} 8 = \frac{3}{4}$ 24. $\log_{27} 9 = \frac{2}{3}$

25. $27^{-2/3} = \frac{1}{9}$ 26. $16^{-3/4} = \frac{1}{8}$ 27. $\log_8 \frac{1}{2} = -\frac{1}{3}$ 28. $\log_4 \frac{1}{2} = -\frac{1}{2}$

29. $\log_{1/2} 4 = -2$ 30. $\log_{1/3} 27 = -3$ 31. $3^0 = 1$ 32. $\log_3 1 = 0$

33. $\log_7 1 = 0$ 34. $7^0 = 1$ 35. $\log_6 \sqrt{6} = \frac{1}{2}$ 36. $\log_2 \sqrt[3]{2} = \frac{1}{3}$

37. $6^{1/2} = \sqrt{6}$ 38. $5^{1/4} = \sqrt[4]{5}$

In Exercises 39–64, evaluate the given logarithm.

39. $\log_2 8$ 40. $\log_3 81$ 41. $\log_9 81$ 42. $\log_{10} 1{,}000$

43. $\log_4 \frac{1}{4}$ 44. $\log_6 \frac{1}{36}$ 45. $\log_5 \frac{1}{125}$ 46. $\log_2 \frac{1}{16}$

47. $\log_4 \frac{1}{2}$ 48. $\log_9 \frac{1}{3}$ 49. $\log_8 4$ 50. $\log_4 8$

51. $\log_9(-27)$ 52. $\log_{27} 0$ 53. $\log_4 \frac{1}{8}$ 54. $\log_8 \frac{1}{4}$

55. $\log_6 \sqrt{6}$ 56. $\log_{10} \sqrt[8]{10}$ 57. $\log_5 \sqrt[3]{25}$ 58. $\log_3 \sqrt{243}$

59. $\log_5(\log_3 243)$ 60. $\log_4(\log_5 625)$ 61. $\log_8(\log_7 7)$ 62. $\log_2(\log_9 9)$

63. $5^{\log_5 7}$ 64. $3^{\log_3 6}$

In Exercises 65–80, solve the given equation for x, y, or b.

65. $\log_5 x = 3$

66. $\log_2 x = 5$

67. $y = \log_{10} 1{,}000$

68. $y = \log_4 \dfrac{1}{64}$

69. $\log_b 64 = 3$

70. $\log_b 64 = 2$

71. $\log_6 x = -2$

72. $\log_3 x = -4$

73. $\log_4 x = \dfrac{3}{2}$

74. $\log_{27} x = \dfrac{2}{3}$

75. $y = \log_8 32$

76. $y = \log_{32} \dfrac{1}{8}$

77. $\log_b \dfrac{1}{8} = -3$

78. $\log_b 4 = -2$

79. $\log_5 x = 0$

80. $\log_b 1 = 0$

81. On the same set of coordinate axes, sketch the graphs of $y = \log_2 x$ and $y = \log_5 x$.

82. On the same set of coordinate axes, sketch the graphs of $y = \log_{1/2} x$ and $y = \log_{1/5} x$.

? QUESTIONS FOR THOUGHT

83. Using what you know about the graph of $y = \log_3 x$, sketch the graph of $y = \log_3(x - 1)$.

84. Repeat Exercise 83 for $y = \log_3(x + 1)$.

85. Repeat Exercise 83 for $y = \log_3 x - 1$.

86. Repeat Exercise 83 for $y = \log_3 x + 1$.

 MINI-REVIEW

87. *Perform operations, simplify, and express your answer using positive exponents only:*

$$\left(\frac{x^{1/3}y^{1/4}}{xy^2}\right)^{12}$$

88. Write in radical form. $3x^{1/5}$

89. Given $f(x) = \sqrt{x - 3}$ and $g(x) = 5x^2 - 2$, find $g[f(x)]$.

90. Given x varies directly as y and inversely as z. If $x = 5$ when $y = 3$ and $z = 4$, find x when $y = 5$ and $z = 2$.

12.3

Properties of Logarithms

Historically, logarithms were developed to simplify complex numerical computations. The availability of inexpensive hand-held calculators has made this use of logarithms virtually obsolete. Nevertheless, the properties of logarithms, which we discuss in this section, serve as the basis for using logarithms for both numerical and nonnumerical purposes.

As we pointed out in the last section, a logarithm is just an exponent:

$$x = b^y \quad \text{is equivalent to} \quad y = \log_b x$$

Consequently, it seems only natural to expect that logarithms will "inherit" the properties of exponents. The following box contains the three basic properties of logarithms. Again, we assume $b > 0$, $u > 0$, and $v > 0$.

Properties of Logarithms	**1.** $\log_b(uv) = \log_b u + \log_b v$ *Product rule*
	2. $\log_b\left(\dfrac{u}{v}\right) = \log_b u - \log_b v$ *Quotient rule*
	3. $\log_b u^r = r\log_b u$ *Power rule*

In words, these properties can be remembered as follows:

1. The product rule says that *"the log of a product is equal to the sum of the logs."*

2. The quotient rule says that *"the log of a quotient is equal to the difference of the logs."*

3. The power rule says that *"the log of a power is the exponent times the log."*

We will prove the product rule for logarithms and leave the proofs of the other two properties as exercises for the student.

PROOF OF THE PRODUCT RULE: Rule 1 for exponents (the "product rule" for exponents) says that

$$b^m \cdot b^n = b^{m+n}$$

Let $u = b^m$ and $v = b^n$ and let's write these exponential statements in logarithmic form:

$$u = b^m \quad \text{is equivalent to} \quad \log_b u = m$$
$$v = b^n \quad \text{is equivalent to} \quad \log_b v = n$$

Thus, we have

$$u \cdot v = b^m \cdot b^n = b^{m+n} \quad \text{which is equivalent to} \quad \log_b(uv) = m + n$$

If we now substitute $m = \log_b u$ and $n = \log_b v$ into the last equation, we get

$$\log_b(uv) = \log_b u + \log_b v$$

which is exactly what the product rule states.

Just as the product rule for logarithms is the logarithmic form of exponent rule 1, so too the quotient and power rules for logarithms are the logarithmic forms of exponent rules 4 and 2, respectively.

These three properties of logarithms enable us to take the logarithm of a complicated expression which involves products, quotients, powers, and roots, and to write it as a sum of logarithms of much simpler expressions.

Several examples will illustrate this process.

EXAMPLE 1

Express as a sum of simpler logarithms. $\log_b(x^2y)$

Solution

We begin by using the product rule:

$$\log_b(x^2y) = \log_b x^2 + \log_b y \qquad \textit{Now use the power rule.}$$

$$= \boxed{2 \log_b x + \log_b y}$$

EXAMPLE 2

Express as a sum of simpler logarithms. $\log_b \dfrac{\sqrt{x}}{z^5}$

Solution

As we have mentioned before, since logarithms are exponents, when we are working with logarithms we generally write radical expressions in exponential form. Thus, we write $\sqrt{x} = x^{1/2}$.

$$\log_b \frac{\sqrt{x}}{z^5} = \log_b \frac{x^{1/2}}{z^5} \qquad \textit{Using the quotient rule, we get}$$

$$= \log_b x^{1/2} - \log_b z^5 \qquad \textit{Now use the power rule.}$$

$$= \boxed{\frac{1}{2} \log_b x - 5 \log_b z}$$

As the expressions get more complex, we must be very careful using the properties of logarithms—particularly the quotient rule.

EXAMPLE 3

Express as a sum of simpler logarithms. $\log_8 \dfrac{\sqrt[3]{x^3y^5}}{8z^4}$

Solution

As we did in Example 2, we first write the radical as a fractional exponent.

$$\log_8 \frac{\sqrt[3]{x^3y^5}}{8z^4} = \log_8 \frac{(x^3y^5)^{1/3}}{8z^4} \qquad \textit{Using the quotient rule, we get}$$

$$= \log_8(x^3y^5)^{1/3} - \log_8 8z^4 \qquad \textit{Using the power rule, we get}$$

$$= \frac{1}{3} \log_8 x^3y^5 - \log_8 8z^4 \qquad \textit{Using the product rule, we get}$$

$$= \frac{1}{3}(\log_8 x^3 + \log_8 y^5) - (\log_8 8 + \log_8 z^4)$$

Note that both sets of parentheses are essential. Now use the power rule again.

$$= \frac{1}{3}(3 \log_8 x + 5 \log_8 y) - (\log_8 8 + 4 \log_8 z)$$

Now we remove parentheses; also, $\log_8 8 = 1$.

$$= \boxed{\log_8 x + \frac{5}{3} \log_8 y - 1 - 4 \log_8 z}$$

It is also important to point out that an expression such as

$$\log_b(x^3 + y^5)$$

cannot be expressed as the sum of simpler logs. The properties of logarithms *do not* apply to the logarithm of a sum. Thus,

$$\log_b(x^3 + y^5) \neq 3 \log_b x + 5 \log_b y$$

Similarly,

$$\frac{\log_b x}{\log_b y} \neq \log_b x - \log_b y$$

because the quotient rule applies to the log of a quotient, *not* to a quotient of logs.

EXAMPLE 4

Write as a single logarithm. $3 \log_b x + \dfrac{1}{2} \log_b y - 6 \log_b z$

Solution

We are going to use the properties of logarithms in the reverse direction, by first noting that the power rule allows us to write the coefficients of the logarithms as exponents of the variables.

$$3 \log_b x + \frac{1}{2} \log_b y - 6 \log_b z = \log_b x^3 + \log_b y^{1/2} - \log_b z^6 \qquad \text{\textit{Now use the product rule.}}$$

$$= \log_b(x^3 y^{1/2}) - \log_b z^6 \qquad \text{\textit{Now use the quotient rule.}}$$

$$= \boxed{\log_b \frac{x^3 y^{1/2}}{z^6}}$$

EXAMPLE 5

Given $\log_b 2 = 1.2$ and $\log_b 3 = 1.38$, find each of the following:

(a) $\log_b 6$ **(b)** $\log_b 81$ **(c)** $\log_b \dfrac{3}{2}$ **(d)** $\log_b 48$

Solution

(a) $\log_b 6$ *First we rewrite 6 as a product or quotient of 2 and/or 3: $6 = 2 \cdot 3$*

$= \log_b(2 \cdot 3)$ *Then use the product rule.*

$= \log_b 2 + \log_b 3$ *Since $\log_b 2 = 1.2$ and $\log_b 3 = 1.38$,*

$= 1.2 + 1.38$

$= \boxed{2.58}$

(b) $\log_b 81$ *Rewrite 81 as 3^4.*

$= \log_b 3^4$ *Then use the power rule.*

$= 4 \log_b 3$ *Since $\log_b 3 = 1.38$,*

$= 4(1.38)$

$= \boxed{5.52}$

(c) $\log_b \dfrac{3}{2}$ *Use the quotient rule.*

$$= \log_b 3 - \log_b 2$$

$$= 1.38 - 1.2$$

$$= \boxed{0.18}$$

(d) $\log_b 48$ *Rewrite 48 as a product of factors of 2 and 3: $48 = 2^4 \cdot 3$*

$$= \log_b(2^4 \cdot 3) \qquad \textit{Use the product rule.}$$

$$= \log_b 2^4 + \log_b 3 \qquad \textit{Use the power rule.}$$

$$= 4 \log_b 2 + \log_b 3$$

$$= 4(1.2) + 1.38$$

$$= \boxed{6.18}$$

Exercises 12.3

In Exercises 1–26, use the properties of logarithms to express the given logarithm as a sum of simpler ones (wherever possible). Assume that all variables represent positive quantities.

1. $\log_5 xyz$ **2.** $\log_2 rst$ **3.** $\log_7 \dfrac{2}{3}$ **4.** $\log_6 \dfrac{4}{7}$

5. $\log_3 x^3$ **6.** $\log_4 y^6$ **7.** $\log_b a^{2/3}$ **8.** $\log_b t^{3/4}$

9. $\log_b b^8$ **10.** $\log_c c^5$ **11.** $\log_s s^{-1/4}$ **12.** $\log_n n^{-5/6}$

13. $\log_b x^2 y^3$ **14.** $\log_b uv^2 w^7$ **15.** $\log_b \dfrac{m^4}{n^2}$ **16.** $\log_b \dfrac{r^3 s}{t^5}$

17. $\log_b \sqrt{xy}$ **18.** $\log_b \sqrt[3]{mn}$ **19.** $\log_2 \sqrt[5]{\dfrac{x^2 y}{z^3}}$ **20.** $\log_7 \sqrt[4]{\dfrac{x^3 y^5}{z}}$

21. $\log_b(xy + z^2)$ **22.** $\log_b(m^2 - n^3)$ **23.** $\log_b \dfrac{x^2}{yz}$ **24.** $\log_b \dfrac{u^3}{4v^2}$

25. $\log_6 \sqrt{\dfrac{6m^2 n}{p^5 q}}$ **26.** $\log_3 \sqrt{\dfrac{27r^3 s}{t^5}}$

In Exercises 27–40, write the given expressions as a single logarithm.

27. $\log_b x + \log_b y$ **28.** $\log_b x - \log_b y$ **29.** $2 \log_b m - 3 \log_b n$

30. $4 \log_b u + 7 \log_b v$ **31.** $4 \log_b 2 + \log_b 5$ **32.** $2 \log_b 3 + \log_b 4$

33. $\dfrac{1}{3} \log_b x + \dfrac{1}{4} \log_b y - \dfrac{1}{5} \log_b z$ **34.** $\log_b \dfrac{x}{3} + \log_b \dfrac{y}{4} - \log_b \dfrac{z}{5}$ **35.** $\dfrac{1}{2}(\log_b x + \log_b y) - 2 \log_b z$

36. $\dfrac{1}{3}(\log_b m + \log_b n) - 6 \log_b p$ **37.** $2 \log_b x - (\log_b y + 3 \log_b z)$ **38.** $5 \log_p u - \left(\dfrac{1}{2} \log_p v + \log_p w\right)$

39. $\dfrac{2}{3} \log_p x + \dfrac{4}{3} \log_p y - \dfrac{3}{7} \log_p z$ **40.** $\dfrac{1}{4} \log_n x - \log_n y - \log_n z$

In Exercises 41–52, evaluate the logarithm given $\log_b 2 = 1.2$, $\log_b 3 = 1.42$, and $\log_b 5 = 2.1$.

41. $\log_b 10$ **42.** $\log_b 15$ **43.** $\log_b \dfrac{2}{5}$ **44.** $\log_b \dfrac{3}{5}$

45. $\log_b \dfrac{1}{3}$ **46.** $\log_b \dfrac{1}{5}$ **47.** $\log_b 32$ **48.** $\log_b 125$

49. $\log_b 100$ **50.** $\log_b 1,000$ **51.** $\log_b \sqrt{20}$ **52.** $\log_b \sqrt{18}$

In Exercises 53–58, let $\log_b x = A$, $\log_b y = B$, *and* $\log_b z = C$. *Express each logarithm in terms of A, B, and C.*

53. $\log_b \sqrt[3]{x^2}$ **54.** $\log_b \dfrac{1}{\sqrt[5]{x}}$ **55.** $\log_b \dfrac{x^3 y^2}{z}$ **56.** $\log_b \dfrac{z}{x^3 y^2}$

57. $\log_b \sqrt{\dfrac{x^5 y}{z^3}}$ **58.** $\log_b \dfrac{\sqrt{x}}{\sqrt[3]{yz}}$

? QUESTIONS FOR THOUGHT

59. Describe what is *wrong* (if anything) with each of the following:

(a) $\log_b(x^3 + y^4) \overset{?}{=} 3 \log_b x + 4 \log_b y$

(b) $\dfrac{\log_b x^3}{\log_b y^2} \overset{?}{=} 3 \log_b x - 2 \log_b y$

60. Use the outline for the proof of the product rule for logarithms to prove the quotient and power rules for logarithms.

12.4

Common Logarithms and Change of Base

As we mentioned before, logarithms were invented to make complex numerical calculations easier to do. While calculators have made this use of logarithms obsolete, logarithms do still have widespread applications in the physical and social sciences. The material we develop in this section will enable us to do some actual computations with logarithms, which in turn gives us an opportunity to practice using the properties of logarithms we developed in the last section. (If your instructor is teaching the topic of logarithms with the use of a calculator he or she may instruct you to skip some material in the next few sections.)

Our development of the properties of logarithms in the last section was independent of the choice of the base b (as long as $b > 0$ and $b \neq 1$). Since we use the decimal number system (that is, a number system using the base 10), one of the most frequently used bases for logarithms is 10.

DEFINITION

Logarithms with base 10 are called *common logarithms*. In other words, $\log_{10} x$ is called a common logarithm.

Normally we write

$$\log_{10} x \quad \text{as} \quad \log x$$

That is, the base 10 is understood. (This is similar to the convention that the index 2 is understood when we write $\sqrt{x}$.)

As we have seen on numerous occasions,

$$\log_b b^r = r \quad \text{which is just the logarithmic form of} \quad b^r = b^r$$

Thus,

$$\log_{10} 1{,}000 = \log_{10} 10^3 \ = 3$$
$$\log_{10} 100 = \log_{10} 10^2 \ = 2$$
$$\log_{10} 10 = \log_{10} 10^1 \ = 1$$
$$\log_{10} 1 = \log_{10} 10^0 \ = 0$$
$$\log_{10} 0.1 = \log_{10} 10^{-1} = -1$$
$$\log_{10} 0.01 = \log_{10} 10^{-2} = -2$$
$$\log_{10} 0.001 = \log_{10} 10^{-3} = -3$$

For numbers which are not easily expressible as powers of 10, we will need the idea of scientific notation discussed in Section 5.3, together with a table of common logarithms. For example, to find log 495 we proceed as follows (remember that log means $\log_{10}$):

log 495	*Write 495 in scientific notation.*
$= \log(4.95 \times 10^2)$	*Use the product rule for logarithms.*
$= \log 4.95 + \log 10^2$	*Since* $\log 10^2 = 2$ *we get*
$= \log 4.95 + 2$	

In order to find log 4.95 we look in the table of common logarithms given in Appendix D. We look down the first column headed N to the row containing 4.9, (the first two digits of 4.95) and then go across to the column with 5 (the third digit of 4.95) at the top (or bottom). The entry we find in row 4.9 and column 5 is 0.6946.

N	0	1	2	3	4	5	6	7	8	9
1.0										
1.1										
1.2										
.										
.										
4.8										
4.9						→ 0.6946				
5.0										
.										
.										

Thus, we have

$$\log 495 = 0.6946 + 2$$
$$= \boxed{2.6946}$$

Note that since $100 < 495 < 1{,}000$, we would expect $2 < \log 495 < 3$.

The decimal part of the logarithm (.6946) is called the ***mantissa***; the integer part of the logarithm (2) is called the ***characteristic***.

It should be noted that the mantissa which we obtain from the table is only a four-decimal approximation. (On a calculator we would obtain the seven-decimal approximation 2.6946052.) The actual value for log 4.95 is an irrational number. Thus, we should actually write $\log 495 \approx 2.6946$; however, it is accepted custom to write $\log 495 = 2.6946$.

EXAMPLE 1

Find the following logarithms.

(a) log 28,300   **(b)** log 0.0749

Solution

(a) $\log 28{,}300 = \log(2.83 \times 10^4)$
$= \log 2.83 + \log 10^4$
$= \log 2.83 + 4$ *The characteristic is* 4; *look up*
$= 0.4518 + 4$ $\log 2.83$ *in the table.*
$= \boxed{4.4518}$

(b) $\log 0.0749 = \log(7.49 \times 10^{-2})$
$= \log 7.49 + \log 10^{-2}$
$= 0.8745 + (-2)$
$= \boxed{0.8745 - 2}$

Based on these two examples, we can easily generalize that the characteristic of a common logarithm is simply the exponent of 10 when the number is written in scientific notation. ∎

If we find log 0.0749 on a calculator and round off we get -1.1255. This is the same value we would obtain if we compute $0.8745 - 2$ from Example 1. (We *do not* get -2.8745. That is, we do not simply attach the mantissa to a negative characteristic.) The calculator value of -1.1255 means $-1 - 0.1255$, which has a *negative* mantissa. However, our table contains only positive decimals; therefore, if we want to keep the association of the decimal 0.8745 with log 7.49, we must keep the mantissa *positive*. Thus, we may leave our answer as $0.8745 - 2$.

Alternatively, we may sometimes find it convenient to write a logarithm with a negative characteristic in other equivalent forms. For example,

$$0.8745 - 2 = 0.8745 + 8 - 10 = 8.8745 - 10$$
$$0.8745 - 2 = 0.8745 + 18 - 20 = 18.8745 - 20$$
$$0.8745 - 2 = 0.8745 + 28 - 30 = 28.8745 - 30$$

In other words, we may write the characteristic of -2 as $8 - 10$, $18 - 20$, $28 - 30$, or in a variety of other forms depending on the situation. In the next section we will look at an example in which one of these alternate forms is useful.

The table of common logarithms can also be used the other way around—to find a number whose logarithm is known.

EXAMPLE 2

Find N if $\log N = 3.9299$.

Solution

We look at $\log N = 3.9299$ as $\log N = 0.9299 + 3$ so we can see that the characteristic is 3. Looking in the body of the table, we find that the mantissa 0.9299 is located in row 8.5 and the column headed by 1:

N	0	1	2	3	4	5	6	7	8	9
.		↑								
.										
.										
.										
8.4										
8.5	←— 0.9299									
8.6										
.										
.										
.										

Thus, $\log 8.51 = 0.9299$. Summarizing, we have

$$\log N = 3.9299$$
$$= 0.9299 + 3 \qquad \textit{3 is the characteristic; 0.9299 is the mantissa.}$$
$$N = 8.51 \times 10^3$$

$$\boxed{N = 8{,}510}$$

The process demonstrated in Example 2 is called finding the *antilogarithm* of 3.9299. Thus, Example 2 could have been worded as "Find antilog 3.9299."

EXAMPLE 3

Find antilog$(6.5391 - 10)$.

Solution | This example is asking us to find N given that $\log N = 6.5391 - 10$.

$\log N = 6.5391 - 10$ *The characteristic is $6 - 10 = -4$; the mantissa is 0.5391.*

$\log N = 0.5391 - 4$ *We find 0.5391 in row 3.4 and in the column headed with 6.*

$N = 3.46 \times 10^{-4}$

$\boxed{N = 0.000346}$ ■

Change of Base

Having a table of logarithms available makes it possible for us to compute logarithms to other bases as well. For example, if we need to compute $\log_8 15$ and we have available only a table of common logarithms, we can proceed as follows:

$$\text{Let } \log_8 15 = y \quad \textit{Write in exponential form.}$$

$8^y = 15$ *We can take the log of both sides of the equation. (We will discuss this step in more detail in Section 12.5.)*

$\log 8^y = \log 15$ *Use the power rule.*

$y \log 8 = \log 15$

$y = \dfrac{\log 15}{\log 8}$ *But $y = \log_8 15$; therefore, we have*

$\log_8 15 = \dfrac{\log 15}{\log 8}$ *We find $\log 15$ and $\log 8$ in the table.*

$= \dfrac{1.1761}{0.9031}$

$= \boxed{1.3023}$

In general, we find that we can prove the following formula for change of base in a logarithmic computation:

Change of Base Formula	$\log_a x = \dfrac{\log_b x}{\log_b a}$

EXAMPLE 4 | Convert $\log_3 x$ into a logarithm using base 9.

Solution | According to the change of base formula,

$\log_3 x = \dfrac{\log_9 x}{\log_9 3}$ *Since $\log_9 3 = \dfrac{1}{2}$,*

$= \dfrac{\log_9 x}{\dfrac{1}{2}} = \boxed{2 \log_9 x}$

In the next section we will see how we can make use of this ability to change bases in a computation. ■

There is a particular number, designated by the letter e, which plays a very prominent role in the sciences. The number e is an irrational number and its value (correct to seven places) is $e = 2.7182818$.

The number e comes up so frequently that it is called the *natural* base, and $\log_e$ is called the *natural logarithm*. It is also given a special notation:

$$\boxed{\log_e \text{ is written as ln}}$$

Using linear interpolation (discussed below) and the value $e = 2.7183$, we will find that

$$\boxed{\log e = 0.4343}$$

In the next section we will also make use of this fact.

Linear Interpolation

The table of common logarithms allows us to look up three-digit numbers. What do we do if we need the logarithm of a number with more than three digits? The fact that the logarithm function increases very slowly allows us to predict that very small changes in x will cause very small changes in $\log x$. In fact, for small changes in x the graph of $y = \log x$ can be closely approximated by a straight line (hence the name *linear* interpolation). Since a straight line has a constant slope, we can set up a proportion to find the logarithm of a number with more than three digits.

EXAMPLE 5

Find $\log 2.837$.

Solution

$2.837 = 2.837 \times 10^0$ so its characteristic is 0. Since 2.837 has four digits it does not appear in the table given in Appendix D. However, since

$$2.830 < \quad 2.837 < \quad 2.840$$
$$\log 2.830 < \log 2.837 < \log 2.840$$

We find $\log 2.830$ and $\log 2.840$ in the table and arrange our information as follows:

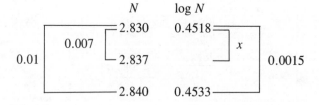

Now we set up the following proportion:

$$\frac{0.007}{0.01} = \frac{x}{0.0015}$$

$$\frac{(0.007)(0.0015)}{0.01} = x$$

$$0.00105 = x \qquad \textit{Rounding off to four places, we get}$$

$$0.0011 = x$$

Therefore, we add 0.0011 to 0.4518 and get

$$\log 2.837 = 0.4518 + 0.0011 = \boxed{0.4529} \qquad \blacksquare$$

We can apply the same procedure to finding antilogarithms.

EXAMPLE 6   Find antilog 2.9318.

Solution   We see that the characteristic is 2 and the mantissa is 0.9318. When we look up 0.9318 in the body of the table we do not find it. However, we find that 0.9318 falls between the two table values 0.9315 and 0.9320. We find the antilog of 0.9315 and 0.9320 in the table and arrange our information as follows:

$$
\begin{array}{ccc}
 & N & \log N \\
 & 8.540 & 0.9315 \\
x & & 0.0003 \\
0.01 & & 0.9318 \quad 0.0005 \\
 & 8.550 & 0.9320
\end{array}
$$

We set up the following proportion:

$$\frac{x}{0.01} = \frac{0.0003}{0.0005}$$

$$x = \frac{(0.01)(0.0003)}{0.0005}$$

$$x = 0.006$$

We add 0.006 to 8.54 and get 8.546.

Thus, antilog $2.9318 = 8.546 \times 10^2 = \boxed{854.6} \qquad \blacksquare$

Exercises 12.4

In Exercises 1–14, find the given logarithm.

1. $\log 584$

2. $\log 0.0584$

3. $\log 0.00371$

4. $\log 37,100$

5. log 280,000   **6.** log 3,600,000   **7.** log 0.0000553   **8.** log 0.000939

9. log 6.21   **10.** log 8.44   **11.** log 0.837   **12.** log 0.102

13. log 8   **14.** log 0.6

In Exercises 15–28, find the given antilogarithm.

15. antilog 2.8420   **16.** antilog 3.6263   **17.** antilog(0.7308 − 3)   **18.** antilog(0.7604 − 5)

19. antilog 4.1875   **20.** antilog(0.8506 − 1)   **21.** antilog(8.9768 − 10)   **22.** antilog 6.2068

23. antilog 5.8733   **24.** antilog(7.0334 − 10)   **25.** antilog 0.7803   **26.** antilog 0.7202

27. antilog 1.5551   **28.** antilog(9.5551 − 10)

In Exercises 29–36, find the given logarithm or antilogarithm.

29. log 0.941   **30.** antilog 0.941   **31.** antilog 0.683   **32.** log 6.83

33. log 776   **34.** antilog 0.7760   **35.** antilog 4.85   **36.** log 4.85

In Exercises 37–40, use the change of base formula to convert the following into logarithms using base 10.

37. $\log_5 x$ **38.** $\log_6 4$ **39.** $\log_7 8$ **40.** $\log_5 10$

41. Convert $\log_5 8$ into a logarithm using base 25.   **42.** Convert $\log_4 5$ into a logarithm using base 2.

In Exercises 43–52, use linear interpolation to find the given logarithm or antilogarithm.

43. log 2,847   **44.** log 0.003896   **45.** antilog 4.6070   **46.** antilog 2.1539

47. log 54,810   **48.** antilog(0.8428 − 3)   **49.** antilog 0.9453   **50.** log 2.164

51. log 0.0006478   **52.** antilog(6.3880 − 10)

? QUESTION FOR THOUGHT

53. Let $y = \log x$. Compute the corresponding values for $x = 3.53, 3.54, 3.55,$ and 3.56. Plot the points (x, y) obtained. Do they look like they fall on a straight line? Do they in fact all fall on a straight line? [*Hint:* Check the slopes between some of the pairs of points you obtained.]

 MINI-REVIEW

Solve for x:

54. $\dfrac{3}{x-2} + 1 = \dfrac{5}{2}$ **55.** $|5x - 2| > 3$ **56.** $\sqrt{2x+3} + 1 = 4$ **57.** $\dfrac{3}{x-2} + 1 = \dfrac{10}{x}$

12.5

Exponential and Logarithmic Equations

In Section 12.2 we solved simple logarithmic equations using the definition of a logarithm. We can use the properties of logarithms to solve more complicated logarithmic equations. For example, in order to solve the logarithmic equation

$$\log_3 x - \log_3 9 = 2$$

we proceed as follows:

$$\log_3 x - \log_3 9 = 2 \qquad \text{\textit{Use the quotient rule to write the}}$$
$$\text{\textit{left-hand side as a single logarithm.}}$$
$$\log_3\left(\frac{x}{9}\right) = 2 \qquad \text{\textit{Write this equation in exponential form.}}$$
$$3^2 = \frac{x}{9}$$
$$9 = \frac{x}{9}$$
$$\boxed{81 = x}$$

CHECK $x = 81$:
$$\log_3 x - \log_3 9 = 2$$
$$\log_3 81 - \log_3 9 \overset{?}{=} 2$$
$$4 - 2 \overset{\checkmark}{=} 2$$

Looking at this solution, we can see that our basic strategy in solving a logarithmic equation is to rewrite the equation so that it involves a single logarithm, and then translate it into exponential form.

EXAMPLE 1

Solve the following equation:

$$\log_6 x + \log_6(x + 1) = 1$$

Solution

$$\log_6 x + \log_6(x + 1) = 1 \qquad \text{\textit{Use the product rule.}}$$
$$\log_6 x(x + 1) = 1 \qquad \text{\textit{Write this equation in exponential form.}}$$
$$x(x + 1) = 6^1$$
$$x^2 + x = 6$$
$$x^2 + x - 6 = 0 \qquad \text{\textit{This is a quadratic equation.}}$$
$$(x + 3)(x - 2) = 0$$
$$x = -3 \quad \text{or} \quad x = 2$$

CHECK $x = -3$:
$$\log_6 x + \log_6(x + 1) = 1$$
$$\log_6(-3) + \log_6(-3 + 1) \overset{?}{=} 1$$

Since the domain for the logarithm function is restricted to positive numbers, $x = -3$ is not an allowable value.

CHECK $x = 2$:
$$\log_6 x + \log_6(x + 1) = 1$$
$$\log_6 2 + \log_6(2 + 1) \overset{?}{=} 1 \qquad \text{\textit{Use the product rule.}}$$
$$\log_6(2 \cdot 3) \overset{?}{=} 1$$
$$\log_6 6 \overset{\checkmark}{=} 1$$

Thus, the solution is $\boxed{x = 2}$

If logarithmic expressions appear on both sides of the equation, we first use the properties of equality to collect all log terms on one side of the equation and all terms without logs on the other side. Then we rewrite the equation as a single logarithm, as illustrated in the next example.

EXAMPLE 2

Solve the following equation for x:

$$\log x = 2 + \log(x - 1)$$

Solution

$\log x = 2 + \log(x - 1)$ *Collect all log terms on one side and "non-log" terms on the other side.*

$\log x - \log(x - 1) = 2$ *Use the quotient rule.*

$\log\left(\dfrac{x}{x - 1}\right) = 2$ *Write this equation in exponential form.*

$\dfrac{x}{x - 1} = 10^2$ *Remember that $\log\left(\dfrac{x}{x + 1}\right)$ has base 10. Solve the equation.*

$x = 10^2(x - 1)$

$x = 100(x - 1)$

$x = 100x - 100$

$99x = 100$

$$x = \boxed{\dfrac{100}{99}}$$

Check to verify that this solution is valid. ■

The log function is a one-to-one function. Thus,

$$\boxed{\text{If } \log_b r = \log_b s, \text{ then } r = s}$$

We can use this fact to solve equations which consist entirely of log terms, as shown in the next example.

EXAMPLE 3

Solve for a. $2 \log_p a = \log_p 3$

Solution

As with the previous examples, we could start by collecting all log terms on one side of the equation and non-log terms on the other side, and then use the log properties. Instead, we will just use the power rule on the left-hand side of the equation.

$\log_p a^2 = \log_p 3$ *Now we have it in the form $\log_b r = \log_b s$ and because the log function is one-to-one,*

$a^2 = 3$ *Now solve for a.*

$a = \pm\sqrt{3}$

CHECK: When we check $a = -\sqrt{3}$, we get

$$2 \log_p(-\sqrt{3}) = \log_p 3$$

Since the domain of the log function is restricted to positive numbers, we eliminate $a = -\sqrt{3}$. The student should check that the value $\boxed{a = \sqrt{3}}$ is a valid solution. ∎

In Section 12.1 we mentioned that the exponential function is a one-to-one function. Hence, if $b^r = b^t$, then we can conclude that $r = t$. This allowed us to solve certain types of exponential equations. We can solve other types of exponential equations using the fact that log is a well-defined function. That is,

> If r and s are positive and $r = s$, then
> $$\log_b r = \log_b s$$

When we use this fact, we will say that we are "taking the log of both sides of the equation."

EXAMPLE 4 | *Solve for x.* $3^{2x} = 5$

Solution | Note that in this equation, the variable for which we are solving appears in the exponent. We cannot rewrite 5 and 3 as powers with the same base as we did in Section 12.1. Let's take the log of both sides and see what happens. We will take the log base 10 so we may use the table in Appendix D.

$3^{2x} = 5$ — *Take the log of both sides of the equation.*

$\log 3^{2x} = \log 5$ — *Use the power rule on the left-hand side.*

$2x(\log 3) = \log 5$ — *Now we can solve for x. Keep in mind that both $\log 3$ and $\log 5$ are constants.*

$x = \dfrac{\log 5}{2 \log 3}$ — *Which, when evaluated, is*

$x = \dfrac{0.6990}{2(0.4771)}$

$x \approx \boxed{0.733}$ ∎

EXAMPLE 5 | *Solve for y.* $2^y = 3^{y+4}$

Solution | $2^y = 3^{y+4}$ — *Since we cannot express both 2 and 3 as powers of the same base, we take the log of both sides.*

$\log 2^y = \log 3^{y+4}$ — *Use the power rule on both sides.*

$y \log 2 = (y + 4) \log 3$ — *Solve for y. To isolate y, first multiply out the right-hand side.*

$$y \log 2 = y \log 3 + 4 \log 3$$

Collect all y terms on one side and all non-y terms on the other side.

$$y \log 2 - y \log 3 = 4 \log 3$$

Factor y from the left-hand side.

$$y(\log 2 - \log 3) = 4 \log 3$$

Divide both sides by the constant multiplier of y, $\log 2 - \log 3$.

$$y = \frac{4 \log 3}{\log 2 - \log 3}$$

Which, when evaluated, is

$$y = \frac{4(0.4771)}{0.3010 - 0.4771}$$

$$y \approx \boxed{-10.84}$$

Exercises 12.5

In Exercises 1–48, solve the equations.

1. $\log_3 5 + \log_3 x = 2$

2. $\log_4 x + \log_4 5 = 1$

3. $\log_2 x = 2 + \log_2 3$

4. $\log_5 x = 2 - \log_5 3$

5. $2 \log_5 x = \log_5 36$

6. $3 \log_4 x = \log_4 125$

7. $\log_3 x + \log_3(x - 8) = 2$

8. $\log_6 x + \log_6(x - 5) = 1$

9. $\log_2 a + \log_2(a + 2) = 3$

10. $\log_3 a + \log_3(a - 2) = 1$

11. $\log_2 y - \log_2(y - 2) = 3$

12. $\log_2 x - \log_2(x + 3) = 2$

13. $\log_3 x - \log_3(x + 3) = 5$

14. $\log_3 2x - \log_3(x - 2) = 4$

15. $\log_b 5 + \log_b x = \log_b 10$

16. $\log_b 6 + \log_b y = \log_b 18$

17. $\log_p x - \log_p 2 = \log_p 7$

18. $\log_p 2 - \log_p x = \log_p 8$

19. $\log_5 x + \log_5(x + 1) = \log_5 2$

20. $\log_3 y + \log_3(y - 2) = \log_3 3$

21. $\log_3 x - \log_3(x - 2) = \log_3 4$

22. $\log_3 a + \log_3(a + 2) = \log_3 15$

23. $\log_4 x - \log_4(x - 4) = \log_4(x - 6)$

24. $\log_9(2x + 7) - \log_9(x - 1) = \log_9(x - 7)$

25. $2 \log_2 x = \log_2(2x - 1)$

26. $2 \log_4 y = \log_4(y + 2)$

27. $\frac{1}{2} \log_3 x = \log_3(x - 6)$

28. $\frac{1}{2} \log_4 x = \log_4(2x + 1)$

29. $2 \log_b x = \log_b(6x - 5)$

30. $2 \log_b x = \log_b(2x + 1)$

31. $2^x = 5$

32. $3^x = 7$

33. $2^{x+1} = 6$

34. $5^{x+1} = 6$

35. $4^{2x+3} = 5$

36. $5^{2x+1} = 9$

37. $7^{y+1} = 3^y$

38. $6^{y+2} = 5^y$

39. $6^{2x+1} = 5^{x+2}$

40. $5^{2x-1} = 3^{x-3}$

41. $8^{3x-2} = 9^{x+2}$

42. $10^{3x+2} = 5^{x+3}$

43. $3^x = 5 \cdot 2^x$

44. $5^x = 7 \cdot 3^x$

45. $2^y 5^y = 3$

46. $3^a 5^a = 10$

47. $4^a 3^{a+1} = 2$

48. $6^a 5^{a-1} = 8$

MINI-REVIEW

49. The sum of three consecutive odd integers is -27. Find them.

50. How long will it take a car traveling at 60 mph to catch up with a car traveling at 45 mph with an hour head start?

51. The sum of a number and twice its reciprocal is $\frac{11}{3}$. Find the numbers.

52. Anne makes two investments. She invests $5,000 more at 7% than she invests at 6.5%. If the total yearly interest from both investments is $1,430, how much is invested at each rate?

53. The distance an object falls is directly proportional to the square of the length of the time it falls. If an object falls 64 feet in 2 seconds, how far does it fall in 5 seconds?

12.6

Applications

In this section we will see how a number of the ideas discussed in this chapter can be used to deal with a variety of applications. (Many of the examples and exercises in this section can be done using a hand-held scientific calculator. Your instructor may have you work this section with the aid of a calculator. However, we will work out these examples using logarithms in order to practice working with logarithms and make use of their properties.)

EXAMPLE 1

Compute to the nearest thousandth using logarithms. $\sqrt[3]{0.0852}$

Solution

The difficulty in computing a cube root directly can be circumvented by applying logarithms as follows:

Let $N = \sqrt[3]{0.0852}$ *Write the radical as a fractional exponent.*

$N = 0.0852^{1/3}$ *Take the logarithm of both sides of the equation.*

$\log N = \log(0.0852)^{1/3}$ *Use the power rule for logarithms.*

$\qquad = \frac{1}{3} \log 0.0852$

$\qquad = \frac{1}{3} \log(8.52 \times 10^{-2})$

$\qquad = \frac{1}{3}(\log 8.52 - 2)$ *Look up* $\log 8.52$.

$\qquad = \frac{1}{3}(0.9304 - 2)$

In order to multiply by $\frac{1}{3}$ and still keep the logarithm in a form in which it is easy to recognize the characteristic and mantissa, we write the characteristic of -2 as $28 - 30$:

$\log N = \frac{1}{3}(28.9304 - 30)$ *Multiply out.*

$\qquad = 9.6435 - 10$

Thus far, we have log $N = 9.6435 - 10$ and so

$N = $ antilog$(9.6435 - 10)$ *We get the antilog from the table.*

$\quad = \boxed{0.440}$

If we multiply out $(0.440)^3$ we get 0.08518, which is quite close to the required value of 0.0852.

Note that the use of logarithms simplified the computation because of the power rule for logarithms which changes a power (the cube root) into a single multiplication (multiplying by $\frac{1}{3}$). ∎

Compound Interest

Formula for Compound Interest	If P dollars is invested at an annual interest rate of $r\%$ compounded n times per year, then A, the amount of money present after t years, is given by the formula $$A = P\left(1 + \frac{r}{n}\right)^{nt}$$ where r is the percentage written as a decimal.

EXAMPLE 2

If \$8,000 is invested in an account which pays an annual rate of 9.2% and the interest is compounded quarterly (4 times per year), find the amount of money in the account after 5 years.

Solution

We use the compound interest formula and substitute $P = 8,000$, $r = 0.092$, $n = 4$, and $t = 5$. Thus, we get

$$A = 8,000\left(1 + \frac{0.092}{4}\right)^{4\cdot5}$$
$$= 8,000(1 + 0.023)^{20}$$
$$= 8,000(1.023)^{20} \quad \text{\textit{Take the logarithm of both sides of the equation.}}$$

$\log A = \log[8,000(1.023)^{20}]$ *Using the properties of logarithms, we get*
$\quad = \log 8,000 + \log(1.023)^{20}$
$\quad = \log 8,000 + 20 \log 1.023$ *We find* $\log 1.023$ *by interpolation.*
$\quad = 3.9031 + 20(0.0099)$
$\quad = 3.9031 + 0.198$
$\quad = 4.1011$

Therefore,

$A = $ antilog 4.1011 *Again we interpolate to find* antilog 4.1011.

$\boxed{A = \$12,620}$

The actual amount is $\$12,606.74$.   ∎

Growth and Decay

There are many relationships in the "real world" which can be fairly accurately described by exponential equations. In such a case, we often say that the equation is a *model* of the situation.

We will now describe two models which involve the number e, the base for natural logarithms mentioned in the last section. First, if we start with a population P_0 which is growing at a continuous rate of $r\%$ per year, then after t years the population will have grown to P, where

$$P = P_0 e^{rt}$$ where r is written as a decimal.

This is called an ***exponential growth model***.

Similarly, if we start with a certain amount of radioactive material A_0 which is decaying at the continuous rate of $r\%$ per year, then after t years the amount of radioactive material still present is A, where

$$A = A_0 e^{-rt}$$ where r is written as a decimal.

This is called an ***exponential decay model***.

EXAMPLE 3

A certain radioactive substance decays at the rate of 6% per year. How long will it take until 50 grams of this substance has decayed, leaving only 25 grams of radioactive substance remaining?

Solution

We use the formula $A = A_0 e^{-rt}$, in which we are being asked to find t when $A_0 = 50$, $A = 25$, and $r = 0.06$.

$25 = 50 e^{-0.06t}$ *Divide both sides of the equation by 50.*

$0.5 = e^{-0.06t}$ *Take the logarithm of both sides of the equation.*

$\log 0.5 = \log e^{-0.06t}$

$\log 0.5 = -0.06t \log e$ *In the last section we saw that* $\log e = 0.4343$.

$0.6990 - 1 = -0.06t(0.4343)$

$-0.3010 = -0.0261t$

$\dfrac{-0.3010}{-0.0261} = t$

$\boxed{11.53 = t}$

Thus, it takes approximately $\boxed{11\dfrac{1}{2}}$ years until 25 grams (or half the original amount) is left.

The amount of time it takes for half of a radioactive substance to decay is called the *half-life* of the substance. ∎

pH

Chemists have defined the pH (which stands for *hydrogen potential*) of a solution to be

$$pH = -\log[H_3O^+]$$

where $[H_3O^+]$ stands for the concentration of the hydronium ion in the solution (in moles per liter). The pH is a measure of the acidity or alkalinity of a solution. Water, which is neutral, has a pH of 7. Solutions with a pH below 7 are acidic, while those with a pH above 7 are alkaline.

EXAMPLE 4

What is the pH of a glass of orange juice if its hydronium ion concentration is 5.43×10^{-4}?

Solution

According to the definition of pH, we have

$$
\begin{aligned}
pH &= -\log[H_3O^+] \\
&= -\log(5.43 \times 10^{-4}) \\
&= -(\log 5.43 + \log 10^{-4}) \\
&= -(\log 5.43 - 4) \\
&= -\log 5.43 + 4 \\
&= -0.7348 + 4 \\
&= 3.2652
\end{aligned}
$$

pH values are customarily rounded off to the nearest tenth. Thus, our answer is $\boxed{pH = 3.3}$ ∎

Notice that in Examples 3 and 4 we were not going to compute an antilogarithm. Therefore, there was no reason not to complete the arithmetic which gave us a negative decimal to work with.

Exercises 12.6

In Exercises 1–10, use logarithms to compute the value of the given expression.

1. $\sqrt{513}$

2. $\sqrt[3]{0.841}$

3. $\sqrt[4]{2.79}$

4. $\sqrt[5]{28,300}$

5. $\sqrt[17]{356}$

6. $\sqrt[11]{93.1}$

7. $(1.31)^{40}$

8. $(0.97)^{50}$

9. $\dfrac{(2.4)^{12}}{(3.5)^8}$

10. $\dfrac{(0.045)^{15}}{(0.27)^{10}}$

11. If \$6,000 is invested in an account which pays 8% per year compounded semiannually, how much money will there be in the account in 5 years?

12. Repeat Exercise 11 if the interest is compounded quarterly.

13. Repeat Exercise 11 if the interest is compounded monthly.

14. Repeat Exercise 11 if the interest is 10% per year compounded semiannually.

15. If \$10,000 is invested at 7.3% per year compounded quarterly, how long will it take for the money to double?

16. Repeat Exercise 15 if the investment pays 14.6%.

17. The population of the earth is estimated to be 4 billion people and is growing at the continuous rate of approximately 2% per year. At this rate, how long will it take for the earth's population to grow to 5 billion? (Use the exponential growth model.)

18. Use the information in Exercise 17 to determine how long it will take for the earth's population to double.

19. The formula for the radioactive decay of radium is $A = A_0 e^{-0.0004t}$. Use this formula to determine the half-life of radium. [*Hint:* Find the value for t for which $A = \frac{1}{2}A_0$.]

20. Find the half-life for carbon-14, whose decay formula is $A = A_0 e^{-0.0001t}$.

21. Radon has a half-life of approximately 4 days and satisfies the exponential decay model. Find the value of r in the formula for radon. (Keep in mind that the value for r is usually given as an *annual* rate of decay. However, the formula applies to any unit of time.)

22. Certain man-made radioactive materials have extremely short half-lives. Repeat Exercise 21 for a substance whose half-life is 3 minutes.

23. The Richter scale is used in measuring the intensity of an earthquake. On the Richter scale, the magnitude of an earthquake, R, is defined to be

$$R = \log \frac{I}{I_0}$$

where I is the intensity of the earthquake and I_0 is a standard intensity. For the sake of simplicity we will assume that $I_0 = 1$. Compare the intensities I_1 and I_2 of two earthquakes whose numbers on the Richter scale are $R_1 = 3.6$ and $R_2 = 7.2$.

24. What is the intensity I of an earthquake whose number on the Richter scale is 6.85?

25. Compute the pH of a solution for which $[H_3O^+] = 8.63 \times 10^{-9}$.

26. What is the hydronium ion concentration of a solution whose pH is 4.7?

27. What is the hydronium ion concentration of water, which has a pH of 7?

28. Compute the pH of a solution for which $[H_3O^+] = 4.82 \times 10^{-3}$.

29. The unit of measurement frequently used to measure sound levels is the *decibel*. Decibels are measured on a logarithmic scale. The number of decibels, N, of a sound with intensity I (usually measured in watts per square centimeter) is defined to be

$$N = 10 \log I + 160$$

What is the intensity of sound of 200 decibels?

30. What is the decibel rating of a jet plane which is emitting a sound whose intensity I is 10^{-1} watt per square centimeter?

CHAPTER 12   Summary

After having completed this chapter, you should be able to:

1. Sketch the graph of an exponential function (Section 12.1).

    For example: Figure 12.6 shows the graph of $y = 4^x$.

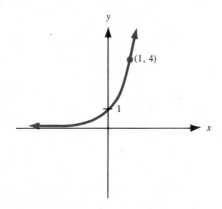

Figure 12.6   Graph of $y = 4^x$

2. Sketch the graph of a logarithmic function (Section 12.2).

    For example: Figure 12.7 shows the graph of $y = \log_4 x$.

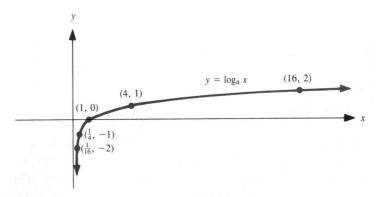

Figure 12.7   Graph of $y = \log_4 x$

3. Translate logarithmic statements into exponential form and vice versa (Section 12.2).

    For example:

    (a) $\log_3 \dfrac{1}{9} = -2$ is equivalent to $3^{-2} = \dfrac{1}{9}$.

    (b) $2^5 = 32$ is equivalent to $\log_2 32 = 5$.

4. Evaluate certain logarithms (Section 12.2).

For example: Find $\log_8 \frac{1}{16}$.

Solution:

Let $\log_8 \frac{1}{16} = t$ *Write in exponential form.*

$8^t = \frac{1}{16}$ *Express both sides in terms of the same base.*

$(2^3)^t = \frac{1}{2^4}$

$2^{3t} = 2^{-4}$

$3t = -4$

$t = -\frac{4}{3}$

Therefore, $\log_8 \frac{1}{16} = -\frac{4}{3}$

5. Use the properties of logarithms to rewrite logarithmic expressions (Section 12.3).

For example:

$\log_b \frac{\sqrt{xy}}{z^3} = \log_b \frac{(xy)^{1/2}}{z^3}$ *Use the quotient rule.*

$= \log_b (xy)^{1/2} - \log_b z^3$ *Now use the power rule.*

$= \frac{1}{2} \log_b xy - 3 \log_b z$ *Now use the product rule.*

$= \frac{1}{2}(\log_b x + \log_b y) - 3 \log_b z$

$= \frac{1}{2} \log_b x + \frac{1}{2} \log_b y - 3 \log_b z$

6. Use the logarithm table to compute logarithms (Section 12.4).

For example: $\log 37{,}500 = \log(3.75 \times 10^4)$

$= \log 3.75 + \log 10^4$

$= \log 3.75 + 4$

$= 0.5740 + 4$

$= 4.5740$

7. Solve exponential and logarithmic equations (Sections 12.1 and 12.5).

For example: Solve the following equations:

(a) $81^x = \dfrac{1}{9}$

$(9^2)^x = 9^{-1}$

$9^{2x} = 9^{-1}$

$2x = -1$

$x = -\dfrac{1}{2}$

(b) $\log_2(x + 5) + \log_2(x - 1) = 4$ *Use the product rule.*

$\log_2(x + 5)(x - 1) = 4$ *Write in exponential form.*

$(x + 5)(x - 1) = 2^4$

$x^2 + 4x - 5 = 16$

$x^2 + 4x - 21 = 0$

$(x + 7)(x - 3) = 0$

$x = -7 \quad \text{or} \quad x = 3$

CHECK $x = -7$: $\log_2(x + 5) + \log_2(x - 1) = 4$

$\log_2(-7 + 5) + \log_2(-7 - 1) \overset{?}{=} 4$

$\log_2(-2) + \log_2(-8) \overset{?}{=} 4$

Since the logarithm of a negative number is undefined, $x = -7$ is not a solution.

CHECK $x = 3$: $\log_2(x + 5) + \log_2(x - 1) = 4$

$\log_2(3 + 5) + \log_2(3 - 1) \overset{?}{=} 4$

$\log_2(8) + \log_2(2) \overset{?}{=} 4$

$3 + 1 \overset{\checkmark}{=} 4$

Thus, the solution is $x = 3$.

(c) $7^{y+5} = 5^{2y}$ *Take the log (base 10) of both sides.*

$\log 7^{y+5} = \log 5^{2y}$ *Use the power rule.*

$(y + 5) \log 7 = (2y) \log 5$ *Solve for y.*

$y \log 7 + 5 \log 7 = 2y \log 5$

$y \log 7 - 2y \log 5 = -5 \log 7$

$y(\log 7 - 2 \log 5) = -5 \log 7$

$y = \dfrac{-5 \log 7}{\log 7 - 2 \log 5}$ *Which, when evaluated, is*

$y = \dfrac{-5(0.8451)}{0.8451 - 2(0.6990)}$

$y = 7.64$

8.   Apply logarithms to solve problems involving exponential and logarithmic expressions (Section 12.6).

CHAPTER 12 Review Exercises

In Exercises 1–6, sketch the graph of the given function.

1. $y = 6^x$

2. $y = \log_6 x$

3. $y = \log_{10} x$

4. $y = \left(\dfrac{2}{3}\right)^x$

5. $y = 2^{-x}$

6. $y = \left(\dfrac{1}{4}\right)^{-x}$

In Exercises 7–18, write exponential statements in logarithmic form and logarithmic statements in exponential form.

7. $\log_3 81 = 4$

8. $6^3 = 216$

9. $4^{-3} = \dfrac{1}{64}$

10. $\log_2 \dfrac{1}{4} = -2$

11. $\log_8 4 = \dfrac{2}{3}$

12. $10^{-3} = 0.001$

13. $25^{1/2} = 5$

14. $\left(\dfrac{1}{3}\right)^{-2} = 9$

15. $\log_7 \sqrt{7} = \dfrac{1}{2}$

16. $\log_{11} \sqrt[3]{11} = \dfrac{1}{3}$

17. $\log_6 1 = 0$

18. $12^1 = 12$

In Exercises 19–32, evaluate the given logarithm.

19. $\log_{10} 1{,}000$

20. $\log_{10} 0.0001$

21. $\log_3 \dfrac{1}{9}$

22. $\log_2 64$

23. $\log_2 \dfrac{1}{4}$

24. $\log_7 \dfrac{1}{7}$

25. $\log_{1/3} 9$

26. $\log_{1/10} 0.01$

27. $\log_3 \dfrac{1}{9}$

28. $\log_4 \dfrac{1}{8}$

29. $\log_b \sqrt{b}$

30. $\log_p \sqrt[3]{p}$

31. $\log_{16} 32$

32. $\log_{32} 16$

In Exercises 33–42, write the given logarithm as a sum of simpler logarithms where possible.

33. $\log_b x^3 y^7$

34. $\log_b \dfrac{x^3}{y^7}$

35. $\log_b \dfrac{u^2 v^5}{w^3}$

36. $\log_b \dfrac{u^2}{v^5 w^3}$

37. $\log_b \sqrt[3]{xy}$

38. $\log_b \sqrt{\dfrac{x}{y}}$

39. $\log_b(x^3 + y^4)$

40. $\log_b \sqrt{\dfrac{x^3 y^5}{z^7}}$

41. $\log_b \sqrt[4]{\dfrac{x^6 y^2}{z^2}}$

42. $\dfrac{\log_b x^4}{\log_b y^9}$

In Exercises 43–44, evaluate the logarithm given $\log_b 7 = 1.32$ and $\log_b 2 = 1.1$.

43. $\log_b 28$

44. $\log_b \left(\dfrac{2}{49}\right)$

In Exercises 45–56, solve the given exponential or logarithmic equation.

45. $9^x = \dfrac{1}{81}$

46. $\left(\dfrac{1}{2}\right)^x = 4$

47. $16^x = 32$

48. $32^x = 16$

49. $5^{x+1} = 3$

50. $2^x = 7^{x+4}$

51. $\log(x + 10) - \log(x + 1) = 1$

52. $\log_2 x + \log_2(x + 1) = 1$

53. $\log_2(t + 1) + \log_2(t - 1) = 3$

54. $\log_3(2x + 1) + \log_3(x - 1) = 3$

55. $\log_b 3x + \log_b(x + 2) = \log_b 9$

56. $\log_p x - \log_p(x + 1) = \log_p 2$

In Exercises 57–64, use the table of common logarithms in Appendix D to compute the given logarithm or antilogarithm.

57. log 783

58. log 0.584

59. antilog(7.7818 − 10)

60. antilog 3.6571

61. log 0.00499

62. antilog 0.4843

63. antilog 2.49

64. log 263,000,000

Use logarithms to solve each of the following exercises.

65. If $6,000 is invested in an account at 8.2% compounded semiannually, how much money will there be in the account after 8 years?

66. How long will it take a sum of money invested at 7.2% and compounded quarterly to triple?

67. If a radioactive substance decays according to the formula $A = A_0 e^{-0.045t}$, how long will it take for 100 grams of radioactive material to decay to 25 grams of radioactive material?

68. If the population of a small town satisfies the exponential growth model with $r = 0.015$, how long will it take the town's population to increase from 8,500 to 15,000?

69. What is the pH of a solution for which $[H_3O^+] = 6.21 \times 10^{-9}$?

70. What is the hydronium ion concentration of a solution whose pH is 8.4?

CHAPTER 12 Practice Test

1. Sketch the graph of $y = \left(\dfrac{1}{3}\right)^x$.

2. Sketch the graph of $y = \log_7 x$.

3. Write in exponential form:

(a) $\log_2 16 = 4$ (b) $\log_9 \dfrac{1}{3} = -\dfrac{1}{2}$

4. Write in logarithmic form:

(a) $8^{2/3} = 4$ (b) $10^{-2} = 0.01$

5. Evaluate each of the following logarithms:

(a) $\log_3 \dfrac{1}{3}$ (b) $\log_{81} 9$ (c) $\log_8 32$

6. Write as a sum of simpler logarithms:

(a) $\log_b x^3 y^5 z$ (b) $\log_4 \dfrac{64\sqrt{x}}{yz}$

7. Use the table of common logarithms to find each of the following:

(a) log 27,900 (b) antilog(0.8549 − 4)

(c) log 0.0783 (d) antilog 6.1072

8. Solve each of the following equations:

(a) $3^x = \dfrac{1}{81}$ (b) $\log_2(x + 4) + \log_2(x - 2) = 4$

(c) $4^{5x} = 32^{3x-4}$ (d) $\log 5x - \log(x - 5) = 1$

(e) $9^{x+3} = 5$

9. If $5,000 is invested in an account paying 8.4% compounded quarterly, how much money will there be in the account after 7 years?

10. Assuming the exponential decay model $A = A_0 e^{-0.04t}$, how long will it take for 100 grams of a radioactive material to decay to 25 grams of radioactive material?

In Exercises 1–4, find the distance and midpoint between P and Q.

1. $P(2, 5)$, $Q(5, 9)$

2. $P(2, 3)$, $Q(-2, -3)$

3. $P(2, -4)$, $Q(6, 3)$

4. $P(5, 8)$, $Q(4, 9)$

In Exercises 5–8, graph the circle and indicate its center and radius.

5. $(x - 2)^2 + (y - 3)^2 = 25$

6. $(x + 1)^2 + (y - 4)^2 = 8$

7. $x^2 + y^2 - 6x + 2y = -1$

8. $x^2 + y^2 + 14x + 6y + 53 = 0$

In Exercises 9–14, graph the parabola. Indicate its vertex, axis of symmetry, and intercepts (if they exist).

9. $y = 3x^2$

10. $y = \dfrac{2}{3}x^2 - 4$

11. $y = x^2 + 2x$

12. $y = x^2 + 2x - 3$

13. $y = -2x^2 - 4x + 6$

14. $x = 4y^2 + 1$

In Exercises 15–17, graph the ellipse. Indicate its vertices.

15. $\dfrac{x^2}{16} + \dfrac{y^2}{5} = 1$

16. $x^2 + 36y^2 = 36$

17. $9x^2 + y^2 = 1$

In Exercises 18–20, graph the hyperbola. Indicate its vertices.

18. $\dfrac{x^2}{24} - \dfrac{y^2}{25} = 1$

19. $4x^2 - 49y^2 = 196$

20. $y^2 - \dfrac{x^2}{9} = 1$

In Exercises 21–26, identify the figure and, if possible, sketch a graph of its equation. Label the important aspects of the figure.

21. $\dfrac{x^2}{9} - \dfrac{y^2}{49} = 1$

22. $x = 3y^2 + 4$

23. $\dfrac{x}{3} + \dfrac{y}{4} = 1$

24. $4x^2 + 4y^2 = 1$

25. $x^2 + y^2 + 4x - 6y = -12$

26. $5x^2 - 25y^2 = 1$

27. The profit (P) made on a 15-minute sideshow is related to the price (p) of a ticket in the following way:

$$P = -200p^2 + 500p$$

where P is the profit in dollars and p is the price of a ticket in dollars. What ticket price would produce the maximum profit? What is the maximum profit?

28. Find the dimensions of a rectangle if its area is 168 sq. in. and its length is 2 in. greater than its width.

(continued)

In Exercises 29–36, solve the given system of equations.

29. $\begin{cases} x + y = 1 \\ x^2 + y^2 = 5 \end{cases}$

30. $\begin{cases} x - 2y = 2 \\ x^2 + 4y^2 = 20 \end{cases}$

31. $\begin{cases} y - x^2 = 4 \\ x^2 + y = 1 \end{cases}$

32. $\begin{cases} y = x^2 - 2x + 1 \\ y = x - x^2 \end{cases}$

33. $\begin{cases} x^2 + y^2 = 10 \\ 3x^2 - 4y^2 = 23 \end{cases}$

34. $\begin{cases} x^2 + 4y^2 = 40 \\ 4x^2 + y^2 = 25 \end{cases}$

35. $\begin{cases} x - y^2 = 3 \\ 3x - 2y = 9 \end{cases}$

36. $\begin{cases} x^2 - y^2 = 5 \\ y^2 - x = 7 \end{cases}$

In Exercises 37–40, find the domain and range of each relation and determine whether it is a function.

37. $\{(2, 3), (4, -1), (7, 5)\}$

38. $\{(1, 3), (-2, 3), (5, 6)\}$

39. $\{(4, 2), (3, 9), (4, 7)\}$

40. $\{(-1, 1), (-2, 4), (1, 1), (2, 4)\}$

In Exercises 41–44, sketch the graph of the given relation or function and determine its domain and range.

41. $x^2 + 4y^2 = 4$

42. $y = x^2 - 4x$

43. $x + 4y = 4$

44. $x^2 - 4y^2 = 4$

In Exercises 45–48, determine the domain of the given function.

45. $y = x^2 - x + 3$

46. $y = \sqrt{x - 3}$

47. $y = \dfrac{x}{x^2 - x - 6}$

48. $y = \dfrac{\sqrt{x + 4}}{x - 3}$

In Exercises 49–52, determine which are graphs of functions.

49.

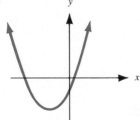

50.

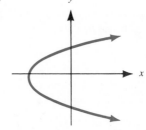

51.

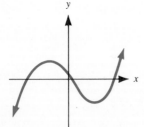

52.

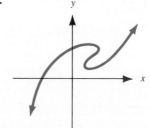

In Exercises 53–68, if $f(x) = 2x^2 - 3x - 4$, $g(x) = \dfrac{1}{x}$, and $h(x) = 3x + 1$, find:

53. $f(-3)$

54. $g\left(\dfrac{3}{4}\right)$

55. $h(x^2)$

56. $[h(x)]^2$

57. $f(a + 3)$

58. $f(a) + 3$

59. $f(3a)$

60. $3f(a)$

61. $f(1) + g(-1)$

62. $h[h(3)]$

63. $f[h(0)]$

64. $h[f(0)]$

65. $(f + g)(2)$

66. $(fg)(2)$

67. $\dfrac{f(x + 3) - f(x)}{3}$

68. $\dfrac{h(x + r) - h(x)}{r}$

In Exercises 69–72, find the inverse of the given function if it exists.

69. $y = x^3 - 1$

70. $y = \dfrac{2x - 5}{3}$

71. $y = \dfrac{x + 1}{x}$

72. $y = x^4$

In Exercises 73–80, sketch the graph of the given function.

73. $f(x) = \sqrt{x + 4}$

74. $f(x) = \sqrt{x} + 4$

75. $f(x) = x^2 - 4x - 5$

76. $f(x) = |x - 3|$

77. $f(x) = 3^{x+1}$

78. $f(x) = \left(\dfrac{1}{2}\right)^x - 1$

79. $f(x) = \log_5 x$

80. $f(x) = \log_2 x + 1$

In Exercises 81–86, translate logarithmic statements into exponential form and vice versa.

81. $\log_2 64 = 6$

82. $3^{-4} = \dfrac{1}{81}$

83. $\sqrt[3]{125} = 5$

84. $\log_{10} 0.01 = -2$

85. $\log_{27} 81 = \dfrac{4}{3}$

86. $4^{-1/2} = \dfrac{1}{2}$

In Exercises 87–94, find the given logarithm.

87. $\log_3 81$

88. $\log_8 2$

89. $\log_4 \dfrac{1}{16}$

90. $\log_{16} 32$

91. $\log_9 \dfrac{1}{3}$

92. $\log_7 7^{2/3}$

93. $\log_b 1$

94. $\log_b b$

In Exercises 95–98, write each logarithm as a sum of simpler logarithms, if possible.

95. $\log_b \sqrt[3]{5xy}$

96. $\log_b \dfrac{x^3 y}{z^5}$

(continued)

97. $\log_3 \dfrac{x^2\sqrt{y}}{9wz}$

98. $\dfrac{\log_b u^8}{\log_b v^4}$

In Exercises 99–104, use the table of common logarithms in Appendix D to find the given logarithm or antilogarithm.

99. log 73,600

100. antilog(7.8745 − 10)

101. antilog 0.6085

102. log 0.0248

103. log 0.4830

104. antilog 3.4014

In Exercises 105–118, solve the given equation.

105. $\dfrac{1}{2} \log_8 x = \log_8 5$

106. $2 \log_b x = \log_b(6x − 9)$

107. $5^x = \dfrac{1}{25}$

108. $16^x = \dfrac{1}{2}$

109. $\log_6 x + \log_6 4 = 3$

110. $\log_4 x + \log_4(x − 6) = 2$

111. $\dfrac{4^{x^2}}{2^x} = 64$

112. $9^{\sqrt{x}} = 3^{3x−8}$

113. $9^x = 7^{x+3}$

114. $\log_b(x + 4) − \log_b(x − 4) = \log_b 5$

115. If y varies directly as x and $y = 8$ when $x = 6$, find y when $x = 20$.

116. If y varies inversely as x and $y = 8$ when $x = 6$, find y when $x = 20$.

117. If x varies jointly as y and z and $x = 10$ when $y = 4$ and $z = 15$, find x when $y = 6$ and $z = 20$.

118. If x varies directly as y and inversely as the square of z, and $x = 2$ when $y = 12$ and $z = 6$, find x when $y = 20$ and $z = 5$.

119. How long will it take $3,000 invested at an annual rate of 8% and compounded quarterly to grow to $5,000?

120. Based on the exponential growth model, how long will it take an initial population of 30,000 growing at 3% per year to double?

121. Based on the exponential decay model with $r = 0.002$, how long will it take 20 grams of radioactive material to decay to 2 grams of radioactive material?

122. What is the pH of a solution with a hydronium ion concentration of 3.6×10^{-4}?

123. What is the hydronium ion concentration of a solution whose pH is 8.2?

CHAPTERS 10–12 CUMULATIVE PRACTICE TEST

1. Find the distance between points $(2, -3)$ and $(5, 1)$, and find the midpoint between the two points.

2. Graph the circle $x^2 + y^2 − 4x + 4y = 4$.

3. Graph the parabola $y = 2x^2 + 4x − 16$. Identify its intercepts (if they exist) and its vertex.

4. Identify and graph the following figures. Label the important aspects of each figure.

   (a) $\dfrac{x^2}{9} + y^2 = 1$

   (b) $\dfrac{x^2}{36} - \dfrac{y^2}{25} = 1$

5. Abel throws a ball into the air. The equation $s = -16t^2 + 48t + 6$ gives the distance (in feet) that the ball is above the ground at t seconds after he throws it. What is the maximum height of the ball?

6. *Solve the system of equations:*

$$\begin{cases} x^2 - y^2 = 7 \\ x - y = 1 \end{cases}$$

7. Does the relation $R = \{(0, 1), (3, 7), (-2, 5), (4, 7)\}$ define a function? Explain your answer. What are its domain and range?

8. Sketch the graph of each of the following relations and determine its domain and range. Is the relation a function? Explain your answer.

   (a) $4x^2 + y^2 = 36$ (b) $y = x^2 - 6x + 9$

9. *Find the domain of each of the following functions:*

   (a) $f(x) = \dfrac{x}{x^2 - 9}$ (b) $g(x) = \sqrt{2x - 3}$

10. Given $f(x) = x^2 - 3x + 2$ and $g(x) = 4x - 1$, find:

   (a) $f(-5)$ (b) $g(x^2)$
   (c) $f(x + 2)$ (d) $f(x) + 2$
   (e) $g(5x)$ (f) $5g(x)$
   (g) $f[g(0)]$ (h) $g[f(0)]$
   (i) $\left(\dfrac{f}{g}\right)(1)$ (j) $\dfrac{f(x + 3) - f(x)}{3}$

11. Find the inverse of the function for $f(x) = \dfrac{x - 3}{7}$.

12. *Sketch the graphs of the following equations:*

   (a) $y = \left(\dfrac{1}{4}\right)^x$ (b) $y = |x + 5|$

13. *Write in exponential form:* $\log_8 \dfrac{1}{4} = -\dfrac{2}{3}$

14. *Find each of the following logarithms:*

   (a) $\log_3 \dfrac{1}{27}$ (b) $\log_4 8$

15. *Express as a sum of simpler logarithms, if possible:*

   (a) $\log_b x\sqrt[3]{y}$ (b) $\log_b \dfrac{x^3}{\sqrt{xy}}$

16. *Use the table of common logarithms in Appendix D to find:*

   (a) $\log 0.00637$ (b) antilog 3.9263

17. Solve the following equations:

   (a) $2^{x^2} = 8^{3x}$ (b) $\log_6 x + \log_6(x + 5) = 2$

18. If y varies inversely as the square of x, and $y = 4$ when $x = 5$, find y when $x = 6$.

19. How long must $5,000 be invested so that if the interest is compounded quarterly at 10%, the $5,000 will have grown to $10,000?

20. What is the pH of a solution with a hydronium ion concentration of 3.98×10^{-8}?

A

The Binomial Theorem

Early in this text we discussed finding powers of the binomial $(x + y)^n$ for small n. For $(x + y)^2$ we pointed out that if we knew the pattern (which we called the *perfect square* of the sum), we could write out the product in simplified form quickly, without using the distributive property.

To find products $(x + y)^3$, we needed to find $(x + y)^2$ first and then multiply the result by $(x + y)$:

$$
\begin{aligned}
(x + y)^3 &= (x + y)(x + y)^2 \\
&= (x + y)(x^2 + 2xy + y^2) \\
&= x^3 + 2x^2y + xy^2 + yx^2 + 2xy^2 + y^3 \\
&= x^3 + 3x^2y + 3xy^2 + y^3
\end{aligned}
$$

To find $(x + y)^4$ we needed to find $(x + y)^3$ first, and then multiply the result by $(x + y)$:

$$
\begin{aligned}
(x + y)^4 &= (x + y)(x + y)^3 \\
&= (x + y)(x^3 + 3x^2y + 3xy^2 + y^3) \\
&= x^4 + 4x^3y + 6x^2y^2 + 4xy^3 + y^4
\end{aligned}
$$

In this appendix we will demonstrate how to find perfect nth powers of the binomial $(x + y)^n$ for any n without multiplying out the binomials. As with the perfect square, we will examine the patterns and then generalize these patterns.

First we examine the product of binomials $(x + y)^n$ multiplied out. When the binomial $(x + y)^n$ is multiplied out and simplified (terms are combined where possible) we call it the *expanded form* of the binomial $(x + y)^n$. We will expand $(x + y)^n$ for $n = 0$ through $n = 5$. We will assume that neither x nor y is 0.

$$
\begin{array}{lll}
n = 0 & (x + y)^0 = & 1 \\
n = 1 & (x + y)^1 = & x + y \\
n = 2 & (x + y)^2 = & x^2 + 2xy + y^2 \\
n = 3 & (x + y)^3 = & x^3 + 3x^2y + 3xy^2 + y^3 \\
n = 4 & (x + y)^4 = & x^4 + 4x^3y + 6x^2y^2 + 4xy^3 + y^4 \\
n = 5 & (x + y)^5 = & x^5 + 5x^4y + 10x^3y^2 + 10x^2y^3 + 5xy^4 + y^5
\end{array}
$$

Let's first ignore the coefficients of the terms and look at the powers of the variables for the terms in each expansion. Note the pattern of the exponents for each expansion:

$$
\begin{array}{ll}
n = 1 & x, \quad y \\
n = 2 & x^2, \quad xy, \quad y^2 \\
n = 3 & x^3, \quad x^2y, \quad xy^2, \quad y^3 \\
n = 4 & x^4, \quad x^3y, \quad x^2y^2, \quad xy^3, \quad y^4 \\
n = 5 & x^5, \quad x^4y, \quad x^3y^2, \quad x^2y^3, \quad xy^4, \quad y^5
\end{array}
$$

First notice that for each n, the expansion starts out with x^n and ends with y^n.

Now observe the terms on the line where $n = 5$. The leftmost x term has an exponent of 5 (which is n), and as we move to the right, the exponent of x decreases (for successive terms) by exactly 1. If we write the last term, y^5, as x^0y^5 (remember that $x^0 = 1$), then the pattern is consistent: Each term has a power of x.

On the other hand, the pattern is reversed for the powers of y in each term. The first term, x^5, can be written as x^5y^0 and hence each subsequent term to the immediate right contains a power of y increased by 1, until the last term contains a fifth (or nth) power of y.

Observe the same pattern for $n = 1$ through $n = 4$: Moving to the right, each term contains consecutively decreasing powers of x and consecutively increasing powers of y.

In general, this pattern of descending powers of x and ascending powers of y for the terms in the expansion of $(x + y)^n$ holds true for all n.

Notice as well that the sum of the exponents of each term in the expansion of $(x + y)^5$ is 5. Again, this is true in general for n: In the expansion of $(x + y)^n$, the sum of the exponents of each term is n. Hence, in the expansion of $(x + y)^n$, the powers of x and y in the terms will have the following pattern:

$$
x^n, \quad x^{n-1}y, \quad x^{n-2}y^2, \quad x^{n-3}y^3, \quad \ldots, \quad x^3y^{n-3}, \quad x^2y^{n-2}, \quad xy^{n-1}, \quad y^n
$$

Note that the exponents in each term add up to n.

We can now write the powers of the variables of the binomial expansion of $(x + y)^8$ with this pattern in mind:

1. The first term is x^8.

2. The exponent of x is decreased by 1 in each subsequent term to the right until we reach x^0.

3. Starting with y^0 in the first term, the exponent of y is increased by 1 in each subsequent term to the right until we reach y^8. Consequently, we get

$$x^8, \quad x^7y, \quad x^6y^2, \quad x^5y^3, \quad x^4y^4, \quad x^3y^5, \quad x^2y^6, \quad xy^7, \quad y^8$$

4. Note that the exponents in each term sum to 8.

We now know what the powers of the variables used in each term will be when we expand the binomial $(x + y)^n$. Let's look at the coefficients.

First we examine the coefficients of the terms in the binomial expansions of $(x + y)^n$ for $n = 0$ through $n = 5$. We can write them out in the form of a triangle:

*This triangle of coefficients is known as **Pascal's triangle**, named after the 17th century philosopher and mathematician, Blaise Pascal.*

One pattern we can identify (indicated by the dashed triangles within Pascal's triangle) is that, other than the end coefficients of 1, each coefficient is the sum of the two coefficients immediately above it. Thus, we can figure out the next two lines of coefficients for the expansion of $(x + y)^6$ and $(x + y)^7$ by adding neighboring coefficients to produce the coefficients in the line below:

This is one way of finding the coefficients of the terms for an expanded binomial. However, this approach is not very practical for finding the coefficients where n is large. There is another pattern in the coefficients we can discern, but before we discuss this pattern, we introduce a new notation to make the pattern more convenient to express (and remember).

We define $n!$, called **n factorial**, as follows:

If $n > 0$, then

$$n! = n \cdot (n - 1) \cdot (n - 2) \cdot (n - 3) \cdot \cdots \cdot 3 \cdot 2 \cdot 1$$

Thus, *n factorial* is the product of *n* consecutively decreasing integers from *n* to 1. For example,

$$3! = 3 \cdot 2 \cdot 1 \qquad \text{\textit{which, when evaluated, is}}$$
$$= 6$$

$$5! = 5 \cdot 4 \cdot 3 \cdot 2 \cdot 1 \qquad \text{\textit{which, when evaluated, is}}$$
$$= 120$$

We also define

$$\boxed{0! = 1}$$

EXAMPLE 1

Evaluate the following:

(a) $6!$ **(b)** $(5 - 1)!$ **(c)** $5! - 1$ **(d)** $\dfrac{8!}{6!}$ **(e)** $\dfrac{6!}{2!4!}$

Solution

(a) $6!$ *6! is the product of the six consecutively decreasing*

$$= 6 \cdot 5 \cdot 4 \cdot 3 \cdot 2 \cdot 1 \qquad \text{\textit{positive integers from 6 to 1.}}$$

$$= \boxed{720}$$

(b) $(5 - 1)!$ *Perform the operation in parentheses first.*

$$= 4! \qquad \text{\textit{4! is the product of the four consecutively decreasing}}$$
$$= 4 \cdot 3 \cdot 2 \cdot 1 \qquad \text{\textit{positive integers from 4 to 1.}}$$

$$= \boxed{24}$$

(c) $5! - 1$ *Multiply (perform factorial operation) first.*

$$= (5 \cdot 4 \cdot 3 \cdot 2 \cdot 1) - 1 \qquad \text{\textit{Multiply.}}$$

$$= 120 - 1 \qquad \text{\textit{Then subtract.}}$$

$$= \boxed{119}$$

Note the differences between parts **(b)** and **(c)**.

(d) $\dfrac{8!}{6!}$ *Do not try to reduce yet; this is not 8 divided by 6. By definition of factorial, we have*

$$= \frac{8 \cdot 7 \cdot 6 \cdot 5 \cdot 4 \cdot 3 \cdot 2 \cdot 1}{6 \cdot 5 \cdot 4 \cdot 3 \cdot 2 \cdot 1} \qquad \text{\textit{Now reduce.}}$$

$$= \frac{8 \cdot 7 \cdot \cancel{6} \cdot \cancel{5} \cdot \cancel{4} \cdot \cancel{3} \cdot \cancel{2} \cdot \cancel{1}}{\cancel{6} \cdot \cancel{5} \cdot \cancel{4} \cdot \cancel{3} \cdot \cancel{2} \cdot \cancel{1}}$$

$$= 8 \cdot 7 = \boxed{56}$$

(e) $\dfrac{6!}{2!4!}$ *By definition of factorial notation, we have*

$$= \dfrac{6 \cdot 5 \cdot 4 \cdot 3 \cdot 2 \cdot 1}{2 \cdot 1 \cdot 4 \cdot 3 \cdot 2 \cdot 1}$$ *Then reduce.*

$$= \dfrac{\overset{3}{\cancel{6}} \cdot 5 \cdot \cancel{4} \cdot \cancel{3} \cdot \cancel{2} \cdot \cancel{1}}{\cancel{2} \cdot 1 \cdot \cancel{4} \cdot \cancel{3} \cdot \cancel{2} \cdot \cancel{1}}$$

$$= \boxed{15}$$

∎

With factorial notation in hand, we can now reexamine the coefficients of the terms of the binomial expansion of $(x + y)^n$, where $n = 5$ and $n = 4$. First we expand $(x + y)^5$:

$$(x + y)^5 = x^5 + 5x^4y + 10x^3y^2 + 10x^2y^3 + 5xy^4 + y^5$$

We will demonstrate that the coefficients of the terms are related to the exponents of the terms.

The third term in the expansion of $(x + y)^5$ is $10x^3y^2$. If we write the expression

$$\dfrac{5!}{3!2!} \quad \text{we get} \quad \dfrac{5 \cdot \overset{2}{\cancel{4}} \cdot \cancel{3} \cdot \cancel{2} \cdot \cancel{1}}{\cancel{3} \cdot \cancel{2} \cdot \cancel{1} \cdot \cancel{2} \cdot 1} = 10$$

The fifth term is $5xy^4$. If we write

$$\dfrac{5!}{1!4!} \quad \text{we get} \quad \dfrac{5 \cdot \cancel{4} \cdot \cancel{3} \cdot \cancel{2} \cdot \cancel{1}}{1 \cdot \cancel{4} \cdot \cancel{3} \cdot \cancel{2} \cdot \cancel{1}} = 5$$

Notice that the denominator is the product of the factorials of the exponents of the variables of the term $5xy^4$.

Let's look at $(x + y)^4 = x^4 + 4x^3y + 6x^2y^2 + 4xy^3 + y^4$. The third term of this expansion is $6x^2y^2$. If we write the expression

$$\dfrac{4!}{2!2!} \quad \text{we get} \quad \dfrac{\overset{2}{\cancel{4}} \cdot 3 \cdot \cancel{2} \cdot \cancel{1}}{\cancel{2} \cdot 1 \cdot \cancel{2} \cdot \cancel{1}} = 6$$

The numerator is the factorial of n and the denominator is the product of the factorials of the exponents of the variables x and y.

In general, we have the result given in the box.

In the expansion of $(x + y)^n$, the term containing the expression $x^{n-k}y^k$ has coefficient $\dfrac{n!}{(n - k)!k!}$, $1 \le n \le k$.

Note that $n - k$ and k, the exponents of x and y, add up to n.

For example, in the expansion of $(x + y)^6$, the term x^2y^4 has coefficient 15, since

$$\frac{n!}{(\text{Exponent of } x)!(\text{Exponent of } y)!} = \frac{n!}{(n - k)!k!}$$

$$= \frac{6!}{2!4!}$$

$$= \frac{\overset{3}{\cancel{6}} \cdot 5 \cdot \cancel{4} \cdot \cancel{3} \cdot \cancel{2} \cdot \cancel{1}}{\cancel{2} \cdot 1 \cdot \cancel{4} \cdot \cancel{3} \cdot \cancel{2} \cdot \cancel{1}}$$

$$= 15$$

We can now give the formula for the expansion of $(x + y)^n$:

| **The Binomial Theorem** | For $n > 0$, $$(x + y)^n = x^n + \frac{n!}{(n - 1)!1!}x^{n-1}y + \frac{n!}{(n - 2)!2!}x^{n-2}y^2$$ $$+ \frac{n!}{(n - 3)!3!}x^{n-3}y^3 + \cdots + \frac{n!}{1!(n - 1)!}xy^{n-1} + y^n$$ |

EXAMPLE 2 Expand the following:

(a) $(x + y)^8$ **(b)** $(2a + 3b)^4$ **(c)** $\left(\dfrac{m}{2} - n\right)^5$

Solution

(a) $(x + y)^8$ *Using the binomial theorem with $n = 8$, we get*

$$= x^8 + \frac{8!}{7!1!}x^7y + \frac{8!}{6!2!}x^6y^2 + \frac{8!}{5!3!}x^5y^3 + \frac{8!}{4!4!}x^4y^4$$

$$+ \frac{8!}{3!5!}x^3y^5 + \frac{8!}{2!6!}x^2y^6 + \frac{8!}{1!7!}xy^7 + y^8$$

Notice that the exponents of the variables appear in the denominators of the coefficients.

$$= \boxed{x^8 + 8x^7y + 28x^6y^2 + 56x^5y^3 + 70x^4y^4 + 56x^3y^5 + 28x^2y^6 + 8xy^7 + y^8}$$

(b) $(2a + 3b)^4$ *We rewrite this as*

$= [(2a) + (3b)]^4$ *Then using $x = 2a$, $y = 3b$, and $n = 4$ in the binomial theorem, we have*

$$= (2a)^4 + \frac{4!}{3!1!}(2a)^3(3b) + \frac{4!}{2!2!}(2a)^2(3b)^2 + \frac{4!}{1!3!}(2a)(3b)^3 + (3b)^4$$

$$= 16a^4 + 4(8a^3)(3b) + 6(4a^2)(9b^2) + 4(2a)(27b^3) + 81b^4$$

$$= \boxed{16a^4 + 96a^3b + 216a^2b^2 + 216ab^3 + 81b^4}$$

(c) $\left(\dfrac{m}{2} - n\right)^5 = \left[\dfrac{m}{2} + (-n)\right]^5 = \left(\dfrac{m}{2}\right)^5 + \dfrac{5!}{4!1!}\left(\dfrac{m}{2}\right)^4(-n)$

$$+ \dfrac{5!}{2!3!}\left(\dfrac{m}{2}\right)^2(-n)^3 + \dfrac{5!}{1!4!}\left(\dfrac{m}{2}\right)(-n)^4 + (-n)^5$$

$$= \dfrac{m^5}{32} - 5\left(\dfrac{m^4}{16}\right)n + 10\left(\dfrac{m^3}{8}\right)n^2 - 10\left(\dfrac{m^2}{4}\right)n^3 + 5\left(\dfrac{m}{2}\right)n^4 - n^5$$

$$= \boxed{\dfrac{1}{32}m^5 - \dfrac{5}{16}m^4n + \dfrac{5}{4}m^3n^2 - \dfrac{5}{2}m^2n^3 + \dfrac{5}{2}mn^4 - n^5} \qquad \blacksquare$$

A general term in the expansion of $(x + y)^n$ can be written as

$$\dfrac{n!}{(n - k)!k!}x^{n-k}y^k$$

Look at the first five terms of the expansion of $(x + y)^9$:

$$x^9 + \dfrac{9!}{8!1!}x^8y + \dfrac{9!}{7!2!}x^7y^2 + \dfrac{9!}{6!3!}x^6y^3 + \dfrac{9!}{5!4!}x^5y^4 + \cdots$$

We now compare these terms with $\dfrac{n!}{(n - k)!k!}x^{n-k}y^k$:

In the 3rd term, $\dfrac{9!}{7!2!}x^7y^2$, k is 2.

In the 4th term, $\dfrac{9!}{6!3!}x^6y^3$, k is 3.

In the 5th term, k is 4.

In general, in the rth term, $k = r - 1$.

Thus, if we want to find the rth term of the expansion of $(x + y)^n$, we substitute $r - 1$ for k in the expression

$$\dfrac{n!}{(n - k)!k!}x^{n-k}y^k$$

to get

$$\dfrac{n!}{[n - (r - 1)]!(r - 1)!}x^{n-(r-1)}y^{r-1}$$

which, when we simplify, becomes the following:

$$\dfrac{n!}{(n - r + 1)!(r - 1)!}x^{n-r+1}y^{r-1}$$

is the rth term in the expansion of $(x + y)^n$.

EXAMPLE 3 | Find the following terms:
(a) The third term in the expansion of $(x + y)^7$
(b) The fourth term in the expansion of $(2a - 5)^6$

Solution

(a) We substitute $n = 7$ and $r = 3$ in the formula

$$\frac{n!}{(n - r + 1)!(r - 1)!}x^{n-r+1}y^{r-1}$$

to get

$$\frac{7!}{(7 - 3 + 1)!(3 - 1)!}x^{7-3+1}y^{3-1} = \frac{7!}{5!2!}x^5y^2$$

$$= \frac{7 \cdot \cancel{6}^3 \cdot \cancel{5} \cdot \cancel{4} \cdot \cancel{3} \cdot \cancel{2} \cdot \cancel{1}}{\cancel{5} \cdot \cancel{4} \cdot \cancel{3} \cdot \cancel{2} \cdot \cancel{1} \cdot 2 \cdot 1}x^5y^2$$

$$= \boxed{21x^5y^2}$$

(b) We substitute 6 for n, 4 for r, $2a$ for x, and -5 for y in

$$\frac{n!}{(n - r + 1)!(r - 1)!}x^{n-r+1}y^{r-1}$$

to get

$$\frac{6!}{(6 - 4 + 1)!(4 - 1)!}(2a)^{6-4+1}(-5)^{4-1} = \frac{6!}{3!3!}(2a)^3(-5)^3$$

$$= \frac{6 \cdot 5 \cdot 4 \cdot \cancel{3} \cdot \cancel{2} \cdot \cancel{1}}{\cancel{3} \cdot \cancel{2} \cdot 1 \cdot \cancel{3} \cdot \cancel{2} \cdot \cancel{1}}(8a^3)(-125)$$

$$= 20(8a^3)(-125)$$

$$= \boxed{-20,000a^3}$$

Appendix A Exercises

In Exercises 1–20, evaluate the expression.

1. $6!$
2. $2!$
3. $4!$
4. $7!$
5. $0!$
6. $1!$
7. $(7 - 2)!$
8. $(8 - 2)!$
9. $7! - 2$
10. $8! - 2$
11. $(2 \cdot 3)!$
12. $(3 \cdot 4)!$
13. $2 \cdot 3!$
14. $3 \cdot 4!$
15. $\frac{5!}{1!4!}$
16. $\frac{7!}{4!3!}$
17. $\frac{8!}{5!3!}$
18. $\frac{9!}{6!3!}$
19. $\frac{12!}{3!4!5!}$
20. $\frac{12!}{4!6!2!}$

In Exercises 21–36, expand the binomial.

21. $(a + b)^6$

22. $(x + y)^7$

23. $(a - b)^6$

24. $(x - y)^7$

25. $(2a + 1)^5$

26. $(3a + 1)^4$

27. $(2a - 1)^5$

28. $(3a - 1)^4$

29. $(1 - 2a)^4$

30. $(1 - 3a)^4$

31. $(a^2 + 2b)^5$

32. $(x + 2y^2)^5$

33. $(2x^2 - 3y^2)^4$

34. $(3x^2 + 2y^2)^4$

35. $\left(\dfrac{x}{3} + 2\right)^4$

36. $\left(\dfrac{a}{2} - 4\right)^4$

In Exercises 37–42, give the first four terms of the binomial expansion.

37. $(a^2 - b^4)^6$

38. $(x^4 - y^2)^6$

39. $(2a + 3b)^8$

40. $(3a - 2b)^7$

41. $(3a^2 - 2)^6$

42. $(2a^2 - 3)^8$

In Exercises 43–52, find the given terms.

43. The third term in the expansion of $(x + y)^8$

44. The fifth term in the expansion of $(x + y)^9$

45. The fourth term in the expansion of $(a - 2b)^6$

46. The fourth term in the expansion of $(2x - b)^6$

47. The third term in the expansion of $(3a^2 - 2)^7$

48. The fourth term in the expansion of $(2r^2 - 3s)^8$

49. The sixth term in the expansion of $(3a^2 - 2)^7$

50. The sixth term in the expansion of $(2r^2 - 3s)^8$

51. The ninth term in the expansion of $(3a^2 - 2)^8$

52. The tenth term in the expansion of $(2a^2 - 1)^9$

? QUESTIONS FOR THOUGHT

53. In words, state the relationship among the exponents of the variables in the terms of the expansion of $(x + y)^n$, the coefficients of the terms of the expansion of $(x + y)^n$, and n.

54. Compute $(1.02)^7$ to four places using the binomial theorem. [*Hint:* Rewrite 1.02 as $1 + 0.02$.] Why do you not need to compute all the terms in the expansion?

55. The expression $\dfrac{n!}{r!(n - r)!}$ is sometimes written in the form $\binom{n}{r}$.

Hence, $\dbinom{7}{2} = \dfrac{7!}{2!5!} = \dfrac{7 \cdot \overset{3}{\cancel{6}} \cdot \cancel{5} \cdot \cancel{4} \cdot \cancel{3} \cdot \cancel{2} \cdot \cancel{1}}{2 \cdot 1 \cdot \cancel{5} \cdot \cancel{4} \cdot \cancel{3} \cdot \cancel{2} \cdot \cancel{1}} = 21.$

Compute the following:

(a) $\dbinom{6}{2}$ **(b)** $\dbinom{5}{0}$ **(c)** $\dbinom{9}{3}$ **(d)** $\dbinom{9}{6}$

B

The Ellipse and Hyperbola: Centered at (h, k)

In Sections 10.3 and 10.4 we discussed the ellipse and hyperbola centered at the origin. We will now examine the more general cases.

The Ellipse

The equation

$$\frac{(x - h)^2}{a^2} + \frac{(y - k)^2}{b^2} = 1$$

is the standard form of the equation of an ellipse centered at (h, k), with vertices $(h + a, k)$, $(h - a, k)$, $(h, k + b)$, and $(h, k - b)$, where a and b are positive constants. The equations of the axes are $x = h$ and $y = k$. (See the accompanying figure.)

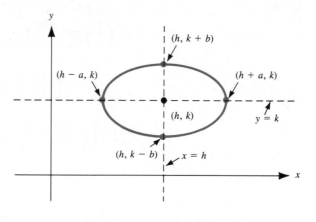

Compare the form of the ellipse given in the box with the form of the ellipse centered at the origin. Note the similarities and differences. The ellipse

$$\frac{(x - h)^2}{a^2} + \frac{(y - k)^2}{b^2} = 1$$

has the same shape as the ellipse

$$\frac{x^2}{a^2} + \frac{y^2}{b^2} = 1$$

but its center is moved h units horizontally and k units vertically.

Instead of memorizing the new vertices and axes of the general form, we can use the fact that both forms have the same shape and graph the new form by *drawing a new set of coordinate axes through the point (h, k) and graphing the form*

$$\frac{x^2}{a^2} + \frac{y^2}{b^2} = 1$$

on the new set of axes. Thus, we just count a units right and left on the new x-axis, and b units up and down on the new y-axis to locate the new vertices. We demonstrate by example.

EXAMPLE 1

Graph the following ellipses:

(a) $\dfrac{(x - 6)^2}{16} + \dfrac{(y - 3)^2}{25} = 1$ **(b)** $\dfrac{(x + 3)^2}{25} + \dfrac{(y - 1)^2}{9} = 1$

Solution

(a)   Comparing with the standard form, we find that

$$h = 6, \quad k = 3, \quad a = 4, \quad \text{and} \quad b = 5.$$

Hence, the center of the ellipse is (h, k), or $(6, 3)$. We draw a new set of coordinate axes centered at $(6, 3)$.

We start at the origin of our new axes and move right $a = 4$ units to find the right vertex and left 4 units to get the left vertex.

We then start at the origin of our new axes and move up $b = 5$ units to find the top vertex and down 5 units to find the bottom vertex.

Now draw the ellipse (see Figure B.1). The vertices are at $(10, 3)$, $(2, 3)$, $(6, -2)$, and $(6, 8)$. The axes of the ellipse are the lines $x = 6$ and $y = 3$.

Figure B.1

Graph of

$\dfrac{(x - 6)^2}{16} + \dfrac{(y - 3)^2}{25} = 1$

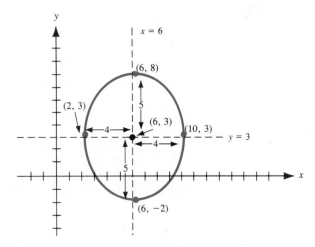

(b)   Comparing this equation with standard form, we note that the expression within the parentheses of the numerator of the first fraction is a sum and not a difference. Consequently,

rewrite $(x + 3)$ as $[x - (-3)]$.

Thus, $h = -3, \quad k = 1, \quad a = 5, \quad \text{and} \quad b = 3.$

Draw a new set of coordinate axes centered at (h, k), which is $(-3, 1)$.

Since $a = 5$, count 5 units right from the origin of the new axes to find the right vertex and 5 units left to find the left vertex.

Count $b = 3$ units up to find the top vertex and 3 units down to find the bottom vertex.

Now draw the ellipse (see Figure B.2 on page 640). The vertices are $(2, 1)$, $(-8, 1)$, $(-3, 4)$, and $(-3, -2)$. The equations of the axes of the ellipse are $y = 1$ and $x = -3$.

Figure B.2

Graph of

$$\frac{(x + 3)^2}{25} + \frac{(y - 1)^2}{9} = 1$$

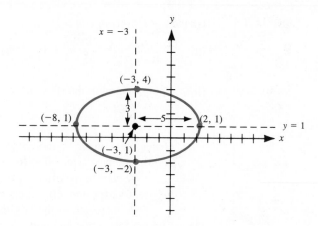

The last case to investigate is that in which the equations are not in standard form. As with circles, we will be required to complete the squares in order to get the equation in standard form.

EXAMPLE 2

Put the equation

$$16x^2 + 9y^2 - 64x - 18y = 71$$

into standard form and sketch its graph.

Solution

We separate the x and y terms:

$$(16x^2 - 64x \quad) + (9y^2 - 18y \quad) = 71$$

Now we note that the square terms have coefficients, and our previous method of completing the square required the square term coefficients to be 1. We get around this by factoring out the squared term coefficient from each group:

$$(16x^2 - 64x \quad) + (9y^2 - 18y \quad) = 71$$
$$16(x^2 - 4x \quad) + 9(y^2 - 2y \quad) = 71 \qquad \textit{We factored out 16 from } 16x^2 - 64x$$
$$\textit{and 9 from } 9y^2 - 18y.$$

We complete the square *for the expressions in parentheses*:

$$16(x^2 - 4x + 4) + 9(y^2 - 2y + 1)$$

Add 4 in here   *Add 1 in here*

Now the tricky part is that we are *not* simply adding 4 and 1 to the left-hand side of the equation, for if you multiply out the left-hand side, you get

$$16(x^2 - 4x + 4) + 9(y^2 - 2y + 1) = 16x^2 - 64x + 64 + 9y^2 - 18y + 9$$

Compared to the original equation you will see that we really added 64 and 9 to the left-hand side, because of the factors 16 and 9. So our equation

$$16(x^2 - 4x \quad) + 9(y^2 - 2y \quad) = 71$$

becomes

$$16(x^2 - 4x + 4) + 9(y^2 - 2y + 1) = 71 + 64 + 9$$
$$16 \cdot 4 = +64 \qquad 9 \cdot 1 = +9$$

Now we write the quadratic expressions in factored form

$$16(x - 2)^2 + 9(y - 1)^2 = 144$$

and divide both sides of the equation by 144 to put the equation in standard form:

$$\frac{16(x - 2)^2 + 9(y - 1)^2}{144} = \frac{144}{144}$$

$$\frac{16(x - 2)^2}{144} + \frac{9(y - 1)^2}{144} = 1 \qquad \textit{Simplify fractions.}$$

$$\frac{(x - 2)^2}{9} + \frac{(y - 1)^2}{16} = 1$$

Once we have the equation in standard form, we can graph it (see Figure B.3). We have the general form of an ellipse with $h = 2$, $k = 1$, $a = 3$, and $b = 4$, which gives us an ellipse with center $(2, 1)$. The vertices are $(-1, 1)$, $(5, 1)$, $(2, -3)$, and $(2, 5)$. The axes are $y = 1$ and $x = 2$.

Figure B.3
Graph of
$$\frac{(x - 2)^2}{9} + \frac{(y - 1)^2}{16} = 1$$

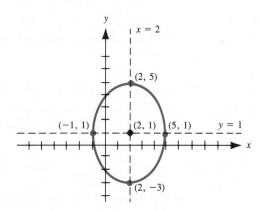

EXAMPLE 3

Graph the following ellipse:

$$9x^2 + y^2 + 6y = 0$$

Solution

Put in standard form:

$$9x^2 \qquad + y^2 + 6y \qquad = 0 \qquad \textit{Separate x and y terms.}$$
$$9x^2 \qquad + y^2 + 6y + 9 = 0 + 9 \qquad \textit{Complete the square for the y terms.}$$
$$9x^2 + (y + 3)^2 = 9 \qquad \textit{Divide both sides by 9.}$$
$$\frac{9x^2}{9} + \frac{(y + 3)^2}{9} = \frac{9}{9}$$
$$x^2 + \frac{(y + 3)^2}{9} = 1$$

Now the equation is in standard form with $h = 0$, $k = -3$ (watch the signs), $a = 1$, and $b = 3$, which gives us an ellipse with center $(0, -3)$ (see Figure B.4). The vertices are $(1, -3)$, $(-1, -3)$, $(0, 0)$, and $(0, -6)$. The axes are $x = 0$ and $y = -3$.

Figure B.4

Graph of $x^2 + \dfrac{(y + 3)^2}{9} = 1$

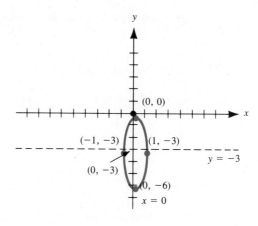

Our next step is to investigate the graph of the hyperbola which is not centered at the origin.

The Hyperbola

As with the ellipse, there is no need to memorize a lot of new information. The hyperbola

$$\frac{(x - h)^2}{a^2} - \frac{(y - k)^2}{b^2} = 1$$

has the same shape as the hyperbola

$$\frac{x^2}{a^2} - \frac{y^2}{b^2} = 1$$

but it is shifted horizontally h units and vertically k units.

As we did for the ellipse, we locate the point (h, k), draw a set of axes with (h, k) as the origin, and graph our hyperbola on the new axes as indicated in the box.

The equation

$$\frac{(x - h)^2}{a^2} - \frac{(y - k)^2}{b^2} = 1$$

is the standard form of the equation of a hyperbola centered at (h, k) with vertices $(h - a, k)$ and $(h + a, k)$, where a and b are positive constants. The equations of the axes are $x = h$ and $y = k$. (See the next figure.)

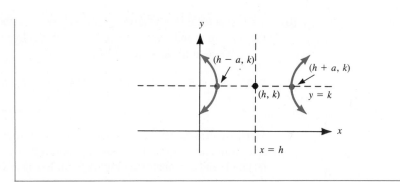

EXAMPLE 4

Graph each of the following:

(a) $\dfrac{(x - 3)^2}{25} - \dfrac{(y - 4)^2}{4} = 1$

(b) $\dfrac{(y - 2)^2}{9} - \dfrac{(x + 5)^2}{64} = 1$

Solution

(a) Since the equation is in standard form, we can easily identify h, k, a, and b:

$$h = 3, \quad k = 4, \quad a = 5, \quad \text{and} \quad b = 2$$

Thus, we locate the point $(h, k) = (3, 4)$ and draw a new set of coordinate axes centered at $(3, 4)$ parallel to the x- and y-coordinate axes.

We count ($a =$) 5 units left and right of $(3, 4)$ on our new x-axis for the vertices, and ($b =$) 2 units up and down along the new y-axis. Draw the rectangle and its diagonals. Then draw the hyperbola (see Figure B.5). The vertices are $(-2, 4)$ and $(8, 4)$. The axes of the hyperbola have equations $x = 3$ and $y = 4$.

Figure B.5

Graph of

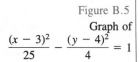

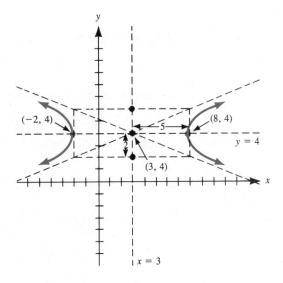

(b)   We first note that the x^2 term is being subtracted rather than the y^2 term. Therefore, the roles of x and y are reversed (as with the hyperbola centered at the origin). We identify h, k, a, and b:

$$h = -5, \quad k = 2, \quad a = 8, \quad b = 3$$

Observe that h and a are still associated with the x^2 term and k and b are still associated with the y^2 term.

Thus, we draw a new set of coordinate axes through $(-5, 2)$. On the new axes, start at the new origin and count left and right 8, up and down 3, and draw the rectangle and diagonals (see Figure B.6). Because the x^2 term is being subtracted, the hyperbola opens up and down and the vertices are $(-5, 5)$ and $(-5, -1)$. The axes are $y = 2$ and $x = -5$.

Figure B.6
Graph of
$$\frac{(y - 2)^2}{9} - \frac{(x + 5)^2}{64} = 1$$

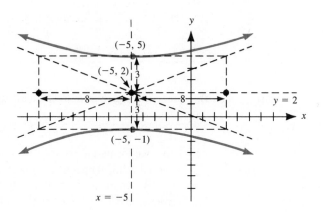

Finally, if the equation is not in standard form, we use the method of completing the square to put our equations in standard form, as with ellipses and circles, as demonstrated in the next example.

EXAMPLE 5

Graph the following equation:

$$4x^2 - 25y^2 - 16x + 150y = 309$$

Solution

We will use the method of completing the square to put the equation in standard form:

$$4x^2 - 16x - 25y^2 + 150y = 309 \qquad \textit{First separate x and y terms.}$$
$$4(x^2 - 4x \quad) - 25(y^2 - 6y \quad) = 309 \qquad \textit{Factor the coefficient of the squared terms from each quadratic factor.}$$

Notice that -25 is factored from $-25y^2 + 150y$, yielding

$$-25(y^2 - 6y \quad)$$
$$\uparrow$$

Note that this sign is changed.

Complete the square for each quadratic expression:

$$4(x^2 - 4x + 4) - 25(y^2 - 6y + 9) = 309 + 16 - 225$$

$4 \cdot 4 = +16 \qquad -25 \cdot 9 = -225$ *Note that we have added $+16$ and also -225. Thus, we must add 16 and add -225 to the right-hand side.*

Represent the quadratic expression in factored form.

$$4(x - 2)^2 - 25(y - 3)^2 = 100 \qquad \text{*Divide both sides by* } 100.$$

$$\frac{4(x - 2)^2}{100} - \frac{25(y - 3)^2}{100} = \frac{100}{100}$$

$$\frac{(x - 2)^2}{25} - \frac{(y - 3)^2}{4} = 1$$

Now the equation is in standard form with $h = 2$, $k = 3$, $a = 5$, and $b = 2$. We plot $(2, 3)$, draw a set of axes, and graph the hyperbola (see Figure B.7). The vertices are $(-3, 3)$ and $(7, 3)$; the axes are $x = 2$ and $y = 3$.

Figure B.7

Graph of

$$\frac{(x - 2)^2}{25} - \frac{(y - 3)^2}{4} = 1$$

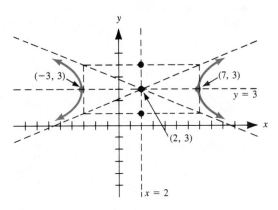

We again point out the differences in form between the hyperbola and the ellipse: The coefficients of the squared terms in the hyperbola have opposite signs, while the coefficients of the squared terms of the ellipse have the same signs.

Appendix B Exercises

In Exercises 1–18, graph the ellipse. Indicate the center, vertices, and equations of the axes.

1. $\dfrac{(x - 1)^2}{9} + \dfrac{(y - 3)^2}{16} = 1$

2. $\dfrac{(x - 3)^2}{25} + \dfrac{(y - 2)^2}{16} = 1$

3. $\dfrac{(x - 3)^2}{9} + \dfrac{(y - 1)^2}{16} = 1$

4. $\dfrac{(x - 2)^2}{25} + \dfrac{(y - 3)^2}{16} = 1$

5. $\dfrac{(x - 2)^2}{25} + \dfrac{(y - 5)^2}{81} = 1$

6. $\dfrac{(x - 4)^2}{81} + \dfrac{(y - 4)^2}{9} = 1$

7. $\dfrac{(x - 3)^2}{9} + \dfrac{(y - 1)^2}{4} = 1$

8. $\dfrac{(x - 5)^2}{25} + \dfrac{(y - 1)^2}{16} = 1$

9. $\dfrac{(x + 3)^2}{9} + \dfrac{(y - 1)^2}{4} = 1$

10. $\dfrac{(x + 2)^2}{9} + \dfrac{(y - 3)^2}{4} = 1$

11. $\dfrac{(x + 3)^2}{9} + \dfrac{(y + 1)^2}{4} = 1$

12. $\dfrac{(x + 3)^2}{25} + \dfrac{(y + 2)^2}{36} = 1$

13. $9(x - 2)^2 + 4(y + 5)^2 = 36$

14. $25(x + 2)^2 + 4(y - 3)^2 = 100$

15. $16(x + 1)^2 + 64(y - 2)^2 = 64$

16. $36(x - 1)^2 + 9(y + 7)^2 = 36$

17. $4(x - 2)^2 + (y - 7)^2 = 1$

18. $5(x - 2)^2 + (y - 3)^2 = 1$

In Exercises 19–36, graph the hyperbola. Indicate the center and vertices.

19. $\dfrac{(x - 1)^2}{16} - \dfrac{(y - 2)^2}{9} = 1$

20. $\dfrac{(x - 3)^2}{25} - \dfrac{(y - 1)^2}{36} = 1$

21. $\dfrac{(y - 2)^2}{9} - \dfrac{(x - 1)^2}{16} = 1$

22. $\dfrac{(y - 3)^2}{36} - \dfrac{(x - 1)^2}{25} = 1$

23. $\dfrac{(x - 3)^2}{25} - \dfrac{(y - 4)^2}{16} = 1$

24. $\dfrac{(x - 5)^2}{25} - \dfrac{(y - 4)^2}{16} = 1$

25. $\dfrac{(x - 4)^2}{9} - \dfrac{(y + 1)^2}{36} = 1$

26. $\dfrac{(x + 2)^2}{9} - (y - 3)^2 = 1$

27. $\dfrac{(y + 2)^2}{16} - \dfrac{(x - 1)^2}{25} = 1$

28. $\dfrac{(y + 5)^2}{36} - \dfrac{(x - 3)^2}{16} = 1$

29. $\dfrac{(x + 3)^2}{36} - \dfrac{(y + 2)^2}{100} = 1$

30. $\dfrac{(x + 4)^2}{9} - \dfrac{(y + 2)^2}{81} = 1$

31. $4(x - 2)^2 - 25(y - 4)^2 = 100$

32. $25(x - 1)^2 - 4(y - 4)^2 = 100$

33. $36(x + 1)^2 - 9(y - 2)^2 = 36$

34. $64(x + 3)^2 - 16(y - 2)^2 = 64$

35. $16(x - 5)^2 - (y + 4)^2 = 1$

36. $25(x + 1)^2 - (y - 3)^2 = 1$

In Exercises 37–52, identify and graph the figure.

37. $4x^2 - 9y^2 - 8x + 54y = 113$

38. $25x^2 + 16y^2 - 150x - 32y = 159$

39. $4x^2 + 9y^2 - 8x - 54y = -49$

40. $25x^2 - 16y^2 - 150x + 32y = 191$

41. $25x^2 + 16y^2 - 250x - 64y = -289$

42. $36x^2 - 49y^2 - 504x - 294y = 441$

43. $25x^2 - 16y^2 - 250x + 64y = -161$

44. $36x^2 + 49y^2 - 504x + 294y = -441$

45. $16y^2 - 9x^2 - 96y + 36x = 36$

46. $36y^2 - 16x^2 + 360y + 128x = -68$

47. $x^2 - 4y^2 + 4x + 8y = 16$

48. $8x^2 - y^2 + 48x - 6y = -47$

49. $36x^2 + 25y^2 + 216x + 100y = 476$

50. $25x^2 - 16y^2 + 150x - 32y = 191$

51. $4x^2 - y^2 + 24x - 4y = -31$

52. $16x^2 + y^2 - 96x - 4y = -147$

?   QUESTION FOR THOUGHT

53. The asymptotes of the hyperbola

$$\frac{(x - h)^2}{a^2} - \frac{(y - k)^2}{b^2} = 1$$

pass through the point (h, k) and have slopes $\pm\dfrac{b}{a}$. Use the point–slope formula to find the equations for the asymptotes.

C

Table of Squares and Square Roots

Using the Table

For example, to find $\sqrt{210}$ we look down the column headed n to line 21, then over to the column headed $\sqrt{10n} = \sqrt{10(21)} = \sqrt{210}$. Thus,

$$\sqrt{210} = 14.491$$

Squares and Square Roots

n	n^2	$\sqrt{n}$	$\sqrt{10n}$	n	n^2	$\sqrt{n}$	$\sqrt{10n}$
1	1	1.000	3.162	26	676	5.099	16.125
2	4	1.414	4.472	27	729	5.196	16.432
3	9	1.732	5.477	28	784	5.292	16.733
4	16	2.000	6.325	29	841	5.385	17.029
5	25	2.236	7.071	30	900	5.477	17.321
6	36	2.449	7.746	31	961	5.568	17.607
7	49	2.646	8.367	32	1,024	5.657	17.889
8	64	2.828	8.944	33	1,089	5.745	18.166
9	81	3.000	9.487	34	1,156	5.831	18.439
10	100	3.162	10.000	35	1,225	5.916	18.708
11	121	3.317	10.488	36	1,296	6.000	18.974
12	144	3.464	10.954	37	1,369	6.083	19.235
13	169	3.606	11.402	38	1,444	6.164	19.494
14	196	3.742	11.832	39	1,521	6.245	19.748
15	225	3.873	12.247	40	1,600	6.325	20.000
16	256	4.000	12.649	41	1,681	6.403	20.248
17	289	4.123	13.038	42	1,764	6.481	20.494
18	324	4.243	13.416	43	1,849	6.557	20.736
19	361	4.359	13.784	44	1,936	6.633	20.976
20	400	4.472	14.142	45	2,025	6.708	21.213
21	441	4.583	14.491	46	2,116	6.782	21.448
22	484	4.690	14.832	47	2,209	6.856	21.679
23	529	4.796	15.166	48	2,304	6.928	21.909
24	576	4.899	15.492	49	2,401	7.000	22.136
25	625	5.000	15.811	50	2,500	7.071	22.361

Squares and Square Roots
(*continued*)

n	n^2	$\sqrt{n}$	$\sqrt{10n}$	n	n^2	$\sqrt{n}$	$\sqrt{10n}$
51	2,601	7.141	22.583	76	5,776	8.718	27.568
52	2,704	7.211	22.804	77	5,929	8.775	27.749
53	2,809	7.280	23.022	78	6,084	8.832	27.928
54	2,916	7.348	23.238	79	6,241	8.888	28.107
55	3,025	7.416	23.452	80	6,400	8.944	28.284
56	3,136	7.483	23.664	81	6,561	9.000	28.460
57	3,249	7.550	23.875	82	6,724	9.055	28.636
58	3,364	7.616	24.083	83	6,889	9.110	28.810
59	3,481	7.681	24.290	84	7,056	9.165	28.983
60	3,600	7.746	24.495	85	7,225	9.220	29.155
61	3,721	7.810	24.698	86	7,396	9.274	29.326
62	3,844	7.874	24.900	87	7,569	9.327	29.496
63	3,969	7.937	25.100	88	7,744	9.381	29.665
64	4,096	8.000	25.298	89	7,921	9.434	29.833
65	4,225	8.062	25.495	90	8,100	9.487	30.000
66	4,356	8.124	25.690	91	8,281	9.539	30.166
67	4,489	8.185	25.884	92	8,464	9.592	30.332
68	4,624	8.246	26.077	93	8,649	9.644	30.496
69	4,761	8.307	26.268	94	8,836	9.695	30.659
70	4,900	8.367	26.458	95	9,025	9.747	30.822
71	5,041	8.426	26.646	96	9,216	9.798	30.984
72	5,184	8.485	26.833	97	9,409	9.849	31.145
73	5,329	8.544	27.019	98	9,604	9.899	31.305
74	5,476	8.602	27.203	99	9,801	9.950	31.464
75	5,625	8.660	27.386	100	10,000	10.000	31.623

D

Table of
Common Logarithms

Common Logarithms

N	0	1	2	3	4	5	6	7	8	9
1.0	0.0000	0.0043	0.0086	0.0128	0.0170	0.0212	0.0253	0.0294	0.0334	0.0374
1.1	0.0414	0.0453	0.0492	0.0531	0.0569	0.0607	0.0645	0.0682	0.0719	0.0755
1.2	0.0792	0.0828	0.0864	0.0899	0.0934	0.0969	0.1004	0.1038	0.1072	0.1106
1.3	0.1139	0.1173	0.1206	0.1239	0.1271	0.1303	0.1335	0.1367	0.1399	0.1430
1.4	0.1461	0.1492	0.1523	0.1553	0.1584	0.1614	0.1644	0.1673	0.1703	0.1732
1.5	0.1761	0.1790	0.1818	0.1847	0.1875	0.1903	0.1931	0.1959	0.1987	0.2014
1.6	0.2041	0.2068	0.2095	0.2122	0.2148	0.2175	0.2201	0.2227	0.2253	0.2279
1.7	0.2304	0.2330	0.2355	0.2380	0.2405	0.2430	0.2455	0.2480	0.2504	0.2529
1.8	0.2553	0.2577	0.2601	0.2625	0.2648	0.2672	0.2695	0.2718	0.2742	0.2765
1.9	0.2788	0.2810	0.2833	0.2856	0.2878	0.2900	0.2923	0.2945	0.2967	0.2989
2.0	0.3010	0.3032	0.3054	0.3075	0.3096	0.3118	0.3139	0.3160	0.3181	0.3201
2.1	0.3222	0.3243	0.3263	0.3284	0.3304	0.3324	0.3345	0.3365	0.3385	0.3404
2.2	0.3424	0.3444	0.3464	0.3483	0.3502	0.3522	0.3541	0.3560	0.3579	0.3598
2.3	0.3617	0.3636	0.3655	0.3674	0.3692	0.3711	0.3729	0.3747	0.3766	0.3784
2.4	0.3802	0.3820	0.3838	0.3856	0.3874	0.3892	0.3909	0.3927	0.3945	0.3962
2.5	0.3979	0.3997	0.4014	0.4031	0.4048	0.4065	0.4082	0.4099	0.4116	0.4133
2.6	0.4150	0.4166	0.4183	0.4200	0.4216	0.4232	0.4249	0.4265	0.4281	0.4298
2.7	0.4314	0.4330	0.4346	0.4362	0.4378	0.4393	0.4409	0.4425	0.4440	0.4456
2.8	0.4472	0.4487	0.4502	0.4518	0.4533	0.4548	0.4564	0.4579	0.4594	0.4609
2.9	0.4624	0.4639	0.4654	0.4669	0.4683	0.4698	0.4713	0.4728	0.4742	0.4757
3.0	0.4771	0.4786	0.4800	0.4814	0.4829	0.4843	0.4857	0.4871	0.4886	0.4900
3.1	0.4914	0.4928	0.4942	0.4955	0.4969	0.4983	0.4997	0.5011	0.5024	0.5038
3.2	0.5051	0.5065	0.5079	0.5092	0.5105	0.5119	0.5132	0.5145	0.5159	0.5172
3.3	0.5185	0.5198	0.5211	0.5224	0.5237	0.5250	0.5263	0.5276	0.5289	0.5302
3.4	0.5315	0.5328	0.5340	0.5353	0.5366	0.5378	0.5391	0.5403	0.5416	0.5428
3.5	0.5441	0.5453	0.5465	0.5478	0.5490	0.5502	0.5514	0.5527	0.5539	0.5551
3.6	0.5563	0.5575	0.5587	0.5599	0.5611	0.5623	0.5635	0.5647	0.5658	0.5670
3.7	0.5682	0.5694	0.5705	0.5717	0.5729	0.5740	0.5752	0.5763	0.5775	0.5786
3.8	0.5798	0.5809	0.5821	0.5832	0.5843	0.5855	0.5866	0.5877	0.5888	0.5899
3.9	0.5911	0.5922	0.5933	0.5944	0.5955	0.5966	0.5977	0.5988	0.5999	0.6010
N	0	1	2	3	4	5	6	7	8	9

(continued)

Common Logarithms

N	0	1	2	3	4	5	6	7	8	9
4.0	0.6021	0.6031	0.6042	0.6053	0.6064	0.6075	0.6085	0.6096	0.6107	0.6117
4.1	0.6128	0.6138	0.6149	0.6160	0.6170	0.6180	0.6191	0.6201	0.6212	0.6222
4.2	0.6232	0.6243	0.6253	0.6263	0.6274	0.6284	0.6294	0.6304	0.6314	0.6325
4.3	0.6335	0.6345	0.6355	0.6365	0.6375	0.6385	0.6395	0.6405	0.6415	0.6425
4.4	0.6435	0.6444	0.6454	0.6464	0.6474	0.6484	0.6493	0.6503	0.6513	0.6522
4.5	0.6532	0.6542	0.6551	0.6561	0.6571	0.6580	0.6590	0.6599	0.6609	0.6618
4.6	0.6628	0.6637	0.6646	0.6656	0.6665	0.6675	0.6684	0.6693	0.6702	0.6712
4.7	0.6721	0.6730	0.6739	0.6749	0.6758	0.6767	0.6776	0.6785	0.6794	0.6803
4.8	0.6812	0.6821	0.6830	0.6839	0.6848	0.6857	0.6866	0.6875	0.6884	0.6893
4.9	0.6902	0.6911	0.6920	0.6928	0.6937	0.6946	0.6955	0.6964	0.6972	0.6981
5.0	0.6990	0.6998	0.7007	0.7016	0.7024	0.7033	0.7042	0.7050	0.7059	0.7067
5.1	0.7076	0.7084	0.7093	0.7101	0.7110	0.7118	0.7126	0.7135	0.7143	0.7152
5.2	0.7160	0.7168	0.7177	0.7185	0.7193	0.7202	0.7210	0.7218	0.7226	0.7235
5.3	0.7243	0.7251	0.7259	0.7267	0.7275	0.7284	0.7292	0.7300	0.7308	0.7316
5.4	0.7324	0.7332	0.7340	0.7348	0.7356	0.7364	0.7372	0.7380	0.7388	0.7396
5.5	0.7404	0.7412	0.7419	0.7427	0.7435	0.7443	0.7451	0.7459	0.7466	0.7474
5.6	0.7482	0.7490	0.7497	0.7505	0.7513	0.7520	0.7528	0.7536	0.7543	0.7551
5.7	0.7559	0.7566	0.7574	0.7582	0.7589	0.7597	0.7604	0.7612	0.7619	0.7627
5.8	0.7634	0.7642	0.7649	0.7657	0.7664	0.7672	0.7679	0.7686	0.7694	0.7701
5.9	0.7709	0.7716	0.7723	0.7731	0.7738	0.7745	0.7752	0.7760	0.7767	0.7774
6.0	0.7782	0.7789	0.7796	0.7803	0.7810	0.7818	0.7825	0.7832	0.7839	0.7846
6.1	0.7853	0.7860	0.7868	0.7875	0.7882	0.7889	0.7896	0.7903	0.7910	0.7917
6.2	0.7924	0.7931	0.7938	0.7945	0.7952	0.7959	0.7966	0.7973	0.7980	0.7987
6.3	0.7993	0.8000	0.8007	0.8014	0.8021	0.8028	0.8035	0.8041	0.8048	0.8055
6.4	0.8062	0.8069	0.8075	0.8082	0.8089	0.8096	0.8102	0.8109	0.8116	0.8122
6.5	0.8129	0.8136	0.8142	0.8149	0.8156	0.8162	0.8169	0.8176	0.8182	0.8189
6.6	0.8195	0.8202	0.8209	0.8215	0.8222	0.8228	0.8235	0.8241	0.8248	0.8254
6.7	0.8261	0.8267	0.8274	0.8280	0.8287	0.8293	0.8299	0.8306	0.8312	0.8319
6.8	0.8325	0.8331	0.8338	0.8344	0.8351	0.8357	0.8363	0.8370	0.8376	0.8382
6.9	0.8388	0.8395	0.8401	0.8407	0.8414	0.8420	0.8426	0.8432	0.8439	0.8445
N	0	1	2	3	4	5	6	7	8	9

(continued)

Common Logarithms

N	0	1	2	3	4	5	6	7	8	9
7.0	0.8451	0.8457	0.8463	0.8470	0.8476	0.8482	0.8488	0.8494	0.8500	0.8506
7.1	0.8513	0.8519	0.8525	0.8531	0.8537	0.8543	0.8549	0.8555	0.8561	0.8567
7.2	0.8573	0.8579	0.8585	0.8591	0.8597	0.8603	0.8609	0.8615	0.8621	0.8627
7.3	0.8633	0.8639	0.8645	0.8651	0.8657	0.8663	0.8669	0.8675	0.8681	0.8686
7.4	0.8692	0.8698	0.8704	0.8710	0.8716	0.8722	0.8727	0.8733	0.8739	0.8745
7.5	0.8751	0.8756	0.8762	0.8768	0.8774	0.8779	0.8785	0.8791	0.8797	0.8802
7.6	0.8808	0.8814	0.8820	0.8825	0.8831	0.8837	0.8842	0.8848	0.8854	0.8859
7.7	0.8865	0.8871	0.8876	0.8882	0.8887	0.8893	0.8899	0.8904	0.8910	0.8915
7.8	0.8921	0.8927	0.8932	0.8938	0.8943	0.8949	0.8954	0.8960	0.8965	0.8971
7.9	0.8976	0.8982	0.8987	0.8993	0.8998	0.9004	0.9009	0.9015	0.9020	0.9025
8.0	0.9031	0.9036	0.9042	0.9047	0.9053	0.9058	0.9063	0.9069	0.9074	0.9079
8.1	0.9085	0.9090	0.9096	0.9101	0.9106	0.9112	0.9117	0.9122	0.9128	0.9133
8.2	0.9138	0.9143	0.9149	0.9154	0.9159	0.9165	0.9170	0.9175	0.9180	0.9186
8.3	0.9191	0.9196	0.9201	0.9206	0.9212	0.9217	0.9222	0.9227	0.9232	0.9238
8.4	0.9243	0.9248	0.9253	0.9258	0.9263	0.9269	0.9274	0.9279	0.9284	0.9289
8.5	0.9294	0.9299	0.9304	0.9309	0.9315	0.9320	0.9325	0.9330	0.9335	0.9340
8.6	0.9345	0.9350	0.9355	0.9360	0.9365	0.9370	0.9375	0.9380	0.9385	0.9390
8.7	0.9395	0.9400	0.9405	0.9410	0.9415	0.9420	0.9425	0.9430	0.9435	0.9440
8.8	0.9445	0.9450	0.9455	0.9460	0.9465	0.9469	0.9474	0.9479	0.9484	0.9489
8.9	0.9494	0.9499	0.9504	0.9509	0.9513	0.9518	0.9523	0.9528	0.9533	0.9538
9.0	0.9542	0.9547	0.9552	0.9557	0.9562	0.9566	0.9571	0.9576	0.9581	0.9586
9.1	0.9590	0.9595	0.9600	0.9605	0.9609	0.9614	0.9619	0.9624	0.9628	0.9633
9.2	0.9638	0.9643	0.9647	0.9652	0.9657	0.9661	0.9666	0.9671	0.9675	0.9680
9.3	0.9685	0.9689	0.9694	0.9699	0.9703	0.9708	0.9713	0.9717	0.9722	0.9727
9.4	0.9731	0.9736	0.9741	0.9745	0.9750	0.9754	0.9759	0.9763	0.9768	0.9773
9.5	0.9777	0.9782	0.9786	0.9791	0.9795	0.9800	0.9805	0.9809	0.9814	0.9818
9.6	0.9823	0.9827	0.9832	0.9836	0.9841	0.9845	0.9850	0.9854	0.9859	0.9863
9.7	0.9868	0.9872	0.9877	0.9881	0.9886	0.9890	0.9894	0.9899	0.9903	0.9908
9.8	0.9912	0.9917	0.9921	0.9926	0.9930	0.9934	0.9939	0.9943	0.9948	0.9952
9.9	0.9956	0.9961	0.9965	0.9969	0.9974	0.9978	0.9983	0.9987	0.9991	0.9996
N	0	1	2	3	4	5	6	7	8	9

Answers to Selected
Exercises and Chapter Tests

Exercises 1.1

1. True   **3.** False   **5.** True   **7.** True   **9.** True   **11.** True   **13.** $\{1, 2, 3, 4, 5, 6, 7, 8, 9, 10, 11\}$
15. $\{7, 8, 9, 10, 11, 12\}$ **17.** $\{3, 4, 5, 6, 7, 8, 9, 10, 11, 12, 13\}$ **19.** $\varnothing$ **21.** $\{41, 43, 47\}$ **23.** $\varnothing$
25. $\{0, 6, 12, 18, \ldots\}$ **27.** $\{0, 8, 16, 24, \ldots\}$ **29.** $\{1, 2, 3, 4, 6, 9, 12, 18, 36\}$ **31.** $\{0, 3, 6\}$
33. $\{0, 1, 2, 3, 4, 5, 6, 9, 12, 15, 18, 21, 24, 27, 30, 33\}$ **35.** $\{7, 11, 13\}$ **37.** $\varnothing$ **39.** $2 \cdot 3 \cdot 11$
41. $2 \cdot 2 \cdot 2 \cdot 2 \cdot 2 \cdot 2 \cdot 2$ **43.** Prime number   **45.** $7 \cdot 13$

Exercises 1.2

1. True   **3.** True   **5.** True   **7.** False   **9.** True   **11.** False   **13.** True   **15.** False   **17.** True
19. $<, \leq, \neq$ **21.** $>, \geq, \neq$ **23.** $<, \leq, \neq$

25.

27.

29.

31.

33.

35.

37.

39.

41.

43.

45. No solution   **47.**

49. $C \cup D = \{x \mid -4 \leq x < 9, x \in Z\}$

51. $A \cap D = \{x \mid 1 \leq x < 9, x \in Z\}$

53. $C \cap D = \{x \mid 1 \leq x \leq 6, x \in Z\}$

55. $A \cup B = \{x \mid x \in R\}$ **57.** $C \cap D = \{x \mid 1 \leq x \leq 6\}$

59. $A \cup C = \{x \mid x \geq -4\}$ **61.** $B \cap D = \varnothing$

Exercises 1.3

1. Associative property of addition   **3.** Commutative property of addition   **5.** Distributive property   **7.** False
9. Distributive property   **11.** Multiplicative identity   **13.** False   **15.** Associative property of multiplication
17. Associative property of multiplication   **19.** False   **21.** Additive inverse property   **23.** False
25. Commutative property of addition   **27.** Distributive property   **29.** Distributive property
31. Closure property of multiplication   **33.** False

Exercises 1.4

1. 5   **3.** -11 **5.** 24   **7.** -12 **9.** -23 **11.** -60 **13.** -27 **15.** -11 **17.** 3   **19.** -16
21. 4   **23.** $\frac{16}{5}$ **25.** 8   **27.** 4   **29.** 1   **31.** -12 **33.** -4 **35.** -3 **37.** 45   **39.** 17   **41.** 10
43. -16 **45.** 2   **47.** 2   **49.** -8 **51.** 36   **53.** 29   **55.** 50   **57.** -81 **59.** -40 **61.** -360
63. -46 **65.** $\frac{50}{7}$ **67.** -13 **69.** 111   **71.** -6 **73.** 10   **75.** 0   **77.** 24   **79.** 0   **81.** 30
83. 6   **85.** 1   **87.** Undefined   **89.** 6.481   **91.** .87   **93.** -4

Exercises 1.5

1. $8x$ **3.** $12x^2$ **5.** $-4x$ **7.** $-12x^2$ **9.** $-6m$ **11.** $60m^3$ **13.** $-9t^2$ **15.** $-24t^6$
17. $2x + 3y + 5z$ **19.** $30xyz$ **21.** $x^3 + x^2 + 2x$ **23.** $2x^6$ **25.** $-17x^2y$ **27.** $30x^4y^2$
29. $x^2 + 2x - 6$ **31.** $11x^2y - 7xy^2$ **33.** $9m + 12n$ **35.** $2a - 16b$ **37.** $6c - 16d$ **39.** $x^2 - 2xy + y^2$
41. 0   **43.** $40a^3b^4c^3$ **45.** $72x^5$ **47.** $18x^5$ **49.** $-32x^{11}$ **51.** 0   **53.** $-b + 10$ **55.** $17t + 12$
57. $13a - 64$ **59.** $12b - 15$ **61.** $4x^2 - 8x$ **63.** $10x - 7y$ **65.** $-8y^3 + 2xy^2 + 2xy + 3x$ **67.** $12s^2$
69. $-7.6x - 15.9y$ **71.** $27.99x - 35.9$

Exercises 1.6

1. Let $x = $ number; $x + 8$ **3.** Let $x = $ number; $2x - 3$ **5.** Let $x = $ number; $3x + 4 = x - 7$
7. Let x and y be the two numbers; $x + y = xy + 1$ **9.** Let the smaller number be x; the second number $= 2x + 5$
11. Let $x = $ smallest number; the other two numbers are $3x$ and $3x + 12$ **13.** x and $x + 1$ **15.** $x, x + 2, x + 4$
17. $x^3 + (x + 2)^3$ **19.** Let $x = $ one number; the other number is $40 - x$
21. Let $x = $ first number; second number $= 2x$; third number $= 100 - 3x$
23. Let $x = $ width; area $= (x)(3x) = 3x^2$; perimeter $= 2x + 2(3x) = 8x$
25. Let $x = $ second side; perimeter $= 2x + x + x + 4 = 4x + 4$ **27. (a)** 31 coins   **(b)** value of coins $= \$4$
29. (a) Number of coins $= n + d + q$ **(b)** value of coins (in cents) $= 5n + 10d + 25q$ cents
31. (a) $2w$ meters   **(b)** $4w$ dollars   **(c)** $6w$ **(d)** $30w$ dollars   **(e)** $34w$ dollars
33. Let $x = $ nickels, then $20 - x = $ dimes; value of coins (in cents) $= 5x + 10(20 - x) = 200 - 5x$ cents

Chapter 1 Review Exercises

1. $A = \{1, 2, 3, 4\}$ **2.** $\{6, 7, 8, 9, \ldots\}$ **3.** $C \cap D = \{b\}$ **4.** $C \cup D = \{a, b, c, e, f, g, r, s\}$
5. $A \cup B = \{1, 2, 3, 4, 6, 7, \ldots\}$ or $A \cup B = \{x \mid x \in N, x \neq 5\}$ **6.** $\varnothing$
7. $A = \{1, 2, 3, 4, 6, 12\}$ **8.** $B = \{0, 12, 24, 36, 48, \ldots\}$ **9.** $A \cap B = \{12\}$
10. $A \cup B = \{0, 1, 2, 3, 4, 6, 12, 24, 36, 48, 60, \ldots, 12n, \ldots\}$, $n = 6, 7, 8, \ldots$
11. $B \cap C = \{x \mid x \in W$ and x is a multiple of $12\} = \{0, 12, 24, 36, \ldots\}$ **12.** $A \cap C = \{6, 12\}$
13. $A \cap B = \{r \mid 3 \leq r \leq 4, r \in Z\} = \{3, 4\}$ **14.** $A \cup B = \{-1, 0, 1, 2, 3, 4, 5, 6, 7, 8, 9, 10, 11, 12\}$

15.   **16.**   **17.**

18.   **19.**   **20.**

21. False   **22.** True   **23.** True   **24.** True   **25.** False   **26.** True   **27.** True   **28.** False

29. Commutative property of addition   **30.** Commutative property of multiplication   **31.** Distributive property
32. Distributive property   **33.** Multiplicative inverse property   **34.** Additive identity property   **35.** False
36. False   **37.** -6 **38.** -11 **39.** -8 **40.** -9 **41.** -30 **42.** -210 **43.** 64 **44.** -64
45. -11 **46.** 3 **47.** 63 **48.** -15 **49.** -34 **50.** 16 **51.** -61 **52.** -136 **53.** -17
54. Undefined   **55.** 1 **56.** 11 **57.** 0 **58.** 4 **59.** Not defined   **60.** 0 **61.** $-6x^3y - 3x^2y^2$
62. $-36a^3b^4$ **63.** $36x^4y^4$ **64.** $54r^8s^6$ **65.** $3y - 4x$ **66.** $-8a - 2b$ **67.** $-5r^2s + rs^2$
68. $-3x^2y^3 - 2xy^2$ **69.** $-x - 3$ **70.** -1 **71.** $3a^2 - 3ab + 3ac$ **72.** $10r^2s + 15rs^2$ **73.** $-x + 12$
74. $7y - 18x - 9$ **75.** $4a - 9$ **76.** $-2r - 3s + 36$ **77.** $-6x + 30$ **78.** $-21y + 31$
79. Let the two numbers be x and y; $xy - 5 = x + y + 3$ **80.** Let x = number; $2x + 8 = x^2 - 3$
81. Let x = first odd integer; $(x) + (x + 2) = (x + 4) - 5$
82. Let x = first even number; $(x + 2)(x + 4) = 10x + 8$
83. Let width = x; area = $(4x - 5)x = 4x^2 - 5x$; perimeter = $2x + 2(4x - 5) = 10x - 10$
84. Let w = width; area = $(3w + 5)(w) = 3w^2 + 5w$; perimeter = $2(3w + 5) + 2w = 8w + 10$
85. Let x and y be the numbers; $x^2 + y^2 = xy + 8$ **86.** Let x and y be the two numbers; $(x + y)^2 = xy + 8$
87. Let x = number of dimes, then $40 - x$ = number of nickels; value (in cents) = $10x + 5(40 - x) = 5x + 200$ cents
88. $12x + 8(30 - x) = 4x + 240$ dollars

Chapter 1 Practice Test

1. (a) $A \cap B = \{2, 3, 5, 7\}$ (b) $A \cup B = \{2, 3, 5, 7, 11, 13, 17, 19, 23\}$
2. (a) False   (b) True   (c) False
3. (a)   (b)

4. (a) False   (b) Commutative property of addition   **5.** (a) 8 (b) 85 (c) 1 (d) 6
6 (a) 1 (b) $\frac{5}{17}$ **7.** (a) $10x^6y^5$ (b) $-rs^2 - 5r^2s - 7rs$ (c) 0 (d) $9r - 6s - 3$
8. Let x = width; perimeter = $2x + 2(3x - 8) = 8x - 16$
9. Number of nickels = $34 - x$; value of coins (in cents) = $10x + 5(34 - x) = 5x + 170$ cents

Exercises 2.1

1. Contradiction   **3.** Identity   **5.** Identity   **7.** Contradiction   **9.** $x = 0$ does not satisfy; $x = 5$ satisfies
11. $a = -3$ does not satisfy; $a = 0$ does not satisfy   **13.** Both $x = -1$ and $x = 5$ satisfy
15. $a = -1$ does not satisfy; $a = 1$ satisfies   **17.** $x = 6$ **19.** $y = -\frac{3}{4}$ **21.** $m = 0$ **23.** Identity
25. $t = 4$ **27.** $y = -9$ **29.** $s = -\frac{9}{2}$ **31.** $x = \frac{7}{8}$ **33.** $x = 4$ **35.** $x = 0$ **37.** $t = 2$
39. No solution   **41.** No solution   **43.** $t = \frac{5}{7}$ **45.** $x = -1$ **47.** No solution   **49.** Identity   **51.** $x = 0$
53. $x = -\frac{2}{11}$ **55.** $x = 23$ **57.** $t = -\frac{1}{2}$ **59.** $x = -\frac{37}{8}$ **61.** $-\frac{11}{9} = -1.\overline{2}$ **63.** $\frac{152}{3} = 50.\overline{6}$

Exercises 2.2

1. Let x = first number; $x + (4x + 3) = 43$; two numbers are 8 and 35
3. Let x = first number; $x + (5x - 8) = -20$; two numbers are -2 and -18
5. Let x = first number; $x + (x + 1) + (x + 2) = 66$; numbers are 21, 22, 23
7. Let x = first number; $(x + 2) + (x + 4) = 2x + 6$; all even integers
9. Let x = first number; $x + (x + 2) + (x + 4) + (x + 6) = 56$; numbers are 11, 13, 15, 17
11. Let x = width; $42 = 2x + 2(2x)$; width is 7 meters; length is 14 meters
13. Let x = width; $54 = 2x + 2(3x + 7)$; width = 5 cm; length = 22 cm
15. Let x = second side; $33 = (x - 5) + x + 2(x - 5)$; sides are 7, 12, 14
17. Let x = original width; $2(x + 2) + 2[2(3x + 1)] = 5(3x + 1) - 3$; original width = 6; original length = 19
19. Let x = nickels; $5x + 10(x + 2) + 25(x + 2 + 4) = 530$; 9 nickels, 11 dimes, 15 quarters
21. Let x = number of \$5 bills; $1[25 - (2x + 1)] + 5x + 10(x + 1) = 164$; four \$1 bills, ten \$5 bills, eleven \$10 bills
23. Let x = number of 10¢ stamps; $5(74 - 3x) + 15(2x) + 10(x) = 670$; thirty-eight 5¢ stamps, twelve 10¢ stamps, twenty-four 15¢ stamps

25. Let x = number of 20-lb packages; $20x + 25(50 - x) = 1,075$; number of 20-lb packages = 35; number of 25-lb packages = 15

27. Let x = number of pairs of shoes sold; $137 = 90 + 2(30 - x) + 1(x)$; 13 pairs of shoes

29. Let x = quantity of $2/lb coffee; $2x + 30(3) = 2.6(x + 30)$; 20 lb of $2/lb coffee

31. Let x = number of orchestra seats; $24x + 14(56 - x) = 1,164$; 38 orchestra seats

33. Let x = number of hours the plumber worked; $22x + 13(x + 2) = 236$; the plumber worked 6 hours

35. $345 = 55t + 60t$; $t = 3$; at 6:00 P.M.   **37.** $595 = 35t + 50t$ where t = time of travel; 7 hours

39. $17t = 7(t + 3)$ where t = time it takes to overtake; $t = 2.1$ hours (2 hours and 6 minutes)

41. Let t = length of time to go to convention; $48t = 54(17 - t)$; $t = 9$ hours; distance to convention = $48 \cdot 9 = 432$ km

43. Let older model work for t minutes; $35t + 50(110 - t) = 5,125$; $t = 25$ minutes; older machine makes $35 \cdot 25 = 875$ copies

45. Let x = number of hours; $60x + 30x = 6,750$; 75 hours

47. Let x = number of hours by experienced worker; $60(x) + 30(x - 3) = 6,750$; 76 hours to complete the job

49. Let x = hours the trainee is working; $80(x - 3) + 48x = 656$; $x = 7$ hours; finishing time = 5:00 P.M.

Exercises 2.3

1. Identity   **3.** Contradiction   **5.** Identity   **7.** Identity   **9.** $u = 2$ does not satisfy; $u = 3$ satisfies

11. $y = -1$ does not satisfy; $y = -2$ does not satisfy   **13.** $z = -2$ satisfies; $z = 4$ does not satisfy

15. Makes sense   **17.** Makes sense ($-8 \leq w < -6$)   **19.** Does not make sense   **21.** Makes sense

23. Makes sense ($-3 < x < 2$)   **25.** $x < 3$ **27.** $y \geq \frac{1}{4}$ **29.** $x \leq -9$ **31.** $x \geq -9$ **33.** $y > -9$

35. $a < \frac{8}{3}$;   **37.** $a < 0$;

39. $t < 4$;   **41.** $y \geq -\frac{10}{3}$;

43. $a < -5$,   **45.** $a > 7$;

47. $t \geq 3$;   **49.** Identity

51. $x \leq -1$;   **53.** $-2 \leq c < 2$;

55. $\frac{7}{2} < k \leq 6$;   **57.** $1 < t < 5$;

59. $-\frac{4}{3} \leq t \leq \frac{7}{3}$ **61.** $-5 < x < 0$ **63.** $3 < x \leq 6$ **65.** $4 < x \leq 7$ **67.** $-\frac{8}{5} \leq z \leq \frac{2}{5}$

69. $2.625 \leq x < 13.\overline{3}$;

Exercises 2.4

1. Let x = number; $3x - 4 < 17$; $x < 7$ **3.** Let x = number; $6x + 12 > 3x$; $x > -4$

5. Let x = number; $4x - 3 \geq 7x - 3$; $x \leq 0$ **7.** Let s = length of a side; $4s \leq 72$; maximum length = 18 feet

9. Let l = length; $16 + 2l \geq 80$; length ≥ 32 cm

11. Let w = width; $50 \leq 2w + 36 \leq 70$; the range of values for the width w is $7 \leq w \leq 17$ inches

13. Let w = width; $100 < 2w + 2(3w) < 200$; $12.5 < w < 25$ feet

15. $106 \leq$ perimeter ≤ 138 inches   **17.** Let x = shortest side; $30 \leq x + 2x + (x + 2) \leq 50$; $7 \leq x \leq 12$ cm

19. Let x = price of reserved ticket; $300x + 150(x - 2) \geq 3,750$; minimum price is $9

21. Let x = number of quarters; $25x + 10(50 - x) \geq 980$; minimum number is 32

23. Let x = number of dimes; $10x + 5(40 - x) \leq 285$; maximum number is 17

25. Let x = number of elephants; $25x + 20(24 - x) \geq 575$; at least 19
27. Let x = number of hours of tutoring time; $6x + 4(30 - x) \geq 190$; $x \geq 35$; he cannot make \$190 working 30 hours
29. Let x = number of superdogs; $45x + 25(100 - x) \geq 3{,}860$; at least 68
31. Let x = number of shares of stock B; $2(1{,}000 - x) + 3x \geq 2{,}400$; at least 400 shares

Exercises 2.5

1. $x = 4$ or $x = -4$ **3.** $-4 < x < 4$ **5.** $x > 4$ or $x < -4$ **7.** $-4 \leq x \leq 4$ **9.** $x \geq 4$ or $x \leq -4$
11. No solution   **13.** Identity   **15.** No solution   **17.** $t = 5$ or $t = 1$ **19.** $t = 5$ or $t = -5$
21. $n = 4$ or $n = 6$ **23.** $2 < a < 8$ **25.** $a \geq 3$ or $a \leq -1$ **27.** No solution   **29.** $-2 < a < 2$
31. $x = 2$ or $x = -\frac{2}{3}$ **33.** No solution   **35.** $x = 0$ or $x = -5$ **37.** $1 \leq a \leq 5$ **39.** $a < 2$ or $a > 3$
41. $-8 < x < 0$ **43.** $x = 6$ or $x = -4$;   **45.** $1 \leq x \leq 5$;

47. $x > -3$ or $x < -4$;   **49.** $\frac{1}{2} < x < 2$

51. $\frac{1}{2} < x < 1$;   **53.** $t = 4$ or $t = -\frac{2}{9}$ **55.** $r = 6$ or $r = -1$ **57.** $a = \frac{7}{2}$

59. $x = -1$ or $x = 1$ **61.** $x = 0$

Chapter 2 Review Exercises

1. $x = 0$ **2.** $x = 4$ **3.** $x = 11$ **4.** $y = \frac{4}{5}$ **5.** $x = -1$ **6.** $x = -2$ **7.** $a = \frac{7}{3}$ **8.** $b = -\frac{1}{2}$
9. No solution   **10.** Identity   **11.** $x = 0$ **12.** $a = 0$ **13.** $a = \frac{62}{11}$ **14.** Identity   **15.** $x = \frac{65}{18}$
16. $x = -\frac{15}{4}$ **17.** $x = \frac{23}{9}$ **18.** $x = \frac{7}{2}$ **19.** $x = 3$ or $x = -3$ **20.** No solution   **21.** No solution
22. $y = 7$ or $y = -7$ **23.** $x = 4$ or $x = -4$ **24.** $x = \frac{5}{3}$ or $x = -\frac{5}{3}$ **25.** $a = 3$ or $a = -5$
26. $a = 8$ or $a = -2$ **27.** $z = -\frac{5}{4}$ **28.** $x = \frac{4}{7}$ or $x = -\frac{8}{7}$ **29.** $y = 5$ or $y = 0$ **30.** $x = \frac{11}{4}$ or $x = \frac{7}{4}$
31. $x = 3$ **32.** $a = -2$ **33.** $t = 2$ **34.** $a = 2$ or $a = \frac{8}{3}$ **35.** $x \leq 3$ **36.** $x \leq -3$ **37.** $x \leq \frac{4}{3}$
38. $x < 2$ **39.** $z > -\frac{5}{3}$ **40.** $y > \frac{5}{4}$ **41.** $s < 4$ **42.** No solution   **43.** Identity   **44.** No solution
45. $-1 \leq x \leq 3$ **46.** $1 \leq x \leq 6$ **47.** $-\frac{1}{2} \leq x \leq \frac{11}{2}$ **48.** $-9 \leq x \leq -1$ **49.** $3 \leq x \leq 6$
50. $2 \leq x \leq \frac{21}{5}$ **51.** $a < -3$;   **52.** No solution   **53.** No solution

54. $x > \frac{2}{3}$;   **55.** $q \leq \frac{12}{5}$

56. $x > -2$;   **57.** $-4 < x < 4$;

58. $x > 4$ or $x < -4$;   **59.** $s \geq 5$ or $s \leq -5$;

60. $-5 \leq x \leq 5$;   **61.** No solution   **62.** $t = 0$;

63. $-1 < t < 3$;   **64.** $t > 3$ or $t < -1$;

65. $a \geq 9$ or $a \leq 3$;   **66.** $3 \leq a \leq 9$;

67. $-13 \leq r \leq -5$ **68.** $r \leq 5$ or $r \geq 13$

69. $x \geq \frac{3}{2}$ or $x \leq -\frac{1}{2}$;   **70.** $x > \frac{3}{2}$ or $x < -\frac{1}{2}$;

71. $-\frac{2}{3} < x < 2$;   **72.** $x > 2$ or $x < -\frac{2}{3}$;

73. $-1 \leq x \leq 4$ **74.** $x \geq 4$ or $x \leq -1$ **75.** $x > 0$ or $x < -\frac{10}{3}$ **76.** $-\frac{10}{3} < x < 0$ **77.** No solution
78. $-\frac{9}{2} \leq x \leq \frac{15}{2}$ **79.** Let x = number; $3x = 4x - 4$; $x = 4$ **80.** Let x = number; $5x - 5 = x + 3$; $x = 2$
81. Let x = number; $5(x + 6) = x - 2$; $x = -8$ **82.** Let x = number; $5x + 6 = x - 2$; $x = -2$
83. Let x = number of nickels; $5x + 10(35 - x) = 345$; 1 nickel and 34 dimes
84. Let x = number of nickels; $5x + 25(70 - x) = 1,250$; 25 nickels and 45 quarters
85. Let x = number of nickels; $5x + 25(42 - x) = 750$; 15 nickels and 27 quarters
86. Let x = number of quarters; $25x + 10(45 - x) = 915$; 31 quarters and 14 dimes
87. Let x = number of quarters; $5(x + 3) + 10(67 - 2x) + 25x = 715$; 3 quarters, 6 nickels, 61 dimes
88. Let x = number of quarters; $25x + 5(x + 7) + 10(82 - x - x - 7) = 985$; 20 quarters, 27 nickels, 35 dimes
89. Let x = number of packages weighing 8 lb each; $8x + 5(30 - x) = 186$; twelve 8-lb packages and eighteen 5-lb packages
90. Let x = number of packages weighing 8 lb; $8x + 5(45 - x) = 276$; seventeen 8-lb packages and twenty-eight 5-lb packages
91. Let x = number of TVs; $15x + 10(23 - x) = 295$; 13 TVs and 10 radios
92. Let x = number of hours the plumber worked; $25x + 10(7 - x) + 27 = 134.50$; plumber worked 2.5 hours; assistant worked 4.5 hours
93. Let x = number of hours of tutoring; $12(30 - x) + 20x \geq 456$; at least 12 hours
94. Let x = number of hours tutoring; $20x + 12(40 - x) \leq 680$; no more than 25 hours tutoring
95. No solution—the most he can have is $20.

Chapter 2 Practice Test

1. $x = \frac{11}{2}$ **2.** No solution **3.** $a = -9$ **4.** $x \geq 4$ **5.** $x < \frac{1}{5}$ **6.** $\frac{1}{2} < x < 4$
7. $x = 7$ or $x = 0$;   **8.** $-\frac{1}{3} < x < 3$;

9. $x \geq \frac{9}{5}$ or $x \leq 1$;

10. Let x = number of boxes weighing 35 kg; $35x + 45(93 - x) = 3,465$; seventy-two 35-kg boxes and twenty-one 45-kg boxes
11. $8x = 3(x + 2)$, where x = time of jogger; $x = \frac{6}{5} = 1\frac{1}{5}$ hours to catch up
12. Let x = number of dimes; $10x + 5(32 - x) \geq 265$; at least 21 dimes

Exercises 3.1

1. (a) Binomial **(b)** 2 **(c)** 1 **3. (a)** Monomial **(b)** 0 **(c)** None **5. (a)** Not a polynomial
7. (a) Monomial **(b)** 11 **(c)** 3 **9.** Not a polynomial **11. (a)** Trinomial **(b)** 12 **(c)** 3
13. $5x^2 - 9x + 9$ **15.** $-x^2 - 3xy - 4y^2 + 3x - 2$ **17.** $6a^2 - 5ab + 3b^2$ **19.** $ab + 5b^2$ **21.** $-5b^2 - ab$
23. $-5y^2 + 3xy - 3y + 4$ **25.** $-11x^2 - 3xy + 2y^2$ **27.** $-5a^2 + ab - 4b^2$ **29.** $3x^4y - 2x^3y^2 + 2x^2y^3$
31. $3x^4y - 2x^3y^2 + 2x^2y^3$ **33.** $x^2 + 7x + 12$ **35.** $6x^2 + x - 1$ **37.** $6x^2 - x - 1$ **39.** $3a^2 + 2ab - b^2$
41. $4r^2 - 4rs + s^2$ **43.** $4r^2 - s^2$ **45.** $9x^2 - 4y^2$ **47.** $9x^2 - 12xy + 4y^2$ **49.** $ax - bx - ay + by$
51. $4ar + 6br + 6as + 9bs$ **53.** $2y^4 - y^2 - 3$ **55.** $5x^3 + 12x^2 - 5x + 12$ **57.** $27a^3 + 18a^2 - 12a - 5$
59. $a^3 + b^3$ **61.** $x^2 + y^2 + z^2 - 2xy - 2xz + 2yz$ **63.** $x^3 + 8$ **65.** $a^2 + b^2 + 2ab + ac + ad + bc + bd$
67. $P = 2w + 2(3w + 5) = 8w + 10$: $A = w(3w + 5) = 3w^2 + 5w$
69. $A = \pi(8 + x)^2 - \pi(8)^2 = \pi x^2 + 16\pi x$ sq. ft **71.** $A = (20 + 2x)(60 + 2x) - (20)(60) = 4x^2 + 160x$ sq. ft
73. $19.344x^3 - 32.448x^2 + 37.44x$

Exercises 3.2

1. $x^2 + 9x + 20$ **3.** $x^2 - 4x - 21$ **5.** $x^2 - 19x + 88$ **7.** $2x^2 - 7x - 4$ **9.** $25a^2 - 16$
11. $9z^2 + 30z + 25$ **13.** $9z^2 - 30z + 25$ **15.** $9z^2 - 25$ **17.** $15r^3 + 25r^2s + 9rs + 15s^2$
19. $9s^2 - 12sy + 4y^2$ **21.** $9y^2 + 60yz + 100z^2$ **23.** $9y^2 - 100z^2$ **25.** $9a^2 + 18ab + 8b^2$
27. $64a^2 - 16a + 1$ **29.** $25r^2 - 20rs + 4s^2$ **31.** $56x^2 - 113x + 56$ **33.** $25t^2 - 15s^2t + 15st - 9s^3$
35. $9y^6 - 24xy^3 + 16x^2$ **37.** $9y^6 - 16x^2$ **39.** $4r^2s^2t^2 - 28rstxyz + 49x^2y^2z^2$ **41.** $4r^2s^2t^2 - 49x^2y^2z^2$
43. $2a^{12} - 7a^6b^3 + 3b^6$ **45.** $2x^4 - 9x^2y^2 + 9y^4$ **47.** $-9x^2 - 5x - 5$ **49.** $-4ab$
51. $-30a^3 - 5a^2 + 10a$ **53.** $-10a^2 + 8a + 2$ **55.** $3rs^2$ **57.** $-y^2 + 19y - 7$
59. $125a^3 - 225a^2b + 135ab^2 - 27b^3$ **61.** $-2x^3 - x^2 + x$ **63.** $a^2 + 2ab + b^2 - 1$
65. $a^2 - 4ab + 4b^2 + 10az - 20bz + 25z^2$ **67.** $a^2 - 4ab + 4b^2 - 25z^2$ **69.** $a^2 + 2ab + b^2 - 4x^2 - 4x - 1$
71. $a^{2n} - 9$

Exercises 3.3

1. $2x(2x + 1)$ **3.** $x(x + 1)$ **5.** $3xy^2(1 - 2xy)$ **7.** $3x^2(2x^2 + 3x - 7)$ **9.** $5x^2y^3z(7x^2y - 3xy^2 + 2z)$
11. $6r^2s^3(4rs - 3rs^2 - 1)$ **13.** $7ab(5ab^2 - 3a^2b + 1)$ **15.** $(3x + 5)(x + 2)$ **17.** $(3x - 5)(x + 2)$
19. $(2x + 5y)(x + 3y)$ **21.** $(x + 4y)(3x - 5y)$ **23.** $2(r + 3)(a - 2)$ **25.** $2(a - 3)^2(x + 1)$
27. $4a(b - 4)(4ab - 16a - 1)$ **29.** $(2x + 3)(x - 4)$ **31.** $(3x + 5y)(x - 4y)$ **33.** $(7x + 3y)(a - b)$
35. $(7x - 3y)(a + b)$ **37.** $(7x - 3y)(a - b)$ **39.** Not factorable **41.** $(2r - s)(r + s)$ **43.** Not factorable
45. $(5a - 2b)(a - b)$ **47.** $(3a - 1)(a - 2)$ **49.** $(3a + 1)(a - 2)$ **51.** $(3a - 1)(a + 2)$
53. $(a + 2)(a^2 + 4)$

Exercises 3.4

1. $(x + 9)(x - 5)$ **3.** $(y - 5)(y + 2)$ **5.** $(x - 5y)(x + 3y)$ **7.** $(r + 9)(r - 9)$ **9.** $(x - 3y)(x + 2y)$
11. Not factorable **13.** $(r - 3s)(r - 4s)$ **15.** $(r - 4s)(r + 3s)$ **17.** $(3x + 7y)(3x - 7y)$
19. $(5x + 4)(3x + 1)$ **21.** $(5x + 3y)(3x - 2y)$ **23.** $(5x - 3y)(3x + 2y)$ **25.** $(3x - 5y)(x + 5y)$
27. Not factorable **29.** $(5a + 2b)(2a - 5b)$ **31.** $(5a - 2b)(2a + 5b)$ **33.** $(5 + y)(5 - 2y)$
35. Not factorable **37.** $2x(x + 4)(x - 2)$ **39.** $3xy(x + 5)(x - 3)$ **41.** $3ab(3x - 2)(2x + 3)$
43. $x(x + 7)(x - 4)$ **45.** $(x - 2)(3x + 1)$ **47.** $3ab(2 - 3x)(3 + 2x)$ **49.** $6y^2(5y - 3)(3y - 2)$
51. $5a^2b(5a + 2b)(5a - 3b)$ **53.** $4rs(3r - 2s)(2r + 3s)$ **55.** $(2r^2 + 1)(3r^2 - 2)$ **57.** $3(x + 3)(x - 3)$
59. $2xy(10y^4 + xy^2 - 4x^2)$ **61.** $3xy(6x - y^2)(6x + 5y^2)$ **63.** $ab(3a^3 - 2)(4a^3 + 3)$ **65.** $(5a^4 - 2)(3a^4 + 5)$

Exercises 3.5

1. $(4x + 3y)(4x - 3y)$ **3.** $(x + 2y)^2$ **5.** $(x - 2y)^2$ **7.** Not factorable **9.** $(3xy - 1)(2xy - 1)$
11. $(9rs - 4)(9rs + 4)$ **13.** $(7xy - 3)^2$ **15.** $(3xy + 2z)(3xy - 2z)$ **17.** $a^2(5 - 2b)(5 + 2b)$
19. $(8x^2 - y)^2$ **21.** $(1 - 7a)^2$ **23.** $(2a - b)(4a^2 + 2ab + b^2)$ **25.** $(x + 5y)(x^2 - 5xy + 25y^2)$
27. Not factorable **29.** $2(2y^2 - 10y + 5)$ **31.** $2a(3a + 2b)^2$ **33.** $3ac(2a + 3c)^2$ **35.** $(3a^3 + b^3)(3a^3 - b^3)$
37. $(2x^2 + 9y^2)(2x^2 - 9y^2)$ **39.** $(2x^3 + 7y^3)^2$ **41.** $(2x^3 + 3y^3)(2x^3 - 3y^3)$ **43.** $(5a^3 + b)^2$
45. $6x^2y(2x + 3y)(2x - 3y)$ **47.** $3y^2(4y^4 + 9)$ **49.** $(a + b + 2)(a + b - 2)$ **51.** $4a(2a + b)$
53. $(2x - 1)(2x + 1)(2x^2 - 3)$ **55.** $(3a + b)(9a^2 - 3ab + b^2)(a + b)(a^2 - ab + b^2)$
57. $3ab(3a^2 - b^2)(a + b)(a - b)$ **59.** $5ab(a + 2b)(a - 2b)(a^2 + b^2)$
61. $(x + y - a - b)(x^2 + 2xy + y^2 + ax + bx + ay + by + a^2 + 2ab + b^2)$ **63.** $(a - b + 4)(a - b - 4)$
65. $(x + 3 + r)(x + 3 - r)$ **67.** $(a + 1)(a + 2)(a - 2)$ **69.** $(x + 1)(x - 1)(x^2 + x + 1)$
71. $(a + 1)^2(a^2 - a + 1)(a - 1)$

Exercises 3.6

1. $x + 4$ **3.** $a + 2$ **5.** $7z - 2, R -14$ **7.** $a^2 + 2a + 3, R -5$ **9.** $3x^2 + x - 1, R 8$
11. $2z - 3, R -6$ **13.** $x^3 + 3x^2 + 2x + 1$ **15.** $3y^3 - 2y^2 + 5, R -5$ **17.** $2a^3 - 5, R 16$
19. $y^2 - y + 1, R -2$ **21.** $4a^2 + 2a + 1, R 2$ **23.** $2y^3 - 3y^2 - 2y + 2, R 3y + 2$
25. $3z^3 - 6z^2 + 11z - 25, R 73z - 35$ **27.** $7x^4 - 3, R x - 5$

Chapter 3 Review Exercises

1. $2x^3 - 7x^2 + 3x - 5$ **2.** $6x^2 - 8x + 15$ **3.** $6x^3y - 9x^2y + 3x^2$ **4.** $15r^3s^3 - 10r^2s^4 - 15rs^2$ **5.** $-ab^2$

6. $-4x^2y - 11xy^2 + 18xy$ **7.** $-2x + 16$ **8.** $21x - 28$ **9.** $x^2 - x - 6$ **10.** $x^2 - 13x + 40$

11. $6y^2 - 7y + 2$ **12.** $25y^2 - 9a^2$ **13.** $9x^2 - 24xy + 16y^2$ **14.** $5a^2 - 7ab - 6b^2$

15. $16x^4 - 40x^2y + 25y^2$ **16.** $16a^2 - 9b^4$ **17.** $49x^4 - 25y^6$ **18.** $6x^3 - 19x^2 + 12x - 5$

19. $10y^3 - 21y^2 + 23y - 21$ **20.** $21x^3 - 5x^2y - 12xy^2 + 4y^3$ **21.** $4x^2 + 12xy + 9y^2 - 2x - 3y - 20$

22. $a^2 + 2ab + b^2 - x^2 + 2xy - y^2$ **23.** $x^2 - 2xy + y^2 + 10x - 10y + 25$ **24.** $x^2 - 6x + 9 - y^2$

25. $a^2 + 2ab + b^2 - 8a - 8b + 16$ **26.** $x^2 + 2xy + y^2 + 10x + 10y + 25$ **27.** $3xy(2x - 4y + 3)$

28. $5ab^2(3a - 2b^2 + 1)$ **29.** $(3x - 2)(a + b)$ **30.** $(y - 1)(5y + 3)$ **31.** $(a - b)(2a - 2b + 3)$

32. $(a - b)(4a - 4b + 7)$ **33.** $(x - 1)(5a + 3b)$ **34.** $(a + b)(2x + 3)$ **35.** $(5y + 3)(y - 1)$

36. $(2a - b)(7a + 3b)$ **37.** $2a(a - 5)$ **38.** $t(t + 3)(t - 2)$ **39.** Not factorable   **40.** Not factorable

41. $(x - 7)(x + 5)$ **42.** $(a + 9)(a - 4)$ **43.** $(a + 7b)(a - 2b)$ **44.** $(y^2 + 3x^2)(y^2 - 2x^2)$

45. $(7a + 2b)(5a + b)$ **46.** $(x - 3y)^2$ **47.** $(a - 3b^2)^2$ **48.** $2x(x^2 + 3x - 27)$ **49.** $3a(a - 5)(a - 2)$

50. $5ab(a + 3b)(a - 2b)$ **51.** $2x(x + 5y)(x - 5y)$ **52.** $3y(y + 4)^2$ **53.** $(3x - 2)(2x + 3)$

54. $(5y - 4)(5y + 3)$ **55.** $(2x + 5y)(4x^2 - 10xy + 25y^2)$ **56.** $(x - 3)(x^2 + 3x + 9)$ **57.** $(6a + b)(a - 3b)$

58. $(6x + 5y^2)(2x + y^2)$ **59.** $(7a^2 + 2b^2)(3a^2 + 5b^2)$ **60.** $2a(3a - 2b)(9a^2 + 6ab + 4b^2)$ **61.** $(5x^2 - 4y^2)^2$

62. $3x(2x + 3y)(3x - 2y)$ **63.** $5xy(2x - 3y)^2$ **64.** $(2a^2 + 3b^2)(2a^2 - 3b^2)$ **65.** $2a^2b(3a + 2b)(a - 2b)$

66. $7x^2y(2x + 3y)(2x - 3y)$ **67.** $2x(x^2 - 2)(3x^2 + 1)$ **68.** $5xy(3x^2 - y)(2x^2 - 5y)$

69. $(a - b + 2)(a - b - 2)$ **70.** $(3x + 5y + 1)(3x - 5y - 5)$ **71.** $(3y + 5 + 3x)(3y + 5 - 3x)$

72. $(5x^2 + y + 4)(5x^2 - y - 4)$ **73.** $3x^2 + 2x + 11, R\ 17$ **74.** $x^3 - x^2 + x - 2, R\ 1$

75. $4a^2 + 6a + 9$ **76.** $y^4 + y, R\ 1$

Chapter 3 Practice Test

1. $27a^3 - 8b^3$ **2.** $6x - 18$ **3.** $x^2 - 2xy + y^2 - 4$ **4.** $9x^2 + 6xy^2 + y^4$ **5.** $2xy(5x^2y - 3xy^2 + 1)$

6. $(3a - b)(2x + 5)$ **7.** $(x + 15y)(x - 4y)$ **8.** $(5x + 2)(2x - 3)$ **9.** Not factorable

10. $3ab(a + b)(a - b)$ **11.** $rs(r - 5s)^2$ **12.** $(2a - 1)(4a^2 + 2a + 1)$ **13.** $(2x + y + 8)(2x + y - 8)$

14. $(a + 4)(a + 1)(a - 1)$ **15.** $x^2 + 3x + 16, R\ 40$

Cumulative Review: Chapters 1–3

The number in parentheses after the answer indicates the section in which the material is discussed.

1. $\{11, 13, 17, 19, 23, 29, 31, 37\}$ (§1.1)   **2.** $\varnothing$ (§1.1)   **3.** $\varnothing$ (§1.1)

4. $\{10, 11, 13, 15, 17, 19, 20, 23, 25, 29, 30, 31, 35, 37\}$ (§1.1)   **5.** True (§1.2)   **6.** False (§1.2)

7. (§1.2)   **8.** (§1.2)

9. Closure property for addition (§1.3)   **10.** Multiplicative inverse property (§1.3)   **11.** -6 (§1.4)

12. -17 (§1.4)   **13.** 36 (§1.4)   **14.** 43 (§1.4)   **15.** 0 (§1.4)   **16.** Undefined (§1.4)

17. $-3x^3 + 6x^2 - 9x$ (§1.5)   **18.** $-12x^4y^3$ (§1.5)   **19.** Let $x =$ the number; $3 + 8x$ (§1.6)

20. Let $x =$ the first number; $x(x + 1) - (x + 2)$ (§1.6)

21. Let $w =$ the width; $A = w(4w - 3)$; $P = 10w - 6$ (§1.6)

22. Let $n =$ number of nickels; $5n + 10(35 - n)$ (in cents) (§1.6)   **23.** $x = -11$ (§2.1)   **24.** No solution (§2.1)

25. $x = 2$ (§2.1)   **26.** $x = -\frac{1}{3}$ (§2.1)   **27.** $x = \frac{4}{19}$ (§2.1)   **28.** $x = -\frac{4}{3}$ (§2.1)   **29.** $x = 4, -5$ (§2.5)

30. $x = \frac{1}{2}, \frac{3}{4}$ (§2.5)   **31.** $x \leq -\frac{3}{2}$ (§2.3);

32. $x < -\frac{1}{4}$ (§2.3);   **33.** Identity (§2.3);

34. $-3 \leq x < 4$ (§2.3);   **35.** $1 < x < 9$ (§2.5);

36. $x < -4$ or $x > 3$ (§2.5);   **37.** $x < -\frac{1}{2}$ or $x > 3$ (§2.5);

38. $0 \le x \le \frac{16}{3}$ (§2.5)

$-2 \quad 0 \quad 2 \quad 4 \; \frac{16}{3} \, 6 \quad 8$

39. Let x = number; $4x = 2x - 3$; $-\frac{3}{2}$ (§2.2)

40. Let x = number of dimes; $10x + 25(25 - x) = 325$; 20 dimes, 5 quarters (§2.2)

41. Let p = number of packages with 100 plates; $1 \cdot p + 2(28 - p) = 33$; 23 packages of 100 plates, 5 packages of 300 plates; 3,800 plates (§2.2)

42. Let x = number of hours assistant painted; $6x + 9(58 - x) = 417$; 35 hours (§2.2)

43. Let t = time for VF-44 printer to do the job; $(30)(60)(80) = x(60)(120)$; 20 minutes (§2.2)

44. Let t = time new machine ran during the job; $90t + 60(120 - t) = 9,750$; $t = 85$ minutes; $(90)(85) = 7,650$ copies (§2.2)

45. Let w = width; perimeter = $8w + 4$; 44 ft to 100 ft (§2.4)

46. Let t = number of hours working as a mechanic; $10t + 6(30 - t) \ge 232$; at least 13 hours (§2.4)　　**47.** 5 (§3.1)

48. 0 (§3.1)　　**49.** $x^2 + 5x - 9$ (§3.1)　　**50.** $-6x^2 + 15xy - 12x$ (§3.1)　　**51.** $x^2 - 4xy + 3y^2$ (§3.1)

52. $3x^2 - 5xy + 2y^2$ (§3.1)　　**53.** $x^2 + 2xy + y^2 - 2x - 2y$ (§3.1)　　**54.** $8a^3 + b^3$ (§3.1)

55. $9x^2 - 30xy + 25y^2$ (§3.2)　　**56.** $4m^2 - 9n^2$ (§3.2)　　**57.** $9x^2 - 25y^2$ (§3.2)　　**58.** $4m^2 + 12mn + 9n^2$ (§3.2)

59. $4x^2 + 4xy + y^2 - 9$ (§3.2)　　**60.** $x^2 - 4xy + 4y^2 + 6x - 12y + 9$ (§3.2)　　**61.** $(x - 8)(x + 3)$ (§3.4)

62. $(y - 2x)(y + 2x)$ (§3.4)　　**63.** $(y - 7x)(y - 5x)$ (§3.4)　　**64.** $(5a + b)(2a - b)$ (§3.4)

65. $(2y + 3z)(2y + 5z)$ (§3.4)　　**66.** $(3x - 5z)(3x + 5z)$ (§3.5)　　**67.** $(5a + 2b)^2$ (§3.5)　　**68.** Not factorable (§3.5)

69. $9(2x - 1)(2x + 1)$ (§3.5)　　**70.** $(5x - 3z)^2$ (§3.5)　　**71.** Not factorable (§3.5)　　**72.** $y(6y + 1)(2y - 3)$ (§3.5)

73. $y(3y - 1)(y + 2)$ (§3.5)　　**74.** $(5a^2 - 2b^2)(5a^2 + 4b^2)$ (§3.5)　　**75.** $(7a^2 - 3z)(7a^2 + z)$ (§3.5)

76. $ab(6a^2 + b)(3a^2 - 2b)$ (§3.5)　　**77.** $(x - y - 4)(x - y + 4)$ (§3.5)　　**78.** $(a - 2b - 5)(a - 2b + 5)$ (§3.5)

79. $(4a^2 + b^2)(2a - b)(2a + b)$ (§3.5)　　**80.** $(x + y + 2)(x + y + 1)$ (§3.5)

81. $(2a + 5b)(4a^2 - 10ab + 25b^2)$ (§3.5)　　**82.** $(x - 1)(x + 5)(x - 5)$ (§3.5)　　**83.** $x - 2$, R 11 (§3.6)

84. $x + 3$, R -4 (§3.6)　　**85.** $2x^2 - 3x + 6$, R -17 (§3.6)　　**86.** $x^2 - 2x + 3$, R $-4x - 1$ (§3.6)

Cumulative Practice Test: Chapters 1–3

1. (a) $\{4, 8, 12, 16, 20\} = A$　　**(b)** $\{2, 4, 6, 8, 10, 12, 14, 16, 18, 20, 22\} = B$　　**2.** Distributive property
3. (a) -4　　**(b)** -18　　**4.** 6　　**5. (a)** $x = -3$　　**(b)** No solution　　**(c)** $x = 11, -5$

6. (a) $x > \frac{9}{4}$;

$0 \quad 1 \quad 2\frac{9}{4} \quad 3$

(b) $-1 < x < \frac{7}{3}$;

$-2 \; -1 \quad 0 \quad 1 \quad 2\frac{7}{3} \, 3$

(c) $x \le 1$ or $x \ge \frac{3}{2}$;

$-1 \quad 0 \quad 1\frac{3}{2} \, 2 \quad 3$

7. Let p = number of packages weighing 30 lb; $30p + 10(170 - p) = 3,140$; seventy-two 30-lb packages, ninety-eight 10-lb packages

8. Let t = time the faster car travels; $55t = 40(t + 1)$; $2\frac{2}{3}$ hours

9. (a) $-3x^2 - xy + 3y^2$　　**(b)** $15a^2 - ab - 6b^2$　　**(c)** $4x^4 - y^2$　　**(d)** $9y^2 - 12yz + 4z^2$
(e) $x^2 + 2xy + y^2 - 6x - 6y + 9$

10. (a) $(a - 7)(a - 2)$　　**(b)** $(3r + 2s)(2r - 3s)$　　**(c)** $(5a^2 + 3y)(2a^2 - 3y)$　　**(d)** $(2a^2 - 3b^2)^2$
(e) $(a + 2b - 5)(a + 2b + 5)$　　**(f)** $(5x - 1)(25x^2 + 5x + 1)$　　**11.** $2x^2 - 4x + 14$, R -23

Exercises 4.1

1. $x \ne 2$　　**3.** x and y can be any number　　**5.** $x \ne 0$　　**7.** $3x \ne y$　　**9.** $\dfrac{x}{5}$　　**11.** $\dfrac{8x}{9ya^2}$　　**13.** $\dfrac{x - 5}{x - 3}$

15. $\dfrac{(2x - 3)(x - 5)}{(2x + 3)(x + 3)}$　　**17.** -1　　**19.** $\dfrac{x - 1}{2(x + 3)}$　　**21.** $\dfrac{x + 3}{x - 3}$　　**23.** $\dfrac{(w - 7z)(w - z)}{(w + 7z)(w + z)}$　　**25.** $x + 4$

27. $-(x + 4)$　　**29.** $\dfrac{a - 6}{5a + 1}$　　**31.** $\dfrac{2(x - 3)(x + 2)}{3(x - 2)}$　　**33.** $\dfrac{2x^2}{3(x^2 - 2)}$　　**35.** $-\dfrac{3a + 2}{1 + 2a}$　　**37.** -1

39. $\dfrac{a^2 - ab + b^2}{(a + b)^2}$ **41.** $-\dfrac{a + b}{2x}$ **43.** $a + b - x - y$ **45.** $6x^3$ **47.** $25a^2b$ **49.** $3x(x + 5)$

51. $(a - b)(x + y)$ **53.** $-(y - 2)(y + 4)$ **55.** $(x - y)(x + y)$ **58.** $\frac{19}{7}$ **59.** $12xy - 18y^2$ **60.** $x = 1$

61. $(x + 1)^2(x - 1)$ **62.** Let $x = \#$ nickels; $5x + 10(28 - x) = 250$; 6 nickels, 22 dimes

Exercises 4.2

1. $\dfrac{6y}{a^2b^6}$ **3.** $\dfrac{b^3}{4}$ **5.** $\dfrac{128x^3ab^4}{3}$ **7.** $\dfrac{12r^2s^5}{a^2}$ **9.** $\dfrac{16r^2s}{27a^2b^2}$ **11.** $\dfrac{2(x + 1)^2(x - 2)}{3(x + 3)^2}$ **13.** $\dfrac{3(x - 2)}{2}$

15. $\dfrac{5a(a - b)}{3}$ **17.** $\dfrac{5a(a - b)}{3(a + b)^2}$ **19.** -1 **21.** $\dfrac{2a + 3}{a + 1}$ **23.** $\dfrac{1}{x - y}$ **25.** $3x$ **27.** $\dfrac{2(y - x)}{(x + y)^2}$ **29.** 1

31. $\dfrac{2(a - b)(2x - 3)}{a + b}$ **33.** $\dfrac{(2a + b)(3a + b)}{(3a + 2b)(2a - b)}$ **35.** $\dfrac{1}{3a + b}$ **37.** x **39.** $\dfrac{3x(2x - y)}{2x + y}$ **41.** 0

42. $x \geq \dfrac{-7}{3}$ **43.** The Distributive Property **44.** $x = 0$ or $x = 5$

45. Let $t =$ the time Bobby was running; $6t = 5(t + 1)$; it takes Bobby 5 hours.

46. Let $t =$ the time it takes for both to complete the job; $32t + 40t = 3{,}960$; it takes 55 minutes.

Exercises 4.3

1. 2 **3.** 10 **5.** $\dfrac{1}{a + 1}$ **7.** $\dfrac{x - y}{x + y}$ **9.** $\dfrac{12(x + 1)}{x + 3}$ **11.** $\dfrac{27x - 8x^2y}{18y^2}$ **13.** $\dfrac{21y^3 - 16x^4}{18x^2y^2}$ **15.** $\dfrac{28}{27y}$

17. $\dfrac{252a^2c^2 + 168b - 3bc^2}{7b^2c^3}$ **19.** $\dfrac{x^2 + 3x + 6}{x(x + 2)}$ **21.** $\dfrac{a^2 - ab + b^2}{a(a - b)}$ **23.** $\dfrac{3x - 29}{(x + 7)(x - 3)}$

25. $\dfrac{3(x^2 - 6x - 1)}{(x - 7)(x + 2)}$ **27.** $\dfrac{r^2 + 6rs - s^2}{(r - s)(r + s)}$ **29.** -1 **31.** 1 **33.** $-\dfrac{(a + 3)}{(a - 3)}$ **35.** $\dfrac{a^2 + ab - b}{a^2 - b^2}$

37. $\dfrac{a(a + b)}{b}$ **39.** $\dfrac{y^3 + 10y^2 + 19y + 3}{(y + 2)(y + 3)}$ **41.** $\dfrac{5x + 3}{x + 2}$ **43.** $\dfrac{5x + 1}{2}$ **45.** $\dfrac{2a^2 + 4a + 9}{a^2 + a - 12}$ **47.** $\dfrac{-1}{y - 3}$

49. $\dfrac{(3x + 4)(x - 2)}{(x - 5)(x + 2)}$ **51.** $\dfrac{2a^2 - 15a - 2}{(a - 2)(a + 3)(a - 5)}$ **53.** $\dfrac{75a^3 + 51a^2 + 2a - 3}{(3a - 1)(5a + 2)}$ **55.** $2 + x$ **57.** $\dfrac{1}{r - s}$

59. $\dfrac{3s^2 - 3st + 2s + t}{s^3 - t^3}$ **61. (a)** $\dfrac{x + 4}{4}$ **(b)** $\dfrac{x}{4} + 1$ **63. (a)** $3x - 2y$ **(b)** $3x - 2y$

65. (a) $\dfrac{6m - 4m^2n - 9}{15n}$ **(b)** $\dfrac{2m}{5n} - \dfrac{4m^2}{15} - \dfrac{3}{5n}$ **67.** $2x - 3$, $R\ 5$

68. Let $x = \#$ 25-lb packages; $25x + 30(60 - x) = 1{,}680$; twenty-four 25-lb packages, and thirty-six 30-lb

packages. **69.** $-1 \leq x \leq 4$ **70.** $\dfrac{1}{y(x + 1)}$ **71.** $\dfrac{x^2y^3(2x - 3)^2(x + 1)}{(x - 4)^2}$

Exercises 4.4

1. $\dfrac{x}{5y}$ **3.** $\dfrac{9}{(x - y)^2}$ **5.** $\dfrac{a}{3a + 1}$ **7.** $\dfrac{x^2y^2 + 2}{y^2(1 + 2x)}$ **9.** $x + 2$ **11.** $\dfrac{1}{y + 8}$ **13.** $z(z - 2)$

15. $\dfrac{2x - 3y}{2x + 3y}$ **17.** $\dfrac{2x + 5y}{y(3x - y)}$ **19.** $\dfrac{6x - 2}{11x - 5}$ **21.** $\dfrac{x + 2}{x - 3}$ **23.** $\dfrac{2x - 2y - xy}{x^2}$ **25.** $\dfrac{2x(x + 2)}{x^2 + 6}$

27. $\dfrac{x - 3}{x - 5}$ **29.** $\dfrac{(a - 8)(a - 3)}{a - 2}$ **31.** $\dfrac{3(3a^2 - 2)}{a(a - 4)}$ **33.** $\dfrac{x^2 - 15x + 27}{9 - 4x}$

Exercises 4.5

1. $x = 12$ **3.** $t < -30$ **5.** $a = 4$ **7.** $y = 7$ **9.** $y \leq 17$ **11.** $x = -17$ **13.** $x \leq -17$

15. $x = -3$ **17.** $x > -\dfrac{177}{11}$ **19.** $x = 4$ **21.** $\dfrac{75 - 2x}{10x}$ **23.** $t = \dfrac{3}{4}$ **25.** $a = 5$ **27.** No solution

29. $y = \dfrac{38}{13}$　**31.** $\dfrac{3y^2 + y + 6}{6y(y + 3)}$　**33.** $x = 2$　**35.** $x = 3$　**37.** No solution　**39.** $\dfrac{3n^2 - 8n + 2}{(3n + 2)(3n - 2)}$

41. $\left\{ n \mid n \neq \dfrac{4}{3},\, n \neq -\dfrac{4}{3} \right\}$　**43.** $x = -\dfrac{5}{8}$　**45.** No solution　**47.** $x = -\dfrac{22}{3}$　**49.** $x = \dfrac{5}{18}$

51. $(x - y - 3)(x - y + 3)$　**52.** No solution　**53.** $\dfrac{-3(x - 3)(x + 3)}{xy}$

54. Let $x = $ first consecutive even integer; $x + (x + 2) + (x + 4) = 174$; 56, 58, and 60

Exercises 4.6

1. $x = \dfrac{4 - 7y}{5}$　**3.** $y = \dfrac{2x - 11}{9}$　**5.** $w = 5z - 4$　**7.** $r > \dfrac{15}{2}t$　**9.** $n = \dfrac{3m + 4p + 8}{9}$　**11.** $a = \dfrac{-5b}{9}$

13. $x = \dfrac{21}{19}y$　**15.** $x = \dfrac{d - b}{a - c}$　**17.** $x = \dfrac{by - 2y + 6}{3 - a}$　**19.** $x = \dfrac{a}{y + 7} - 3 = \dfrac{a - 3y - 21}{y + 7}$　**21.** $u = \dfrac{-y - 1}{y - 1}$

23. $t = \dfrac{2x - 3}{3x - 2}$　**25.** $b = \dfrac{2A}{h}$　**27.** $b_1 = \dfrac{2A}{h} - b_2$　**29.** $r = \dfrac{A - P}{Pt}$　**31.** $F = \dfrac{9}{5}C + 32$

33. $P_2 = \dfrac{P_1 V_2}{V_1}$　**35.** $g = \dfrac{2}{t^2}(S - s_0 - v_0 t)$　**37.** $x < \mu + 1.96s$　**39.** $f_1 = \dfrac{ff_2}{f_2 - f}$　**41.** $h = \dfrac{S - 2\pi r^2}{2\pi r}$

43. $x < -1$ or $x > \dfrac{7}{3}$　**44.** $8(x - 2y)(x^2 + 2xy + 4y^2)$　**45.** $\dfrac{x + 5}{x + 6}$

46. Let $t = $ the time travelled by both cars; $45t + 55t = 325$; $3\dfrac{1}{4}$ hrs

47. Let $x = $ the width of the rectangle; $2x + 2(3x + 5) = 42$; 4 by 17 inches

Exercises 4.7

1. Let $x = $ number; $\dfrac{3}{4}x = \dfrac{2}{5}x - 7$; $x = -20$　**3.** Let $x = $ # of men; $\dfrac{7}{9} = \dfrac{x}{810}$; 630 men

5. Let $x = $ the larger number; $\dfrac{5}{12} = \dfrac{x - 21}{x}$; two numbers are 36 and 15

7. Let $x = $ # inches in 52 cm; $\dfrac{1}{2.54} = \dfrac{x}{52}$; 20.47 in.

9. Let $x = $ # of dribbles; $\dfrac{x}{28} = \dfrac{25}{7} \cdot \dfrac{4}{5}$; 80 dribbles

11. Let $x = $ second side; $P = 22 = \dfrac{1}{2}x + x + x + 2$; lengths are 4, 8, and 10 cm

13. Let $x = $ width; $50 = 2(2.5x + x)$; $x = \dfrac{50}{7}$ cm (width), length $= \dfrac{125}{7}$ cm

15. Let $x = $ distance from home; $\dfrac{3x}{4} = 6$; 8 miles

17. Let $x = $ distance of home from school; $x - \dfrac{1}{5}x - \dfrac{1}{4}\left(\dfrac{4}{5}x\right) = 2$; $3\dfrac{1}{3}$ blocks　**19.** $\dfrac{4}{5} = \dfrac{1}{2} + \dfrac{1}{5} + \dfrac{1}{R_3}$; $R_3 = 10$ ohms

21. Let $\$x$ be invested in bond; $0.12x + 0.11(x + 3{,}000) = 905$; \$2,500 in bond; \$5,500 in CD

23. Let $\$x$ be invested at $5\dfrac{1}{2}\%$; $0.055x + 0.07(25{,}000 - x) = 1{,}465$; \$19,000 at $5\dfrac{1}{2}\%$, \$6,000 at 7%

25. Let $\$x$ be invested at $8\dfrac{1}{2}\%$; $0.085x + 0.11(18{,}000 - x) = 0.10(18{,}000)$; \$7,200 at $8\dfrac{1}{2}\%$, \$10,800 at 11%

27. Let $x = $ number of hours working together; $\dfrac{x}{3} + \dfrac{x}{5} = 1$; $1\dfrac{7}{8}$ hours

29. Let $x = $ number of hours working together; $\dfrac{x}{30} + \dfrac{x}{20} = 1$; 12 hours

31. Let x = time working together; $\dfrac{x}{2\frac{2}{3}} + \dfrac{x}{5} = 1$; $1\dfrac{17}{23}$ days.

33. Let x = # hours for the Super Quickie Service to finish cleaning; $\dfrac{10}{30} + \dfrac{x}{20} = 1$; $x = \dfrac{40}{3}$; Total time = $23\frac{1}{3}$ hours

35. Let x = number of minutes to overflow; $\dfrac{x}{10} - \dfrac{x}{15} = 1$; 30 minutes   **37.** Let x = Bill's rate; $\dfrac{10}{x} = \dfrac{15}{x + 10}$; 20 kph

39. Let x = number of miles at slow speed; $\dfrac{600 - x}{50} + \dfrac{x}{20} = 14$; $66\dfrac{2}{3}$ miles

41. Let x = number of ounces of 20% alcohol; $0.20x + 0.50(5) = 0.30(x + 5)$; 10 ounces of 20% alcohol

43. Let x = ml of 30% solution; $0.30x + 0.75(80 - x) = 0.50(80)$; $44\frac{4}{9} \approx 44.44$ ml of 30% solution, $35\frac{5}{9} \approx 35.56$ ml of 75% solution

45. Let x = tons of 40% iron; $0.40x + 0.60(80 - x) = 0.55(80)$; $x = 20$ tons of 40% iron; 60 tons of 60% iron

47. Let x = liters of pure alcohol; $0.60(2) + 1(x) = 0.80(x + 2)$; 2 liters of pure alcohol

49. Let x = gallons drained off = water added; $0.30(3 - x) + 0(x) = 0.20(3)$; 1 gallon

51. Let x = number of advance tickets sold; $7.50x + 10.25(3,600 - x) = 30,850.00$; 2,200 advance tickets

53. Let x = number of nickels; $0.05x + 0.10(x + 5) + 0.25(2x) = 20.00$; 30 nickels, 35 dimes, 60 quarters

55. Let x = number of orchestra tickets; $12.50x + 8(2x) + 10.25(900 - 3x) = 8,662.50$; 250 orchestra tickets, 500 general admission, 150 balcony

57. Let x = score on final exam; $0.20(85) + 0.20(65) + 0.20(72) + 0.40(x) \geq 80$; he needs at least an 89.

60. $-5x + 23$ **61.** $3xy(x - y)^2$ **62.** $(x - a - 3)(x - a + 3)$

63. $-9 \leq x \leq 1$

$$-10\ -8\ -6\ -4\ -2\ \ 0\ \ 2\ \ 4\ \ 6\ \ 8\ \ 10$$

Chapter 4 Review Exercises

1. $\dfrac{x}{4y^2}$ **2.** $\dfrac{3b^3}{4a^3c^2}$ **3.** $\dfrac{x + 4}{x + 5}$ **4.** $\dfrac{3x + 1}{2x - 3}$ **5.** $x^2 - 2x + 3$ **6.** $\dfrac{6a^2b^4 - 8b^3 + 9}{2ab}$ **7.** $\dfrac{5x - 7}{3x - 2}$

8. $2a - b$ **9.** $\dfrac{xy^3z^2}{2}$ **10.** $\dfrac{2xz^2}{9y^5}$ **11.** $\dfrac{2}{x^2y}$ **12.** 0 **13.** $\dfrac{3(x + 1)}{x - 1}$ **14.** 2 **15.** 2 **16.** 3

17. $\dfrac{5b^3 - 6a}{3a^2b^4}$ **18.** $\dfrac{4y + 35}{15xy^2}$ **19.** $\dfrac{-6x^2 + 19x + 5}{10x^2}$ **20.** $\dfrac{14x^2 + 3x + 2}{6x^2}$ **21.** $\dfrac{2(x + 5)}{x - 5}$ **22.** $\dfrac{a - 2}{a - 4}$

23. $2x$ **24.** $\dfrac{a - 3}{2}$ **25.** $\dfrac{1}{a - b}$ **26.** $\dfrac{r + s}{3}$ **27.** $\dfrac{3x^2 - 22x - 15}{(2x + 3)(x - 4)}$ **28.** $\dfrac{13a^2 + 2a}{(a - 1)(2a + 1)}$

29. $\dfrac{16x^2 - 44x - 8}{(2x - 7)(2x - 3)}$ **30.** $\dfrac{5x^2 - 22x - 11}{(x - 2)(x + 3)}$ **31.** $\dfrac{10a^3 + 10a^2 + a - 3}{2a^2(a + 1)(a - 3)}$ **32.** $\dfrac{-2x - 17}{(x + 4)(x - 2)(x + 1)}$

33. $\dfrac{8x + 4}{(x - 2)^2(x + 2)}$ **34.** $\dfrac{-2a + 24}{(a + 3)(a - 3)^2}$ **35.** $\dfrac{x(x + 3)}{7(x - 3)(x + 1)}$ **36.** $\dfrac{5x + 5y - 3}{x^2 - y^2}$ **37.** $\dfrac{8x^2 + 2x - 3}{x^2 - 4}$

38. $\dfrac{-x^2 + 26x + 1}{x^2 - 25}$ **39.** $\dfrac{1}{y}$ **40.** b^2 **41.** $\dfrac{3(x^2 + 5x + 5)}{(x + 3)(x - 2)(x + 1)}$ **42.** $\dfrac{x^2 - 9x + 2}{(2x + 3)(x + 5)(x - 3)}$

43. $\dfrac{5x + 3}{(x - 2)(x + 1)}$ **44.** $\dfrac{2r^2 + 2s^2 + 3r - 6s}{(r - 2s)(r + s)(r - s)}$ **45.** $\dfrac{3x}{2x - 3}$ **46.** $\dfrac{3x}{2x + 3}$ **47.** $\dfrac{2x + 3y}{2}$ **48.** $\dfrac{2}{2x + 3y}$

49. $\dfrac{(x^2 + 6)(x + 1)}{2x^2}$ **50.** xy **51.** $\dfrac{8axy}{3b}$ **52.** $\dfrac{x}{y}$ **53.** $\dfrac{1}{5}$ **54.** $\dfrac{1}{2 - x}$ **55.** $\dfrac{(b - 1)(2b + 5)}{(b + 1)(b^2 - b + 2)}$

56. $\dfrac{z^2 + 2z - 2}{3z}$ **57.** $x = 2$ **58.** $x = 5$ **59.** $x > -5$ **60.** $x \geq 5$ **61.** $x = 4$ **62.** $x = \frac{11}{10}$

63. $x < -22$ **64.** $x \geq -\frac{28}{3}$ **65.** $x = \frac{1}{2}$ **66.** $x = \frac{3}{2}$ **67.** $x = 7$ **68.** $x = 4$ **69.** $x > \frac{8}{3}$

70. $x \leq \frac{6}{11}$ **71.** No solution   **72.** No solution   **73.** $x = 4$ **74.** No solution   **75.** $x = \dfrac{10y}{3}$

76. $y = \dfrac{3}{10}x$ **77.** $y = \dfrac{4}{x}$ **78.** $b = \dfrac{a}{a - 1}$ **79.** $y = \dfrac{2x + 1}{x}$ **80.** $a = \dfrac{b}{c - x}$ **81.** $x = \dfrac{dy - b}{a - cy}$

82. $s = \dfrac{3 + 2r}{3r - 2}$ **83.** $b = \dfrac{adc}{ac - cd - ad}$ **84.** $c = \dfrac{ad}{b + de}$

85. Let x = # of inches in 1 cm; $\dfrac{2.54 \text{ cm}}{1 \text{ in.}} = \dfrac{1}{x}$, $x = \dfrac{1}{2.54} \approx 0.39$ in.

86. Let x = # of Democrats; $\dfrac{4}{3} = \dfrac{x}{1{,}890 - x}$; 1,080 Democrats

87. Let total distance = x; $\dfrac{1}{2}x + \dfrac{1}{3}\left(\dfrac{1}{2}x\right) + \dfrac{3}{2} = x$; 4.5 miles

88. Let r = Carlos' rate; $\dfrac{3}{r-5} = \dfrac{5}{r}$; $12\dfrac{1}{2}$ mph

89. Let x = number of days required by both; $\dfrac{x}{2\frac{1}{2}} + \dfrac{x}{2\frac{1}{3}} = 1$; $1\dfrac{6}{29}$ days

90. Let x = amount of time to clean kitchen together; $\dfrac{x}{\frac{1}{3}} + \dfrac{x}{\frac{2}{3}} = 1$; $\dfrac{2}{9}$ hour

91. Let x = amount of 35% alcohol; $0.70(5) + 0.35x = 0.60(x + 5)$; 2 liters
92. Let x = quantity of pure water; $0(x) + 0.65(3) = 0.30(x + 3)$; $x = 3.5$ liters
93. Let x = number of children's tickets; $1.50x + 4.25(980 - x) = 3{,}010.00$; 420 children's tickets, 560 adult tickets
94. Let x = number of dimes; $0.05(4x) + 0.10x + 0.01(82 - 5x) = 3.32$; 40 nickels, 10 dimes, and 32 pennies

Chapter 4 Practice Test

1. (a) $\dfrac{3y^3}{8x}$ **(b)** $\dfrac{x+3}{x-3}$ **(c)** $\dfrac{3(2x+1)}{5x}$

2. (a) $\dfrac{3}{4x^3y^2ab^4}$ **(b)** $\dfrac{4x^3 + 15y}{24x^2y^2}$ **(c)** $2(r+s)$ **(d)** 2 **(e)** $\dfrac{x^2 - x + 8}{(x+2)(x-2)(x-3)}$ **(f)** $\dfrac{x+3}{x}$

3. $\dfrac{x(1-2x)}{(x+1)(x+5)}$ **4. (a)** $x > \dfrac{-15}{4}$ **(b)** No solution **(c)** $x = 2$ **5.** $x = \dfrac{-y-2}{2y-1}$

6. Let x = amount of 30% alcohol solution; $0.30(x) + 0.45(8) = 0.42(x + 8)$; 2 liters

7. Let x = time required working together; $\dfrac{x}{3\frac{1}{2}} + \dfrac{x}{2} = 1$; $x = 1\dfrac{3}{11}$ hours

Exercises 5.1

1. x^{11} **3.** $-6a^7b^{10}$ **5.** a^{10} **7.** $\dfrac{4}{9}$ **9.** $x^2 + 2xy^3 + y^6$ **11.** x^2y^6 **13.** $2^6 \cdot 3^4$ **15.** $x^{26}y^{19}$

17. $-8a^{10}b^8$ **19.** $16r^{10}s^{11}t^7$ **21.** $144a^8b^6$ **23.** x^3 **25.** $\dfrac{x^2}{y^2}$ **27.** $\dfrac{1}{4}$ **29.** $\dfrac{ab^8}{c^4}$ **31.** $-\dfrac{1}{y}$ **33.** $\dfrac{1}{18}$

35. x^7y^8 **37.** $\dfrac{1}{y^9}$ **39.** y^{15} **41.** $\dfrac{4r^8s^9}{3}$ **43.** $\dfrac{x^{15}}{y^{10}}$ **45.** $\dfrac{x^5y^4}{2}$ **47.** $-1{,}728a^3b^9$ **49.** $\dfrac{27}{8}r^6s^9$

51. $11{,}664x^{12}y^8$ **53.** $-2{,}304x^8y^9$ **55.** $r^{26}s^{28}$ **57.** $\dfrac{b^6}{2^{14}a^2}$ **63.** 4 **64.** -6 **65.** -2

66. $\dfrac{1}{x^2 - 2x + 4}$

67. Let x = the length of the second side; $2x + x + 24 = 75$; first side = 34 in., second side = 17 in.

68. Let x = # kg in 42 lb; $\dfrac{1}{2.2} = \dfrac{x}{42}$; 19.09 kg

Exercises 5.2

1. $\dfrac{1}{x}$ **3.** $\dfrac{x^2}{y^2}$ **5.** 3^6 **7.** $\dfrac{1}{a^4 b^6}$ **9.** $\dfrac{r^6}{s^8}$ **11.** $-\dfrac{1}{4}$ **13.** $\dfrac{1}{9}$ **15.** $\dfrac{2^6}{3^6}$ **17.** $\dfrac{s^{18}}{3^{12}}$ **19.** $\dfrac{r^6}{4s^2}$

21. $-\dfrac{y^9}{864 x^{10}}$ **23.** x^6 **25.** $y^2 x^2$ **27.** 2 **29.** $-\dfrac{125}{27}$ **31.** $\dfrac{1}{x^4}$ **33.** $\dfrac{a^5}{b^4}$ **35.** x^6 **37.** 81 **39.** -1

41. $\dfrac{4y^6}{9x^6}$ **43.** $\dfrac{1}{x^6 y^9}$ **45.** $\dfrac{9b^3}{a^4}$ **47.** $\dfrac{r^{11}}{s^8}$ **49.** $\dfrac{32 s^{15}}{r^{11}}$ **51.** $\dfrac{x^2}{y^3}$ **53.** $\dfrac{x^2 y^3 + 1}{y^3}$ **55.** $\dfrac{y^6}{x^4}$ **57.** $\dfrac{y^6}{(x^2 y^3 + 1)^2}$

59. $\dfrac{x^4 y^4}{(x^2 + y^2)^2}$ **61.** 1 **63.** $\dfrac{x + y}{x^2}$ **65.** $\dfrac{s(s + r^2)}{r(s^2 + r)}$ **67.** $\dfrac{a(2b^2 + a)}{b^2(1 + a^2 b)}$ **69.** $\dfrac{(x^2 + y^2)(x + y)^2}{x^2 y^2}$

73. $2(9ay - 4)(9ay + 4)$ **74.** $(x - y)^2(x + y)$ **75.** $\dfrac{(x - 1)(2x + 5)}{(x + 1)(x^2 - x + 2)}$

76. Let t = the time the carpenter works; $30t + 18(t + 3) = 270$; 4.5 hr

77. Let x = amount of 30% solution; $0.30x + 0.70(6) = 0.60(x + 6)$; 2 oz of 30% solution

Exercises 5.3

1. 1,000   **3.** 1,000,000,000   **5.** 10   **7.** 0.000001   **9.** 1,000,000   **11.** 16.2   **13.** 760,000,000

15. 0.000000851   **17.** 6,000   **19.** 8.24×10^2 **21.** 5.0×10^0 **23.** 9.3×10^{-3} **25.** 8.27546×10^8

27. 7.2×10^{-7} **29.** 7.932×10^1 **31.** $0.16\overline{6}$ or 0.17   **33.** 0.000001   **35.** 6,000,000,000,000   **37.** 2,000

39. 5 hr. 28 min. 51 sec.   **41.** 6.048×10^{13} tons in one day, 2.20752×10^{16} tons in one year

43. There are 10^4 Å in 1 micron   **45.** 0.66×10^{-8} cm $= 6.6 \times 10^{-9}$ cm

47. $\dfrac{1 \text{ mile}}{1.6 \text{ km}} = \dfrac{5.86 \times 10^{12} \text{ miles}}{x \text{ km}}$; 9.376×10^{12} km   **49.** 60×10^{-4} cm $= 6.0 \times 10^{-3}$ cm

51. Mass $= 75{,}000 \times 10^{-23} = 7.5 \times 10^{-19}$ gm   **55.** $24y$ **56.** 1, the multiplicative identity   **57.** $9x^4 y^6$

58. $\dfrac{x^4}{y^4}$ **59.** $\dfrac{x^2}{y(x^2 + 2x + 4)}$

Exercises 5.4

1. 2 **3.** -2 **5.** No real solution   **7.** -9 **9.** 16 **11.** 16 **13.** -8 **15.** $\frac{1}{2}$ **17.** $-\frac{1}{2}$

19. $-\frac{1}{9}$ **21.** $\frac{1}{16}$ **23.** No real solution   **25.** $-\dfrac{1}{16}$ **27.** $\dfrac{2^{18}}{3^9}$ **29.** $\dfrac{2}{3}$ **31.** 16 **33.** $x^{7/6}$ **35.** $x^{17/60}$

37. $a^{3/8}$ **39.** 1 **41.** $\dfrac{1}{r}$ **43.** $\dfrac{r^{1/3}}{s}$ **45.** $\dfrac{r}{s^{1/3}}$ **47.** $\dfrac{1}{x^{1/6}}$ **49.** $\dfrac{b^{2/15}}{a^{3/4}}$ **51.** x^5 **53.** $\dfrac{x^{5/12}}{y^{1/15}}$ **55.** $-\dfrac{1}{2}$

57. $\dfrac{1}{x^6 y^{47/12}}$ **59.** $x + x^{1/2} y$ **61.** $a - b$ **63.** $\dfrac{x^2 - 4x + 4}{x}$ **65.** $\sqrt[3]{x^2}$ **67.** $\sqrt[3]{49}$ **69.** $6\sqrt[3]{x}$

71. $6\sqrt[5]{x^2}$ **77.** No solution   **78.** $1 \le a \le 5$ **79.** $a \le 1$ or $a \ge 5$ **80.** No solution   **81.** $a = \dfrac{bc}{b + c}$

Chapter 5 Review Exercises

1. x^{12} **2.** $x^{11} y^9$ **3.** $-6x^4 y^5$ **4.** $-30 x^3 y^6$ **5.** a^{12} **6.** b^{20} **7.** $a^{14} b^{21}$ **8.** $r^{16} s^{24}$ **9.** $a^{10} b^9$

10. $x^7 y^{22}$ **11.** $2^{10} 3^{10}$ **12.** -32 **13.** $a^7 b^8 c^7$ **14.** $-162 x^8 y^5$ **15.** $\dfrac{1}{a}$ **16.** x^5 **17.** 1 **18.** y^2

19. $\dfrac{1}{x^{16} y^{14}}$ **20.** $\dfrac{1}{a^2 b^6}$ **21.** a^4 **22.** $\dfrac{a^4 b^4}{c^4}$ **23.** $9a^4 x^4$ **24.** $-3x^6 y^2$ **25.** $-\dfrac{72 y^8}{x^2}$ **26.** $\dfrac{-72 b^5}{125 a^5}$

27. $18 y^3$ **28.** $5{,}184 a^{16} b^{10}$ **29.** $\dfrac{9 b^6 c^4}{4}$ **30.** $\dfrac{1}{4 r^4 s^8}$ **31.** $\dfrac{1}{a^2}$ **32.** $\dfrac{1}{x^9}$ **33.** $\dfrac{x^8}{y^{20}}$ **34.** $\dfrac{1}{x^4}$ **35.** $\dfrac{a^{12}}{b^{14}}$

36. $\dfrac{1}{4 a^4 b^2}$ **37.** $-\dfrac{1}{162}$ **38.** 1 **39.** $\dfrac{x^{16}}{3^8 y^5}$ **40.** $\dfrac{1}{a^2}$ **41.** 1 **42.** 1 **43.** $\dfrac{1}{x^4}$ **44.** $\dfrac{y}{x^4}$ **45.** $\dfrac{1}{x^9 y^2}$

46. $\dfrac{1}{x^{13}}$ **47.** x^8 **48.** a^8b^{12} **49.** $\dfrac{r^{11}}{s^5}$ **50.** $-\dfrac{s^{11}}{3r^5}$ **51.** $\dfrac{25}{4}$ **52.** $\dfrac{256}{81}$ **53.** $\dfrac{27y^8}{4xz^2}$ **54.** $\dfrac{8x}{9yz^8}$

55. $\dfrac{3^4 2^4}{x^{12}}$ **56.** $\dfrac{256}{y^{20}}$ **57.** $\dfrac{x^2 - y^2}{xy}$ **58.** $\dfrac{y^4 + 4xy^2 + 4x^2}{x^2 y^4}$ **59.** $\dfrac{y^3 + x}{y^5}$ **60.** $\dfrac{b(b + a^2)}{a}$

61. $\dfrac{y(y - x^2)}{x(x + y^2)}$ **62.** $\dfrac{ab(b^2 + a)}{b^3 - a^2}$ **63.** 28,300 **64.** 6.29 **65.** 0.0000796 **66.** 0.0000008264

67. 7.936×10^0 **68.** 9.259×10^1 **69.** 5.78×10^{-3} **70.** 8.0×10^0 **71.** 6.25897×10^5

72. 7.3×10^{-6} **73.** 0.00000001 **74.** 100 **75.** 5 **76.** 16 **77.** $-\dfrac{1}{27}$ **78.** $-\dfrac{1}{6}$ **79.** $\dfrac{4}{3}$ **80.** $\dfrac{4}{5}$

81. $-\dfrac{5}{3}$ **82.** No real solution **83.** $x^{5/6}$ **84.** $\dfrac{1}{y^{1/6}}$ **85.** x **86.** $y^{1/12}$ **87.** $x^{13/30}$ **88.** $a^{1/2}$

89. $\dfrac{s^{1/3}}{r^{5/6}}$ **90.** $\dfrac{1}{r^{1/5}s^{4/3}}$ **91.** $a^5 b^3$ **92.** $\dfrac{y^3}{x}$ **93.** $\dfrac{x^{5/6}}{y^{7/6}}$ **94.** $\dfrac{1}{a^{1/6}}$ **95.** 2^{10} **96.** 1

97. $\dfrac{1}{a^{1/6}} - \dfrac{b^{1/2}}{a^{2/3}} = \dfrac{a^{1/2} - b^{1/2}}{a^{2/3}}$ **98.** $a^{2/3} - b^{2/3}$ **99.** $\dfrac{1}{a} + \dfrac{4b^{1/2}}{a^{1/2}} + 4b$

100. $\dfrac{4}{a} + \dfrac{12}{a^{1/2}b^{1/2}} + \dfrac{9}{b} = \dfrac{4b + 12a^{1/2}b^{1/2} + 9a}{ab}$ **101.** $\sqrt[5]{x^2}$ **102.** $\sqrt[3]{x}$ **103.** $3\sqrt[5]{y^2}$ **104.** $\sqrt[5]{(3y)^2}$

Chapter 5 Practice Test

1. $a^4 b^{11}$ **2.** $-200x^8 y^3$ **3.** $-\dfrac{4b^2 y^2}{3ax^2}$ **4.** $-\dfrac{2x^4}{9y^5}$ **5.** $\dfrac{9r^4 s^{10}}{25}$ **6.** $\dfrac{x^{14}}{y^4}$ **7.** $\dfrac{1}{3}$ **8.** $x^{7/3}$ **9.** $\dfrac{1 + x^2}{xy}$

10. $x - x^{4/3}$ **11.** $\dfrac{1 + 6x + 9x^2}{x}$ **12.** (a) -5 (b) $-\dfrac{1}{8}$ **13.** $3\sqrt[3]{a^2}$

14. In one hour it travels $186{,}000 \times 60 \times 60 = 6.696 \times 10^8$ miles.

Exercises 6.1

1. 4 **3.** 3 **5.** Not a real number **7.** 1 **9.** -7 **11.** -6 **13.** 2 **15.** 5 **17.** 3 **19.** -5
21. Not a real number **23.** 4 **25.** 27 **27.** 9 **29.** 16 **31.** Not a real number **33.** -16 **35.** 9
37. $(xy)^{1/3}$ **39.** $(x^2 + y^2)^{1/2}$ **41.** $(5a^2 b^3)^{1/5}$ **43.** $2(3xyz^4)^{1/3}$ **45.** $5(x - y)^{2/3}$ **47.** $(x^n - y^n)^{1/n}$
49. $(x^{5n+1}y^{2n-1})^{1/n}$ **51.** $\sqrt[3]{x}$ **53.** $m\sqrt[3]{n}$ **55.** $\sqrt[3]{(-a)^2}$ **57.** $-\sqrt[3]{a^2}$ **59.** $\sqrt[3]{a^2 b}$ **61.** $\sqrt{x^2 + y^2}$

63. $\sqrt{x^n - y^n}$ **65.** 6.16 **67.** 10.95 **69.** -25.88 **74.** $\dfrac{1}{4}$ **75.** $\dfrac{x^4}{y^6}$ **76.** $a^{32/15}b^2$ **77.** $y = \dfrac{2}{3x + 1}$

78. Let $t =$ the time it takes for them to complete the job working together; $\dfrac{x}{\frac{1}{2}} + \dfrac{x}{\frac{3}{4}} = 1$; 18 min.

79. 2.976×10^5 km/sec

Exercises 6.2

1. $2\sqrt{14}$ **3.** $4\sqrt{3}$ **5.** $2\sqrt[5]{2}$ **7.** 12 **9.** $8x^4$ **11.** $3x^3$ **13.** $8x^{30}\sqrt{2}$ **15.** $2x^{15}\sqrt[4]{8}$ **17.** $2x^{12}\sqrt[5]{4}$
19. xy^2 **21.** $4ab^2\sqrt{2}$ **23.** $a^7 b^{15}$ **25.** $x^2 y^2$ **27.** $\dfrac{\sqrt{2}}{2}$ **29.** $\dfrac{\sqrt{5x}}{5}$ **31.** $\dfrac{3\sqrt{5}}{2}$ **33.** $\dfrac{\sqrt{3}}{15}$
35. $8x^2 y^4\sqrt{x}$ **37.** $3x^2 y^2\sqrt[3]{3x^2 y}$ **39.** $6y\sqrt[4]{2y^2}$ **41.** $x + y^2$ **43.** $\sqrt[4]{x^4 - y^4}$ **45.** $30st\sqrt{2st}$ **47.** $12a^3 b^2$
49. $\dfrac{xy^2}{2}$ **51.** $\dfrac{2x^2\sqrt[4]{2x}}{y^3}$ **53.** $3\sqrt{3}$ **55.** $\dfrac{y^4}{x^3}$ **57.** $\dfrac{\sqrt{15x}}{5x}$ **59.** $\dfrac{\sqrt[3]{3xy}}{xy^2}$ **61.** $\dfrac{\sqrt[3]{12}}{2}$ **63.** $\dfrac{\sqrt[3]{18}}{2}$
65. $\dfrac{\sqrt[4]{36}}{2}$ **67.** $\dfrac{3y}{2x}\sqrt[3]{12x^2}$ **69.** $\dfrac{x^2}{3a^5}$ **71.** $-\dfrac{6s^3\sqrt{sr}}{r}$ **73.** $\sqrt{a}$ **75.** $x\sqrt[12]{x}$ **77.** $\sqrt[6]{a}$ **79.** $x^5 y^3$
82. $\dfrac{x}{1 + 2x}$ **83.** Distributive property **84.** $\dfrac{3x^2 - 5x + 4}{(x + 3)(x - 1)(x - 2)}$

85. Let $x =$ the number; $\dfrac{2}{3}x = \dfrac{3}{5}x + 5$; 75

Exercises 6.3

1. $4\sqrt{3}$ **3.** $-3\sqrt{5}$ **5.** $4\sqrt{3} + 2\sqrt{6}$ **7.** $\sqrt{3} - \sqrt{5}$ **9.** 0 **11.** $(7a - 3a^3)\sqrt{b}$ **13.** $(3x + 4)\sqrt[3]{x^2}$
15. $1 - 2\sqrt[3]{a}$ **17.** $-\sqrt{3}$ **19.** $5\sqrt{6} - 3\sqrt{3}$ **21.** $6\sqrt{3} - 36$ **23.** $5\sqrt{6} - 20\sqrt{3}$ **25.** $6\sqrt[3]{3} - 11\sqrt[3]{6}$
27. $ab(2\sqrt{a} - 3\sqrt{b} - \sqrt{ab})$ **29.** 0 **31.** $4x^4y^4\sqrt{5x}$ **33.** $9x\sqrt[3]{9x^2} - 3x\sqrt[3]{x^2}$ **35.** $(8 - 4x)\sqrt[4]{x}$
37. $\dfrac{\sqrt{5} + 10}{5}$ **39.** 0 **41.** $\dfrac{3\sqrt{2}}{2}$ **43.** $\dfrac{7\sqrt{10}}{10}$ **45.** $\dfrac{5\sqrt{7} - 21\sqrt{5}}{35}$ **47.** $-\dfrac{11\sqrt[3]{4}}{2}$ **49.** $\dfrac{y\sqrt{x} + x\sqrt{y}}{xy}$
51. $\dfrac{\sqrt[3]{3} - 3\sqrt[3]{9}}{3}$ **53.** $\dfrac{3\sqrt{2} + 21\sqrt{7}}{7}$ **55.** $\dfrac{14\sqrt{10}}{5}$ **57.** $4\sqrt[3]{25}$ **59.** $\dfrac{y^5}{x^8}$ **60.** $\dfrac{y^3}{x^4}$ **61.** $\dfrac{7x + 7}{x^2 + x - 6}$
62. $x^3 + 10x^2 + 50x + 125$ **63.** Let t = the time it takes for the car to catch up; $50t = 40(t + 1)$; 4 hr
64. Let n = the number of nickels; $5n + 10(32 - n) = 200$; 24 nickels and 8 dimes

Exercises 6.4

1. $5\sqrt{5} - 15$ **3.** $-2\sqrt{3} - 6\sqrt{5}$ **5.** $a + \sqrt{ab}$ **7.** $\sqrt{10} + 2$ **9.** $6\sqrt{15} - 60$ **11.** $-5\sqrt{6} + 14$
13. $\sqrt{15} + \sqrt{5} - 2\sqrt{3} - 2$ **15.** 2 **17.** $8 - 2\sqrt{15}$ **19.** 3 **21.** 5 **23.** $4a - 4\sqrt{ab} + b$
25. $9x - 4y$ **27.** $x - 6\sqrt{x} + 9$ **29.** $x - 3$ **31.** $-2\sqrt{x}$ **33.** -1 **35.** $2\sqrt{2} - 3\sqrt{3}$
37. $\dfrac{5\sqrt{2} - \sqrt{7}}{4}$ **39.** $3\sqrt{2} + \sqrt{5}$ **41.** $\dfrac{2 + \sqrt{7}}{2}$ **43.** $-\dfrac{\sqrt{2} + 3}{7}$ **45.** $\dfrac{5\sqrt{5} - 5}{2}$ **47.** $\dfrac{2\sqrt{3} + 2\sqrt{a}}{3 - a}$
49. $\dfrac{x + \sqrt{xy}}{x - y}$ **51.** $\dfrac{\sqrt{10} + 2}{3}$ **53.** $\dfrac{2\sqrt{10} + 2}{9}$ **55.** $\dfrac{\sqrt{10} - 1}{2}$ **57.** $5 + 2\sqrt{6}$ **59.** $\dfrac{18 + 5\sqrt{10}}{2}$
61. $\sqrt{x} + \sqrt{y}$ **63.** $(x + 1)(\sqrt{x} + \sqrt{2})$ **65.** 12 **67.** $9\sqrt{7} - 5\sqrt{3}$ **70.** $(4 - x - y)(4 + x + y)$
71. $-\dfrac{2}{3} \le x < \dfrac{7}{3}$ **72.** $\dfrac{x^6}{y^4}$ **73.** $\dfrac{1}{8x^{1/2}y^{1/10}}$

Exercises 6.5

1. $a = 49$ **3.** $x = 28$ **5.** $x = 1$ **7.** $a = \frac{1}{3}$ **9.** $a = 7$ **11.** $y = 78$ **13.** $x = \frac{10}{3}$ **15.** $y = -\frac{7}{3}$
17. $x = 4$ **19.** $a = \frac{81}{16}$ **21.** $x = \frac{225}{16}$ **23.** $y = \frac{81}{25}$ **25.** $x = \frac{37}{2}$ **27.** $x = 17$ **29.** $y = \frac{21}{4}$
31. $x = 8$ **33.** No solution **35.** $a = \frac{3}{2}$ **37.** $x = 64$ **39.** $y = 81$ **41.** $s = -243$ **43.** $x = 344$
45. $x = -\frac{67}{2}$ **47.** $x = 125$ **49.** $x = -8$ **51.** $a = 57$ **53.** $x = \frac{1}{256}$ **55.** $x = -11$
58. $\dfrac{3x^2 + 3x + 2}{(x - 1)(x + 1)^2}$ **59.** $3 + \sqrt{6}$
60. Let x = amount invested in the 6% bond; $0.06x + 0.075(50{,}000 - x) = 3{,}150$; \$40,000 invested at 6% and \$10,000 invested at $7\frac{1}{2}$%

Exercises 6.6

1. $-i$ **3.** $-i$ **5.** $-i$ **7.** 1 **9.** $-i$ **11.** -1 **13.** $i\sqrt{5}$ **15.** $3 - i\sqrt{2}$ **17.** $-2\sqrt{7} + 4i$
19. $\dfrac{3}{5} - i\dfrac{\sqrt{2}}{5}$ **21.** $2 - i\dfrac{\sqrt{3}}{3}$ **23.** $1 - i$ **25.** 4 **27.** $15 - 35i$ **29.** $-4 + 2i$ **31.** -15 **33.** -9
35. $-4 + 6i$ **37.** $10 + 6i$ **39.** 0 **41.** $25 - 5i$ **43.** 10 **45.** 53 **47.** $2i$ **49.** -5 **51.** -1
53. $2 - \frac{5}{3}i$ **55.** $-1 - 3i$ **57.** $-1 - \frac{5}{2}i$ **59.** $-\frac{4}{29} + \frac{10}{29}i$ **61.** $\frac{10}{29} + \frac{4}{29}i$ **63.** $\frac{3}{5} + \frac{4}{5}i$
65. $-\frac{21}{29} - \frac{20}{29}i$ **67.** $\frac{1}{29} - \frac{41}{29}i$ **69.** $\frac{5}{34} - \frac{31}{34}i$

Chapter 6 Review Exercises

1. $\sqrt{x}$ **2.** $\sqrt[3]{x}$ **3.** $x\sqrt{y}$ **4.** $\sqrt{xy}$ **5.** $\sqrt[3]{m^2}$ **6.** $\sqrt{m^3}$ **7.** $\sqrt[4]{(5x)^3}$ **8.** $5\sqrt[4]{x^3}$ **9.** $\dfrac{1}{\sqrt[5]{s^4}} = \dfrac{\sqrt[5]{s}}{s}$

10. $\dfrac{1}{\sqrt[4]{s^5}} = \dfrac{\sqrt[4]{s^3}}{s^2}$ **11.** $a^{1/3}$ **12.** $t^{1/5}$ **13.** $-n^{4/5}$ **14.** $-n^{5/4}$ **15.** $t^{5/3}$ **16.** $t^{5/3}$ **17.** $t^{-7/5}$

18. $t^{-5/7}$ **19.** $3\sqrt{6}$ **20.** $3\sqrt[3]{2}$ **21.** x^{30} **22.** x^{15} **23.** x^{12} **24.** x^{20} **25.** $2xy^2\sqrt[3]{6xy^2}$

26. $2x^4y^6\sqrt{7xy}$ **27.** $3x^2y^2\sqrt[4]{xy^2}$ **28.** $2z\sqrt[5]{y^4z}$ **29.** $15x\sqrt{y}$ **30.** $24ab$ **31.** $2x^4y^2\sqrt{y}$ **32.** $6x^7y^2$

33. $\frac{2}{3}$ **34.** $\frac{2}{3}$ **35.** $\frac{\sqrt{y}}{x}$ **36.** $\frac{\sqrt{xy}}{x}$ **37.** $\frac{4\sqrt{ab}}{a^2b}$ **38.** $\frac{9y\sqrt{2x}}{2x}$ **39.** $\frac{x\sqrt{2x}}{y}$ **40.** $6\sqrt{6}$ **41.** $\frac{\sqrt{5a}}{a}$

42. $\frac{\sqrt[3]{5a^2}}{a}$ **43.** $\frac{2\sqrt[3]{4a^2}}{a}$ **44.** $\frac{2\sqrt[3]{4a}}{a}$ **45.** $\sqrt[3]{x^2}$ **46.** $\sqrt[4]{x^3}$ **47.** $\sqrt[6]{2^5}$ **48.** $\sqrt[6]{2}$ **49.** $-\sqrt{3}$

50. 0 **51.** $33\sqrt{6}$ **52.** 0 **53.** $2ab\sqrt{ab}$ **54.** $2s^2t^2\sqrt{s} - 3s^2t^2\sqrt{t}$ **55.** $\dfrac{3\sqrt{6} + 2\sqrt{15}}{6}$

56. $\dfrac{3\sqrt{35} + 7\sqrt{6}}{21}$ **57.** $-\dfrac{7\sqrt[3]{3}}{3}$ **58.** $\dfrac{5\sqrt[3]{35} + 14\sqrt[3]{7}}{7}$ **59.** 0 **60.** $-\sqrt{10}$ **61.** $27 - 7\sqrt{21}$

62. $6x + 4\sqrt{x} - 3\sqrt{5x} - 2\sqrt{5}$ **63.** $x - 10\sqrt{x} + 25$ **64.** $10 - 2\sqrt{21}$ **65.** $-14\sqrt{a} - 42$

66. $12 - 6\sqrt{b}$ **67.** $-10\sqrt{3} - 10\sqrt{5}$ **68.** $6\sqrt{5} - 6\sqrt{3}$ **69.** $\sqrt{m} + n$ **70.** $4\sqrt{2x} + 4\sqrt{5y}$

71. $15 - 5\sqrt{5}$ **72.** $-3\sqrt{2}$ **73.** 3 **74.** $s + t$ **75.** $x = 72$ **76.** $x = \frac{4}{3}$ **77.** $x = 27$ **78.** $x = \frac{14}{3}$

79. $x = 85$ **80.** $x = 5$ **81.** $x = 16$ **82.** $x = 80$ **83.** $-i$ **84.** i **85.** 1

86. -1 **87.** $9 - i$ **88.** $1 - 5i$ **89.** $8 - 25i$ **90.** 41 **91.** $-6 - 8i$ **92.** $-5 - 10i$

93. $5 - 12i$ **94.** $-5 - 12i$ **95.** -37 **96.** $34 - 6i$ **97.** $\frac{9}{10} - \frac{13}{10}i$ **98.** $-i$

Chapter 6 Practice Test

1. $3x^2y^3$ **2.** $2x\sqrt[3]{y^2}$ **3.** $6x^4y$ **4.** $\frac{7}{2}$ **5.** $\frac{\sqrt{35}}{7}$ **6.** $\frac{2\sqrt[3]{3}}{3}$ **7.** $\frac{3xy^2\sqrt{2}}{2}$ **8.** $\sqrt[4]{5^3}$ **9.** $5\sqrt{2}$

10. $x - 6\sqrt{x} + 9$ **11.** $\frac{2\sqrt{6}}{3}$ **12.** $3 + \sqrt{6}$ **13.** $x = 19$ **14.** $x = 121$ **15.** $x = -5$ **16.** $x = 625$

17. $-i$ **18.** $-5 - 5i$ **19.** $-\frac{4}{5} - \frac{7}{5}i$

Cumulative Review: Chapters 4–6

1. $\frac{9b}{8a^2}$ (§4.1) **2.** $\frac{x + 3y}{x - 3y}$ (§4.1) **3.** $\frac{2a + b}{a(a + b)}$ (§4.1) **4.** $\frac{3x - y}{2}$ (§4.1) **5.** $\frac{5yb^2}{7a^2x^2}$ (§§4.2–4.3)

6. $(2x + y)(x - y)$ (§§4.2–4.3) **7.** $\frac{9x^2 + 4y^2}{6xy}$ (§§4.2–4.3) **8.** -3 (§§4.2–4.3) **9.** $\frac{x - 4}{2}$ (§§4.2–4.3)

10. $3a - b$ (§§4.2–4.3) **11.** $\frac{-x + 11}{(x - 5)(x - 2)}$ (§§4.2–4.3) **12.** $\frac{-7x - 2y}{(x - 3y)(x + y)(2x + y)}$ (§§4.2–4.3)

13. $\frac{x}{x + 3y}$ (§§4.2–4.3) **14.** $\frac{x - 2y}{x - 5y}$ (§§4.2–4.3) **15.** $\frac{2x^2 - 4}{x - 2x^2}$ (§4.4) **16.** $\frac{(x - 2y)(x + 2y)}{(2x - y)(x + 4y)}$ (§4.4)

17. $x = \frac{6}{5}$ (§4.5) **18.** $x = -2$ (§4.5) **19.** $x > 6$ (§4.5) **20.** No solution (§4.5) **21.** $x = 1$ (§4.5)

22. $x \geq -\frac{1}{5}$ (§4.5) **23.** $a = \frac{b}{3}$ (§4.6) **24.** $y = \frac{5x}{3x - 5}$ (§4.6) **25.** $y = \frac{x}{x + 1}$ (§4.6)

26. $x = \frac{2y + 3}{y - 2}$ (§4.6) **27.** Let x = total # of cars; $\frac{5}{6} = \frac{x - 1,200}{1,200}$; 2,200 cars (§4.7)

28. Let x = amount of 20% solution; $0.20x + 0.35(8) = 0.30(8 + x)$; 4 liters (§4.7)

29. Let t = time they would take to work together; $\frac{t}{4} + \frac{t}{4\frac{1}{2}} = 1$; $2\frac{2}{17}$ hours (§4.7)

30. Let x = number of general admission tickets sold; $3.50x + 4.25(505 - x) = 1,861.25$; 380 general admission tickets, 125 reserved seat tickets (§4.7)

31. $54x^5y^7$ (§5.1) **32.** $\frac{y^4}{2x^2}$ (§5.1) **33.** $\frac{1}{4x^2y^2}$ (§5.1) **34.** $-6a^2b^3$ (§5.1) **35.** $\frac{1}{x^{12}y}$ (§5.2)

36. $\frac{x^6}{4y^2}$ (§5.2) **37.** $\frac{1}{rs^2}$ (§5.2) **38.** $-\frac{27}{8}$ (§5.2) **39.** $-\frac{8}{rs^7}$ (§5.2) **40.** $\frac{5y^3}{x^3}$ (§5.2) **41.** $\frac{y^2 - x^2}{xy}$ (§5.2)

42. $\dfrac{b^2 + a}{b^2 + ab^2}$ (§5.2)   **43.** 5.642932×10^4 (§5.3)   **44.** 7.52×10^{-5} (§5.3)

45. $(1.86 \times 10^5)(60)(60)(24) = 1.60704 \times 10^{10}$ miles (§5.3)   **46.** 0.16 (§5.3)   **47.** -10 (§5.4)   **48.** $\frac{16}{9}$ (§5.4)

49. $a^{1/6}$ (§5.4)   **50.** $x^{1/15}$ (§5.4)   **51.** $\frac{1}{9}$ (§5.4)   **52.** $\dfrac{1}{x^{1/2}}$ (§5.4)   **53.** $\sqrt[4]{x^3}$ (§§6.1–6.2)

54. $2\sqrt{x}$ (§§6.1–6.2)   **55.** $4a^2b^4$ (§§6.1–6.2)   **56.** $2ab^5\sqrt{2a}$ (§§6.1–6.2)

57. $24x^3y^2\sqrt{x}$ (§§6.1–6.2)   **58.** $3xy\sqrt[3]{3xy^2}$ (§§6.1–6.2)   **59.** 4 (§§6.1–6.2)   **60.** $\dfrac{\sqrt{3x}}{xy}$ (§§6.1–6.2)

61. $\sqrt{x}$ (§§6.1–6.2)   **62.** $\dfrac{\sqrt[3]{50x^2}}{5x}$ (§§6.1–6.2)   **63.** $7\sqrt{5}$ (§6.3)   **64.** 0 (§6.3)   **65.** $2a^2\sqrt[3]{a}$ (§6.3)

66. $-\dfrac{\sqrt{6}}{6}$ (§6.3)   **67.** $2 + \sqrt{2}$ (§6.4)   **68.** $8 - 2\sqrt{15}$ (§6.4)   **69.** 2 (§6.4)   **70.** $-8\sqrt{x} - 12$ (§6.4)

71. $4\sqrt{3} - 5\sqrt{2}$ (§6.4)   **72.** $\sqrt{15} - 3$ (§6.4)   **73.** $x = \frac{47}{3}$ (§6.5)   **74.** $x = \frac{25}{3}$ (§6.5)   **75.** $x = -7$ (§6.5)
76. $x = -1$ (§6.5)   **77.** $-i$ (§6.6)   **78.** -1 (§6.6)   **79.** $17 - 7i$ (§6.6)   **80.** $\frac{7}{5} + \frac{1}{5}i$ (§6.6)

Cumulative Practice Test: Chapters 4–6

1. (a) $5x + 3y$ **(b)** -2 **(c)** $\dfrac{(3x - 14)(x - 1)}{(x - 5)(x - 3)(x - 7)}$ **2.** $\dfrac{-2x^2 - x + 1}{3x^2 + 4x}$ **3. (a)** $x = \dfrac{9}{7}$ **(b)** $x < \dfrac{45}{2}$

(c) No solution   **4.** $a = \dfrac{y}{1 - y}$ **5.** Let $t = $ time it takes to process 80 forms together; $\dfrac{t}{6} + \dfrac{t}{5\frac{1}{2}} = 1; 2\frac{20}{23}$ hours

6. (a) $-72x^7y^{11}$ **(b)** $-\dfrac{18x^3}{y^3}$ **(c)** $\dfrac{9x^6}{y^4}$ **(d)** $\dfrac{b}{b + a}$ **7.** 3.4×10^{-5} **8.** $3{,}000{,}000$ **9.** $-\frac{1}{8}$

10. (a) $x^{2/3}$ **(b)** $\dfrac{1}{x^{7/6}y^{3/4}}$ **11. (a)** $2xy^2\sqrt{6y}$ **(b)** $2a^3b^3\sqrt{2ab}$ **(c)** $\dfrac{\sqrt[3]{5a^2}}{a}$

12. (a) 0 **(b)** $\dfrac{2\sqrt{xy} - y\sqrt{2x}}{2y}$ **(c)** $4\sqrt{3} - 7$ **(d)** $\dfrac{5\sqrt{10} + 10}{3}$ **13.** $x = \dfrac{1}{2}$ **14.** $\dfrac{7}{10} - \dfrac{1}{10}i$

Exercises 7.1

1. $x = -2, 3$ **3.** $y = \frac{1}{2}, 4$ **5.** $x = \pm 5$ **7.** $x = 0, 4$ **9.** $x = 6, -2$ **11.** $y = 0, 7$ **13.** $x = \pm 4$
15. $c = \pm\frac{4}{3}$ **17.** $a = \pm\frac{3}{2}$ **19.** $x = 3, -2$ **21.** $y = 1, \frac{1}{2}$ **23.** $x = \frac{7}{2}, -4$ **25.** $x = \pm\sqrt{6}$

27. $a = \pm\dfrac{\sqrt{133}}{7}$ **29.** $x = 5, -2$ **31.** $a = -\frac{1}{3}, \frac{1}{2}$ **33.** $y = \frac{3}{2}, -\frac{2}{3}$ **35.** $x = \pm\dfrac{i\sqrt{15}}{3}$ **37.** $s = -\frac{5}{6}, 2$

39. $x = 3, -7$ **41.** $x = 5, 1$ **43.** $a = \pm\dfrac{\sqrt{14}}{2}$ **45.** $x = -9$ **47.** $x = \pm i\sqrt{2}$ **49.** $x = 8 \pm 2\sqrt{2}$

51. $y = 5 \pm 4i$ **53.** $x = 2, 3$ **55.** $a = \pm\sqrt{13}$ **57.** $x = -\dfrac{1}{3}$ **59.** $x = -\dfrac{4}{3}$ **61.** $x = 1$

63. $x = 5$ **65.** $x = \dfrac{14}{3}, -1$ **67.** $a = \pm\dfrac{\sqrt{b}}{2}$ **69.** $y = \pm\dfrac{\sqrt{(9 - 5x^2)(7)}}{7}$ **71.** $r = \dfrac{\sqrt{6V\pi}}{2\pi}$

73. $a = \pm 2b$ **75.** $x = 3y, -2y$

Exercises 7.2

1. $x = 3 \pm \sqrt{10}$ **3.** $c = 1 \pm \sqrt{6}$ **5.** $y = \dfrac{-5 \pm \sqrt{33}}{2}$ **7.** $x = \dfrac{-3 \pm \sqrt{5}}{2}$ **9.** $a = \dfrac{1 \pm \sqrt{17}}{2}$

11. $a = -1 \pm i$ **13.** $x = -1 \pm \sqrt{3}$ **15.** $a = \dfrac{1 \pm \sqrt{33}}{2}$ **17.** $x = \dfrac{5 \pm i\sqrt{19}}{2}$ **19.** $x = \dfrac{-5 \pm \sqrt{195}}{5}$

21. $t = -\dfrac{3}{2}, 1$ **23.** $x = 2 \pm \sqrt{15}$ **25.** $x = \dfrac{1}{4}, 3$ **27.** $a = \dfrac{1 \pm i\sqrt{3}}{2}$ **29.** $y = \dfrac{-1 \pm 3i}{2}$

31. $n = \dfrac{1 \pm i\sqrt{7}}{2}$ **33.** $t = \dfrac{-1 \pm \sqrt{7}}{2}$ **35.** $x = \dfrac{-2 \pm \sqrt{19}}{5}$ **41.** -267 **42.** $\dfrac{-3x^2 - 7x + 1}{x^2 - 9}$

43. $\dfrac{-2x - 1}{3x^2 + 9x}$ **44.** $q = \dfrac{-1}{3a + 10}$ **45. (a)** $\dfrac{4 + \sqrt{7}}{5}$ **(b)** $\dfrac{\sqrt{30} + \sqrt{10}}{2}$

46. Let l = the length of the rectangle; $2l + 2\left(\dfrac{3}{4}l\right) = 100$; $28\dfrac{4}{7}$ feet by $21\dfrac{3}{7}$ feet

Exercises 7.3

1. $x = 5, -1$ **3.** $a = \dfrac{3 \pm \sqrt{41}}{4}$ **5.** $y = \dfrac{2 \pm \sqrt{2}}{2}$ **7.** $y = \dfrac{-7 \pm \sqrt{77}}{2}$ **9.** $x = \dfrac{3 \pm \sqrt{65}}{4}$

11. $a = \dfrac{-1 \pm i\sqrt{23}}{6}$ **13.** $s = -1 \pm 3i$ **15.** Roots are not real. **17.** Roots are real and equal.

19. Roots are real and distinct. **21.** Roots are real and distinct. **23.** Roots are real and equal.

25. Roots are real and distinct. **27.** $a = 4, -1$ **29.** $y = \pm \dfrac{\sqrt{6}}{4}$ **31.** $x = \dfrac{4}{5}, \dfrac{3}{2}$ **33.** $x = \dfrac{5}{2}, -\dfrac{1}{5}$

35. $a = 0, -1$ **37.** $y = \pm\sqrt{21}$ **39.** $x = 0, -1$ **41.** $x = \dfrac{3 \pm i\sqrt{11}}{2}$ **43.** $s = \pm\sqrt{6}$

45. $x = \dfrac{-1 \pm i\sqrt{39}}{2}$ **47.** $x = \dfrac{19}{4}$ **49.** $y = 4$ **51.** $a = \dfrac{2 \pm i\sqrt{2}}{3}$ **53.** $x = \dfrac{5 \pm \sqrt{57}}{2}$ **55.** $a = 1, \dfrac{1}{2}$

57. $z = \dfrac{7}{6}$ **59.** $x = \pm\sqrt{7}$ **61.** $y = \dfrac{6}{5}$ **63.** $y = -\dfrac{4}{7}$ **65.** $a = \dfrac{1 \pm \sqrt{97}}{4}$ **67.** $a = \dfrac{-1 \pm i\sqrt{15}}{2}$

69. $x = .3, 5$

Exercises 7.4

1. Let x = number; $x^2 - 5 = 5x + 1$; $x = 6$ **3.** Let x = number; $x^2 + 64 = 68$; $x = 2$

5. Let x = number; $(x + 6)^2 = 169$; $x = -19$ or $x = 7$ **7.** Let x = number; $x + \dfrac{1}{x} = \dfrac{13}{6}$; $x = \dfrac{2}{3}$ or $x = \dfrac{3}{2}$

9. Let x = number; $x + \dfrac{2}{x} = 3$; $x = 2$ or $x = 1$ **11. (a)** Profit = \$0 **(b)** Price = \$6

13. (a) $s = 24$ ft **(b)** $t = \dfrac{\sqrt{10}}{2}$ sec ≈ 1.58 sec **(c)** 40 ft

15. Let x = width; $(x + 2)(x) = 80$; length = 10 ft, width = 8 ft

17. Let side of square = x inches; $x^2 = 60$; $2\sqrt{15} \approx 7.75$ inches (each side)

19. Let x = larger number; $(x)(x - 2) = 120$; $x = 12$ and $x = -10$; two sets of numbers: 12 and 10, -12 and -10

21. Let x = one number; $x(3x + 2) = 85$; two sets of numbers: 5 and 17, $-\dfrac{17}{3}$ and -15

23. Let x = number; $x(x - 3) = 40$; two sets of numbers: 8 and 5, -5 and -8

25. Hypotenuse = $\sqrt{73} \approx 8.54''$ **27.** Length of other leg = $\sqrt{33} \approx 5.74''$ **29.** Dimensions are $8''$, $2\frac{1}{3}''$, and $8\frac{1}{3}''$

31. Length of diagonal = $\sqrt{193} \approx 13.9''$ **33.** Length of diagonal = $5\sqrt{2} \approx 7.07''$

35. Length of side = $4\sqrt{2} \approx 5.66''$ **37.** It reaches $2\sqrt{209} \approx 28.9$ ft high **39.** Area = 32 sq. in.

41. Area = $5\sqrt{119} \approx 54.54$ sq. in.

43. Let x = width of picture; $(x + 2)(2x + 2) = 60$; dimensions of picture are 4 in. by 8 in.

45. Let x = width of walkway; $(20 + 2x)(55 + 2x) - (20)(55) = 400$; width of walkway is $2\frac{1}{2}$ feet.

47. Let x = width of path; $\pi(10 + x)^2 - \pi(10)^2 = 44\pi$; width of path is 2 feet.

49. Let r = speed of boat in still water; $\dfrac{20}{r - 5} + \dfrac{10}{r + 5} = \dfrac{4}{3}$; 25 kph **51.** $(5y^2 - 1)(25y^4 + 5y^2 + 1)$ **52.** 34

53. $\dfrac{6}{5} + \dfrac{3}{5}i$ **54.** $x = 5$ **55.** $x = 625$

Exercises 7.5

1. $x = \frac{9}{4}$ **3.** $x = 4$ **5.** $a = 13$ **7.** $a = 8$ **9.** $x = 8$ **11.** $a = 1$ **13.** $y = 1$ **15.** $a = 2$

17. $s = 9$ **19.** $a = -2$ **21.** $x = (b - a)^2$ **23.** $L = \frac{g^2 T^2}{\pi}$ **25.** $x = \frac{b^2 + 11b + 36}{5}$

27. $x = 0, 5, -3$ **29.** $a = 0, \frac{2}{3}, -\frac{1}{2}$ **31.** $y = \pm 4, \pm 1$ **33.** $a = \pm 2, \pm\sqrt{2}$ **35.** $b = \pm 4, \pm\sqrt{7}$
37. $x = \pm 2\sqrt{2}, \pm 1$ **39.** $x = \pm 1$ **41.** $x = \pm 8$ **43.** $x = 4$ **45.** $y = 125, y = -1$ **47.** No solution
49. $x = \frac{1}{2}, \frac{1}{3}$ **51.** $x = -2, 3$ **53.** $x = \pm\frac{1}{3}, \pm\frac{1}{2}$ **55.** $a = 81$ **57.** $x = 625$ **59.** $a = -7$

61. $x = -7, -1$ **63.** $x = \frac{2}{3}, -\frac{1}{2}$ **65.** $x = -2, 1, x = \frac{-1 \pm i\sqrt{7}}{2}$ **67.** $a = 10, -1, 5, -2$ **70.** $a \leq \frac{5}{18}$

71. No solution **72.** $-2 < x < \frac{8}{5}$ **73.** $x \leq -2$ or $x \geq \frac{8}{5}$

Exercises 7.6

1. $-4 < x < 2$; **3.** $x < -4$ or $x > 2$;

5. $-2 \leq x \leq 5$; **7.** $x \leq \frac{1}{2}$ or $x \geq 3$;

9. $-4 < a < 5$ **11.** $x \leq -4$ or $x \geq 3$ **13.** $-\frac{1}{2} \leq a \leq 5$ **15.** $y < -\frac{1}{3}$ or $y > \frac{1}{2}$ **17.** $-5 \leq x \leq \frac{2}{3}$
19. All real numbers **21.** No solution **23.** $x < \frac{3}{2}$ or $x > 5$ **25.** $y \leq -\frac{1}{3}$ or $y \geq 2$ **27.** $x = -1$
29. $-1 < x < 2$ **31.** $y \leq -3$ or $y > 5$
33. $-4 < a < 6$ **35.** $y > 4$ **37.** $y < 1$ or $y > 4$;

39. $-\frac{5}{2} < y < -2$; **41.** $y < 1$;

43. $y > \frac{1}{3}$ or $y < -1$; **45.** $1 < x \leq 3$;

47. $x < -3$; **49.** $-3 < x < 2$;

52. $\dfrac{3x^2 + 8x - 3}{5x^3 + 13x^2 - 19x + 1}$ **53.** $\dfrac{2y^{3/4}}{3x^{9/4}}$ **54.** $9 - \sqrt{6}$ **55.** $30 + 2\sqrt{3} - \sqrt{6}$

Chapter 7 Review Exercises

1. $x = -7, 4$ **2.** $x = \frac{1}{2}, -3$ **3.** $y = -\frac{1}{2}, 1$ **4.** $a = -\frac{2}{3}, 5$ **5.** $x = -\frac{4}{3}, 7$ **6.** $a = \frac{1}{2}, -\frac{1}{5}$

7. $a = \pm 9$ **8.** $y = \pm\sqrt{65}$ **9.** $z = \pm i\sqrt{5}$ **10.** $r = \pm\frac{i\sqrt{10}}{2}$ **11.** $x = 3, 3$ **12.** $a = 2, -1$

13. $a = 5, -2$ **14.** $y = 3, -\frac{1}{2}$ **15.** $x = \pm\sqrt{5}$ **16.** $a = 6, 4$ **17.** $x = -6, 3$ **18.** $x = -1, -\frac{11}{3}$

19. $x = -1 \pm \sqrt{5}$ **20.** $x = 1 \pm \sqrt{5}$ **21.** $y = \frac{-2 \pm \sqrt{10}}{2}$ **22.** $y = \frac{-2 \pm i\sqrt{2}}{2}$ **23.** $a = \frac{-3 \pm 2\sqrt{6}}{3}$

24. $a = \frac{3 \pm i\sqrt{6}}{3}$ **25.** $a = \frac{4 \pm \sqrt{43}}{2}$ **26.** $a = \pm\frac{\sqrt{22}}{2}$ **27.** $a = -\frac{1}{3}, \frac{5}{2}$ **28.** $x = 4, -\frac{2}{5}$

29. $a = -\frac{2}{3}, 3$ **30.** $y = 1, -\frac{7}{2}$ **31.** $a = \pm 1$ **32.** $a = 5, -3$ **33.** $x = 2 \pm \sqrt{6}$ **34.** $y = 1 \pm \sqrt{39}$

35. $t = -4$ **36.** $u = \frac{5}{3}$ **37.** $x = \pm\frac{\sqrt{6}}{2}$ **38.** $x = \pm\frac{\sqrt{57}}{3}$ **39.** $x = 0$ **40.** $y = \pm\sqrt{5}$

41. $x = 2 \pm \sqrt{2}$ **42.** $z = 3 \pm i\sqrt{2}$ **43.** $z = 1 \pm \sqrt{10}$ **44.** $z = 3, -1$ **45.** $\dfrac{-16 \pm \sqrt{6}}{5}$

46. $x = \dfrac{-7 \pm \sqrt{201}}{4}$ **47.** $x = \dfrac{4 \pm \sqrt{22}}{2}$ **48.** $x = 9, -1$ **49.** $r = \dfrac{\sqrt{\pi A h}}{\pi h}$ **50.** $t = \dfrac{\sqrt{2lg}}{g}$

51. $x = \dfrac{-3y}{2}, x = y$ **52.** $y = 5x, y = -x$ **53.** $a = 3$ **54.** $a = 4, 1$ **55.** $a = 5$ **56.** $y = 0, 3$

57. $x = 84$ **58.** $x = 16$ **59.** $x = 7, 4$ **60.** $x = 5$ **61.** $y = \dfrac{x^2 - z}{3}$ **62.** $y = \dfrac{x^2 + z}{5}$

63. $y = \dfrac{(x - z)^2}{3}$ **64.** $y = \dfrac{(x + z)^2}{5}$ **65.** $x = 0, 5, -3$ **66.** $x = 0, 0, 7, -5$ **67.** $x = 0, 3, -\frac{1}{2}$

68. $x = 0, -1, \frac{2}{3}$ **69.** $a = \pm 4, \pm 1$ **70.** $y = \pm 3$ **71.** $y = \pm 2, \pm i$ **72.** $a = \pm 3, \pm 2i$

73. $z = \pm \sqrt{5}, \pm 1$ **74.** $z = \pm 3, \pm \sqrt{2}$ **75.** $a = 81$ **76.** $a = 27, -8$ **77.** $x = 27, -\frac{1}{8}$ **78.** $x = 81$

79. $x = 625$ **80.** $x = 16, 81$ **81.** $x = -1, \frac{3}{2}$ **82.** $x = -\frac{5}{3}, 1$ **83.** $x < -1$ or $x > 2$

84. $a \leq -5$ or $a \geq 1$ **85.** $-\frac{1}{3} \leq x \leq 2$ **86.** $-\frac{1}{2} < y < 6$

87. $y < 1$ or $y > 4$;   **88.** $4 \leq a \leq 9$;

89. $-9 < a < 9$;   **90.** $a > 9$ or $a < -9$;

91. $s \leq -\frac{2}{5}$ or $s \geq 4$;   **92.** $-3 \leq a \leq 5$;

93. $-2 < x < 3$;   **94.** $x < -4$ or $x > 2$;

95. $x < -2$ or $x \geq 3$;   **96.** $-4 \leq x < 2$;

97. $x < 3$ **98.** $x < 2$ or $x \geq \frac{11}{4}$ **99.** $-4 < x \leq -\frac{11}{4}$

100. $x < -5$ or $x > -\frac{8}{3}$ **101.** Let $x = $ number; $x^2 + 4 = 36$; $x = \pm 4\sqrt{2}$

102. Let $x = $ number; $(x + 4)^2 = 36$; $x = 2$ or -10

103. Let $x = $ number; $x + \dfrac{1}{x} = \frac{53}{14}$; $x = \frac{2}{7}$ or $x = \frac{7}{2}$ **104.** Let $x = $ number; $x - \dfrac{1}{x} = \frac{40}{21}$; $x = \frac{7}{3}$ or $-\frac{3}{7}$

105. Let width $= x$; $50 = x(2x)$; $x = 5$ ft (width); length $= 10$ ft

106. Let $w = $ width; $w(3w + 2) = 85$; width $= 5''$, length $= 17''$

107. Let width of frame $= x$; $(5 + 2x)(8 + 2x) - 40 = 114$; $x = 3''$

108. Let $x = $ width of walkway; $(2x + 20)(2x + 30) - 600 = 216$; 2 feet wide

109. Hypotenuse $= 5\sqrt{10} \approx 15.81''$ **110.** Length of leg $= 15\sqrt{3} \approx 25.98'$

111. Length of diagonal $= \sqrt{41} \approx 6.40''$ **112.** Length of side $= 10\sqrt{2} \approx 14.14''$

113. Let rate of wind $= x$ mph; $\dfrac{300}{200 - x} + \dfrac{300}{200 + x} = \dfrac{25}{8}$; $x = 40$ mph

Chapter 7 Practice Test

1. (a) $z = 3, -\frac{1}{2}$ **(b)** $x = \pm \sqrt{5}$ **2. (a)** $y = -2 \pm \sqrt{5}$ **(b)** $a = \dfrac{-1 \pm \sqrt{17}}{4}$ **(c)** $x = 5, -4$

3. Roots are distinct and imaginary. **4.** $r = \dfrac{\sqrt{2V\pi h}}{2\pi h}$ **5.** $x = 12$ **6.** $x = 625$

7. (a) $x \leq -4$ or $x \geq 9$;   **(b)** $\frac{2}{3} < x < 6$;

8. Length of diagonal $= 9\sqrt{2} \approx 12.73'$

9. Let speed of boat in still water $= x$ mph; $\dfrac{15}{x-3} + \dfrac{12}{x+3} = \dfrac{9}{4}$; $x = 13$ mph

Exercises 8.1

1. No   **3.** No   **5.** Yes   **7.** Yes   **9.** $(-1, 9)$, $(0, 8)$, $(1, 7)$, $(10, -2)$, $(8, 0)$, $(4, 4)$

11. $\left(-2, \frac{15}{2}\right)$, $(0, 5)$, $(4, 0)$, $(8, -5)$, $(4, 0)$, $\left(\frac{4}{5}, 4\right)$ **13.** $(-3, 8)$, $(0, 4)$, $(3, 0)$, $(6, -4)$, $(3, 0)$, $(0, 4)$

15. x-intercept: 6; y-intercept: 6   **17.** x-intercept: 6; y-intercept: -6 **19.** x-intercept: -6; y-intercept: 6

21. x-intercept: 6; y-intercept: 3   **23.** x-intercept: 3; y-intercept: 4   **25.** x-intercept: 5; y-intercept: -3

27. x-intercept: $-\frac{7}{3}$; y-intercept: $\frac{7}{2}$

29.

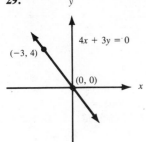

31.

33.

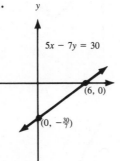

35.

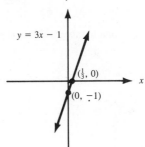

37.

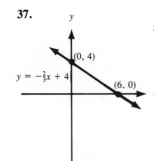

39.

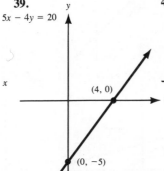

41.

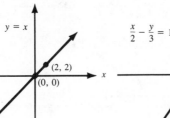

43.

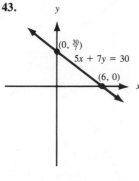

45.

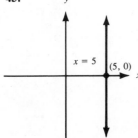

47.

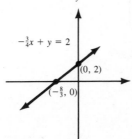

49.

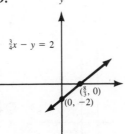

51.

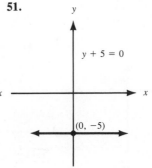

53.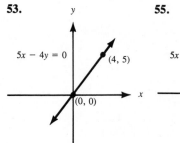

55.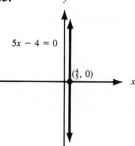

61. $2x(x + 4)(x - 3)$ **62.** $\dfrac{x^3 - 4x + 3}{x^2 - 4}$ **63.** $\dfrac{y^4}{x^2y^4 + 2xy^2 + 1}$ **64.** $\dfrac{3y^2\sqrt[3]{4x}}{2x}$ **65.** $\dfrac{-1 \pm \sqrt{11}}{2}$

66. Let x = one number; $x(23 - x) = 132$; 11, 12

Exercises 8.2

1.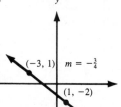

3.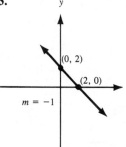

5.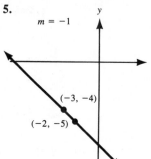

7. Slope = 0
9. Undefined   **11.** 1
13. $b + a$ **15.** $\frac{3}{5}$

17.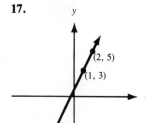

19.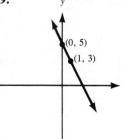

21.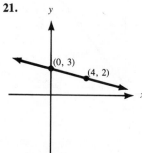

23.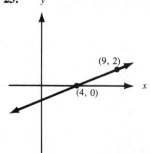

25.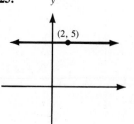

27.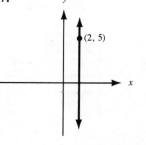

29. Parallel **31.** Neither **33.** Perpendicular **35.** Perpendicular **37.** $h = 13$ **39.** $n = 4, n = -3$

41. No real solutions for c; $c = \dfrac{3 \pm i\sqrt{7}}{4}$ **51.** $\dfrac{1}{32}$ **52.** $6\sqrt{3}$ **53.** $\dfrac{5\sqrt{2} + \sqrt{5}}{3}$ **54.** $x = -6, 5$

55. $2\sqrt{61}$

Exercises 8.3

1. $y = 5x - 8$ **3.** $y = -3x - 13$ **5.** $y = \frac{2}{3}x - 3$ **7.** $y = -\frac{1}{2}x + 2$ **9.** $y = \frac{3}{4}x + 5$ **11.** $y = -4$

13. $x = -4$ **15.** $y = \frac{4}{3}x + \frac{1}{3}$ **17.** $y = 4x + 7$ **19.** $y = -x + 5$ **21.** $y = 4x + 6$ **23.** $y = -2x - 6$

25. $y = 3$ **27.** $x = -2$ **29.** $y = \frac{2}{3}x + 2$ **31.** $y = 3x - 4$ **33.** $y = \frac{3}{2}x + \frac{3}{2}$ **35.** $y = -x$

37. $y = \frac{3}{2}x - \frac{1}{2}$ **39.** $y = -\frac{3}{4}x - \frac{9}{4}$ **41.** $y = -\frac{5}{8}x - 4$ **43.** $y = -2x + 4$ **45.** $y = 3$ **47.** $x = -1$

49. Parallel **51.** Neither **53.** Perpendicular **55.** Parallel **57.** $P = 30x - 340$; if $x = 200$, $P = \$5,660$

59. Test B score: 85 **61.** $V = -1,000E + 105,000$; $E = 90$ when $V = 15,000$ **65.** $0 \le x < \dfrac{7}{3}$ **66.** $\dfrac{y^{1/2}}{x^{4/3}}$

67. $x = 1$ **68.** $x < -1$ or $x > 4$ **69.** $x \le -2$ or $x \ge 7$;

Exercises 8.4

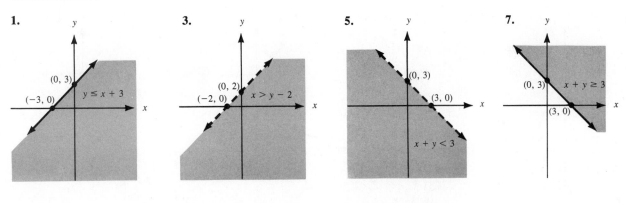

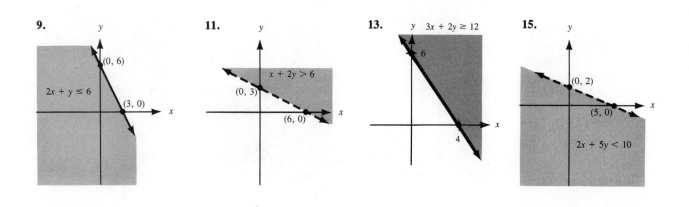

17.

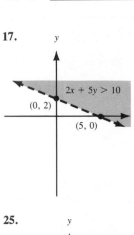

$2x + 5y > 10$

$(0, 2)$

$(5, 0)$

19.

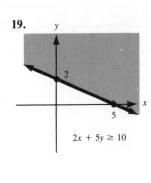

$2x + 5y \geq 10$

21.

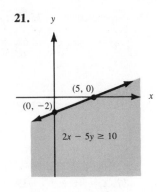

$(5, 0)$

$(0, -2)$

$2x - 5y \geq 10$

23.

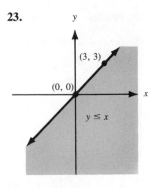

$(3, 3)$

$(0, 0)$

$y \leq x$

25.

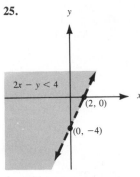

$2x - y < 4$

$(2, 0)$

$(0, -4)$

27.

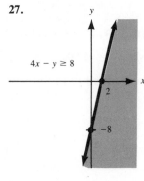

$4x - y \geq 8$

2

-8

29.

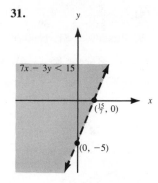

$(4, 0)$

$(0, -3)$

$3x - 4y > 12$

31.

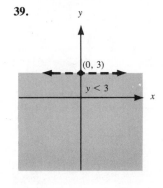

$7x - 3y < 15$

$\left(\frac{15}{7}, 0\right)$

$(0, -5)$

33.

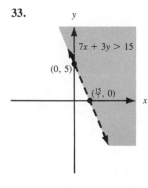

$7x + 3y > 15$

$(0, 5)$

$\left(\frac{15}{7}, 0\right)$

35.

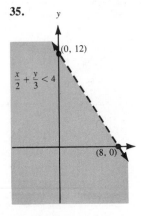

$(0, 12)$

$\frac{x}{2} + \frac{y}{3} < 4$

$(8, 0)$

37.

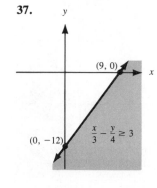

$(9, 0)$

$(0, -12)$

$\frac{x}{3} - \frac{y}{4} \geq 3$

39.

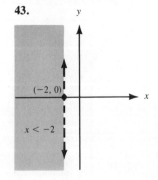

$(0, 3)$

$y < 3$

41.

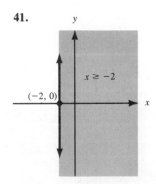

$x \geq -2$

$(-2, 0)$

43.

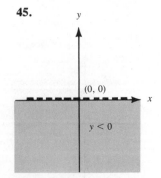

$(-2, 0)$

$x < -2$

45.

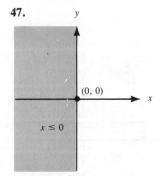

$(0, 0)$

$y < 0$

47.

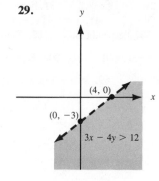

$(0, 0)$

$x \leq 0$

49.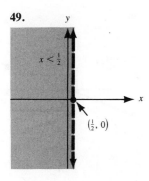

51.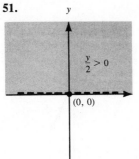

54. $\dfrac{4x(x + 2)}{(2x + 3)(x - 1)}$ **55.** 7.28×10^2 **56.** $-\dfrac{5}{13} + \dfrac{12}{13}i$

57. Let $x =$ the length of the rectangle; $2(12) + 2x \geq 100$; the length has to be at least 38 cm.

Chapter 8 Review Exercises

1.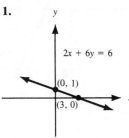

2.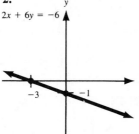

3.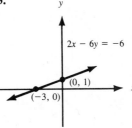

4.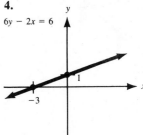

5.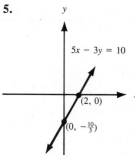

6.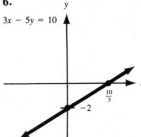

7.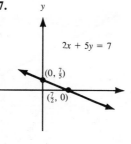

8.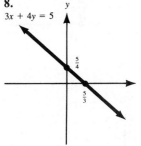

9.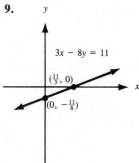

10.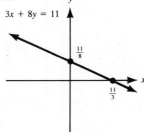

11.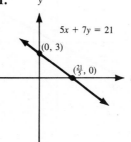

12.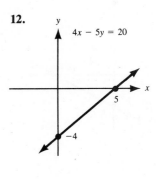

13.

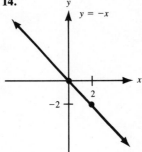

$y = x$

$(3, 3)$

$(0, 0)$

14.

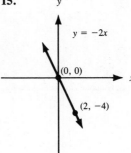

$y = -x$

-2 2

15.

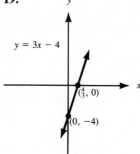

$y = -2x$

$(0, 0)$

$(2, -4)$

16.

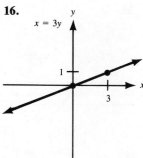

$x = 3y$

1

3

17.

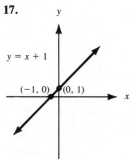

$y = x + 1$

$(-1, 0)$ $(0, 1)$

18.

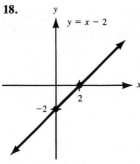

$y = x - 2$

-2 2

19.

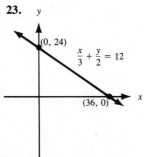

$y = 3x - 4$

$(\frac{4}{3}, 0)$

$(0, -4)$

20.

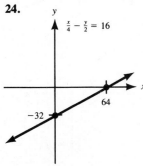

$y = -2x + 5$

5

$\frac{5}{2}$

21.

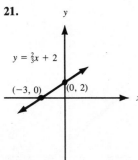

$y = \frac{2}{3}x + 2$

$(-3, 0)$ $(0, 2)$

22.

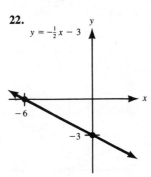

$y = -\frac{1}{2}x - 3$

-6

-3

23.

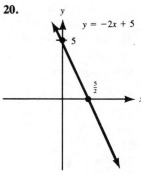

$(0, 24)$

$\frac{x}{3} + \frac{y}{2} = 12$

$(36, 0)$

24.

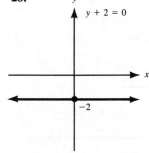

$\frac{x}{4} - \frac{y}{2} = 16$

-32 64

25.

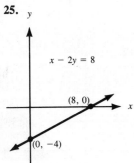

$x - 2y = 8$

$(8, 0)$

$(0, -4)$

26.

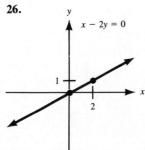

$x - 2y = 0$

1

2

27.

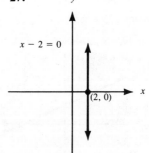

$x - 2 = 0$

$(2, 0)$

28.

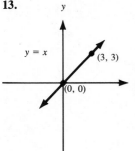

$y + 2 = 0$

-2

29.

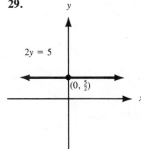

30.

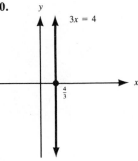

31. $-\frac{1}{2}$ **32.** $-\frac{3}{5}$ **33.** $\frac{4}{5}$ **34.** $-\frac{4}{5}$ **35.** 3 **36.** $\frac{2}{3}$ **37.** $\frac{3}{4}$ **38.** $\frac{1}{3}$ **39.** $\frac{1}{2}$ **40.** $-\frac{4}{3}$ **41.** 0

42. -2 **43.** 0 **44.** Undefined **45.** 3 **46.** $-\frac{1}{5}$ **47.** $-\frac{3}{5}$ **48.** $\frac{3}{2}$ **49.** Undefined

50. 0 **51.** -1 **52.** a **53.** $a = 14$ **54.** $a = 0$ **55.** $a = -8$ **56.** $a = 5, -4$

57. No real solutions: $a = \dfrac{3 \pm 5i\sqrt{3}}{2}$ **58.** $a = 3, 1$ **59.** $y = -\frac{7}{3}x - \frac{5}{3}$ **60.** $y = 2x - 2$ **61.** $y = \frac{5}{3}x$

62. $y = -3x$ **63.** $y = \frac{2}{5}x + \frac{21}{5}$ **64.** $y = -4x + 4$ **65.** $y = 5x - 13$ **66.** $y = -\frac{3}{4}x + \frac{41}{4}$

67. $y = 5x + 3$ **68.** $y = -3x - 4$ **69.** $y = 3$ **70.** $x = 2$ **71.** $x = -3$ **72.** $y = -4$

73. $y = \frac{3}{2}x$ **74.** $y = \frac{1}{5}x + \frac{3}{5}$ **75.** $y = -\frac{2}{5}x + 6$ **76.** $y = \frac{6}{7}x + \frac{5}{7}$ **77.** $y = \frac{5}{3}x$

78. $y = -\frac{3}{5}x$ **79.** $y = \frac{5}{3}x - 5$ **80.** $y = -\frac{8}{5}x + 8$ **81.** $y = \frac{5}{2}x + 5$ **82.** $y = 2x - 2$

83. $y = \frac{3}{2}x + \frac{3}{5}$ **84.** $y = -\frac{2}{5}x - 2$

85.

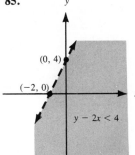

86.

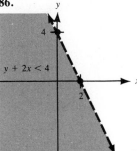

87.

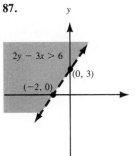

88.

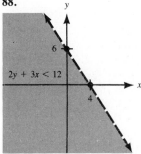

89.

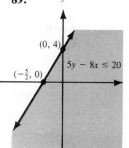

90.

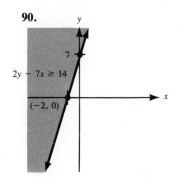

91.

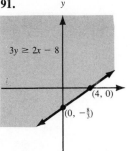

92.

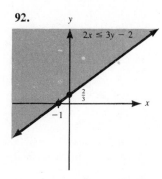

93.

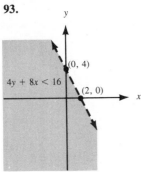

94.

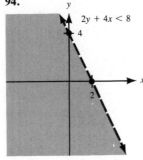

95.

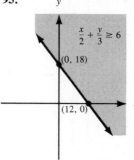

96.

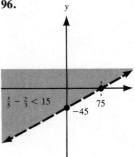

97.

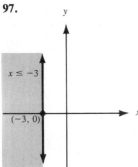

98.

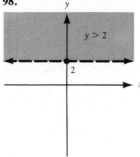

99.

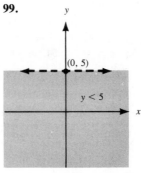

100.

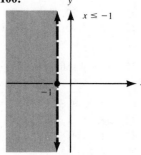

101. $P = 160x - 28,000$; $P = \$36,000$ when $x = 400$ gadgets

102. Jake's score on test B: 86; Charles' score on test A: 37

Chapter 8 Practice Test

1. (a)

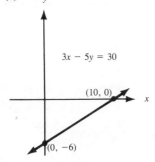

(b)

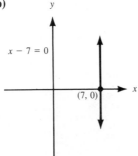

2. (a) 1 **(b)** $\frac{3}{2}$ **3.** $a = \frac{1}{2}$

4. (a) $y = 8x - 19$ **(b)** $y = -4x + 4$ **(c)** $y = x + 3$ **(d)** $y = -\dfrac{x}{3} - \dfrac{7}{3}$ **(e)** $y = 3x - 9$ **(f)** $y = -1$

5.

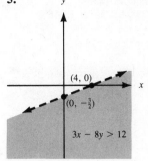

(4, 0)

$(0, -\frac{3}{2})$

$3x - 8y > 12$

6. 165 on test B

Exercises 9.1

1. $x = 5, y = 2$ **3.** $x = 4, y = 3$ **5.** $x = 1, y = 3$ **7.** $x = 6, y = 2$ **9.** $x = 5, y = -4$
11. Dependent **13.** Inconsistent **15.** $x = 5, y = 0$ **17.** $x = -1, y = 1$ **19.** $a = \frac{13}{27}, b = \frac{17}{27}$
21. $s = \frac{5}{2}, t = \frac{5}{2}$ **23.** $m = \frac{8}{7}, n = -\frac{16}{7}$ **25.** $p = 3, q = 1$ **27.** $u = \frac{27}{5}, v = \frac{4}{5}$ **29.** $w = 12, z = 6$
31. $x = 3, y = 2$ **33.** $x = 2, y = 6$ **35.** $x = 3, y = 7$ **37.** $x = \frac{1}{2}, y = 2$
39. Let $x =$ amount invested at 9% and $y =$ amount at 11%; $x + y = 14,000, 0.09x + 0.11y = 1,350$; $9,500 at 9%
and $4,500 at 11%
41. Let $x =$ amount at 8% and $y =$ amount at 10%; $x + y = 10,000, 0.08x = 0.10y$; $5,555.56 at 8% and
$4,444.44 at 10%
43. Let $L =$ length and $W =$ width; $L = W + 2, 2L + 2W = 36$; 8 cm by 10 cm
45. Let $x =$ amount of 30% solution and $y =$ amount of 70% solution; $x + y = 100, 0.30x + 0.70y = .54(100)$; 40 ml
of 30% solution and 60 ml of 70% solution
47. Let $T =$ cost of 35-mm roll and $M =$ cost of movie roll; $5T + 3M = 35.60, 3T + 5M = 43.60$; cost of movie
roll = $6.95 each; cost of 35-mm roll = $2.95 each
49. Let $c =$ cost of cream-filled donut and $j =$ cost of jelly donut; $7c + 5j = 3.16, 4c + 8j = 3.04$; single cream
donut = 28¢; single jelly donut = 24¢
51. Let $E = $ # of expensive models and $L = $ # of less expensive models; $6E + 5L = 730, 3E + 2L = 340$; 80 of the
more expensive type and 50 of the less expensive type
53. Let $f = $ # of five dollar bills and $t = $ # of ten dollar bills; $f + t = 43, 5f + 10t = 340$; eighteen $5 bills and
twenty-five $10 bills
55. Let $f =$ flat rate and $r =$ mileage rate; $f + 85r = 44.30, f + 125r = 51.50$; $29 flat rate and 18¢ per mile
57. Let $x =$ speed of plane and $y =$ wind speed; $6(x + y) = 2,310, 6(x - y) = 1,530$; speed of plane = 320 mph,
speed of wind = 65 mph
59. $\frac{y - 3}{x - 2} = -1; \frac{y + 2}{x - 1} = 2; x = 3, y = 2$ **64.** $(x - y + 5)(x - y - 5)$ **65.** $\frac{-27}{x^2 - x - 20}$
66. $x = 2, -1$ **67.** $\frac{-21\sqrt{2}}{2}$
68. Let $t =$ the time it takes both to complete the job; $\frac{x}{3} + \frac{x}{1} = 1; \frac{3}{4}$ day

Exercises 9.2

1. $x = 6, y = 3, z = 0$ **3.** $x = 2, y = 1, z = 0$ **5.** $x = 2, y = \frac{-3}{2}, z = \frac{-1}{2}$
7. $\{(x, y, z) \mid x + 2y + 3z = 1\}$ **9.** $x = 1, y = -2, z = -1$ **11.** $a = \frac{1}{2}, b = -1, c = 2$
13. $x = 1, y = 0, z = 1$ **15.** $s = 6, t = -6, u = 2$ **17.** $p = 1, q = 2, r = 3$ **19.** $x = 1, y = -1, z = 1$
21. $a = 3, b = -3, c = 6$ **23.** No solutions **25.** $x = 1,000, y = 2,000, z = 3,000$
27. Let $d = $ # of dimes, $q = $ # of quarters, and $h = $ # of half-dollars; $h + q + d = 48, d = q + h - 2$,
$0.10d + 0.25q + 0.50h = 10.55$; 23 dimes, 17 quarters, 8 half-dollars

29. Let a = amount in the bank account, b = amount in bonds, and c = amount in stocks; $a + b + c = 12,000$, $0.087a + 0.093b + 0.1266c = 1,266$, $0.1266c = 0.087a + 0.093b$; $3,000 at 8.7\%, $4,000 at 9.3\%, $5,000 at 12.66\%
31. Let r = # of orchestra tickets, m = # of mezzanine tickets, and b = # of balcony tickets; $r + m + b = 750$, $12r + 8m + 6b = 7,290$, $m + b = r - 100$; 425 orchestra tickets, 120 mezzanine tickets, and 205 balcony tickets
33. Let A = # of model A's, B = # of model B's, and C = # of model C's; $2.1A + 2.8B + 3.2C = 721$, $3.2A + 3.6B + 4C = 974$, $0.5A + 0.6B + 0.8C = 168$; Model A: 110, model B: 95, model C: 70

35.

36. $\dfrac{\sqrt[3]{63}}{3}$ **37.** $\dfrac{8y^{3/5}}{27x^{21/4}}$ **38.** $-\dfrac{9}{5}$

[Graph showing line $3x - 2y = 12$ with intercepts $(4, 0)$ and $(0, -6)$]

39. Let x = the amount of pure water to be added; $0x + 0.60(5) = 0.40(x + 5)$; $2\frac{1}{2}$ quarts

Exercises 9.3

1. $\begin{bmatrix} 3 & -2 & | & 5 \\ 1 & -1 & | & 8 \end{bmatrix}$ **3.** $\begin{bmatrix} 1 & -2 & 3 & | & 4 \\ 0 & 1 & -1 & | & -3 \\ 2 & 3 & 0 & | & 8 \end{bmatrix}$ **5.** $(3, -2)$ **7.** $(0, -2)$ **9.** $(-4, 0)$

11. $\left(\frac{1}{2}, 3\right)$ **13.** $\left(\frac{2}{3}, \frac{1}{2}\right)$ **15.** $(2, 1, 3)$ **17.** $(-2, 1, -3)$ **19.** $(3, -2, 0)$ **21.** $\left(\frac{1}{2}, 2, 3\right)$
23. $\{(x, y, 3) \mid 2x + 3y = -2\}$ **25.** $(1, -2, 0, 3)$ **27.** $\begin{bmatrix} 1 & 0 & | & 4 \\ 0 & 1 & | & -1 \end{bmatrix}$
29. $\begin{bmatrix} 1 & 0 & 0 & | & 2 \\ 0 & 1 & 0 & | & 1 \\ 0 & 0 & 1 & | & -1 \end{bmatrix}$ **31.** $(2, -1)$ **33.** $(1, 3, 1)$ **35.** $(2, 3, -1)$ **37.** $(0, 2, 2, 3)$

Exercises 9.4

1. -2 **3.** 22 **5.** 8 **7.** 4 **9.** 0 **11.** 11 **13.** 0 **15.** -92 **17.** 1 **19.** $x = 3$
21. $x = 7$ or $x = -2$ **23.** $x = -5, x = 3$ **25.** $x = -1, y = 5$ **27.** No unique solution
29. $x = 4, y = 2$ **31.** $x = -2, y = 0$ **33.** $x = -\frac{53}{9}, y = -\frac{17}{3}$ **35.** $s = 2, t = 0$ **37.** $u = -3, v = 3$
39. No unique solution   **41.** $x = -\frac{1}{2}, y = \frac{1}{2}, z = 3$ **43.** $x = 1, y = 0, z = -1$ **45.** No unique solution
47. $x = 2, y = 2, z = 2$ **49.** No unique solution   **51.** No unique solution   **53.** $a = \frac{49}{11}, b = -\frac{10}{11}, c = \frac{4}{11}$

56. $a = \dfrac{3}{r + 3}$ **57.** $x = \dfrac{3 \pm \sqrt{73}}{2}$ **58.** $y = \dfrac{3}{2}x + 11$

59.

60. $2\sqrt{2} \approx 2.83$ in.

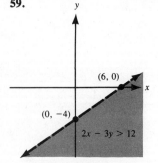

[Graph showing shaded region with boundary line through $(6, 0)$ and $(0, -4)$, labeled $2x - 3y > 12$]

Exercises 9.5

1.

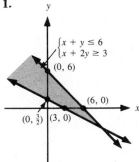

$$\begin{cases} x + y \le 6 \\ x + 2y \ge 3 \end{cases}$$

(0, 6)

(6, 0)

$(0, \frac{3}{2})$ (3, 0)

3.

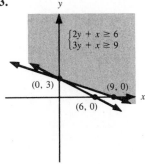

$$\begin{cases} 2y + x \ge 6 \\ 3y + x \ge 9 \end{cases}$$

(0, 3) (9, 0)

(6, 0)

5. No solution

7.

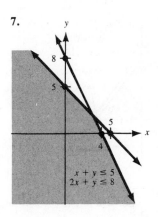

8

5

5

4

$$x + y \le 5$$
$$2x + y \le 8$$

9.

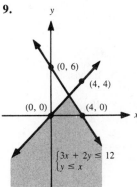

(0, 6)

(4, 4)

(0, 0) (4, 0)

$$\begin{cases} 3x + 2y \le 12 \\ y \le x \end{cases}$$

11.

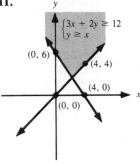

$$\begin{cases} 3x + 2y \ge 12 \\ y \ge x \end{cases}$$

(0, 6)

(4, 4)

(4, 0)

(0, 0)

13.

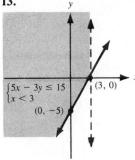

$$\begin{cases} 5x - 3y \le 15 \\ x < 3 \end{cases}$$

(3, 0)

(0, -5)

15.

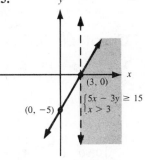

(3, 0)

$$\begin{cases} 5x - 3y \ge 15 \\ x > 3 \end{cases}$$

(0, -5)

17.

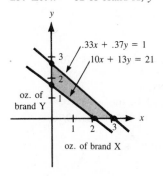

(0, 2)

(2, 0) (10, 0)

(0, -5)

$$x - 2y < 10$$
$$x \ge 2$$
$$y \le 2$$

19.

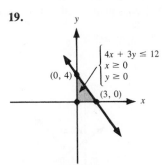

$$4x + 3y \le 12$$
$$x \ge 0$$
$$y \ge 0$$

(0, 4)

(3, 0)

21.

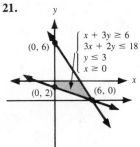

$$x + 3y \ge 6$$
$$3x + 2y \le 18$$
$$y \le 3$$
$$x \ge 0$$

(0, 6)

(0, 2) (6, 0)

23.

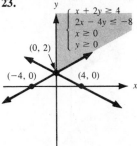

$$\begin{cases} x + 2y \ge 4 \\ 2x - 4y \le -8 \\ x \ge 0 \\ y \ge 0 \end{cases}$$

(0, 2)

(-4, 0) (4, 0)

25. Let x = oz of brand X, y = oz of brand Y; $10x + 13y \ge 21$, $0.33x + 0.37y \le 1$, $x \ge 0$, $y \ge 0$

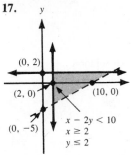

.33x + .37y = 1

10x + 13y = 21

3

2

oz. of
brand Y 1

1 2 3

oz. of brand X

27. Let x = men's shoes, y = ladies' shoes; $1x + \frac{3}{4}y \le 500$, $\frac{1}{3}x + \frac{1}{4}y \le 150$

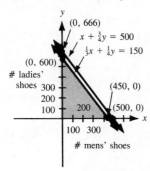

29. $x = \dfrac{11}{2}$ **30.** $4(x - 1)(x^2 + x + 1)$ **31.** $y = \dfrac{x}{3}, -\dfrac{x}{2}$ **32.** $y = -\dfrac{3}{2}x$

Chapter 9 Review Exercises

1. $x = 5, y = 1$ **2.** $x = 3, y = -2$ **3.** $x = 8, y = 0$ **4.** $x = \frac{1}{2}, y = \frac{1}{3}$ **5.** Dependent   **6.** Inconsistent
7. $x = -\frac{1}{5}, y = \frac{7}{5}$ **8.** $s = \frac{64}{23}, t = -\frac{2}{23}$ **9.** $x = 1, y = 1, z = 1$ **10.** $x = 2, y = 0, z = -1$

11. $x = -1, y = -2, z = -3$ **12.** $a = 0, b = 0, c = 1$ **13.** $\begin{bmatrix} 1 & 0 & | & 10 \\ 0 & 1 & | & -2 \end{bmatrix}$ **14.** $\begin{bmatrix} 1 & 0 & | & -10 \\ 0 & 1 & | & 8 \end{bmatrix}$

15. $\begin{bmatrix} 1 & 0 & 0 & | & 2 \\ 0 & 1 & 0 & | & -3 \\ 0 & 0 & 1 & | & 1 \end{bmatrix}$ **16.** $\begin{bmatrix} 1 & 0 & 0 & | & 0 \\ 0 & 1 & 0 & | & 0 \\ 0 & 0 & 1 & | & 3 \end{bmatrix}$ **17.** $(3, 1)$ **18.** $(1, 0)$ **19.** $(2, 1, -1)$

20. $(-1, 2, -1)$ **21.** $(2, 1)$ **22.** $(-1, 4, -1)$ **23.** -14 **24.** 22 **25.** 98
26. 54 **27.** $x = -\frac{2}{23}, y = \frac{14}{23}$ **28.** $x = \frac{19}{23}, y = \frac{13}{23}$ **29.** Inconsistent   **30.** Dependent

31. $s = \frac{32}{53}, t = \frac{16}{53}, u = -\frac{1}{53}$ **32.** $u = \dfrac{-95}{107}, v = \dfrac{-156}{107}, w = \dfrac{86}{107}$

33.   **34.**   **35.**   **36.** No solution

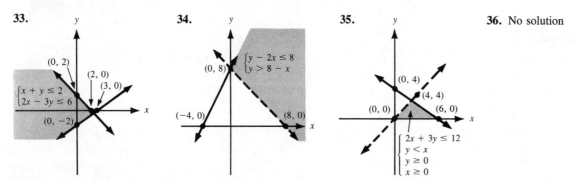

37. Let \$$x$ be deposited at 8.75% and \$$y$ be deposited at 9.65%; $0.0875x + 0.0965y = 768.95$, $x + y = 8{,}500$; \$5,700 at 8.75% and \$2,800 at 9.65%
38. Let \$$x$ be invested at 7.7%, \$$y$ be invested at 8.6%, and \$$z$ be invested at 9.8%; $y + z = 3{,}000 + x$, $x + y + z = 20{,}000$, $0.077x + 0.086y + 0.098z = 1{,}719.10$; \$8,500 at 7.7%, \$5,200 at 8.6%, and \$6,300 at 9.8%
39. Let price per lb of bread = \$$x$ and price per lb of cookies = \$$y$; $3x + 5y = 22.02$, $2x + 3y = 13.43$; $x = \$1.09/\text{lb}$, $y = \$3.75/\text{lb}$
40. Let s = price of a sweater, b = price of a blouse, and j = price of jeans; $s + 2b + j = 28.50$, $2s + b + j = 30$, $2s + 3b + 2j = 52$; \$6.50 for each sweater, \$5 for each blouse, and \$12 for each pair of jeans

Chapter 9 Practice Test

1. **(a)** $x = 8, y = -5$ **(b)** $a = -3, b = 6$ **(c)** $x = 3, y = 0, z = -1$ **2.** $(3, 1, 2)$ **3. (a)** 11   **(b)** 23

4. (a) $x = \frac{19}{4}, y = \frac{7}{4}$ **(b)** $x = 2, y = -1, z = 3$

5. Let f = # of \$5 bills, t = # of \$10 bills, w = # of \$20 bills; $2w = t, f + t + w = 40, 5f + 10t + 20w = 500$; four \$5 bills, twenty-four \$10 bills, twelve \$20 bills

6. Let x = amount invested at $8\frac{1}{2}\%$ and y = amount invested at 9%; $x + y = 3{,}500, 0.085x + 0.09y = 309$; \$1,200 at $8\frac{1}{2}\%$, \$2,300 at 9%

7.

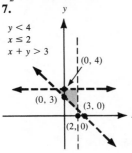

Cumulative Review: Chapters 7–9

1. $a = 5, -3$ (§7.1)   **2.** $a = 6, 1$ (§7.1)   **3.** $x = \pm\dfrac{\sqrt{38}}{2}$ (§7.1)   **4.** $x = 5 \pm 2\sqrt{7}$ (§7.1)

5. $y = -3 \pm \sqrt{10}$ (§7.2)   **6.** $x = \dfrac{-3 \pm \sqrt{15}}{3}$ (§7.2)   **7.** $a = \dfrac{1 \pm \sqrt{7}}{3}$ (§7.3)   **8.** $x = 2 \pm i\sqrt{3}$ (§7.3)

9. $x = -1, -3$ (§7.3)   **10.** $y = -4, 3$ (§7.3)   **11.** $x = 3, -\frac{9}{2}$ (§7.3)   **12.** $b = 5$ (§7.3)

13. $x = 2 \pm \sqrt{7}$ (§7.3)   **14.** $\dfrac{25 \pm \sqrt{865}}{10}$ (§7.3)   **15.** $z = \pm\dfrac{\sqrt{xys}}{s}$ (§7.1)   **16.** $b = \dfrac{a}{5}, b = -a$ (§7.1)

17. $x = 1, 2$ (§7.5)   **18.** $x = 3$ (§7.5)   **19.** $x = \dfrac{4y^2 + 1}{5}$ (§7.5)   **20.** $x = \dfrac{4y^2 + 4y + 1}{5}$ (§7.5)

21. $x = \pm3, \pm3i$ (§7.5)   **22.** $a = 625$ (§7.5)

23. $5 < x < 8$ (§7.6);   **24.** $x \leq -\frac{1}{2}$ or $x \geq 5$ (§7.6);

25. $x < -3$ or $x > 2$ (§7.6);   **26.** $-\frac{1}{2} < x \leq 0$ (§7.6);

27. Let x = the number; $x - \dfrac{1}{x} = \dfrac{5}{6}; \dfrac{3}{2}$ or $\dfrac{-2}{3}$ (§7.6)

28. Let w = the width; $w(w + 4) = 21$; width is 3, length is 7 (§7.4)

29. Let s = side of square; $28^2 = s^2 + s^2$; $s = 14\sqrt{2} \approx 19.8$ inches (§7.4)

30. Let x = width of walkway; $(20 + 2x)(15 + 2x) - 15 \cdot 20 = 294$; $3\frac{1}{2}$ feet (§7.4)

31. (§8.1)   **32.** (§8.1)   **33.** (§8.1)   **34.** (§8.1)

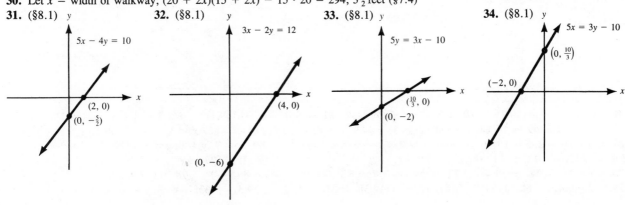

35. (§8.1)

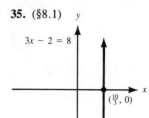

$3x - 2 = 8$

$(\frac{10}{3}, 0)$

36. (§8.1)

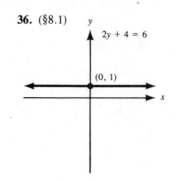

$2y + 4 = 6$

$(0, 1)$

37. $m = 8$ (§8.2)   **38.** $m = -1$ (§8.2)   **39.** $m = 5$ (§8.2)   **40.** $m = \frac{3}{2}$ (§8.2)   **41.** $m = \frac{5}{4}$ (§8.2)
42. $m = -\frac{2}{5}$ (§8.2)   **43.** $a = 1$ (§8.2)   **44.** $a = \frac{1}{2}$ (§8.2)   **45.** $y = 3x + 13$ (§8.3)
46. $y = -2x + 9$ (§8.3)   **47.** $y = -6x + 19$ (§8.3)   **48.** $y = 5$ (§8.3)   **49.** $y = 4x + 2$ (§8.3)
50. $y = -x + 3$ (§8.3)   **51.** $y = -\frac{3}{5}x - \frac{9}{5}$ (§8.3)   **52.** $y = -\frac{1}{2}x - 2$ (§8.3)

53. (§8.4)

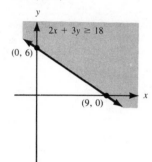

$2x + 3y \geq 18$

$(0, 6)$

$(9, 0)$

54. (§8.4)

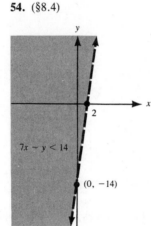

$7x - y < 14$

2

$(0, -14)$

55. (§8.4)

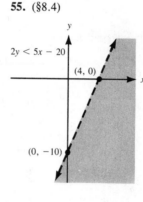

$2y < 5x - 20$

$(4, 0)$

$(0, -10)$

56. (§8.4)

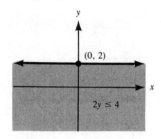

$(0, 2)$

$2y \leq 4$

57. (§8.4)

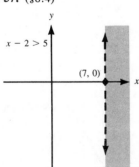

$x - 2 > 5$

$(7, 0)$

58. $P = \dfrac{125}{6}T - \dfrac{8,050}{6}$; $533.33 (§8.3)

59. $x = 2$, $y = -1$ (§9.1)   **60.** $x = \frac{1}{2}$, $y = 0$ (§9.1)   **61.** $u = 4$, $v = -1$ (§9.1)   **62.** $s = \frac{1}{4}$, $t = -\frac{1}{2}$ (§9.1)
63. Inconsistent (§9.1)   **64.** Dependent (§9.1)   **65.** $x = 8$, $y = 15$ (§9.1)   **66.** $x = 3$, $y = -2$ (§9.1)
67. $x = 3$, $y = 2$, $z = 1$ (§9.2)   **68.** $x = 0$, $y = -2$, $z = 4$ (§9.2)   **69.** $x = 7$, $y = 0$, $z = -4$ (§9.2)
70. $x = 12$, $y = 12$, $z = 12$ (§9.2)   **71.** $\{(x, y, z) \mid x - 2y + 3z = 4\}$ (§9.2)   **72.** $x = \frac{1}{6}$, $y = \frac{1}{2}$, $z = \frac{1}{3}$ (§9.2)
73. 7 (§9.4)   **74.** 5 (§9.4)   **75.** 0 (§9.4)   **76.** -21 (§9.4)   **77.** -8 (§9.4)   **78.** 0 (§9.4)
79. -2 (§9.4)   **80.** 0 (§9.4)   **81.** (2, -1) (§9.3)   **82.** (-1, 2, 3) (§9.3)   **83.** $x = \frac{67}{55}$, $y = -\frac{13}{55}$ (§9.4)

84. $x = \frac{31}{53}$, $y = -\frac{19}{106}$ (§9.4) **85.** $x = 0$, $y = 1$, $z = 0$ (§9.4) **86.** $x = \frac{19}{32}$, $y = -\frac{79}{48}$, $z = \frac{17}{4}$ (§9.4)

87. (§9.5) **88.** (§9.5) **89.** (§9.5) **90.** (§9.5)

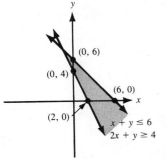
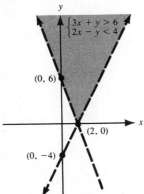
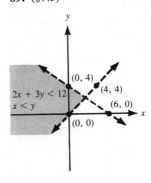
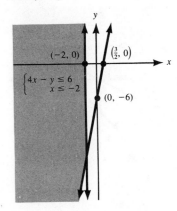

91. (§9.5) **92.** (§9.5)

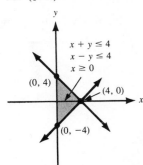
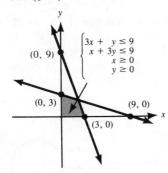

Cumulative Practice Test: Chapters 7–9

1. (a) $x = -\frac{3}{2}, 1$ **(b)** $x = \pm 2\sqrt{3}$ **(c)** $x = \dfrac{-2 \pm \sqrt{10}}{2}$ **(d)** $x = \dfrac{7 \pm \sqrt{217}}{6}$ **2.** $a = \pm \dfrac{\sqrt{3(b^2 - 2)}}{3}$

3. (a) $x = 4$ **(b)** $x = 0, -3, 2$ **4.** $x \leq -\frac{1}{2}$ or $x \geq 4$;

5. Let r = rate of stream; $\dfrac{4}{7 - r} + \dfrac{4}{7 + r} = \dfrac{7}{5}$; 3 mph

6.   **7.** $m = -1$ **8. (a)** $y = \frac{2}{3}x - \frac{13}{3}$ **(b)** $y = -\frac{3}{2}x$ **9.**

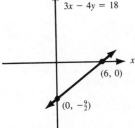
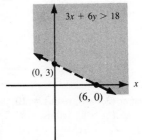

10. $B = 28$ **11.** (a) $x = 2, y = -3$ (b) $x = 5, y = -1, z = 2$

12.

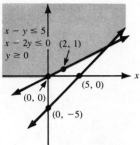

13. $(0, 2, -2)$ **14.** (a). $x = -2, y = 5$ (b) $x = 3, y = -1, z = 0$

Exercises 10.1

1. 5   **3.** $3\sqrt{2}$ **5.** $6\sqrt{13}$ **7.** 1   **9.** $\dfrac{\sqrt{145}}{6}$ **11.** 0.5   **13.** $(0, 6)$ **15.** $(0, 1)$ **17.** $(0, 0)$

19. $\left(\dfrac{11}{30}, \dfrac{11}{8}\right)$ **21.** Yes   **23.** No   **25.** Yes; both diagonals have midpoint $(7, 5)$.   **27.** No   **29.** $x^2 + y^2 = 1$

31. $(x - 1)^2 + y^2 = 49$ **33.** $(x - 2)^2 + (y - 5)^2 = 36$ **35.** $(x - 6)^2 + (y + 2)^2 = 25$

37. $(x + 3)^2 + (y + 2)^2 = 1$ **39.** Center $(0, 0)$; radius $= 4$ **41.** Center $(0, 0)$; radius $= 2\sqrt{6}$

43. Center $(3, 0)$; radius $= 4$ **45.** Center $(2, 1)$; radius $= 1$ **47.** Center $(-1, 3)$; radius $= 5$

49.   **51.**   **53.**

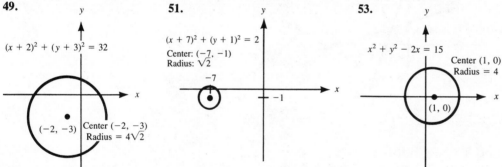

55.   **57.**

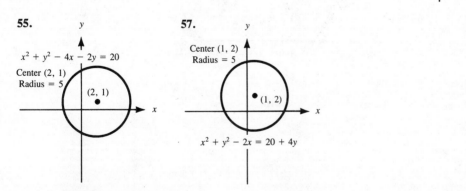

59. Center $(2, -5)$; radius $= 10$ **61.** Center $(-3, 1)$; radius $= 2\sqrt{2}$ **63.** Center $(1, -1)$; radius $= \sqrt{13}$

65. Center $\left(\tfrac{1}{2}, -1\right)$; radius $= 4$

67.

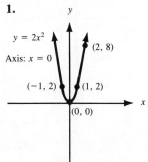

Circle
Center (0, 0)
Radius = 4

$x^2 + y^2 = 16$

(0, 0)

69.

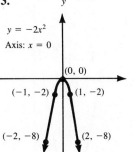

$x + y = 4$
Straight line

(0, 4)

(4, 0)

74. $\dfrac{x^2 + x - 2}{3x + 15}$ **75.** $\dfrac{x^9}{27y^{18}}$ **76.** $\sqrt{5} + 2$ **77.** $m = -\dfrac{2}{5}$ **78.** $\sqrt{65}$

Exercises 10.2

1.

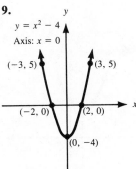

$y = 2x^2$

Axis: $x = 0$

(2, 8)

(−1, 2) (1, 2)

(0, 0)

3.

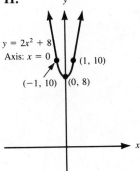

$y = -2x^2$

Axis: $x = 0$

(0, 0)

(−1, −2) (1, −2)

(−2, −8) (2, −8)

5.

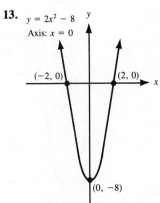

$y = \frac{1}{2}x^2$

Axis: $x = 0$

(−4, 8) (4, 8)

(−2, 2) (2, 2)

(0, 0)

7.

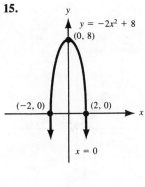

$y = -5x^2$

$x = 0$

(0, 0)

−1 1

−5

9.

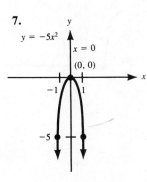

$y = x^2 - 4$

Axis: $x = 0$

(−3, 5) (3, 5)

(−2, 0) (2, 0)

(0, −4)

11.

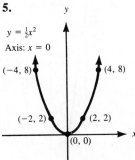

$y = 2x^2 + 8$
Axis: $x = 0$

(1, 10)

(−1, 10) (0, 8)

13.

$y = 2x^2 - 8$
Axis: $x = 0$

(−2, 0) (2, 0)

(0, −8)

15.

$y = -2x^2 + 8$

(0, 8)

(−2, 0) (2, 0)

$x = 0$

17.

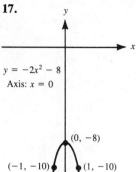

$y = -2x^2 - 8$
Axis: $x = 0$
$(0, -8)$
$(-1, -10)$ $(1, -10)$
$(-2, -16)$ $(2, -16)$

19.

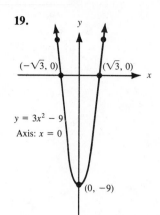

$(-\sqrt{3}, 0)$ $(\sqrt{3}, 0)$
$y = 3x^2 - 9$
Axis: $x = 0$
$(0, -9)$

21.

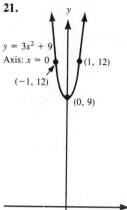

$y = 3x^2 + 9$
Axis: $x = 0$ $(1, 12)$
$(-1, 12)$
$(0, 9)$

23.

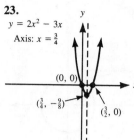

$y = 2x^2 - 3x$
Axis: $x = \frac{3}{4}$
$(0, 0)$
$(\frac{3}{4}, -\frac{9}{8})$ $(\frac{3}{2}, 0)$

25.

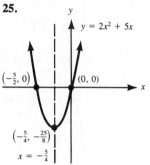

$y = 2x^2 + 5x$
$(-\frac{5}{2}, 0)$ $(0, 0)$
$(-\frac{5}{4}, -\frac{25}{8})$
$x = -\frac{5}{4}$

27.

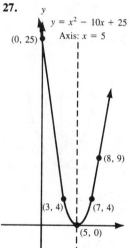

$y = x^2 - 10x + 25$
Axis: $x = 5$
$(0, 25)$
$(8, 9)$
$(3, 4)$ $(7, 4)$
$(5, 0)$

29.

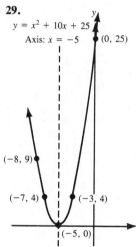

$y = x^2 + 10x + 25$
Axis: $x = -5$ $(0, 25)$
$(-8, 9)$
$(-7, 4)$ $(-3, 4)$
$(-5, 0)$

31.

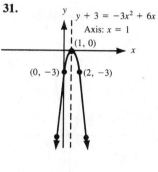

$y + 3 = -3x^2 + 6x$
Axis: $x = 1$
$(1, 0)$
$(0, -3)$ $(2, -3)$

33.

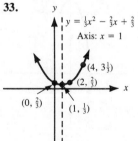

$y = \frac{1}{3}x^2 - \frac{2}{3}x + \frac{2}{3}$
Axis: $x = 1$
$(4, 3\frac{1}{3})$
$(2, \frac{2}{3})$
$(0, \frac{2}{3})$ $(1, \frac{1}{3})$

35.

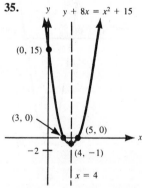

$y + 8x = x^2 + 15$
$(0, 15)$
$(3, 0)$ $(5, 0)$
-2 $(4, -1)$
$x = 4$

37.

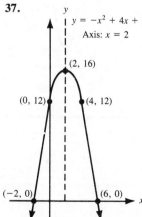

$y = -x^2 + 4x + 12$
Axis: $x = 2$
$(2, 16)$
$(0, 12)$ $(4, 12)$
$(-2, 0)$ $(6, 0)$

39.

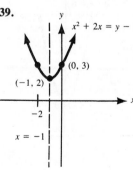

$x^2 + 2x = y - 3$
$(0, 3)$
$(-1, 2)$
-2
$x = -1$

41.

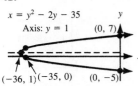
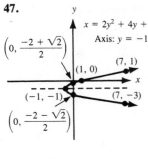

$x = y^2 - 2y - 35$
Axis: $y = 1$

43.

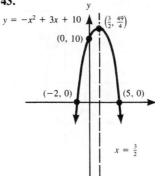

$y = -x^2 + 3x + 10$

45.

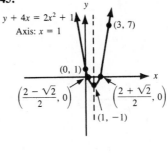

$y + 4x = 2x^2 + 1$
Axis: $x = 1$

47.

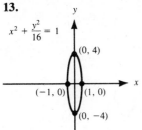

$x = 2y^2 + 4y + 1$
Axis: $y = -1$

49.

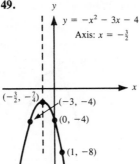

$y = -x^2 - 3x - 4$
Axis: $x = -\frac{3}{2}$

51.

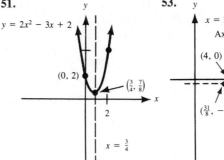

$y = 2x^2 - 3x + 2$

53.

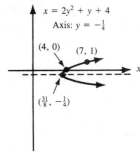

$x = 2y^2 + y + 4$
Axis: $y = -\frac{1}{4}$

55. 35 widgets; $P = \$1,225$ **57.** Number of cases = 56; maximum $P = \$2,601$
59. $t = \frac{25}{2}$ seconds; maximum height = 2,500 ft   **61.** Dimensions: 12.5′ by 25′

Exercises 10.3

1. $(0, 3), (0, -3), (-4, 0), (4, 0)$ **3.** $(0, 4), (0, -4), (-3, 0), (3, 0)$ **5.** $(0, 6), (0, -6), (-5, 0), (5, 0)$
7. $(0, 3), (0, -3), (-2\sqrt{3}, 0), (2\sqrt{3}, 0)$ **9.** $(0, 2\sqrt{5}), (0, -2\sqrt{5}), (-2\sqrt{6}, 0), (2\sqrt{6}, 0)$
11. $(0, 4), (0, -4), (-2\sqrt{3}, 0), (2\sqrt{3}, 0)$

13.

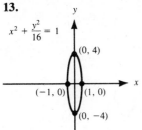

$x^2 + \dfrac{y^2}{16} = 1$

15.

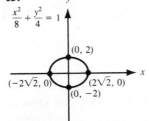

$\dfrac{x^2}{8} + \dfrac{y^2}{4} = 1$

17.

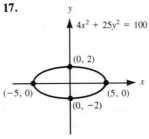

$4x^2 + 25y^2 = 100$

19.

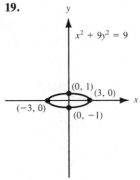

$x^2 + 9y^2 = 9$

21.

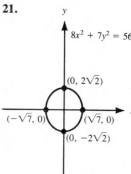

$8x^2 + 7y^2 = 56$
$(0, 2\sqrt{2})$
$(-\sqrt{7}, 0)$ $(\sqrt{7}, 0)$
$(0, -2\sqrt{2})$

23.

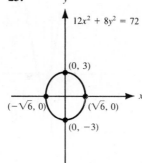

$12x^2 + 8y^2 = 72$
$(0, 3)$
$(-\sqrt{6}, 0)$ $(\sqrt{6}, 0)$
$(0, -3)$

25.

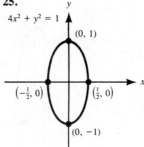

$4x^2 + y^2 = 1$
$(0, 1)$
$(-\frac{1}{2}, 0)$ $(\frac{1}{2}, 0)$
$(0, -1)$

27.

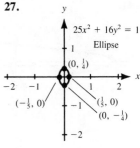

$25x^2 + 16y^2 = 1$
Ellipse
$(0, \frac{1}{4})$
$(-\frac{1}{5}, 0)$ $(\frac{1}{5}, 0)$
$(0, -\frac{1}{4})$

29. Ellipse with vertices $(0, 4)$, $(0, -4)$, $(-2, 0)$, $(2, 0)$ **31.** Straight line with x-intercept 4 and y-intercept 16
33. Circle with center $(0, 0)$ and radius = 4

35.

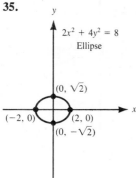

$2x^2 + 4y^2 = 8$
Ellipse
$(0, \sqrt{2})$
$(-2, 0)$ $(2, 0)$
$(0, -\sqrt{2})$

37.

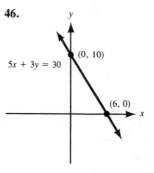

$2x + 4y = 8$
Straight line
$(0, 2)$
$(4, 0)$

39.

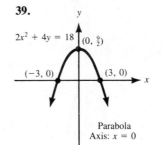

$2x^2 + 4y = 18$
$(0, \frac{9}{2})$
$(-3, 0)$ $(3, 0)$
Parabola
Axis: $x = 0$

44. $x = 63$ **45.** $x = 3$ **46.**

$5x + 3y = 30$
$(0, 10)$
$(6, 0)$

47. Let x = amount invested at 8% and y = amount invested at 10%: $0.08x + 0.10y = 730$, $0.10x + 0.08y = 710$: \$8,000 total was invested.

Exercises 10.4

1. Vertices $(-3, 0)$, $(3, 0)$; asymptotes $y = \pm\frac{4}{3}x$ **3.** Vertices $(-4, 0)$, $(4, 0)$; asymptotes $y = \pm\frac{3}{4}x$
5. Vertices $(0, 3)$, $(0, -3)$; asymptotes $y = \pm\frac{3}{4}x$ **7.** Vertices $(-1, 0)$, $(1, 0)$; asymptotes $y = \pm4x$
9. Vertices $(-2\sqrt{3}, 0)$, $(2\sqrt{3}, 0)$; asymptotes $y = \pm\frac{\sqrt{3}}{3}x$

11.

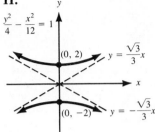

$$\frac{y^2}{4} - \frac{x^2}{12} = 1$$

$(0, 2)$ $y = \frac{\sqrt{3}}{3}x$

$(0, -2)$ $y = -\frac{\sqrt{3}}{3}x$

13.

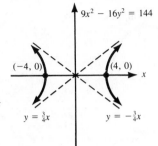

$9x^2 - 16y^2 = 144$

$(-4, 0)$ $(4, 0)$

$y = \frac{3}{4}x$ $y = -\frac{3}{4}x$

15.

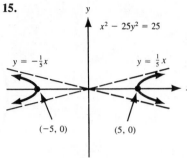

$x^2 - 25y^2 = 25$

$y = -\frac{1}{5}x$ $y = \frac{1}{5}x$

$(-5, 0)$ $(5, 0)$

17.

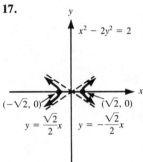

$x^2 - 2y^2 = 2$

$(-\sqrt{2}, 0)$ $(\sqrt{2}, 0)$

$y = \frac{\sqrt{2}}{2}x$ $y = -\frac{\sqrt{2}}{2}x$

19.

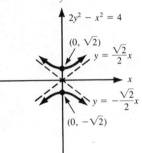

$2y^2 - x^2 = 4$

$(0, \sqrt{2})$ $y = \frac{\sqrt{2}}{2}x$

$y = -\frac{\sqrt{2}}{2}x$

$(0, -\sqrt{2})$

21.

$12y^2 - 5x^2 = 60$

$(0, \sqrt{5})$ $y = \frac{\sqrt{15}}{6}x$

$y = -\frac{\sqrt{15}}{6}x$

$(0, -\sqrt{5})$

23.

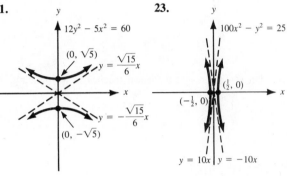

$100x^2 - y^2 = 25$

$(\frac{1}{2}, 0)$

$(-\frac{1}{2}, 0)$

$y = 10x$ $y = -10x$

25.

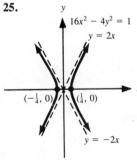

$16x^2 - 4y^2 = 1$

$y = 2x$

$(-\frac{1}{4}, 0)$ $(\frac{1}{4}, 0)$

$y = -2x$

27. Hyperbola; vertices $(-5, 0)$, $(5, 0)$; asymptotes $y = \pm 2x$ **29.** Ellipse; vertices $(0, 5)$, $(0, -5)$, $(-10, 0)$, $(10, 0)$
31. Ellipse; vertices $(0, 4)$, $(0, -4)$, $(-3, 0)$, $(3, 0)$ **33.** Hyperbola; vertices $(0, \sqrt{2})$, $(0, -\sqrt{2})$; asymptotes $y = \pm x$
35. Straight line; y-intercept 4; x-intercept 4

37.

Parabola
Axis: $x = 0$

$x^2 + y = 4$

$(0, 4)$

$(-2, 0)$ $(2, 0)$

39.

Parabola
Axis: $y = 0$

$x + y^2 = 4$

$(0, 2)$

$(4, 0)$

$(0, -2)$

41.

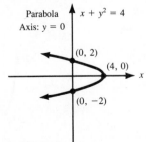

$4x^2 + y^2 = 16$

Ellipse

$(0, 4)$

$(-2, 0)$ $(2, 0)$

$(0, -4)$

43.

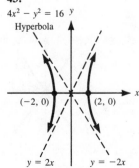

$4x^2 - y^2 = 16$

Hyperbola

$(-2, 0)$ $(2, 0)$

$y = 2x$ $y = -2x$

45.

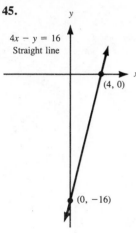

$4x - y = 16$
Straight line
$(4, 0)$
$(0, -16)$

51. $y = \frac{5}{2}, -\frac{4}{3}$

52.

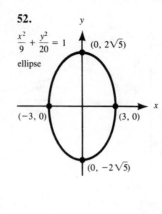

$\frac{x^2}{9} + \frac{y^2}{20} = 1$
ellipse
$(0, 2\sqrt{5})$
$(-3, 0)$
$(3, 0)$
$(0, -2\sqrt{5})$

53. Let $x =$ the height the ladder reaches on the building; $40^2 - 6^2 = x^2$: $x = 2\sqrt{391} \approx 39.55'$ **54.** $-2 < x \leq 3$

Exercises 10.5

1. Circle   **3.** Ellipse   **5.** Parabola   **7.** Straight line   **9.** Hyperbola   **11.** Hyperbola   **13.** Circle
15. Circle   **17.** Hyperbola   **19.** Parabola

21.

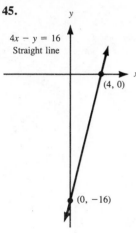

$x^2 + y^2 = 16$
Circle: Center $(0, 0)$,
Radius $= 4$

23.

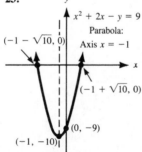

$x^2 + 2x - y = 9$
Parabola:
Axis $x = -1$
$(-1 - \sqrt{10}, 0)$
$(-1 + \sqrt{10}, 0)$
$(0, -9)$
$(-1, -10)$

25.

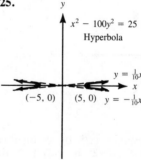

$x^2 - 100y^2 = 25$
Hyperbola
$y = \frac{1}{10}x$
$(-5, 0)$
$(5, 0)$ $y = -\frac{1}{10}x$

27.

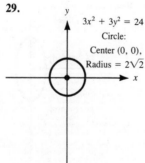

Degenerate hyperbola;
two intersecting
straight lines
$16x^2 - 9y^2 = 0$
$y = \frac{4}{3}x$
$(3, 4)$
$(-3, 4)$ $(0, 0)$
$y = -\frac{4}{3}x$

29.

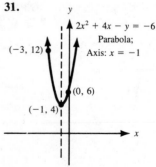

$3x^2 + 3y^2 = 24$
Circle:
Center $(0, 0)$,
Radius $= 2\sqrt{2}$

31.

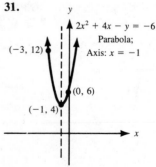

$2x^2 + 4x - y = -6$
Parabola;
Axis: $x = -1$
$(-3, 12)$
$(0, 6)$
$(-1, 4)$

33.

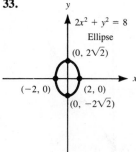

$2x^2 + y^2 = 8$
Ellipse
$(0, 2\sqrt{2})$
$(-2, 0)$ $(2, 0)$
$(0, -2\sqrt{2})$

35.

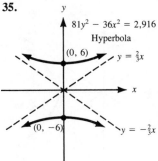

$81y^2 - 36x^2 = 2{,}916$
Hyperbola
$(0, 6)$ $y = \frac{2}{3}x$
$(0, -6)$ $y = -\frac{2}{3}x$

Exercises 10.6

1. $(3, 1)$, $(3, -1)$, $(-3, 1)$, $(-3, -1)$ **3.** $(5, 0)$, $(0, -5)$ **5.** $(\sqrt{5}, 2)$, $(\sqrt{5}, -2)$, $(-\sqrt{5}, 2)$, $(-\sqrt{5}, -2)$

7. $(0, -3)$, $\left(\frac{12}{5}, \frac{9}{5}\right)$ **9.** $(2, 3)$, $(2, -3)$, $(-2, 3)$, $(-2, -3)$ **11.** $\left(\dfrac{1 + \sqrt{13}}{2}, \dfrac{7 + \sqrt{13}}{2}\right)$, $\left(\dfrac{1 - \sqrt{13}}{2}, \dfrac{7 - \sqrt{13}}{2}\right)$

13. No real solutions   **15.** $(3, -9)$, $(4, -8)$ **17.** $(0, 4)$, $(0, -4)$, $(-1, \sqrt{15})$, $(-1, -\sqrt{15})$

19. $(3, 4)$, $(4, 3)$ **21.** $\left(\dfrac{2\sqrt{15}}{3}, \dfrac{2\sqrt{6}}{3}\right)$, $\left(\dfrac{2\sqrt{15}}{3}, \dfrac{-2\sqrt{6}}{3}\right)$, $\left(\dfrac{-2\sqrt{15}}{3}, \dfrac{2\sqrt{6}}{3}\right)$, $\left(\dfrac{-2\sqrt{15}}{3}, \dfrac{-2\sqrt{6}}{3}\right)$

23. $(1, 0)$, $(0, 3)$ **25.** $\left(-\frac{1}{2}, \frac{35}{4}\right)$, $(2, 5)$ **27.** $(13, -4)$, $(6, 3)$ **29.** $(2, 1)$, $\left(-\frac{6}{5}, \frac{37}{5}\right)$

31. $(2\sqrt{2}, \sqrt{2})$, $(-2\sqrt{2}, -\sqrt{2})$, $(\sqrt{2}, 2\sqrt{2})$, $(-\sqrt{2}, -2\sqrt{2})$ **33.** $(\sqrt{6}, \sqrt{2})$, $(-\sqrt{6}, -\sqrt{2})$

35. $(4, 3)$, $(-4, -3)$, $(3, 4)$, $(-3, -4)$

37. $(5, 2)$, $(-2, -5)$ **39.** $(4, \sqrt{13})$, $(4, -\sqrt{13})$, $(-4, \sqrt{13})$, $(-4, -\sqrt{13})$ **41.** No solution (graphs never intersect)

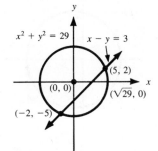

$x^2 + y^2 = 29$ $x - y = 3$
$(5, 2)$
$(0, 0)$ $(\sqrt{29}, 0)$
$(-2, -5)$

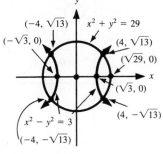

$(-4, \sqrt{13})$ $x^2 + y^2 = 29$
$(-\sqrt{3}, 0)$ $(4, \sqrt{13})$
$(\sqrt{29}, 0)$
$(\sqrt{3}, 0)$
$x^2 - y^2 = 3$ $(4, -\sqrt{13})$
$(-4, -\sqrt{13})$

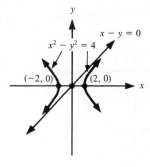

$x - y = 0$
$x^2 - y^2 = 4$
$(-2, 0)$ $(2, 0)$

43. $(2, 0)$, $(-2, 0)$ **45.** $(-5, 0)$, $(5, 0)$

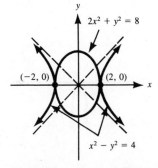

$2x^2 + y^2 = 8$
$(-2, 0)$ $(2, 0)$
$x^2 - y^2 = 4$

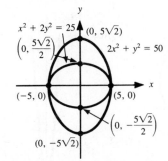

$x^2 + 2y^2 = 25$ $(0, 5\sqrt{2})$
$\left(0, \dfrac{5\sqrt{2}}{2}\right)$ $2x^2 + y^2 = 50$
$(-5, 0)$ $(5, 0)$
$\left(0, -\dfrac{5\sqrt{2}}{2}\right)$
$(0, -5\sqrt{2})$

47. Let L = length and W = width; $2L + 2W = 59$, $LW = 210$; 17.5 in. by 12 in.

49. Let r = rate going and t = time going; $60 = rt$, $60 = (r + 12)\left(t - \frac{1}{4}\right)$; rate going = 48 mph and time = 1.25 hr, rate returning = 60 mph and time = 1 hr

52. $\dfrac{x-3}{(x+3)(x-1)}$ **53.** No solution   **54.**

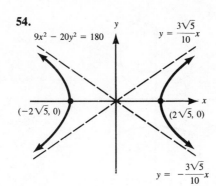

55.

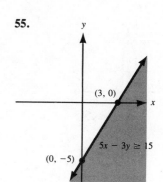

Chapter 10 Review Exercises

1. $2\sqrt{10}$; $(1, 3)$ **2.** $2\sqrt{5}$; $(3, 10)$ **3.** 4; $(0, 5)$ **4.** 16; $(3, 0)$ **5.** $2\sqrt{2}$; $(5, -5)$ **6.** $2\sqrt{58}$; $(0, 0)$
7. $2\sqrt{29}$; $(0, 0)$ **8.** $2\sqrt{29}$; $(0, 0)$ **9.** $\sqrt{26}$; $\left(\frac{11}{2}, -\frac{7}{2}\right)$ **10.** $7\sqrt{2}$; $\left(-\frac{1}{2}, -\frac{1}{2}\right)$ **11.** Yes   **12.** No
13.

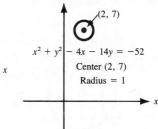

14.

15.

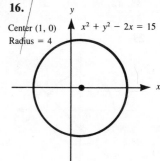

16.

17.

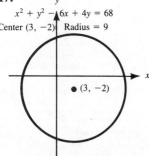

18.

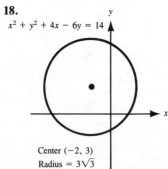

19.

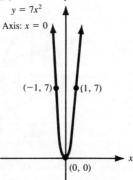

20.

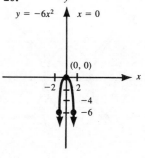

21.

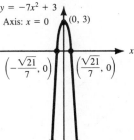

$y = -7x^2 + 3$
Axis: $x = 0$
$(0, 3)$
$\left(-\dfrac{\sqrt{21}}{7}, 0\right)$ $\left(\dfrac{\sqrt{21}}{7}, 0\right)$

22.

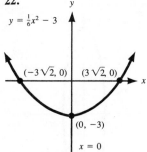

$y = \dfrac{1}{6}x^2 - 3$
$(-3\sqrt{2}, 0)$ $(3\sqrt{2}, 0)$
$(0, -3)$
$x = 0$

23.

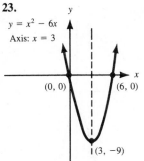

$y = x^2 - 6x$
Axis: $x = 3$
$(0, 0)$ $(6, 0)$
$(3, -9)$

24.

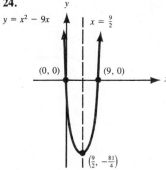

$y = x^2 - 9x$
$x = \dfrac{9}{2}$
$(0, 0)$ $(9, 0)$
$\left(\dfrac{9}{2}, -\dfrac{81}{4}\right)$

25.

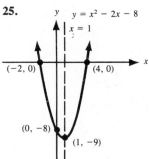

$y = x^2 - 2x - 8$
$x = 1$
$(-2, 0)$ $(4, 0)$
$(0, -8)$ $(1, -9)$

26.

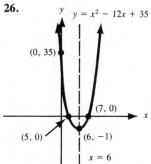

$y = x^2 - 12x + 35$
$(0, 35)$
$(7, 0)$
$(5, 0)$ $(6, -1)$
$x = 6$

27.

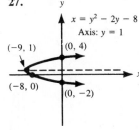

$x = y^2 - 2y - 8$
Axis: $y = 1$
$(-9, 1)$ $(0, 4)$
$(-8, 0)$
$(0, -2)$

28.

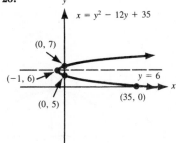

$x = y^2 - 12y + 35$
$(0, 7)$
$(-1, 6)$ $y = 6$
$(0, 5)$ $(35, 0)$

29.

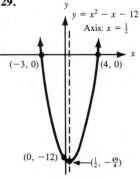

$y = x^2 - x - 12$
Axis: $x = \dfrac{1}{2}$
$(-3, 0)$ $(4, 0)$
$(0, -12)$ $\left(\dfrac{1}{2}, -\dfrac{49}{4}\right)$

30.

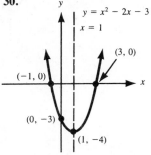

$y = x^2 - 2x - 3$
$x = 1$
$(3, 0)$
$(-1, 0)$
$(0, -3)$
$(1, -4)$

31.

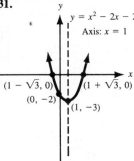

$y = x^2 - 2x - 2$
Axis: $x = 1$
$(1 - \sqrt{3}, 0)$ $(1 + \sqrt{3}, 0)$
$(0, -2)$ $(1, -3)$

32.

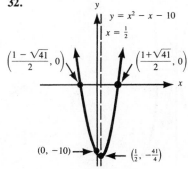

$y = x^2 - x - 10$
$x = \dfrac{1}{2}$
$\left(\dfrac{1 - \sqrt{41}}{2}, 0\right)$ $\left(\dfrac{1 + \sqrt{41}}{2}, 0\right)$
$(0, -10)$ $\left(\dfrac{1}{2}, -\dfrac{41}{4}\right)$

33.

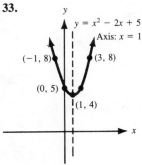

$y = x^2 - 2x + 5$
Axis: $x = 1$
$(-1, 8)$ $(3, 8)$
$(0, 5)$
$(1, 4)$

34.

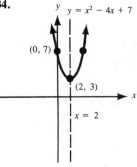

$y = x^2 - 4x + 7$
$(0, 7)$
$(2, 3)$
$x = 2$

35.

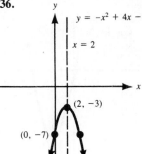

$y = -x^2 + 2x - 5$
Axis: $x = 1$
$(1, -4)$
$(0, -5)$
$(-1, -8)$ $(3, -8)$

36.

$y = -x^2 + 4x - 7$
$x = 2$
$(2, -3)$
$(0, -7)$

37.

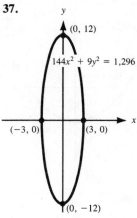

$(0, 12)$
$144x^2 + 9y^2 = 1{,}296$
$(-3, 0)$ $(3, 0)$
$(0, -12)$

38.

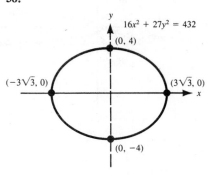

$16x^2 + 27y^2 = 432$
$(0, 4)$
$(-3\sqrt{3}, 0)$ $(3\sqrt{3}, 0)$
$(0, -4)$

39.

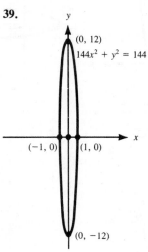

$(0, 12)$
$144x^2 + y^2 = 144$
$(-1, 0)$ $(1, 0)$
$(0, -12)$

40.

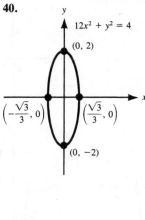

$12x^2 + y^2 = 4$
$(0, 2)$
$\left(-\dfrac{\sqrt{3}}{3}, 0\right)$ $\left(\dfrac{\sqrt{3}}{3}, 0\right)$
$(0, -2)$

41.

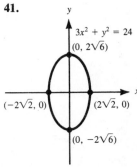

$3x^2 + y^2 = 24$
$(0, 2\sqrt{6})$
$(-2\sqrt{2}, 0)$ $(2\sqrt{2}, 0)$
$(0, -2\sqrt{6})$

42.

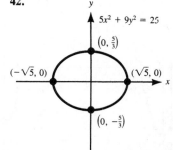

$5x^2 + 9y^2 = 25$
$\left(0, \frac{5}{3}\right)$
$(-\sqrt{5}, 0)$ $(\sqrt{5}, 0)$
$\left(0, -\frac{5}{3}\right)$

43.

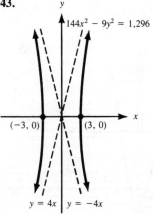

$144x^2 - 9y^2 = 1{,}296$
$(-3, 0)$ $(3, 0)$
$y = 4x$ $y = -4x$

44.

45.

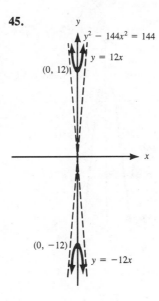

46.

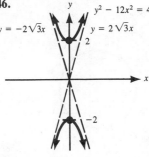

47.

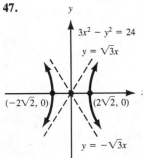

48.

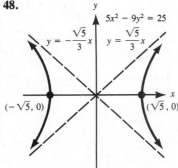

49. Ellipse; vertices $(0, 4)$, $(0, -4)$, $(-2, 0)$, $(2, 0)$ **50.** Hyperbola; vertices $(-1, 0)$, $(1, 0)$; asymptotes $y = \pm 2x$

51. Parabola; vertex $(5, 0)$; axis $x = 5$ **52.** Ellipse; vertices $(-\sqrt{6}, 0)$, $(\sqrt{6}, 0)$, $(0, 2)$, $(0, -2)$

53. Circle; center $(0, 0)$; radius $6\sqrt{3}$ **54.** Hyperbola; vertices $(-4\sqrt{2}, 0)$, $(4\sqrt{2}, 0)$; asymptotes $y = \pm \dfrac{5\sqrt{2}}{8}x$

55. Straight line; x-intercept 9; y-intercept 9 **56.** Straight line; x-intercept 32; y-intercept -25

57. Ellipse; vertices $(-4\sqrt{2}, 0)$, $(4\sqrt{2}, 0)$, $(0, 5)$, $(0, -5)$ **58.** Parabola; vertex $(-1, -26)$; axis $x = -1$

59. Hyperbola; vertices $(-\sqrt{6}, 0)$, $(\sqrt{6}, 0)$; asymptotes $y = \pm \dfrac{\sqrt{6}}{3}x$

60. Hyperbola; vertices $(-\sqrt{10}, 0)$, $(\sqrt{10}, 0)$; asymptotes $y = \pm x$

61. Parabola; vertex $(-26, -1)$; axis $y = -1$ **62.** Circle; center $(-7, 3)$; radius $2\sqrt{2}$

63. Circle; center $(7, 3)$; radius $2\sqrt{2}$ **64.** Parabola; vertex $(3, -26)$; axis $x = 3$

65. Circle; center $(-2, -2)$; radius 2 **66.** Circle; center $(-2, -2)$; radius $2\sqrt{3}$

67. Hyperbola; vertices $(-4, 0)$, $(4, 0)$; asymptotes $y = \pm x$

68. Ellipse; vertices $\left(-\dfrac{\sqrt{3}}{3}, 0\right)$, $\left(\dfrac{\sqrt{3}}{3}, 0\right)$, $\left(0, \dfrac{\sqrt{2}}{2}\right)$, $\left(0, -\dfrac{\sqrt{2}}{2}\right)$

69. Straight line; x-intercept 16; y-intercept -16 **70.** Straight line; x-intercept $\frac{1}{3}$; y-intercept $\frac{1}{2}$

71. Degenerate circle; point $(0, 0)$ **72.** Parabola; vertex $(0, 0)$; axis $x = 0$ **73.** Impossible

74.

$x^2 + y^2 + 2x - 6y = 14$
Circle

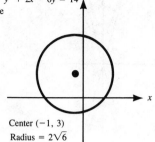

Center $(-1, 3)$
Radius $= 2\sqrt{6}$

75.

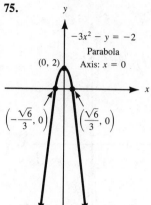

$-3x^2 - y = -2$
Parabola
Axis: $x = 0$

$(0, 2)$

$\left(-\dfrac{\sqrt{6}}{3}, 0\right)$ $\left(\dfrac{\sqrt{6}}{3}, 0\right)$

76.

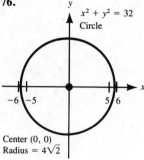

$x^2 + y^2 = 32$
Circle

-6 -5 5 6

Center $(0, 0)$
Radius $= 4\sqrt{2}$

77.

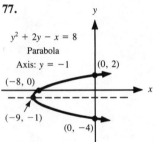

$y^2 + 2y - x = 8$
Parabola
Axis: $y = -1$ $(0, 2)$
$(-8, 0)$

$(-9, -1)$ $(0, -4)$

78.

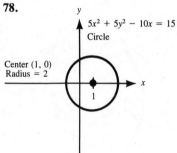

$5x^2 + 5y^2 - 10x = 15$
Circle

Center $(1, 0)$
Radius $= 2$

1

79.

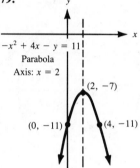

$-x^2 + 4x - y = 11$
Parabola
Axis: $x = 2$

$(2, -7)$

$(0, -11)$ $(4, -11)$

80.

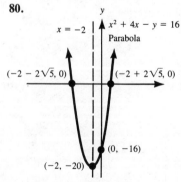

$x = -2$ $x^2 + 4x - y = 16$
Parabola

$(-2 - 2\sqrt{5}, 0)$ $(-2 + 2\sqrt{5}, 0)$

$(0, -16)$

$(-2, -20)$

81.

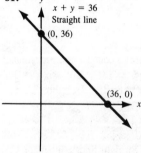

$x + y = 36$
Straight line
$(0, 36)$

$(36, 0)$

82.

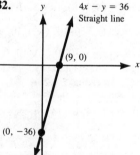

$4x - y = 36$
Straight line

$(9, 0)$

$(0, -36)$

83.

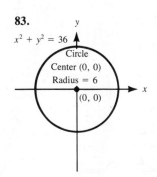

$x^2 + y^2 = 36$

Circle
Center $(0, 0)$
Radius = 6
$(0, 0)$

84.

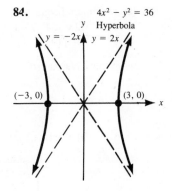

$4x^2 - y^2 = 36$
Hyperbola
$y = -2x$ $y = 2x$
$(-3, 0)$ $(3, 0)$

85.

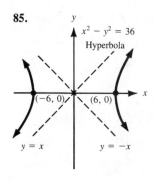

$x^2 - y^2 = 36$
Hyperbola
$(-6, 0)$ $(6, 0)$
$y = x$ $y = -x$

86.

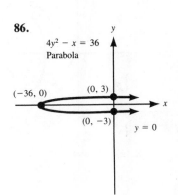

$4y^2 - x = 36$
Parabola
$(-36, 0)$ $(0, 3)$
$(0, -3)$
$y = 0$

87.

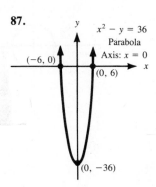

$x^2 - y = 36$
Parabola
Axis: $x = 0$
$(-6, 0)$ $(0, 6)$
$(0, -36)$

88.

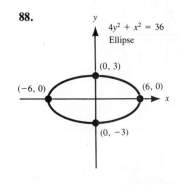

$4y^2 + x^2 = 36$
Ellipse
$(0, 3)$
$(-6, 0)$ $(6, 0)$
$(0, -3)$

89.

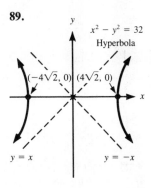

$x^2 - y^2 = 32$
Hyperbola
$(-4\sqrt{2}, 0)$ $(4\sqrt{2}, 0)$
$y = x$ $y = -x$

90.

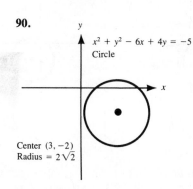

$x^2 + y^2 - 6x + 4y = -5$
Circle
Center $(3, -2)$
Radius = $2\sqrt{2}$

91.

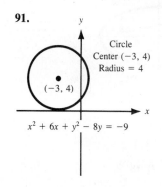

Circle
Center $(-3, 4)$
Radius = 4
$(-3, 4)$
$x^2 + 6x + y^2 - 8y = -9$

92.

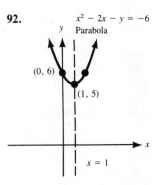

$x^2 - 2x - y = -6$
Parabola
$(0, 6)$
$(1, 5)$
$x = 1$

93. 500 widgets; \$1,000　　**94.** Ticket price = \$6　　**95. (a)** Maximum height = $\frac{65}{8}$ ft　　**(b)** 5 feet

96. $x = 2, y = 2$　　**97.** $\left(\frac{2\sqrt{154}}{11}, \frac{2\sqrt{11}}{11}\right), \left(\frac{2\sqrt{154}}{11}, -\frac{2\sqrt{11}}{11}\right), \left(-\frac{2\sqrt{154}}{11}, \frac{2\sqrt{11}}{11}\right), \left(-\frac{2\sqrt{154}}{11}, -\frac{2\sqrt{11}}{11}\right)$

98. $(5, 6), \left(\frac{-19}{7}, \frac{6}{7}\right)$　　**99.** $(0, 3), (\sqrt{5}, -2), (-\sqrt{5}, -2)$　　**100.** No real solutions　　**101.** No real solutions

Chapter 10 Practice Test

1. $5\sqrt{2}$　　**2.** $\left(4, \frac{11}{2}\right)$　　**3. (a)** Center $(3, -2)$; radius $= 2\sqrt{3}$　　**(b)** Center $(2, 3)$; radius $= 4$

4. (a)　**(b)**　**5.**　**6.**

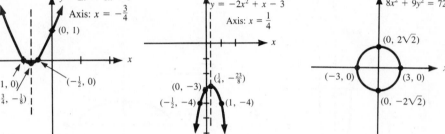

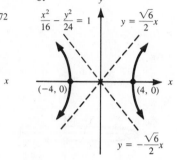

7. (a)　**(b)**　**(c)**　　**8.** $x = 2, y = -1$

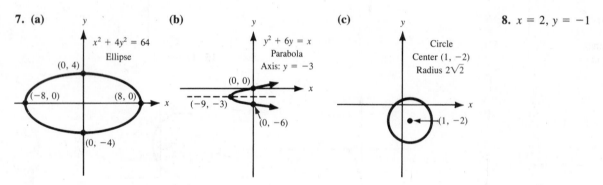

Exercises 11.1

1. $\{(3, 9), (3, 7), (8, 2), (7, 2)\}$　　**3.** $\{(3, a), (3, b), (8, b), (-1, b), (-1, c)\}$　　**5.** Domain: $\{3, 4, 5\}$; range: $\{2, 3\}$

7. Domain: $\{3, -2, 4\}$; range: $\{-2, 3, -1\}$　　**9.** $\{x \mid x \neq 0\}$　　**11.** All reals　　**13.** $\{x \mid x \neq -\frac{3}{2}\}$　　**15.** All reals

17. $\{x \mid x \geq 4\}$　　**19.** All reals　　**21.** $\{x \mid x \leq \frac{5}{4}\}$　　**23.** $\{x \mid x \geq 0\}$　　**25.** $\{x \mid x > 3\}$

27. Domain: $\{x \mid -3 \leq x \leq 3\}$; range: $\{y \mid -3 \leq y \leq 3\}$　　**29.** Domain: all reals; range: $\{y \mid y \geq -4\}$

31. Domain: $\{x \mid x \geq -4\}$; range: all reals　　**33.** Domain: all reals; range: all reals

35. Domain $\{x \mid -3 \leq x \leq 3\}$; range: $\{y \mid -4 \leq y \leq 4\}$　　**37.** Domain: $\{x \mid x \leq -2$ or $x \geq 2\}$; range: all reals

39. Function　　**41.** Not a function　　**43.** Function　　**45.** Not a function　　**47.** Function　　**49.** Function

51. Not a function　　**53.** Function　　**55.** Not a function　　**57.** Not a function　　**59.** Function

61. Not a function　　**63.** Function　　**65.** Not a function

67. (a) Yes　　**(b)** Domain: $\{r \mid \$200 \leq r \leq \$400\}$; range: $\{P \mid \$35,000 \leq P \leq \$50,000\}$　　**(c)** \$300　　**(d)** \$50,000

Exercises 11.2

1. -3　　**3.** 11　　**5.** 15　　**7.** $2\sqrt{2}$　　**9.** $\sqrt{2}$　　**11.** $\sqrt{a + 5}$　　**13.** $x^2 + 2x + 3$　　**15.** $2x - 7$　　**17.** $2x + 1$

19. $4x^2 + 2$　　**21.** $6x + 1$　　**23.** $x^2 + 4x$　　**25.** $x^2 + 2x - 2$　　**27.** $x^2 + 2x - 3$

29. $9x^2 + 6x - 3$　　**31.** $3x^2 + 6x - 9$　　**33.** $\sqrt{9x + 5}$　　**35.** $3\sqrt{3x + 2} + 1$　　**37.** $\sqrt{3x + 3h + 2}$

39. $\sqrt{3x + 2} + \sqrt{3h + 2}$ **41.** $\sqrt{3x + 2} + h$ **43.** $\sqrt{3ax + 2}$ **45.** $a\sqrt{3x + 2}$ **47.** $2x + 5$

49. $2x + h + 3$ **51.** 0 **53.** $\sqrt{6}$ **55.** $x - 3$ **57.** $\sqrt{x^2 - 3}$ **59.** $\dfrac{2\sqrt{3}}{3}$ **61.** $-\dfrac{5}{2}$ **63.** $\dfrac{\sqrt{x}(1 + x)}{x}$

65. -4 **67.** 3 **69.** 0 **71.** Undefined $\left(3 \text{ is not in the domain of } \dfrac{g}{f}\right)$ **73.** $\dfrac{(x - 3)(x + 1)}{(x - 1)}, x \neq 1$

76. $6 + 4\sqrt{3}$ **77.** $y = \dfrac{1}{2}x + 4$ **78.** **79.** $x = 0, y = 1, z = -1$

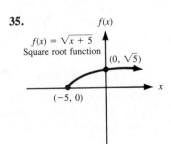

$y = x^2 - 4x - 5$
Parabola
Axis: $x = 2$

$(-1, 0)$ $(5, 0)$

$(0, -5)$

$(2, -9)$

80. Let $x = $ # of five-dollar bills; $5x + 20(50 - x) = 430$; 38 five-dollar bills and 12 twenty-dollar bills.

Exercises 11.3

1. Linear function **3.** Quadratic function **5.** Polynomial function **7.** Quadratic function
9. Square root function **11.** Absolute value function **13.** Linear function **15.** Quadratic function
17. $9, 5, 1, 3, 7$ **19.** $7, 3, -1, 3, 7$
21. $9, 5, 1, 5, 9$ **23.** $7, 3, 1, 5, 9$
25. **27.** **29.** **31.**

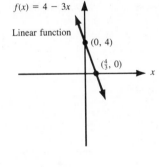

$f(x)$

$f(x) = 4 - 3x$

Linear function

$(0, 4)$

$\left(\frac{4}{3}, 0\right)$

x

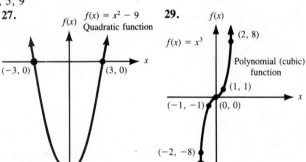

$f(x)$ $f(x) = x^2 - 9$
Quadratic function

$(-3, 0)$ $(3, 0)$ x

$(0, -9)$

$f(x)$

$f(x) = x^3$

Polynomial (cubic) function

$(2, 8)$

$(1, 1)$

$(-1, -1)$ $(0, 0)$ x

$(-2, -8)$

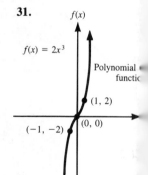

$f(x)$

$f(x) = 2x^3$

Polynomial function

$(1, 2)$

$(-1, -2)$ $(0, 0)$

33. **35.** **37.**

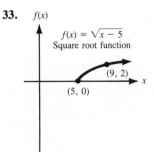

$f(x)$

$f(x) = \sqrt{x - 5}$
Square root function

$(9, 2)$

$(5, 0)$ x

$f(x)$

$f(x) = \sqrt{x + 5}$
Square root function

$(0, \sqrt{5})$

$(-5, 0)$ x

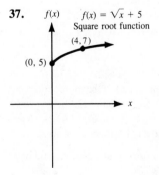

$f(x)$ $f(x) = \sqrt{x} + 5$
Square root function

$(4, 7)$

$(0, 5)$

x

$f(x) = \sqrt{x} - 5$
uare root function

ntercept is 25)

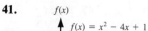

−3)

41.

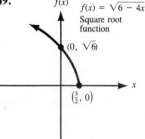

$f(x)$

$f(x) = x^2 - 4x + 1$
Quadratic function

$(0, 1)$
$(2 - \sqrt{3}, 0)$
$(2 + \sqrt{3}, 0)$
x
$(2, -3)$

43.

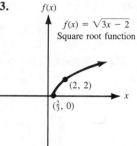

$f(x)$

$f(x) = \sqrt{3x - 2}$
Square root function

$(2, 2)$
$(\frac{2}{3}, 0)$
x

45.

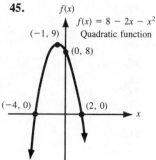

$f(x)$

$f(x) = 8 - 2x - x^2$
Quadratic function

$(-1, 9)$
$(0, 8)$
$(-4, 0)$
$(2, 0)$
x

$(x) = \sqrt{8 - 2x}$
iare root function

$\sqrt{2})$

x
4, 0)

49.

$f(x)$

$f(x) = \sqrt{6 - 4x}$
Square root function

$(0, \sqrt{6})$
$(\frac{3}{2}, 0)$
x

51.

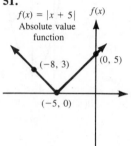

$f(x) = |x + 5|$
Absolute value function

$f(x)$

$(-8, 3)$
$(0, 5)$
x
$(-5, 0)$

53.

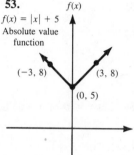

$f(x)$

$f(x) = |x| + 5$
Absolute value function

$(-3, 8)$
$(3, 8)$
$(0, 5)$
x

$f(x) = x + 5$
Linear function

, 5)

ubic)

x

57.

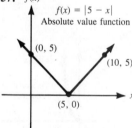

$f(x)$

$f(x) = |5 - x|$
Absolute value function

$(0, 5)$
$(10, 5)$
$(5, 0)$
x

63. $x = \frac{1}{2}, 5$ **64.** $\left(\frac{1}{2}, \frac{9}{2}\right)$ **65.** $(x - 3)^2 + (y + 4)^2 = 25$ **66.** 58

4

3. $\{(2, 3), (-1, 3), (3, -1)\}$ **5.** $\{(5, 3), (-5, 2), (5, 2), (6, 2)\}$ **7.** $R^{-1}: y = \dfrac{x + 4}{3}$

11. $R^{-1}: y = \dfrac{\sqrt[3]{4x}}{2}$ **13.** $R^{-1}: y = \pm\sqrt{x - 4}$ **15.** Function **17.** One-to-one function

Function **23.** One-to-one function **25.** One-to-one function **27.** Function **29.** Neither

ction **33.** Domain: $\{2, 3\}$; inverse: $\{(-2, 3), (-3, 2)\}$; domain of inverse: $\{-2, -3\}$

2, 6}; inverse: $\{(-3, 6), (-4, 2), (6, -3)\}$; domain of inverse: $\{-4, -3, 6\}$

4}; inverse not a function

; domain of $f(x)$: all reals; domain of $f^{-1}(x)$: all reals

41. $g^{-1}(x) = \dfrac{x + 3}{2}$; domain of $g(x)$; all reals; domain of $g^{-1}(x)$: all reals

43. $h^{-1}(x) = \dfrac{4 - x}{5}$; domain of $h(x)$: all reals; domain of $h^{-1}(x)$: all reals

45. Domain of $f(x)$: all reals; inverse not a function

47. $f^{-1}(x) = \sqrt[3]{x - 4}$; domain of $f(x)$: all reals; domain of $f^{-1}(x)$: all reals

49. $g^{-1}(x) = \dfrac{1}{x}$; domain of $g(x)$: $\{x \mid x \neq 0\}$; domain of $g^{-1}(x)$: $\{x \mid x \neq 0\}$

51. $g^{-1}(x) = \dfrac{2 - 3x}{x}$; domain of $g(x)$: $\{x \mid x \neq -3\}$; domain of $g^{-1}(x)$: $\{x \mid x \neq 0\}$

53. $g^{-1}(x) = \dfrac{1}{1 - x}$; domain of $g(x)$; $\{x \mid x \neq 0\}$; domain of $g^{-1}(x)$: $\{x \mid x \neq 1\}$

55. $h^{-1}(x) = \dfrac{x + 2}{x - 1}$; domain of $h(x)$: $\{x \mid x \neq 1\}$; domain of $h^{-1}(x)$: $\{x \mid x \neq 1\}$

57.　　　　　　　　　　　　**59.** Graph is its own inverse (same graph).　　**63.** $19 - 25i$　　**64.** $-5 < x < 2$

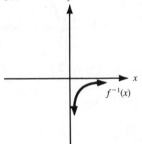

65. $-1 < x < 5$　　**66.** Hyperbola　　**67.**

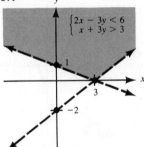

$\begin{cases} 2x - 3y < 6 \\ x + 3y > 3 \end{cases}$

Exercises 11.5

1. $y = 6$　　**3.** $y = \frac{40}{3}$　　**5.** $x = \frac{15}{22}$　　**7.** $a = 36$　　**9.** $r = \frac{243}{4}$　　**11.** $y = 10$　　**13.** $x = 28$　　**15.** $x = \frac{125}{6}$
17. $a = \frac{3}{256}$　　**19.** $a = \frac{32}{3}$　　**21.** $z = 15$　　**23.** $z = \frac{50}{3}$　　**25.** $d = \frac{5}{4}$　　**27.** $z = 40$　　**29.** $z = \frac{160}{9}$
31. $z = 160$　　**33.** $z = \dfrac{32}{3}$　　**35.** $V = \dfrac{1,000}{3}\text{m}^3$　　**37.** $V = \dfrac{256\pi}{3}\text{cm}^3$　　**39.** $5\sqrt{2} \approx 7.07$ seconds

41. $E = 6.25$ footcandles　　**43.** $6\pi\text{m}^3$　　**45.** 3.75 ohms　　**47.** $\dfrac{2x + 5y}{3x - y}$　　**48.** $y = \dfrac{7}{3}x - \dfrac{14}{3}$　　**49.** $5x^2y^3 \sqrt[3]{2x}$

50. $(-2, -3)$

Chapter 11 Review Exercises

1. Domain $= \{2, 3\}$; range $= \{4, 5\}$　　**2.** Domain $= \{-2, -7, -3\}$; range $= \{6, -3, -7\}$
3. Domain: all reals; range: $\{y \mid y \geq -1\}$　　**4.** Domain: $\{x \mid x \geq -1\}$; range: all reals

5. Domain: $\{x \mid -6 \le x \le 6\}$; range: $\{y \mid -6 \le y \le 6\}$ **6.** Domain: $\{x \mid -5 \le x \le 5\}$; range: $\{y \mid -2 \le y \le 2\}$
7. All reals   **8.** All reals   **9.** $x \le 4$ **10.** $x \ge -3$ **11.** $\{x \mid x \ne -2\}$ **12.** $\{x \mid x \ne \frac{1}{2}\}$ **13.** Function
14. Not a function   **15.** Not a function   **16.** Function   **17.** Function   **18.** Function   **19.** Function
20. Not a function   **21.** Function   **22.** Function   **23.** Not a function   **24.** Function   **25.** 2, 5, 8, 11
26. 9, 5, 1, -3 **27.** 7, 2, 1, 4   **28.** $-8, -3, 2, 25$ **29.** 1, 0, not a real number [4 not in the domain of $h(x)$]
30. $\sqrt{2}$, not a real number [2 not in the domain of $g(x)$], $2\sqrt{2}$ **31.** 0, $\frac{1}{3}$, undefined [-3 not in the domain of $h(x)$]
32. 0, undefined [0 not in the domain of $h(x)$], $\frac{5}{4}$ **33.** $2a^2 + 4a - 1$, $2z^2 + 4z - 1$
34. $2x^2 - 4x + 2$, $2z^2 - 4z + 2$
35. $5x + 12$ **36.** $5x + 17$ **37.** $5x + 4$ **38.** $5x + 7$ **39.** $5x + 14$ **40.** $5x + 19$ **41.** $\sqrt{-1 - 2x}$
42. $\sqrt{-2x - 3}$ **43.** $\sqrt{3 - 4x}$ **44.** $\sqrt{3 - 6x}$ **45.** $2\sqrt{3 - 2x}$ **46.** $3\sqrt{3 - 2x}$
47. $5h$ **48.** $\sqrt{3 - 2x - 2h} - \sqrt{3 - 2x}$ **49.** $5\sqrt{3 - 2x} + 2$ **50.** $\sqrt{-1 - 10x}$

51.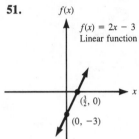

52.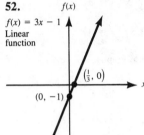

53.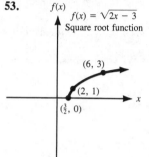

54.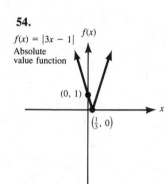

55.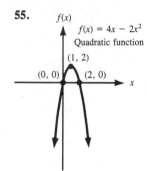

56.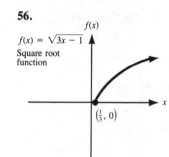

57.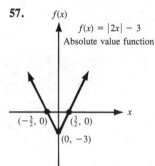

58.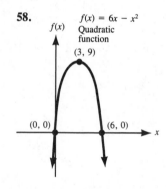

59. $\{(-1, 0), (0, 2), (-1, 3)\}$ **60.** $\{(0, 2), (3, -1), (2, -1)\}$ **61.** $R^{-1}: y = \pm\sqrt{x + 3}$ **62.** $R^{-1}: y = \sqrt[3]{x + 3}$
63. Function   **64.** Neither   **65.** One-to-one function   **66.** One-to-one function   **67.** One-to-one function
68. Function   **69.** Not a function   **70.** One-to-one function   **71.** $\{(3, 2), (4, 3)\}$ **72.** $\{(-8, 5), (7, 3), (6, 2)\}$
73. $y = \dfrac{x - 8}{3}$ **74.** $y = \dfrac{x + 1}{5}$ **75.** $y = \sqrt[3]{x}$ **76.** $y = \sqrt[5]{x}$ **77.** $y = \dfrac{3 - x}{x}$ **78.** $y = \dfrac{3x}{2 - x}$

79.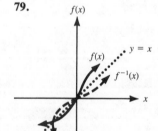

80.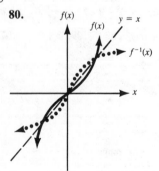

81. $x = \frac{16}{3}$ **82.** $x = 12$ **83.** $s = \pm 2\sqrt{2}$ **84.** $s = \sqrt[3]{2}$ **85.** $z = \frac{96}{7}$ **86.** $z = 36$ **87.** 288π in.^{3}
88. 144.63 lb   **89.** 60 m^3 **90.** $\$11,400$

Chapter 11 Practice Test

1. (a) Domain: $\{2, 3, 4\}$; range: $\{-3, 5, 6\}$ **(b)**

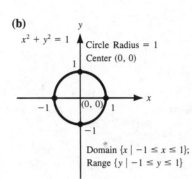

2. (a) $\{x \mid x \geq 4\}$ **(b)** $\{x \mid x \neq \frac{4}{3}\}$
3. (a) Not a function   **(b)** Function   **(c)** Not a function   **(d)** Function
4. (a) $\sqrt{5}$ **(b)** 23 **(c)** $3x^2 - 12x + 8$ **(d)** 23
5.

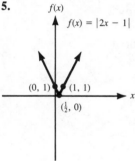

6. $\{(-3, 2), (8, 5), (8, 3)\}$ **7.** No   **8. (a)** $f^{-1}(x) = \dfrac{x + 4}{2}$ **(b)** $f^{-1}(x) = \dfrac{3x}{1 - x}$ **9.** $y = \frac{8}{3}$ **10.** $z = 405$

Exercises 12.1

1.

3.

5.

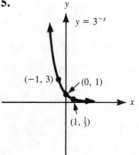

7.

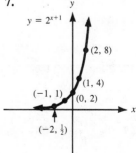

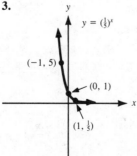

9.

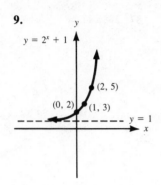

11. $x = 1$ **13.** $x = 2$ **15.** $x = -1$ **17.** $x = 2$ **19.** $x = 2, x = -1$ **21.** $x = 9$

23. $x = 1, x = -\frac{1}{4}$ **25.** $5{,}000, 10{,}000, 80{,}000, 2{,}500(2^{t/10})$ **27.** $40{,}000, 80{,}000, 20{,}000(2^{(y-1985)/14})$

29. $2{,}500, 1{,}250, 10{,}000(\frac{1}{2})^t$ **31.** $\dfrac{xy^5}{2}$ **32.** $\dfrac{5\sqrt{3} - \sqrt{30}}{3}$ **33.** $x = 5, 1$

34. $f(-2) = 21, f(x - 2) = 3x^2 - 14x + 21$ **35.** $f^{-1}(x) = \sqrt[3]{\dfrac{x + 2}{3}}$

Exercises 12.2

1. $7^2 = 49$ **3.** $\log_3 81 = 4$ **5.** $10^4 = 10{,}000$ **7.** $\log_{10} 1{,}000 = 3$ **9.** $9^2 = 81$ **11.** $81^{1/2} = 9$

13. $\log_6(\frac{1}{36}) = -2$ **15.** $3^{-1} = \frac{1}{3}$ **17.** $\log_{25} 5 = \frac{1}{2}$ **19.** $8^1 = 8$ **21.** $\log_8 1 = 0$ **23.** $16^{3/4} = 8$

25. $\log_{27}(\frac{1}{9}) = -\frac{2}{3}$ **27.** $8^{-1/3} = \frac{1}{2}$ **29.** $(\frac{1}{2})^{-2} = 4$ **31.** $\log_3 1 = 0$ **33.** $7^0 = 1$ **35.** $6^{1/2} = \sqrt{6}$

37. $\log_6 \sqrt{6} = \frac{1}{2}$ **39.** 3 **41.** 2 **43.** -1 **45.** -3 **47.** $-\frac{1}{2}$ **49.** $\frac{2}{3}$ **51.** Not defined

53. $-\frac{3}{2}$ **55.** $\frac{1}{2}$ **57.** $\frac{2}{3}$ **59.** 1 **61.** 0 **63.** 7 **65.** $x = 125$ **67.** $y = 3$ **69.** $b = 4$

71. $x = \frac{1}{36}$ **73.** $x = 8$ **75.** $y = \dfrac{5}{3}$ **77.** $b = 2$ **79.** $x = 1$

81. **87.** $\dfrac{1}{x^8 y^{21}}$ **88.** $3\sqrt[5]{x}$ **89.** $g[f(x)] = 5x - 17$ **90.** $x = \dfrac{50}{3}$

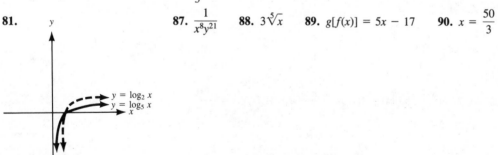

Exercises 12.3

1. $\log_5 x + \log_5 y + \log_5 z$ **3.** $\log_7 2 - \log_7 3$ **5.** $3 \log_3 x$ **7.** $\left(\frac{2}{3}\right) \log_b a$ **9.** 8 **11.** $-\frac{1}{4}$

13. $2 \log_b x + 3 \log_b y$ **15.** $4 \log_b m - 2 \log_b n$ **17.** $\dfrac{\log_b x}{2} + \dfrac{\log_b y}{2}$ **19.** $\dfrac{2 \log_2 x}{5} + \dfrac{\log_2 y}{5} - \dfrac{3 \log_2 z}{5}$

21. $\log_b(xy + z^2)$ **23.** $2 \log_b x - \log_b y - \log_b z$ **25.** $\dfrac{1}{2} + \log_6 m + \dfrac{\log_6 n}{2} - \dfrac{5 \log_6 p}{2} - \dfrac{\log_6 q}{2}$ **27.** $\log_b(xy)$

29. $\log_b\left(\dfrac{m^2}{n^3}\right)$ **31.** $\log_b 80$ **33.** $\log_b\left(\dfrac{x^{1/3} y^{1/4}}{z^{1/5}}\right)$ **35.** $\log_b \dfrac{\sqrt{xy}}{z^2}$ **37.** $\log_b\left(\dfrac{x^2}{yz^3}\right)$ **39.** $\log_p\left(\dfrac{x^{2/3} y^{4/3}}{z^{3/7}}\right)$

41. 3.3   **43.** -0.9 **45.** -1.42 **47.** 6   **49.** 6.6   **51.** 2.25   **53.** $\dfrac{2A}{3}$ **55.** $3A + 2B - C$

57. $\dfrac{5A}{2} + \dfrac{B}{2} - \dfrac{3C}{2}$

Exercises 12.4

1. 2.7664   **3.** -2.4306 or $0.5694 - 3$ **5.** 5.4472   **7.** -4.2573 or $0.7427 - 5$ **9.** 0.7931
11. -0.0773 or $0.9227 - 1$ **13.** 0.9031   **15.** 695.0   **17.** 0.00538   **19.** 15,400   **21.** 0.0948
23. 747,000   **25.** 6.030   **27.** 35.90   **29.** -0.0264 or $0.9736 - 1$ **31.** 4.820   **33.** 2.8899
35. 70,790   **37.** $\dfrac{\log x}{\log 5}$ **39.** $\dfrac{\log 8}{\log 7}$ **41.** $2 \log_{25} 8$ **43.** 3.4544   **45.** 40,450   **47.** 4.7389
49. 8.816   **51.** -3.1885 or $.8115 - 4$ **54.** $x = 4$ **55.** $x < -\frac{1}{5}$ or $x > 1$ **56.** $x = 3$ **57.** $x = 4, 5$

Exercises 12.5

1. $x = \frac{9}{5}$ **3.** $x = 12$ **5.** $x = 6$ **7.** $x = 9$ **9.** $a = 2$ **11.** $y = \frac{16}{7}$ **13.** No solution   **15.** $x = 2$
17. $x = 14$ **19.** $x = 1$ **21.** $x = \frac{8}{3}$ **23.** $x = 8$ **25.** $x = 1$ **27.** $x = 9$ **29.** $x = 5, x = 1$

31. $x = \dfrac{\log 5}{\log 2} \approx 2.322$ **33.** $x = \dfrac{\log 6 - \log 2}{\log 2} = \dfrac{\log 3}{\log 2} \approx 1.585$ **35.** $x = \dfrac{\log 5 - 3 \log 4}{2 \log 4} \approx -0.9195$

37. $y = \dfrac{\log 7}{\log 3 - \log 7} \approx -2.296$ **39.** $x = \dfrac{2 \log 5 - \log 6}{2 \log 6 - \log 5} \approx 0.723$ **41.** $x = \dfrac{2 \log 8 + 2 \log 9}{3 \log 8 - \log 9} \approx 2.116$

43. $x = \dfrac{\log 5}{\log 3 - \log 2} \approx 3.969$ **45.** $y = \dfrac{\log 3}{\log 2 + \log 5} = \dfrac{\log 3}{\log 10} = \log 3 \approx 0.4771$

47. $a = \dfrac{\log 2 - \log 3}{\log 4 + \log 3} \approx -0.1632$ **49.** Let $x =$ the first odd integer; $x + (x + 2) + (x + 4) = -27$; $-11, -9, -7$

50. Let $t =$ the time it takes for the car to catch up; $60t = 45(t + 1)$; 3 hr

51. Let $x =$ the number; $x + 2\left(\dfrac{1}{x}\right) = \dfrac{11}{3}$; $3, \dfrac{2}{3}$

52. Let $x =$ the amount invested at 6.5%; $0.065x + 0.07(5,000 + x) = 1,430$; $8,000 at 6.5% and $13,000 at 7%
53. 400 ft

Exercises 12.6

1. 22.65   **3.** 1.292   **5.** 1.413   **7.** 49,074.04 (49,204 using interpolation)
9. 1.622 (1.620 using interpolation)   **11.** $8,881.47 ($8,876 using interpolation)
13. $8,939.07 ($8,957 using interpolation)   **15.** 9.58 years (9.77 using interpolation)
17. 11.16 years   **19.** $t = 1,732.67$ years   **21.** Rate $= 17.33\%$ per day   **23.** $I_1 = 3,981.07$, $I_2 = 15,848,932$
25. 8.1   **27.** 1×10^{-7} **29.** $I = 10,000$ watts/cm^2

Chapter 12 Review Exercises

1.   **2.**

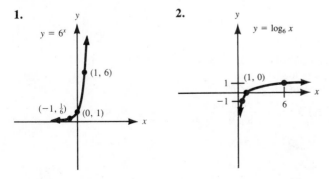

3.

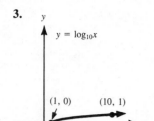

4.

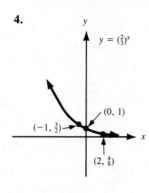

5.

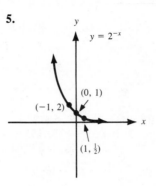

6.

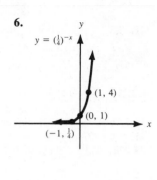

7. $3^4 = 81$ **8.** $\log_6 216 = 3$ **9.** $\log_4\left(\frac{1}{64}\right) = -3$ **10.** $2^{-2} = \frac{1}{4}$ **11.** $8^{2/3} = 4$ **12.** $\log_{10}(0.001) = -3$
13. $\log_{25} 5 = \frac{1}{2}$ **14.** $\log_{1/3} 9 = -2$ **15.** $7^{1/2} = \sqrt{7}$ **16.** $11^{1/3} = \sqrt[3]{11}$ **17.** $6^0 = 1$
18. $\log_{12} 12 = 1$ **19.** 3 **20.** -4 **21.** -2 **22.** 6 **23.** -2 **24.** -1 **25.** -2 **26.** 2
27. -2 **28.** $-\frac{3}{2}$ **29.** $\frac{1}{2}$ **30.** $\frac{1}{3}$ **31.** $\frac{5}{4}$ **32.** $\frac{4}{5}$ **33.** $3 \log_b x + 7 \log_b y$
34. $3 \log_b x - 7 \log_b y$ **35.** $2 \log_b u + 5 \log_b v - 3 \log_b w$ **36.** $2 \log_b u - 5 \log_b v - 3 \log_b w$
37. $\dfrac{\log_b x}{3} + \dfrac{\log_b y}{3}$ **38.** $\dfrac{\log_b x}{2} - \dfrac{\log_b y}{2}$ **39.** $\log_b(x^3 + y^4)$ **40.** $\dfrac{3 \log_b x}{2} + \dfrac{5 \log_b y}{2} - \dfrac{7 \log_b z}{2}$
41. $\dfrac{3 \log_b x}{2} + \dfrac{\log_b y}{2} - \dfrac{\log_b z}{2}$ **42.** $\dfrac{4 \log_b x}{9 \log_b y}$ **43.** 3.52 **44.** -1.54 **45.** $x = -2$ **46.** $x = -2$
47. $x = \frac{5}{4}$ **48.** $x = \frac{4}{5}$ **49.** $\dfrac{\log 3 - \log 5}{\log 5} \approx -0.3173$ **50.** $\dfrac{4 \log 7}{\log 2 - \log 7} \approx -6.213$ **51.** $x = 0$
52. $x = 1$ **53.** $t = 3$ **54.** $x = 4$ **55.** $x = 1$ **56.** No solution **57.** 2.8938
58. -0.2336 or $0.7664 - 1$ **59.** 0.006051 **60.** 4,540 **61.** -2.3019 or $0.6981 - 3$ **62.** 3.050
63. 309 **64.** 8.4200 **65.** \$11,412.03 **66.** 15.4 years **67.** $t = 30.81$ years **68.** 37.87 years
69. 8.2 **70.** 3.981×10^{-9}

Chapter 12 Practice Test

1.

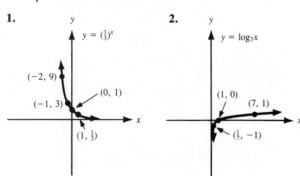

2.

3. (a) $2^4 = 16$ (b) $9^{-1/2} = \frac{1}{3}$ **4.** (a) $\log_8 4 = \frac{2}{3}$ (b) $\log_{10}(0.01) = -2$ **5.** (a) -1 (b) $\frac{1}{2}$ (c) $\frac{5}{3}$
6. (a) $3 \log_b x + 5 \log_b y + \log_b z$ (b) $3 + \frac{1}{2}\log_4 x - \log_4 y - \log_4 z$
7. (a) 4.4456 (b) 0.000716 (c) -1.1062 or $0.8938 - 2$ (d) 1.28×10^6

8. (a) $x = -4$ (b) $x = 4$ (c) $x = 4$ (d) $x = 10$ (e) $x = \dfrac{\log 5 - 3 \log 9}{\log 9} \approx -2.268$ **9.** \$8,947

10. ≈ 34.7 years

Cumulative Review: Chapters 10–12

1. $5; \left(\frac{7}{2}, 7\right)$ (§10.1)   **2.** $2\sqrt{13}; (0, 0)$ (§10.1)   **3.** $\sqrt{65}; \left(4, -\frac{1}{2}\right)$ (§10.1)   **4.** $\sqrt{2}; \left(\frac{9}{2}, \frac{17}{2}\right)$ (§9.1)

5. (§10.1)

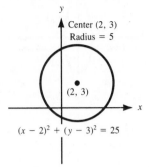

Center (2, 3)
Radius = 5

(2, 3)

$(x - 2)^2 + (y - 3)^2 = 25$

6. (§10.1)

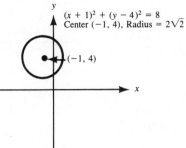

$(x + 1)^2 + (y - 4)^2 = 8$
Center $(-1, 4)$, Radius = $2\sqrt{2}$

$(-1, 4)$

7. (§10.1)

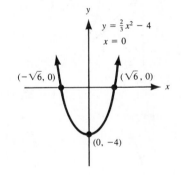

Center (3, −1)
Radius = 3

(3, −1)

$x^2 + y^2 - 6x + 2y = -1$

8. (§10.1)

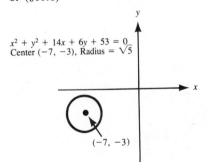

$x^2 + y^2 + 14x + 6y + 53 = 0$
Center $(-7, -3)$, Radius = $\sqrt{5}$

$(-7, -3)$

9. (§10.2)

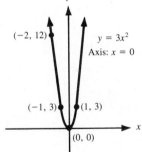

$(-2, 12)$

$y = 3x^2$
Axis: $x = 0$

$(-1, 3)$ $(1, 3)$

$(0, 0)$

10. (§10.2)

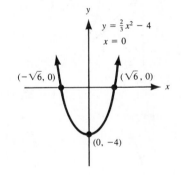

$y = \frac{2}{3}x^2 - 4$
$x = 0$

$(-\sqrt{6}, 0)$ $(\sqrt{6}, 0)$

$(0, -4)$

11. (§10.2)

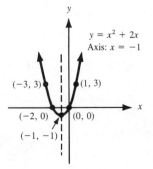

$y = x^2 + 2x$
Axis: $x = -1$

$(-3, 3)$ $(1, 3)$

$(-2, 0)$ $(0, 0)$
$(-1, -1)$

12. (§10.2)

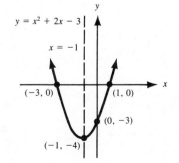

$y = x^2 + 2x - 3$

$x = -1$

$(-3, 0)$ $(1, 0)$

$(0, -3)$

$(-1, -4)$

13. (§10.2)

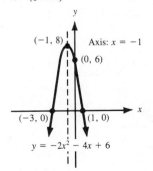

$(-1, 8)$ Axis: $x = -1$
$(0, 6)$

$(-3, 0)$ $(1, 0)$

$y = -2x^2 - 4x + 6$

14. (§10.2)

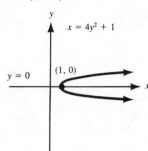

$x = 4y^2 + 1$

$y = 0$ (1, 0)

15. (§10.3)

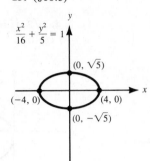

$\dfrac{x^2}{16} + \dfrac{y^2}{5} = 1$

$(0, \sqrt{5})$

$(-4, 0)$ (4, 0)

$(0, -\sqrt{5})$

16. (§10.3)

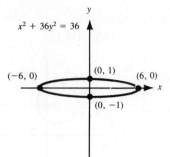

$x^2 + 36y^2 = 36$

$(-6, 0)$ (0, 1)   (6, 0)

$(0, -1)$

17. (§10.3)

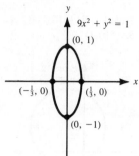

$9x^2 + y^2 = 1$

(0, 1)

$\left(-\tfrac{1}{3}, 0\right)$ $\left(\tfrac{1}{3}, 0\right)$

$(0, -1)$

18. (§10.4)

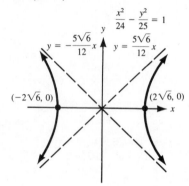

$\dfrac{x^2}{24} - \dfrac{y^2}{25} = 1$

$y = -\dfrac{5\sqrt{6}}{12}x$ $y = \dfrac{5\sqrt{6}}{12}x$

$(-2\sqrt{6}, 0)$ $(2\sqrt{6}, 0)$

19. (§10.4)

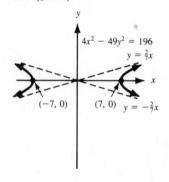

$4x^2 - 49y^2 = 196$

$y = \tfrac{2}{7}x$

$(-7, 0)$ (7, 0) $y = -\tfrac{2}{7}x$

20. (§10.4)

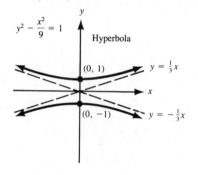

$y^2 - \dfrac{x^2}{9} = 1$ Hyperbola

(0, 1)

$y = \tfrac{1}{3}x$

$(0, -1)$ $y = -\tfrac{1}{3}x$

21. (§10.5)

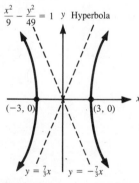

$\dfrac{x^2}{9} - \dfrac{y^2}{49} = 1$ Hyperbola

$(-3, 0)$ (3, 0)

$y = \tfrac{7}{3}x$ $y = -\tfrac{7}{3}x$

22. (§10.5)

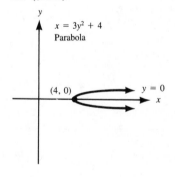

$x = 3y^2 + 4$
Parabola

(4, 0) $y = 0$

23. (§10.5)

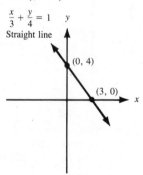

$\dfrac{x}{3} + \dfrac{y}{4} = 1$
Straight line

(0, 4)

(3, 0)

24. (§10.5)

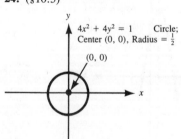

$4x^2 + 4y^2 = 1$ Circle;
Center (0, 0), Radius = $\tfrac{1}{2}$

(0, 0)

25. (§10.5)

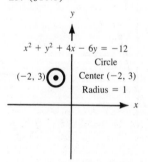

$x^2 + y^2 + 4x - 6y = -12$
Circle
$(-2, 3)$ Center $(-2, 3)$
Radius = 1

26. (§10.5)

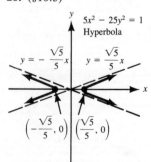

$5x^2 - 25y^2 = 1$
Hyperbola

$y = -\dfrac{\sqrt{5}}{5}x$ $y = \dfrac{\sqrt{5}}{5}x$

$\left(-\dfrac{\sqrt{5}}{5}, 0\right)$ $\left(\dfrac{\sqrt{5}}{5}, 0\right)$

27. $1.25; $312.50 (§10.2)   **28.** Let l = length and w = width; $lw = 168$, $l = w + 2$: 12″ by 14″ (§10.6)

29. $(2, -1), (-1, 2)$ (§10.6)   **30.** $(-2, -2), (4, 1)$ (§10.6)   **31.** No solution (§10.6)   **32.** $(1, 0), \left(\frac{1}{2}, \frac{1}{4}\right)$ (§10.6)

33. $(3, 1), (3, -1), (-3, 1), (-3, -1)$ (§10.6)   **34.** $(2, 3), (2, -3), (-2, 3), (-2, -3)$ (§10.6)

35. $(3, 0), \left(\frac{31}{9}, \frac{2}{3}\right)$ (§10.6)   **36.** $(4, \sqrt{11}), (4, -\sqrt{11}), (-3, 2), (-3, -2)$ (§10.6)

37. Domain: $\{2, 4, 7\}$; range: $\{-1, 3, 5\}$; function (§11.1)   **38.** Domain: $\{-2, 1, 5\}$; range: $\{3, 6\}$; function (§11.1)

39. Domain: $\{3, 4\}$; range: $\{2, 9, 7\}$; not a function (§11.1)

40. Domain: $\{-2, -1, 1, 2\}$; range: $\{1, 4\}$; function (§11.1)

41. Domain $\{x \mid -2 \le x \le 2\}$; range: $\{y \mid -1 \le y \le 1\}$ (§11.1)   **42.** Domain: all reals; range: $\{y \mid y \ge -4\}$ (§11.1)

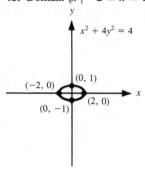

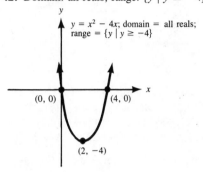

43. Domain: all reals; range: all reals (§11.1)   **44.** Domain: $\{x \mid x \le -2 \text{ or } x \ge 2\}$; range: all reals (§11.1)

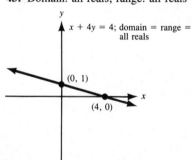

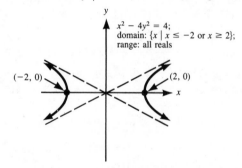

45. All reals (§11.1)   **46.** $\{x \mid x \ge 3\}$ (§11.1)   **47.** $\{x \mid x \ne 3, x \ne -2\}$ (§11.1)

48. $\{x \mid x \ge -4, x \ne 3\}$ (§11.1)   **49.** Is a function (§11.1)   **50.** Not a function (§11.1)

51. Is a function (§11.1)   **52.** Not a function (§11.1)   **53.** 23 (§11.2)   **54.** $\frac{4}{3}$ (§11.2)   **55.** $3x^2 + 1$ (§11.2)

56. $9x^2 + 6x + 1$ (§11.2)   **57.** $2a^2 + 9a + 5$ (§11.2)   **58.** $2a^2 - 3a - 1$ (§11.2)   **59.** $18a^2 - 9a - 4$ (§11.2)

60. $6a^2 - 9a - 12$ (§11.2)   **61.** -6 (§11.2)   **62.** 31 (§11.2)   **63.** -5 (§11.2)   **64.** -11 (§11.2)

65. $-\frac{3}{2}$ (§11.2)   **66.** -1 (§11.2)   **67.** $4x + 3$ (§11.2)   **68.** 3 (§11.2)   **69.** $y = \sqrt[3]{x + 1}$ (§11.4)

70. $y = \dfrac{3x + 5}{2}$ (§11.4)   **71.** $y = \dfrac{1}{x - 1}$ (§11.4)   **72.** Inverse is not a function (§11.4)

73. (§11.4)   **74.** (§11.4)   **75.** (§11.4)   **76.** (§11.4)

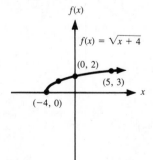

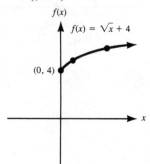

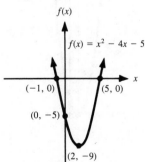

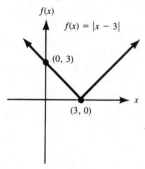

77. (§11.4)

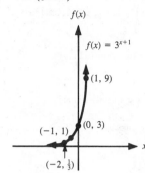

78. (§11.4)

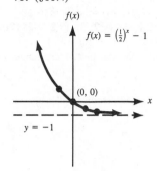

79. (§11.4)

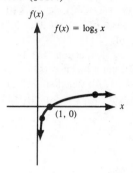

80. (§11.4)

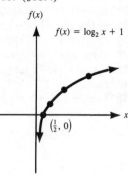

81. $2^6 = 64$ (§12.2)   **82.** $\log_3\left(\frac{1}{81}\right) = -4$ (§12.2)   **83.** $\log_{125} 5 = \frac{1}{3}$ (§12.2)   **84.** $10^{-2} = 0.01$ (§12.2)

85. $27^{4/3} = 81$ (§12.2)   **86.** $\log_4\left(\frac{1}{2}\right) = -\frac{1}{2}$ (§12.2)   **87.** 4 (§12.2)   **88.** $\frac{1}{3}$ (§12.2)   **89.** -2 (§12.2)

90. $\frac{5}{4}$ (§12.2)   **91.** $-\frac{1}{2}$ (§12.2)   **92.** $\frac{2}{3}$ (§12.2)   **93.** 0 (§12.2)   **94.** 1 (§12.2)

95. $\frac{1}{3}\log_b 5 + \frac{1}{3}\log_b x + \frac{1}{3}\log_b y$ (§12.3)   **96.** $3\log_b x + \log_b y - 5\log_b z$ (§12.3)

97. $2\log_3 x + \frac{1}{2}\log_3 y - 2 - \log_3 w - \log_3 z$ (§12.3)   **98.** $\dfrac{2\log_b u}{\log_b v}$ (§12.3)   **99.** 4.8669 (§12.4)

100. 7.49×10^{-3} (§12.4)   **101.** 4.060 (§12.4)   **102.** $0.3945 - 2 = -1.6055$ (§12.4)

103. $0.6839 - 1 = -0.3161$ (§12.4)   **104.** 2.52×10^3 (§12.4)   **105.** $x = 25$ (§§12.1, 12.4)

106. $x = 3$ (§§12.1, 12.4)   **107.** $x = -2$ (§§12.1, 12.4)   **108.** $x = -\frac{1}{4}$ (§§12.1, 12.4)

109. $x = 54$ (§§12.1, 12.4)   **110.** $x = 8$ (§§12.1, 12.4)   **111.** $x = 2, -\frac{3}{2}$ (§§12.1, 12.4)

112. $x = 4$ (§§12.1, 12.4)   **113.** $x = \dfrac{3\log 7}{\log 9 - \log 7} \approx 23.24$ (§§12.1, 12.4)   **114.** $x = 6$ (§§12.1, 12.4)

115. $y = \frac{80}{3}$ (§11.5)   **116.** $y = \frac{12}{5}$ (§11.5)   **117.** $x = 20$ (§11.5)   **118.** $x = \frac{24}{5}$ (§11.5)

119. ≈ 6.448 yr (§12.5)   **120.** ≈ 23.1 yr (§12.5)   **121.** $\approx 1{,}151$ yr (§12.5)   **122.** 3.4 (§12.5)

123. 6.31×10^{-9} (§12.5)

Cumulative Practice Test: Chapters 10–12

1. $5; \left(\frac{7}{2}, -1\right)$

2.

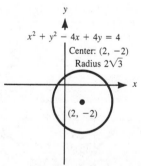

3.

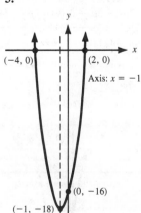

4. (a)

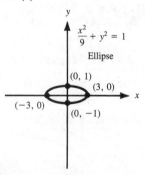

(b)

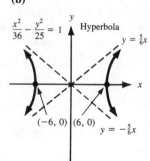

5. 42 feet   **6.** $x = 4, y = 3$

7. Yes; each value of x is assigned no more than one value of y; domain: $\{-2, 0, 3, 4\}$; range: $\{1, 5, 7\}$

8. (a) Domain: $\{x \mid -3 \le x \le 3\}$; range: $\{y \mid -6 \le y \le 6\}$; not a function: line $x = 0$ intersects graph at two points;

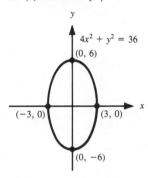

(b) Domain: all reals; range: $\{y \mid y \ge 0\}$; is a function: any vertical line will intersect the graph at exactly one point

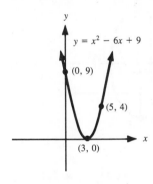

9. (a) $\{x \mid x \ne 3, x \ne -3\}$ **(b)** $\{x \mid x \ge \frac{3}{2}\}$

10. (a) 42   **(b)** $4x^2 - 1$ **(c)** $x^2 + x$ **(d)** $x^2 - 3x + 4$ **(e)** $20x - 1$ **(f)** $20x - 5$ **(g)** 6

(h) 7   **(i)** 0   **(j)** $2x$ **11.** $f^{-1}(x) = 7x + 3$

12. (a)   **(b)**

13. $8^{-2/3} = \frac{1}{4}$ **14. (a)** -3 **(b)** $\frac{3}{2}$ **15. (a)** $\log_b x + \frac{1}{3}\log_b y$ **(b)** $\frac{5}{2}\log_b x - \frac{1}{2}\log_b y$

16. (a) $0.8041 - 3 = -2.1959$ **(b)** 8.439×10^3 **17. (a)** $x = 0, 9$ **(b)** $x = 4$ **18.** $y = \frac{25}{9}$

19. ≈ 7.032 yr   **20.** 7.4

Appendix A

1. 720   **3.** 24   **5.** 1   **7.** 120   **9.** 5,038   **11.** 720   **13.** 12   **15.** 5   **17.** 56   **19.** 27,720

21. $a^6 + 6a^5b + 15a^4b^2 + 20a^3b^3 + 15a^2b^4 + 6ab^5 + b^6$

23. $a^6 - 6a^5b + 15a^4b^2 - 20a^3b^3 + 15a^2b^4 - 6ab^5 + b^6$

25. $32a^5 + 80a^4 + 80a^3 + 40a^2 + 10a + 1$ **27.** $32a^5 - 80a^4 + 80a^3 - 40a^2 + 10a - 1$

29. $1 - 8a + 24a^2 - 32a^3 + 16a^4$ **31.** $a^{10} + 10a^8b + 40a^6b^2 + 80a^4b^3 + 80a^2b^4 + 32b^5$

33. $16x^8 - 96x^6y^2 + 216x^4y^4 - 216x^2y^6 + 81y^8$ **35.** $\dfrac{x^4}{81} + \dfrac{8}{27}x^3 + \dfrac{8}{3}x^2 + \dfrac{32}{3}x + 16$

37. $a^{12} - 6a^{10}b^4 + 15a^8b^8 - 20a^6b^{12}$ **39.** $256a^8 + 3,072a^7b + 16,128a^6b^2 + 48,384a^5b^3$

41. $729a^{12} - 2,916a^{10} + 4,860a^8 - 4,320a^6$ **43.** $28x^6y^2$ **45.** $-160a^3b^3$ **47.** $20,412a^{10}$

49. $-6,048a^4$ **51.** 256

Appendix B

1.

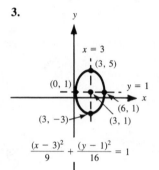

$$\frac{(x - 1)^2}{9} + \frac{(y - 3)^2}{16} = 1$$

3.

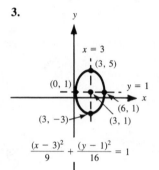

$$\frac{(x - 3)^2}{9} + \frac{(y - 1)^2}{16} = 1$$

5. $\dfrac{(x - 2)^2}{25} + \dfrac{(y - 5)^2}{81} = 1$

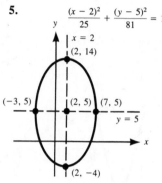

7.

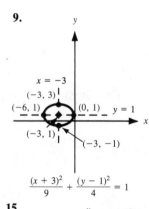

$$\frac{(x - 3)^2}{9} + \frac{(y - 1)^2}{4} = 1$$

9.

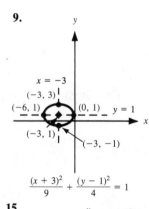

$$\frac{(x + 3)^2}{9} + \frac{(y - 1)^2}{4} = 1$$

11. $\dfrac{(x + 3)^2}{9} + \dfrac{(y + 1)^2}{4} = 1$

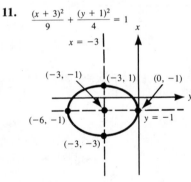

13. $9(x - 2)^2 + 4(y + 5)^2 = 36$

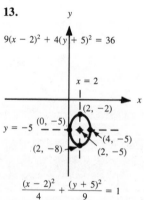

$$\frac{(x - 2)^2}{4} + \frac{(y + 5)^2}{9} = 1$$

15. $16(x + 1)^2 + 64(y - 2)^2 = 64$

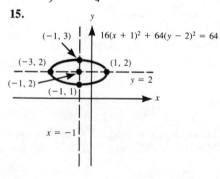

17.

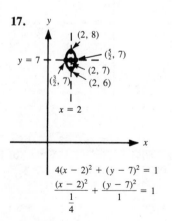

$$4(x - 2)^2 + (y - 7)^2 = 1$$

$$\frac{(x - 2)^2}{\frac{1}{4}} + \frac{(y - 7)^2}{1} = 1$$

19.

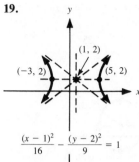

$$\frac{(x-1)^2}{16} - \frac{(y-2)^2}{9} = 1$$

21.

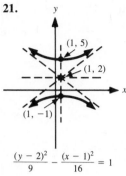

$$\frac{(y-2)^2}{9} - \frac{(x-1)^2}{16} = 1$$

23. $\dfrac{(x-3)^2}{25} - \dfrac{(y-4)^2}{16} = 1$

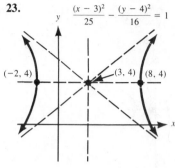

25. $\dfrac{(x-4)^2}{9} - \dfrac{(y+1)^2}{36} = 1$

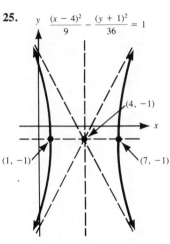

27. $\dfrac{(y+2)^2}{16} - \dfrac{(x-1)^2}{25} = 1$

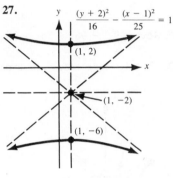

29. $\dfrac{(x+3)^2}{36} - \dfrac{(y+2)^2}{100} = 1$

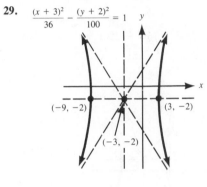

31.

$$4(x-2)^2 - 25(y-4)^2 = 100$$

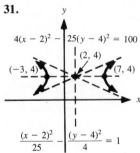

$$\frac{(x-2)^2}{25} - \frac{(y-4)^2}{4} = 1$$

33. $36(x+1)^2 - 9(y-2)^2 = 36$

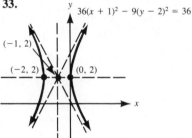

35.

$$16(x-5)^2 - (y+4)^2 = 1$$

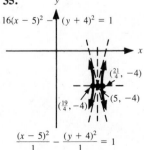

$$\frac{(x-5)^2}{\frac{1}{16}} - \frac{(y+4)^2}{1} = 1$$

37. $4x^2 - 9y^2 - 8x + 54y = 113$;
hyperbola: $\dfrac{(x-1)^2}{9} - \dfrac{(y-3)^2}{4} = 1$;

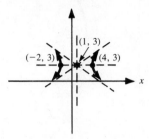

39. $4x^2 + 9y^2 - 8x - 54y = -49$;
ellipse: $\dfrac{(x-1)^2}{9} + \dfrac{(y-3)^2}{4} = 1$

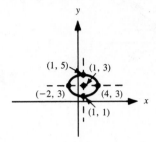

41. $25x^2 + 16y^2 - 250x - 64y = -289$;
ellipse: $\dfrac{(x-5)^2}{16} + \dfrac{(y-2)^2}{25} = 1$

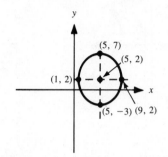

43. $25x^2 - 16y^2 - 250x + 64y = -161$;
hyperbola: $\dfrac{(x-5)^2}{16} - \dfrac{(y-2)^2}{25} = 1$

45. $16y^2 - 9x^2 - 96y + 36x = 36$; hyperbola $\dfrac{(y-3)^2}{9} - \dfrac{(x-2)^2}{16} = 1$

47. $x^2 - 4y^2 + 4x + 8y = 16$; hyperbola $\dfrac{(x+2)^2}{16} - \dfrac{(y-1)^2}{4} = 1$;

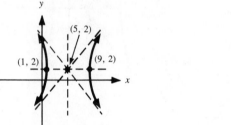

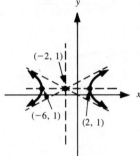

49. $36x^2 + 25y^2 + 216x + 100y = 476$; ellipse $\dfrac{(x+3)^2}{25} + \dfrac{(y+2)^2}{36} = 1$

51. $4x^2 - y^2 + 24x - 4y = -31$; hyperbola: $\dfrac{(x+3)^2}{\frac{1}{4}} - (y+2)^2 = 1$

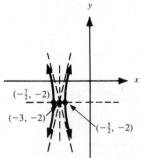

Index

Key Definitions, Symbols, and Properties

$\in$ is an element of

$\notin$ is not an element of

$=$ is equal to

$\neq$ is not equal to

$\subset$ is a subset of

$A \cap B = \{x \mid x \in A \text{ and } x \in B\}$ *intersection*

$A \cup B = \{x \mid x \in A \text{ or } x \in B\}$ *union*

$<$ is less than

$\leq$ is less than or equal to

$>$ is greater than

$\geq$ is greater than or equal to

$\approx$ is approximately equal to

$|x| = \begin{cases} x & \text{if } x \geq 0 \\ -x & \text{if } x < 0 \end{cases}$ *(absolute value)*

$a - b = a + (-b)$ *(subtraction)*

$a^n = \underbrace{a \cdot a \cdot a \cdot \cdots \cdot a}_{n \text{ times}} \left(\begin{array}{c}\text{natural number} \\ \text{exponent}\end{array}\right)$

$a^0 = 1$ if $a \neq 0$ *(zero exponent)*

$a^{-n} = \dfrac{1}{a^n}$ if $a \neq 0$ *(negative exponent)*

$a^{1/n} = \begin{cases} b & \text{if } n \text{ is odd and } b^n = a \quad (\textbf{nth root}) \\ |b| & \text{if } n \text{ is even, } a \geq 0 \text{ and } b^n = a \end{cases}$

$a^{m/n} = (a^{1/n})^m = (a^m)^{1/n}$ *(rational exponent)*
if m/n is reduced to lowest terms
and $a^{1/n}$ is a real number

$\sqrt[n]{a} = a^{1/n}$ *(radical notation)*

$i = \sqrt{-1}$ *(imaginary unit)*

(x, y) ordered pair

$\begin{bmatrix} a & b \\ c & d \end{bmatrix}$ matrix $\begin{vmatrix} a & b \\ c & d \end{vmatrix}$ determinant

$f(x)$ f evaluated at x *(function notation)*

$f^{-1}(x)$ inverse function of $f(x)$

$e \approx 2.71828$ *(natural log base)*

$\log_b y = x$ means $b^x = y$ *(log base b)*

$\ln y = x$ means $e^x = y$ *natural logarithm (log base e)*

$\log y = x$ means $10^x = y$ *common logarithm*
(log base 10)

Properties of Exponents

$x^n x^m = x^{n+m}$

$(x^n)^m = x^{nm}$

$(xy)^n = x^n y^n$

$\dfrac{x^n}{x^m} = x^{n-m}$ $(x \neq 0)$

$\left(\dfrac{x}{y}\right)^n = \dfrac{x^n}{y^n}$ $(y \neq 0)$

Properties of Radicals

$\sqrt[n]{xy} = \sqrt[n]{x}\,\sqrt[n]{y}$

$\sqrt[n]{\dfrac{x}{y}} = \dfrac{\sqrt[n]{x}}{\sqrt[n]{y}}$

$\sqrt[np]{x^{mp}} = \sqrt[n]{x^m}$ $(x > 0)$

Properties of Logarithms

$\log_b(xy) = \log_b x + \log_b y$

$\log_b x^r = r \log_b x$

$\log_b \dfrac{x}{y} = \log_b x - \log_b y$

Properties of Equality

If $a = b$ then:

$a + c = b + c$

$a - c = b - c$

$ac = bc$

$\dfrac{a}{c} = \dfrac{b}{c}$ $(c \neq 0)$

Zero-Product Rule

If $ab = 0$ then $a = 0$ or $b = 0$.

Square Root Theorem

If $x^2 = d$ then $x = \pm\sqrt{d}$.